D1387708

An Inclusive Environment

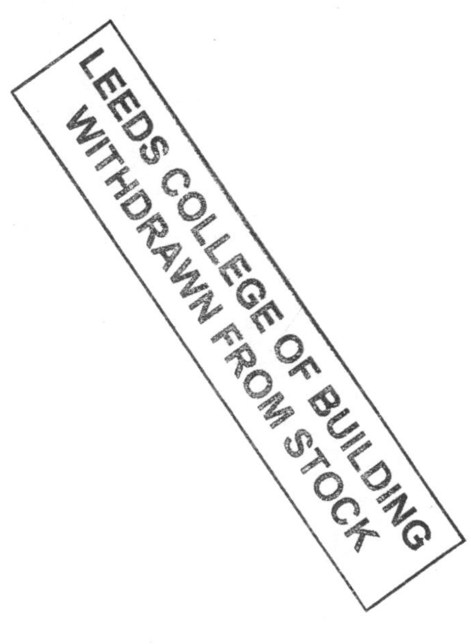

To David Chappell, who averted disaster

An Inclusive Environment

An A–Z guide to legislation, policies and products

Maritz Vandenberg

AMSTERDAM • BOSTON • HEIDELBERG • LONDON • NEW YORK • OXFORD
PARIS • SAN DIEGO • SAN FRANCISCO • SYDNEY • TOKYO
Butterworth-Heinemann is an imprint of Elsevier

Butterworth-Heinemann is an imprint of Elsevier
Linacre House, Jordan Hill, Oxford OX2 8DP, UK
30 Corporate Drive, Suite 400, Burlington, MA 01803, USA

First published 2008

British Library Cataloguing-in-Publication Data
A catalogue record for this title is available from the British Library

Library of Congress Cataloging-in-Publication Data
A catalog record for this title is available from the Library of Congress

ISBN: 978-0-7506-8456-9

For information on all Butterworth-Heinemann publications visit our website at www.books.elsevier.com

All illustrations by the author

Typeset by Charon Tec Ltd., A Macmillan Company. (www.macmillansolutions.com)

Printed and bound in Great Britain

08 09 10 11   10 9 8 7 6 5 4 3 2 1

Contents

Foreword

An inclusive environment is one that considers everyone. It is the result of a simple but novel design philosophy, inclusive design, which I have defined as follows. Inclusive design:

- places people at the heart of the design process
- responds to human diversity and difference
- offers dignity, autonomy and choice
- provides for flexibility in use.

This definition embraces all of the human population, not just disabled people, and its purpose is social – inclusion by design. It articulates an aspiration in the form of a set of precepts, but it is not prescriptive technically. Inclusive design is concerned with the culture that informs the decision-making of designers. It should be informed by research-based technical data, but it is not in itself a set of rules.

The idea of inclusive design is so simple, so potent, one wonders why it took so long to emerge. The answer, I think, lies with the usually benign welfare state and its sometimes perverse consequences. In 1970 the Chronically Sick and Disabled Persons Act introduced a series of entitlements for disabled people. In order to qualify, they needed to meet the legal definition of a disabled person. Inevitably, this was couched in largely negative terms, and the 'medical model of disability' was born. In accordance with this model disabled people were perceived as being different from able-bodied people (whoever they might be), and as having 'special needs'.

No-one did well out of this. A 'disabled person' was defined by drawing a line arbitrarily across the continuum of ability/disability. A false antithesis was established between able-bodied people, who had general needs (as in housing), and disabled people, whose needs were not seen as coming within the generality of need and were therefore different and special. The idea that people came in essentially two varieties, disabled or able-bodied, requiring different treatment in the built environment, is, when examined critically, clearly quite bizarre, albeit bureaucratically neat. Under this regime society was the poorer for denying much of the built environment to disabled people. I also believe that this comfortable but misleading antithesis held back for decades research into how people in their infinite variety interact with the built environment. It is only with the emergence of the inclusive design philosophy that we are starting to see whole population research, yielding important information about people's decision-making and preferences in the built environment.

Fortunately, in the 1980s, academics and activists within the disability movement threw their weight behind a new model of disability, the social model. This identified the way society is organised (including the design, regulation and management of the built environment) as the means of oppression and denial of opportunity to disabled people.

The social model of disability has proved to be a powerful and influential idea. It paved the way, eventually, for the first inclusive revision of Part M of the Building Regulations in 2004. This now avoids all attempts at categorising building users, and calls, essentially, for buildings to be suitable for the people who use them.

Architects have on the whole been slow to appreciate the potential of inclusive design. So ingrained is the belief that disabled people form a self-contained and separate building user population, that it is all too easy to interpret inclusive design as the new euphemism for design for special needs. Some of the most inclusive work to date has been carried out in the context of historic buildings, which resist standard solutions and where the creative problem-solving skills of our great architects are stretched to the full.

For the truth is that the inclusive design philosophy is heady stuff. It prompts new curiosity about how people use the built environment. It positively invites more research, not so much on discrete user groups, but based on whole populations, to explore more widely people's strategies for engaging with the built environment. Wayfinding, for example, emerges as a particular issue.

Inclusive design, in short, is concerned with freeing problem-solving skills and inciting dialogue with users, not in a sterile obligatory consultation mode, but in a spirit of collaborative inquiry as to what works best and why.

Inclusive design is good design. It is the result of good manners and concern for others. It is absolutely not about designing an exclusive building for some people to use, and then sticking on the 'disability access ramp' or the all too often literally 'disabled WC'. It is about new ways of satisfying our thirst for beautiful buildings that are a joy to use.

For all those embarked on the quest for inclusive environments Maritz's remarkable book is an essential companion. It is replete with legal and technical detail, spiced with irreverent commentary on the human foible and inconsistency that attend the political journey towards a better society. Like all proper reference books it invites browsing; it is both impeccably organised and wonderfully discursive. It is an unusual and useful work.

Sarah Langton-Lockton OBE

Sarah Langton-Lockton was Chief Executive of the Centre for Accessible Environments from 1979 until her retirement in December 2006. She received an OBE in the Birthday Honours List of 2000 'for services to disabled people'.

Introduction

An inclusive environment is one in which everyone (or virtually everyone) has dignified and easy access to all the good things that civilised life has to offer. These may be summarised as safety, comfort, a rich cultural life, and freedom – and without freedom the rest are not only unlikely, but emptied of meaning.

People can be excluded from these goods by many factors including age, disability, poverty, crime or the fear of crime, lack of education, unfair discrimination, unresponsive bureaucracies, and – in the case of freedom – governments that act as though the people belong to them rather than vice versa. Most of these matters are to some degree interrelated. This book tries to cover all of them, and to show the links where such exist.

Where legal matters are touched upon the present work only gives simplified summaries of original documents and does not purport to provide full, authoritative, and up-to-date statements of the law.

Corrections, comments, and suggestions for improvement would be welcomed. Please send them to maritz2000@aol.com.

Maritz Vandenberg
May 2008

A

AbilityNet	1 The *charity* (qv) AbilityNet is the UK's leading source of advice on computer use by *disabled people* (qv). It provides advice, support, equipment, and guidance on *accessible website design* (qv). 2 For more information visit www.abilitynet.org.uk/ See also: *Accessible websites;* and *Reading aids.*
Abortion	1 Prenatal diagnosis techniques, which are becoming constantly more effective, enable parents to find out whether a foetus has abnormalities and to decide whether or not to terminate the pregnancy. Late terminations are extremely undesirable: in January 2008 a leading foetal pain expert, Professor K J S Anand, told a House of Commons meeting that the part of the human brain that feels pain develops at around 20 weeks, and that foetuses aborted after that date 'experience excruciating pain' from the physical dismemberment or crushing of body parts that accompany abortion. Quite clearly early screening of foetuses should be universal practice. 2 For a well-informed and thoughtful discussion refer to pp 118–121 of DISABILITY RIGHTS AND WRONGS by the *disabled* writer Tom Shakespeare – see the bibliography. See also: *Babies: premature.*
Academics for Academic Freedom (AFAF)	1 AFAF was formed in December 2006 by 60 UK academics who had become so concerned at the growing reluctance of academics to voice controversial opinions that they felt it necessary to call for a change in the law to guarantee the freedom to question and test received wisdom and put forward unpopular views – for more on this see: *Speech: freedom of.* 2 For more information visit www.afaf.org.uk/ See also: *Political correctness: speech codes;* and *Speech: freedom of.*
Access	See: *Accessible.*
Access action plan	1 A methodical plan by a building owner or facilities manager to make their premises more *accessible* (qv). This essentially means bringing the premises into conformity with the requirements of the Disability Discrimination Act (qv). 2 The drawing up of an access action plan should normally commence with the commissioning of an *access audit* (qv).
Access All Areas	See: *Access awards.*
Access appraisals	1 An access appraisal is very similar to an *access audit* (qv), except that 'appraisals' are carried out at design stage, whereas 'audits' are carried out on completed buildings. 2 See: *Access audits.*
Access Association	1 The Access Association, formerly the Access Officers' Association, was formed in 1991 to represent *access* professionals working in organisations of all kinds, with the aim of helping to ensure the proper implementation of the Disability Discrimination Act (qv). Its quarterly periodical THE ACCESS JOURNAL is a useful information source for all professionals with an interest in *inclusive design* (qv) – see the bibliography. 2 For more information visit www.access-association.org.uk/

Access auditors

See: *National Register of Access Consultants.*

Access audits

1 An access audit is a formal evaluation of the *accessibility* (qv) of a building against some recognised standard, carried out by a qualified *access consultant* (qv) on behalf of the building owner or management. The audit will usually consist of a detailed examination of the building, leading to a written report that establishes three things:

- What are the barriers to accessibility in or around that building?
- What needs to be done to remove those barriers?
- How can these improvements be implemented as a planned programme carried out over a stated period, and fitting in (if possible) with the normal maintenance and investment programme to achieve maximum economy and efficiency?

2 As of 2008, useful references include the following. They are given in alphabetical order. For publication details see the bibliography.

- ACCESS AND THE DDA: A SURVEYOR'S HANDBOOK by Mark Ratcliff.
- ACCESS AUDIT HANDBOOK by Alison Grant, published by the Centre for Accessible Environments (qv), is the essential guide to all access auditing. It includes a comprehensive list of further references.
- ACCESS AUDIT PRICE GUIDE, published by the Building Cost Information Service (BCIS) of the Royal Institution of Chartered Surveyors.
- ACCESS AUDITS: A PLANNING TOOL FOR BUSINESS is a DVD/CD ROM sponsored by Barclays PLC and published by the Centre for Accessible Environments.
- ACCESS AUDITS OF PRIMARY HEALTHCARE FACILITIES. Health Facilities Note 20.
- ACCESS FOR DISABLED PEOPLE is a design guidance note from Sport England, and includes a series of checklists for auditing sports buildings.

See also: *Disability Discrimination Act.*

Access audits: live

1 In contrast with the structured nature of a conventional *access audit* (qv), a live audit is an informal evaluation of the *accessibility* (qv) of a building, or of an environment such as a town centre, carried out by a suitably chosen group of people who explore the area together and record their thoughts and comments as they go. The aim also is less formal: to bridge the gap between users and deciders (planners, architects, interior designers and the like) by helping each side to see the point of view of the others.

2 For an example of such an exploration (in this case a London walk from the footbridge opposite the Tate Modern Gallery to the piazza of Somerset House) refer to pp 11–13 of issue 85 of ACCESS BY DESIGN (qv), the journal of the Centre for Accessible Environments (qv) – see the bibliography.

See also: *Access audits;* and *Urban design.*

Access awards

1 The purpose of an *access award* (also called an *inclusive design award*, or by other similar names) is to publicly recognise and record the effort that has been made by an owner and/or designer to improve the *accessibility* (qv) of a public space or building.

2 Such schemes are to be welcomed, but RADAR (qv) recommends that awarding bodies should always make a clear record of:

- What exactly the access award is for.
- What criteria were used.
- The date of the award.

Such caution will help avoid the risk of (for example) a *disabled person* (qv) bringing a valid complaint of *discrimination* (qv) against an *employer or service provider* (qv), and of the defendant producing some previously received access award to counter the charge, even though the award in question may have been based on dubious criteria, or the building radically altered since receiving the award.

Access awards case examples

1 ACCESS ALL AREAS AWARDS are run by the Department for Work and Pensions (qv) and give *service providers* (qv) with fewer than 100 employees a chance to demonstrate how they have made changes to their business to help *disabled people* (qv) use their services.

2003 award winners were:

London and Southeast	Wildfowl and Wetlands Trust, London.
Southwest	Brewhouse Theatre and Arts Centre.
West Midlands	Mouzer Employment Consultancy.
East Midlands	Willow Farm Riding School.
Northwest	Council for Voluntary Services.

2004 award winners included:

Entertainment & leisure	Salisbury and South Wiltshire Museum, Wiltshire.
Health & beauty	Janet Maitland Hair Excellence, Durham.
Hotels and bars etc.	D. Fecci & Sons fish and chip shop, Tenby.
Charities	Holton Lee holiday cottages, Poole, Dorset.

---

2 ADAPT TRUST AWARDS are made by the Adapt Trust (qv) to existing *arts* and *heritage* buildings which have been altered to make them more accessible to disabled people. Awards made under the Trust's auspices in 2003 were as follows:

Adapt Trust:
National Portrait Gallery, London.
Dundee Contemporary Arts.
Manchester Art Gallery.
Royal Lyceum Theatre, Edinburgh.
The Space, Dundee College.
York Theatre Royal.

Barry Foster Memorial Awards:
Hampstead Theatre, London.
Chicken Shed Theatre Company.

John S Cohen Foundation Awards:
Highland Folk Museum, Newton more site.
LSO – St Luke's.
Dorman Museum, Middllesbrough.
FACT (Foundation for Art & Creative Technology) Liverpool.
National Centre for Early Music, York.

Highly commended:
Ikon Gallery, Birmingham.
Millennium Forum, Derry/Londonderry.
Royal Centre, Nottingham.

---

3 AUTOMOBILE ASSOCIATION ACCESSIBLE HOTEL OF THE YEAR AWARDS are made by the AA – visit www.theaa.com/

2001	De Vere White Hotel in Reebok Stadium, Horwich, Manchester.
2002	Old Government House Hotel, St Peter Port, Guernsey.
2003	Castle House Hotel, Hereford.
2004	De Vere Daresbury Park, Warrington.

4 CIVIC TRUST ACCESS AWARDS are made by the Civic Trust and sponsored by English Heritage. They are given annually to the scheme which is judged to best reconcile the access needs of *disabled people* (qv) with the conservation requirements of a *historic building* or *site* (qv). Recent award-winning schemes are:

2000 The Point, Doncaster.
2001 Great Dunmow Maltings, Essex.
2002 Royal Court Theatre, Kensington & Chelsea, London.
2003 Whitby Abbey Headland, Yorkshire.
2004 LSO St Luke's, Islington, London.

5 DN/SCOPE ACCESSIBLE HIGH STREETS AWARDS are made by the monthly periodical Disability Now (qv) and the charity Scope.

6 RIBA INCLUSIVE DESIGN AWARDS are made by the Royal Institute of British Architects (qv) in association with the Centre for Accessible Environments (qv) and the architectural ironmonger Allgood. Recent winners are:

2004 City of Manchester Stadium, Manchester.
2005 The Sage, Gateshead.
2006 Idea Store, Whitechapel, London.

See also: *Concert halls; Museums;* and *Theatres*.

Access by Design See: *Centre for Accessible Environments*.

Access champion 1 An individual or organisation concerned with promoting the principles of *inclusive design* (qv) when a building or facility is commissioned, planned, designed, or occupied. For a note on their duties refer to section 3 of PLANNING, BUILDINGS, STREETS AND DISABILITY EQUALITY – see the bibliography.
2 For more information visit www.dptac.gov.uk/inclusive/guide/07.htm

Access consultant See: *National Register of Access Consultants*.

Access controls to public buildings: complying with the Disability Discrimination Act 1 Buildings used by the public must in general comply with the Disability Discrimination Act (qv). In the case of access controls, *authoritative practical guides* (qv) include DESIGNING FOR ACCESSIBILITY (qv) and in addition the following documents. For publication details see the bibliography.

• APPROVED DOCUMENT M (qv), the provisions of which must be heeded in order to comply with the Building Regulations (qv) for England and Wales. See paras 2.14–2.21.
• BS 8300, the provisions of which apply to all buildings in the UK. It is an advisory document. See paras 6.5.1–6.5.6.
• INCLUSIVE MOBILITY (qv), which is the most authoritative reference for *inclusive design* (qv) of *pedestrian infrastructure* (qv) and for *transport-related buildings* (qv) such as bus and coach stations, railway stations, air terminals, and transport interchanges. It is an advisory document. See para 8.2.

While the above advisory documents do not have the force of law, conformity with their recommendations will make it easier to demonstrate that the requirements of the Disability Discrimination Act have been met.
2 The dimensions shown on the diagram on the next page will in general satisfy the provisions of AD M and BS 8300, though each particular case should be checked in detail before decisions are finalised.

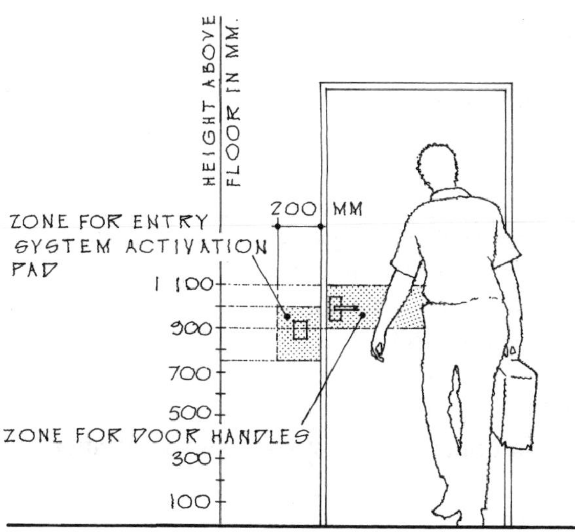

3 For places other than England and Wales the following regulatory documents should be consulted:

• In Northern Ireland: TECHNICAL BOOKLETS E, H, and R of the Building Regulations (Northern Ireland) (qv).

• In Scotland: the building (SCOTLAND) REGULATIONS and the associated TECHNICAL HANDBOOKS (qv).

4 In the USA, where the Americans with Disabilities Act applies, refer to the ADA AND ABA ACCESSIBILITY GUIDELINES (qv).

See also: *Doors.*

Access controls to public buildings: other information sources

1 As of 2008, other useful references include the following. For publication details see the bibliography.

2 From the British Security Industry Association:

• ACCESS CONTROL USER GUIDES.

• A GUIDE TO ASSIST IN COMPLIANCE WITH THE DISABILITY DISCRIMINATION ACT.

3 From the Centre for Accessible Environments (qv):

• ARCHITECTURAL IRONMONGERY.

• AUTOMATIC DOOR SYSTEMS.

4 Useful websites include the following, given in alphabetical order.

• British Security Industry Association www.bsia.co.uk/

• Disabled Living Foundation www.dlf.org.uk/

Access Directory See: *DPTAC Access Directory.*

Access for All See: *Rail travel: accessibility.*

Access groups See: *Local access groups.*

Access Journal See: *Access Association;* and *Authoritative practical guides.*

Access knowledge map See: *DPTAC Access Directory.*

Access: legal advice	See: *Disabled access: legal advice.*
Access: legislation, regulations and standards	See: *Authoritative practical guides.*
Access officers	1 Access Officers are employed by *local authorities* (qv), probably in the social services, planning, or building control departments to look after the interests of *disabled persons* (qv) in the local area. Their membership organisation is the *Access Association* (qv). For a note on their duties refer to section 3 of PLANNING, BUILDINGS, STREETS AND DISABILITY EQUALITY – see the bibliography. 2 For more information visit www.access-association.org.uk/
Access officers' Association	See: *Access Association.*
Access plans	See: *Accessibility plans.*
Access routes to buildings in general	1 Access routes are *footways* (qv) enabling pedestrians (including *wheelchair-users* (qv) and people using *walking and standing aids* (qv) such as walking sticks and crutches to comfortably make their way: • Between the site entrance or car park and the principal entrance of the building; and • Between the entrances of the various buildings on the site. 2 Footways alongside and across streets, in pedestrian shopping precincts, and other situations of that kind are covered by the term *pedestrian infrastructure* (qv). They serve a more public role and are subject to more stringent requirements. 3 As of 2008, useful general references include the METRIC HANDBOOK – see the bibliography. For more specific references see the next two entries.
Access routes to dwellings	1 The following references contain authoritative recommendations on access routes to dwellings They are given in alphabetical order. For publication details see the bibliography. • Section 6 of APPROVED DOCUMENT M (qv). The provisions in AD M must be heeded in order to comply with the Building Regulations (qv) in England and Wales. In the case of private dwellings the requirement is, in brief, a firm and smooth surface with a minimum clear width of 900 mm. If the footway is steeper than 1:20 it should be designed as a *ramp* (qv). For housing estates see the next entry. • THE HABINTEG HOUSING ASSOCIATION DESIGN GUIDE (qv) applies most specifically to *housing association* (qv) schemes, but ought to be by the designer's side in the design of all housing developments and individual dwellings. • THE HOUSING DESIGN GUIDE is a best-practice guide for housing in London. • WHEELCHAIR ACCESSIBLE HOUSING: DESIGNING HOMES THAT CAN BE EASILY ADAPTED FOR RESIDENTS WHO ARE WHEELCHAIR-USERS is a best-practice guide for housing in London, and is meant to be used in conjunction with the next title. • THE WHEELCHAIR HOUSING DESIGN GUIDE (qv) applies most specifically to *housing association* (qv) schemes, but its usefulness extends beyond that special remit.

Access routes to public buildings: complying with the Discrimination Act

1 Buildings used by the public must in general comply with the Disability Discrimination Act (qv). In the case of access routes to public buildings, *authoritative practical guides* (qv) include DESIGNING FOR ACCESSIBILITY (qv) plus the following:

- Section 1 of APPROVED DOCUMENT M (qv). The provisions in AD M must be heeded in order to comply with the Building Regulations (qv) in England and Wales.
- Section 5 of BS 8300 (qv) for all buildings in general. BS 8300 is an advisory document.
- Sections 3, 4, and 10 of INCLUSIVE MOBILITY (qv) for *transport-related buildings* such as bus and coach stations, railway stations, air terminals, and transport interchanges. INCLUSIVE MOBILITY is an advisory document.

The advisory documents above do not have the force of law, but conformity with their recommendations will help to demonstrate that the requirements of the Disability Discrimination Act have been met.

2 Surfaces should be firm and smooth, and if the gradient is steeper than 1:20 the footway should be designed as a *ramp* (qv). The widths shown on the diagram below will generally comply with the above guidance documents, though each particular case should be checked in detail before decisions are finalised.

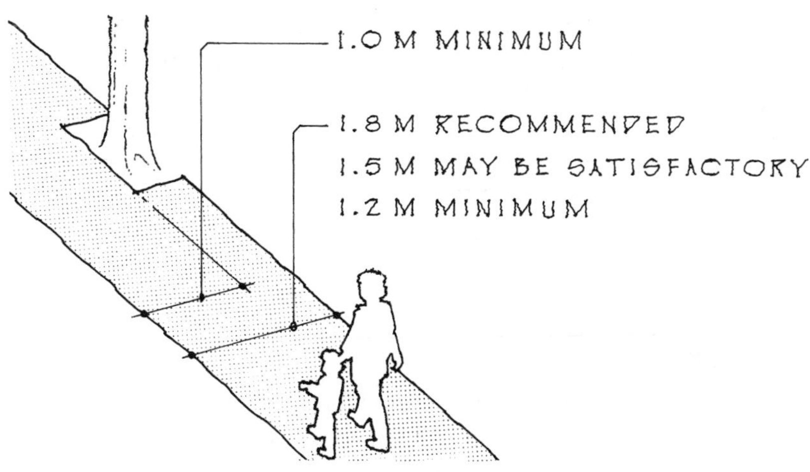

3 For places other than England and Wales the following regulatory documents should be consulted:

- In Northern Ireland: TECHNICAL BOOKLET R of the Building Regulations (Northern Ireland) (qv).
- In Scotland: the BUILDING (SCOTLAND) REGULATIONS and the associated TECHNICAL HANDBOOKS (qv).
- In the USA, where the Americans with Disabilities Act applies, refer to section 4 of the ADA AND ABA ACCESSIBILITY GUIDELINES (qv).

See also: *Cobbled surfaces; Gravel surfaces;* and *Pedestrian environment.*

Access routes: recommendations for historic sites

1 Para 13.10 of BS 8300 (qv) states that 'historic buildings should be made accessible to *disabled people*, wherever possible, without compromising conservation and heritage issues'. Where such sites have *cobbled* (qv) or *gravel* (qv) access routes, which present a barrier to *wheelchair-users* (qv) or to visitors with walking difficulties or ones pushing baby buggies, acceptable solutions could include the following:

- Laying a smooth path in a sympathetic material such as York stone along the centre of the existing path.

• Creating a smooth *by-pass* route (also termed a *diversion route*) as an alternative to the cobbled or gravel route.
Examples are shown in the publications below.
2 As of 2008, useful references include the following. For publication details see the bibliography.
• EASY ACCESS TO HISTORIC BUILDINGS.
• EASY ACCESS TO HISTORIC LANDSCAPES.
These informative and beautifully illustrated twin publications from English Heritage give excellent guidance.

See also: *Cobbles;* and *Gravel.*

Access statements See: *Design and access statements.*

Access strategies See: *Accessibility strategies.*

Access to information 1 The Freedom of Information Act 2000 and Environmental Information Regulations 2004 give citizens the right to request official information held by all public bodies including *government* (qv) departments, *local authorities* (qv), and *quangos* (qv). Certain kinds of information cannot be given for security and other reasons, but on the whole requests must be honoured. Inquirers who cannot get satisfaction can make a complaint to the Information Officer's Office (qv) after all other avenues have been exhausted.
2 As of 2008, useful websites include the following. They are given in alphabetical order

• DEFRA	www.defra.gov.uk/
• Info4local	www.info4local.gov.uk/
• Information Officer's Office	www.ico.gov.uk/
• Office of Public Sector Information	www.opsi.gov.uk/acts.htm

See also: *Data Protection Act; Freedom of Information Act;* and *Information Commissioner's Office.*

Accessible 1 A dictionary definition of *accessible* is 'within reach; approachable; easily understood'. The following notes discuss what this may mean in the context of inclusive environments:
2 *Accessible buildings* are ones which everyone – or virtually everyone – regardless of body size, fitness, disability, age, or gender can safely and easily enter, use, and leave. See in this connection section 0 of APPROVED DOCUMENT M (qv) and section 3 of BS 8300 (qv).
3 In the case of *public transport vehicles* (qv) such as taxis, buses/coaches, and rail vehicles, accessibility means that everyone – or virtually everyone – should be able to (a) get in and out of the vehicle safely and easily, and (b) travel within it safely and in comfort. See in this connection part 5 of the Disability Discrimination Act (qv).
4 In the case of *education* (qv); *employment* (qv), *housing* (qv), and other public services, accessibility means that no tenant, applicant, student, employee, etc., should receive less favourable treatment than another because of his or her age, colour, disability, ethnic grouping, gender, marital status, race, or religion. For more detail see: *Discrimination;* and *Equality and Human Rights Commission.*
5 In the area of education, culture, and the arts the question of what is meant by *accessibility* is complicated by a fundamental but as yet unresolved issue that underlies many of the policies now being implemented by the Department for Children, Schools and Families (qv), the Department for Culture, Media and

Sport (qv), and the Qualifications and Curriculum Authority (qv) in the name of *access* and *inclusion* (qv). The issue is that of standards, and may be put thus. If the great pinnacles of intellectual achievement (for instance, the works of Shakespeare) be represented as Mount Everest, does 'access' mean helping and encouraging as many people as possible to undertake the demanding climb to the summit, but accepting the inescapable fact that a few will make it to the top, most will reach some intermediate level, and some will be unable or unwilling to get beyond the foothills? Or does it mean bringing the summit down to a level so low that everyone (or just about everyone) can stroll to the top and then be officially assured that they have been given 'access' to a great and rarefied experience? Ministers in the above departments constantly profess to respect excellence, which implies the first approach, but in fact government policies – particularly in education – have moved virtually all the way towards the second. If examples be required, here is one – and readers will readily find many more in the daily news.

• In May 2004 we read of the exam, devised by the Qualifications and Curriculum Authority (qv) and sat by 630 000 teenage pupils, in which their knowledge and grasp of Shakespeare were tested by the following question: 'In Macbeth, Banquo warns Macbeth about the witches' influence. You give advice in a magazine for young people. You receive this request: 'Please advise me. I have recently moved school and made some new friends. I like spending time with them but my form tutor thinks my work is suffering. What should I do? Sam'. Write your advice to be published in the magazine. The National Association for the Teaching of English, and many individual teachers, expressed outrage at the shallowness and philistinism of the exam, and the fact that a question on the works of Shakespeare could have been readily answered by pupils who had never read a word written by him. A spokesman for the QCA unrepentantly defended the question for the 'accessible' approach it brought to the subject.

At the opposite pole to this approach stands THE USES OF LITERACY, a book in which Richard Hoggart writes movingly of his childhood in the slums of Leeds in the 1920s and 1930s, and his liberation into the best things that civilised life has to offer by a state education system that strove to stretch the minds of children from even the humblest background to the highest standards. Which of these two paths do we wish our current education system to take? This is a matter of fundamental importance, which deserves a thorough national debate.

See also: *Aimhigher*; *Department for Education and Skills; Education;* and *Qualifications and Curriculum Authority.*

Accessible books

See: *Reading aids; Right to Read alliance;* and *Talking books.*

Accessible buses

See: *Buses and coaches.*

Accessible cars

1 Cars that can be easily used by *disabled people* are a form of *mobility aid* (qv), and an excellent example of *assistive technology* (qv). Here the term 'accessibility' has two meanings.

2 First, being able to easily enter and leave the vehicle. Aids in this context include ramps, hoists, and other devices. Rear-entry ramps as shown below are quite simple, and both inbuilt and portable models are widely available.

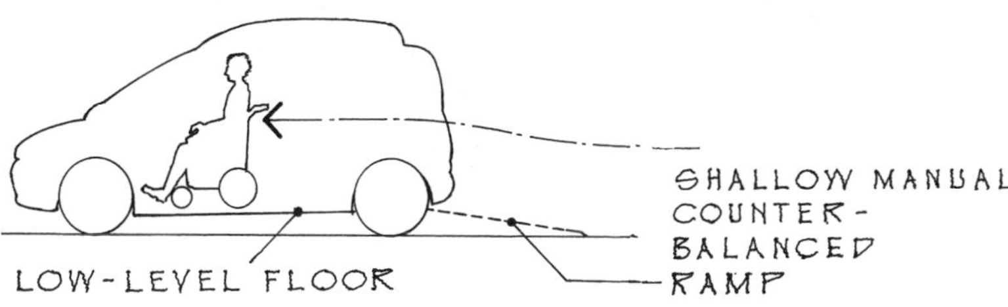

Side-entry ramps are more complex, but also readily available. The figure below shows two types.

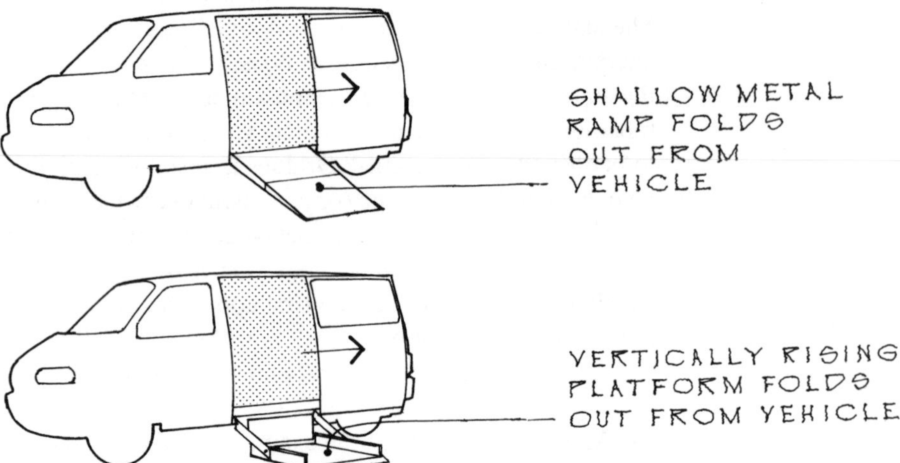

3 Second, being able to drive the vehicle. Hand-operated levers to replace accelerator and brake pedals are a relatively simple example.

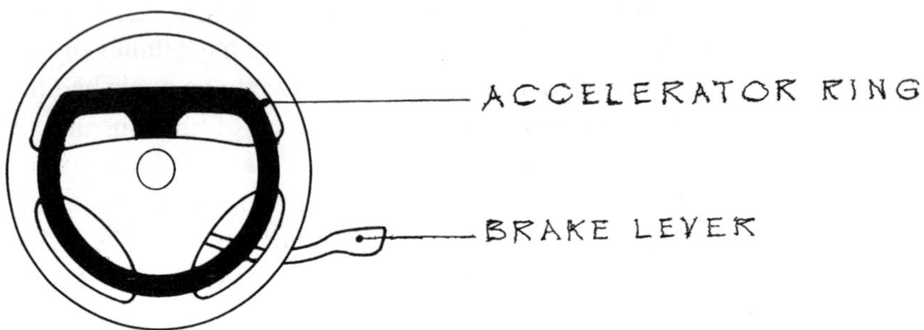

The diagram below shows a more *high-tech* example, consisting of a spinner (to help someone with a weak grip hold the wheel) combined with an electronic keypad that enables a severely disabled person to operate the horn, lights, indicators, windshield wipers, etc., by fingertip control.

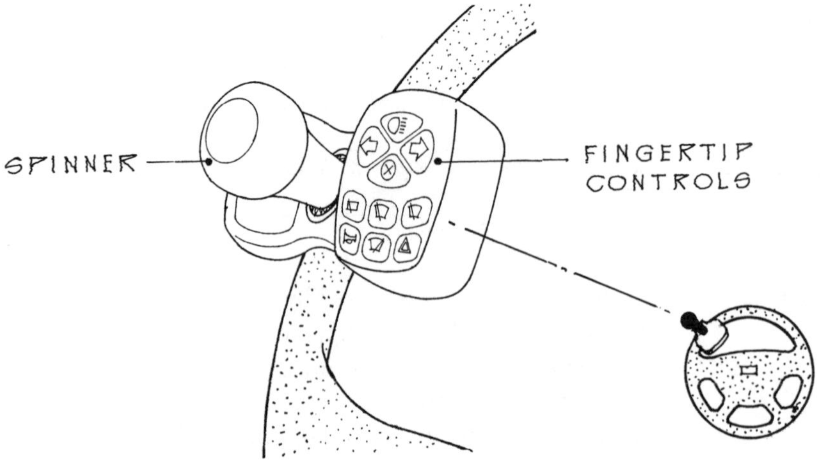

4 Useful guides to the above matters include the following, all from the *charity* (qv) Ricability (qv). For publication details see the bibliography.

- THE INS AND OUTS OF CHOOSING A CAR.
- GETTING A WHEELCHAIR INTO A CAR.
- CAR CONTROLS.

• MOTORING AFTER A STROKE.
• MOTORING WITH ARTHRITIS.
• MOTORING WITH MULTIPLE SCLEROSIS.

The above booklets also contain comprehensive lists of useful organisations, publications, exhibitions, and companies that supply cars and other vehicles that have been adapted for disabled people, together with contact details and websites.

5 As of 2008, useful websites include the following. They are given in alphabetical order.

• Directgov: disabled people www.disability.gov.uk/
• Disabled Living Foundation www.dlf.org.uk/
• Mavis www.dft.gov.uk/
• Mobilise www.mobilise.info/
• Motability www.motability.co.uk/
• Ricability www.ricability.org.uk/

Accessible computers See: *Computers: accessible.*

Accessible design See: *Inclusive design.*

Accessible entrances See: *Entrances to historic buildings.*

Accessible housing See: *Housing: accessibility.*

Accessible lavatories See: *Lavatories.*

Accessible maps 1 Conventional maps, such as Ordnance Survey Maps, are not satisfactory for people with *impaired vision* (qv), who may need larger print and better colour contrast or for people with *impaired mobility* (qv), who will need more detailed information on the steepness of paths, what the surfaces are like, whether there are barriers such as stiles, and the location of accessible lavatories.

2 The Fieldfare Trust has produced a series of self-guiding walk leaflets that give careful attention to these issues. An interesting aspect of these publications is the use of special symbols that give more information about the terrain than conventional maps, using images that can be understood at a glance.

3 As of 2008, useful websites include the following. They are given in alphabetical order.

• Fieldfare Trust www.fieldfare.org.uk/
• Fieldsman Trails www.fieldsmanbookshop.co.uk
• International Cartographic Association's Commission on Maps & Graphics for Blind & Partially Sighted people www.surrey.ac.uk/~pss1su/intact/

See also: *Symbols.*

Accessible parking See: *Accessible roads; Blue badge schemes;* and *Car parking.*

Accessible print 1 CLEAR PRINT GUIDELINES, published by the RNIB (qv), gives advice on all aspects of printwork for easy readability by people with less than perfect vision. The guide covers type size and font; colour contrast; word spacing and alignment; line lengths; and just about everything else. Examples of its recommendations include the following:

• PAPER: Matt paper with no show-through from the reverse side is best. High-gloss and thin or transparent papers create problems.

• CONTRAST: Black type on white or yellow paper is ideal. For all other colour combinations the printing ink should be as dark, and the background as light as possible. Pale ink on a coloured background is difficult to read, and yellow printing ink is just about the worst possible choice.

• TYPE SIZE: General readers are well suited by a height of approximately 2mm for the letter x (for comparison the present text, in 11pt Times New Roman, is around 1.8mm). In Arial a 2mm letter height equates to 11 point type, and in Times New Roman to 12 or 13 point type – but fonts vary and heights should be checked. Partially sighted readers may require 14 or even 16 point type. Here are some examples of type sizes:

This sentence is set in 10pt Times New Roman serif font.
This sentence is set in 11pt Times New Roman serif font.
This sentence is set in 12pt Times New Roman serif font.
This sentence is set in 13pt Times New Roman serif font.
This sentence is set in 14pt Times New Roman serif font.
This sentence is set in 10pt Arial sans serif font.
This sentence is set in 11pt Arial sans serif font.
This sentence is set in 12pt Arial sans serif font.
This sentence is set in 13pt Arial sans serif font.

• TYPE WEIGHT AND TYPEFACE: Light (i.e. thin) typefaces are difficult to read, especially in small sizes. Most of the fonts used in books and newspapers are easy to read; and in all cases well-established typefaces which have stood the test of time are best. Bizarre 'novelty' typefaces are generally difficult to read and should be avoided.

• CAPITAL LETTERS: Capitals throughout are more difficult to read than the conventional mix of upper and lower case, and the latter is recommended.

• SPACING AND JUSTIFICATION: An even word-spacing is clearest, and stretching or compressing lines of type or single words to fit a line length is not recommended. Unjustified right-hand margins (as in the present text) make for easier reading.

• LINE LENGTH: 50 to 65 characters per line are the recommended maximum, and partially sighted readers may prefer even shorter lines. For comparison, the present text has lines of around 75 characters.

2 The above are summaries only. The original document should be consulted. For publication details see the bibliography.

Accessibility

See: *Accessible.*

Accessibility plans

1 As of 2008, the 'planning duties framework', which derives from the Special Educational Needs and Disability Act or SENDA (qv), requires each *Local Education Authority* (qv) and school in England to create and implement:

• An 'accessibility strategy', drawn up by the LEA and covering its entire stock of schools; and

• An 'accessibility plan', drawn up by each individual school and covering that school.

2 These strategies and plans are aimed at improving *access* (qv) to education in schools. In general three issues must be addressed:

• The ability of *disabled* (qv) pupils to participate in the school curriculum;
• The provision to disabled children of information in formats that suit them;
• The physical *accessibility* (qv) of premises.

3 As of 2008, useful websites for latest information include the following.

• Department for Education and Skills www.dfes.gov.uk/
• Teachernet www.teachernet.gov.uk/

Accessible roads

1 GOWRINGS MOBILITY UK ROAD ATLAS gives UK-wide guidance on accessible car parks; Blue Badge (qv) parking schemes; accessible *service stations* (qv); accessible beaches; and tourist information.
2 For publication details see the bibliography.

Accessible stadia

1 ACCESSIBLE STADIA: A GOOD PRACTICE GUIDE TO THE DESIGN OF FACILITIES TO MEET THE NEEDS OF DISABLED SPECTATORS AND OTHER USERS is, as of 2008, the most authoritative general reference for the *inclusive* design of sports and recreation facilities. The document is advisory and does not have the force of law, but conformity with its recommendations will make it easier to demonstrate that the requirements of the Disability Discrimination Act (qv) have been met.
2 For publication details see the bibliography.

See also: *Recreation and sports facilities.*

Accessible thresholds

See: *Thresholds: level.*

Accessible toilets

See: *Lavatories.*

Accessible websites

1 Under the Disability Discrimination Act (qv) all organisations offering goods and services on the web – see for instance *Service providers* (qv) – have a duty to make their websites *accessible* (qv) to all readers, including, within reason, those who have mental or physical *impairments* (qv).
2 Tools that will help web providers and designers achieve the required standards include the following:
• The Web Accessibility Initiative (WAI) is internationally respected and issues guidelines, tools, and online validators that can be used to check websites for compliance with standards. Visit www.rnib.org.uk/
• Bobby is another highly respected and widely-used online testing tool to help developers assess web sites for accessibility. Visit www.rnib.org.uk/or www.cast.org/bobby/
• The Society for Information Technology Management (Socitm) conducts an annual 'Better Connected' survey of the websites of all 468 local authorities and assesses the performance of each on a range of criteria. Visit www.rnib.org.uk/or www.socitm.gov.uk/socitm/
3 Useful references include the following. For publication details see the bibliography.
• PAS 78 – A GUIDE TO GOOD PRACTICE IN COMMISSIONING ACCESSIBLE WEBSITES is a Publicly Available Specification – i.e. a near-equivalent to a British Standard (qv), but introduced more quickly and updated every two years.
4 Useful websites include the following. They are given in alphabetical order.

• AbilityNet	www.abilitynet.co.uk/
• Browsealoud	www.browsealoud.com/
• RNIB	www.rnib.org.uk/xpedio/groups/public/documents/code/public_rnib008789.hcsp
• SiteMorse	www.sitemorse.co.uk/
• WebXACT	www.cast.org/bobby/
• World Wide Web consortium	www.w3c.org/

See also: *Disability Discrimination Act; Communication aids;* and *Impaired vision.*

Accessories for daily living

See: *Activities of daily living.*

Accidents

1 The Royal Society for the Prevention of Accidents (qv) reported a total of around 13 000 deaths by accident in the UK in 2000. Of these some 4 000 occurred in homes, 3 500 on roads, and 200 at work.
2 These figures represent a rate of about 22 accidental deaths a year per 100 000 people. Among people under 24 the incidence fell by nearly half over ten years, from around 2 700 in 1991 to 1 500 in 2000; among people aged between 25 and 64 it remained roughly constant, at about 4 500; and among people over 65 it rose from about 6 150 to 7 000 – perhaps in part because the total number of older people increased over that period.
3 The subject of accidents is closely bound up with that of *risk management* (qv).
4 As of 2008, useful references include the following. For publication details see the bibliography.
• PLANNING YOUR HOME FOR SAFETY AND CONVENIENCE: PRACTICAL ADVICE FOR DISABLED AND OLDER PEOPLE.
5 Useful websites include the following. They are given in alphabetical order.

• Department of Health	www.dh.gov.uk/
• Health and Safety Executive	www.hse.gov.uk/
• Office for National Statistics	www.statistics.gov.uk/
• Parliamentary Advisory Council	www.pacts.org.uk/
• Royal Society for the Prevention of Accidents	www.rospa.co.uk/

See also: *Compensation culture; Product packaging; Risk*; and *Transport safety*.

Accidents: cycling
Accidents: railways
Accidents: roads

See: *Transport safety.*

Acoustics

1 Acoustics are one aspect of the *sensory environment* (qv), the others being *colour* (qv), *lighting* (qv), and *touch* (qv). Proper attention to all of these is essential if people – and in particular those with mental or physical *impairments* (qv) of one kind or another – are to navigate their way safely and comfortably through buildings and open spaces. In acoustic design four matters need attention:
2 Reduction of incoming noise into buildings and rooms. Noise reduction methods include:
• Noise reduction at source, eg quieter aeroplanes, trains, trucks, and road-breaking machinery, etc.
• Designing the building envelope (roofs, walls, and windows) to resist noise entry from the outside. In technical terminology, these elements should have high sound insulation values.
• Designing internal building divisions (floors, partitions, and doors) to resist the passage of sound from one room to another – again a matter of high sound insulation values.
Relevant standards on the above matters include:
• APPROVED DOCUMENT E (qv) of the Building Regulations.
• BS 8300 (qv), section 9.
• INCLUSIVE MOBILITY (qv), para 10.1.8.
3 Reduction of the noise that is generated within rooms and spaces. This is essentially a matter of providing enough sound-absorbent surfaces, within

noise-generating spaces such as restaurants and busy waiting rooms, to absorb the sounds of people speaking, cutlery clattering on tables, etc., instead of reflecting the sounds back into the room. Carpeted floors, acoustic tiled ceilings, and thick tapestries on walls are useful sound-absorbing devices. Bare clay-tiled floors, concrete ceilings, and concrete or brickwork walls (all of which are currently fashionable) should be used with extreme care.
4 In public places, installation of *hearing enhancement systems* (qv) to assist people with *impaired hearing* (qv).
5 All of these are specialised subjects, and the above notes are intended only to highlight some key issues, not to give technical guidance.
6 As of 2008 useful publications include the following. They are given in alphabetical sequence. For publication details see the bibliography.

- ACOUSTIC GUIDES, a set of free leaflets from the manufacturer Armstrong, gives clear simple guidance on definitions, and the contribution that correct ceiling design can make to good acoustics.
- BB93 THE ACOUSTIC DESIGN OF SCHOOLS is the official guidance document on this specialist subject.
- THE BUILDING REGULATIONS EXPLAINED AND ILLUSTRATED explains how the provisions of APPROVED DOCUMENT E of the Building Regulations (qv), which deals with sound insulation, may be satisfied.
- THE METRIC HANDBOOK gives a simple summary of basic principles. Though some of the specific data may have been superseded by more recent standards and regulations – a matter which should be checked – the general explanations may aid background understanding.

7 Useful websites include the following. They are given in alphabetical order.

- Armstrong — www.armstrong-ceilings.co.uk/
- Hodgson & Hodgson — www.acoustic.co.uk/
- Lorient — www.norseal.co.uk/
- Oscar Acoustics — www.oscar-acoustics.co.uk/
- Sound Reduction Systems — www.soundreduction.co.uk/
- Soundsorba — www.soundsorba.com/

See also: *Impaired hearing;* and *Sensory environments*.

Active Community Unit (ACU)

1 A unit in the Office of the Third Sector (qv) in the Home Office, which aims to promote the development of the *voluntary and community sector* (qv) and encourage people to become more actively involved in their local communities, particularly in *areas of deprivation* (qv).
2 For more detail visit:

- Office of the Third Sector — www.cabinetoffice.gov.uk/thirdsector/
- Info4local — www.info4local.gov.uk/
- Neighbourhood Renewal Unit — www.neighbourhood.gov.uk/

See also: *Neighbourhood renewal*.

Activities of daily living (ADL)

1 This is a term used in the caring professions for self-care activities such as washing and bathing. It is commonly abbreviated to ADLs. For *aids and equipment* (qv) that might be helpful in this context (also referred to as 'aids for daily living') consult the Disabled Living Foundation (qv).
2 Useful websites include the following. They are given in alphabetical order.

- Coopers Healthcare Services — www.sunrisemedical.co.uk/
- Disabled Living Foundation — www.dlf.org.uk/public/factsheets.html.
- Nottingham Rehab — www.nrs-uk.co.uk/

See also: *Products for easier living.*

Acts of Parliament

See: *Statutes*

Acute Trusts

See: *NHS structure.*

ADA and ABA Assessibility Guidelines

1 The ADA AND ABA ACCESSIBILITY GUIDELINES FOR BUILDINGS AND FACILITIES give guidance on how to comply with the *accessibility* (qv) regulations that are issued by USA federal agencies under the Americans with Disabilities Act. The latter is, very broadly speaking, the USA's counterpart to the UK's Disability Discrimination Act (qv).
2 As of 2007, useful websites include the following.

- The Act www.usdoj.gov/crt/ada/
- The Guidelines www.access-board.gov/ada-aba/final.htm

Adapt Trust

1 During its lifetime this Trust, which closed in July 2008, advised the owners and managers of *arts* and *heritage* (qv) venues on how to create effective *access* (qv) for everyone – see: *Inclusive design* (qv).
2 One of its most useful activities was giving annual awards for excellence. See: *Access awards.*

Adaptation: ocular

See: *Lighting: transitional.*

ADHD

See: *Attention-deficit/hyperactivity disorder.*

Adjustable furniture

1 To most people, going to bed or getting up from a chair may seem the easiest thing in the world, but there are many *ageing* (qv), *disabled* (qv), or otherwise frail people for whom these actions are painful or impossible. They can be greatly helped by height-adjustable beds, chairs, bathroom fittings, desks and tables, and kitchen fittings.
2 Many such products are now readily available. As of 2008, useful websites include the following. They are given in alphabetical order.

- Astor Bannerman www.astorbannerman.co.uk/
- Bakare www.bakare.co.uk/
- DCS Joncare www.dcsjoncare.freeserve.co.uk/
- Disabled Living Foundation www.dlf.org.uk/public/factsheets.html
- OTDirect www.otdirect.co.uk/link-equ.html#Platform
- Phlexicare www.nichollsandclarke.com/

See also: *Assistive technology; Hoists; Powered lifters;* and *Products for easier living.*

Adjustments under the Disability Discrimination Act

1 Under the Disability Discrimination Act (qv), which effectively outlaws *discrimination* (qv) against *disabled persons* (qv) in public life, *employers* (qv), *service providers* (qv) and others can avoid or remove discrimination against disabled people, or remove existing discrimination, by making 'reasonable adjustments' to the way they run their business or to their buildings (adjustments to vehicles are discussed separately, further down). As shown by the examples in the next paragraph, such adjustments need not necessarily be complex or expensive.
2 The Act recognises three categories of adjustment:

- Changing the *management practices* of an organisation. For instance, a shop or restaurant with a 'no dogs' policy can change it to admit *guide dogs* (qv) so that *blind* (qv) people can enter and use the facility.
- Providing *auxiliary aids and services*. For instance, installing an *induction loop* (qv), or providing a *sign language* (qv) interpreter at a public meeting, to enable *deaf* (qv) people to participate in the meeting.

• Making helpful physical adjustments to their premises. This could be as simple as moving a workstation to a better position, or as complicated as installing a lift.

These matters are explained in more detail in the entries below, which follow the sequence shown in the diagram below.

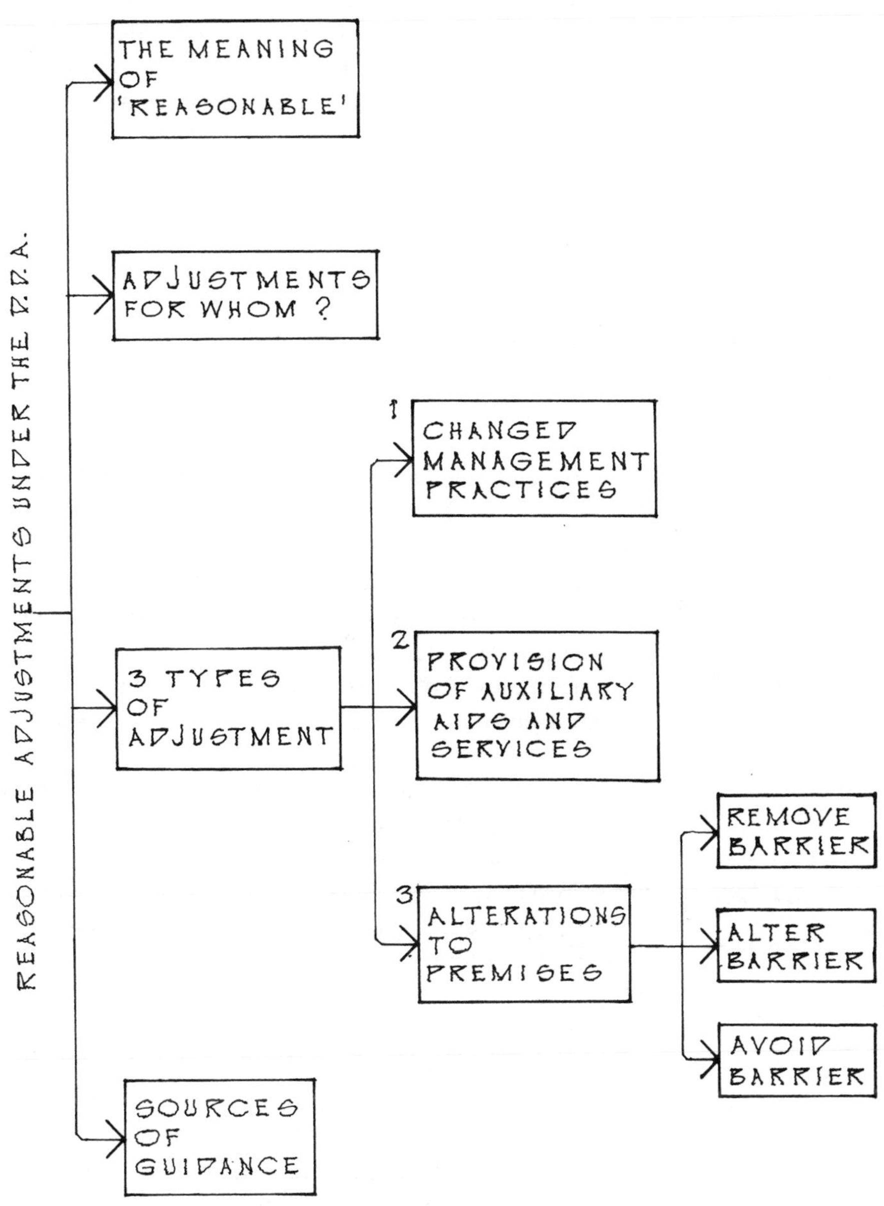

3 Readers are reminded that the present work only gives simplified summaries of original documents and does not purport to provide full, authoritative and up-to-date statements of the law. For latest information refer to DISABILITY DISCRIMINATION: LAW AND PRACTICE (see the bibliography) and the following websites:

• Directgov: disabled people www.disability.gov.uk/
• Equality & Human Rights Commission www.equalityhumanrights.com/

4 Readers are reminded that the present work only gives simplified summaries of original documents and does not purport to provide full, authoritative and up-to-date statements of the law. Useful websites for checking the latest situation include:

• Directgov: disabled people www.disability.gov.uk/
• Equality & Human Rights Commission www.equalityhumanrights.com/

See also: *Disability Discrimination Act.*

Adjustments under the Disability Discrimination Act: adjustments for whom?

1 In the case of employers, *adjustments* to avoid or remove *discrimination* need only address the needs of known employees or job applicants. These needs will best be identified in a formal *workplace assessment* (qv), probably carried out by an *occupational therapist*; and any physical adjustments that are required must be done as soon as practicable.

2 In the case of service providers the situation is different, and adjustments to avoid or remove discrimination must anticipate the needs of *disabled people* at large – see: *Service providers*. These needs will best be assessed by an *access audit* (qv), and any physical adjustments that are required will probably be implemented as a planned sequence of step-by-step improvements carried out in an affordable manner over a reasonable period of time. It would usually be most cost effective if these improvements could be planned to form part of the landlord's normal maintenance and repair programme, and not done as a special project – though no doubt there are exceptions.

Adjustments under the Disability Discrimination Act: authoritative practical guides

1 As of 2008, useful references include the following. For publication details see the bibliography.

2 For a general introduction refer to Section 3.4 and relevant subsequent sections of DISABILITY DISCRIMINATION: LAW AND PRACTICE by Brian Doyle, a commercial publication that states the law as at February 2008.

3 The supreme official source of information is the Equality and Human Rights Commission (qv) at www.cehr.org.uk/, and official guidance documents include the following, given here in alphabetical order:

- CODE OF PRACTICE: EMPLOYMENT AND OCCUPATION explains how *disabled people* (qv) are protected from *discrimination* (qv) if they are in employment or seeking employment. It also gives guidance on the law, which will help lawyers when representing disabled clients and assist courts and *tribunals* (qv).
- CODE OF PRACTICE FOR SCHOOLS – DDA 1995: PART 4 covers the duty of education providers not to *discriminate* (qv) against *disabled* (qv) pupils and prospective pupils in schools.
- CODE OF PRACTICE (REVISED) FOR PROVIDERS OF POST-16 EDUCATION AND RELATED SERVICES covers the duty of education providers not to *discriminate* (qv) against post-16 *disabled* (qv) pupils. It replaces the earlier CODE OF PRACTICE FOR PROVIDERS OF POST-16 EDUCATION AND RELATED SERVICES, which was published in 2002 and is now obsolete.
- CODE OF PRACTICE – RIGHTS OF ACCESS: SERVICES TO THE PUBLIC, PUBLIC AUTHORITY FUNCTIONS, PRIVATE CLUBS AND PREMISES gives guidance on the duty of *service providers* (qv) not to *discriminate* (qv) against *disabled* (qv) customers.
- CODE OF PRACTICE – TRADE ORGANISATIONS AND QUALIFICATIONS BODIES explains how *disabled people* (qv) are protected from *discrimination* (qv) by trade organisations and qualifications bodies.

4 Other useful publications include the following. They are given here in alphabetical order:

- ACCESS AND THE DDA: A SURVEYOR'S HANDBOOK by Mark Ratcliff is meant primarily for surveyors, but will also be useful to other professionals.
- THE ACCESS AUDIT HANDBOOK by Alison Grant helps to establish the link between the diagnosis of access problems in an existing building, and possible solutions. See: *Access audits.*
- THE ACCESS AUDIT PRICING GUIDE gives guidance on the costs of physical adjustments.
- THE ACCESS MANUAL: AUDITING AND MANAGING INCLUSIVE BUILT ENVIRONMENTS by Ann Sawyer and Keith Bright is a short practical guide written by two *access auditors* (qv).
- ACCESS TO BUILDINGS by Stephen Garvin is a BRE Digest that outlines the requirements of the Building Regulations (qv) and the Disability

Discrimination Act for both domestic and non-domestic buildings.
• DESIGNING FOR ACCESSIBILITY, published by the Centre for Accessible Environments (qv), gives simple guidance for the design of public premises to conform with Disability Discrimination Act.
• CLAIMS ABOUT PHYSICAL BARRIERS TO ACCESS: A GUIDE TO THE DISABILITY DISCRIMINATION ACT is an official publication to help people who wish to make a claim under the Disability Discrimination Act.
• INCLUSIVE ACCESSIBLE DESIGN by Adrian Cave provides a route-map through the principal regulatory documents.
• It is nothing less than an outrage that an Act that was intended to benefit disabled people has become instead, as the several thousand pages of dense legal/technical text in the above guidance documents demonstrate, a gold-mine for fee-earning specialist lawyers and consultants. See in this connection also: *Better regulation bodies; Consultants' charter; Over-legislation;* and *Over-regulation.*
5 As of 2007, useful websites include the following. They are given in alphabetical order.

• Directgov	www.direct.gov.uk/en/DisabledPeople/index.htm
• Directgov: disabled people	www.disability.gov.uk/
• Equality & Human Rights Commission	www.equalityhumanrights.com/
• Office of Public Sector Information	www.opsi.gov.uk/acts.htm

See also: *Authoritative practical guides.*

Adjustments under the Disability Discrimination Act: the meaning of 'reasonable'

1 The meaning of *reasonable* in the phrase 'reasonable adjustments' will always be partly subjective. The Act does however make clear that what is regarded as 'reasonable' should be judged against three questions:
2 Would the proposed adjustment be <u>effective</u> in preventing *discrimination*? For instance, in the particular case of employers:
• An employer can probably not reasonably be expected to buy costly equipment for a *disabled* employee if the equipment would give the employee only a small added benefit.
3 Would it be <u>practicable</u> to take that step? For instance:
• An employer can probably not reasonably be expected to make the building entrance more *accessible* for a would-be employee if the work could not realistically be done by the time the employee is needed to start work.
4 Can the organisation <u>afford</u> to take that step? For instance:
• An employer can probably not reasonably be expected to spend substantially more on an adjustment to suit an employee who has become disabled, than he would have to spend on recruiting and training a new person for that job.
• An employer can probably not reasonably be expected to make an adjustment to suit a disabled job applicant if it will disrupt the work of other employees so severely as to prevent them doing their job.
• The sum of money any organisation can reasonably be expected to spend on adjustments is related to its overall resources, and would be less for a small low-profit organisation than for a large rich one.
5 These examples are suggestive only. For authoritative guidance refer to CODE OF PRACTICE: EMPLOYMENT AND OCCUPATION and the other *codes of practice* (qv) listed under *Sources of guidance* below.

Adjustments under the Disability Discrimination Act: three types

1 Sections 6, 15, and 21 of the Disability Discrimination Act mention three types of *adjustment* that can be made to avoid or overcome *discrimination* against *disabled people*. They are discussed below.
2 Throughout this discussion, readers are reminded that the present work only gives simplified summaries of original documents and does not purport to provide full, authoritative, and up-to-date statements of the law.

Adjustments type 1: changes to management practices

1 In the case of an employer, the following might be examples of *adjustments* carried out to avoid *discrimination* against a *disabled* employee:

- An employee suffering from *dyslexia* is excused from having to write the occasional letter, if this is not an essential feature of the job, and the task is given to someone else.
- The work station of a *wheelchair-user* is moved from the third floor of the building to a more *accessible* place on the ground floor.
- An employee who develops a condition of *impaired mobility* is moved from an outdoor job to a desk job.

These examples are suggestive only. For authoritative guidance refer to CODE OF PRACTICE: EMPLOYMENT AND OCCUPATION – see: *Sources of guidance* below.

2 In the case of a service provider, the following might be examples of adjustments to avoid discrimination against disabled customers:

- A shop that has a 'no-dogs' policy relaxes this rule for *blind* people accompanied by *guide dogs*.
- A local insurance broker's reception desk is too high for a wheelchair customer to use in filling in a form, and it is impracticable to alter the desk; therefore the customer is given a clipboard and is courteously assisted by the receptionist.

These examples are suggestive only. For authoritative guidance refer to CODE OF PRACTICE – RIGHTS OF ACCESS: SERVICES TO THE PUBLIC, PUBLIC AUTHORITY FUNCTIONS, PRIVATE CLUBS AND PREMISES – see: *Sources of guidance* below.

Adjustments type 2: provision of auxiliary aids and services

1 In the case of an employer, the following might be examples of *adjustments* to avoid *discrimination* against a *disabled* employee:

- A disabled applicant for a job that involves typing would be the best person for the job, except that her typing speed is slowed by *arthritis*. The employer investigates, and finds that purchase of a special keyboard would solve the problem. As her typing speed could not, in these circumstances, in itself be used as a *justification* (qv) for refusing to employ her, the employer hires the applicant and buys the equipment.
- Similarly an employee (whether existing or about to be hired) suffering from *impaired hearing* might be provided with a *textphone,* and one suffering from *impaired vision* with a *big-button telephone*, in order to 'level the playing field' between themselves and their able-bodied colleagues and help them perform to their full potential.

These examples are suggestive only. For authoritative guidance refer to CODE OF PRACTICE: EMPLOYMENT AND OCCUPATION – see: *Sources of guidance* below.

2 In the case of a service provider, the following might be examples of adjustments to avoid *discrimination* against *disabled* customers:

- A conference organiser provides *induction loops* (= an 'auxiliary aid') and/or *British Sign Language* interpreters (= an 'auxiliary service') to convey what is being said to deaf people in the audience.
- A railway station helps travellers with *impaired hearing* by installing *induction loops* at all ticket counters and a *textphone* alongside the public telephone.
- A public building loans *tactile plans* of each floor of the building to blind or near-blind visitors, free of charge.
- A large shopping centre provides a *shopmobility* facility for customers with *impaired mobility;* large-print and *braille* information leaflets for customers

with *impaired vision*; and *induction loops* and hosts trained in *British Sign Language* for customers with *impaired hearing.*
These examples are suggestive only. For authoritative guidance refer to CODE OF PRACTICE – RIGHTS OF ACCESS: SERVICES TO THE PUBLIC, PUBLIC AUTHORITY FUNCTIONS, PRIVATE CLUBS AND PREMISES – see: *Sources of guidance* below.

Adjustments type 3: physical alterations to premises

1 By 'physical features' the Disability Discrimination Act means kerbs, outdoor pavements, outdoor benches, parking areas, gates, approach paths, steps, building entrances and exits (including *emergency escape* routes), lifts, escalators, stairways, lavatories, washing facilities, lighting and ventilation, public facilities such as telephones and service counters – in short, just about everything that forms part of the physical environment.
2 Under the Act a physical feature that creates problems for *disabled people* can be (a) removed, (b) altered, or (c) avoided in order to prevent *discrimination*. Thus, taking the fairly common case of an obstructive step in an entrance doorway, this obstruction can be:
• Removed by raising the pavement outside the doorway to threshold level;
• Altered by converting the step into a ramp; or
• Avoided by creating a separate, step-free entrance nearby.
For examples of how this principle might be applied in practice see: *Entrances to historic buildings.*
3 Looking at some other possibilities, showing that 'physical adjustments' need not necessarily involve bricks, mortar, and a lot of money:
• In the case of an employer, a *wheelchair-user* applies for a job which he could do perfectly well, except that the workstation is not *accessible* to *wheelchairs*. The employer finds that he can solve the problem by an easily affordable rearrangement of furniture.
• In the case of a service provider, a shop owner makes life easier for disabled customers (and incidentally everyone else) by smoothing the pavings and cutting back dangerous overhanging branches over the approach to the shop entrance. He also removes a loose entrance mat that constitutes a tripping *hazard,* and marks the edges all stairs in a contrasting colour to make them more visible – see: *Colour and tonal contrast.*
Other simple but effective measures of this kind include:
• Fitting *support rails* and improving the *lighting* at those hazardous spots in the building which cannot be removed or radically altered.
• Providing clearer *wayfinding* and other *signs* on approach routes and in buildings.
4 These examples are suggestive only. For authoritative guidance refer to CODE OF PRACTICE: EMPLOYMENT AND OCCUPATION; TO CODE OF PRACTICE – RIGHTS OF ACCESS: SERVICES TO THE PUBLIC, PUBLIC AUTHORITY FUNCTIONS, PRIVATE CLUBS AND PREMISES; and to the other *codes of practice* (qv) listed under *Sources of guidance* below.

Adjustments: vehicles

1 The above entries focus mainly on adjustments to buildings. Authoritative guidance on adjustments to vehicles is given in sections 5 to 7 of PROVISION AND USE OF TRANSPORT VEHICLES, STATUTORY CODE OF PRACTICE, SUPPLEMENT TO PART 3 CODE OF PRACTICE.
2 See in this context also: *Transport: complying with the Disability Discrimination Act.*

ADLs

See: *Activities of daily living.*

Administrative and commercial buildings: design: complying with the Disability Discrimination Act

1 Buildings used by the public must in general comply with the Disability Discrimination Act (qv). In the case of offices, shops, courts of law, and police stations, *authoritative practical guides* (qv) include DESIGNING FOR ACCESSIBILITY (qv) plus the following:
2 APPROVED DOCUMENT M (qv), the provisions of which must be heeded in order to comply with the Building Regulations (qv) for England and Wales. For places other than England and Wales the following regulatory documents should be consulted:
• In Northern Ireland: TECHNICAL BOOKLET R of the Building Regulations (Northern Ireland) (qv).
• In Scotland: the TECHNICAL HANDBOOKS of the Building (Scotland) Regulations (qv).
3 BS 8300, the provisions of which apply to all buildings in the UK. Para 13.3 gives brief design recommendations for all the building types listed above. BS 8300 does not have the force of law, but conformity with its recommendations will make it easier to demonstrate that the requirements of the Disability Discrimination Act have been met.
4 In the USA, where the Americans with Disabilities Act applies, refer to the ADA AND ABA ACCESSIBILITY GUIDELINES (qv).
5 For further publications, case studies, and commentary on design quality visit www.cabe.org.uk/default.aspx?contentitemid=40

Administrative and commercial buildings: management: complying with the Disability Discrimination Act

See: *Service providers*.

Administrative and commercial buildings: basic planning data

1 Distilled planning data for most of the major building types is given in the METRIC HANDBOOK: PLANNING AND DESIGN DATA. For publication details see the bibliography.
2 For general design guidance, commentary, and case examples visit:
• CABE www.cabe.org.uk/default.aspx?contentitemid=40

Adult learning

Visit: www.direct.gov.uk/EducationAndLearning/fs/en

Adult Learning Inspectorate

See: *Office for Standards in Education*.

Advocacy research

1 Research that is carried out not to discover new truths, but to produce data that will support a pre-existing conclusion. The practice is widespread and is often accompanied by outrageous misuse of statistical methods. The major culprits include commercial interest groups, political parties or groups, and *pressure groups* (qv). For an example of the influence of the latter see: *Tactile pavings*.
2 THE TIGER THAT ISN'T: SEEING THROUGH A WORLD OF NUMBERS, by Michael Blastland and Andrew Dilnot, explains to lay people how they can make sense of numbers and avoid having the wool pulled over their eyes. For publication details see the bibliography.

See also: *Medicalisation; Health scares*; and *Pressure groups*.

Affordable housing

See: *Housing: affordable*.

Affordable Rural Housing Commission

1 This body was set up in 2005 to look into the shortage of *affordable housing* (qv) in rural England and Wales.
2 For more information, and the Report which was subsequently published, visit www.defra.gov.uk/rural/housing/commission/default.htm

See also: *Housing: affordable.*

Age discrimination

1 *Discrimination* (qv) on the basis of age is outlawed in the UK by anti-discrimination legislation – for details see: *Discrimination.*
2 For more information refer to AGE DISCRIMINATION: THE NEW LAW (see the bibliography) and visit www.cehr.org.uk/
See also: *Employment Equality (Age) Regulations.*

Age Positive

1 Age Positive is a team working within the Department for Work and Pensions (qv) to promote the benefits of employing a mixed-age workforce that includes older and younger people, not just those in the middle.
2 For more information visit www.agepositive.gov.uk/

Ageing

1 For the changing age structure of the UK population see: *Population: average age.*
2 On current trends the proportion of the population aged 65 or over is expected to rise from under just 16% in 1993 to around 24% by 2050. The buildings and environments we create now should therefore be designed with these future users prominently in mind. Such a policy would in fact be immediately beneficial to everyone, because environments and products that have been designed to suit older persons (from less steep stairs to easy-twist jar tops) will also be safer and more convenient for most other people. As Professor Bernard Isaacs, founding Director of the Centre for Applied Gerontology at Birmingham University has put it: 'Design for the young and you exclude the old; design for the old and you include the young.'
3 As of 2008, useful references include the following. They are given in alphabetical order. For publication details see the bibliography.

- LIVING WELL IN LATER LIFE, a study undertaken jointly by the Healthcare Commission (qv), the Audit Commission (qv), and the Commission for Social Care Inspection (qv), and published by the Healthcare Commission, argues that old people are being failed by the NHS and other public services.
- SECURING GOOD SOCIAL CARE FOR OLDER PEOPLE: TAKING A LONG TERM VIEW by Sir Derek Wanless was named 'think tank publication of the year' in 2006 by Prospect magazine.
- SOCIAL TRENDS (qv), an annual publication of the Office for National Statistics (qv), gives latest statistical data plus background explanation and comment. SOCIAL TRENDS 34 additionally contained a special extended essay titled 'Ageing and gender: diversity and change'.

4 Websites relating to the needs of older people include:

- Age Concern — www.ageconcern.org.uk/
- Directgov — www.direct.gov.uk/
- Disabled Living Foundation — www.dlf.org.uk/
- Equality & Human Rights Commission — www.equalityhumanrights.com/
- NHS Direct — www.nhsdirect.nhs.uk/
- Opportunity Age — www.dwp.gov.uk/opportunity_age/
- RADAR — www.radar.org.uk/
- Ricability — www.ricability.org.uk/

See also: *Impaired dexterity; Opportunity Age; Products for easier living*; and Social care.

Agencies and public bodies

See: *Quangos*.

Aids

1 This is an acronym for 'Acquired Immune Deficiency Syndrome', which is a condition brought about by the Human Immunodeficiency Virus (HIV) which causes the body's immune system to become deficient, leaving the victim very vulnerable to infection, leading possibly (though not necessarily) to death.
2 Persons suffering from Aids qualify as *disabled people* (qv) under the Disability Discrimination Act (qv).
3 As of 2008, useful websites include the following. They are given in alphabetical order.

- Directgov www.direct.gov.uk/
- Directgov: disabled people www.disability.gov.uk/
- Equality & Human Rights Commission www.equalityhumanrights.com/
- NHS Direct www.nhsdirect.nhs.uk/

Aids and equipment

1 We all use aids and equipment in our daily lives, either to make normal activities easier or to help us do things that would otherwise be impossible; and this applies with particular force to *ageing* (qv) or *disabled* (qv) people – see: *Independence.*
2 The appropriate aids for people with *special needs* (qv) may be more easily identified if two questions be asked:

- What is the condition (eg having a *disability* such as *impaired dexterity)* that needs to be catered for ?
- What is the activity (eg cooking a meal) that is to be assisted ?

3 For devices that are appropriate to particular conditions see the following entries:

- Impaired dexterity.
- Impaired hearing.
- Impaired mobility.
- Impaired understanding.
- Impaired vision.
- Limb deficiency.

For devices that are appropriate to particular types of activity see the following entries:

- Activities of daily living.
- Bathing aids.
- Communication aids.
- Driving aids.
- Eating and drinking aids.
- Exercise and rehabilitation aids.
- Food preparation aids.
- Gardening aids.
- Gripping, reaching and turning aids.
- Hearing aids.
- Home adaptations.
- Learning and development aids.
- Mobility aids.
- Patient lifting and moving aids.
- Reading aids.
- Recreation and sports aids.
- Speech aids.

- Telephoning aids.
- Toileting aids.
- Walking and standing aids.
- Way-finding aids.
- Writing aids.

4 As of 2008, useful websites include:

• Directgov	www.direct.gov.uk/
• Directgov: disabled people	www.disability.gov.uk/
• Disabled Living Foundation	www.dlf.org.uk/public/factsheets.html
• RADAR	www.radar.org.uk/
• Ricability	www.ricability.org.uk/

See also: *Assistive technology;* and *Products for easier living.*

Aids and equipment: entitlement

1 *Disabled people* (qv) and their *carers* (qv) are legally entitled to an assessment of their need for assistive aids and equipment such as *hoists* (qv), *stairlifts* (qv), or *through-floor lifts* (qv).
2 Under various Acts (qv) and regulations (qv), as may be found by visiting www.direct.gov.uk/, disabled children and adults have a right to an assessment of their needs by the *local authority* (qv); and if a need is demonstrated then the authority has a statutory duty to arrange for the necessary *aids and equipment* (qv) and/or *home improvements* (qv).
3 In October 2004 it was announced that a consortium of three charities hoped to take over some of the responsibility for providing community equipment to disabled people – the RNIB (qv), the RNID (qv), and the British Red Cross (qv).
4 As of 2008, useful websites include the following. They are given in alphabetical order.

• Directgov	www.direct.gov.uk/
• Directgov: disabled people	www.disability.gov.uk/
• Equality & Human Rights Commission	www.equalityhumanrights.com/
• Office of Public Sector Information	www.opsi.gov.uk/acts.htm

See also: *Disabled Facilities Grants; Home adaptations; Occupational therapists*; and *Social services departments.*

Aids for daily living

See: *Activities of daily living.*

Aimhigher

1 This is a *quango* (qv) with responsibility for encouraging more young people into higher education, particularly those from disadvantaged social and economic backgrounds, those from some ethnic minorities, and those who are *disabled* (qv). The programme was launched by the then DfES (qv) in 2001 under the name Excellence Challenge. In 2004 it joined with Partnerships for Progression, run by the Higher Education Funding Council for England (qv) and the Learning and Skills Council (qv), to become Aimhigher.
2 The programme has not been particularly successful, and some of its ideas are bizarre. In October 2006 its suggestion that more than 40 points should be added to the A-level scores of pupils thought to have been disadvantaged (for instance by domestic troubles or having attended a bad school) attracted sharp criticism. Such a scheme could mean that a C-grade earned by a pupil considered to be from a deprived background would automatically be treated as equal to an A-grade earned by one who was not considered thus. The ill-effects of such a policy would include:

- Pretending that something is what it is not, which is seldom the best approach to difficult problems.

• Unfairness to the millions of students who gained their A-grades by hard work and competence.
• Perhaps worst of all, reducing the pressure upon bad schools to improve. If their badly-educated pupils can sail past the competition into university places simply by virtue of having attended a bad school, why should such schools bother to raise their standards ?
3 For more information on Aimhigher visit: www.aimhigher.ac.uk/

See also: *Accessible; Education: tertiary;* and *Qualifications and Curriculum Authority.*

Air travel: complying with the Disability Discrimination Act

1 The Disability Discrimination Act (qv) effectively outlaws *discrimination* against *disabled persons* in public life. The implications for air travel are broadly as follows:
2 Airport buildings are covered by the Act. For guidance see: *Airport design*; and *Transport related buildings*.
3 Airport services and booking services are also covered by the Act. For guidance see: *Service providers*.
4 Air travel itself is exempt from the provisions of the Act. The Department for Transport (qv) and the Disabled Persons' Transport Advisory Committee, better known as DPTAC (qv), have however published several guidance documents aimed at developing and improving services. They include:
• ACCESS TO AIR TRAVEL: GUIDANCE FOR DISABLED AND LESS MOBILE PASSENGERS. This is a voluntary Code of Practice (qv) that applies only to UK companies.
• ACCESS TO AIR TRAVEL FOR DISABLED PEOPLE: CODE OF PRACTICE. This is a companion to the document above and is aimed at people involved in the air travel industry including travel agents, tour operators, UK airlines, and UK airports. It applies only to UK companies, and includes relevant information for airport operators who are legally bound by the Disability Discrimination Act as *service providers* (qv).
• CIVIL AVIATION AUTHORITY GUIDANCE ON SEATING RESTRICTIONS explains the reasons for safety restrictions on aircraft.
For publication details of the above documents see the bibliography.
5 In 2005 the application of the Disability Discrimination Act to public *transport* (qv) was extended to include the management of transport services, a matter that should be checked by all transport operators.
6 The situation will be constantly developing in coming years, and designers and managers must keep up with current information. As of 2007, useful websites for checking the latest situation include the following, given in alphabetical order:
• Department for Transport
www.dft.gov.uk/transportforyou/access/aviationshipping/
• DPTAC Access Directory www.dptac.gov.uk/adnotes.htm
• DPTAC planes www.dptac.gov.uk/planes.htm
• DPTAC Publications www.dptac.gov.uk/pubs.htm
• Office of Public Sector Information www.opsi.gov.uk/acts.htm

See also: *Flight Rights;* and *Transport: complying with the Disability Discrimination Act*.

Air travel: guidance for disabled people

See: *Door to Door*.

Airport design: complying with the Disability Discrimination Act

1 Buildings used by the public must in general comply with the Disability Discrimination Act (qv). In the case of airports, *authoritative practical guides* (qv) include DESIGNING FOR ACCESSIBILITY (qv) and in addition the following documents. For publication details see the bibliography.
• BS 8300 (qv). Para 13.1 states brief design requirements for all transport-related buildings.
• INCLUSIVE MOBILITY (qv) gives extensive and detailed guidance for all transport-related buildings.
These guides do not have the force of law, but conformity with their recommendations will make it easier to demonstrate that the requirements of the Disability Discrimination Act DDA have been met.
2 As of 2008, useful websites include the following. They are given in alphabetical order.
• DPTAC Access Directory www.dptac.gov.uk/adnotes.htm
• DPTAC Publications www.dptac.gov.uk/pubs.htm

See also: *Transport-related buildings.*

Alarms: community

1 Community alarms (also called *warden alarms*) are systems installed in groups of dwellings, such as *sheltered housing schemes*, allowing residents to call the warden or emergency services (probably provided by the *local authority*) by pulling a cord or pressing a portable trigger.
2 As of 2008, useful websites include the following. They are given in alphabetical order.
• Ricability www.ricability.org.uk/
• Telecare Services Association www.asap-uk.org/

See also: *Sheltered housing.*

Alarms: fire

1 *Authoritative practical guides* (qv) to the provision of fire alarms in buildings, with special reference to the needs of *disabled people*, include the following. Publication details are given in the bibliography:
• APPROVED DOCUMENT B (qv) of the Building Regulations. Refer in particular to paras 0.14; 1.1-1.32; and 5.18.
• APPROVED DOCUMENT M (qv) of the Building Regulations. Refer in particular to paras 3.10; 4.24; and 5.4.
• BS 5588-8: CODE OF PRACTICE FOR MEANS OF ESCAPE FOR DISABLED PEOPLE (qv). Refer in particular to para 18.1.
• BS 8300 (qv). Refer in particular to para 9.37.
• FIRE SAFETY: AN EMPLOYER'S GUIDE.
• SAFETY, SECURITY AND ENVIRONMENTAL CONTROLS.
This list is not comprehensive, but the two last-named documents contain useful bibliographies.
2 With respect to alerters for people with *impaired hearing*, useful websites include the following:
• Deaf Alerter plc www.deaf-alerter.com/website.htm
• RNID (qv) www.rnid.org.uk/shop/

See also: *Fire safety;* and *Impaired hearing.*

Alarms: personal

1 People who live alone, or are alone for longish periods, may be greatly reassured by having an alarm system that can summon help at any moment. Such a system may be built into the home (as in *sheltered housing* schemes – see: *Community alarms* above) or be portable. There are many systems

available, ranging from simple pull-cords which activate an alarm somewhere else, to autodialler alarms which directly call a 24-hour monitoring station by telephone, to a simple neck-hung device with a push-button.
2 As of 2008, useful websites include the following. They are given in alphabetical order.
• Disabled Living Foundation www.dlf.org.uk/public/factsheets.html.
• Ricability www.ricability.org.uk/
• RNID (qv) www.rnid.org.uk/shop/

Alerting devices

1 For people with *impaired hearing* who cannot hear the ringing of the doorbell or telephone, or the crying of a baby, there are alerting devices such as doorbell, telephone, smoke alarm, and baby-cry transmitters. They alert the deaf person by:
• Vibrating a receiver carried in the pocket or placed under a mattress or pillow;
• Flashing a light; and/or
• Sounding an audible alarm.
Some have a bed-shaking mechanism.
2 As of 2008, useful websites include the following. They are given in alphabetical order.
• Disabled Living Foundation www.dlf.org.uk/.
• Ricability www.ricability.org.uk/
• RNID (qv) www.rnid.org.uk/shop/

See also: *Alarms;* and *Impaired hearing.*

Alternative and augmentative communication (AAC)

1 AAC are methods of communicating which supplement or replace speech and handwriting. Such methods aim to help non-speaking people convey what they are trying to say as fully and naturally as possible by other means. There are two variants:
• Unaided communication does not involve equipment and relies on facial expression, gesture, manual signs, or eye-pointing. See for instance: *British Sign Language.*
• Aided communication makes use of technical aids and equipment. These can either be low-tech (word and phrase lists, picture charts, etc.) or high-tech (electronic devices of various kinds). High-tech aids are glamorous and can do impressive things, but they will seldom be the whole answer; and most non-speaking people are best helped by combinations of low-tech and high-tech systems which give them different tools for different situations. See: *Speech aids.*
2 A key aspect of AACs is the need for readily usable means of representing objects, ideas, and meanings. The traditional methods are of course spoken and written language, but some children and adults cannot manage these; and in such cases pictures, symbols, and signs must be used. Symbol and sign systems include *British Sign Language* (qv), *Makaton vocubulary* (qv), *Paget-Gorman signed speech;* and *Signed English* (qv).
3 Both users (or 'speakers') and communication partners (or 'listeners') must be thoroughly trained in using the systems they are adopting – even the simplest ones. This is particularly important if the speakers have *impaired understanding* as well as *impaired speech*, which is quite frequently the case.
Without such training the whole exercise is very likely to end up as a frustrating failure for all concerned.
4 As of 2008, useful websites include the following. They are given in alphabetical order.
• RNID www.rnid.org.uk/

• Royal College of Speech and Language Therapists www.rcslt.org/

See also: *Aphasia; Communications aids; Sign and symbol systems;* and *Speech aids*.

Ambulant disabled persons

1 *Ambulant* disabled persons, as distinct from *wheelchair-users*, are people who suffer from *impaired mobility* but are able to get about satisfactorily on their legs plus simple aids, with no need for *wheelchairs*.
2 The *aids and equipment* that are available to them include calipers, crutches, walking sticks, and walking frames – for more detail see: *Walking and standing aids*.

Americans with Disabilities Act 1990

See: *Ada and Aba Accessibility Guidelines*.

Anthropometrics

1 *Anthropometrics* is the science of body measurement. The things measured include body size, shape, strength, and working capacity; and *anthropometry* includes matching the physical form and dimensions of a workspace or product to those of the user.

Anthropometric data

1 It may occasionally be possible to design clothing, furniture, fittings, building spaces, and other products to suit a particular individual (called 'bespoke design'), but in most cases a limited range of sizes must perforce suit the whole of the population, including people big and small, tall and short, fat and thin, strong and weak. Thus in designing some building part or product to be used by men, it may be necessary to select one particular body height to represent all adult British males, when in fact men range in stature from very roughly 1 500 mm to 1 930 mm.
2 In some cases (for instance clothing) it is possible to manufacture products in a range of standard sizes, so that several heights could be selected between, say, 1 550 mm and 1 925 mm. In others (for instance office chairs) it is possible to build in some degree of adjustment, so that one manufactured size can be adapted after purchase to suit users of all body heights. But in many cases (for instance tables and chairs, desks and worktops, and door heights) most products must be manufactured in one single size to suit the whole of the population.
3 Tables of anthropometric data have therefore been developed to give designers the best possible 'average' figures for a range of human characteristics such as body height, eye-height, reach, and many others. As of 2007, useful references include the following. Publication details are given in the bibliography.
• BODYSPACE: ANTHROPOMETRY, ERGONOMICS AND THE DESIGN OF WORK by Stephen Pheasant is a respected and widely used reference, written in a clear and lively style. It gives guidance on longterm basic principles, not on the latest official recommendations, which are subject to change. The main respect in which some of its basic data may need revisiting is that both children and adults have become fatter since it was written – see: *Obesity*.
• THE METRIC HANDBOOK, which contains distilled information on a wide range of design-related matters, gives useful summary data.

Anthropometric data: application

1 Great care is required in using tables of anthropometric data such as those in the publications above. For a useful discussion see pp 15–44 of BODYSPACE; but the points at issue include the following:
• In some cases (seat height, table/desk height, or notice board height) the average figure – more accurately called the 50th *percentile* – is the one to use.

The number of people who find the height too high or too low will be roughly equal; most will find it acceptable; and very few will find it impossible.
• In other cases (eg door height, clear space under tables and desks, and seat width) a dimension that suits the largest people will also suit everyone else, while an 'average' dimension will cause severe problems for everyone much larger than the average. For these situations designers should therefore choose not the 50th percentile but – say – the 95th, a figure which includes the whole population except the 5% largest.
• In yet other situations (eg *reach* distances to hand or foot controls) a dimension that suits the smallest people will also suit most others, while an 'average' dimension will mean that users who are much smaller than the average may well be unable to reach the control. For these situations designers should therefore choose not the 50th percentile but – say – the 5th, which comprises the smallest 5% of the population.
2 A further complication is the fact that few buildings, spaces, or fittings are used exclusively by the neatly specified groups of people in anthropometric tables, which tend to have headings such as 'British adults aged 19 to 65'; 'British 15-year-old males'; or 'British elderly women'. The ultimate users may well include unpredictable varieties of people, including children with their very different body sizes, making nonsense of some designer's careful selection of body dimensions from an appropriate-seeming table.
3 Better sources of information, where these are available, are therefore those dimensional recommendations issued by authoritative bodies like the British Standards Institution (qv) for specific applications such as lift button heights, seat heights, or notice board heights. All practicalities of the above kinds have been taken into account in these documents, and a considered conclusion reached. For such sources see the next entry.

See also: *Normal distribution; Obesity;* and *Percentiles.*

Anthropometric data: complying with the Disability Discrimination Act

1 The Disability Discrimination Act (qv) offers no design guidance of its own. Instead, the Disability Rights Commission (qv) advises designers to consult *authoritative practical guides* (qv).
2 BS 8300 (qv) is the most authoritative general reference, and applies to all buildings in the UK. The document is advisory and does not have the force of law, but conformity with its recommendations will make it easier to demonstrate that the requirements of the DDA have been met. Annex D (pp 151 to 162) gives tables of recommended design dimensions specially researched to establish the needs of *disabled persons* (qv). Other dimensional recommendations are scattered throughout BS 8300. See for example the following entries in the present work:
• Access controls.
• Automatic teller machines.
• Controls and switches.
• Corridors.
• Counters and reception desks.
• Handrails.
• Seats.
• Windows.
• Worktops.
3 INCLUSIVE MOBILITY (qv) applies particularly to the needs of *disabled persons* in pedestrian and transport infrastructure. The 'Summary of dimensions' that is included as a loose insert gives a most useful and very wide range of officially recommended dimensions.

Anti Poverty Strategies 1 A governmental attempt at a co-ordinated approach to tackling *poverty* (qv), including programmes to help people claim benefits, manage debt, have *access* to low interest small loans, and to social work and *housing* services. Visit www.neighbourhood.gov.uk/glossary.asp

2 See also: *Neighbourhood renewal.*

Anti-social behaviour 1 Anti-social behaviour, particularly on *housing estates* (qv), has become one of the worst blights on the quality of life in Britain. The problem remains essentially unsolved, but a scattering of encouraging results have been achieved by application of the Crime and Disorder Act 1998 (qv); the Anti-social Behaviour Act 2003; and by governmental promotion of good practice, especially through multi-agency approaches such as *Crime and Disorder Reduction Partnerships* (qv) with the active involvement of the local community. These tools have helped some *social landlords* (qv) get a degree of control on problems such as rooting out nuisance tenants who make their neighbours' lives a misery.

2 But huge problems remain. Mr Frank Field (qv) a much-respected Labour MP and a member of the advisory board of the *think tank* (qv) Reform, has argued (a) that parental and teacher authority, backed up by the informal authority that was traditionally exercised by the whole of the surrounding community, have decayed to such a degree that society must find new ways of protecting itself against disorder; and (b) that the present approach of centrally-directed policing initiatives, including no fewer than 42 *Asbo* (qv) type powers,
is absolutely the wrong way of going about things. In his view local communities must be given direct power to police their own neighbourhoods, and he has proposed the following legal reforms:

- A local community should have the right to ask magistrates to bring before them an offending yob.
- The magistrate should be able to rule that a request for immediate action is a public matter.
- The police would then be required to enforce the warrant.

The outcome of these measures must be that the perpetrators of disorder can be brought before the court that very day.

Mr Field further proposes that the police should have the power to issue warnings, and if these are ignored to impose a restriction on bad behaviour there and then. The offender would of course have the freedom to go to court, but in many cases anti-social behaviour would be nipped in the bud, and most young offenders would be kept out of the criminal justice system.

3 As of 2008, useful publications include the following. For publication details see the bibliography.

- ANTI-SOCIAL BEHAVIOUR LAW by Paul Greatorex and Damian Falkowski.
- TACKLING ANTI-SOCIAL BEHAVIOUR, published in December 2006 by the National Audit Office (qv).

4 Useful websites include the following. They are given in alphabetical order.

• Civitas	www.civitas.org.uk/
• Directgov	www.direct.gov.uk/CrimeJusticeAndTheLaw/
• Home Office	www.crimereduction.gov.uk/
• Home Office	www.homeoffice.gov.uk/justice/what-happens-at-court/sentencing/
• The Met	www.met.police.uk/askthemet/vandalism.htm
• Nacro	www.nacro.org.uk/
• Policy Research Bureau	www.prb.org.uk/

See also: *Crime: youth;* and *Housing estates.*

Anti-social behaviour orders (Asbos)

1 Asbos were introduced by Prime Minister Tony Blair as a method of combating yobbish behaviour, particularly on *housing estates* (qv) and in town centres. They are court orders which prohibit offenders from committing specific anti-social acts or entering defined areas, and are imposed for a term of two to five years. They most commonly include bans on causing harassment, alarm or distress, bans on entering particular places or parts of town, and bans on mixing with other named individuals; and they have been used particularly against young offenders, the aim being to restrain their bad behaviour without putting them in prison.

2 Over 16 000 of these orders were issued between April 1999 and 1 January 2007. But there was a strong decline towards the end of that period, and in May 2008 the Government was beginning to talk about Asbos in the past tense. They had caused much controversy for criminalising behaviour that was undesirable but otherwise lawful, and critics complained that they were a blunt instrument that did not address the underlying problems causing anti-social behaviour. But the main problem was that these orders were simply being ignored – in early 2008 around half were being disregarded, with a breach rate of 61% among juveniles and 43% among adults.

3 Alternative measures promoted by the Government include acceptable behaviour contracts (ABCs), which are written agreements between a young person, the local housing office or registered social landlord, and the local police, in which the offender agrees not to do a series of identifiable things which have been defined as anti-social. But as ABCs were breached by almost two thirds of under-18-year-olds, critics were not impressed.

4 While punitive responses cannot be the only answer to the yobbish behaviour that inflicts misery upon many of Britain's housing estates, the following Home Office initiatives are promising:

• First, Operation Leopard, which was trialled on an estate in Basildon. The police claim that it dramatically reduced offending and proved highly popular with residents. Under this scheme, which was created by Inspector Jon Burgess and Sergeant Gavin Brock, local police targeted 14 people in their teens and early 20s who were well known to the force, having built up criminal records for offences such as intimidation, burglary, criminal damage, anti-social behaviour and vehicle crime. The police knocked on their doors, followed them on the estate and repeatedly searched them, warning them that local people had identified them as lawbreakers.

• Second, making regular checks against a hard core of troublemakers, including whether their vehicles were licensed and whether they had paid their council tax and possessed television licenses.

5 As of 2008, useful references include the following. They are given in date order. For publication details see the bibliography.

• ANTI-SOCIAL BEHAVIOUR ORDERS: A SUMMARY OF RESEARCH INTO ANTISOCIAL BEHAVIOUR ORDERS GIVEN TO YOUNG PEOPLE BETWEEN JANUARY 2004 AND JANUARY 2005 was published in November 2006 by the Policy Research Bureau and Nacro, the crime reduction charity.

• TACKLING ANTI-SOCIAL BEHAVIOUR was published in December 2006 by the National Audit Office (qv).

• THE ASBO: WRONG TURNING, DEAD END by Detective Chief Superintendent Neil Wain, head of the Stockport division of Greater Manchester police, and Elizabeth Burney, was published in 2007. It questioned the success of ASBOs in preventing crime and anti-social behaviour, and called for a pause while researchers test growing evidence that they may be more of a political stunt than an effective tool.

6 For useful websites see the previous entry.

Anti-Terrorism, Crime and Security Act 2001

See: *Civil liberties.*

Aphasia

1 An inability to understand or produce speech, as a result of damage to those parts of the brain that are concerned with the use and meaning of words. Such damage is very commonly the result of *stroke*. The problem affects the understanding and production of language rather than the purely physical aspects of speech such as the movement of lips and tongue. There are usually, but not necessarily, associated difficulties in reading and writing.
2 As of 2008, useful websites include the following. They are given in alphabetical order.

- NHS Direct www.nhsdirect.nhs.uk/
- Speakability www.speakability.org.uk/
- The Stroke Association www.stroke.org.uk/

See also: *Alternative and augmentative communication; Communications aids; Impaired communication;* and *Speech aids.*

Appliances: domestic

1 Domestic appliances that are designed to be easily usable by most people, and not only those with sharp eye-sight, good grip, and ready understanding, will generally possess the following characteristics.
2 With respect to general design:

- Comfortable height, not obliging people with stiff joints to bend double in order to operate or clean the appliance.
- No risk of burning users. For instance, ovens with downward-folding doors have badly burnt many *wheelchair-users* (qv) as they reached across the scorching surface to reach the inside.
- Large, sensibly shaped handles that can be easily gripped by weak, stiff, or shaky hands.

3 With respect to controls:

- Large knobs with a non-circular shape, that can be easily gripped and turned by weak, stiff, or shaky hands.
- Big, brightly coloured buttons.
- Large-lettered labels that are easily readable in bad light, even by people with poor eyesight who have left their glasses in another room. See: *Accessible print.*

4 With respect to instructions:

- Easily understood instructions in simple language.
- Written instructions supplemented by spoken instructions and warnings – see: *Talking household appliances.*

5 As of 2008, useful websites include the following. They are given in alphabetical order.

- Disabled Living Foundation www.dlf.org.uk/public/factsheets.html.
- RADAR www.radar.org.uk/
- Ricability www.ricability.org.uk/

See also: *Assistive technology;* and *Products for easier living.*

Approaches

See: *Access routes.*

Approved Documents (ADs)

1 The Building Regulations (qv) are brief statements of requirements, generally to do with the health and safety of building users, but giving no detail on how these requirements may be met in practice. The latter are given separately in Approved Documents, abbreviated as ADs. Thus Part M of the Building

Regulations requires that reasonable provision be made to ensure that buildings are accessible and usable by all those people who could be expected to use them; and APPROVED DOCUMENT M shows how building designers may meet this requirement (though they are free to use other methods, provided their chosen solution can be shown to be soundly based).

2 For a complete list of ADs, and more detailed explanations, visit www.planningportal.gov.uk/

3 As of 2008, useful guides to the Building Regulations for England and Wales include the following. For publication details see the bibliography.

- BUILDING REGULATIONS EXPLAINED AND ILLUSTRATED by M J Billington, M W Simons and J R Waters.
- BUILDING REGULATIONS EXPLANATORY BOOOKLET.

4 As of 2008, useful websites include the following. They are given in alphabetical order.

- Building Regulations www.thebuildingregs.com/
- Dept for Communities & Local Government www.communities.gov.uk/
- Planning Portal www.planningportal.gov.uk/

5 In parts of the UK other than England and Wales the following documents fulfil approximately similar functions to Approved Documents:

- In Northern Ireland: the TECHNICAL BOOKLETS.
- In Scotland: the DOMESTIC TECHNICAL HANDBOOK and the NON-DOMESTIC TECHNICAL HANDBOOK.

See the bibliography for particulars of these publications.

Architectural disablement

See: *Disability: other definitions.*

Architecture

1 Architecture is the art of creating buildings, streets and towns that are convenient, safe, and beautiful – or, in the more elegant words of Sir Henry Wotton four centuries ago: 'Well building hath three conditions: commodity, firmness, and delight'. That is what citizens expect; and as they are the users of the built environment, bear most of the cost of creating that environment, and largely pick up the bill for training the architects who design it, they are entitled to see their reasonable wishes respected.

2 Today's buildings are safer and more convenient than ever before, so two of the above conditions are met; but the third, in the view of very many people, is not. Every survey of 'most loved buildings' and 'most hated buildings' shows most of Britain's best-loved buildings to have been built before 1950, and most of its best-hated ones after 1950. People's everyday behaviour confirms their responses to opinion poll: each year millions of them visit old houses, old villages, and old town and city centres for the specific purpose of enjoying their beauty, while the number who visit modern buildings or towns for that reason is – to put it kindly – modest. Popular buying preferences point in the same direction: Cotswold villages, Georgian and Regency terraces and squares, and period country houses are at or near the top of the housing price pyramid. In short, while modern architects do produce isolated examples of visual delight, the profession is failing to meet society's deep and legitimate yearning for entire streets, suburbs or towns that may be treasured for their charm or beauty – words which to most people imply ornamented, pre-modern buildings of one kind or another.

3 Architects like to argue that asking them to create buildings in pre-modern styles is inappropriate or impractical. Are modern functions not impossible to house satisfactorily in buildings of – using the term in its broadest sense – the 'classical' styles of architecture? Do modern ways of working and living, and modern methods of construction, not require modern-style buildings?

4 The answer to these questions is 'no', as will be shown below. Architects could quite easily give people houses, streets and towns in the building styles they love and desire, and ought to be doing so.

Architecture: what went wrong?

1 The root cause of the gracelessness if not downright ugliness of most modern buildings and townscapes (the accuracy of these descriptions can be verified from any train window approaching Waterloo station or Brighton, Bath and a dozen other cities and towns) may be traced to two changes in architectural training and practice, instituted in the 1950s.

• First, the most supreme aim of architects and master builders in all ages and cultures – the pursuit of beauty as a goal for its own sake – was summarily banished from architectural training courses and replaced by the newfangled pursuits of 'functionality' and 'honesty'. Beauty – if the thinkers of the modern movement could be bothered to use that old-fashioned word – was imperiously redefined as something that would result if a building, a chair, or any other designed element 'honestly' expressed its function, the nature of its structure, and the nature of its materials. In other words, what used to be a value in its own right was now relegated to being no more than a by-product of other, supposedly more important concerns.

• Having effectively defined the traditional ideal of beauty out of existence, modernists proceeded also to eliminate from architectural training courses the practical toolkit that had for centuries enabled building designers to achieve that all-important goal.

2 The toolkit in question was – again using the term in its broadest sense – the classical style of architecture. This architectural language has for centuries given building designers at all aesthetic levels, from cathedral down to cottage, a rich vocabulary of parts (façades, towers, roofs and pinnacles, surface mouldings, balustrades, urns, and statuary), and a grammar for the assemblage of those parts, that could be applied to almost the entire spectrum of building types. And while the results were eminently practical, as will be shown in the next entry, the ultimate aim went deeper. It was – to give a formal definition of what is meant by architectural beauty – to create 'appearances that give pleasure when contemplated'. The creation of such appearances was what people expected of architects, and what architects most prided themselves upon delivering. And not only architects: uneducated builders, working from pattern-books, could successfully produce cottages and terraced dwellings which used the language of classical design at a more modest level. In some cases beauty was achieved; in other cases merely charm, or at an even less assuming level the simple but important quality of a building being agreeable, of soothing the eye and making onlookers happy to be in its presence. These mostly ordinary buildings, created mostly by oridinary builders for ordinary people, managed to achieve a pleasent civility that delights us to this day.

3 In the 1950s that entire approach was tossed aside and replaced by the dogmas of functionalism. The unloved consequences may be seen, as already noted, in Brighton, Bath and a score of other towns and city centres across Britain. And the time has come to draw the obvious conclusion. After fifty years of enduring successive failed architectural experiments, from functionalism to the shoddy banalities of post-modernism and deconstructivism, citizens are finally entitled to ask the architectural profession to stop playing private games and revert to what used to be its core skill – designing building façades (for the primary of these in architectural design see: *Urban design*) and interiors that are practical, durable, and will delight the eyes of users and passers-by now, and for generations to come.

4 The practicalities of such an approach are considered in the next entry.

Is a revival of the classical vernacular feasible?

1 Proponents of modernism like to argue that modern uses cannot be efficiently accommodated in classical-style buildings. But consider the following examples – all of them, for the sake of convenience, in London:

• Museums. Behind a noble and strikingly beautiful 120 year-old Romanesque façade, the iron-framed interior spaces of the Natural History Museum (1881) in Cromwell Road readily accommodates the very latest, state-of-the-art displays.

• Department stores. Behind a grand neoclassical façade that rates as one of the most satisfying examples of street architecture in the world, the steel-framed interior spaces of Selfridges (1928) in Oxford Street meet the most demanding requirements of 21st century retailing.

• Town halls. Behind its virtually unmodified 'Edwardian Renaissance' exterior the old LCC County Hall (1933) at Westminster Bridge very satisfactorily accommodates new luxury apartments, two modern hotels, an entertainment centre, an aquarium, a museum, and several restaurants, all of them to 21st century standards.

• Hospitals. Behind a handsome and virtually unmodified classical exterior, St George's Hospital (1829) at Hyde Park Corner has been converted into one of London's most luxurious hotels.

• Churches. With very little modification to its main interior space, and virtually none to its exterior, the fine baroque church of St John's (1728) in Smith Square, Westminster, now functions as a popular and acoustically excellent concert venue.

• Houses. All over London, Georgian and Regency streets and squares built 200 or 300 years ago are among the most sought-after dwellings for 21st century residents. In Holborn and Mayfair these erstwhile dwellings command even higher prices as prestige legal or commercial offices.

2 A second objection, that modern construction methods demand modern building designs, may be true in some cases but does not hold as a general rule. The neo-Romanesque Natural History Museum in London (1881) and the 57-storey neo-Gothic Woolworth Building in New York (1913) consists, respectively, of cast iron and steel frames clad with machine-produced terracotta panels.

3 A third factor, the now vital need for sustainable design, quite demonstrably favours traditional building design over modern. Georgian and Regency buildings have remained in use for two or three centuries, while steel, glass and plastic buildings are unlikely to last one century. Can anyone doubt that Mr Quinlan Terry's fine row of neo-classical office buildings at 20–36 Baker Street – between Blandford and George streets – completed in 2002, will still be in use when the modernist Lord (Norman) Foster's London City Hall at Tooley Street, SE1, termed a 'gas guzzler' after a damning Building Research Establishment report on its energy consumption; and his 'Gherkin' Tower at 20 St Mary Axe, EC3 (one of whose giant windows fell out and crashed onto the pavement 400 feet below), have become scrap metal? Readers should inspect these buildings and come to their own conclusions.

4 A final objection, that there aren't – and cannot be – enough talented designers to lovingly create beautiful townscapes of the kind that most citizens yearn for turns out to be the reverse of the truth when one inspects the methods by which 'functionalist' and 'traditional' buildings are designed.

• The doctrine of functionalism requires each building to be designed as an individual creation, starting from an intricate analysis of the functions to be accommodated within, and laboriously working outwards to a final building form that, as architects like to say, 'expresses' those functions – thus obeying the precept, beloved of modernists, that 'form follows function'. This process requires college-trained professionals, a lot of talent, is extremely time-consuming, and cannot possibly be used for designing more than a tiny fraction of our total habitat.

• Against that, the classical tradition does not expect each building to be a unique invention – which is just as well, as geniuses are scarce. Instead, it provides a range of well-tested type solutions (the basilica, the palazzo, the villa, etc.) the loose-fit plans of which can be adapted to become grand country houses, modest suburban dwellings, office blocks, department stores, museums, libraries, hospitals, schools, railway terminals, warehouses, electrical substations, or public lavatories. And once constructed these buildings can be readily adapted to successive new uses – nicely illustrated by the little pitched-roof Victorian building on the corner of Fulham Palace Road and Lillie Road, SW6, which started life as a public lavatory and has since been re-used as an architect's office, then a shop, then a health centre.

5 In short, the successful building is the one that is least attached to its initial function, and most able to survive a change in function. And what encourages us to save such a building is that we like the look of it. On both counts it is a matter of record that buildings designed in the classical vernacular outscore those designed in the modern style. The way forward, for architecture, is a revival of the thoroughly proven and much-loved classical vernacular style for most of our buildings, leaving experimental modern styles for those particular clients and users who actually want them.

Architecture: sources of further information

1 As of 2008 useful references include the following. For publication details see the bibliography.

• THE CLASSICAL VERNACULAR: ARCHITECTURAL PRINCIPLES IN AN AGE OF NIHILISM by Roger Scruton. The above notes are greatly indebted to Professor Scruton's civilised essays.

• GENTLE REGRETS: THOUGHTS FROM A LIFE by Roger Scruton. This is not primarily a book about architecture, but Chapter 11 'Returning Home' (pp 197-217) contains wiser thoughts, better expressed, than the pretentious gobbledygook that fills many of today's professional journals.

• FORM FOLLOWS FIASCO: WHY MODERN ARCHITECTURE HASN'T WORKED by Peter Blake. After writing about and practising modern architecture for 20 years, the author of the standard work THE MASTER BUILDERS came to the conclusion that the modern movement has been a dreadful mistake. This mea culpa by an insider who knew, and in many cases worked with, almost all the major figures in modern architecture from Frank Lloyd Wright and le Corbusier to Louis Kahn and Robert Venturi is devastating.

• NO PLACE LIKE UTOPIA: MODERN ARCHITECTURE AND THE COMPANY WE KEPT by Peter Blake. An autobiographical account of the above-mentioned career.

• RADICAL CLASSICISM: THE ARCHITECTURE OF QUINLAN TERRY by David Watkin. A comprehensive, well-illustrated record of the neo-classical buildings of Quinlan Terry. It includes a foreword by HRH the Prince of Wales and a scholarly introduction by Professor Watkin. This beautiful book will be enjoyed by everyone who loves architecture.

• LEAVING AN ENVIRONMENTALLY SOUND, ATTRACTIVE LEGACY by Quinlan Terry. This 7-page essay argues that buildings constructed in the traditional way are more environmentally responsible than ones made of steel, glass, and plastic. The 80-page booklet of which the essay forms part, titled the RICHARD H DRIEHAUS PRIZE BOOK 2005, contains full-colour photographs of several of Mr Terry's neo-classical buildings, all of which promise to look as handsome a century hence as they do today.

See also: *Le Corbusier;* and *Urban design.*

Area based regeneration

1 These are government initiatives for reviving communities which are based on tackling all the problems in a neighbourhood, rather than just one or two in isolation. The aim is to deal with physical, economic and social decline in the round.
2 As of 2008, useful websites include the following. They are given in alphabetical order.

- Dept for Communities & Local Government www.communities.gov.uk/
- Info4local www.info4local.gov.uk/
- Joseph Rowntree Foundation www.jrf.org.uk/
- Neighbourhood Renewal Unit www.neighbourhood.gov.uk/

See also: *Neighbourhoods; Regeneration;* and *Social exclusion.*

Areas for regeneration

1 These are defined in the Mayor's *London Plan* (qv) as the *wards* (qv) in greatest socio-economic need, on the basis of being the 20% most deprived wards in the London Index – the latter being the Greater London Assembly's index of deprivation.
2 For more information visit www.lda.gov.uk/

See also: *London Plan.*

Areas of deprivation

See: *Neighbourhood renewal.*

Arm's length bodies

See: *Quangos.*

Art galleries

See: *Exhibition design*; and *Exhibition management.*

Arthritis

1 This is a general term for a number of painful conditions of the joints and bones. It is an all too common condition, and a major cause of *disablement* (qv) and the need to use *wheelchairs* (qv).
2 Arthritis is formally recognised as a *disability* (qv) under the Disability Discrimination Act (qv) if the effect is severe enough.
3 As of 2008, useful websites include the following. They are given in alphabetical order.

- Directgov www.direct.gov.uk/
- Directgov: disabled people www.disability.gov.uk/
- Equality & Human Rights Commission www.equalityhumanrights.com/
- NHS Direct www.nhsdirect.nhs.uk/

See also: *Impaired dexterity; and Products for easier living.*

Artificial limbs and appliances

See: *Limb deficiency*; and *Prosthesis.*

Arts Council England

1 Arts Council England is a *quango* (qv) that is responsible for funding arts projects in England, both from taxation (qv) and the proceeds of the National Lottery (qv). It is a *quango* (qv), and behaves like one, being apparently less interested in nourishing the arts than in meeting *politically correct* (qv) *targets* (qv). In April 2008 we learnt (a) that the Council announced drastic cuts to hundreds of theatres and orchestras while sitting on £152 million of unspent funds, and (b) that it now asks grant applicants to state how many of their Board members are bisexual, homosexual, heterosexual, lesbian, or whose inclinations are 'not known'. This impudent question, which according to the form had to be answered, was publicly criticised by the actors Sir Ian McKellen, Vanessa Redgrave, and Simon Callow; the playwrights Christopher Hampton ('It's bureaucracy and political correctness gone mad') and Michael Frayn; the

gallery director Julian Spalding; the painter Maggi Hambling, and others from across the political spectrum.
2 For more information visit www.artscouncil.org.uk/

ASAP See: *Association of Social and Community Alarm Providers.*

Asbestos and health risk

1 The word 'asbestos' confusingly covers two entirely different groups of minerals:
• Blue and brown asbestos, which consist of fibres that resemble tiny shards of broken glass and can pose an extremely severe risk to health if inhaled.
• White asbestos, which consists of soft, silky fibres that dissolve in the lungs within a few days. It has been very widely used in the form of asbestos cement products such as roof tiles, ceiling panels, and ironing boards, etc.; and in this form it is a particularly safe material since its fibres are not only soft (very like talc), but so securely locked in the cement that they cannot escape in respirable form. White asbestos cement products constitute around 90% of all asbestos in the UK, and one of the most comprehensive reviews yet conducted of the scientific literature (Hoskins and Lange 2004) found that they pose 'no measurable risk to health'.
2 Unscrupulous contractors, lawyers, and surveyors in the asbestos-removal trade often blur the distinction between these different types of asbestos, and frighten building owners into spending millions of pounds a year for needless removal work. Building owners can get sensible advice on such matters from the Asbestos Watchdog at www.asbestowatchdog.co.uk/
3 Other useful websites include www.hse.gov.uk/asbestos/regulations.htm

See also: *Hazards; Health scares; Risk;* and *Risk assessments: basic procedure.*

Assembly and waiting areas: complying with the Disability Discrimination Act

1 Buildings used by the public must in general comply with the Disability Discrimination Act (qv). In the case of assembly and waiting areas in public buildings, *authoritative practical guides* (qv) include DESIGNING FOR ACCESSIBILITY (qv) and in addition the following documents. For publication details see the bibliography.
• BS 8300 (qv). Sections 11 and 13.1.6 state brief design requirements for assembly areas and waiting rooms.
• INCLUSIVE MOBILITY (qv). Section 9 gives guidance for waiting areas in all transport-related buildings such as bus and coach stations, railway stations, air terminals, and transport interchanges.
These guides do not have the force of law, but conformity with their recommendations will make it easier to demonstrate that the requirements of the Disability Discrimination Act have been met.
2 As of 2008, useful websites include the following. They are given in alphabetical order.
• DPTAC Access Directory www.dptac.gov.uk/adnotes.htm
• DPTAC Publications www.dptac.gov.uk/pubs.htm

See also: *Auditoria; Seating*; and *Transport-related buildings.*

Assimilationism See: *Diversity: alternative models.*

Assistance dogs

1 Three kinds of assistance dog are available to help *disabled people* (qv) get about and lead reasonably *independent* (qv) lives:
• 'Guide dogs' help people who are *blind* (qv) or near-blind. Most owners are elderly and have lost their sight with old age. The dogs are taught to lead their owner round *hazards* (qv). They will skirt obstructions such as lamp posts; but

for the rest will simply take a straight line from edge to edge, stopping at each edge such as an upstand *kerb*, a flight of *steps*, or a hole in the ground, to await further instructions. They cannot respond to changes of texture or colour underfoot, such as *tactile paving surfaces* (qv), or give warning of above-knee obstacles. They are therefore liable to lead their owners straight into injury from high-level hazards such as open windows, horizontal or oblique scaffolding poles, or overhanging branches. Holly can do terrible damage to peoples' eyes. See: *Hazards;* and *Streetworks.*
• 'Hearing dogs' help people who are *deaf* (qv) or near-deaf by alerting them to sounds such as a door knock or doorbell, alarm clock, oven buzzer, telephone, baby cry, name call, or smoke alarm. The dogs touch their owners with a paw, then lead them to the sound.
• 'Service dogs', of which there are at present very few in the UK, help people with physical *impairments* (qv) by retrieving objects that are out of their reach, pulling wheelchairs, opening and closing doors, turning light switches off and on, barking for alert, finding another person, assisting ambulatory persons to walk by providing balance and counterbalance, and other tasks. They can also be trained to alert owners of an imminent epileptic seizure.
2 As of 2008, useful websites include the following. They are given in alphabetical order.

• Assistance Dogs International Inc	www.adionline.org/
• Canine Partners	www.c-p-i.org.uk/
• Dogs for the Disabled	www.dogsforthedisabled.org/
• Guide Dogs for the Blind Association	www.gdba.org.uk/
• Hearing Dogs for Deaf people	www.hearing-dogs.co.uk/
• Support Dogs	www.support-dogs.org.uk/

3 As most assistance dogs are basically, and probably exclusively, *guide dogs*, this widely-used and well-understood expression is an acceptable everyday term for the whole group.

See also: *Canes for blind people; Hazards*; and *Impaired vision.*

Assisted suicide

See: *Self-deliverance.*

Assistive technology (AT)

1 The term refers to products or services that help to give people with physical, sensory, or mental *impairments* (qv) a greater degree of *independence* (qv). Examples range from devices as simple as a *crutch* (qv) or *walking stick* (qv), to ones as advanced as electronic *speech aids* (qv). Many are a blend of the old and the new – for instance a traditional door opened or closed by automatic controls, or a traditional clock that also speaks the day, date and time for people who cannot see well – see: *Talking household appliances.*
2 For convenience the present work divides such products into three practical categories – low-tech, medium-tech, and high-tech.

See also: *Aids and equipment.*

Assistive technology: high-tech

1 The term 'high-tech' refers broadly to products and devices of a sophisticated *electronic* kind. They are also called 'Electronic Assistive Technology' or EAT. Such products can be conveniently divided into two categories:
INSTALLATIONS IN PUBLIC BUILDINGS OR SPACES INCLUDE:
• Alarms.
• Induction and infrared communication systems.
• Talking signs.
PRODUCTS FOR MAINLY PERSONAL AND DOMESTIC USE INCLUDE:
• Alerting devices.

• Communication aids (a generic term that includes some of the others below).
• Domestic induction loop systems.
• Environmental control systems.
• Hearing aids.
• Hearing enhancement systems.
• Listening devices.
• Talking household appliances.
• Talking signs.
• Special telephones.
• Speech aids.
For more detail see the individual entries for most of the above items.
2 As of 2008, useful publications include the following. For publication details see the bibliography.
• SAFETY, SECURITY AND ENVIRONMENTAL CONTROLS.
3 Useful websites include the following. They are given in alphabetical order.

• Ricability	www.ricability.org.uk/
• RNIB	www.rnib.org.uk/
• RNID	www.rnid.org.uk/
• Possum Controls Ltd	www.possum.co.uk/
• RSL Steeper	www.rslsteeper.co.uk/

See also: *Environmental control systems*.

Assistive technology: low-tech

1 The term 'low-tech' refers broadly to *manual* devices – i.e. ones that are worked and controlled mainly by hand. Examples include:
• Adjustable furniture (qv), bathroom fittings, and kitchen fittings which are height-adjusted by hand.
• Manual wheelchairs (qv).
• Walking frames (qv).
2 As of 2008, useful websites include the following. They are given in alphabetical order.

• Astor Bannerman	www.astorbannerman.co.uk/
• Bakare	www.bakare.co.uk/
• DCS Joncare	www.dcsjoncare.freeserve.co.uk/
• Disabled Living Foundation	www.dlf.org.uk/public/factsheets.html
• OTDirect	www.otdirect.co.uk/link-equ.html#Platform
• Phlexicare	www.nichollsandclarke.com/

Assistive technology: medium-tech

1 The term 'medium-tech' refers broadly to products and devices of a traditional kind, but *power-operated* for greater ease. Examples include:
• Powered doors.
• Powered adjustable furniture, bathroom fittings, etc.
• Hoists.
• Platform lifts.
• Powered door and window operators.
• Powered wheelchairs.
• Stairlifts.
2 For more detail see the individual entries for most of the above items. Useful websites are as for the previous entry.

AT

See: *Assistive technology*.

ATM

See: *Automatic teller machines*.

Attention deficit/ hyperactivity disorder (ADHD)

1 Children affected by ADHD are disturbingly active and restless; seem unable to regulate their behaviour to fit the situation; cannot pay attention when required, and lack a capacity for delaying gratification. According to some estimates, up to 5% of school-age children in England and Wales have ADHD – i.e. perhaps two children in a class of 30.

2 ADHD may qualify as a *disability* (qv) under the Disability Discrimination Act (qv), and children thought to have the syndrome may qualify as having *special educational needs* (qv), with all that that entails.

3 Some people suspect that many children labelled as 'ADHD' are simply somewhat livelier than their parents or teachers have the patience to deal with. In her column in *The Times* of 30 August 2005, the broadcaster Libby Purves recalled her interview with a mother who said her four-year-old definitely had ADHD and needed Ritalin, because he was 'always asking questions, then when you answer he thinks of another, it goes on all day'. Ms Purves commented: 'She got her drugs, and her child got the message that there's no point in asking'. Similarly, in March 2007 the psychiatrist who first identified ADHD, Professor Robert Spitzer of Columbia University in New York, told *The Times Educational Supplement* that 20% to 30% of youngsters classified as suffering from ADHD might simply be showing perfectly normal signs of being happy or sad. He thought that there was probably a lot of over-prescription going on, but was less concerned by normal children being needlessly put on the prescribed drugs, as there were usually no serious side-effects, than by failure to treat genuine cases. A fair conclusion from all the above might be that more accurate and reliable diagnostic techniques are needed – see also in this context *dyslexia.*

4 As of 2008, useful websites include the following. They are given in alphabetical order.

• Directgov	www.direct.gov.uk/
• Directgov: disabled people	www.disability.gov.uk/
• Equality & Human Rights Commission	www.equalityhumanrights.com/
• Dyslexia Research Trust	www.dyslexic.org.uk/
• Healthmatters	www.healthmatters.org.uk/issue52/adhd
• NHS Direct	www.nhsdirect.nhs.uk/

See also: *Dyslexia; Medicalisation;* and *Special educational needs.*

Audience seating: complying with the Disability Discrimination Act

1 Public buildings and spaces must in general comply with the Disability Discrimination Act (qv). In the design of audience seating areas in lecture/conference facilities; theatres, concert halls and cinemas; and sports stadia, *authoritative practical guides* (qv) include DESIGNING FOR ACCESSIBILITY (qv) plus the following:

• APPROVED DOCUMENT M (qv) of the Building Regulations. Paras 4.5 to 4.12 contain provisions on *wheelchair access* that must be heeded in order to comply with the Building Regulations (qv).

• BS 8300 (qv). Sections 11.3 and 11.4 state brief design requirements for assembly areas and waiting rooms.

• INCLUSIVE MOBILITY (qv). Section 9 gives guidance for waiting areas in all transport-related buildings such as bus and coach stations, railway stations, air terminals, and transport interchanges.

• ACCESSIBLE STADIA (qv). Pages 36–49 give guidance for viewing areas in sports stadia.

Except for AD M, the above guides do not have the force of law, but conformity with their recommendations will make it easier to demonstrate that the requirements of the Disability Discrimination Act have been met.

2 For places other than England and Wales the following regulatory documents

should be consulted:
• In Northern Ireland: TECHNICAL BOOKLET R of the Building Regulations (Northern Ireland) (qv).
• In Scotland: the NON-DOMESTIC TECHNICAL HANDBOOK of the Building (Scotland) Regulations (qv).
• In the USA, where the Americans with Disabilities Act applies: para 802 of the ADA AND ABA ACCESSIBILITY GUIDELINES (qv).
3 As of 2008, useful websites include the following. They are given in alphabetical order.

• DPTAC Access Directory	www.dptac.gov.uk/adnotes.htm
• DPTAC Publications	www.dptac.gov.uk/pubs.htm

See also: *Assembly and waiting rooms*; and *Auditorium design.*

Audio description

1 Audio description is a service for people with *impaired vision* (qv), helping them to enjoy leisure activities more fully. The service consists of a running commentary that can be added to television programmes; video and DVD; films; museum and gallery displays or events; theatrical performances; and sports events. The commentary gives spoken descriptions of body language, facial expression, scenery, action, costumes – in fact, of any aspect that might be important in conveying to hearers the storyline, event, or image.
2 In the UK a main provider of audio descriptions, and the most reliable source of expertise, is the RNIB (qv).
3 As of 2008, useful websites include the following. They are given in alphabetical order.

• Centre for Translation Studies	www.surrey.ac.uk/lcts/cts/
• Ofcom	www.ofcom.org.uk/
• RNIB	www.rnib.org.uk/

See also: *Vocaleyes.*

Audioguides

1 These are neck-hung devices with earphones, that visitors may carry around an art gallery or museum to hear commentaries on the displays. In addition to the standard packages available to the public at large, many galleries now have special audioguide packages for visitors with *impaired vision* (qv), and the best of these evoke the qualities of paintings and sculptures very skilfully. The equipment can be used with *hearing aids* (qv). The RNIB promotes and develops such schemes through its *Vocaleyes* (qv) project.
2 Under the Disability Discrimination Act (qv) audioguides and similar services come under the heading of *auxiliary aids and services* (qv). The proper provision of such services will make it easier for gallery owners and managers, in their role as *service providers* (qv), to demonstrate that the requirements of the DDA have been met.
3 For commendable case examples visit the websites given under: *Exhibition spaces: accessible*.

See also: *Touch tours; Virtual tours*; and *Vocaleyes*.

Audio signs

See: *Talking signs.*

Audit Commission

1 The Audit Commission is a *quango* (qv) under the Department for Communities and Local Government (qv). It is in turn audited by the National Audit Office (qv), which also audits central *government* (qv) departments.
2 As of 2008, useful websites include the following. They are given in

alphabetical order.

- Audit Commission www.audit-commission.gov.uk/
- Dept for Communities &c www.communities.gov.uk/
- Info4local www.info4local.gov.uk/
- National Audit Office www.nao.org.uk/

Auditorium design: complying with the Disability Discrimination Act

1 Buildings used by the public must in general comply with the Disability Discrimination Act (qv). In the case of theatres, cinemas, concert halls, conference facilities, lecture theatres, and teaching spaces *authoritative practical guides* (qv) include DESIGNING FOR ACCESSIBILITY (qv) plus the following:

2 APPROVED DOCUMENT M (qv), the provisions of which must be heeded in order to comply with the Building Regulations (qv) for England and Wales. Paras 4.10–4.12 and 4.35–4.36 contain important provisions on *wheelchair access.* For places other than England and Wales the following regulatory documents should be consulted:

- In Northern Ireland: TECHNICAL BOOKLET R of the Building Regulations (Northern Ireland) (qv).
- In Scotland: the TECHNICAL HANDBOOKS of the Building (Scotland) Regulations (qv).

3 BS 8300, the provisions of which apply to all buildings in the UK. Section 11.3 states brief design requirements for conference facilities, lecture theatres, and teaching spaces; and section 13.6 does the same for cinemas, concert halls, and theatres. BS 8300 does not have the force of law, but conformity with its recommendations will make it easier to demonstrate that the requirements of the Disability Discrimination Act have been met.

4 In the USA, where the Americans with Disabilities Act applies, refer to the ADA AND ABA ACCESSIBILITY GUIDELINES (qv).

5 As of 2008, other useful publications include the following. For publication details see the bibliography.

- BUILDINGS FOR THE PERFORMING ARTS by Ian Appleton.

See also: *Cinemas; Concert halls; Entertainment-related buildings;* and *Theatres*.

Auditorium design: emergency escape

1 Because *cinemas* (qv), *concert halls* (qv), *theatres* (qv), and other such assembly areas are occupied by large numbers of people, many of whom are unfamiliar with the building layout, the design and management of *means of escape* (qv) in such premises is a particularly difficult problem. The authorities have in the past placed severe restrictions on access for disabled persons to buildings of this kind. But such discrimination is no longer acceptable, and the following references give guidance on making equal or near-equal provision for *disabled people* (qv), including *wheelchair-users* (qv).

- BS 5588 FIRE PRECAUTIONS IN THE DESIGN, CONSTRUCTION AND USE OF BUILDINGS – PART 8: CODE OF PRACTICE FOR MEANS OF ESCAPE FOR DISABLED PEOPLE is an essential reference in the design of auditoria.
- In sports stadia, which may accommodate 80 000 or more spectators, the problem is especially acute. For authoritative guidance refer to GUIDE TO SAFETY AT SPORTS GROUNDS, and to the more recent and highly authoritative ACCESSIBLE STADIA.

2 For publication details of the above references see the bibliography.

See also: *BS 5588;* and *Means of Escape.*

ugmentative and lternative ommunication	See: *Alternative and augmentative communication.*
uthoritative practical uides to complying ith the Disability iscrimination Act: wellings	1 Private dwellings are not subject to the provisions of the Disability Discrimination Ac (qv). 2 For general design guides in this area see: *Housing: design.*
uthoritative practical uides to complying ith the Disability iscrimination Act: uildings used by the ublic	1 While this is not an explicit requirement of the Disability Discrimination Act (qv), the manner in which the Act is being interpreted means that the great majority of public premises (but not necessarily private ones such as private homes and some smaller clubs) must be physically *accessible* (qv) to *disabled people* (qv) – meaning in brief that such people should be able to enter and use those premises without being placed at an avoidable disadvantage in comparison with others. For more information on the Act and its application visit www.direct.gov.uk/disabilityand www.equalityhumanrights.com/ 2 The Act offers no specific design guidance on how to comply with its provisions. Pages 309–310 of CODE OF PRACTICE – RIGHTS OF ACCESS: SERVICES TO THE PUBLIC, PUBLIC AUTHORITY FUNCTIONS, PRIVATE CLUBS AND PREMISES (see the bibliography) vaguely direct designers to 'the wealth of published advice on the principles and practice of inclusive design' without specifically identifying the documents concerned; and other official sources advise designers to consult 'authoritative practical guides', again without identifying them. 3 However, the following publications are published by officially recognised bodies, are mentioned in official documents, and may reasonably be presumed to qualify. They are extensively quoted in the present work, and architects and other designers are strongly recommended to acquire all those titles that are relevant to their particular area of work for their reference shelf. Publication details are given in the bibliography. 4 First guides, which will help building designers to quickly get most of their early decisions broadly right when designing buildings that will be used by the public, include the following. They are given in alphabetical order. • ACCESS TO BUILDINGS by Stephen Garvin is a 12-page BRE Digest that outlines the requirements of the Building Regulations (qv) and the Disability Discrimination Act for both domestic and non-domestic buildings. • DESIGNING FOR ACCESSIBILITY, published by the Centre for Accessible Environments (qv), gives simple guidance for the design of public premises. It should be constantly at the designer's side. • INCLUSIVE ACCESSIBLE DESIGN by Adrian Cave provides a route-map through the principal regulatory documents, so that designers can see at a glance which clause they must consult when designing any particular element. • OPEN FOR BUSINESS, published by the Employers' Forum on Disability (qv) is similar to DESIGNING FOR ACCESSIBILITY, but focuses specifically on business premises. It should be used in addition to, not instead of, that reference. 5 Core references, which are more detailed essential references to augment the simple guides given above. • APPROVED DOCUMENT M (qv) of the Building Regulations. Outside England and Wales its approximate counterparts are TECHNICAL BOOKLET R of the Building Regulations (Northern Ireland); and the two TECHNICAL HANDBOOKS of the Building (Scotland) Regulations. • BS8300:2001, DESIGN OF BUILDINGS AND THEIR APPROACHES TO MEET THE NEEDS OF DISABLED PEOPLE, CODE OF PRACTICE (qv). This British Standard

• BS8300:2001, DESIGN OF BUILDINGS AND THEIR APPROACHES TO MEET THE NEEDS OF DISABLED PEOPLE, CODE OF PRACTICE (qv). This British Standard (qv) applies throughout the UK. Unlike AD M it does not have the force of law, but conformity with its recommendations will make it easier to demonstrate that the requirements of the Disability Discrimination Act have been met. This 164-page BS is exceedingly difficult to use owing to the lack of an index. See the entry for BS 8300 in the present work for a detailed contents list.
• INCLUSIVE MOBILITY (qv). This 164-page guidance document is free, and its excellent and clearly-written recommendations on the design of *pedestrian environments* (qv) and *transport-related buildings* (qv) apply throughout the UK.
6 Highly recommended references include the following.
• THE GOOD LOO DESIGN GUIDE (qv), published by the Centre for Accessible Environments (qv), gives admirably clear and sensible guidance on an important subject.
• THE WHEELCHAIR HOUSING DESIGN GUIDE (qv), from the Habinteg Housing Association, refers specifically to housing design, but may well be found useful in other situations as well.
7 Background documents include the following.
• ACCESS AND THE DDA: A SURVEYOR'S HANDBOOK by Mark Ratcliff was published in May 2007 and is therefore more up-to-date than some of the other references below. It is meant primarily for surveyors, but will also be useful to other professionals.
• CODE OF PRACTICE – RIGHTS OF ACCESS: SERVICES TO THE PUBLIC, PUBLIC AUTHORITY FUNCTIONS, PRIVATE CLUBS AND PREMISES, published in 2006, is a highly authoritative guide.
• ENGLISH PARTNERSHIPS GUIDANCE NOTE: INCLUSIVE DESIGN by Rita Newton and Marcus Ormerod, and published by English Partnerships, lists all relevant official documents as of 2007. The stupefying plethora of references shown here adds up to an impenetrable informational thicket that urgently needs simplification – see: *Better regulation bodies; Consultants' charter*; and *Over-regulation.*
• PLANNING AND ACCESS FOR DISABLED PEOPLE: A GOOD PRACTICE GUIDE (qv) by Driver Jonas, published by the Department for Communities and Local Government (qv).
8 Periodicals for keeping up to date with key issues include:
• ACCESS BY DESIGN
• ACCESS JOURNAL
9 Specialised documents will be required by professionals such as lighting designers, or for special building types such as schools and hospitals; but for the general run of buildings the dozen or so guides listed above ought – in a rational world – to be adequate basis for designing a genuinely *accessible* (qv) and *inclusive* (qv) built environment. If in doubt, designers should consult the Equality and Human Rights Commission (qv), which has overall responsibility in this area.
10 Readers are reminded that the present work only gives simplified summaries of original documents and does not purport to provide full, authoritative, and up-to-date statements of the law. As of 2008, useful websites for checking the latest situation include:
• Equality & Human Rights Commission www.equalityhumanrights.com/
• NRAC www.nrac.org.uk/Resource_list.html

See also: *Better regulation bodies; Consultants' charter; Disability Discrimination Act; Over-legislation; and Over-regulation.*

Authoritative practical guides to related legislation

1 The Disability Discrimination Act (qv) does not stand in isolation, but is surrounded by other legislation and standards that regulate the health, safety, and *accessibility* (qv) of the environment. As of 2007, useful guidance will be found in the following references. They are given in alphabetical order.

- THE ACCESS MANUAL: AUDITING AND MANAGING INCLUSIVE BUILT ENVIRONMENTS by Ann Sawyer and Keith Bright. Pages 46–85 give brief guidance to Acts etc. relating to *accessibility* (qv), *fire safety* (qv), and occupier liability.
- THE BUILDING REGULATIONS EXPLAINED AND ILLUSTRATED by M J Billington, M W Simons and J R Waters. Section 5 (13 pages) gives brief guidance to Acts etc. relating to *fire safety* (qv) and *health and safety* (qv).
- ENGLISH PARTNERSHIPS GUIDANCE NOTE: INCLUSIVE DESIGN by Rita Newton and Marcus Ormerod is a 30-page listing of virtually all current official standards and guidance documents to do with *inclusive design* (qv). See the comment in the previous entry.

2 For publication details of the above documents see the bibliography.

Autistic spectrum disorders (ASD)

1 A condition broadly linked to *Attention-deficit/hyperactivity disorder* (qv), *dyslexia* (qv), and *dyspraxia* (qv). Children affected by ASD have marked problems with communication and social interaction and exhibit a restricted, stereotyped range of behaviours. They have difficulty in initiating conversations, replying appropriately, volunteering information, or forming relationships with other children.

2 These developmental problems are said to affect up to 20% of school-age children, which is probably an exaggeration. Simon Baron-Cohen, the psychologist who revolutionised thinking about autism, has been quoted as saying that doctors are now putting on the autistic spectrum children they would once have dismissed as shy and a little odd. Be that as it may, they account for the majority of children with *Special Educational Needs* (qv) or SEN.

3 ASD may qualify as a *disability* (qv) under the Disability Discrimination Act (qv), and children thought to have the syndrome may qualify as having SEN. If they do, they are a particularly unsuitable category of children to accommodate in *mainstream schools* (qv), into which officialdom is increasingly trying to force them whether they like it or not. B IS FOR BULLIED (see the bibliography), a report published in 2006 by the National Autistic Society (NAS) drew attention to a huge incidence of bullying of autistic children, and even more so of children with Asperger's syndrome. Parents who think their autistic children should be in the protected environment of a *special school* (qv) should fight to have them placed in one. That is their constitutional right, and they are better placed to know what is best for their children than administrators. For more on this see: *Special Educational Needs*.

4 As of 2008, useful websites include the following. They are given in alphabetical order.

- Department for Children etc. — www.dcsf.gov.uk/
- Directgov — www.direct.gov.uk/
- Directgov: disabled people — www.disability.gov.uk/
- Disability Rights Commission — www.drc-gb.org/
- Dyslexia Research Trust — www.dyslexic.org.uk/
- Gloucester Special Schools Protection League — www.gsspl.org.uk/
- National Autistic Society — www.nas.org.uk/
- NHS Direct — www.nhsdirect.nhs.uk/
- Teachernet — www.teachernet.gov.uk/

See also: *Dyslexia; Medicalisation;* and *Special educational needs.*

Automatic control

See: *Environmental control systems;* and *Smart homes.*

Automatic doors

1 Few inventions have done more to make life easier for people who are burdened with luggage or push chairs, are frail or weak, or have *disabilities* – in a word: most of us, at one time or another – than the automatically-opening entrance door. It is an excellent example of *assistive technology* (qv) and of *inclusive design* (qv) – a design feature that raises the quality of life for just about the whole population.

2 An automatic powered door, preferably sliding, should be installed almost as a matter of course in busy new public building entrances, and retro-fitted wherever possible to existing ones.

3 Buildings used by the public must in general comply with the Disability Discrimination Act (qv). In the case of automatic, powered entrance doors *authoritative practical guides* (qv) include DESIGNING FOR ACCESSIBILITY (qv) and in addition the following documents. For publication details see the bibliography.

• APPROVED DOCUMENT M (qv): the provisions in section 2 must be heeded in order to comply with the Building Regulations for England and Wales.

• BS 8300 (qv): the recommendations in Section 6.3 apply to all buildings in the UK.

• INCLUSIVE MOBILITY (qv): the recommendations in Section 8.2 apply to *transport-related buildings* such as bus and coach stations, railway stations, air terminals, and transport interchanges.

Except for AD M, the above guides do not have the force of law, but conformity with their recommendations will make it easier to demonstrate that the requirements of the Disability Discrimination Act have been met.

For places other than England and Wales the following regulatory documents should be consulted:

• In Northern Ireland: TECHNICAL BOOKLET R of the Building Regulations (Northern Ireland) (qv).

• In Scotland: the TECHNICAL HANDBOOKS of the Building (Scotland) Regulations (qv).

• In the USA, where the Americans with Disabilities Act applies: refer to the ADA AND ABA ACCESSIBILITY GUIDELINES (qv).

4 As of 2008, other useful publications include the following. For publication details see the bibliography.

• AUTOMATIC DOOR SYSTEMS, published by the Centre for Accessible Environments (qv), is a highly authoritative general guide.

5 Useful websites include:

• Automated Door Systems www.automateddoorsystems.co.uk/

Automatic teller machines (ATMs)

1 This is the technical term for the cash-dispensing machines found inside and outside banks, in high streets and in many other public places. Important design considerations include:

• Correct height and other dimensions to suit all users, whether upstanding adults or sitting *wheelchair-users*.

• Protection from the rain, and screening from the sun, so that the text on the screen can be easily seen by day and night.

• Careful location, so that vulnerable people drawing money at any time during the day or night are not at risk from criminals. Pensioners drawing their pension or social security are particularly vulnerable.

2 Facilities used by the public must in general comply with the Disability

Discrimination Act (qv). In the case of ATMs, *authoritative practical guides* (qv) include the following documents. For publication details see the bibliography.

- BS 8300 (qv). Paragraphs 5.7.4 and 10.2 state basic standards.
- ACCESS TO ATMS: UK DESIGN GUIDELINES, published by the Centre for Accessible Environments (qv), is an authoritative guide.

Though these documents do not have the force of law, conformity with their recommendations will make it easier to demonstrate that the requirements of the Disability Discrimination Act have been met.
3 In the USA, where the Americans with Disabilities Act applies: refer to the ADA AND ABA ACCESSIBILITY GUIDELINES (qv).

---

utomatic window ontrols See: *Windows: controls.*

---

utonomy See: *Independent living.*

---

utowalks See: *Travelators.*

---

uxiliary aids and ervices See: *Adjustments under the Disability Discrimination Act.*

---

verage 1 This is an indispensable statistical concept, without which we would be unable to talk meaningfully about – for instance – the tallness of British people, their incomes, their longevity, and much else. But it is easily abused and must be used with care. To quote a familiar analogy: can a man sitting with one foot in a bucketful of ice, and the other in a bucket of boiling water, claim to be 'on average, quite comfortable'? Quite so.
2 See in this context also the following entries: *Anthropometric data; Median income; Normal distribution; Percentiles*; and *Statistics*.

---

verage wage See: Incomes.

B

Babies: premature

1 Each year approximately 800 babies are born in NHS maternity units under 25 weeks old, compared with the full term of 40 weeks. At 24 weeks 39% survive, and at 23 weeks only 17%.
2 For many of those children who do survive the prospects are grim. The EPICure study of babies born at 25 weeks or less, led by researchers at Nottingham University, found that by the age of six, 22% of the surviving children were severely *disabled,* some being blind and unable to walk; 34% had lesser problems such as a squint; and only 20% had no impairments. For these reasons Dutch doctors recently took the humane policy decision no longer to help babies born at 25 weeks to survive, and to permit mercy killing for a range of incurable conditions including an absolutely horrific skin condition called epidermolysis bullosa.
3 In Britain Baroness Warnock (see: *Special educational needs*) has called for doctors to stop competing for the technical 'triumph' of keeping babies alive at increasingly young ages, despite knowing that many of the survivors would be agonisingly disabled, and in November 2006 the Royal College of Obstetricians and Gynaecology proposed that a policy of 'active euthanasia' should be considered in the UK. The college's submission was welcomed by Dr Richard Nicholson, editor of the *Bulletin of Medical Ethics*, who spoke about his own experience of the agonies suffered by severely handicapped babies, and by John Harris, a member of the government's Human Genetics Commission and professor of bioethics at Manchester University.

See also: *Abortion*.

Balustrade

A balustrade normally consists of:
• A *guarding* – defined as a screen to stop people falling off the edge of a stair, ramp, landing, balcony, etc. Such a screen should be high enough to prevent people falling over (probably between 900 mm and 1 100 mm above floor surface or stair *pitch line*); be structurally resistant to horizontal forces; and where children under the age of 5 may be present the guarding should not be easily climbable nor have apertures which allow passage of a 100 mm diameter sphere.
• A *handrail* – defined as a rail that people may grip for guidance or support. Such a rail should be easy to grip (probably between 40 mm and 50 mm in diameter); smooth-surfaced; and mounted at a convenient height (900 mm to 1 000 mm above floor surface or stair pitch line).

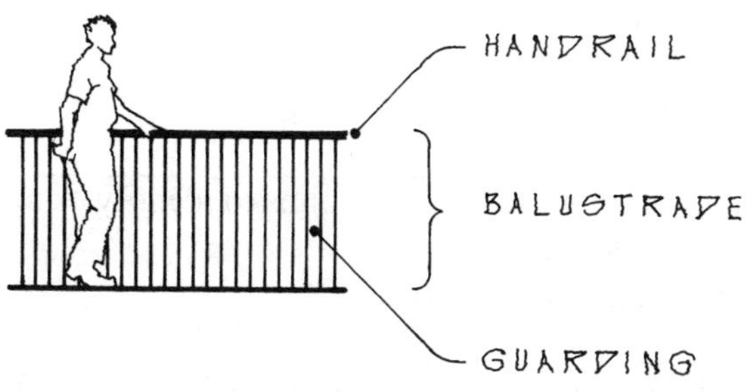

2 For more detail see: *Guardings; Handrails; Ramps*; and *Stairs*.

Basic Skills Agency

1 An agency (see: *Quangos*) under the Department for Children, Schools and Families (qv). Its remit is to help people of all ages who struggle with words and numbers.
2 For more information visit www.basic-skills.co.uk/

Bathing aids

See: *Products for easier living.*

Bathrooms in dwellings

1 The following references contain authoritative guidance:

- THE HABINTEG HOUSING ASSOCIATION DESIGN GUIDE (qv) applies most specifically to *housing association* (qv) schemes, but ought to be by the designer's side in the design of all housing developments and individual dwellings.
- THE WHEELCHAIR HOUSING DESIGN GUIDE (qv), also from the Habinteg Housing Association, is another illustrated manual whose usefulness extends far beyond the special remit implied by the title.
- THE HOUSING DESIGN GUIDE is a best-practice guide for housing in London.

For publication details of the above references see the bibliography.

2 As of 2008, useful websites include:

- Disabled Living Foundation www.dlf.org.uk/public/factsheets.html
- Disabled Living Foundation http://gus.dlf.org.uk/
- Disabled Living Online www.disabledliving.co.uk/

Bathrooms in public buildings: complying with the Disability Discrimination Act

1 Buildings used by the public must in general comply with the Disability Discrimination Act (qv). In the case of bathrooms in public buildings, *authoritative practical guides* (qv) include DESIGNING FOR ACCESSIBILITY (qv) and in addition the following documents. For publication details see the bibliography.

- APPROVED DOCUMENT M (qv), the provisions of which must be heeded in order to comply with the Building Regulations (qv) in England and Wales. Refer to paras 4.19, 4.24, and 5.19–5.21. These provisions apply to hotels, motels, student accommodation, etc.
- BS 8300 (qv) applies to all buildings in the UK. Refer to section 12.2, which makes recommendations for hotels, motels, nursing or *residential homes* (qv), hostels, halls of residence, *day centres* (qv), *hospitals* (qv), and *sports-related buildings* (qv). BS 8300 does not have the force of law, but conformity with its recommendations will make it easier to demonstrate that the requirements of the Disability Discrimination Act have been met.

For places other than England and Wales the following regulatory documents should be consulted:

- In Northern Ireland: TECHNICAL BOOKLET R of the Building Regulations (Northern Ireland) (qv).
- In Scotland: the TECHNICAL HANDBOOKS of the Building (Scotland) Regulations (qv).

2 In the USA, where the Americans with Disabilities Act applies, refer to the ADA AND ABA ACCESSIBILITY GUIDELINES (qv).

See also: *Disability Discrimination Act.*

Beacon Councils

1 This is a government scheme that aims to identify excellence and innovation in *local government* (qv).

2 Visit www.idea.gov.uk/beacons/

Becta

1 The British Education Communications and Technology Agency is an agency (see: *Quangos*) that supports all four education departments in the UK in their Information and Communications Technology (ICT) developments, with the aim of improving education through technology.

2 For more information visit www.becta.org.uk/

See also: *Children: fingerprinting.*

Bedrooms in dwellings 1 The following references contain authoritative guidance:
• THE HABINTEG HOUSING ASSOCIATION DESIGN GUIDE (qv) applies most specifically to *housing association* (qv) schemes, but ought to be by the designer's side in the design of all housing developments and individual dwellings.
• THE WHEELCHAIR HOUSING DESIGN GUIDE (qv), also from the Habinteg Housing Association, is another illustrated manual whose usefulness extends far beyond the special remit implied by the title.
• THE HOUSING DESIGN GUIDE is a best-practice guide for housing in London.
2 For publication details of the above see the bibliography.

Bedrooms in public buildings: complying with the Disability Discrimination Act 1 Buildings used by the public must in general comply with the Disability Discrimination Act (qv). In the case of bedrooms in public buildings, *authoritative practical guides* (qv) include DESIGNING FOR ACCESSIBILITY (qv) and in addition the following documents. For publication details see the bibliography.
• APPROVED DOCUMENT M (qv), the provisions of which must be heeded in order to comply with the Building Regulations (qv) in England and Wales. Refer to paras 3.10, 4.4 and 4.17–4.24. These provisions apply to hotels, motels, student accommodation, etc.
• BS 8300 (qv), which applies to all buildings in the UK. Refer to section 12.5, which makes recommendations for hotels, motels, nursing or *residential homes* (qv), hostels, college and university halls of residence, day centres, visitors' accommodation in hospitals, and *sports-related buildings* (qv). BS 8300 does not have the force of law, but conformity with its recommendations will make it easier to demonstrate that the requirements of the Disability Discrimination Act have been met.
For places other than England and Wales the following regulatory documents should be consulted:
• In Northern Ireland: TECHNICAL BOOKLETS E, H, and R of the Building Regulations (Northern Ireland) (qv).
• In Scotland: the BUILDING (SCOTLAND) REGULATIONS and the associated TECHNICAL HANDBOOKS (qv).
2 In the USA, where the Americans with Disabilities Act applies, refer to the ADA AND ABA ACCESSIBILITY GUIDELINES (qv).

See also: *Disability Discrimination Act.*

Beds 1 To able-bodied people, going to bed or getting up may seem easy; but many *ageing* (qv), *disabled* (qv) or sick people have great difficulty getting into or out of bed, turning over, sitting up, etc. They and their *carers* (qv) can be helped by variable-posture beds, tilting beds, height-adjustable beds, and *hoists* (qv).
2 For useful websites see: *Adjustable furniture.*

See also: *Adjustable furniture;* and *Products for easier living.*

Bending main programmes 1 Officialese for 'tackling deprivation by focusing local agency and government department spending more specifically on the most deprived areas'.
2 See also: *Mainstreaming.*

Benefit 1 For more detail see the specific entries: *Child benefit; Council tax benefit; Housing benefit; Incapacity benefit*; and *Unemployment benefit.*
2 As of 2008, useful websites include:
• Directgov www.direct.gov.uk/

Benefit reform

1 Britain's benefit system needs to be hugely simplified. The report BENEFITS SIMPLIFICATION, produced in July 2007 by the House of Commons' Work and Pensions select committee (see the bibliography), called the current system 'stunningly complicated'. The Department for Work and Pensions (qv) alone administers around 40 benefits, each with different rules. The Treasury-managed *tax credits* (qv) system adds an additional layer of bafflement. Such is the overall complexity, according to the Select Committee, that the taxpayer now loses more money to official and claimant error than to fraud.
2 See *Child benefit; Council tax benefit; Housing benefit; Incapacity benefit; Unemployment benefit,* and *Welfare reform.*

Best Value (BV)

1 This is a system that requires *local authorities* (qv) to:
• Continuously improve their performance;
• Undergo regular inspections under the Best Value Inspection Service, and
• Be awarded two marks ranging between zero and three. The first mark is for performance, the second for the likelihood of improved performance. The system is well-meant but excessively bureaucratic see: *Better regulation bodies* and *Over-regulation.*
2 For more information visit www.bvpi.gov.uk/pages/Index.asp

Better regulation bodies

1 Since 1997 the Government has launched a series of 'better regulation' initiatives in a half-hearted attempt to cut the mountain of *red tape* (qv) that is suffocating British businesses and *voluntary organisations* (qv). Each of these has failed spectacularly. On the one hand the Government creates over 70 000 pages of new legislation each year; on the other the much-trumpeted Regulatory Reform Act 2006, which was meant to cut excessive legislation, has not (according to Lord Bach in the House of Lords on 9 December 2007) succeeded in abolishing one single rule or regulation in the 18 months from May 2006 to December 2007. The mountain continues to grow apace.
2 Prime Minister Tony Blair commenced his attack on excessive regulation by creating the Better Regulation Task Force in 1997. In October 2004 the BRTF published a seminal report titled AVOIDING REGULATORY CREEP, which sets out five 'Principles of good regulation'. The latter are still officially recognised in 2007 as the basis of sensible regulation-making. They are:
• Proportionality.
• Accountability.
• Consistency.
• Targeting.
• Transparency.
3 Mr Blair immediately asked the BRTF to follow up its initial report with a set of practical recommendations, and from that point on the history of this ill-fated programme is as follows.
• In March 2005 the BRTF produced a set of practical proposals under the title REGULATION – LESS IS MORE: REDUCING BURDENS, IMPROVING OUTCOMES.
• On 18 July 2005 Mr Blair wrote to the BRTF confirming that the Government had 'accepted in full' (sic) its recommendations.
• The BRTF was then dissolved and replaced by a plethora of new bodies including the 'Better regulation bodies' which as of 2007 include (in alphabetical order) the Better Regulation Commission; the Better Regulation Executive; the Better Regulation Review Group; the Bureaucracy Reduction Group; the Cutting Burdens on Teachers initiative; the Higher Education Regulation Review Group; the Implementation Review Unit; the Lifting Burdens Task Force; the Reducing Burdens programme; the Regulatory Impact Unit; and the DTI's Cutting Red Tape programme and Local Better Regulation Office – see these individual entries for more detail.

• In May 2006 the Government passed the Legislative and Regulatory Reform Act (qv), to help it carry through a programme of regulatory simplification – for details visit www.cabinetoffice.gov.uk/regulation/reform/bill/
• In February 2007 a group of 19 government departments and *quangos* (qv) published their respective 'Simplification Plans'. These are strategies for cutting the administrative burdens on businesses and voluntary societies by at least one quarter by 2010.
• In April 2007 the Cabinet Office (qv) announced that Regulatory Impact Assessments (qv) would be replaced forthwith by Impact Assessments 'to ensure that all new regulation is necessary and carried out with minimum burdens'. The new Impact Assessments would, it was claimed, 'mean clearer guidelines for policy makers' to help them in the task of not generating unnecessary bureaucracy. For more on Impact Assessments visit www.cabinetoffice.gov.uk/regulation/news/2007/070402_ia.asp and www.cabinetoffice.gov.uk/regulation/ria/index.asp
• On 9 December 2007 Lord Bach had to admit to the House of Lords that the Government had in the 18 months since May 2006 not managed to abolish a single rule or regulation under its much-vaunted Legislative and Regulatory Reform Act (qv). To fully grasp the implications of this admission, it must be must be set alongside the commentator Ross Clark's finding, on pp 5-6 of HOW TO LABEL A GOAT (published in 2006 – see the bibliography) that the Government is imposing over 70 000 pages of new legislation upon Britain each year.

4 This is an abysmal record. Yet official websites publications such as SIMPLIFICATION PLAN 2007: PROMOTING BUSINESS AND ENTERPRISE THROUGH BETTER REGULATION, published in December 2007 by the Department for Business, Enterprise and Regulatory Reform (qv), which is currently in charge of red tape reduction, speak of nothing but success and the prospects of more success.

5 Such optimistic words are contradicted by the everyday experience of businessmen, professionals, and voluntary organisations. The bureaucratic load is not reducing, but increasing, and one key reason is that the Government reneged on Mr Blair's 'acceptance in full' (see above) of REGULATION – LESS IS MORE: REDUCING BURDENS, IMPROVING OUTCOMES. The cornerstone of that document, which was regarded as so fundamental that the authors highlighted it no fewer than 11 times in the 12 pages of the Executive Summary, was the proposed 'One in, One out' rule, whereby one existing provision must be removed each time a new regulation is made. Here was a red tape-busting measure that is simple, concrete, and (judging by the Dutch experience) effective. But Mr Blair's Government quietly ditched this key proposal and decided to place its entire reliance on a forest of new bureaucracy to fight existing bureaucracy.

6 The Government no longer has any realistic hope of fulfilling its pledge in early 2007 to cut the burden of red tape by one quarter, in real terms, by the year 2010. It is time for Ministers to abandon their make-believe policies and take three practical measures:
• Repeal the European Communities Act 1972 (qv), under which between 50% and 80% of UK laws arrive from Brussels readymade and pass virtually automatically onto our lawbook.
• Stop trying to control every aspect of our lives and drastically cut the amount of home-grown legislation. See: *Over-legislation; Over-regulation;* and *Privacy.*
• Implement the 'One in, One out' rule.

7 As of 2008, useful references include the following. They are listed in date order. For publication details see the bibliography.

• AVOIDING REGULATORY CREEP by the Better Regulation Task Force, was published in October 2004.

• REGULATION – LESS IS MORE: REDUCING BURDENS, IMPROVING OUTCOMES by the Better Regulation Task Force, was published in March 2005.

• SIMPLIFICATION PLANS: A SUMMARY by the Better Regulation Commission and the Better Regulation Executive, summarised progress up to December 2006.

• HOW TO LABEL A GOAT: THE SILLY RULES AND REGULATIONS THAT ARE STRANGLING BRITAIN by Ross Clark, was published in 2006.

• SIMPLIFICATION PLAN 2007: PROMOTING, BUSINESS AND ENTERPRISE THROUGH BETTER REGULATION by the Department for Business, Enterprise and Regulatory Reform (qv), was published in December 2007.

8 Useful websites include the following. They are listed in alphabetical order.

• Better Regulation www.betterregulation.gov.uk/
• Better Regulation Commission www.brc.gov.uk/
• Better Regulation Executive www.cabinetoffice.gov.uk/regulation/
• BDR publications http://publications.brc.gov.uk/
• Cabinet Office www.cabinetoffice.gov.uk/better_regulation/
• DWP www.dwp.gov.uk/aboutus/better_regulation/
• Directgov www.direct.gov.uk/

See also: *Contract of compliance; Over-legislation; Over-regulation;* and *Post-democracy.*

Better Regulation Commission

See the previous entry.

Better Regulation Review Group (BRRG)

1 The BRRG was established in 2003, and was superseded in 2004 by the Higher Education Regulation Review Group (qv).

2 For a short history of the BRRG, and its final report, visit www.hefce.ac.uk/lgm/account/brrg.asp

Bill of Rights

See: *Constitution;* and *Human Rights Act.*

Biometric data

See: *Children: fingerprinting; National DNA Database;* and *Privacy.*

Blindness

1 Blindness is a substantial and permanent, but not necessarily total lack of sight. Most blind people can distinguish between light and dark, and many can recognise a friend at arms length. The number of users of public buildings and spaces who are blind is impossible to estimate accurately, but may be of the order of 1 in 5 000, according to a 1992 survey of shopping centres quoted on p 372 of DESIGNING FOR THE DISABLED: THE NEW PARADIGM by Selwyn Goldsmith – see the bibliography.

2 Visit www.rnib.org.uk/

See also: *Braille;* and *Impaired vision.*

Blissymbolics

1 Blissymbolics is a *symbol system* (see: *Speech aids*) for people with *impaired communication* (qv). Some of its symbols are 'pictographs' which look like the objects they represent; others are 'ideographs' which represent ideas; and yet others are arbitrary. It is said to be cognitively more demanding, but visually less busy than purely pictographic systems.

2 Visit www.blissymbolics.org/

See also: *Alternative and augmentative communication; Sign and symbol systems;* and *Speech aids.*

Blister surfaces

1 Blister surfaces, consisting of parallel rows of flat-topped blisters, are one of six types of *tactile pavings* (qv). Their function is to warn *blind* (qv) people that they are about to step onto a carriageway. They are used at *pedestrian crossings* where the *kerb* (qv), which would normally be the warning signal, has been dropped to be flush with the carriageway. They must not be applied to other situations, as consistency of use is essential if potentially fatal misunderstandings are to be avoided.

2 The diagram below shows a common example, but the rules for correct use in particular situations are very precise, and designers must in all cases consult Chapter 1 of GUIDANCE ON THE USE OF TACTILE PAVING SURFACES (qv) and para 4.1 of INCLUSIVE MOBILITY (qv) – see the bibliography.

3 For details of the six types of tactile paving, including blister surfaces, see: *Tactile pavings*.

Blue Badge schemes

1 A European-wide arrangement that provides parking concessions to people with severely *impaired mobility* (qv). Useful references include the LONDON BLUE BADGE PARKING GUIDE – see the bibliography.

2 As of 2008, useful websites include:

- Dept for Transport www.dft.gov.uk/transportforyou/access/bluebadge/
- DirectGov www.directgov.gov.uk/

See also: *Accessible roads;* and *Car parking.*

Braille

1 Braille is one of two systems of *tactile text* (qv) that enable *blind* (qv) people to read by touch, the other being *Moon* (qv). Braille uses varying patterns of six raised dots which can give 63 different distinguishable combinations. While it may be used for very short messages in those buildings that are likely to be frequented by blind people, Braille cannot be relied upon as the sole method of signage for partially sighted and blind people, as fewer than 50 000 of Britain's 1.8 million people with severely impaired vision are regular users of the system. Signs in large print, audio messages, and conventional lettering or symbols embossed onto signs, lift control buttons, etc., would help far more people – see: *Signs*.

2 Where Braille is used the recommendations given in para 9.2.4 of BS 8300 (qv) should be followed.

3 As of 2008, useful websites include:
- Braille Signs www.braille-signs.co.uk/
- RNIB www.rnib.org.uk/
- Tactile Group signs www.tactilesigneurope.com/

See also: *Blindness*, *Impaired vision; Moon; Tactile displays;* and *Tactile signs.*

Bridges

See: *Footbridges.*

British constitution

See: *Constitution.*

British Crime Survey

See: *Crime.*

British Educational and Communications and Technology Agency

See: *Becta*

British Sign Language (BSL)

1 BSL is a standard sign language used to communicate with people who are *deaf* (qv) or have *impaired hearing* (qv). The RNID (qv) estimates that around 50 000 people in the UK use BSL to communicate, and that it is the first language of many of them. Services offered include:
- BSL interpreters are employed at conferences for simultaneous translation of speeches. Many major museums and art galleries have regular BSL interpreted lectures, talks, and tours.
- The RNID offers deaf people the services of BSL interpreters via a videophone.

2 In March 2003 the Government announced that BSL would henceforth be recognised as an official language, alongside other minority languages such as Welsh. The RNID has predicted that *tribunals (qv)* hearing cases of *discrimination* (qv) brought under the Disability Discrimination Act (qv) will in future be more likely to regard provision of BSL interpreting as a reasonable *adjustment* (qv) that could be expected in situations such as job interviews and going to see one's GP. The National Deaf Children's Society similarly predicted a revolution in education, with schools having to offer deaf children the opportunity of using BSL and studying it as a GCSE or A-level subject.
3 As of 2008 useful websites include:
- Association of Sign Language Interpreters www.asli.org.uk/
- British Sign Language Com http://web.ukonline.co.uk/p.mortlock/
- RNID www.rnid.org.uk/

See also: *Impaired hearing;* and *Sign and symbol systems.*

British Social Attitudes Survey (BSAS)

1 The BRITISH SOCIAL ATTITUDES SURVEY series has been published annually since 1983. It is conducted by the National Centre for Social Research (formerly Social and Community Planning Research) and is designed to produce annual measures of attitudinal movements to complement large-scale government surveys such as the General Household Survey and the Labour Force Survey. The subjects covered include housing and home ownership; work and unemployment; health and social care; education; business and industry; social security and dependency; tax and spending; the welfare state; transport; the environment and countryside; constitutional reform; law and order; civil liberties; moral issues and sexual mores; racism and sexism; social inequality; religion; politics and governance.
2 For publication details see the bibliography.

British Standards (BSs)

1 A 'BS' is a published specification of some material, product, method, or service, setting minimum standards for the matters at issue. BSs do not have the force of law unless some *law* (qv) or *regulation* (qv) specifically makes compliance with a particular Standard compulsory. But professionals should familiarise themselves with all British Standards that are relevant to the task upon which they are engaged. Those that are likely to be particularly relevant to *inclusive design* (qv) include:

- BS 5588-8: FIRE PRECAUTIONS IN THE DESIGN, CONSTRUCTION AND USE OF BUILDINGS – CODE OF PRACTICE FOR MEANS OF ESCAPE FOR DISABLED PEOPLE (qv).
- BS 8300 DESIGN OF BUILDINGS AND THEIR APPROACHES TO MEET THE NEEDS OF DISABLED PEOPLE – CODE OF PRACTICE (qv).

2 BSs are, like DINs in Germany and AFNOR in France, being superseded by the EU-wide CE mark, which is a declaration by manufacturers that their product meets all the appropriate European Directives – for more information visit www.dti.gov.uk/innovation/strd/cemark/page11646.html

3 For more information on BSs in general visit www.bsi-global.com/en/Standards-and-Publications/

British Toilet Association (BTA)

1 The BTA represents the interests of (a) 'away from home' toilet providers, and (b) toilet users of all types, and works to improve standards. It is unglamorous work that deserves the highest commendation – especially in view of the reprehensible abandonment by *local authorities* (qv) of their civic duty to provide a sufficiency of good, clean public lavatories in all public places. For more on this see: *Lavatories: public.*

2 Visit www.britloos.co.uk/

See also: *Lavatories.*

Broadcasting Advertising Clearance Centre (BACC)

1 If there were an annual Nanny State award (alas, there isn't) the BACC might have won it in 2007 with its ban on re-runs of Tony Hancock's 1950s commercials using the slogan 'Go to work on an egg', on the grounds that these advertisements would 'encourage an unhealthy lifestyle'.

2 Visit www.bacc.org.uk/

See also: *Health scares; Political correctness;* and *Risk.*

BS 5588

1 BS 5588: FIRE PRECAUTIONS IN THE DESIGN, CONSTRUCTION AND USE OF BUILDINGS has several parts. Part 8, which is most directly relevant to users of the present work, is noted below.

2 Visit www.bsonline.bsi-global.com/server/index.jsp

BS 5588: Part 8

1 BS 5588-8: FIRE PRECAUTIONS IN THE DESIGN, CONSTRUCTION AND USE OF BUILDINGS – CODE OF PRACTICE FOR MEANS OF ESCAPE FOR DISABLED PEOPLE was produced to help meet the needs of *disabled people* (qv), whose particular problems in trying to escape from a fire in a building were not adequately addressed in previous codes of practice and other documents.

2 It is based on the following important concepts:

- That *means of escape* (qv) routes in multi-level premises should be planned in such a way that disabled people can move as far away from the fire as possible without assistance, travelling along a strictly horizontal route;
- That they will find at the end of that horizontal route a safe *refuge* (qv) in a *place of relative safety* (qv) such as a *protected* lift or stair lobby, where they

can await the arrival of managers or the fire brigade to help them negotiate the vertical part of the escape route to a *place of safety* (qv).
• That managers are required to augment these design measures by clear *evacuation* (qv) plans which ensure that every staff member and disabled person knows exactly what to do in case of fire.
The above principles are schematically shown below.

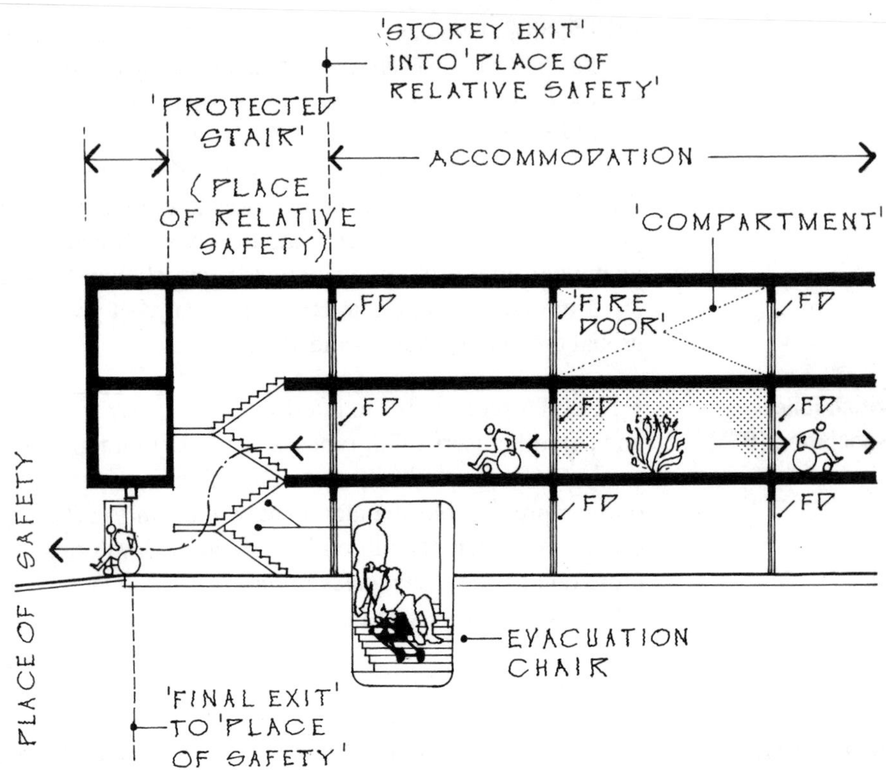

3 Some commentators doubt that disabled people fleeing from a fire will patiently await assistance in a refuge, and suspect that they are more likely to do something dangerous like trying to get down the stair. This may be true, but against that the number of disabled people injured or killed by fire in buildings each year is so vanishingly small that the cost of implementing more radical and ambitious design measures in all public buildings could be better spent in other ways.
4 The principles set out in BS 5588-8 should be followed. While the document does not have the force of law, conformity with its recommendations will make it easier to demonstrate that the requirements of the Disability Discrimination Act (qv) have been met

See also: *Evacuation; Means of escape; Place of relative safety; Place of safety*, and *Statutory instruments and Codes of practice.*

BS 8300

1 BS 8300 DESIGN OF BUILDINGS AND THEIR APPROACHES TO MEET THE NEEDS OF DISABLED PEOPLE – CODE OF PRACTICE is, after APPROVED DOCUMENT M (qv) of the Building Regulations, the most authoritative source of design guidance for complying with the Disability Discrimination Act (qv).
2 Unlike AD M, BS 8300 does not have the force of law, but conformity with its recommendations will make it easier to demonstrate that the requirements of the Disability Discrimination Act have been met. This 164-page document is expensive, and exceedingly difficult to use owing to the lack of a detailed index. But designers are very strongly advised to acquire a copy and not to depart from its recommendations without good reason and on an informed basis.

3 For publication details of BS 8300 see the bibliography.

See also: *Authoritative practical guides*; and *Statutory instruments and Codes of practice.*

BS 8300: contents

The following list of clauses in BS 8300 (the left-hand column) are shown alongside their approximate counterparts in APPROVED DOCUMENT M (the two central columns) so that designers can check both sets of recommendations for any particular building feature. Sections 1 to 3 of BS 8300 are introductory.

BS 8300	AD M: (non-dwellings)	AD M: (dwellings)	DESIGN ASPECT
4	1.1–1.18	6.1	CAR PARKING, SETTING-DOWN POINTS, AND GARAGING
4.1			Car parking, garaging, enclosed parking
4.1.1			On-street parking bays
4.1.2			Provision of designated parking bays
4.1.2.2			Workplaces
4.1.2.3			Shopping, recreation & leisure facilities
4.1.2.4			Railway car parks
4.1.2.5			Churches
4.1.2.6			Crematoria and cemetary chapels
4.1.3			Off-street garages, enclosed parking
4.1.3.1			Uncovered parking areas
4.1.3.2	1.18		Designated off-street parking spaces
4.1.3.3			Multi-storey car parks
4.1.3.4			Garaging and enclosed parking
4.1.4			Entrances to car parks
4.1.4.1			Signs to designated parking spaces
4.1.4.2			Barrier control systems
4.1.4.3			Vehicle heights barriers
4.1.5			Parking meters, controls, ticket dispensers
4.1.5.1			Information
4.1.5.2	1.16–1.18		Pay-and-display systems
4.1.5.3	1.16–1.18		Ticket dispensers
4.2	1.17–1.18		Setting-down points
5	1.1–1.39	6.1–6.18	ACCESS ROUTES TO AND AROUND BUILDINGS
5.1	1.1–1.25	6.1–6.12	Location and general design
5.2	1.13	6.9–6.10	Width and height
5.3	1.13		Passing places
5.4	1.19–1.26	6.11	Gradients
5.5	1.13	6.9–6.11	Footway & footpath surfaces
5.5.1	1.13	6.9–6.11	General recommendations
5.5.2	1.13		Tactile pavings
5.6			Drainage gratings
5.7			Barriers, restrictions, hazards
5.7.1			Street furniture
5.7.1.1			Location of street furniture
5.7.1.2			Free-standing posts & columns
5.7.2	1.38–1.39		Hazard protection
5.7.3			Free-standing stairs & ramps
5.7.4			Automatic teller machines
5.8	1.19–1.39	6.14–6.15	Ramped access
5.8.1	1.19–1.25		General recommendations

BS 8300	AD M: (non-dwellings)	AD M: (dwellings)	DESIGN ASPECT
5.8.2	1.26	6.14	Gradients
5.8.3	1.26	6.15	Width
5.8.4	1.26	6.15	Landings
5.8.5	1.34–1.37		Handrails
5.8.6	1.26		Surface materials
5.8.7	1.26		Protection against falling
5.8.8	1.25		Lighting
5.8.9			Portable or temporary ramps
5.9	1.27–1.39	6.16–6.17	Stepped access
5.9.1			Protection against weather
5.9.2	1.33	6.17	Design of steps
5.9.3	1.33	6.17	Rise of a flight
5.9.4	1.33	6.17	Width
5.9.4	1.33		Identification of nosings
5.9.6	1.33	6.17	Landings
5.9.7	1.34–1.37	6.17	Handrails
5.9.7.1		6.17	Where to provide
5.9.7.2	1.37	6.17	Height and extension
5.9.8	1.32		Lighting
5.10	1.34–1.37	6.17	Handrail design
5.10.1	1.37	6.17	Profile, colour, detailing of terminations
5.10.2	1.33	6.17	Dimensions
5.9.6	1.33		Tactile warning surfaces
6	2.1–2.29	6.19–6.23	ENTERING A BUILDING
6.1	2.1–2.7		The principal entrance
6.1.1	2.1–2.3		General principles
6.1.2	2.7		Visual clarity
6.1.3			Protection against weather
6.2		6.21	Threshold
6.3	2. 8–2.24	6.22–6.23	Principal entrance door
6.3.2	2.14–2.17		Self-closing single swing-doors
6.3.3	2.18–2.22		Power-operated (automatic) doors
6.3.4			Low energy swing-doors
6.3.5	2.18–2.22		Power-operated revolving doors
6.3.6	2.25–2.29		Entrance lobbies
6.3.6.1	2.25–2.28		General principles
6.3.6.2	2.29		Dimensions
6.3.6.3	2.29		Glazing
6.3.6.4	2.29		Projections
6.4			DOOR DESIGN
6.4	2.13–3.10	6.22–6.23	External door design
6.4			Internal door design
6.4.1	2.13	6.23	Effective clear width
6.4.2	2.15–2.17		Location and minimum clear space at side
6.4.3	2.13		Vision panels
6.4.4	2.23, 2.14		Glass doors
6.5	2.16, 2.17		Door furniture
6.5.1	2.17		Handles
6.5.2			Hinges
6.5.3			Securing of entrance doors
6.5.3.2			Locks & latches
6.5.3.3			Keys
6.5.3.4			Door bolts

BS 8300	AD M: (non-dwellings)	AD M: (dwellings)	DESIGN ASPECT
6.5.4			Door entry systems
6.5.5			Entryphones
6.5.6			Wide-angle viewing lenses
7	3.1–3.38	7.1–7.5	HORIZONTAL CIRCULATION
7.1	3.2–3.6		Entrance hall & reception area
7.1.2	3.6		Indentification of reception point
7.1.3	3.3		Accessible route
7.2	3.11–3.14	7.2–7.5	Corridors & passageways
7.2.1	3.14	7.5	Projections into corridors
7.2.2.	3.14	7.5	Dimensions of corridors
7.2.3	3.14	–	Floors in corridors
7.2.4	3.12	–	Lighting in corridors
7.2.5	3.14	–	Doors leading into corridors
7.2.6	3.10	–	Doors across corridors
7.3	3.10	–	Doors fitted with closing devices
7.3.1	3.10	–	General principles
7.3.2	3.10	–	Fire doors fitted with closing devices
7.3.3		–	Fire doors in residential buildings
–	3.15–3.16	–	Internal lobbies
8	3.17–3.20	7.6–7.7	VERTICAL CIRCULATION
8.1	3.50–3.51	7.6–9.5	Internal steps & stairs
8.1.1	–	7.7–9.5	Design of steps & stairs
8.1.2	–	–	Rise of a flight
8.1.3	–	7.7	Stair clear width
8.1.4	–	–	Identification of nosing & riser
8.1.5	–	–	Landings
8.1.6	3.54–3.55	7.7–9.5	Handrails to steps & stairs
8.1.6.1	–	9.5	Handrail provision
8.1.6.2	–	–	Height & extension of handrails
8.1.7	–	–	Artificial lighting
8.1.8	–	–	Surface finishes
8.2	3.52–53	–	Internal ramps
8.2.1	–	–	General principles
8.2.2	–	–	Ramp gradients
8.2.3.	–	–	Ramp widths
8.2.4	–	–	Landings
8.2.5	3.54–3.55		Handrails to ramps
8.2.6	–	–	Surface materials
8.2.7	–	–	Protection against falling
8.2.8	–	–	Lighting
8.3	–	–	Handrail design
8.4	3.17–3.49	9.1–9.7	LIFTS
8.4.1	3.17–3.49	9.1–9.7	Provision of lifts
8.4.2	3.17–3.49	–	Access to lifts
8.4.3	3.29–3.34	9.7	Passenger lifts
8.4.4	3.35–3.43	–	Platform lifts
8.4.5	3.44–3.49	–	Stairlifts
8.4.5.1	–	–	Stairlifts for dwellings
8.4.5.2	3.44–3.49	–	Stairlifts for public buildings
8.5	–	–	ESCALATORS & PASSENGER CONVEYORS
9	4.31–4.36	–	SURFACES & COMMUNICATION AIDS

BS 8300	AD M: (non-dwellings)	AD M: (dwellings)	DESIGN ASPECT
9.1	4.32	–	Surface finishes
9.1.1	–	–	Visual characteristics
9.1.2	–	–	Materials acoustic design
9.1.3	3.6, 3.14, 5.18, 5.21	– –	Floor surfaces
9.1.4	3.4, 3,32	–	Wall surfaces
9.1.5	2.22–3.14		Glazed walls & screens
9.2	–	–	SIGNS & INFORMATION
9.2.1	2.5–2.7, 3.5, 3.18	–	Provision of signs & information
9.2.2	–	–	Location and design of signs
9.3	–	–	AUDIBLE COMMUNICATION SYSTEMS
9.3.1	4.36	–	Public address etc. systems
9.3.2	4.33–4.36	–	Hearing enhancement systems
9.3.3	4.36	–	Induction loop systems
9.3.4	4.36	–	Infrared systems
9.3.5	–	–	Radio systems
9.3.6	4.36		Inductive couplers
9.3.7	–	–	Fire alarm systems
9.4	4.34	–	LIGHTING
9.4.1	4.34	–	General principles
9.4.2	4.32–4.34	–	Avoiding glare & shadows
9.4.3	–	–	Colour rendering
9.4.4	–	–	Illumination for lip reading
10	4.1–5.21	–	FACILITIES IN BUILDINGS
10.1	–	–	Storage facilities
10.1.1	–	–	Provision of storage facilities
10.1.2	–	–	Access to storage facilities
10.1.3	4.20	–	Shelving heights
10.2	–	–	ATMS AND OTHER COIN AND CARD-OPERATED DEVICES
10.3	–	–	WINDOWS & WINDOW CONTROLS
10.3.1	–	–	Window sill & transome heights
10.3.2	4.24	–	Window controls
10.4	4.14	–	PUBLIC TELEPHONES
10.4.1	–	–	Provision & location of telephones
10.4.2	–	–	Telephone controls
10.4.3	–	–	Telephone booths
10.5	–	–	ELECTRICAL OUTLETS, SWITCHES & CONTROLS
10.5.1	4.25–4.29	8.1–8.2	General principles
10.5.2	4.30	8.3	Location of outlets, switches & controls
11	–	–	ASSEMBLY AREAS & WAITING ROOMS
11.1	3.2–3.6	–	Counters & reception desks
11.1.1	–	–	Location & access for visitors
11.1.2	–	–	Space in front of counters & desks
11.1.3	–	–	Height of counters & desks

BS 8300	AD M: (non-dwellings)	AD M: (dwellings)	DESIGN ASPECT
11.1.4	–	–	Counters & desks for wheelchair-users
11.1.5	–	–	Space below counters for wheelchair-users
11.1.6	–	–	Work surface heights
11.1.7	–	–	Work surface depths
11.1.8	–	–	Work surface profiles
11.1.9	–	–	Work surface widths for wheelchair-users
11.1.10	4.32	–	Communications at counters & reception desks
11.1.11	4.32	–	Acoustics at counters & reception desks
11.2	–	–	Seating in waiting areas
11.2.1	–	–	Spaces for wheelchair-users
11.2.2	–	–	Access to waiting areas
11.2.3	–	–	Seating layouts
11.2.4	–	–	Seating types
11.2.5	–	–	Visual clarity & orientation
11.2.6	–	–	Acoustics in waiting areas
11.2.7	4.31–4.36	–	Audible communication systems
11.3	–	–	Audience seating
11.3.1	–	–	Spaces for wheelchair-users
11.3.2	–	–	Access to audience seating
11.3.3	–	–	Visual clarity & orientation
11.4	4.2–4.12	–	Lecture & conference facilities
11.4.1	4.12	–	Spaces for wheelchair-users
11.4.2	–	–	Access to lecture & conference facilities
11.4.3	4.12	–	Raked floors
11.4.4	4.9	–	Sight lines
11.4.5	–	–	Seating types
11.4.6	–	–	Ancillary equipment
11.4.7	–	–	Study spaces
11.4.8	–	–	Reading distance & size of lettering
11.4.9	4.9	–	Visual clarity & orientation
11.4.10	–	–	Acoustics in lecture & conference facilities
11.2.7	4.12, 4.31–4.36	–	Sound enhancement systems
12	–	–	INDIVIDUAL ROOMS IN BUILDINGS
12.1	–	–	KITCHEN AREAS
12.1.1	–	–	Accessible routes
12.1.2	–	–	Floor space & knee recesses below worktops
12.1.3	4.16	–	Work surfaces
12.1.3.1	–	–	Work surface layout & work sequence
12.1.3.2	–	–	Dual-height work surfaces
12.1.3.3	–	–	Single-height work surfaces
12.1.3.4	–	–	Variable-height work surfaces
12.1.3.5	–	–	Work surface depth
12.1.3.6	–	–	Work surface accessories
12.1.4	–	–	Storage units accessible to wheelchair-users

BS 8300	AD M: (non-dwellings)	AD M: (dwellings)	DESIGN ASPECT
12.1.5	–	–	Cupboard & drawer handles
12.1.6	–	–	Kitchen sinks & taps
12.1.7	–	–	Kitchen appliances
12.1.7.1	–	–	Hobs & cooking rings
12.1.7.2	–	–	Ovens
12.1.7.3	–	–	Other kitchen & laundry appliances
12.1.8	–	–	Switches, sockets, & controls
12.1.9	–	–	Heat emitters
12.1.10	–	–	Alarms
12.1.11	–	–	Eating/drinking spaces for wheelchair-users
12.1.12	–	–	Surface finishes
12.1.13	–	–	Lighting
12.2	5.19–5.21	–	BATHROOMS
12.2.1	5.19–5.21	–	Accessible bathroom accommodation
12.2.2	–	–	Accessible routes
12.2.3	–	–	Independent-use bathrooms with corner wc
12.2.4	–	–	Independent-use bathrooms with peninsular wc
12.2.5	–	–	En suite bathrooms with ceiling track
12.2.7	–	–	Baths and taps
12.2.9	5.21	–	Bath transfer seat
12.2.10	5.21	–	Bath grab rails
12.2.11	5.21	–	Other sanitary fittings & accessories
12.2.12	5.21	–	Doors & door furniture
12.2.13	–	–	Heat emitters
12.2.14	–	–	Alarms
12.2.15	–	–	Lighting
12.2.16	5.21	–	Surface finishes
12.3	5.15–5.18	–	CHANGING & SHOWER AREAS
12.3.1	5.15–5.17	–	Provision of changing & shower areas
12.3.2	–	–	Accessible routes
12.3.3	5.18	–	Self-contained changing areas
12.3.6	–	–	Controls for self-contained changing areas
12.3.4	5.18	–	Communal changing areas
12.3.7	5.18	–	Controls for communal changing areas
12.3.5	–	–	Lockers
12.3.8	–	–	Shower areas with corner wc
12.3.9	–	–	En suite shower areas with ceiling track
12.3.10	–	–	Sanitary fittings & accessories
12.3.11	–	–	Doors & door furniture
12.3.12	–	–	Lobbies
12.3.13	–	–	Lighting
12.3.14	5.18	–	Alarms
12.3.15	–	–	Heat emitters
12.3.16	5.18	–	Surface finishes
12.4	5.1–5.14	10.1–10.3	LAVATORIES
12.4.1	5.1–5.14	10.3	Provision of lavatory accommodation
12.4.2	5.14	10.3	Accessible routes
12.4.3.1	5.9–5.14	–	Unisex accessible corner wc
12.4.3.2	5.9–5.14	–	Unisex accessible peninsular wc

BS 8300	AD M: (non-dwellings)	AD M: (dwellings)	DESIGN ASPECT
12.4.3.3	5.8	–	Compartments for ambulant disabled people
12.4.4	–	–	Urinals
12.4.5.1	5.4	–	Wash basins and taps
12.4.6	–	–	Lavatory accessories
12.4.6.1	–	–	General principles
12.4.6.2	–	–	Mirrors
12.4.7	5.4	–	Doors & door furniture
12.4.8	–	–	Lobbies
12.4.9	5.4	–	Lighting
12.4.10	5.4	–	Alarms
12.4.11	5.4	–	Heat emitters
12.4.12	5.4	–	Surface finishes
12.5	4.17–4.24	–	BEDROOMS
12.5.1	4.17–4.24	–	Provision of accessible bedrooms
12.5.2	4.18	–	Accessible routes
12.5.3	4.24	–	Space around beds
12.5.4	4.23–4.24	–	Doors & door furniture
12.5.5	4.24	–	Balconies
12.5.6	4.24	–	Windows
12.5.7.1	4.24	–	Beds
12.5.7.2	4.24	–	Wardrobes & storage systems
12.5.7.3	4.24	–	Other furniture & fittings
12.5.8	4.25–4.30	–	Switches, sockets & controls
12.5.9	–	–	Heat emitters
12.5.10	4.24	–	Alarms
12.5.11	4.34	–	Lighting
12.5.12	4.32	–	Surface finishes
	4.36	–	Communication aids
13	–	–	BUILDING TYPES
13.1	–	–	TRANSPORT-RELATED BUILDINGS
13.1.2	–	–	Location & accessible routes
13.1.3	–	–	Ticket sales & information points
13.1.5	–	–	Obstructions
13.1.6	–	–	Waiting areas & seating
13.1.7	–	–	Ticket barriers & gates
13.1.8	–	–	Boarding points & platforms
13.1.9	–	–	Lavatories
13.1.10	–	–	Escalators & passenger conveyors
13.2	–	–	INDUSTRIAL BUILDINGS
13.2.2	–	–	Access routes
13.2.3	–	–	Lavatories
13.2.4	–	–	Storage of hazardous materials
13.2.5	–	–	Equipment
13.3.1	–	–	ADMINISTRATIVE & COMMERCIAL BUILDINGS
13.3.2	–	–	Offices & commercial buildings lighting
13.3.3	–	–	Shops, supermarkets & shopping malls
13.3.3.2	–	–	Arrival
13.3.3.3	–	–	Circulation
13.3.3.4	–	–	Fitting rooms

BS 8300	AD M: (non-dwellings)	AD M: (dwellings)	DESIGN ASPECT
13.3.4	–	–	Law courts, police stations, & prisons
13.4	–	–	HEALTH AND WELFARE BUILDINGS
13.5	4.13–4.16	–	REFRESHMENT BUILDINGS
13.5.3	4.16	–	Counter & table heights
13.6	4.10	–	ENTERTAINMENT BUILDINGS
13.6.2.1	4.12	–	Seating & viewing distances
13.6.2.2	–	–	Access for performers
13.6.3	–	–	Lavatories
13.6.4	–	–	Box office counters
13.6.5	–	–	Sound enhancement systems
13.7	4.11–4.12	–	SPORTS-RELATED BUILDINGS
13.7.2	4.12	–	Seating, sound enhancement, and lavatories
13.7.3	–	–	Facilities for disabled performers
13.7.4	–	–	Swimming pools
13.7.5	–	–	Fitness & exercise studios
13.8	–	–	RELIGIOUS BUILDINGS
13.8.2	–	–	Churches: seating & sound enhancement systems
13.8.3	–	–	Crematoria: access, seating, lavatories & sound enhancement systems
13.9	–	–	EDUCATIONAL, CULTURAL & SCIENTIFIC BUILDINGS
13.9.3	–	–	Display cases
13.9.4	–	–	Libraries & reading carrels
13.9.5	–	–	Seating
13.9.6	–	–	Wayfinding & obstructions
13.9.7	–	–	Sound enhancement systems
13.10	–	–	HISTORIC BUILDINGS
13.11	–	–	TRAVEL AND TOURISM-RELATED BUILDINGS.
Annex B			Space allowances for ambulant disabled people and wheelchair-users on access routes
Annex C			Slip-resistance of floor and stair tread finishes when wet and dry
Annex D			Bodyspace dimensions and space requirements of wheelchair, powerchair and scooter users

It will be seen that the contents of both BS8300 and APPROVED DOCUMENT M are structured in accordance with the *journey sequence* (qv) pioneered by the architects C Wycliffe Noble and Geoffrey Lord.

uggies

1 Buggies are a *mobility aid* (qv), and comprise one of five categories of *small personal vehicles* for frail and *disabled* (qv) people:

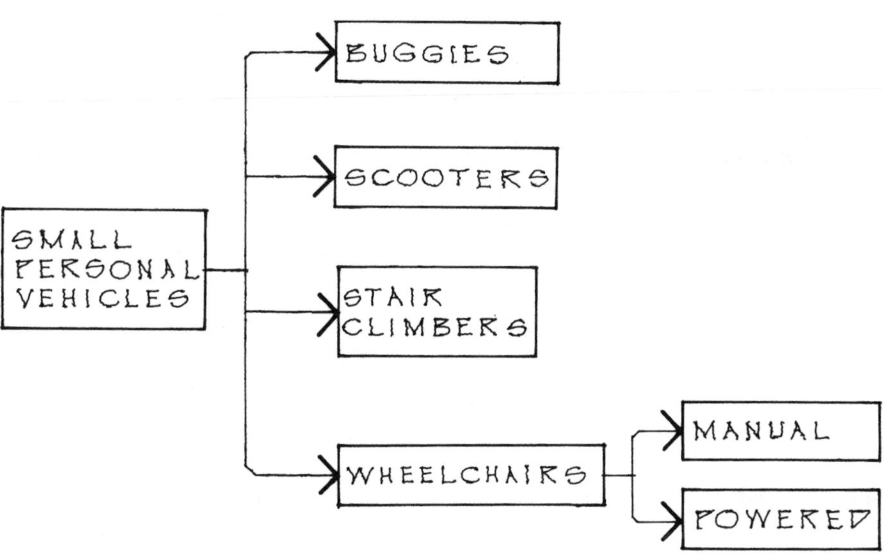

2 They are very like *scooters* (qv), but designed specifically for use on rough, soft, and uneven ground, and are intended essentially to give disabled people access to the countryside, to riversides for fishing or boating, or for other recreational uses. The model below is a Tramper, made by Beamer Ltd.

3 As of 2008, useful references include the following. For publication details see the bibliography:

- CHOOSING A SCOOTER OR BUGGY.
- GET MOBILE – YOUR GUIDE TO BUYING A SCOOTER OR POWERED WHEELCHAIR

4 Useful websites include the following. They are given in alphabetical order.

- Dept for Transport www.dft.gov.uk/transportforyou/access/tipws/
- Disabled Living Foundation www.dlf.org.uk/
- Disabled Living Online www.disabledliving.co.uk/
- KeepAble www.keepable.co.uk/
- OTDirect www.otdirect.co.uk/link-equ.html#Platform
- Patient UK www.patient.co.uk/showdoc/12/
- Ricability www.ricability.org.uk/
- Tramper www.tramper.co.uk/

5 The annual Mobility Roadshow exhibits scooters and other personal vehicles wheelchairs, etc., and allows test drives. For more information visit www.mobilityroadshow.co.uk/
6 For baby buggies, an entirely different subject, see: *Pushchairs*.

See also: *Mobility aids;* and *Scooters*.

Building Regulations

1 The Building Regulations 2000 apply in England and Wales. Useful guides include the following.
• THE BUILDING REGULATIONS EXPLAINED AND ILLUSTRATED by J R Waters, Keith Bright and M J Billington. 13th edition.
• BUILDING REGULATIONS EXPLANATORY BOOKLET.
• THE FUTURE OF BUILDING CONTROL.
Other parts of the UK are covered by the Building Regulations (Northern Ireland) 2000, and Building (Scotland) Regulations 2004 – see the bibliography
2 Visit www.planningportal.gov.uk/

Building Schools for the Future

1 When this £45 billion programme was launched in 2004, Ministers promised that it would result in every state secondary school in England being rebuilt by 2020. Progress to date has been as follows:
• 100 building contracts were scheduled to be signed in 2006; the actual number signed was 5.
• 100 new schools were scheduled to be opened in 2007; the actual number opened was 9.
• 200 new schools were scheduled to be opened in 2008; the actual number now expected to open is 35.
2 In January 2008 Mr Ty Goddard, director of the British Council for School Environments (which represents hundreds of schools, local authorities, architects and building companies) charged that the stupefyingly burdensome bidding process imposed by the Government (and which all participants agree to be responsible for the above delays) was wasting enough taxpayers' money 'to fund a new school in every local authority'.
3 In April 2008 the 2004 plan was scrapped and replaced by vague promises to commence work on some schools, some time, in each area.
4 Taken together with the CABE (qv) report ASSESSING SECONDARY SCHOOL DESIGN QUALITY, which charged in July 2006 that half of the 124 schools built between 2003 and 2005 were 'poorly built, badly designed, and failed to provide inspiring educational environments', the above record of failure casts lethal doubt on the Government's school-building initiatives generally. While billions of pounds are being wasted, we are also creating a generation of buildings that future generations might wish they could demolish.
5 The organisation that is responsible for delivering the Building Schools for the Future programme is Partnerships for Schools – visit www.p4s.org.uk/
6 Useful references include the following. For publication details see the bibliography.
• ASSESSING SECONDARY SCHOOL DESIGN QUALITY: RESEARCH REPORT.
• SMART PFI: RIBA POSITION PAPER.

See also: *Over-regulation*.

Built environment

1 The term 'built environment' refers to those elements of our surroundings which are created artificially, including buildings, transport infrastructure, and landscaping.
2 The diagram on pages 76 and 77 summarises these elements, many of which have entries in the present work.

uilt Environment orking Group

1 The Built Environment Working Group, also known as the Built Environment Group, was located within DPTAC (qv) until 2005. It was then superseded by the Inclusive Environment Group at CABE (qv).
2 For latest information visit www.cabe.org.uk/

ungalows

1 Bungalows are small single-storey houses. Their compactness and lack of stairs mean that they are ideal for *ageing* (qv) and *disabled (qv)* people, or those who suffer from severely *impaired mobility* (qv) owing, for instance, to *arthritis* (qv).
2 They are also, according to a succession of opinion surveys, either the most popular or the second most popular house types in Britain – see: *Happiness and homes.* This preference creates a problem for the Government because single-storey dwellings take up a lot of land, and England is becoming uncomfortably overcrowded. The entry *Housing: affordability* touches upon some of the issues raised by this problem.

ureaucracy eduction Group BRG)

1 For the nature and purpose of the Bureaucracy Reduction Group see: *Cutting Burdens on Teachers.*
2 For more information on the BRG visit www.dfes.gov.uk/furthereducation/index.cfm?fuseaction=content.view&Category

See also: *Better regulation bodies.*

us and coach design: omplying with the isability iscrimination Act

1 The Disability Discrimination Act 1995 (qv) allows the Government to make Public Service Vehicle (or 'PSV') regulations on both the design and operation of buses and coaches. The aim is (a) that *disabled people* (qv) can get in and out of these vehicles safely, and (b) that they can be conveyed safely and in reasonable comfort. For *wheelchair-users* (qv) both of these things should be possible while they are sitting in their wheelchairs.
2 The situation as of 2008 is this:

• Vehicles carrying fewer than nine passengers are not covered by the Disability Discrimination Act.

• Buses and coaches carrying from 9 to 22 passengers – a category informally referred to as 'small buses' – will in time be covered by regulations issued under the Act, but as yet none have yet been made. These vehicles, as may be seen below, are normally thought of as 'minibuses' rather than 'buses', and typically include mini-buses, midi-buses; van conversions; etc.

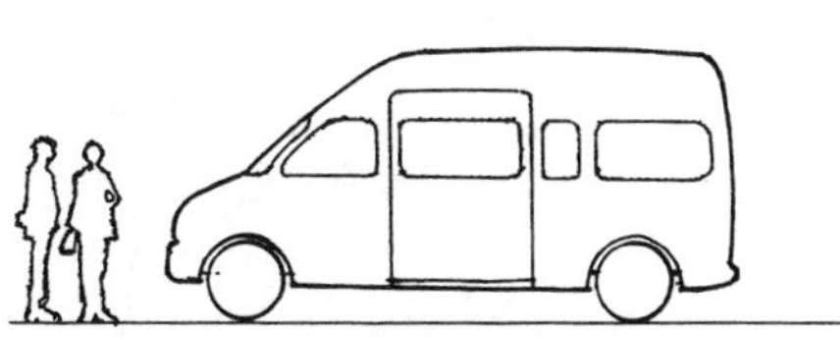

There are probably about 100 000 small buses in the UK, most being very informally operated and many carrying *ageing* (qv) and *disabled* (qv) passengers, and in the interests of public safety DPTAC (qv) has issued a set of good practice details titled ACCESSIBILITY SPECIFICATION FOR SMALL BUSES DESIGNED TO CARRY 9 TO 22 PASSENGERS (see below).

• New buses and coaches that carry more than 22 passengers do (in England and Wales) come under the *Public Service Vehicles Accessibility Regulations 2000* (qv), and coverage will be gradually extended to existing vehicles, so that all vehicles are covered by 2017. These vehicles are informally referred to as 'large buses', and include all double-deck and the vast majority of single-deck

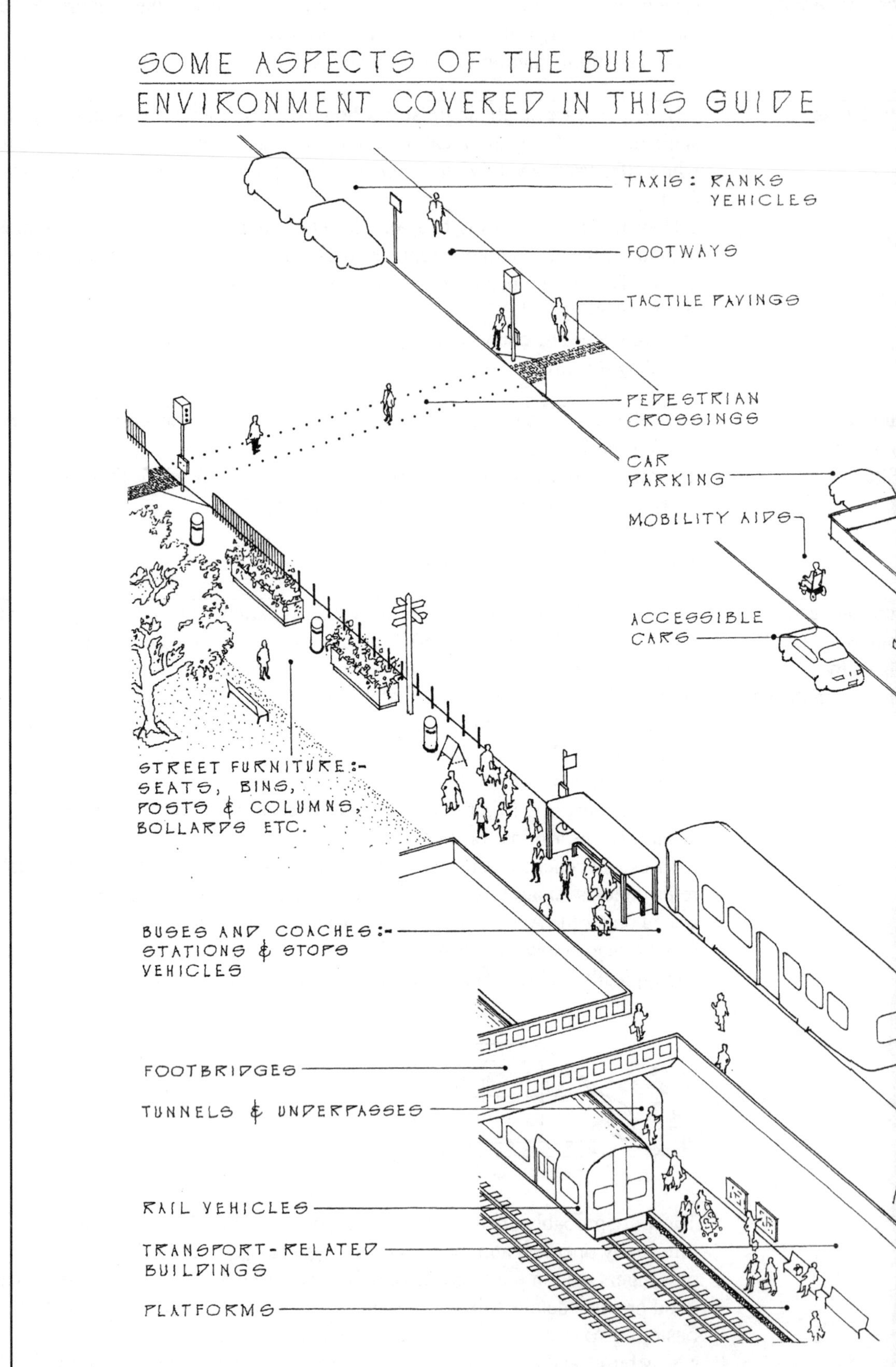
SOME ASPECTS OF THE BUILT ENVIRONMENT COVERED IN THIS GUIDE
TAXIS: RANKS VEHICLES
FOOTWAYS
TACTILE PAVINGS
PEDESTRIAN CROSSINGS
CAR PARKING
MOBILITY AIDS
ACCESSIBLE CARS
STREET FURNITURE:- SEATS, BINS, POSTS & COLUMNS, BOLLARDS ETC.
BUSES AND COACHES:- STATIONS & STOPS VEHICLES
FOOTBRIDGES
TUNNELS & UNDERPASSES
RAIL VEHICLES
TRANSPORT-RELATED BUILDINGS
PLATFORMS

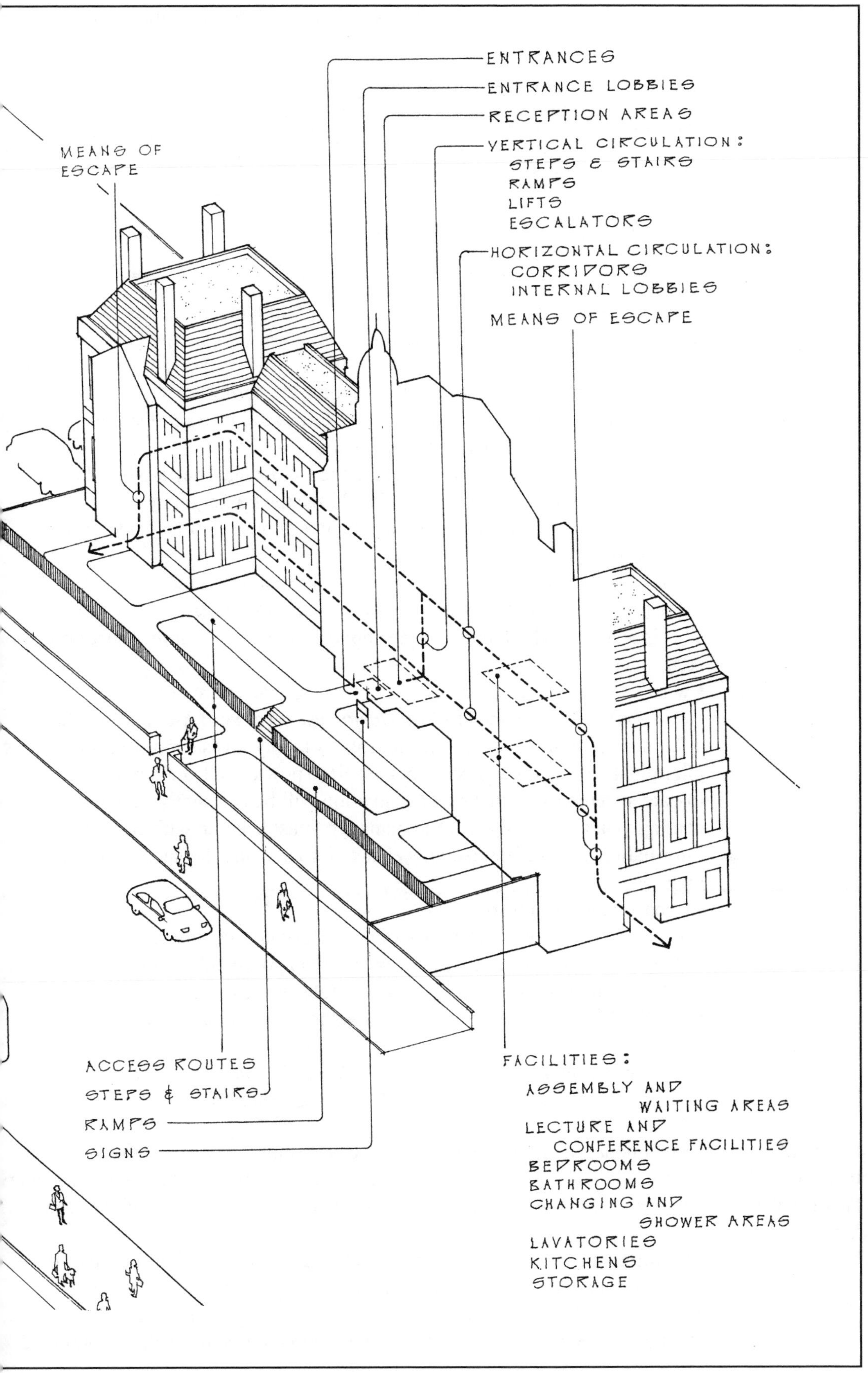
ENTRANCES
ENTRANCE LOBBIES
RECEPTION AREAS
VERTICAL CIRCULATION:
STEPS & STAIRS
RAMPS
LIFTS
ESCALATORS
HORIZONTAL CIRCULATION:
CORRIDORS
INTERNAL LOBBIES
MEANS OF ESCAPE
MEANS OF ESCAPE
ACCESS ROUTES
STEPS & STAIRS
RAMPS
SIGNS
FACILITIES:
ASSEMBLY AND WAITING AREAS
LECTURE AND CONFERENCE FACILITIES
BEDROOMS
BATHROOMS
CHANGING AND SHOWER AREAS
LAVATORIES
KITCHENS
STORAGE

buses used on local and scheduled services. Two typical examples are shown below.

Matters covered by the term *accessibility* (qv) in large buses include wide doors; flat floors; a ramp or lift at one door; a *wheelchair* (qv) parking space measuring at least 1 300 mm long by 750 mm wide, with headroom of at least 1 500 mm; vertical and horizontal handholds at appropriate places; palm-operated bell-pushes; good *colour contrast* (qv) to make key features clearly visible; etc. For seat dimensions see: *Seats*.

3 As of 2008, useful references include the following. They are listed in alphabetical order. For publication details see the bibliography.

- ACCESSIBILITY SPECIFICATION FOR SMALL BUSES DESIGNED TO CARRY 9 TO 22 PASSENGERS.
- CODE OF PRACTICE. PROVISION AND USE OF TRANSPORT VEHICLES: STATUTORY CODE OF PRACTICE: SUPPLEMENT TO PART 3 CODE OF PRACTICE. This is an essential reference in the application of the Disability Discrimination Act.
- WHEELS WITHIN WHEELS: A GUIDE TO USING A WHEELCHAIR ON PUBLIC TRANSPORT gives guidance for *wheelchair-users* (qv) of buses and coaches.

4 Readers are reminded that the present work only gives simplified summaries of original documents and does not purport to provide full, authoritative and up-to-date statements of the law. The situation will be constantly developing in coming years, and designers and managers must keep up with current information. As of 2008, useful websites for checking the latest situation include:

- Department for Transport www.dft.gov.uk/transportforyou/access/buses/
- DPTAC Access Directory www.dptac.gov.uk/adnotes.htm
- DPTAC Publications www.dptac.gov.uk/pubs.htm
- DPTAC buses www.dptac.gov.uk/buses.htm

See also: *Public Service Vehicles Accessibility Regulations 2000;* and *Transport: complying with the Disability Discrimination Act.*

Bus and coach operation: complying with the Disability Discrimination Act

1 The operation of bus and coach services is subject to the Disability Discrimination Act 1995, which creates a *civil duty* (qv) for transport providers not to *discriminate* (qv) against *disabled* (qv) customers or would-be customers in the provision of:

- Facilities such as booking counters, waiting areas, toilet facilities, platforms, and other public areas – for guidance see: *Transport-related buildings*.
- Services such as timetables, booking, and ticketing arrangements.

On the latter point, the kinds of information that bus and coach operators are expected to provide to passengers include:

- Advance information, available to users before they set out, about routes, timetables, and links in both directions. The information should help travellers to identify the *accessible* (qv) facilities and services that are available en route (eg lavatories) so that they can plan a practicable journey.

• Audible and visual information throughout the journey – at bus stops, in the bus, and at interchanges etc. – keeping travellers informed and helping them to plan their onward journey.

Information should be supplied in a format that people with *impaired hearing* (qv), *impaired vision* (qv), or *impaired understanding* (qv) can use and understand, at no extra cost and with no delay, where such provision would make it easier for them to use the service. Partial guidance on these matters will be found in sections 6, 7, 9, and 10 of INCLUSIVE MOBILITY (qv) – see the bibliography.

2 Management duties in transport services were not covered by The Disability Discrimination Act 1995, but this gap was closed by the Disability Discrimination Act 2005 (qv). If for instance a bus driver parks an *accessible* (qv) bus too far from the kerb to give a wheelchair user ready access, the latter now has redress under the Disability Discrimination Act.

3 Readers are reminded that the present work only gives simplified summaries of original documents and does not purport to provide full, authoritative and up-to-date statements of the law. Designers and managers must keep up with current information.

4 As of 2008, useful references include the following. For publication details see the bibliography.

• AVOIDING DISABILITY DISCRIMINATION IN TRANSPORT: A PRACTICAL GUIDE FOR BUSES AND SCHEDULED COACHES.

• AVOIDING DISABILITY DISCRIMINATION IN TRANSPORT: A PRACTICAL GUIDE FOR TOUR COACH OPERATORS.

5 Useful websites include the following. They are given in alphabetical order.

• Dept for Transport www.dft.gov.uk/transportforyou/access/buses/
• DPTAC Access Directory www.dptac.gov.uk/adnotes.htm
• DPTAC Publications www.dptac.gov.uk/pubs.htm

See also: *Transport: complying with the Disability Discrimination Act.*

us and coach :ations

1 For general design guidance see: *Transport-related buildings.*

2 For specific design guidance on bus stations refer to section 8 of INCLUSIVE MOBILITY (qv) – see the bibliography.

3 As of 2008, useful websites include:

• DPTAC Access Directory www.dptac.gov.uk/adnotes.htm
• DPTAC Publications www.dptac.gov.uk/pubs.htm

us and coach :ops

1 For design guidance refer to sections 6 and 10 of INCLUSIVE MOBILITY (qv) – see the bibliography.

2 As of 2008, useful websites include:

• DPTAC Access Directory www.dptac.gov.uk/adnotes.htm
• DPTAC Publications www.dptac.gov.uk/pubs.htm

See also: *Pedestrian environments; Seats; Traffic advisory leaflets;* and *Walking distances.*

us priority and park-nd-ride schemes

See: *Traffic management.*

us travel

1 The figure on the next page shows that the number of passenger journeys made on local buses in Great Britain is fewer than in 1970, but suggests that the decline has halted. At date 'A' council-run bus services were privatised by the 1985 Transport Act, and at date 'B' the railways were split up and privatised by the 1993 Railways Act.

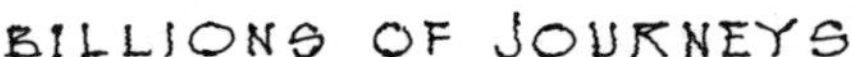

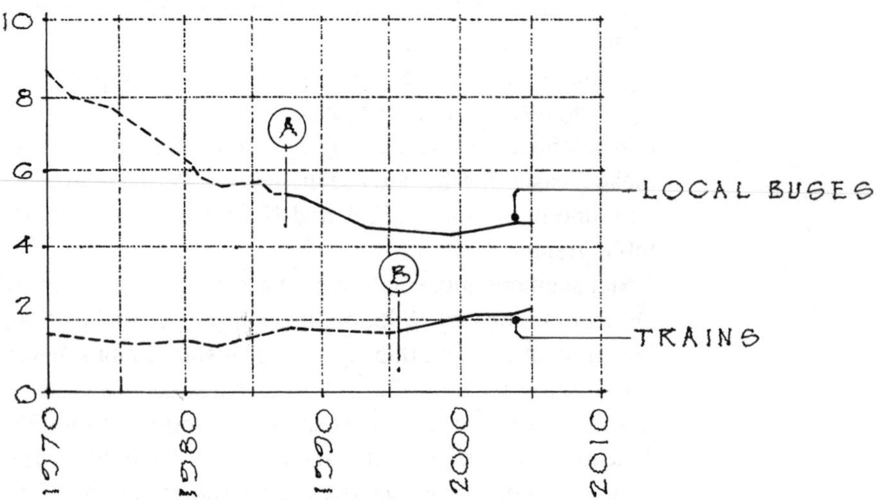

The graph is from www.statistics.gov.uk/cci/nugget_print.asp?ID=1094
For more recent data refer to Figure. 12.14 in SOCIAL TRENDS 37 (qv), which shows that the number of vehicle kilometres increased steeply after 1988.

2 In December 2006 the Transport Secretary, Mr Douglas Alexander, presented proposals with the stated aim of substantially increasing bus travel across England. The intention was to end the free-for-all that began when the industry was privatised in 1986. Thus bus companies would lose their powers to set fares, frequencies, and timetables; they would have to submit detailed information about punctuality; and those that performed poorly would be penalised. Local *charities* (qv) and *voluntary and community* (qv) groups would be allowed to run services in areas that have been neglected. For more information visit www.dft.gov.uk/

3 As of 2008, useful websites include:

- Office for National Statistics — www.statistics.gov.uk/
- Transport Statistics for Great Britain (TSGB) — www.dft.gov.uk

Bus travel: guidance for disabled people

See: *Door to Door*.

C

ABE 1 The Commission for Architecture and the Built Environment (CABE) was created by the Government in 1999 to promote the importance of high-quality architecture and urban design in England.
2 For more information visit www.cabe.org.uk/

ABE Space 1 CABE Space works to bring excellence to the design and management of public spaces in England's towns and cities.
2 For more information visit www.cabe.org.uk/default.aspx?contentitemid=465

abinet Office 1 The Cabinet Office is one of the departments of central *government* (qv). It works alongside the Treasury and the Prime Minister's Office with the aim of making government more effective by providing a strong centre.
2 One of its most important responsibilities is the fight against excessive *red tape* (qv) – see: *Better regulation bodies* (qv); and *Over-regulation* (qv).
3 For more information visit www.cabinetoffice.gov.uk/

See also: *Better regulation bodies*; and *Over-regulation*.

AE See: *Centre for Accessible Environments*.

allipers See: *Walking and standing aids*.

ancer 1 Persons who qualify as *disabled people* (qv) under the Disability Discrimination Act (qv) owing to having cancer, qualify from the time that cancer is diagnosed as being likely to require 'substantial treatment', and not (as is the case with most other illnesses) from the point at which the effects of the condition have actually become 'substantial'.
2 As of 2008, useful websites include the following. They are given in alphabetical order.
- Equality & Human Rights Commission www.equalityhumanrights.com/
- Directgov www.direct.gov.uk/
- Office of Public Sector Information www.opsi.gov.uk/acts.htm

See also: *Disability Discrimination Act*.

andle power See: *Candela*.

andela See: *Lighting*.

anes for blind eople 1 Several kinds of cane may be used as *mobility aids* (qv) by *blind* (qv) and near-blind people. They give warning only of below-knee obstacles, and are liable to lead their users straight into injury from high-level *hazards* (qv) such as overhanging branches, untrimmed hedges (holly can do terrible damage to people's eyes), horizontal or oblique scaffolding poles, open windows, etc. Such hazards should either not exist or, if they cannot be avoided, should be surrounded by warning surfaces or protective railings – see: *Hazards: protection* (qv); and *Streetworks* (qv).
2 The conventional types have recently been augmented by the *high-tech* (qv) UltraCane which incorporates a number of ultrasonic sensors as used by bats for echo-location. These give users of the cane a greatly enhanced spatial 'picture' of what is around them, even at head height, enabling them to move around more confidently and more safely than with conventional canes. The signals reach the user in the form of vibrations in the handle, which they detect by touch and

quickly learn to interpret almost by instinct. For more information visit www.soundforesight.co.uk/

See also: *Impaired vision: mobility aids*.

Captioning

See: *StageText*.

Car parking: dwellings

1 The following references contain authoritative design guidance. For publication details see the bibliography.

• Section 6 of APPROVED DOCUMENT M (qv). The provisions in AD M must be heeded in order to comply with the Building Regulations for England and Wales. In Northern Ireland refer to TECHNICAL BOOKLET R of the Building Regulations (Northern Ireland); and in Scotland to the TECHNICAL HANDBOOKS of the Building (Scotland) Regulations (qv).

• Section 4 of BS 8300 (qv), which applies to all buildings in the UK.

• THE HABINTEG HOUSING ASSOCIATION DESIGN GUIDE (qv), which applies most specifically to *housing association* (qv) schemes, but ought to be by the designer's side in the design of all housing developments and individual dwellings.

• THE WHEELCHAIR HOUSING DESIGN GUIDE (qv), also from the Habinteg Housing Association, is another housing design reference whose usefulness extends well beyond the special remit implied by the title.

Car parking: public: complying with the Disability Discrimination Act

1 Public buildings and spaces must in general comply with the Disability Discrimination Act (qv). In the case of car parking, relevant *authoritative practical guides* (qv) include DESIGNING FOR ACCESSIBILITY (qv) and in addition the following documents. For publication details see the bibliography.

• APPROVED DOCUMENT M (qv): the provisions in section 1 must be heeded in order to comply with the Building Regulations for England and Wales.

• BS 8300 (qv): the recommendations in section 4 apply to all buildings in the UK.

• INCLUSIVE MOBILITY (qv): the recommendations in section 5 apply to *transport-related buildings* such as bus and coach stations, railway stations, air terminals, and transport interchanges.

Except for AD M, the above guides do not have the force of law, but conformity with their recommendations will make it easier to demonstrate that the requirements of the Disability Discrimination Act have been met.

2 For places other than England and Wales the following regulatory documents should be consulted:

• In Northern Ireland: TECHNICAL BOOKLET R of the Building Regulations (Northern Ireland) (qv).

• In Scotland: the TECHNICAL HANDBOOKS of the Building (Scotland) Regulations (qv).

• In the USA, where the Americans with Disabilities Act applies: refer to the ADA AND ABA ACCESSIBILITY GUIDELINES (qv).

3 As of 2008, other useful references include the following. For publication details see the bibliography.

• CAR PARKING: WHAT WORKS WHERE, a 134-page guide published in 2006 by English Partnerships, gives guidance on the number of cars to be provided for and on options such as on-street parking; mechanized parking; *homezones* (qv); car clubs; and others. It also illustrates around two dozen case examples – one unintended result of which is to show how hideous much of current housing developments and townscapes have become – see in this connection: *Architecture*.

4 Useful websites include:
• Department for Transport
www.dft.gov.uk/transportforyou/roads/howparkingismanaged

See also: *Streets; Traffic advisory leaflets*; and *Vehicle barriers*.

r parks: safety

1 Car parks can be lonely and even frightening places, especially for women and especially at night. The following measures can help:
• Good lighting. Dark places are very likely to be dangerous places.
• Surveillance, both formal (see: *Closed circuit television*) and informal. The latter is especially important. The more visible all parts of a car park are to overlooking windows, to neighbours, and to passers-by, the safer it is likely to be. While it may often be necessary – and, from a landscaping point of view, highly desirable – to sink car parking areas below ground level, they should not be completely covered over and hidden from sight except as an absolute last resort – see the sketch below.

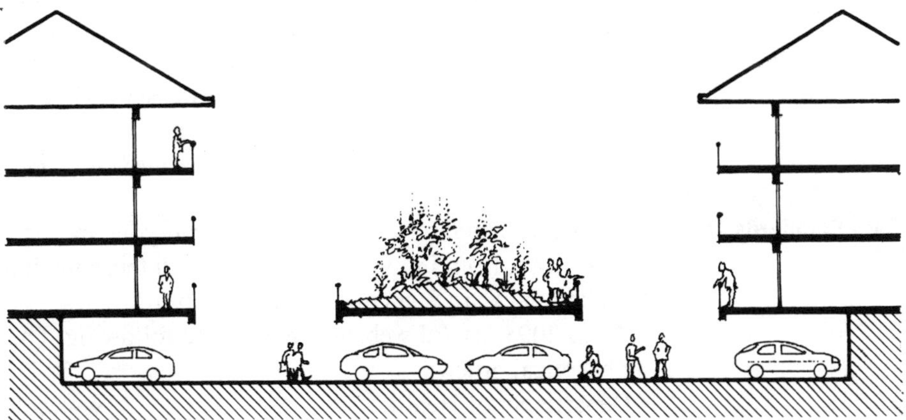

2 As of 2008, useful references include the following. For publication details see the bibliography.
• CAR PARKING: WHAT WORKS WHERE, published by English Heritage gives general advice on car parking provision.
3 Useful websites include:
• British Parking Association www.britishparking.co.uk/
• Park Mark www.securedcarparks.com/

are homes for older ople

1 When older people are no longer able to live *independently* (qv), even with support services as in *sheltered housing* (qv), the typical next stage is moving to a care home.
2 As of 2008, useful references include the following. They are given in alphabetical order. For publication details see the bibliography.
• CHANGES IN COMMUNAL PROVISION FOR ADULT SOCIAL CARE 1991–2001 by Laura Banks, Philip Haynes, Susan Balloch, and Michael Hill.
This report, which was published by the Joseph Rowntree Foundation (qv) in July 2006, examines trends in care home provision. The authors found great inequalities in provision, and raise important questions about the future.
• DESIGN OF RESIDENTIAL CARE AND NURSING HOMES FOR OLDER PEOPLE, also known as Health Facilities Note 19, is published by NHS estates (qv) on behalf of the Centre for Accessible Environments (qv). It is fully illustrated in colour and gives profiles of care homes plus references, legislation, a bibliography, and a list of useful organisations. It was reviewed on p 24 of issue 77 of ACCESS BY DESIGN (qv), the Journal of the Centre for Accessible Environments – see the bibliography.

• FIRE SAFETY IN CARE HOMES FOR OLDER PEOPLE AND CHILDREN is a BSI (qv) publication that provides guidance on advisable fire safety precautions. The book is aimed at designers, owners and managers of homes for older people a children's homes; for those involved in their renovation or refurbishment; and for anyone responsible for enforcing fire safety standards.
• THE STATE OF SOCIAL CARE IN ENGLAND 2005–06 predicts that local authorities will increasingly be forced to deny financial aid for home care to a but the most critical cases. At its launch Minister Ivan Lewis said we had bett get used to the idea that the state will provide less funding for home care for elderly and *disabled* people, and called for a national debate on the issue.
3 Useful websites include the following, given in alphabetical order.
• Commission for Social Care Inspection www.csci.org.uk/
• Directgov www.direct.gov.uk/
• English Community Care Association www.ecca.org.uk/
• Estates and Facilities Management www.dh.gov.uk/
• Joseph Rowntree Foundation www.jrf.org.uk/
• Laing & Buisson www.laingbuisson.co.uk/
• National Care Homes Association www.guide-information.org.u
• National Care Standards Commission www.dh.gov.uk/
• Social Trends www.statistics.gov.uk/StatBase/Product.asp?vlnk = 1367

See also: *Ageing; Care homes; Care Standards Act*; and *Day centres.*

Care Standards Act

1 The Care Standards Act 2000 covers everyone in England and Wales in any type of care, from infants to old people. All these institutions must register wi the National Care Standards Commission (qv).
2 As of 2008, useful websites include the following. They are given in alphabetical order.
• Department of Health www.dh.gov.uk/
• Directgov www.direct.gov.uk/
• National Care Standards Commission www.dh.gov.uk/
• Office of Public Sector Information www.opsi.gov.uk/acts.htm

See also: Commission for Social Care Inspection; and *National Care Standar Commission.*

Care Trusts

See: *NHS structure.*

Carers

1 Carers are people who look after those who are in need of additional care, assistance, or support. The term covers both relatives doing the work unpaid, and paid professionals. The Carers and Disabled Children Act 2000 gave local councils in England and Wales duties to support carers by providing services to them directly, and in the provision of breaks. The Carers (Equal Opportunities Act 2004 enhanced the rights of carers to lead the kinds of lives that others take for granted – for instance, adequate opportunities for studying, or leisure, and in April 2007 more than 2.5 million carers were given the right to request flexible working hours. In Scotland, the Community Care and Health (Scotlan Act 2002 extended carers' rights. Finally, the Disability and Carer's Service (qv) provides a variety of support services for carers.
2 Tensions can arise between carers and their clients, and conflicts between the rights of carers and the rights of *disabled people*. For a useful discussion of these issues refer to pp189–93 of DISABILITY RIGHTS AND WRONGS by the disabled writer Tom Shakespeare – see the bibliography.
3 As of 2008, useful websites include the following. They are given in alphabetical order.

- Carers UK www.carersuk.org/
- Department for Work and Pensions www.dwp.gov.uk/
- Department of Health www.dh.gov.uk/
- Directgov www.direct.gov.uk/carers
- IDeA Knowledge www.idea.gov.uk/
- Office of Public Sector Information www.opsi.gov.uk/acts.htm
- UK Home Care Association www.ukhca.co.uk/

arers' Allowance

1 This is a benefit to help people who are responsible for looking after someone who is *disabled* (qv), whether the parties are related or not.
2 For more information visit www.direct.gov.uk/

ars: accessible

See: *Accessible cars.*

ash machines

See: *Automatic teller machines.*

entral Office of formation (COI)

1 The COI works with government departments and *quangos* (qv) to produce information campaigns on issues that affect the lives of all British citizens.
2 For more information visit www.coi.gov.uk/

entre for Accessible nvironments

1 The CAE provides information and consultancy on how the built environment can be made more *accessible* (qv) and *inclusive* (qv) by design. Its quarterly journal access by design is a useful information source for all professionals with an interest in *inclusive design* (qv) – see the bibliography.
2 For more information visit www.cae.org.uk/

hairlifts

1 Chairlifts are one of two categories of stairlift as shown below.

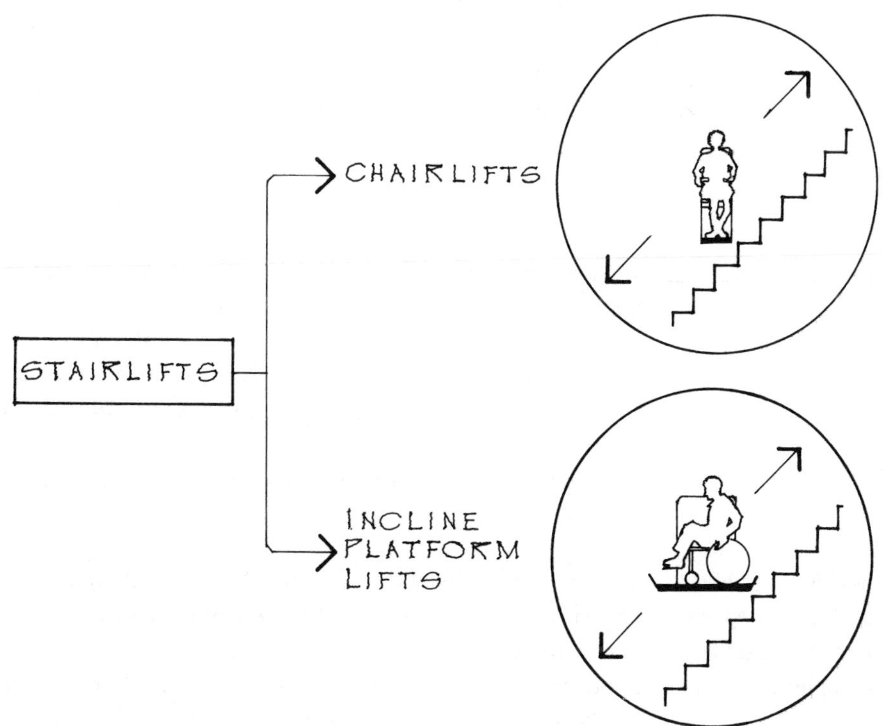

2 There are two main types – 'Sitting' chairlifts for the majority of users, as shown in the upper diagram on the next page; and 'Perching' chairlifts, as shown in the lower diagram, for people who cannot easily bend their knees and prefer to perch in a posture midway between sitting and standing.

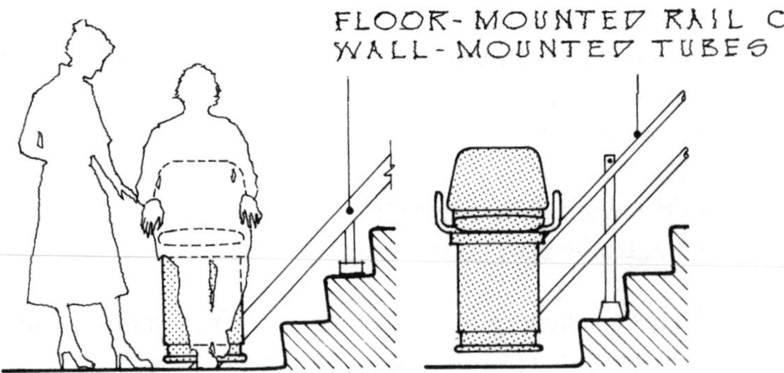

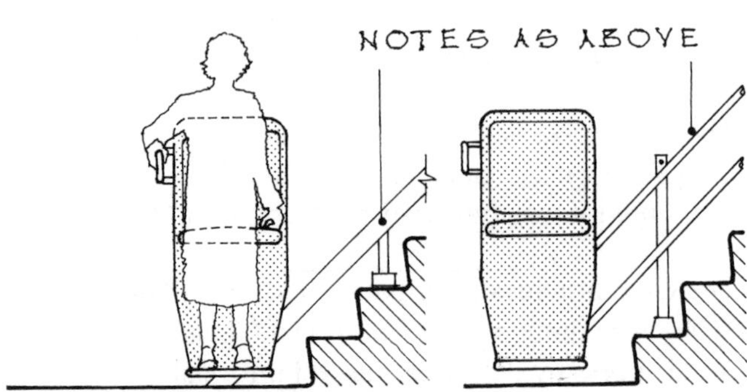

3 As of 2008, useful references include the following. They are given in alphabetical order. For publication details see the bibliography.

- CHOOSING EQUIPMENT TO GET UP AND DOWN STAIRS, published by the Disabled Living Foundation (qv).

4 Useful websites include:

• Disabled Living Foundation	www.dlf.org.uk/public/factsheets.html
• Patient UK	www.patient.co.uk/showdoc/12/
• OTDirect	www.otdirect.co.uk/link-equ.html#Platform
• Ricability	www.ricability.org.uk/

See also: *Incline platform lifts; Lifts*; and *Stairlifts*.

Chairlifts in dwellings

1 With respect to private dwellings: there are no statutory regulations covering the installation of chairlifts, and homeowners may do as they please.
2 With respect to blocks of flats: section 9 of APPROVED DOCUMENT M (qv) contains provisions on *passenger lifts* (qv) that must be heeded in order to comply with the Building Regulations. As of 2008 this text makes no reference to chairlifts.
3 For possible grant aid see: *Aids and equipment: entitlement*; and *Home improvement grants*.

Chairlifts in public buildings: complying with the Disability Discrimination Act

1 Public buildings and spaces must in general comply with the Disability Discrimination Act (qv). In the case of chairlifts, relevant *authoritative practical guides* (qv) include DESIGNING FOR ACCESSIBILITY (qv) and in addition the following documents. For publication details see the bibliography.

- APPROVED DOCUMENT M (qv): the provisions in paras 3.17 to 3.49 must be heeded in order to comply with the Building Regulations for England and Wales.
- BS 8300 (qv): the recommendations in section 8.4 apply to all buildings in the UK. BS 8300 does not have the force of law, but conformity with its recommendations will make it easier to demonstrate that the requirements of the Disability Discrimination Act have been met.

For places other than England and Wales the following regulatory documents should be consulted:
• In Northern Ireland: TECHNICAL BOOKLET R of the Building Regulations (Northern Ireland) (qv).
• In Scotland: the TECHNICAL HANDBOOKS of the Building (Scotland) Regulations (qv).
The general import of these recommendations is that chairlifts are the least appropriate type of lift for public buildings, and that they should not be used except as a last resort. If it is necessary to install a chairlift instead of a *passenger lift* (qv), the decision should be noted and justified in the *Design and Access Statement* (qv) that accompanies the application for *planning permission* or *Building Regulation consent*.
2 In the USA, where the Americans with Disabilities Act applies, refer to the ADA AND ABA ACCESSIBILITY GUIDELINES (qv).

See also: *Lifts*.

---

hairs in public ıildings: complying ith the Disability iscrimination Act

See: *Seats in public buildings: complying with the Disability Discrimination Act*.

---

hanging rooms and ıowers: dwellings

1 The following references contain authoritative recommendations:
• THE HABINTEG HOUSING ASSOCIATION DESIGN GUIDE (qv) applies most specifically to *housing association* (qv) schemes, but ought to be by the designer's side in the design of all housing developments and individual dwellings.
• the wheelchair housing design guide (qv), also from the Habinteg Housing Association, is another illustrated manual whose usefulness extends far beyond the special remit implied by the title.
2 For publication details of the above references see the bibliography.

ıanging rooms and ıowers: public ıildings: complying th the Disability scrimination Act

1 Buildings used by the public, such as hotels and motels, hostels, schools, sports grounds, and transport terminals and interchanges, must in general comply with the Disability Discrimination Act (qv). In the design of changing rooms and showers in such buildings, relevant *authoritative practical guides* (qv) include designing for accessibility (qv) and in addition the following documents. For publication details see the bibliography.
• APPROVED DOCUMENT M (qv): the provisions in paras 5.15 to 5.18 must be heeded in order to comply with the Building Regulations for England and Wales.
• BS 8300 (qv): the recommendations in section 12.3 apply to all buildings in the UK. BS 8300 does not have the force of law, but conformity with its recommendations will make it easier to demonstrate that the requirements of the Disability Discrimination Act have been met.
2 For places other than England and Wales the following regulatory documents should be consulted:
• In Northern Ireland: TECHNICAL BOOKLET R of the Building Regulations (Northern Ireland) (qv).
• In Scotland: the TECHNICAL HANDBOOKS of the Building (Scotland) Regulations (qv).
• In the USA, where the Americans with Disabilities Act applies: the ADA AND ABA ACCESSIBILITY GUIDELINES (qv).

See also: *Bathrooms*; and *Lavatories*.

Charities See: *Voluntary and community sector.*

Charities Act 2006 Visit www.charity-commission.gov.uk/spr/ca2006prov.asp

Charities and disabled people 1 Relations between *disabled people* (qv) and the charities that speak for them – or claim to do so – have been troubled for at least a quarter of a century, and remain so today. One of the burning issues has always been that of 'pity' versu[s] 'rights'.
2 For a balanced account refer to pp 153–166 of DISABILITY RIGHTS AND WRONGS by the disabled writer Tom Shakespeare – see the bibliography.

Charities and the state 1 Charities, which traditionally were funded and run by volunteers selflessly trying to make the world a better place, are receiving increasing proportions o[f] their incomes from the Government.
2 WHO CARES? HOW STATE FUNDING AND POLITICAL ACTIVISM CHANGE CHARITY, a report for the *think tank* (qv) Civitas by Nick Seddons, argues that charities receiving over 70% of their income from government, as some now d[o] have effectively become part of the state, rather than civil society, and should lose their charitable status in order to preserve the integrity of the sector. He therefore proposes that charities be classed in three groups:
• Those receiving less than 30% of their income from the state should continu[e] to benefit from charitable status.
• Those receiving between 30% and 70% should be called state-funded chariti[es] and receive more modest benefits.
• Those receiving over 70% of their income from the state are already *de facto* state agencies, and should be forced to choose between reducing their dependency on statutory funding or losing their charitable status.
3 For publication details of who cares? see the bibliography.

See also: *Commission for the Compact; Office of the Third Sector; Pressure groups*; and *Voluntary and community sector.*

Charity Commission Visit www.charitycommission.gov.uk/

Charter 88 1 A left-of-centre *pressure group* (qv) which holds that in the UK too much power has become concentrated in the hands of too few people.
2 For more information visit www.charter88.org.uk/

See also: *Constitution*; and *Post-democracy.*

Charter for Childhood Visit www.playengland.org.uk/

Charter of basic rights See: *Human Rights Act.*

Child abuse 1 Despite media reports to the contrary, Britain is not facing a major child abus[e] epidemic. Official figures suggest quite the opposite. In 2004–05 just 23 children in every 10 000 were on the child protection register, compared with [2…] in 2001, and 32 in 1995.
2 As of 2008, useful publications include the following. For publication detail[s] see the bibliography.
• REFERRALS, ASSESSMENTS, AND CHILDREN AND YOUNG PEOPLE ON CHILD PROTECTION REGISTERS, ENGLAND – YEAR ENDING 31 MARCH 2005.

See also: *ContactPoint.*

Child benefit

1 Child benefit is a regular payment made by the Government to anyone bringing up children. It is paid for each child, and is not affected by income or savings.
2 As of 2008, useful websites include the following. They are given in alphabetical order.
• Directgov www.direct.gov.uk/
• HM Revenue and Customs www.hmrc.gov.uk/

See also: *Welfare reform*.

Child Maintenance and Enforcement Agency

See: *Child Support Agency*.

Child poverty

1 Children in households on less than 60% of the national *median income* (qv) are officially defined as living in poverty – for a more detailed discussion see: *Poverty*. The graph below, from the Department for Work and Pensions (qv), shows that the number of children in the UK living in poverty by this definition, after housing costs, fell by 800 000 between 1998–99 and 2004–05, and then rose by 200 000 in 2005–06.

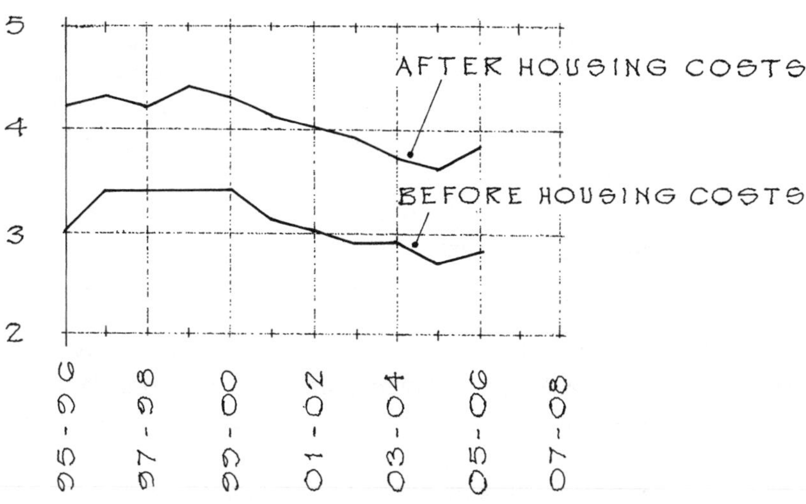

2 Children in households on less than 40% of the national median income are officially defined as living in severe poverty. According to Table 2 of WELFARE ISN'T WORKING: CHILD POVERTY by Frank Field and Ben Cackett (see below) this figure has hardly changed since 1997 – the BHC figure has remained steady at 0.7 million children, and the AHC figure has fallen marginally from 1.4 to 1.3 million. This study from the *think tank* (qv) Reform contains much thoughtful comment and is well worth reading.
3 As of 2008, useful references include the following. They are given in alphabetical order. For publication details see the bibliography.
• CHILD POVERTY IN PERSPECTIVE: AN OVERVIEW OF CHILD WELL-BEING IN RICH COUNTRIES. This UNICEF study assesses the well-being of children in the world's advanced economies on six criteria – material well-being, health and safety, education, peer and family relationships, behaviours and risks, and young people's own subjective opinions. Britain is near bottom of the list.
• THE COST OF EXCLUSION, a study carried out by the London School of Economics for the Prince's Trust, reported in 2007 that around 20% of Britain's young people are *Neets* (qv), and predicted that they are likely to spend their entire future lives in poverty.

• THE MILLENNIUM COHORT STUDY: SECOND SURVEY – A USER'S GUIDE TO INITIAL FINDINGS by Kirstine Hansen and Heather Joshi, is tracking a group of over 15 000 children from their birth to their dying day, and is undoubtedly one of the major social research projects of our time. Its successive stages will be an essential record.
• NARROWING THE GAP, a report by the Fabian Commission on Life Chances and Child Poverty. This final report from the above Commission, published in 2006, claims to provide a comprehensive audit of inequality in Britain and proposes steps which could make life chances more equal. It calls for maternity services, support and special benefits to be refocused and concentrated on the most disadvantaged mothers.
• REDUCING INEQUALITIES, a report published in September 2007 by the National Children's Bureau, the Institute of Education at the University of London, and the Family and Parenting Institute, analysed data from a study that tracked the lives of 17 000 people born in 1970. It found that inequalities in British society are entrenched; social mobility is a myth; and despite billions of pounds of government funding to cut child poverty, the gap between the poorest and richest children is probably wider now than it was 30 years ago.
• SOCIAL TRENDS (qv), an annual publication of the Office for National Statistics (qv), gives latest statistical data plus background explanation and comment. Social trends 32 additionally contained a special extended essay titled 'Children'.
• WELFARE ISN'T WORKING: CHILD POVERTY by Frank Field and Ben Cackett, from the *think tank* (qv) Reform, argues that chancellor Gordon Brown's *tax credit* (qv) policy has failed to help the poorest children, and that it has had the very undesirable side-effect of discriminating against two-parent families. See: *Children: single-parent.*
4 As of 2008, useful websites include the following. They are given in alphabetical order.
• Child poverty action group www.cpag.org.uk/
• Directgov www.direct.gov.uk/
• Frank Field www.frankfield.co.uk/type3.asp?id=20&type=3
• Joseph Rowntree Foundation www.jrf.org.uk/child-poverty/
• Social Trends www.statistics.gov.uk/StatBase/Product.asp?vlnk=13675

See also: *Homelessness; Housing: affordable; Poverty*; and *Social exclusion*.

Child Support Agency (CSA)

1 The Child Support Agency (now replaced by the Child Maintenance and Enforcement Commission) was launched in 1993 with an unusual degree of cross-party support. Its aim was to make absent parents pay a fair share of the costs of raising the children they had left behind.
2 It turned out to be a disaster. In the 13 years of its existence millions of children received little or no maintenance money even though taxpayers paid out almost £50 billion. In June 2006 the House of Commons cross-party Public Accounts Committee issued one of the most damning indictments in its history stating that recent attempts to simplify the CSA's child support system and speed up payments had actually made a bad situation worse on every single measure, and amounted to what the Committee termed 'one of the worst public administration scandals in modern times'. The following month, in July 2006, the Agency was dissolved. It had at that time identified but failed to collect £3.5 billion in maintenance (which a year later was still owed to families and children); new cases had remained unopened for a year owing to a backlog of over 300 000 unprocessed claims; and a third of phone calls from applicants – many of them desperate – were going unanswered. Despite expenditure of £91

million on private consultants, the computerised system still had 500 identified faults.

3 The cautionary lessons to emerge from this disaster are important enough to be worth noting. Various analyses, including those of the House of Commons Public Accounts Committee quoted above, have mentioned the following blunders and shortcomings:

- One that has now become wearisomely familiar in ambitious Government projects – relying on hugely ambitious but inadequately piloted computer systems to make them work, with no fall-back position if an untested system fails. For other examples see: *IT systems*.
- An early decision (in which the present author detects the hand of the Treasury, whose invincible indifference to the way things work in the real world has wrecked numerous well-intentioned government schemes under all political parties) to use the CSA to save government money rather than to achieve its original aim of trying to erase *childhood poverty*. Ministers decided that for each £1 a lone parent received in maintenance, the state would deduct £1 of benefits, so that the CSA became an agency for recovering the meagre state subsidies received by impoverished people living on benefits.
- Over-complexity. When the CSA was created in 1993 it required its staff to deal with 100 pieces of information in order to work out a maintenance payment.
- Ministerial inexperience. Whereas cabinet secretaries in the USA are appointed for their expertise and usually serve at least a four-year term in order to develop a commanding knowledge of their portfolio, their counterparts in the UK generally take responsibility for matters of which they have no prior knowledge whatsoever, and are then moved on before they've had time to truly master their department. The CSA had four secretaries of state in its nine years after the 1997 election: Baroness Hollis from 1997; Mr Alistair Darling from June 2001; Mr Andrew Smith from May 2002; and Mr John Hutton (who was then taking up his third ministerial post in seven years) from November 2005 to the Agency's demise in July 2006. Need one ask why these intelligent people failed to get a grip on their multi-billion pound department?
- Ministerial impatience. Like so many other complex governmental schemes this one was never tested in a pilot project; employees were not adequately trained; and systems were not properly tested in the urge to get it running.
- Ministerial arrogance. The Department for Work and Pensions (qv) ignored numerous warnings of critical defects in the system; introduced a new and complex £800 million *IT system* (qv) at the same time as a departmental restructuring was being carried out; and failed to put in place a contingency plan in the event of anything going wrong. Ministers with little experience of life outside politics appeared to believe that if they demanded firmly enough that a complex system must work, then it would work.

4 It has been announced that the CSA would be replaced by the Child Maintenance and Enforcement Commission. On behalf of the vulnerable children and mothers who are still owed £3.5 billion, and the taxpayers who have so far stumped up around £50 billion with small result, one must hope (without much confidence) that lessons have been learnt from the above failures.

5 As of 2008, useful websites include the following. They are given in alphabetical order.

- Child Support Agency www.csa.gov.uk/
- Directgov www.direct.gov.uk/

See also: *Child poverty*.

Child Tax Credits See: *Tax Credits*.

Childcare

1 In pursuance of two well-intentioned policies (to help women compete with men in the workplace, and to get poorer women off benefit and into work) the Government has for some years been encouraging mothers to go to work, and funding a great expansion of the number of available nursery places.

2 The creation of more nursery places is to be welcomed, but not if based on policies which encourage the separation of young children from their mothers. All published studies agree that the best type of infant care is at home, by the mother; and some suggest that a day nursery may well be the worst option. In a letter titled 'Childcare problems', published in the *Daily Telegraph* on 21 October 2006, eight well-informed people led by Sir Richard Bowlby, President of the Centre for Child Mental Health, called for an urgent national debate before proceeding with current policies.

3 Most mothers do in fact want to look after their own infants, and only hand them over to strangers with great reluctance. They may be driven out to work against their will by many factors, including:

• The exorbitant cost of housing – see: *Housing: affordability*. Impossible house prices lie at the root of so many social problems that a big house-building programme should be a top Government priority.

• Britain's irrational benefit system, which rewards single parents for staying single and punishes them financially if they decide to form a two-parent family – see: *Children: single-parent*.

4 As an indication of current trends, a news release from the DfES in May 2007 stated that all 3 and 4-year-olds are entitled to 12.5 hours early education for 38 weeks of the year, and that many more of them were doing so. The take-up rate in 2007 was said to be 96% of 3-year-olds (up from 87% in 2003), and virtually 100% of 4-year-olds – visit www.dfes.gov.uk/rsgateway/DB/SFR/S000729/index.shtml. Most of these youngsters were enrolled in nursery or infant classes in state primary schools, and Margaret Morrissey of the National Confederation of Parent Teacher Associations found the figures disturbing. She said: 'My 3-year-old grand-daughter goes to pre-school and comes home mentally exhausted. We have been assured that schooling at that age is informal and all about learning through play but that is simply not the case. Thanks to the influence of Government inspectors, tests, and league tables over the past 10 years, lessons even for 3 and 4-year-olds are becoming very rigid'. In similar vein, the annual conference of the NUT in March 2008 expressed concern that under-fives could lose the freedom to play thanks to a 'toddler's curriculum' that imposes 69 learning goals on pre-school youngsters. Speakers feared that the regime could become a box-ticking exercise, with nursery staff and child-minders under pressure to follow it 'to the letter'.

5 There are no simple answers, but Ms Minette Marrin, whose column in the *Sunday Times* is always worth reading, holds that 'it is wrong to pressgang mothers into work with massive tax and benefit incentives; those incentives should be offered in the opposite direction – to mothers (or fathers) who stay at home to bring up children and who take on community work and charity work. Family life would become affordable; wider good works would become possible'.

6 As of 2008, useful references include the following. They are given in alphabetical order. For publication details see the bibliography.

• CHILDREN'S NURSERIES UK MARKET REPORT 2007, the latest edition of an annual report by the specialist analysts and information providers Laing & Buisson. This survey reports that in 2003 there had been 425 000 nursery places with a take-up of 407 000 children; in 2007 there were 725 000 places with a take-up of 580 000 children.

• CHOOSING TO BE DIFFERENT – WOMEN, WORK AND THE FAMILY by Jill Kirby. Mrs Kirby cites research by the Joseph Rowntree Foundation (qv) which found that under-fives whose mothers were in full-time employment had less chance of obtaining qualifications, and were more likely to be unemployed and suffer psychological distress in adulthood, than others.

• FAMILIES, CHILDREN AND CHILDCARE by Penelope Leach, Kathy Sylva and Alan Stein, published by the Institute for the Study of Children, Families & Social Issues at Birkbeck College, the University of London. The authors tracked 1 200 children from the age of three months to four years and found that children looked after by stay-at-home mothers had fared best, followed by those looked after by nannies and childminders in a homely situation, followed by those looked after by grandparents and other relatives. Children cared for in day nurseries had fared worst.

• SHARED CARE: ESTABLISHING A BALANCE BETWEEN HOME AND CHILD CARE SETTINGS. This article in the journal 'Child Development' by Professor Lieselotte Ahnert of the Free University of Berlin, and Professor Michael E Lamb of the National Institute of Child Health and Human Development, reports on evidence of stress among very young children who are being cared for in day nurseries.

• SOCIAL TRENDS (qv) gives latest statistics plus commentary on the above and related matters.

7 As of 2008, useful websites include the following. They are given in alphabetical order.

• Civitas	www.civitas.org.uk/pubs/familyMain.php
• Directgov	www.direct.gov.uk/
• Directgov schoolsfinder	http://schoolsfinder.direct.gov.uk/
• Full Time Mothers	www.fulltimemothers.org/
• Laing & Buisson	www.laingbuisson.co.uk/P
• National Family & Parenting Institute	www.familyandparenting.org/
• Watch	www.jbaassoc.demon.co.uk/watch/
• What about the children?	www.jbaassoc.demon.co.uk/watch/

See also: *Childhood; Children's play; Education: pre-school; Neighbourhood Nursery Initiative; Families*; and *Sure Start*;

---

hildcare: vetting

1 The term 'vetting' refers in this instance to a variety of checks that must, by law, be carried out on adults who come into professional contact with children. The aim is laudable, but the methods being introduced to achieve those aims are seen by many as wildly *disproportionate* (qv) and – given the proven unreliability of large government databases (see: *IT systems*; and *Privacy*) a serious threat to our *civil liberties* (qv).

2 The scale of intrusion is disturbing: in the three years prior to 2006 the number of checks of this kind carried out by the Criminal Records Bureau (qv) added up to 8.9 million, and in 2006 it was estimated that the Safeguarding Vulnerable Groups Bill (qv) would, if enacted, mean that up to 9.5 million adults in England and Wales – more than a third of the adult working population – would be subject to ongoing criminal checks to establish their suitability to work with children. The people obliged to undergo such checks would include lollipop ladies, youth club workers, 16-year-old boys teaching sport to younger children at weekends, and a father who offers to coach his son's team on a Sunday morning. In all of these cases failure to have undergone a formal criminal records check would result in a criminal record and possibly a £5 000 fine.

3 As of 2008 useful references include the following. For publication details see the bibliography.

• THE CASE AGAINST VETTING: HOW THE CHILD PROTECTION INDUSTRY IS POISONING ADULT-CHILD RELATIONS. This is a publication from the Manifesto

Club (qv), which conducts valiant campaigns against vetting and other intrusion upon our *privacy* (qv) and liberty.
4 As of 2008 useful websites include the following. They are given in alphabetical order.

- Directgov www.direct.gov.uk/
- Foundation for Information Policy Research www.fipr.org/
- Manifesto Club www.manifestoclub.com/
- Office of Public Sector Information www.opsi.gov.uk/acts.htm

See also: *ContactPoint; Privacy; Safeguarding Vulnerable Groups Act 2006*; and *Transformational government*.

Children and crime See: *Crime: youth*.

Children and Family Court Advisory Service See: *Office for Standards in Education*.

Children and risk See: *Safeguarding Vulnerable Groups Act*; and *Risk assessments: play areas*.

Children: behaviour 1 Research results published in July 2007 by the left-of-centre *think tank* (qv) Institute for Public Policy Research showed British youngsters to be among the worst behaved in Europe, and that 30% of adults in Britain would definitely not challenge a group of youngsters vandalising a bus shelter compared with 19% in France and 7% in Germany; etc. Visit www.ippr.org.uk/pressreleases/?id=281
2 Some of the reasons may be found in some of the entries preceding and succeeding this one, and in entries such as *Crime: youth; Families* (qv) and *Neets*.

Children born outside marriage 1 The UK has, at 42%, one of the highest levels of births outside marriage in the EU. Figures within the UK are 52% for Wales; 47% for Scotland; 42% for England; and 36% for Northern Ireland.
2 The graph below, from Figure 2.16 in SOCIAL TRENDS 31 (qv), shows the figure for the UK to 1999; since then it has climbed to 42%.

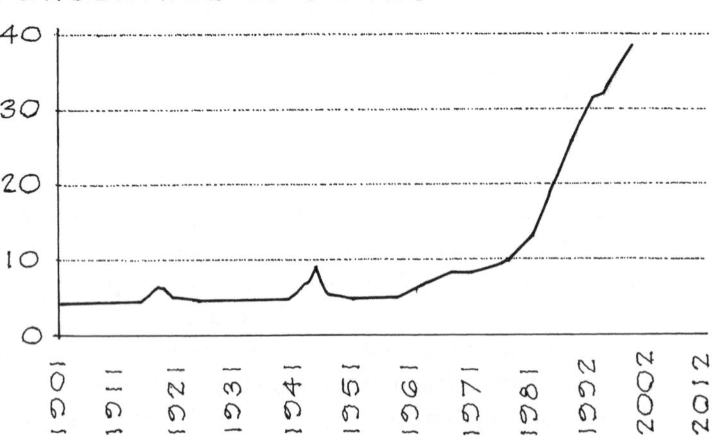

3 As of 2008, useful references include the following. For publication details see the bibliography.

- SOCIAL TRENDS (qv) gives latest statistics plus commentary on the above and related matters.

See also: *Children: single parent*; and *Families*.

hildren: equipment

1 Children, particularly those who are *disabled* (qv) or have *special needs* (qv) of other kinds, need learning equipment, living equipment, mobility equipment, and play equipment that has been specially designed to suit their needs.
2 As of 2008, useful websites include the following. Alphabetical order.

• DCS Joncare	www.dcsjoncare.freeserve.co.uk/
• Disabled Living Foundation	wwwww.dlf.org.uk/public/factsheets.html
• Dragonmobility	www.dragonmobility.com/
• Mission	www.missioncycles.co.uk/
• Tendercare	www.tendercareltd.com/

See also: *Multi-sensory environments*.

hildren: ngerprinting by chools

1 In July 2007 it emerged that schools had been given ministerial permission to take fingerprints from children as young as five without parental permission. They could similarly take *biometric* (qv) data such as retina and iris scans, face shapes, and hand measurements. An investigation earlier in the year showed that 285 schools were already using fingerprint scanners in this way. According to a survey by Mr Greg Mulholland, parental permission had been sought in fewer than 20% of cases, and polls by lobby groups suggested that around 3 500 schools had bought the necessary equipment. These revelations prompt a question posed by Terri Dowty of the children's rights group Action Rights for Children (ARCH) in the entry for ContactPoint: 'Who is bringing children up? Are parents effectively nannies for the state's children, or are children born to families and the state just helps families when they ask for it?'.
2 The Schools Minister Mr Jim Knight defended the new fingerprinting practices, quoting guidance from Becta (qv) that schools should not pass on such data to outside bodies. Civil liberties (qv) campaigners such as Liberty were unimpressed, pointing out that:
• School computers are not secure against hackers. Children are therefore being exposed to the risk of identity fraud at a later stage, quite apart from other forms of misuse.
• The police have the right to enter any database, private or public, and their own management of data has been repeatedly shown to be unreliable – see: *Criminal Records Bureau*.
3 As of 2008, useful websites include the following. They are given in alphabetical order.

• Becta	www.becta.org.uk/
• Children's database	www.arch-ed.org/issues/databases/databases.htm
• GeneWatch UK	www.genewatch.org/
• Liberty	www.liberty-human-rights.org.uk/

See also: *National DNA Database; Post-democracy*; and *Privacy*.

hildren in care

1 If parents cannot properly look after their children, *local authorities* (qv) can take them into care. They are then described as being 'looked after'. There are around 68 000 such children in England, Wales, and Northern Ireland, over two thirds of them in foster homes.
2 For the effects of the UK care system upon the youngsters who pass through it see HANDLE WITH CARE below. For evidence of the attitude that the Government as of 2007 brings to bear upon the problem, consider the following. Britain has a dozen Roman Catholic adoption agencies which handle roughly a third of all adoptions of older and troubled children – ones who are generally very difficult to place. These bodies do excellent work, and their disruption rates are among the lowest in the field. In January 2007 the

Government decided that from April of that year all adoption agencies must be willing to place children with homosexual couples. The Roman Catholic church stated that it accepted the right of homosexual couples to adopt children, and would refer any such couples who approached it to a more suitable agency while declining to make such placements itself. In this way the church could continue to place hundreds of children in stable, loving homes, while other agencies filled their own particular niches. An admirable solution, one would have thought. But the Government, led by spokespersons Harriet Harman MP and Angela Eagle MP, flatly decreed that after December 2008 there would be no exemptions whatever to the Sexual Orientation Regulations, and that agencies such as those mentioned above would cease to be recognised. Thus many vulnerable children who could have been placed in stable families will now go through the kinds of experiences detailed in HANDLE WITH CARE below, and suffer the consequences. In a word, *ideology* (qv) rules and the interests of children come a distant second.

3 For more details on children in care refer to Chapter 8 of the latest edition of SOCIAL TRENDS – see below.

4 As of 2007, useful references include the following. They are given in alphabetical order. For publication details see the bibliography.

- BRITISH JUSTICE: A FAMILY RUINED by Camilla Cavendish, published in the Times, 21 February 2008, is one of a succession of revelations of the tyrannous secrecy with which social services regularly transfer children from parents into the dysfunctional care system. Many of these cases beggar belief.
- CARE MATTERS: TRANSFORMING THE LIVES OF CHILDREN AND YOUNG PEOPLE IN CARE, a *green paper* (qv) published in 2006, sought to start a productive debate on these matters such as the above.
- CARE MATTERS: TIME FOR CHANGE, a *white paper* (qv) published in June 2007, produced proposals that were generally well-received.
- HANDLE WITH CARE, a study that makes shocking reading, reveals that of the roughly 6 000 children who emerge from the care of the state each year, most are likely to leave school without a single qualification; half will end up jobless; the boys are very likely to go to prison; and only 1% of these youngsters will go to university, compared with a national average of around 42%.
- CARE MATTERS: TIME FOR CHANGE, a *white paper* (qv) published in June 2007, responded to criticisms such as the above with proposals that were generally well-received.
- SOCIAL TRENDS (qv) gives latest statistics plus commentary on the above and related matters.

5 Useful websites include the following, given in alphabetical order.

• Barnado's	www.barnardos.org.uk/
• Braintree Children's Trust	www.idea.gov.uk/idk/aio/6441806
• Dept for Children, Schools and Families	www.dcsf.gov.uk/
• Directgov	www.direct.gov.uk/
• Joseph Rowntree Foundation	www.jrf.org.uk/child-poverty/

See also: *Crime; Families*; and *Social exclusion*.

Children: invisible See: *Neets*.

Children: looked after See: *Children in care*.

Children: single-parent 1 All published studies agree that children thrive best if brought up in two-parent families.

2 The bar chart on the next page, from the Office for National Statistics (qv), shows that the proportion of children in Great Britain living in lone-parent families rose steeply until 1997 – perhaps due in part to the steep rise in

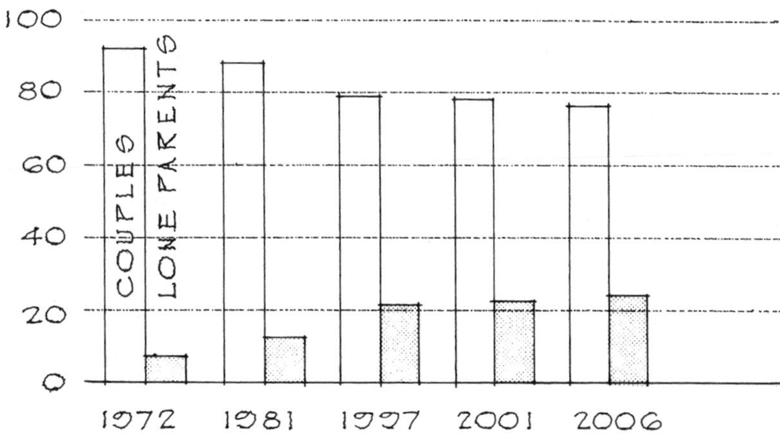

unemployment (qv) at that time, though the ONS does not say this – and then levelled off at just under 24%. For the original set of data visit www.statistics.gov.uk/cci/nugget.asp?id=1748.

3 Changing social attitudes doubtless help to drive the trend, but government policies may also be playing a part. WELFARE ISN'T WORKING: CHILD POVERTY written by the Labour MP Frank Field and the researcher Ben Cackett for the *think tank* (qv) Reform, charges that the *tax credit* (qv) and benefit strategy that is in place in 2007 'brutally discriminates against two-parent families' and actively discourages single parents from forming stable relationships. As an example, a single parent with two children working 16 hours a week will gain a weekly income after tax credit payment of £487. The breadwinner of a two-parent family, also with two children, will have to work 116 hours to get the same income. As Mr Field puts it: 'Such effort is impossible to sustain for more than a week or two', and he adds: 'This is not a plea to take help from single parents (but) to equalise this support to all parents'. The above study recommends that no further general increases in tax credits should occur until proper weighting is given to the second adult in poor households. Hopefully this strategy would not only cut the number of poor children in working households, but would also increase the chances of such children being brought up by two parents, giving them a better start in life.

4 Useful references include the following. They are given in data order. For publication details see the bibliography.

- CHILD POVERTY IN PERSPECTIVE: AN OVERVIEW OF CHILD WELL-BEING IN RICH COUNTRIES, which was published by Unicef in 2007, looked at 40 indicators of child well-being including poverty, family relationships, and health. The UK did worst of 21 industrialised countries.
- WELFARE ISN'T WORKING: CHILD POVERTY by Frank Field and Ben Cackett was published in June 2007 by the *think tank* Reform (qv). See the notes above.
- TAXATION OF MARRIED FAMILIES: HOW THE UK COMPARES INTERNATIONALLY, a study carried out by former Treasury officials and published in January 2008, after the above notes were written, found that the financial penalties for parents who live together are actually getting worse. Ex-chancellor Gordon Brown's taxation policies now ensure that 3 out of 4 ordinary couples would be better living apart, even after the extra costs of running two homes. Families in which one partner works and the other stays at home to bring up one child are likely to gain £95 a week by splitting up.

See also: *Child poverty*; *Children born outside marriage*; and *Families*.

Children's Assessment Framework See: *Common Assessment Framework.*

Children: vulnerable

1 These are officially defined as disadvantaged children who would benefit from extra help from public agencies, to increase their chances of making the most of their opportunites in life. For more detail visit www.dh.gov.uk/PolicyAndGuidance/SocialServicesInspectorate/
2 For child protection more generally, see: *Safeguarding Vulnerable Groups Bill.*

Children with special educational needs

See: *Special educational needs.*

Children's Fund

1 The Children's Fund was launched in November 2000 as part of the Government's commitment to tackle disadvantage among children and young people.
2 As of 2008, useful sources of information include:
• Directgov www.direct.gov.uk/
• Every child matters www.everychildmatters.gov.uk/

Children's Index

See: *ContactPoint.*

Children's play

1 Free, unsupervised play, of which many older people (including the present author) have happy memories, is fast disappearing from the lives of UK children and this is a tragedy. First, British youngsters are thrust into formal education far too soon (under pressure from a nannying state, that threshold is currently heading downward towards the age of 3 whereas in many European countries school begins at the sensible age of 6 or even 7; and second, allowing children to wander around and play without supervision is fast becoming regarded as something close to criminal neglect. But children need to play, and much of that play must be free, explorative, and risk-taking – in a word, unsupervised. That is what children most enjoy, and that is how they learn about the world and about themselves.
2 An essay could be written on this important subject, but we have not the space. For partial notes see: *Childcare; Education: pre-school; Health scares; Risk assessments: play areas; Safeguarding Vulnerable Groups Act;* and *Streets.*
3 As of 2008, useful references include the following. For publication details see the bibliography.
• OVER-PARENTING IS THE CURSE OF OUR TIME by Johann Hari.
• NO FEAR: GROWING UP IN A RISK AVERSE SOCIETY by Tim Gill.
• RISK AND CHILDHOOD by Nichola Madge and John Barker.

Children's Society

Visit www.childrenssociety.org.uk/ and see also: *Childhood.*

Choose and Book

See: *NHS Connecting for Health.*

Cinemas: aids for disabled people

1 *Auxiliary aids* (qv) to give *disabled people* (qv) as near a quality of service to other customers as feasible, include the following:
• 'Audio Description' gives cinema-goers with *impaired vision* (qv) an explanation of what is going on via a cordless headset.
• 'Digital Theatre Systems Cinemas Soft Subtitling' (DTS/CSS) is a sub-titling system.
2 As of 2008, useful websites on the above matters include:
• British Film Institute www.bfi.org.uk/
• RNIB www.rnib.org.uk/

See also: *StageText.*

Cinemas: complying with the Disability Discrimination Act

See: *Auditorium design*; and *Service providers*.

Cinemas: design

Visit: www.cabe.org.uk/default.aspx?contentitemid=42

Civic buildings

1 For guidance on complying with the Disability Discrimination Act (qv) see the following entries:
- Administrative and commercial buildings.
- Educational, cultural and scientific buildings.
- Entertainment-related buildings.
- Health and welfare buildings.
- Offices and commercial buildings.
- Sports-related buildings.

2 For additional references, case studies, and commentary on design quality visit www.cabe.org.uk/default.aspx?contentitemid=40

Civil contingencies Act 2004

See: *Civil liberties*.

Civil duty

See: *Civil proceedings*.

Civil liberties

1 UK citizens enjoy more civil liberties than ever before, and more than most of the world's population today. But particularly since 1997, our freedoms have been eroded by Ministers who talk about liberty (especially when pontificating on 'British values') while ruthlessly pursuing intrusion, oversight and control. For instance:
- Our most personal details are being loaded onto colossal central databases covering up to 50 million people each, to which several hundred thousand officials have access. This is being done without our consent or even – in many cases – our knowledge. See: *Privacy*.
- In the 1950s the authorities had 10 legal powers to enter our homes without our consent; by the 1990s they had taken unto themselves more than 60. Again see: *Privacy*.
- Parliament, which ought to be our watchdog against such abuses of power, has been emasculated. Between 50% and 80% (estimates vary) of our laws are now made in Brussels and virtually nodded through a UK Parliament which, under the 1972 European Communities Act, is powerless to block their passage onto the lawbook. In the case of our own home-made laws, the commentator Mr Henry Porter (whose column in the *Observer* newspaper on Sundays is essential reading on these matters) estimates that the guillotine was used only 80 times in the 94 years between 1881 and 1975, but has been applied 216 times in the 6 years between 1997 and 2003 to cut off parliamentary debate and ram legislation through the House of Commons. See: *Parliament*.

2 Partly as the result of this weakening of Parliament the following laws have been passed, many of them after little (and some after no) debate. They are given here in date order:
- The Protection from Harassment Act 1977 was stated, when passed, to be aimed at defending people from stalkers. But the first three people to be arrested under the Act were peaceful protesters, and since then it has been used repeatedly to prevent the people of Radley village in Oxfordshire from peacefully protesting against a plan to empty Thrupp Lake, which is a treasured local amenity, line it with clay, and fill it with a hideous grey slurry. For more information visit www.saveradleylakes.org.uk/

- The Public Order Act 1986 makes it an offence to stir up hatred against defined groups of people. In May 2005 Mr Sam Brown, an Oxford student out celebrating the end of his finals, asked a mounted police officer: 'Do you realise your horse is gay?'. The policeman radioed for assistance and within minutes two squad cars arrived, disgorging a group of officers who arrested Mr Brown under Section 5 of the Act for making homophobic remarks. A police spokesperson told the student newspaper *Cherwell* that the remarks had been 'not only offensive to the policeman and his horse, but any members of the general public in the area'. The accused spent a night in the cells and was fined £80, which he refused to pay. He was acquitted of the charge after a court hearing in January 2006.
- The Firearms (Amendment) (No. 2) Act 1997, which was passed following the shootings at Dunblane on 13 March 1996, banned the private ownership of all cartridge ammunition handguns, regardless of calibre. No exceptions were allowed, not even the British Olympic shooting team, who have since then been training outside Britain. Yet ten years later the country is awash with illegal guns and gun crime has rocketed. Laws are supposed to be aimed at criminals, but this one robbed law-abiding gun-owners of a traditional civil liberty, while having no effect whatsoever on criminals or on the rate of gun-crime, except possibly an inverse one. For figures from the USA showing that gun crime goes down when citizens are allowed to carry arms, and rises when they are not see the article titled WOULDN'T YOU FEEL SAFER WITH A GUN? by Richard Munday in *The Times* on 8 September 2007.
- The Harassment Act 1997 has been used to charge a woman for sending two e-mails to a company requesting them not to conduct animal experiments. Her offence was to send two e-mails, for in that lies the repeated action that can now held to be illegal.
- The Terrorism Act 2000, which according to the Home Office website was passed 'in response to the changing threat from international terrorism'. In September 2005 Mr Walter Wolfgang, an 82-year old member of the Labour Party, was first frogmarched out of the annual party conference for heckling the Foreign Secretary Jack Straw on the Iraq war, and then prevented from re-entering by police citing the Terrorism Act. After initial denials a Sussex police spokesman admitted that Mr Wolfgang had been issued with a section 44 stop-and-search form under the Terrorism Act.
- The Courts Act 2003 swept away the long-standing principle of British justice that a citizen's home is inviolate, and allows magistrates to appoint a fines officer who may break into a private home to seize goods. These provisions were never debated in Parliament. As noted by the commentator Mr Henry Porter, they were smuggled into law in the Domestic Violence, Crime and Victims Act 2004 (qv), thus avoiding Parliamentary scrutiny on the crucial question of forced entry in civil cases.
- The Civil Contingencies Act 2004 allows Ministers in an emergency, which they only have to believe is about to occur, to make practically any provision they like without reference to Parliament. During such a declared emergency they can pass special legislation which allows the forced evacuation of people, the seizing of property without compensation, and the banning of any assembly, which could conceivably include Parliament itself.
- The Inquiries Act 2005 gave Ministers an unprecedented degree of control over government inquiries, in effect allowing them to scrutinise their own behaviour instead of letting Parliament do so. Under the Act they are able to appoint the members of the inquiry, set its terms, restrict public access, suppress evidence, and shut it down without having to explain themselves to anybody. Inquiry reports are now routinely presented to Ministers and not, as they once were, to *Parliament* (qv).

• The Serious Organised Crime and Police Act 2005 (a) allows police officers to remove DNA from innocent people, using force if this is necessary; (b) bans demonstrations (even if this means standing quietly with a placard) within 1 km (just over half a mile) of Parliament without prior police consent; and (c) allows the new Serious Organised Crime Agency to forcibly enter someone's home, even if the householder is not suspected of a criminal offence, if the agency has received no response to a 'disclosure notice' served upon that person, or if it is believed that it would not be 'practicable' to issue such a notice. As an example of the application of (b) above: in October 2005 Miss Maya Anne Evans was arrested after a peaceful remembrance ceremony at central London's Cenotaph, at which she read out the names of 97 British soldiers killed in Iraq while fellow campaigner Milan Rai read out the names of dead Iraqi civilians, ringing a bell each time a name was read out. Both were arrested and Miss Evans was held for five hours at Charing Cross police station before being charged. She appeared in court in December for demonstrating without police consent. After a 3-hour trial magistrates found her guilty and gave her a conditional discharge. As a second example: this law was also used to charge a mime artist named Neil Goodwin for doing an impersonation of Charlie Chaplin outside Parliament. Mr Goodwin's statement to the court concluded: 'In truth, one of the first things to go under a dictatorship is a good sense of humour'. On a happier note: soon after taking office as Prime Minister in June 2007 Mr Gordon Brown announced that the clause under which Miss Evans was arrested would be revised. We await the outcome with interest.

• The Identity Cards Act 2006 (qv) will in due course oblige every citizen to inform the authorities each time he or she changes their address, or be fined up to £1 000 per offence. For further comment see that entry

• The Safeguarding Vulnerable People Act 2006 (qv) extends criminal checks to around 10 million adults, about one third of the working population. The Bill was so badly drafted that it needed 250 amendments at the end of its passage through Parliament, which almost certainly means that the law was not properly scrutinized and is riddled with imperfections. For notes on this Act, which has not received the public attention it deserves, see that entry.

• The Legislative and Regulatory Reform Act 2006 would, if passed as originally drafted, have virtually set aside the need for Parliamentary debate when our legislators found the latter inconvenient. In the name of reducing the unreasonable burden of regulations that now stifles businesses, charities, schools, hospitals, and local authorities (see: *Better regulation bodies*; and *Over-regulation*), the Government proposed to give itself the power to amend, replace, or repeal any law whatsoever without Parliamentary scrutiny – in short, to rule by diktat. Fortunately a cross-party coalition of outraged MPs, peers, civil libertarian organizations, and media commentators objected so vociferously to this summary abolition of Parliamentary democracy that the Government was forced in May 2006 to announce a radical revision to the Bill, reducing it to what it should always have been – a law that is limited to the reduction of business regulation. For its effectiveness see: *Better regulation bodies*.

• The Racial and Religious Hatred Bill 2006 (qv) would, had it been enacted in its original form, have outlawed the expression of provocative views on anyone's religion, a profound curtailment of free speech. Mercifully the Bill was defeated by a campaign mounted very effectively outside Parliament by the actor Rowan Atkinson and the writers' organisation Pen.

• The Tribunals, Courts & Enforcement Bill going through Parliament in April 2007 proposes to greatly extend the rights of officials to enter our homes, but the final shape of the Act is unknown as this is written.

3 The above laws are disquieting in themselves. Taken together they amount to an assault upon our freedoms that is unprecedented in peacetime Britain, and

against which the present Human Rights Act 1998 (qv) has given us no protection at all. What can be done? Mr Henry Porter, to whom the above notes owe much, has called for three actions:
• A detailed audit, by an independent body, to establish the degree to which British citizens' rights and liberties have been compromised by the above and other legislation.
• The appointment of a commission to investigate the effects of mass surveillance – see: *Privacy*.
• The writing of a new British *constitution* (qv) that enshrines a bill of rights and places our rights and liberties beyond the reach of an increasingly power-hungry Executive and Parliament.
4 As of 2007, useful references include the following. They are given in alphabetical order. For publication details see the bibliography.
• CHARGE OR RELEASE: TERRORISM PRE-CHARGE DETENTION – COMPARATIVE LAW STUDY. This is an international comparison of the periods for which terrorism suspects may be held without charge. In Britain the figure is 28 days; the rest of the democratic world is said to average around 4 days.
• CROSSING THE THRESHOLD: 266 WAYS IN WHICH THE STATE CAN ENTER YOUR HOME by Harry Snook. Under English law the home has traditionally been a privileged space, which the authorities may not enter without the permission of the owner unless authorised to do so by a specific law. In the 1950s the authorities had 10 such legal powers; today they have more than 60 – and in most of these cases force can be used.
• THE GOVERNANCE OF BRITAIN. This *green paper* (qv) was published within days after Prime Minister Gordon Brown took office in June 2007. It proposes a nationwide debate on a new constitutional settlement.
5 Useful websites include the following, given in alphabetical order.

• British Institute of Human Rights	www.bihr.org.uk/
• Democratic Audit	www.democraticaudit.co.uk
• Directgov	www.direct.gov.uk/
• Information Commissioner's Office	www.ico.gov.uk/
• Justice	www.justice.org.uk/
• Liberty	www.liberty-human-rights.org.uk/
• National Identity Scheme	www.identitycards.gov.uk/scheme.asp
• No2ID	www.no2id.net/
• Office of Public Sector Information	www.opsi.gov.uk/acts.htm
• Open Rights Group	www.openrightsgroup.org/

See also: *Constitution; Human rights; Parliament; Post-democracy; Privacy*; and *Transformational government.*

Civil partnerships

1 Civil partnerships, which came into force in December 2005, allow two people of the same sex to register their relationship and have it legally recognised. This entitles them to equal treatment with married couples in a wide range of legal matters.
2 For more detail visit www.dwp.gov.uk/

Civil proceedings

1 Proceedings in UK courts of law are of two kinds:
• Civil proceedings, in which one citizen (which may be a company) brings a case against another. Examples include breaches of contract; trespassing on someone else's property; or negligence. The consequence of being found guilty is not punishment, but an order to compensate the wronged party. Losing a case does not result in a criminal record.

• Criminal proceedings, in which the state prosecutes a citizen for a breach of

law. The consequence of being found guilty is punishment and acquiring a criminal record.
2 The duty under the Disability Discrimination Act (qv) not to *discriminate* (qv) against *disabled persons* (qv), is an example of a civil duty. The onus rests on one citizen (the person who feels that he or she has been discriminated against) to bring a case against another (whoever is accused of doing the discriminating), and the state is not involved. If the court or tribunal finds in favour of the complainant it can apply one or more of the following remedies:
• Compensation to the aggrieved person for actual financial losses incurred. For instance, loss of earnings.
• Damages to the aggrieved person for injury to feelings. As of 2006, the sum awarded often amounts to around £1 000.
• An injunction to the guilty party (for instance, a shop or hotel) not to repeat any discriminatory act in future.
• An order (called a 'mandatory injunction') to the guilty party to carry out a particular action. For instance, a shop or hotel could be ordered to make certain physical *adjustments* (qv) to the building, such as installing a ramp or lift. While this is a possibility, courts traditionally dislike ordering people to carry out tasks, and they seldom do it.
3 Readers are reminded that the present work only gives simplified summaries of original documents and does not purport to provide full, authoritative and up-to-date statements of the law.
4 The situation will be constantly developing in coming years, and designers and managers must keep up with current information. As of 2007, useful websites for checking the latest situation include:
• Commission for Equality and Human Rights www.cehr.org.uk/
• Directgov www.direct.gov.uk/

See also: *Disability Discrimination Act*; and *Tribunals*.

---

ivitas See: *Think tanks.*

---

limbié, Victoria See: *Victoria Climbié.*

---

limate change See: *Kyoto Protocol.*

---

linics See: *Primary healthcare centres.*

---

losed circuit elevision (CCTV)

1 CCTV enables public places such as streets, shops, airport terminals, as well as private premises, to be kept under surveillance and the records to be preserved for as long as appropriate. According to Sheffield University's Centre for Criminological Research, Britain in 2004 had over 4 million CCTV cameras – i.e. one for every 14 people in the UK, which was thought to be the highest national figure in the world.
2 CCTV systems must be operated within legislation such as the Data Protection Act (qv), and in a manner that retains the confidence and support of citizens. So far most people appear to be remarkably unconcerned about the extent to which they are being surveilled in this manner, believing that 'if you are innocent you have nothing to fear'. But they ought to be taking a closer interest – see: *Privacy*.
3 As of 2008, useful references include the following. They are given in alphabetical order. For publication details see the bibliography.
• CRIME PREVENTION EFFECTS OF CLOSED CIRCUIT TELEVISION: A SYSTEMATIC REVIEW by Brandon C Welsh and David P Farrington.
• DILEMMAS OF PRIVACY AND SURVEILLANCE – CHALLENGES OF

TECHNOLOGICAL CHANGE by Professor Nigel Gilbert and others, and published by the Royal Academy of Engineering in March 2007.

• A REPORT ON THE SURVEILLANCE SOCIETY. This study, produced by the Surveillance Studies Network for the Information Commissioner (qv), provides a systematic and unusually thorough survey of the many types of surveillance in operation as of 2006, and offers brief suggestions on regulation.

4 Useful websites include:

• British Security Industry Association	www.bsia.co.uk/index.php
• Liberty	www.liberty-human-rights.org.uk
• Privacy International	www.privacyinternational.org/

See also: *Privacy*.

Coaches

See: *Buses and coaches*.

Cobbled surfaces

1 Cobbled surfaces, which are almost impossible for older or *disabled people* (qv) to traverse, are a feature of some *historic buildings and sites* (qv). One way of making such properties *accessible* (qv) while retaining their character is to form a smooth-surfaced pathway, made of some sympathetic material, within the cobbled expanse, leading for instance from the site entrance to the front door. York stone pavings may be suitable.

2 As of 2008, useful references include the following. For publication details see the bibliography.

• EASY ACCESS TO HISTORIC BUILDINGS.

• EASY ACCESS TO HISTORIC LANDSCAPES.

These twin guides give excellent advice on how to reconcile the requirement for *accessibility* (qv) with the special nature of historic buildings

See also: *Access routes; Gravel surfaces*; and *Historic buildings and sites*.

Codes of practice

See: *Statutory instruments* and *codes of practice*.

College and university design: complying with the Disability Discrimination Act

1 Buildings used by the public must in general comply with the Disability Discrimination Act (qv). In the case of colleges and universities, *authoritative practical guides* (qv) include DESIGNING FOR ACCESSIBILITY (qv) and in addition the following documents. For publication details see the bibliography.

• Para 13.9 of BS 8300 gives brief design recommendations for universities, colleges, libraries, laboratories, and other educational, cultural, and scientific buildings.

BS 8300 does not have the force of law, but conformity with its recommendations will make it easier to demonstrate that the requirements of the Disability Discrimination Act DDA have been met.

2 With respect to university and college halls of residence, see also the following entries:

• Bathrooms: complying with the Disability Discrimination Act.

• Bedrooms: complying with the Disability Discrimination Act.

• Changing and shower areas: complying with the Disability Discrimination Act.

• Lavatories: complying with the Disability Discrimination Act.

3 As of 2008, useful websites include the following. They are given in alphabetical order.

• Commission for Equality and Human Rights	www.cehr.org.uk/
• DPTAC Access Directory	www.dptac.gov.uk/adnotes.htm
• DPTAC Publications	www.dptac.gov.uk/pubs.htm

olour in buildings: omplying with e Disability iscrimination Act

1 Buildings used by the public must in general comply with the Disability Discrimination Act (qv). The Act does not in general apply to private dwellings.

2 In this context, recommendations on the use of colour generally refer to the exploitation of *visual contrast* (a term that is defined in para 0.29 of APPROVED DOCUMENT M (qv) of the Building Regulations) to help people with poor vision identify objects and avoid hazards.

3 Typical examples of the use of colour contrast include the following:

- Lettering on signs should be in *visual contrast* with the background colour – for an illustration see the next entry.
- Stair nosings should be in visual contrast with the main body of the stair – for an illustration see the next entry.
- Stair handrails should be in visual contrast with their surroundings.
- Door handles should be in visual contrast with the door.
- Light switches should be in visual contrast with the wall.
- WC fittings such as soap dispensers should be in visual contrast with the wall.

4 The above examples of visual contrast are common sense. Some of the other recommendations in BS 8300, for instance that wall colours should be 'noticeably different' from those of ceilings and floors, are in the view of the present author based on dubious research, of no practical importance to over 99% of building users, and quite unacceptably prescriptive. Their impracticality is commented upon by Mr Selwyn Goldsmith on pp 195–6 of DESIGNING FOR THE DISABLED: THE NEW PARADIGM. Designers should resist attempts by officialdom to dictate to them on these matters.

5 *Authoritative practical guides* (qv) for complying with the Disability Discrimination Act include DESIGNING FOR ACCESSIBILITY (qv) and in addition the following documents. For publication details see the bibliography.

- approved document m (qv) deals with access to and use of buildings. Its provisions must be heeded in order to comply with the Building Regulations in England and Wales.
- BS 8300 (qv) applies to all buildings in the UK. It is an advisory document.
- INCLUSIVE MOBILITY (qv) is the most authoritative reference for the *inclusive* design of *pedestrian infrastructure* (qv) and for transport-related facilities such as bus and coach stations, railway stations, air terminals, and transport interchanges. It is an advisory document.
- BUILDING SIGHT: A HANDBOOK OF BUILDING AND INTERIOR DESIGN SOLUTIONS is a useful guide for situations in which colour and tonal contrast are judged to be particularly important. It is an advisory document.
- COLOUR AND CONTRAST – A DESIGN GUIDE FOR THE USE OF COLOUR AND CONTRAST TO MEET THE NEEDS OF DISABLED PEOPLE shows colour and/or tonal combinations that will meet the criteria stated in the preceding works. It is an advisory document.

While the advisory documents above do not have the force of law, conformity with their recommendations will make it easier to demonstrate that the requirements of the Disability Discrimination Act have been met. The relevant clauses in the first three documents are tabulated below.

DOCUMENT	RELEVANT PARAGRAPHS
APPROVED DOCUMENT M	0.29; 2.17; 3.8–3.12; 3.18; 3.28; 3.34; 3.43; 4.28; and 5.4
BS 8300	9.1–9.2
INCLUSIVE MOBILITY	10.1.4

For places other than England and Wales the following regulatory documents should be consulted:
• In Northern Ireland: TECHNICAL BOOKLET R of the Building Regulations (Northern Ireland) (qv).
• In Scotland: the TECHNICAL HANDBOOKS of the Building (Scotland) Regulations (qv).

Colour in buildings: examples of visual contrast

1 In the design of *signs* (qv), para 9.2.3.2 of BS 8300 (qv) recommends use of the colour combinations that are tabulated below to achieve satisfactory visual contrast. The preference is for dark letters, symbols, or pictograms on a light-coloured background, not the other way round, though both are possible.

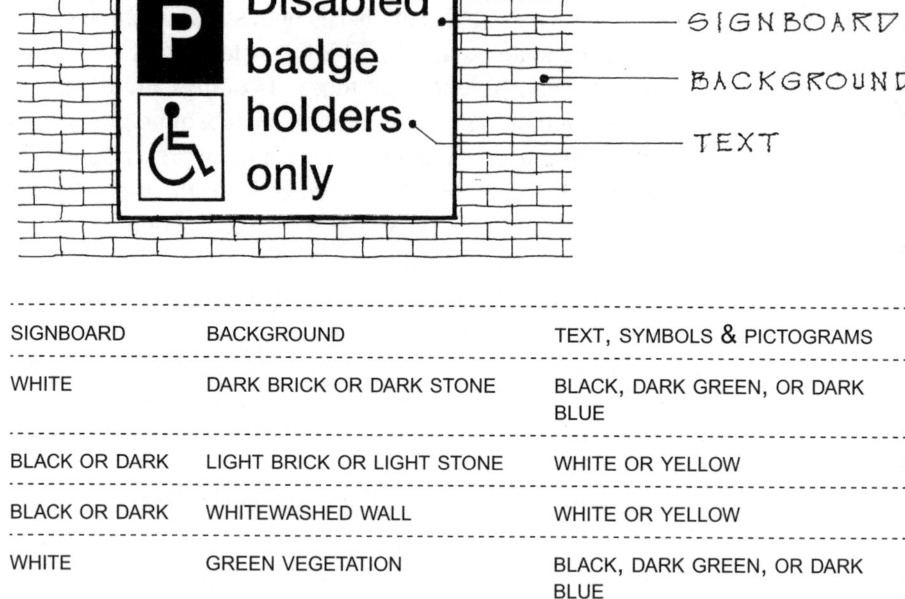

SIGNBOARD	BACKGROUND	TEXT, SYMBOLS & PICTOGRAMS
WHITE	DARK BRICK OR DARK STONE	BLACK, DARK GREEN, OR DARK BLUE
BLACK OR DARK	LIGHT BRICK OR LIGHT STONE	WHITE OR YELLOW
BLACK OR DARK	WHITEWASHED WALL	WHITE OR YELLOW
WHITE	GREEN VEGETATION	BLACK, DARK GREEN, OR DARK BLUE

2 In *stair* (qv) design for non-domestic buildings, para 1.33 of APPROVED DOCUMENT M (qv) of the Building Regulations requires visually contrasting stair nosings as shown below. While such contrast would also be highly desirable in private dwellings, there is no regulation to that effect.

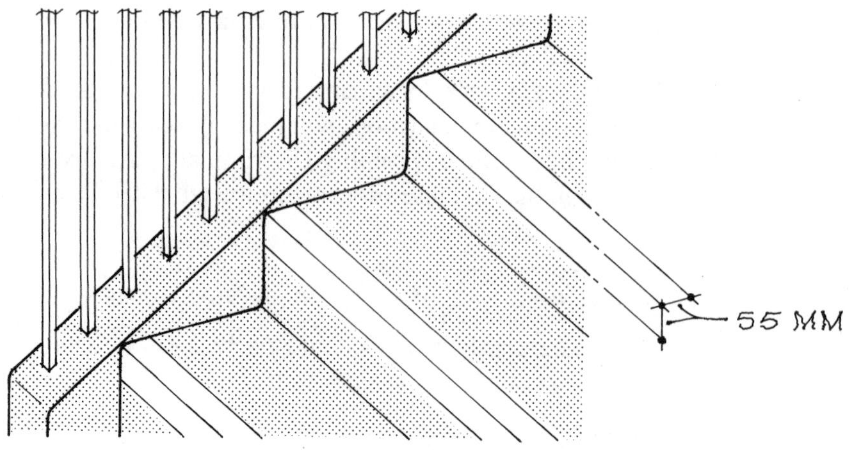

See also: *Signs*; and *Steps and stairs: public buildings*.

mmission for chitecture and the ilt Environment	See: CABE.
mmission for sability Rights	See: *Disability Rights Commission.*
mmission for uality and Human ghts (CEHR)	See: *Equality and Human Rights Commission.*
mmission for Health provement	See: *Healthcare Commission.*
mmission for ealthcare Audit d Inspection	See: *Healthcare Commission.*
mmission for tient and Public volvement in Health	See: *Dental care*; and *NHS structure.*

mmission for Public pointments

1 The Commission was created in 1995 in response to a recommendation by the Nolan Committee on Standards in Public Life. The Commissioner's task is to ensure that all appointments made by Ministers to *quangos* (qv), NHS bodies, and other public bodies are made on merit after fair and open competition.
2 Yet senior appointments to these bodies do in many cases appear to be drawn disproportionately from a pool of politically well-connected people. As examples:
• In November 2006 the shadow Culture Secretary Mr Hugo Swire MP published a dossier listing 20 such individuals, all with close government connections, who have between them been appointed to 83 influential public posts at salaries of up to £120 000 for a two or three-day working week.
• In July 2007 the shadow health spokesman Mr Stephen O'Brien MP used the Freedom of Information Act (qv) to uncover the fact that of 342 people recently appointed to NHS bodies, 312 had declared affiliations with the Labour Party, 77 with the Conservative Party, and 53 with the Liberal Democrats.
Pure coincidence?
3 For more information on the Office of the Commissioner for Public Appointments visit www.ocpa.gov.uk/

See also: *Quangos.*

ommission for Racial quality (CRE)

1 Now replaced by the Equality and Human Rights Commission (qv), this *quango* (qv) was set up under the Race Relations Act 1976 to tackle race discrimination and promote racial equality.
2 The CRE was troubled throughout its life by the difficult problem of how to fight vigorously for the interests of a racially defined group of citizens, without thereby encouraging the dividing up of British society into distinct groups competing with each other for rights and entitlements. It never managed to achieve a really happy balance, and one must hope that its successor body is more successful.

See also: *Equality; Equality and Human Rights Commission*; and *Victimhood.*

Commission for Rural Communities (CRC)

1 The CRC publishes regular reports on social, economic and environmental conditions in local communities in England, and works for improvement in all these areas. For one example of its informational output see: *Housing: affordability*.
2 For more information visit www.ruralcommunities.gov.uk/

Commission for Social Care Inspection (CSCI)

1 This *quango* (qv) is the inspectorate for social care in England. It was create by the Health and Social Care (Community Health and Standards) Act 2003 t bring together the inspection, regulation, and review of all social care services into one organisation. It is responsible for ensuring that National Minimum Standards are met, registers all *care homes* (qv), and does regular checks to ensure that they are complying with these standards.
2 The CSCI will be absorbed in a new Care Quality Commission in 2008. Meanwhile, visit www.csci.org.uk/for more information.

See also: *Care homes; Care Standards Act; General Social Care Council; National Care Standards Commission*; and *Social Care.*

Commission for the Compact

See *Voluntary and community sector.*

Commission for Women

See: *Women's National Commission.*

Commission on Cohesion and Integration

1 This quango (qv), consisting of around a dozen Commissioners, was created in August 2006 by Ms Ruth Kelly, Secretary of State for Communities and Loc Government, and given the task of examining the issues which raise tensions between different groups of citizens, leading to conflict and segregation. It wa also asked to consider how local community and political leadership can break down perceived barriers to cohesion and integration, and how communities can be helped to tackle extremist ideologies. Ms Kelly called for a 'new and honest debate' about integration, cohesion, and multiculturalism, promising in particular to revive the debate on *multiculturalism* (qv) and whether it has lead t increased division within society.
2 The Commission was expected to make recommendations for improving community cohesion and tackling extremism in July 2007.
3 For more information visit www.edf.org.uk/

See also: *Diversity; Equality*; and *Institute for Community Cohesion.*

Committee on Standards in Public Life

1 A body created in October 1994 by Prime Minister John Major to act as a watchdog over public affairs, investigate allegations of wrong-doing, and generally use its powers to ensure the highest standards of probity in British public life.
2 In March 2007 there were reports that the Government was planning to merg five watchdogs (one of them the Committee on Standards in Public Life) that had recently been critical of Ministerial sleaze, government practice, and appointments, into a single new 'super-regulator' on public ethics.
3 For more information visit www.public-standards.gov.uk/

See also: *Electoral Commission*; and *Electoral fraud.*

Common Assessment Framework (CAF)

1 The Common Assessment Framework for Children and Young People is a standardised approach to conducting an assessment of a child's additional needs, and deciding how those needs should be met. Twelve local areas

formally trialled the CAF during 2005–06; the trials were evaluated by the University of East Anglia; and as of 2006 all *local authority* areas are expected to implement the CAF between April 2006 and the end of 2008.
2 For more informationwww.everychildmatters.gov.uk/

Communication aids and systems 1 These comprise one category of *Aids and equipment* (qv) – i.e. devices intended to make normal activities easier, or to help people do things that would otherwise be impossible. They are especially important to older and *disabled* (qv) people.
2 For notes on personal devices see the following entries:
- Computers.
- Hearing aids.
- Lightwriters.
- Listening devices.
- Mobile phones.
- Reading aids.
- Seeing aids.
- Sign and symbol systems.
- Speech aids.
- Speech recognition devices.
- Telephones: private.
- Writing aids.

3 For authoritative recommendations on building installations see the following clauses in BS 8300 (qv):

• Colour in buildings	para 9.1.1.
• Fire alarms	para 9.3.7
• Hearing enhancement systems	paras 9.3.2 to 9.3.7
• Lighting	para 9.4
• Public address systems	para 9.3.1
• Signs	para 9.2
• Telephones: public	para 10.4

Communities England 1 In January 2007 Ms Ruth Kelly, Secretary of State for the Department for Communities and Local Government (qv), announced proposals to create this agency, the task of which will be to expedite and co-ordinate the delivery of new housing and regeneration agency in England.
2 For latest information visit www.communities.gov.uk/index.asp?id=1506002

See also: *Housing: affordability*.

Communities: rural See: *Commission for Rural Communities*.

Community alarms See: *Alarms*.

Community businesses 1 These are organisations which are established to provide services and/or employment in a local community. They claim to be focused on building the community and the local economy, but doing so in a business-like way as independent and self-supporting organisations.
2 For more detail visit www.neighbourhood.gov.uk/

Community care See: *English Community Care Association*.

Community Care Grants 1 Community Care Grants help people who are getting *Income Support* (qv), income-based *Jobseeker's Allowance* (qv), or *Pension Credit* (qv), when moving out of institutional care or a *care home* (qv). The purpose is to help such people live independently in the community.
2 For more information visit www.dwp.gov.uk/

Community centres See: *Village halls*.

Community Champions Fund 1 A fund that makes modest grants to people who wish to become active in the local communities, or want to help others become active. It applies only in England.
2 For more information visit www.dfes.gov.uk/communitychampions/

Community Chest 1 Neighbourhood Renewal Community Chests are administered by voluntary sector 'lead organisations' and offer small grants (typically a few thousand pounds) to community groups for projects to help them renew their own neighbourhoods. As of 2007 it has become part of the Single Community Programme (qv).
2 For more detail visit http://www.neighbourhood.gov.uk/

Community Empowerment Fund (CEF) 1 This fund, which was created to help community and voluntary groups participate in Local Strategic Partnerships (qv) and neighbourhood renewal, is now amalgamated into the Single Community Programme (qv).
2 For more detail visit www.neighbourhood.gov.uk/

Community Forum 1 The Community Forum was launched in January 2002 to act as a sounding board for government ministers and the Neighbourhood Renewal Unit (qv), a to provide a 'grass-roots' perspective on neighbourhood renewal strategies.
2 For more detail visit www.neighbourhood.gov.uk/page.asp?id=521

Community Fund 1 This is the operating name of the National Lottery Charities Board, an independent organisation set up by Parliament in 1994 to distribute money raised by the National Lottery (qv) to support *charities* (qv) and *voluntary and community groups* (qv) throughout the UK, and to UK agencies working abroa
2 For more information visit www.community-fund.org.uk/

See also: *National lottery*.

Community Health Councils (CHCs) See: *NHS structure*.

Community Legal Service Partnerships 1 Local networks of providers of legal services, supported by co-ordinated funding and delivering services to local communities. They claim to work on the basis of 'identified priority need'.
2 For more information visit www.clsdirect.org.uk/index.jsp

See also: *Disability Law Service*; and *Mediation*.

Community planning 1 The process whereby a *local authority* (qv) and partner organisations come together to plan, provide, and promote the well-being of their communities. Their areas of activity include for example health, education, transport, the economy, safety, and the environment.
2 For more detail visit www.neighbourhood.gov.uk/

Community strategies 1 *Local authorities* (qv) are required to prepare these strategies for improving the economic, environmental, and social well being of local areas, and to co-ordinate the actions of local public, private, voluntary, and community organisations to that end. The strategies should be prepared in a way that allows local communities to say what they want.
2 For more detail visit www.neighbourhood.gov.uk/

mpensation	See: *Tribunals: compensation awards for discrimination.*
mpensation culture	1 An affluent and civilised society should of course ensure (a) that its citizens receive compensation for injury (physical or mental) not caused by their own negligence, and (b) that the sums awarded are roughly commensurate with the harm suffered. 2 But currently huge sums are awarded annually to the victims of real or supposed harm – perhaps £50 billion a year in claims, costs, lost earnings and benefits, as an unverifiable guess – with a lack of commensurateness that shames Britain's system of justice. Three examples may suffice: • A nursery teacher who was horribly injured defending her class from a machete-wielding maniac received £57 000. • A woman police officer whose career was damaged by sex discrimination received £500 000. • A female merchant banker who was told she had 'nice waps' received £1.2 million. 3 In short: we have a compensation culture in which the sums awarded depend less on the degree of harm suffered by the victim than on the skill of the plaintiff's lawyers. The situation calls out for a sensible Act of Parliament. 4 As of 2008, useful references include the following. For publication details see the bibliography: • SOUR GRIPES by Simon Carr. The above examples are taken from this eye-opening little book. See also: *Risk,* and *Victimhood.*
mprehensive rformance sessment (CPA)	1 An assessment of the performance of every *local authority* (qv). These have been carried out annually since 2002 by the Audit Commission (qv), using evidence from other review bodies plus the AC's own judgements. 2 For more information visit www.audit-commission.gov.uk/cpa/ See also: *Local authorities*; and *Local government league table.*
mprehensive hools	See: *Education*; and *Social mobility.*
mputers	See: *Accessible websites; Communication aids*; and *Reading aids*
ncert halls: mplying with the sability scrimination Act	See: *Auditorium design*; and *Service providers.*
ncert halls: lected case amples	1 The Adapt Trust (qv) makes regular awards to arts venues considered to be especially meritorious in terms of overall *accessibility.* Recent award-winning (asterisked) and highly commended concert halls include: • Belfast Waterfront Hall, Northern Ireland.* • Snape Maltings, Aldeburgh.* • Djanogly Recital Hall, University of Nottingham.* • Duff House, Banff. 2 As of 2008, useful websites include: • CABE www.cabe.org.uk/default.aspx?contentitemid=42 See also: *Access awards.*

Conciliation See: *Mediation and the Disability Discrimination Act.*

Connecting for Health See: *NHS Care Records.*

Constitution

1 Virtually any organisation, from a tennis club to a great nation, has a docume that sets out the rules, rights, and responsibilities by which members and governors are bound. In the case of nations, constitutions normally define the powers and duties of governments, and in democracies they almost invariably guarantee certain basic rights and freedoms to their citizens. The yearning for such a guarantee prompted the making of the world's first and most influentia written constitution, that of the USA (1787), and it remains a major justificati for any modern constitution.

2 Unlike the American, French, and most other major examples the British constitution is not a single written document, but an informal set of historical papers (starting with the Magna Carta of 1215), unwritten conventions, and act of Parliament. Until British politicians signed the Treaty of Rome in 1972, whereby all legislation from the European Union (qv) and judgments from the European Court of Justice (qv) automatically become part of the UK constitutio this informal constitution grew organically from the bottom up via democratic processes. Being flexible it could continually adapt itself to changing circumstances, and until recently this pragmatic arrangement served us well. From the late 18th century onward, as the modern world took shape, Britain was seen by liberals in Europe and elsewhere as a bastion of freedom; it was the fi country in history to outlaw slavery (1833); and in contrast with just about all o European neighbours (the honourable exceptions including the Netherlands, Scandinavia, and Switzerland) we never even came close to falling for those murderous tyrannies of the 20th century, communism and fascism.

3 But gentlemen's agreements only work while all concerned act as gentlemer and the informal British constitution has been unable to deal with the pressure put upon it since the 1970s. As of 2007, the UK is troubled by at least four constitutional or quasi-constitutional problems that need to be satisfactorily resolved:

• An unprecedented erosion of our *civil liberties* (qv) and *privacy* (qv), partly driven by processes outlined under *post-democracy* (qv).

• An inequitable relationship between England and Scotland. Since 1999 Westminster MPs with Scottish seats have been able to vote on all matters affecting English constituencies, but Westminster MPs with English seats do no have an equivalent power to vote on all matters affecting Scotland – an asymmetry often referred to as the 'West Lothian question'. Thus were highly unpopular tuition fees imposed upon English students by means of the votes of Scottish MPs at Westminster, but not upon Scottish students because the devolved Scottish Parliament preferred to give Scottish students their universi education for free – using, as a finishing touch, almost £2 billion of English taxpayers' money to fund this generosity.

• A loss of control over the making of our laws. Between 50% and 80% (estimates vary) of UK laws now give effect to what is ordered by Brussels, and under the 1972 European Communities Act (qv) our Parliament cannot block their passage onto the lawbook. This lack of debate and scrutiny, or the power say No, is not only undemocratic but is helping to submerge British citizens a organisations under some thoroughly bad law (see as one instance: *Employme Equality (Age) Regulations*), a lot of unwanted law, and far too much law – se *Over-legislation; Over-regulation*; and *Post-democracy*.

• There is a widespread feeling that Britain's governmental machinery has become too centralised – see: *Localism*. This may or may not be true, but the question can no longer be left to politicians to decide.

4 Arising from the above anomalies and injustices, the time has come for a new British constitutional settlement that arises from a national debate in which the views of the people prevail over those of politicians. The new constitution should:

• Be a written one, which according to p 132 of POWER TO THE PEOPLE (see below) is wanted by 70% of people, rising to 80% in 2004.

• Contain a Bill of Rights that is clear enough for everyone to understand. Preferably – though this is a problematic concept – the Bill should entrench the essential freedoms of all UK citizens beyond the reach of Parliament.

5 Such a constitution-making process should start with a wide and deep public debate on the following fundamental matters:

• The proper relationship between citizens and the state. A series of laws that have been smuggled onto the lawbook in virtual secrecy (see: *Civil Liberties*), and privacy-invading schemes such as ContactPoint (qv), the National DNA Database (qv), and the NHS Care Records Service (qv) that are being imposed upon us whether we agree or not, seem to be founded on an assumption that citizens – or in some cases their children – belong to the state rather than the other way round.

• The proper relationship between our elected Parliament, the Executive, and the Judiciary.

• The proper relationship between the lower and the upper houses of the UK Parliament.

• The proper relationship between the four entities within the UK (England, Wales, Scotland, and Northern Ireland).

• The proper relationship between the UK and the EU.

6 Within his first week in office, in July 2007, the new Prime Minister Mr Gordon Brown announced plans for a debate of this kind, but immediately cast doubt upon the whole exercise by ruling out referenda on the West Lothian question (qv) and the Reform Treaty (qv) for establishing a constitution for Europe – two of the most fundamental constitutional issues of our time.

7 As of 2008, useful references include the following. They are given in alphabetical order. For publication details see the bibliography.

• ABC: A BALANCED CONSTITUTION FOR THE 21st CENTURY by Martin Howe. This pamphlet briefly discusses some of the above issues and suggests a way forward.

• A BRITISH BILL OF RIGHTS: INFORMING THE DEBATE. This unusually thorough document sets the factual framework for any discussion on a future British Bill of Rights, without itself taking any particular position.

• THE BRITISH CONSTITUTION by Anthony King is excellent, and up-to-date.

• THE GOVERNANCE OF BRITAIN. This *green paper* (qv) calls for an open debate on a new constitutional settlement. But Prime Minister Gordon Brown's decisions on the West Lothian question (qv) and the Reform Treaty (qv), soon after publication, suggest that we will be consulted on a few matters of second and third-order importance and on none of first-order importance.

• POWER TO THE PEOPLE. THE REPORT OF POWER: AN INDEPENDENT INQUIRY INTO BRITAIN'S DEMOCRACY, published by the Joseph Rowntree Foundation (qv). See for instance p 132.

8 Useful websites include the following. They are given in alphabetical order.

• British Institute of Human Rights	www.bihr.org.uk/
• Democratic Audit	www.democraticaudit.co.uk/
• Department for Constitutional Affairs	www.dca.gov.uk/

- Direct Democracy www.direct-democracy.co.uk/
- openDemocracy www.opendemocracy.net/
- Wikipedia http://en.wikipedia.org/wiki/Constitutio
- Your Rights www.yourrights.org.uk/

See also: *Civil liberties; Parliament; Privacy*; and *Post-democracy*.

Consultants' charter

1 The terms 'consultants' charter' and 'lawyers' charter' refer to regulations or technical documents that are so difficult to understand or apply that perfectly competent managers and professionals are forced to turn to specialist consultan to help them carry out their normal duties. Such consultants are benefiting fro the increasing complexity of regulation in all areas, including – disgracefully – those of *accessibility* (qv) and *inclusion* (qv), which ought to be readily understandable and applicable by everyone.
2 For an example from the field of education, a glance at the 366 pages of the standard reference SPECIAL EDUCATIONAL NEEDS AND THE LAW by Simon Oliver (see the bibliography) reveals the extent to which laws that were intended to benefit children have become, instead, a veritable gold-mine for fee-earning specialist lawyers and consultants while children, parents, and teachers look in baffled, from the outside.
3 For an example from the field of planning and building design visit www.drc gb.org/businessandservices/access.asp, go to 'access statements', and downloa the document titled ACCESS STATEMENTS: ACHIEVING AN INCLUSIVE ENVIRONMENT BY ENSURING CONTINUITY THROUGHOUT THE PLANNING, DESIGN AND MANAGEMENT OF BUILDINGS AND SPACES. The latter is recommended to architects as a guide to the preparation of the *access statement* (qv) that must accompany each planning application to certify that the proposed design will be *accessible* (qv) to *disabled people* (qv). Instead of the simple practical guide th should (and could) have been framed we have here a mind-numbing guidance document that is itself 40 pages long; contains a baffling flowchart replete with numerous boxes, bubbles, and arrows; and gives in Appendix A a list of 30 additional references which designers ignore at their legal peril – with, to cap it all, a note in para 9.3.4 that these are only 'some of the more established sources and that the list 'is by no means exhaustive'. Faced with this daunting docume most architects will be forced to turn to fee-earning consultants to help them fin their way through an informational thicket that need not exist, and was created b consultants for the ultimate benefit of consultants.

See also: *Authoritative practical guides; Better regulation bodies; Over-regulation*; and *Single Equality Act*.

ContactPoint

1 ContactPoint (previously known as the 'Children's Index', and before that as the 'Information Sharing Index') forms part of the Government's *Every child matters* (qv) programme, and its dysfunctional *Transformational Government* strategy (qv). It involves the creation of a vast electronic databank containing personal information about all 11 million children up to the age of 18 in England and Wales. No-one will be permitted to opt out, and around 330 000 officials will have access to this supersensitive information. In November 2007 after the revelation (see: *Privacy*) that another government department had 'los in the post' two discs containing the unencrypted personal details of 11 million children, plus the bank account and NI numbers of their 15 million parents or guardians, the Government announced that it would delay the launch of Contactpoint by 5 months for a 'security review'. As Minister of Stat Beverley Hughes is determined to press ahead with the scheme, it is worth reviewing the scheme as matters stand in October 2007.

2 In December 2006 a report by a panel of experts on child protection, law, and computers, and published by Mr Richard Thomas, Parliament's Information Commissioner (qv), alleged that the proposed scheme would shatter family privacy; undermine parental authority; violate the law; would not be secure; would inevitably contain inaccurate information (with potentially traumatic consequences for falsely accused parents); would waste millions of pounds; and would actually put children at greater risk because the vast amount of data would make it harder to spot those in genuine danger. As Mr Thomas put it: 'When you are looking for a needle in a haystack, don't make the haystack bigger'. Ms Liz Davies, a senior lecturer in social work at London Metropolitan University, wrote in the *Guardian* on 28 February 2007 that 'the new database is in effect a population-surveillance tool. It has nothing to do with protecting children'. On 22 June 2007 the *Guardian* published a letter from representatives of the Independent Schools Council, Action on Rights for Children, the Foundation for Information Policy Research (qv), the Open Rights Group (qv), and Privacy International (qv), opposing the scheme.

3 What justification can there be for this Orwellian project, which parents have never asked for? The official explanation is that it arose partly from the death of Victoria Climbié (qv), a little girl who was tortured to death by her foster parents in 2000. But this argument fails on two grounds:

• Writing in the *Guardian* on 6 April 2004 under the title 'National child database will increase risk', Dr Eileen Munro, reader in social policy at the London School of Economics, pointed out that the idea of such a database predated the Climbié report by at least a year and had nothing to do with the prevention of child abuse. It was first proposed in a report by the Performance and Innovation Unit at the Cabinet Office (qv) titled PRIVACY AND DATA-SHARING: THE WAY FORWARD FOR THE PUBLIC SERVICES, and related to *social exclusion* (qv).

• That apart, the problem in Victoria Climbié's case was not that Government departments did not know about her. She was on the records of four social services departments, two housing authorities, two police child protection teams, two hospitals, and the NSPCC. Staff with full access to the facts were too overworked, too badly managed, or too incompetent to make proper use of that information. The same applies to Jessica Randall, who died at the age of 54 days in November 2005 after systematic and horrific abuse by her parents despite being monitored by 30 health workers in Kettering, Northamptonshire, and being seen by them 10 times in her short life. The same applies to Kimberly Harte and Samuel Duncan, convicted in February 2007 of pouring boiling water on their little girl, breaking her arm, kicking her so violently in the groin as to damage her liver, and forcing her to eat her own excrement, despite the fact that social workers saw her more than 20 times during the 7 weeks she was being tortured. The same applies to 4-year old Leticia Wright, who was battered, burned and tortured to death in 2007 by her mother and her mother's boyfriend while she was being monitored by Kirklees social workers.

4 Soon after the conviction of Harte and Duncan Lord Laming (who had chaired the inquiry into the death of little Victoria Climbié) used the Victoria Climbié Memorial Lecture to complain that social workers and managers were still not following practical management procedures that would help protect children. His remarks suggest that the answer to the problem of abused children lies (a) in a small, tightly controlled database of children known to be at risk, and (b) in commonsense management practices; not in gigantic databases and ever more powers for officials to force themselves into the private lives of the millions of parents who do not abuse their children.

5 For more information on ContactPoint visit www.everychildmatters.gov.uk/deliveringservices/contactpoint/

6 As of 2008, useful references include the following. They are given in alphabetical order. For publication details see the bibliography.
• AFTER CLIMBIÉ, CHILDREN ARE AT EVEN MORE RISK by Jill Kirby argues that reforms in child protection instituted after 2003 make it even less likely that social workers will concentrate on regular home visits and apply the kind of robust common sense that might have saved Victoria.
• CHILDREN'S DATABASES – SAFETY AND PRIVACY: A REPORT FOR THE INFORMATION COMMISSIONERS
• SAFETY FEARS OVER NEW REGISTER OF ALL CHILDREN, a report by Francis Elliott in *The Times* on 27 August 2007, tells of a letter to the Government fro the Association of Children's Services (ACDS) outlining senior social worker 'significant concerns' about the system, which was rushed through Parliamen (qv) without publicity in July despite severe warnings by the House of Lords Select Committee on Merits of Statutory Instruments.
7 As of 2008, useful websites include:

• Children's databases	www.arch-ed.org/issues/databases/databases.ht
• DCSF	www.dcsf.gov.uk/
• Liberty	www.liberty-human-rights.org.uk/

See also: *Child abuse; IT systems; Privacy; and Victoria Climbié.*

Contract of compliance

1 A system that is being increasingly imposed on all companies and charities bidding for government contracts, obliging them to demonstrate their commitment to *diversity* (qv) and *equality* (qv) policies.
2 For an indication of the amount of paperwork that may be involved in such procedures, visit www.mpa.gov.uk/committees/eodb/2005/050110/07.htm This is a document from the Metropolitan Police titled DIVERSITY WITHIN PROCUREMENT SERVICES which requires all suppliers of goods and services to the Met to fill in detailed employment questionnaires, accept site visits and monitoring, and give extensive details of 'how they apply equal opportunities their selection of subcontractors'.
3 Metropolitan Police Officers have commented that the public might be bette served if their scarce time was devoted to reducing London's accelerating incidence of violent *crime* (qv) rather than researching, monitoring, and recording the ethnic, gender, and age profiles of their subcontractors' staff.

See also: *Better regulation bodies; Over-regulation*; and *Voluntary and community sector.*

Controls and switches: dwellings

1 The following references contain authoritative recommendations:
• Section 8 of APPROVED DOCUMENT M (qv). The provisions in AD M must be heeded in order to comply with the Building Regulations (qv) in England and Wales. See the diagram on the next page.
• THE HABINTEG HOUSING ASSOCIATION DESIGN GUIDE (qv) applies most specifically to *housing association* (qv) schemes, but ought to be by the designer's side in the design of all housing developments and individual dwellings.
• THE WHEELCHAIR HOUSING DESIGN GUIDE (qv), also from the Habinteg Housing Association, is another illustrated manual whose usefulness extends f beyond the special remit implied by the title.
• THE HOUSING DESIGN GUIDE is a best-practice guide for housing in London. For publication details of the above references see the bibliography.
2 The dimensions shown on the next page will in general satisfy the provisions c APPROVED DOCUMENT M, though each particular case should be checked in deta before decisions are finalised.

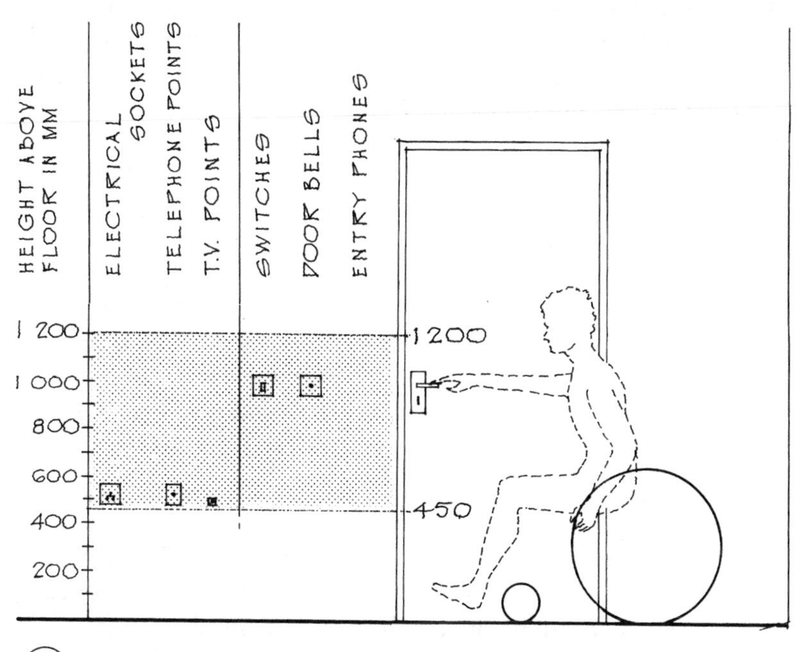
HEIGHT ABOVE FLOOR IN MM
ELECTRICAL SOCKETS
TELEPHONE POINTS
T.V. POINTS
SWITCHES
DOOR BELLS
ENTRY PHONES
1 200
1 000
800
600
400
200
1200
450

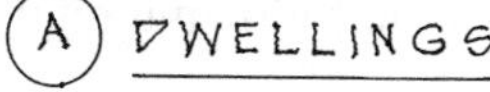
A DWELLINGS

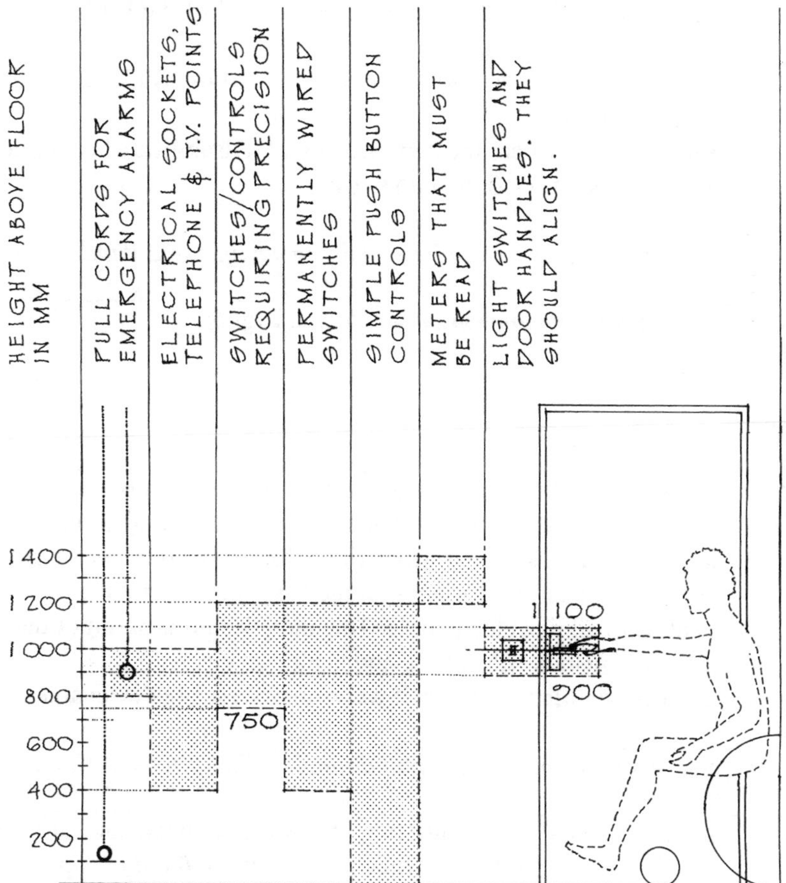
HEIGHT ABOVE FLOOR IN MM
PULL CORDS FOR EMERGENCY ALARMS
ELECTRICAL SOCKETS, TELEPHONE & T.V. POINTS
SWITCHES/CONTROLS REQUIRING PRECISION
PERMANENTLY WIRED SWITCHES
SIMPLE PUSH BUTTON CONTROLS
METERS THAT MUST BE READ
LIGHT SWITCHES AND DOOR HANDLES. THEY SHOULD ALIGN.
1 400
1 200
1 000
800
600
400
200
750
1 100
900

B NON-DOMESTIC BUILDINGS

Controls and switches: public buildings: complying with the Disability Discrimination Act

1 Public buildings and spaces must in general comply with the Disability Discrimination Act (qv). In the case of controls and switches, relevant *authoritative practical guides* (qv) include the following:
- APPROVED DOCUMENT M (qv), which deals with access to and use of buildings. Its provisions must be heeded in order to comply with the Building Regulations in England and Wales.
- BS 8300 (qv), which applies to all buildings in the UK. It is an advisory document. While BS 8300 does not have the force of law, conformity with its recommendations will make it easier to demonstrate that the requirements of the Disability Discrimination Act have been met.

The relevant clauses of the above documents are tabulated below:

DOCUMENT	RELEVANT PARAGRAPHS
APPROVED DOCUMENT M	2.21; 3.27–3.28; 3.43; 3.49; 4.24–4.30; 5.4; and 5.18
BS 8300	10

2 The dimensions shown on the previous page will in general satisfy the provisions of AD M and BS 8300, though each particular case should be checked in detail before decisions are finalised.
3 For places other than England and Wales the following regulatory documents should be consulted:
- In Northern Ireland: TECHNICAL BOOKLET R of the Building Regulations (Northern Ireland) (qv).
- In Scotland: the TECHNICAL HANDBOOKS of the Building (Scotland) Regulations (qv).
- In the USA, where the Americans with Disabilities Act applies: the ADA AND ABA ACCESSIBILITY GUIDELINES (qv).

See also: *Colour in buildings; Passenger lifts*; and *Windows*.

Convention for the Protection of Human Rights and Fundamental Freedoms

See: *European Convention on Human Rights*.

Convention on the Rights of persons with Disabilities

1 The Convention was approved by the United Nations in September 2006. The next step would be signature and ratification by individual member states, including the UK. It enshrines rights for disabled people in international law, and for the first time applies all the basic human rights provided in earlier treaties to all disabled people. This includes prohibition of discrimination and access to justice, education, and health services.
2 For more information visit www.un.org/esa/socdev/enable

See also: *Disability Discrimination Act; Equality*; and *Human rights*.

Corduroy surfaces

1 Corduroy surfaces, consisting of rows of rounded bars, are one of six types of *tactile pavings* (qv). Their function is to warn *blind* (qv) people that they are approaching:
- A flight of external *steps* (qv).
- A tram, LRT, or any other *on-street platform* (qv).

An example of each application is shown on the next page. They must not be applied to other situations, as consistency of use is essential if misunderstandings are to be avoided.

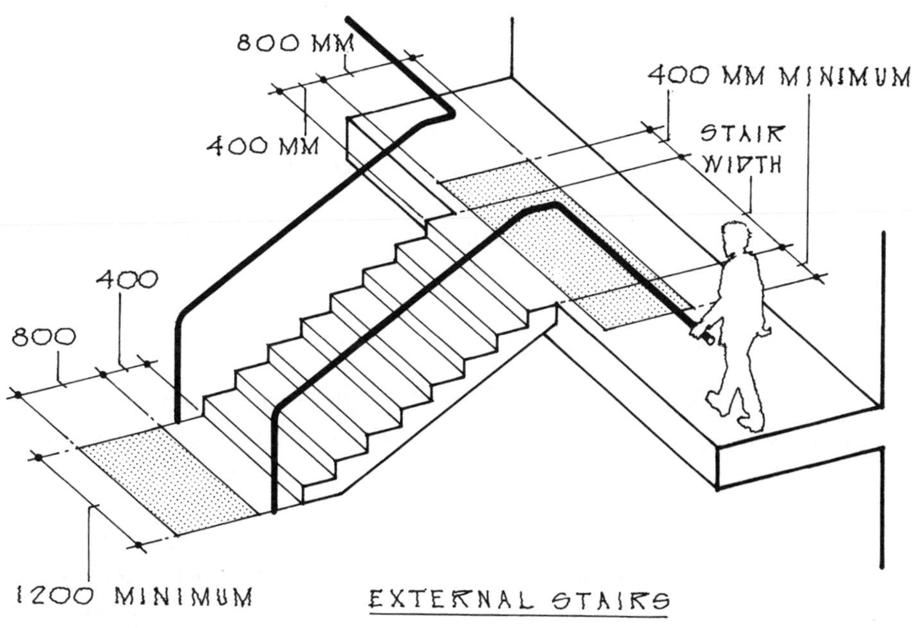

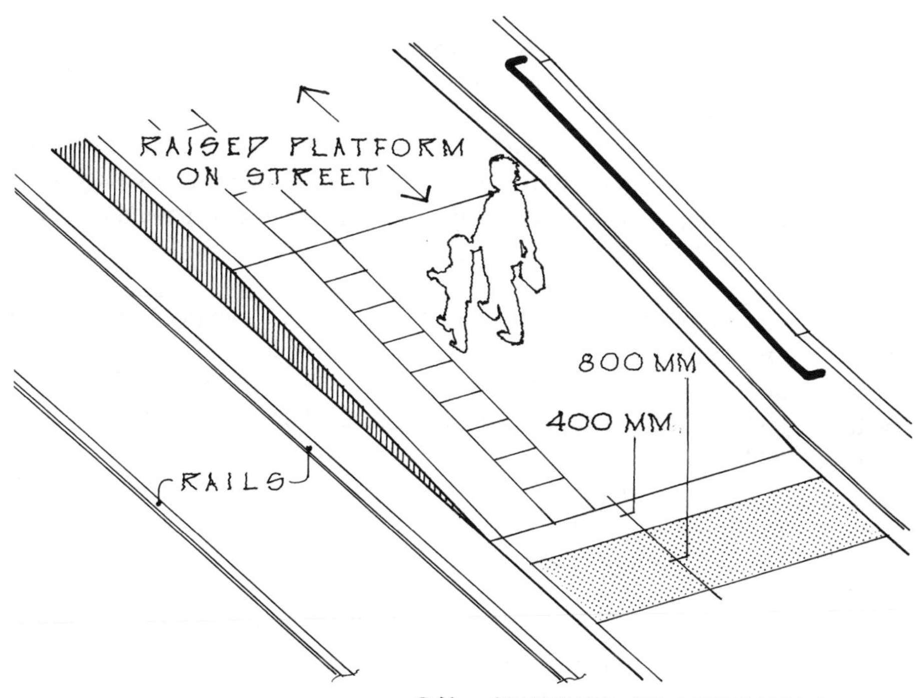

2 The above are typical examples, but the rules for correct use in particular situations are very precise and designers must in all cases consult Chapter 2 of GUIDANCE ON THE USE OF TACTILE PAVING SURFACES (qv) and para 4.2 of INCLUSIVE MOBILITY (qv) – see the bibliography.

3 For details of the six types of tactile paving, including corduroy surfaces, see: *Tactile pavings*.

re poor See: *Poor: core.*

re references See: *Authoritative practical guides.*

rpus Juris 1 Corpus Juris is a European system of criminal justice that is set to replace British Common Law as a consequence of our membership of the European Union (qv). One difference between the two systems is that Corpus Juris lacks the right to trial by jury, whereby accused persons are entitled to be tried by a

panel of jurors, chosen from the ordinary people, who are sovereign to the poi that they can disregard the law if they think it would result in an unjust conviction. Many English laws regarded as cruel or unjust by ordinary citizens have been amended or scrapped because juries would not convict.
2 The scrapping of a British legal system that has developed incrementally and democratically, bottom-up, over a period of centuries, and its replacement by a top-down European system, seems quite likely to come about in the near futur without public debate, consultation, or consent.

See also: *Constitution; European Union*; and *Post-democracy*.

Corridors and passages in dwellings

1 The following references contain authoritative design guidance.
• Section 7 of APPROVED DOCUMENT M (qv). The provisions in AD M must be heeded in order to comply with the Building Regulations (qv) in England and Wales.
• THE HABINTEG HOUSING ASSOCIATION DESIGN GUIDE (qv) applies most specifically to *housing association* (qv) schemes, but ought to be by the designer's side in the design of all housing developments and individual dwellings.
• THE WHEELCHAIR HOUSING DESIGN GUIDE (qv), also from the Habinteg Housing Association, is another illustrated manual whose usefulness extends f. beyond the special remit implied by the title.
For publication details of the above references see the bibliography.
2 The dimensions shown on the next page will in general satisfy the provisions APPROVED DOCUMENT M, though each particular case should be checked in det before decisions are finalised.

Corridors and passages in public buildings: complying with the Disability Discrimination Act

1 Buildings used by the public must in general comply with the Disability Discrimination Act (qv). In the case of corridors in public buildings, *authoritative practical guides* (qv) include DESIGNING FOR ACCESSIBILITY (qv) and in addition the following documents. For publication details see the bibliography.
• APPROVED DOCUMENT M (qv), which deals with access to and use of buildings. Its provisions must be heeded in order to comply with the Building Regulations in England and Wales.
• BS 8300 (qv), which applies to all buildings in the UK.
• INCLUSIVE MOBILITY (qv), which is the most authoritative reference for the *inclusive* design of *pedestrian infrastructure* and for transport-related facilitie such as bus and coach stations, railway stations, air terminals, and transport interchanges.
While the latter two documents above do not have the force of law, conformity with their recommendations will make it easier to demonstrate that the requirements of the Disability Discrimination Act have been met. The relevant clauses in each are tabulated below:

DOCUMENT	RELEVANT PARAGRAPHS
APPROVED DOCUMENT M	3.11–3.14.
BS 8300	7.2.2
INCLUSIVE MOBILITY	8.3

The dimensions shown on the next page will in general satisfy the provisions of AD M and BS 8300, though each particular case should be checked in deta before decisions are finalised. See in this connection also the figure under *Wheelchairs: space requirements*.

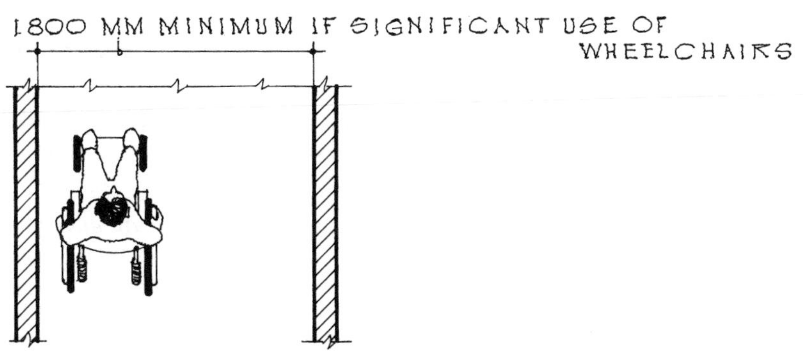

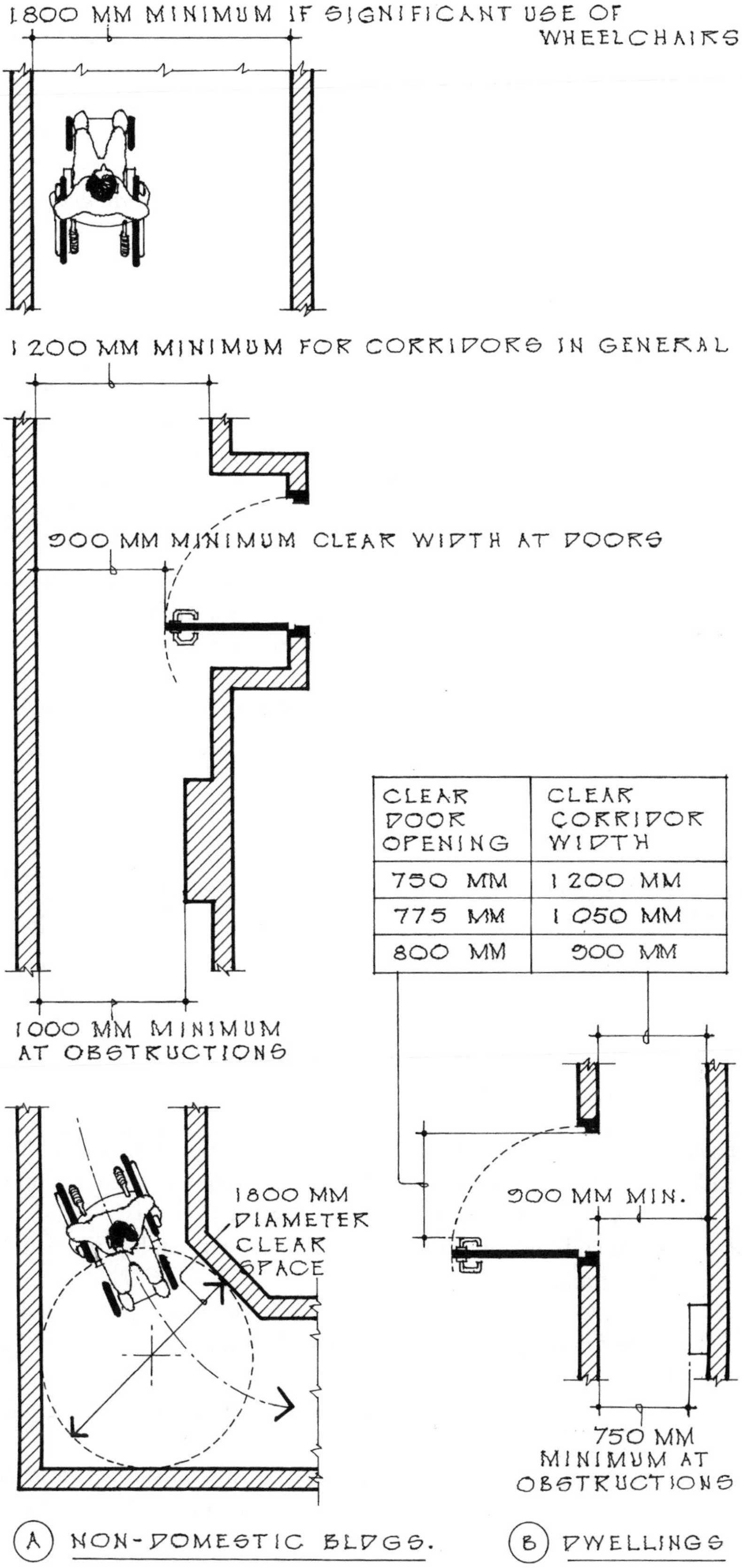

1800 MM MINIMUM IF SIGNIFICANT USE OF WHEELCHAIRS
1 200 MM MINIMUM FOR CORRIDORS IN GENERAL
900 MM MINIMUM CLEAR WIDTH AT DOORS
1000 MM MINIMUM AT OBSTRUCTIONS
1800 MM DIAMETER CLEAR SPACE
900 MM MIN.
750 MM MINIMUM AT OBSTRUCTIONS
(A) NON-DOMESTIC BLDGS.
(B) DWELLINGS

CLEAR DOOR OPENING	CLEAR CORRIDOR WIDTH
750 MM	1 200 MM
775 MM	1 050 MM
800 MM	900 MM

For places other than England and Wales the following regulatory documents should be consulted:
• In Northern Ireland: TECHNICAL BOOKLET R of the Building Regulations (Northern Ireland) (qv).
• In Scotland: the TECHNICAL HANDBOOKS of the Building (Scotland) Regulations (qv).
2 In the USA, where the Americans with Disabilities Act applies, refer to the ADA AND ABA ACCESSIBILITY GUIDELINES (qv).

See also: *Doors; Means of escape*; and *the Pedestrian environment.*

Co-sited schools See: *Special educational needs.*

Council housing See: *Housing: social.*

Council of Europe 1 The Council of the European Union (also called the 'Council of Ministers' or the 'Council') is the EU's main decision-making body. It is composed of the ministers of the Member States and thus constitutes the EU institution in which the governments of the Member States are represented. The Council, together with the European Parliament, acts in a legislative and budgetary capacity. It is also the lead institution for decision-making on the common foreign and security policy (CFSP), and on the coordination of economic policies. Each Member State in turn presides over the Council for six months.
2 For more detail visit www.consilium.europa.eu/

See also: *European Convention on Human Rights*; and *Human Rights Act.*

Council on Tribunals See: Tribunals.

Council tax See: *Local authorities: funding.*

Council tax benefit 1 Council tax benefit is paid as a rebate on recipients' *council tax* (qv) bill. It may be paid to people who rent their homes, own them, or live rent-free. For details visit www.direct.gov.uk/
2 As of 2008, the system is badly in need of reform. Useful references in this regard include REFORMING WELFARE by Nicholas Boys Smith. For publication details see the bibliography.
See also: *Welfare reform.*

Counters and reception desks: complying with the Disability Discrimination Act 1 Public buildings and spaces must in general comply with the Disability Discrimination Act (qv). In the design of counters and reception desks, *authoritative practical guides* (qv) include DESIGNING FOR ACCESSIBILITY (qv) and in addition the following documents. For publication details see the bibliography.
• APPROVED DOCUMENT M (qv): the provisions in paras 3.3 and 3.6 must be heeded in order to comply with the Building Regulations for England and Wales.
• BS 8300 (qv): the recommendations in para 7.1.2 and section 11 apply to all buildings in the UK.
• INCLUSIVE MOBILITY (qv): the recommendations in section 9 apply to transport-related buildings such as bus and coach stations, railway stations, air terminals, and transport interchanges.

Except for AD M, the above guides do not have the force of law, but conformity with their recommendations will make it easier to demonstrate that the requirements of the Disability Discrimination Act have been met.

2 The layouts and dimensions shown below will in general satisfy the provisions of AD M and BS 8300, though each particular case should be checked in detail before decisions are finalised.

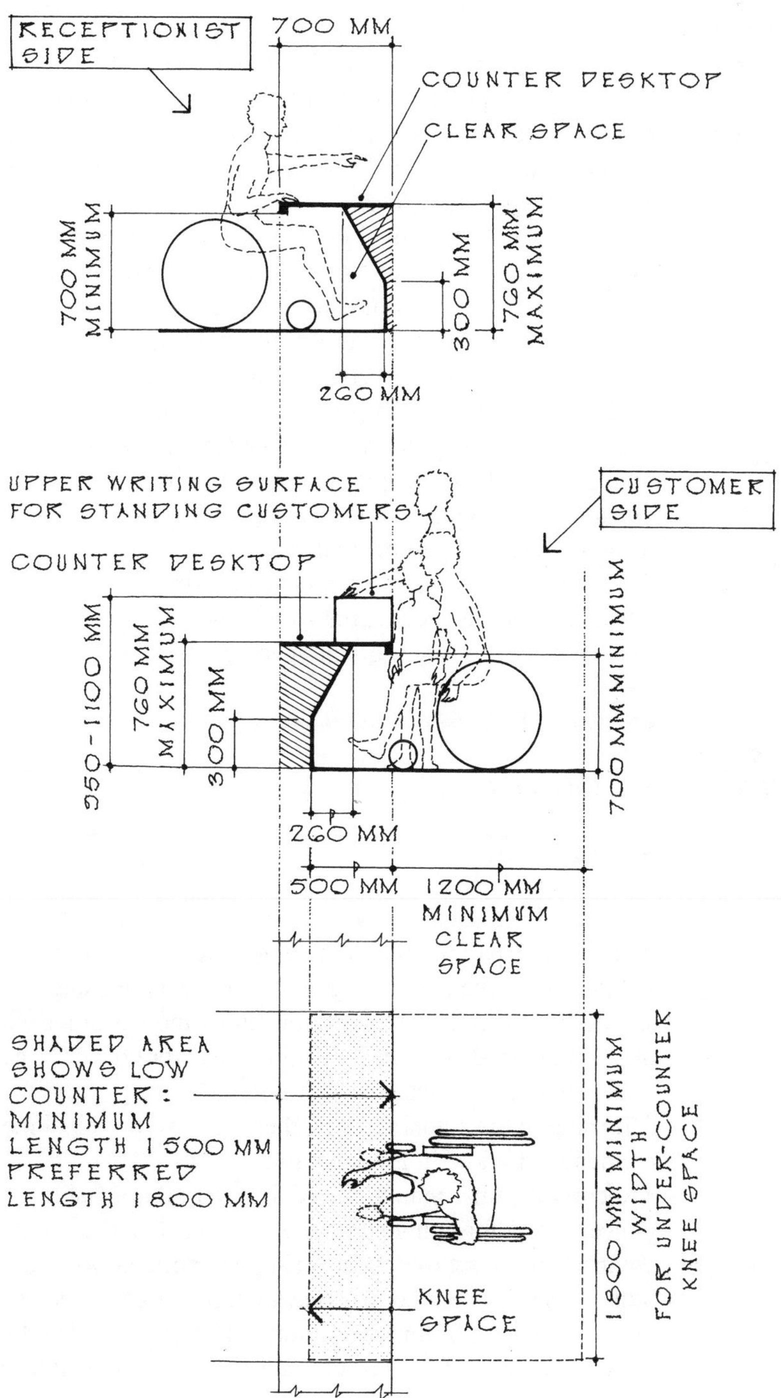

3 For places other than England and Wales the following regulatory documents should be consulted:
• In Northern Ireland: TECHNICAL BOOKLET R of the Building Regulations (Northern Ireland) (qv).
• In Scotland: the TECHNICAL HANDBOOKS of the Building (Scotland) Regulations (qv).

Counters in refreshment facilities

See: *Refreshment facilities.*

Countryside facilities: complying with the Disability Discrimination Act

1 Public spaces must in general comply with the Disability Discrimination Act (qv). As of 2008, relevant *authoritative practical guides* (qv) include the following. For publication details see the bibliography:
• BT COUNTRYSIDE FOR ALL STANDARDS AND GUIDELINES. A detailed good practice guide to disabled people's access to the countryside.
• BY ALL REASONABLE MEANS: INCLUSIVE ACCESS TO THE OUTDOORS FOR DISABLED PEOPLE. Clear and detailed guidance from the Countryside Agency.
• EASY ACCESS TO HISTORIC LANDSCAPES, an informative and beautifully illustrated guide from English Heritage.
• MAKING CONNECTIONS: A GUIDE TO ACCESSIBLE GREENSPACE. Guidance for greenspace managers on effective ways of opening up their sites to a wider audience, including example of good practice.
• SENSE AND ACCESSIBILITY: how to improve access on countryside paths, routes and trails for people with mobility impairments.
The above publications do not have the force of law, but conformity with their recommendations will make it easier to demonstrate that the requirements of the Disability Discrimination Act have been met.
2 Useful websites include the following:
• Greenspace www.green-space.org.uk/

• See also: *Pedestrian environments.*

Courts Act 2003

See: *Civil liberties.*

Courts of law

See: *Law courts.*

Crime

1 In a civilised society women should be able to walk the streets at night without fear; parents should be able to let their children freely roam about without worrying about drug dealers or juvenile gangs; and householders should be able to leave their doors and windows unlocked. Such conditions did once obtain in Britain and it is time they were restored, as they have been in large parts of America.
2 Turning to the actual figures, there are two principal measures of the incidence of crime in England and Wales:
• Recorded Crime Statistics (RCS) cover all notifiable offences recorded by the police, except for minor summary offences. For an extremely detailed and comprehensive set of data covering the entire period from 1898 to 2006, visit http://uk.sitestat.com/homeoffice/homeoffice/s?rds.100yearsxls&ns_type=clickout&ns_url=%5Bhttp://www.homeoffice.gov.uk/rds/pdfs/100years.xls%5D Visit www.homeoffice.gov.uk/rds/recordedcrime1.html for more general information on recorded crime.
• The British Crime Survey (BCS) takes a different approach, regularly questioning a sample of around 40 000 people aged 16 or over about crimes

which they personally have experienced in the past year, and extrapolating a national figure from that. The BCS is more reliable than Recorded Crime Statistics in that (a) its results directly reflect people's experience (b) it includes many crimes which may not have been reported to the police. But against that, its website tells us that it 'does not record crimes which are serious but too small statistically to measure (such as murder and rape) or crimes committed against businesses (such as fraud or shoplifting) or against people under 16 (since it only surveys adults)'. The omission of crime against under-16s is a serious defect (see: *Crime: youth on youth*) that needs to be rectified. The omission of homicides is unavoidable, as the victims cannot be interviewed, but many criminologists regard these crimes as one of the best indicators of the general level of violence in society, and from this point of view the rise in the murder rate from 608 to 820 between 1997 and 2005 is ominous. Note also that according to CRIME IN ENGLAND AND WALES: MORE VIOLENCE AND MORE CHRONIC VICTIMS (see below) the BCS uses a flawed methodology that causes it to understate all crime by 30%, and violent crime by more than 50%. Be that as it may, BCS surveys have been carried out since 1981, and details are given at www.homeoffice.gov.uk/rds/bcs1.html and in the annual publication SOCIAL TRENDS (qv).

3 The graph below, drawn from data contained on the Home Office website http://uk.sitestat.com/homeoffice/homeoffice/s?rds.100yearsxls&ns_type=click out&ns_url=%5Bhttp://www.homeoffice.gov.uk/rds/pdfs/100years.xls%5D shows Recorded Crime Statistics from 1906 to 2006. The line 'A' marks the date when Mr Michael Howard became Home Secretary and started imprisoning more criminals. The break line in 1999 reflects a change in Home Office Counting Rules on 1 April 1998, which had the effect of increasing the number of crimes counted.

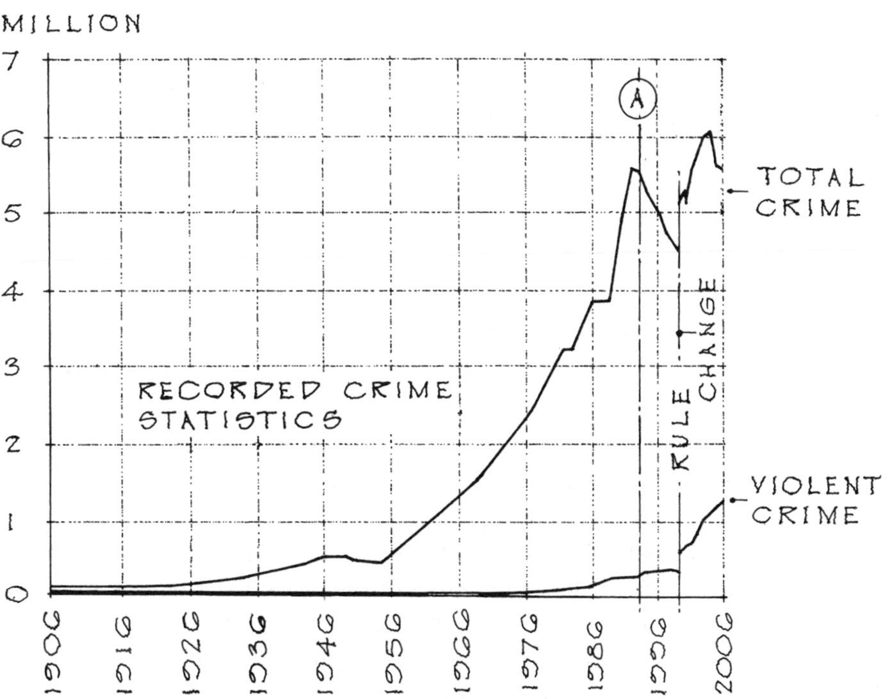

4 The graph on the next page shows both Recorded Crime Statistics and British Crime Survey estimates for England and Wales, starting in 1981 (which is the year in which the BCS commenced). The graph also shows the prison population for England and Wales. As before the line 'A' marks the date when the prison population began to rise sharply, reaching 78 000 in June 2006.

5 The picture that emerges from these graphs shames our nation. The Recordec Crime Statistics show a huge rise in crime over the past half century, and a trul staggering increase over the past century. The BCS estimates, though showing a fall in the particular types of crimes that are covered, still mean that adults in England and Wales ran a 24% risk of being the victim of crime in 2005 and 2006. This is better than the 40% risk in 1995, but remains unacceptable in a civilised society.

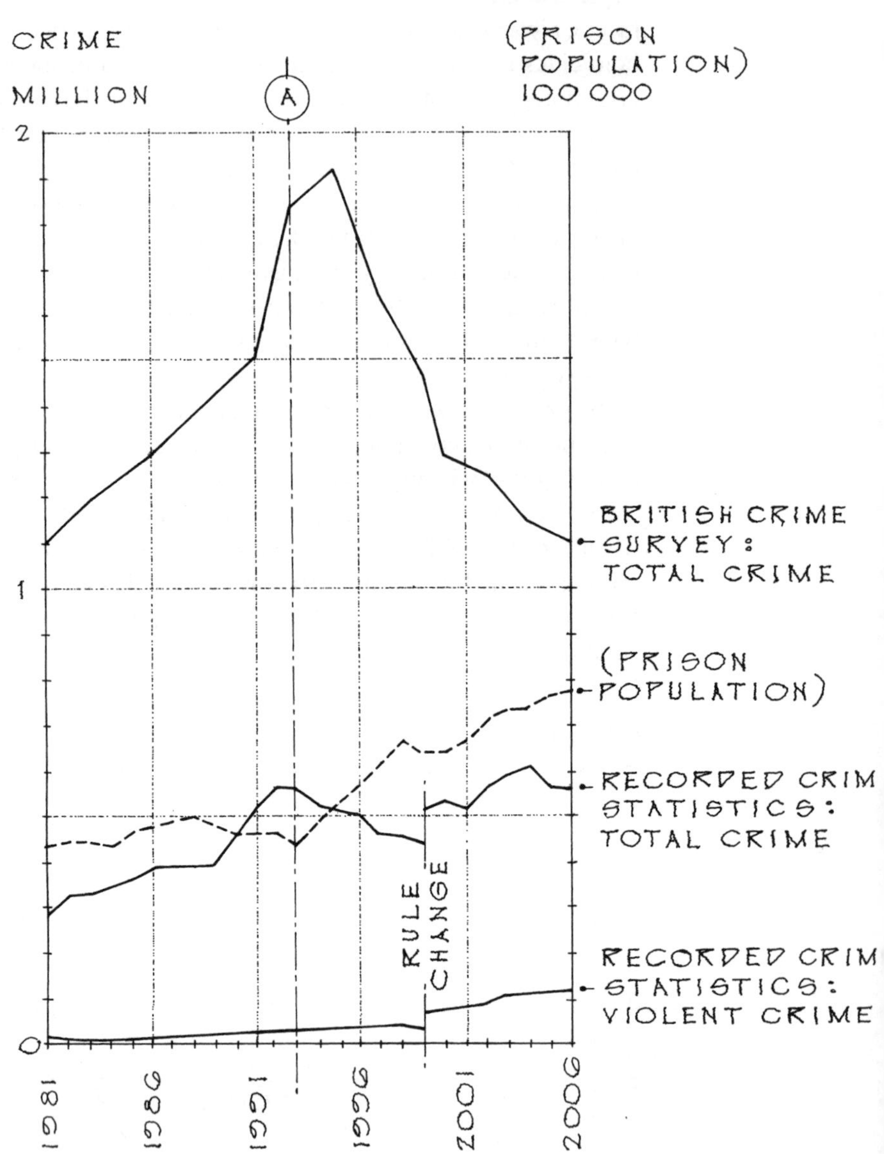

6 Any long-term strategy for cutting the crime rate back to what it was 50 year ago must include several strands:

- Finding ways of helping children in troubled homes. The Home Office has found that nearly half of prisoners had run away from home as youngsters; tha almost as many had a family member with a conviction; and that a third had family members who had spent time in prison.
- Radically reforming Britain's state care system for 'looked-after children'. Youngsters who have passed through 'care' have often been so brutalised by the experience that despite their small overall numbers (around 6 000 emerge from the system each year) they form more than a quarter of Britain's prison population. See: *Children in care*.
- Greatly improving the teaching of literacy in schools – an area, unlike some of those above, that can be directly influenced by government action. Home

Office figures show that four fifths of people in prison lack the literacy skills needed to fill in a job application form. With that handicap it is not surprising that they have drifted into crime. See: *Literacy*.

• Either decriminalising the use of drugs, or evolving effective policies for reducing drug addiction, or both. More than half of those arrested for theft or robbery test positive for heroin or cocaine, and it is reasonable to deduce that they are stealing to fund their drug addiction.

• Instituting effective education and training programmes in penal institutions, to maximise prisoners' chances of making a decent life for themselves after release. See: *Prisons*.

• Moving policemen and women out of their offices and cars onto the streets, and applying zero-tolerance policies. THE GREAT AMERICAN CRIME DECLINE (see below) shows how such policies caused a three-quarters drop in youth violence in New York after 1990, and in Britain similar policies cut stranger killings in Manchester from 37 in 1999 to 5 in 2005.

7 Meanwhile, society must be made safe against crime now. *Asbos* (qv) and community sentences have a role, but it is a limited one (see the Civitas websites listed below), and there is overwhelming and well-founded public support for a policy of locking up persistent criminals in secure institutions. The Government believes that 100 000 criminals commit half of all offences carried out in England and Wales each year. Yet between 1997 and 2007 all proposals from Home Secretaries to build more prison places in order to accommodate a higher percentage of these wrongdoers were rejected by the then chancellor Gordon Brown. The time is overdue for a very large prison-building programme. Britain currently has 81 000 prisoners at a cost of £1.94 billion, or £24 000 per prisoner per annum. In August 2007 the Government committed itself to creating 9 500 additional places by 2012 (but also told the prison service to cut £60 million from its budget in 2007–08). Respected commentators say that even 91 000 prison places are nowhere near enough. According to the Justice Minister Jack Straw, debating with his shadow, Mr Nick Herbert MP, in July 2007, 60 000 additional prison places would be needed if prisoners had to serve their full sentences instead of being released halfway through, as is now common practice. That suggests that Britain needs not far short of 140 000 secure places, even before we start replacing 2 and 3-person cells with single-person cells as we ought to. Bearing in mind that courts might impose shorter sentences if they knew that the full term would be served, a sensible overall figure might therefore be not less than 120 000, and probably more – see: *Prisons*.

8 Even if higher imprisonment rates had no deterrence effect, which is unlikely; the prison-building programme outlined above would have two beneficial effects:

• Decent accommodation and an end to overcrowding would enable us to undertake genuine prisoner rehabilitation programmes.

• As an average prisoner has on his own admission carried out 140 crimes in the previous year, each full year in prison is likely to spare the public at 140 or more crimes per inmate.

9 As of 2008, useful references include the following. They are given in alphabetical order. For publication details see the bibliography.

• A BETTER WAY. In 2002 the *think tank* (qv) Reform established a Commission on the Reform of Public Services with the aim of finding, on a non-party basis, the best way to provide and fund high-quality public services in Britain. This 140-page report, published in 2003, contains the Commission's proposals in the areas of education, healthcare, and crime.

• CRIME AND CIVILISED SOCIETY by David G Green, Emma Grove, and Nadia Martin is an account of how Britain came to have one of the highest crime rates in Europe, and how anti-crime policies have failed.

• CRIME IN ENGLAND AND WALES: MORE VIOLENCE AND MORE CHRONIC VICTIMS, which was published in June 2007, argues that the British Crime Survey uses a flawed methodology and that the overall incidence of crime is 14 million a year, not 10.9 million as stated by the BCS, and the incidence of violent crime 4.4 million, not 2.4 million as stated by the BCS. The report was written by Professor Ken Pease, former acting head of the Home Office's policy research group, and Professor Gary Farrell of Loughborough University. In a subsequent exchange of letters in the *Guardian*, Professor Mike Hough of Kings College, London, who had helped to design the BCS, defended the BCS methodology, but Professors Pease and Farrell refused to budge.
• THE FAILURE OF BRITAIN'S POLICE: LONDON AND NEW YORK COMPARED by Norman Dennis. Since publication of this book Professor Dennis, who is Visiting Fellow in the Department of Psychology at Newcastle University, had an article in the *Sunday Telegraph* on 26 August 2007 under the title 'Fight Crime by Stamping out the Seedbed', in which he stated that the number of robberies in London more than doubled from 22 000 to 46 000 between 1991 and 2007, while over the same period the corresponding figure in New York fell from 99 000 to 24 000.
• THE GREAT AMERICAN CRIME DECLINE: STUDIES IN CRIME AND PUBLIC POLICY by Franklin E Zimring. Professor Zimring analyses the longest and largest fall in crime in the USA since the 1940s. The decline occurred across violent and non-violent crime, all regions, and all demographics.
• A LAND FIT FOR CRIMINALS by David Fraser. The author, who spent 26 years working in the probation and prison services, shows how successive British governments have misled the public on the true nature of their policies on crime and punishment, with traumatic consequences for millions of victims.
• SEND FOR THE SHERIFF. This is one of a set of Localist Papers from the *think tank* (qv) Direct Democracy. It proposes a radical decentralisation and democratisation of justice, with power passed downward from distant international and national bodies to elected local sheriffs.
• SIMPLE JUSTICE by Charles Murray and others. The author argues that the explosion in UK crime since the 1950s is due partly to a system of justice that has become too lenient (in the 1950s one in 3 robberies led to a prison sentence compared with one in 22 today) and too complex.

10 Useful websites include the following. They are given in alphabetical order.

• British Crime Survey www.homeoffice.gov.uk/rds/bcs1.html
• Centre for Crime & Justice Studies www.kcl.ac.uk/depsta/rel/ccjs/
• Civitas www.civitas.org.uk/crime/index.php
• Crime & Society Research Association www.csra.org.uk/
• European Sourcebook of Crime www.europeansourcebook.org/
• Home Office www.homeoffice.gov.uk/justice/what-happens-at-court/sentencing/
• Home Office www.crimereduction.gov.uk/
• Home Office www.crimestatistics.org.uk/output/Page1.asp
• Home Office www.homeoffice.gov.uk/rds/recordedcrime1.html
• Home Office www.homeoffice.gov.uk/rds/bcs1.html
• Office for National Statistics www.statistics.gov.uk/
• Reform www.reform.co.uk/website/crime.aspx
• Social Trends www.statistics.gov.uk/StatBase/Product.asp?vlnk = 1367
• Statistics Commission www.statscom.org.uk/

See also: *Judicial system; National Probation Service*; and *Prisons*.

Crime and Disorder Act 1998 See: *Anti-social behaviour*; and *Discrimination*.

ime and Disorder eduction rtnerships (CDRPs)

1 CDRPs are combinations of police, local authorities, and other organisations and businesses who have banded together to tackle crime and disorder at a local level. As noted under *anti-social behaviour* (qv) they have greatly helped some social landlords (qv) in their struggles with anti-social behaviour.
2 For more information, and an alphabetical list of partnerships, visit www.crimereduction.gov.uk/regions/regionsaz.htm

ime and Society esearch Association

See: *Think tanks*.

ime Concern

1 This is a national crime reduction organisation and registered *charity* (qv). It runs approximately 60 projects across England and Wales.
2 Visit www.crimeconcern.org.uk/

ime: hate

See: *Hate crime*.

ime Reduction rtnerships (CRPs)

1 CRPs are statutory partnerships formed as a consequence of the Crime and Disorder Act 1998, which required police and local authorities to collaborate with each other and with other organizations to reduce crime and disorder within a *local authority* (qv).
2 As of 2008, useful websites include:

- Crimereduction www.crimereduction.gov.uk/
- IDeA Knowledge www.idea.gov.uk/
- Police service gateway www.acpo.police.uk/links.asp

rime reduction by esign

1 As of 2008, useful references include the following. For publication details see the bibliography:

- CRIME OPPORTUNITY PROFILING OF STREETS (COPS): A QUICK CRIME ANALYSIS-RAPID IMPLEMENTATION APPROACH. This is an evaluation of Crime Prevention through Environmental Design (CPTED) strategies.
- CRIME PREVENTION THROUGH ENVIRONMENTAL DESIGN: applications of architectural design and space management concepts by Timothy D Crowe, supported by the National Crime Prevention Institute.
- SECURITY WITHOUT THE SPIKES? A practical resource pack for crime prevention in the public realm.

2 Useful websites include:

- Secured by design www.securedbydesign.com/

rime Reduction rogramme

1 A government-funded programme consisting of several initiatives which, it is claimed, have been effective at reducing crime or the fear of crime.
2 For more information visit www.crimereduction.gov.uk/crpevaluation.htm

rime statistics

1 The Home Office (qv) is traditionally the official source of data on crime and justice in the UK. But the public has become so distrustful of official crime figures (for instance, in October 2007 Mr Norman Brennan, director of the Victims of Crime Trust, told the press that he believed the true crime figure to be five times higher than the published statistics) that there have been two calls for changes to the present system:

- In September 2006 the Statistics Commission (qv), an independent watchdog, called for the Home Office to be stripped of its involvement in publishing crime statistics. Though still collecting data, it would under this arrangement hand over control of the British Crime Survey to the Office for National Statistics (qv). The Commission also recommended changing the definition of violent crime. For details visit www.statistics.gov.uk/

• In November 2006 a panel led by Professor Adrian Smith, a former president of the Royal Statistical Society, published a set of practical recommendations for reforming crime statistics to restore public confidence. The report, which was commissioned in early 2006 by the then Home Secretary Mr Charles Clarke, is titled CRIME STATISTICS: AN INDEPENDENT REVIEW – see the bibliography.

2 In 2008 a new Statistics Board (qv) will try to bring more rigour to UK national statistics, but its brief excludes Home Office figures such as crime.

3 As of 2008, useful websites include:

- Crime in London www.met.police.uk/crimefigures/index.phpLondon
- Home Office www.crimestatistics.org.uk/output/Page1.asp
- Office for National Statistics www.statistics.gov.uk/
- Social Trends www.statistics.gov.uk/socialtrends

See also: *Office for National Statistics*; and *Statistics Board*.

Crime: youth

1 Youth crime, and in particular youth-on-youth crime, is in some ways more troubling than other kinds, not only because it affects the lives of people at a particularly vulnerable stage of life, but also because of its ominous implications for the future.

2 Current figures for this category of crime are alarming. According to Home Office statistics released in March 2007 the number of 10 to 17-year-olds sent for trial on homicide (murder or manslaughter) charges leapt from 44 in 2003 to 10[illegible] in 2005, and the number of people convicted of murder before their 18th birthday from 9 to 26. Whereas only 1in 15 suspected killers in 2003 was a minor, the figure in 2005 was 1 in 7.

3 Underlying much of this is the fast-growing phenomenon of gang culture. Writing in the *Sunday Times* on 19 August 2007 under the title GANGS, ALAS, ARE OFFERING WHAT BOYS NEED Harriet Sergeant, author of HANDLE WITH CARE (see: *Children in care*) quotes Norman Brennan, director of the Victims of Crime Trust, as saying that many young people now 'put a knife in their pockets as routinely as they pull on their trainers in the morning'. She herself comments: 'Drugs and alcohol are merely the symptoms of a deeper problem. Too many young men suffer from an absence of authority at home, in school and on the street. Young boys join gangs, they told me, because they are afraid. There is nobody else to protect them, certainly no responsible adult. "You don't start off as a killer", said a 19-year-old gang leader, "but you get bullied on the street. So you go to the gym and you end up a fighter, a violent person. All you want is for them to leave you alone but they push you and push you". Another boy aged 13 explained that in his area boys "would do anything" to join a gang. If they join a gang with "a big name" people will "look at them differently, be scared of them"'. She concludes: 'We have failed to provide a safe, disciplined and principled environment in which young people can relax, find themselves and channel their best efforts. Instead we have relegated many of them to a ghetto of violence and despair'. Justine Greening, who is MP for an inner London constituency, has aptly likened the situation in the worst areas to that depicted in William Golding's novel LORD OF THE FLIES.

4 The situation is indeed particularly bad in London. On 27 March 2007 Sir Ian Blair, Commissioner of the Metropolitan Police, told a Metropolitan Police Authority meeting that teenage gangs were a menacing 'cloud on the horizon' who must be tackled before they got out of hand. Research suggested that there were around 170 gangs in London with about 5 000 members. Only about 20 of these were currently linked to high-level crime, but it is difficult to be confident that this will remain so.

5 As of 2008, useful references include the following. They are given in alphabetical order. For publication details see the bibliography.

• BRITAIN HAS LOST THE ART OF SOCIALISING THE YOUNG by Peter Wilby.

• CHILDREN AS VICTIMS: CHILD-SIZED CRIMES IN A CHILD-SIZED WORLD, a survey published by the Howard League for Penal Reform in October 2007, found that three-quarters of the children interviewed had been assaulted in the past year, and two-thirds had been the victims of theft.

• EDUCATIONAL FAILURE AND WORKING-CLASS WHITE CHILDREN IN BRITAIN by Gillian Evans, published in 2006, examines middle-class assumptions about the reasons why white working-class boys do so badly at school. The general belief, as she has put it in a separate article, is that boys who fail at school 'come from problem families' or are 'just good for nothing'. Dr Evans discovered, on the contrary, that many of these boys are 'as good as gold at home' and go wrong when they seek the freedom to 'play out', where they meet gangs 'who rule the closely defined territories of the street with ruthless intimidation and violence'. To succeed in one of these gangs, you must yourself become intimidating, and the place where you can most easily start to do that is in a school where adult authority has broken down. Failing schools are places where the staff spend their time and energy containing gang violence instead of teaching.

• FROM PUNISHMENT TO PROBLEM SOLVING – A NEW APPROACH TO CHILDREN IN TROUBLE BY Rob Allen.

• YOUNG PEOPLE AND CRIME: FINDINGS FROM THE 2005 OFFENDING, CRIME AND JUSTICE SURVEY by Debbie Wilson, Clare Sharp and Alison Patterson.

• YOUTH JUSTICE 2004: A REVIEW OF THE REFORMED YOUTH JUSTICE SYSTEM.

6 Useful websites include:

• Connexions North London www.connexions-northlondon.co.uk/
• Crime reduction www.crimereduction.gov.uk/
• Crime Reduction Toolkits www.crimereduction.gov.uk/toolkits/py10.htm
• Frank Field www.frankfield.co.uk/type3.asp?id=20&type=3
• Home Office www.homeoffice.gov.uk/rds/youthjustice1.html
• Home Office www.homeoffice.gov.uk/crime-victims/reducing-crime/youth-crime/
• Metropolitan Police www.met.police.uk/youngpeople/
• Nacro www.nacro.org.uk/services/youthcrime.htm
• SmartJustice www.smartjustice.org/
• Trust for the Study of Adolescence www.studyofadolescence.org.uk/
• Youth Justice Board www.yjb.gov.uk/en-gb/
• Youth Justice system www.yjb.gov.uk/en-gb/yjs/TheSystem/

See also: *Anti-social behaviour*.

riminal Justice Act 994 See: *Civil liberties*; and *DNA database*.

riminal Justice and olice Act 2001 See: *Civil liberties*.

riminal offence See: *Civil proceedings*.

riminal Records ureau (CRB) 1 The CRB acts as a one-stop-shop for organisations wishing to check any data on job applicants, etc, that may be held by the Police National Computer (qv), the Department of Health (qv), and the Department for Education and Skills (qv). It is an Executive Agency (see: *Quangos*) and is used for instance for checking the backgrounds of all people whose work might bring them into contact with children – visit www.crb.gov.uk/

2 In the three years prior to 2006 the number of checks of this kind carried out added up to 8.9 million, and in 2006 it was estimated that the Safeguarding Vulnerable Groups Bill (qv), if enacted, would mean that up to 9.5 million adul in England and Wales – more than a third of the adult working population – would be subject to ongoing criminal checks to establish their suitability to wor with children.

3 As with all databases of personal records, CRB records are riddled with errors. As just one of a series of revelations of innocent people being listed as criminals, the CRB discovered in 2006 that it had wrongly labelled 2 700 innocent people as paedophiles, pornographers and violent criminals – refer to para 9.11.6 of A REPORT ON THE SURVEILLANCE SOCIETY (see below). One cannot help wondering how many more cases of this kind have never come to light, and what consequences they suffered from the undetected error.

4 For a note on problems with other very large UK government-commissioned computer-based systems see: *IT systems*; and for a comment on this and other invasions of our freedom and privacy see: *Civil rights*; and *Privacy*.

5 As of 2008, useful references include the following. For publication details see the bibliography.

- A REPORT ON THE SURVEILLANCE SOCIETY, written for the Information Commissioner (qv) by the Surveillance Studies Network.

See also: *Childcare: vetting; Children: fingerprinting by schools; IT systems; National DNA database; Police National Computer*; and *Privacy*.

Crossings See: *Pedestrian crossings*; and *Footbridges*.

Culture 1 As for societies (see: *Economic success*), so for individual people, the available evidence suggests that the key to success lies in cultural values and attitudes more than in race, gender, or the effects of discrimination.

2 First, as regards educational achievement, the following figures, taken from Table 3.17 in SOCIAL TRENDS 37 (qv), show the proportions of men and women from Great Britain's major ethnic groups who attained a degree or equivalent qualification in 2004.

Group	Percentage
Chinese men	33%.
Indian men	30%.
Chinese women	29%.
White Irish women	25%.
Black African men	24%.
White Irish men	23%.
Indian women	21%.
White British men	18%.
Black African women	17%.
White British women	16%.
Black Caribbean women	15%.
Pakistani men	15%.
Black Caribbean men	11%.
Bangladeshi men	11%.
Pakistani women	10%.
Bangladeshi women	05%.

This ranking fails to show any clear pattern of race or gender, or of being at the receiving end of unfair discrimination, as various *pressure groups* (qv) would have us believe. But it does conform with easily-observed differences in attitude to academic achievement among UK population groups.
3 Second, as regards subsequent career success, the pattern is broadly consistent with the pattern above, but there are discrepancies too. These may be partly due to race or gender prejudice among employers, but again cultural values and priorities appear to be more important. For instance, some cultures encourage women to go out and compete for success at the highest business and professional levels; others discourage such behaviour and prefer women to devote themselves to the home and the family. The results show up very clearly in career achievement, and therefore in incomes.
4 Both in respect of education and career success the important (and reassuring) point is that families have the key to success for their children largely in their own hands, and that few people are doomed to exclusion from the good things in life by race, gender, or unfair discrimination. As with societies, a commitment to high academic standards and a dedicated, disciplined, and purposeful attitude to work are likely to win through – as is demonstrated throughout the world by, for instance, Jewish and Chinese populations. It must unfortunately be added that parents' best efforts to nurture high standards in their children may be sabotaged by the poor quality of education on offer from their local school, but that's a different subject – see: *Social mobility*.

See also: *Economic success; Social mobility*; and *Victimhood.*

---

ulture and leisure uildings

See: *Educational, cultural and scientific buildings*; and *Sports-related buildings.*

SA

See: *Child Support Agency.*

---

utting Burdens on achers

1 This programme was created to cut red tape and reduce bureaucracy in education. It comes under the Department for Education and Skill (qv), and has three 'gatekeeper' groups:
• The Implementation Review Unit (qv) for schools.
• The Bureaucracy Reduction Group (qv) for further education and training.
• The Higher Education Regulation Review Group (qv) for higher education.
It collaborates with the Lifting Burdens Task Force (qv) in the area of children's services.
2 Despite these activities the burden of red tape upon teachers has massively grown rather than diminished. However, in February 2007 a new phase was entered when Cutting Burdens on Teachers published its Simplification Plan, by the implementation of which the Government is committed to cutting the burden of red tape by at least one quarter in three years (i.e. by 2010). We await the results with interest.
3 For more detail on the Cutting Burdens on Teachers programme visit www.dfes.gov.uk/reducingbureaucracy/

See also: *Better regulation bodies*; and *Over-regulation.*

---

utting Red Tape rogramme

1 One of an array of programmes being created by the Government to help it cut red tape – visit www.dti.gov.uk/bbf/better-regulation/red-tape/page23677.html
2 For a comment on other bodies of this kind see: *Better regulation bodies.*

See also: *Better regulation bodies*; and *Over-regulation.*

Cycle track/footway surface See: *Segregated shared cycle track/footway surface.*

D

ily living aids See: *Aids to daily living.*

nger sign See: *Signage: safety.*

ta Protection Act 1 The Data Protection Act 1998 entitles citizens to gain access to information held about them by organisations, and governs how such organisations can use the personal data that they hold. This includes how they acquire the information, share it, or dispose of it. The Act is administered and enforced by the Information Commissioner's Office (qv).
2 As of 2008, useful websites include the following. They are given in alphabetical order.

- Data protection www.foi.gov.uk/datprot.htm
- Directgov www.direct.gov.uk/
- Information Commissioner's Office www.informationcommissioner.gov.uk/
- Office of Public Sector Information www.opsi.gov.uk/

See also: *Access to information; Freedom of Information Act; Notification*; and *Privacy*.

y centres 1 Local authority day centres provide facilities for recreation and training to help *disabled people* (qv) back to work. They may offer such users supported employment, educational and sporting activities, and opportunities to meet socially.
2 For more information contact the *Social services* (qv) department of the *local authority* (qv) concerned.

See also: *Care homes; Health and welfare buildings*; and *Social services departments*.

y centres for ildren See: *Childcare.*

y centres for meless people 1 These are a key service for many homeless and street people, and people living in temporary accommodation or at risk of *homelessness* (qv). They offer daytime shelter, advice, support, and often provide practical services such as food, showers, clothing, or laundry facilities. Such shelters have traditionally been organised mainly by *voluntary* (qv) groups at venues such as church halls or *community centres* (qv), but in recent years some larger day centres have been developed to offer a wider range of essential support and resettlement services. Some have large purpose built premises with teams of paid and specialist staff that provide services including housing, health care, mental health care, alcohol and drugs training, and employment advice.
2 As of 2008, useful websites include the following. They are given in alphabetical order.

- Centrepoint www.centrepoint.org.uk/
- Crash www.crash.org.uk/homelessness/intro.lml
- Directgov www.direct.gov.uk/
- Homeless Link www.homeless.org.uk/
- Homelessness and Housing Support Directorate www.communities.gov.uk/
- Shelter http://england.shelter.org.uk/home/index.cfm

See also: *Homelessness.*

Day centres for older people

1 Loneliness can be a great problem for older people, and day centres (or day care centres) can provide company, meals, activities and entertainment, and in the best cases an enjoyable day out. Providers include *local authorities* (qv) and charities and *voluntary and community sector* (qv) organisations such as Age Concern.
2 For design guidance see: Health and welfare buildings.
3 As of 2008, useful websites include:
• Age concern www.ageconcern.org.uk/
• Joseph Rowntree Foundation www.jrf.org.uk/

DCMS

See: *Department for Culture, Media and Sport.*

Deafblindness

1 This is a combination of *impaired hearing* (qv) and *impaired vision* (qv), often caused by rubella. The condition is sometimes referred to as 'dual sensory impairment' or even 'multi-sensory impairment'. Both terms are inexact, very easily misunderstood, and best avoided – see: *Political correctness: speech codes.*
2 The RNID (qv) estimates that there may be 20 000 to 25 000 deafblind people in the UK. They fall into four broad categories:
• Those who are born with little or no hearing and sight, or who have largely lost these faculties in early life. Quite understandably they have great difficulty in acquiring language.
• Those who are born with impaired hearing and develop sight loss some years later, often as a result of Usher syndrome.
• Those who are born with a visual impairment, and subsequently develop hearing loss. If their ability to hear speech endured for long enough to acquire spoken language, then this may well remain one of their methods of communication.
• Those who are born with full hearing and sight, and develop impairments to both in later life, usually in old age. Communication will almost certainly be by speech.
3 There are very few *aids and equipment* (qv) designed specifically for deafblind people, who will usually be helped by an appropriate combination of the aids listed under *impaired hearing* and *impaired vision.* The exact choice will depend upon the nature of their condition.
4 In 2004 the European Parliament recognised deafblindness as a distinct *disability* (qv).
5 As of 2008, useful websites include the following.
• RNID www.rnid.org.uk/
• Sense www.sense.org.uk/

See also: *Impaired hearing;* and *Impaired vision*

Deafness

See: *Impaired hearing.*

Deafness among children

1 The RNID (qv) estimates that there may be 20 000 to 25 000 children up to the age of 15 in the UK who are permanently *deaf* (qv) or hard of hearing. They therefore have *special educational needs* (qv). Most go to *mainstream schools* (qv), where some are accommodated in units for deaf children where a qualified teacher of the deaf gives extra support. Others attend *special schools* (qv) for deaf children with professionals such as speech and language therapists. The decision on which particular option is best for any individual child belongs to that child's parents, guided by experts, and parents should insist on their right to decide.

2 Turning to older age groups: there are an estimated 4 000 to 5 000 deaf students in further and higher education, most of them in mainstream colleges where they are integrated with hearing students.
3 As of 2008, useful websites include:
• RNID www.rnid.org.uk/

See also: *Impaired hearing*; and *Special Educational Needs*.

cent homes ndard See: *Housing: Decent homes standard.*

D See: *Disability equality duty*.

formation fessionelle 1 This term refers to the tendency of highly-educated academics and professionals to develop a world-view so driven by theory that they are blind to everyday realities which are perfectly visible to ordinary people. When combined with the power that is exerted upon society by theorists, both directly in their daily work and indirectly through their influence upon professional institutions and governments, the effects can be (and often are) very damaging. The following examples are given in alphabetical order.
• In architecture and urban design, much-loved urban forms and building styles which had served society well for centuries, and created environments that function well and delight us to this day, were virtually banished from the menu of available options after the mid-1950s by an architectural establishment in thrall to theorists such as le Corbusier (qv). Traditional ornamented, pitched-roof houses facing onto lively streets were remorselessly replaced by graceless slab and tower-blocks set in lifeless landscapes, unloved by their inhabitants and looking shabbier with every passing year as their flat roofs leaked and their concrete surfaces became disfigured by staining. While ordinary people complained bitterly, the profession disdained their views and bestowed awards on buildings such as Robin Hood Gardens, designed in 1968 by the late Alison and Peter Smithson, on the corner of Cotton Street and Robin Hood Lane in Poplar, London E14. See: *Architecture*.
• In crime, the slow increase in both general crime and violent crime over the previous half century suddenly became precipitous in the mid-1950s, and this socially devastating trend continued unchecked until 1993, when the Home Office finally acquired a Secretary of State self-confident enough to defy the intellectual establishment and start imprisoning large numbers of criminals, as demanded by the general public. The Recorded Crime Statistics immediately fell from 5.526 million in 1993 to 4.598 million in 1997 – the first sustained fall in 50 years. See: *Crime*.
• In education, an estimated 4 million British children who could have been taught to read have emerged from school functionally illiterate, as the direct result of the obstinate enforcement (via teacher training colleges, the Department of Education, and local authority education authorities) of modish but ineffective teaching methods. See: *Literacy*.
2 Other examples include education professionals so obsessed by the *social model of disability* (qv) that they are oblivious to the misery suffered by many disabled children in mainstream schools (see: *Special educational needs*); and teams of child care professionals who were unable to see over a period of weeks what a minicab driver, Mr Salman Pinarbasi, noticed at a glance – that a child under their care was being tortured to death (see: *Victoria Climbié*).

See also: *Architecture; Crime; Experts; Ideology; Literacy; Special educational needs*; and *Victoria Climbié*.

DEFRA See: *Department for Environment, Food and Rural Affairs.*

Democracy See: *Civil liberties; Constitution; Human rights; Post-democracy; and Speech, freedom of.*

Democratic deficit 1 A long-winded way of saying 'undemocratic', but enabling politicians and journalists to sound more thoughtful and learned.

2 See also: *Jargon.*

Dench and Gavron See: *Housing policy and social friction.*

Dental care 1 In March 2007 the 'Gaps to Fill' survey conducted on behalf of Citizens Advice, the umbrella group for Citizens Advice Bureaux, revealed that one ye after the Government's reforms of NHS dentistry were introduced, two millio people could not get treatment. Two thirds said they were going without denta treatment owing to 'huge problems accessing NHS dentistry and not being ab to pay for private treatments'; and more than three quarters complained that none of their local dentists was accepting NHS patients. Some were making round trips as long as 120 miles to reach a NHS dentist. For the press release visit www.citizensadvice.org.uk/index/pressoffice/press_index.htm
2 In the same month the British Dental Association (BDA) stated that 95% of dentists surveyed felt less confident about the future of NHS dentistry than th did two years ago. Susie Sanderson, chairwoman of the BDA's executive boar blamed a 'reductive, target-driven system' for these failures. For details visit t media centre at /www.bda.org/
3 In June 2007 it was reported that despite dentists' average earnings rising by 11% over the previous year to £86 000 per annum, it was harder than ever to g treatment by an NHS dentist because many of them were so unhappy with the new contracts (the announced purpose of which had been to improve access to dentists) that they were accepting private patients only.
4 DENTISTRY WATCH: NATIONAL SURVEY OF THE NHS DENTISTRY SYSTEM WITH VIEWS FROM BOTH PATIENTS AND DENTISTS, published in October 2007 by the Commission for Patient and Public Involvement in Health (qv), revealed an even worse situation – see the bibliography. The report was released under the heading 'Patients pulling out of rotting NHS dental system'.
5 It is safe to assume that the above trends and events are hitting poor people hardest.

See also: *NHS*; and *Public services: quality.*

Dentists' surgeries See: *Health and welfare buildings.*

Department for Business, Enterprise and Regulatory Reform (Berr) 1 In June 2007, immediately after Mr Gordon Brown became Prime Minister, the Department for Trade and Industry (qv) was abolished and most of its functions taken over by the newly created Department for Business, Enterprise and Regulatory Reform.
2 Berr also took over the very important area of regulatory reform, *better regulation bodies* (qv) and the reduction of *red tape* (qv) from the Cabinet Office (qv).
3 For more information visit:

• Berr	www.berr.gov.uk/
• Guide to Government	www.number-10.gov.uk/output/Page30.asp

See also: *Government.*

epartment for hildren, Schools and amilies (DCSF)

1 At the same time as the above, the Department for Education and Skills (qv) was abolished and its responsibilities divided between the newly-created Department for Children, Schools and Families (which assumed a wide range of responsibilities for the younger age groups), and Department of Innovation, Universities and Skills (ditto for the older age groups).
2 Among other things the DCSF took control of the Respect agenda from the Home Office (qv) and shares several joint areas of responsibility with other departments. They include youth justice with the new Ministry of Justice (qv); *child poverty* with the Treasury (qv) and Department for Work and Pensions (qv); children's health with the Department of Health (qv); and youth sport with the Department for Culture, Media and Sport (qv).
3 A particularly interesting move was the shifting of responsibility for Asbos (qv) from the Home Office to the new DCSF – in other words, from punishment to rehabilitation. This is an encouraging initiative, and we must hope for good results – see: *Crime: youth.*
4 For more information visit:

- DCSF www.dcsf.gov.uk/
- Guide to Government www.number-10.gov.uk/output/Page30.asp

See also: *Children; Education; Literacy;* and *Social mobility.*

partment for mmunities and cal Government CLG)

1 The Department for Communities (as it is usually called) was created in May 2006 with the remit of promoting community cohesion. It replaced the now-dissolved Office of the Deputy Prime Minister. In addition to inheriting the ODPM's responsibility for housing, planning, and the building regulations it also took over several functions from the Home Office (qv) so as to create a department that is in charge of the social as well as the physical aspects of communities.
2 This recasting follows so many previous short-lived reorganisations that confused readers may be helped by a résumé of recent history – particularly as some of the previous departmental imprints still survive on documents that are in current use.
3 Ten years ago there was the Department of the Environment (DoE), which had a long life during which it did good work and gained much respect. Then we entered a period during which government departments were created and replaced as casually as a child builds and destroys sandcastles:

- Following the general election of 1997, the DoE and the Department of Transport (DoT) were merged to form the new Department of the Environment, Transport and the Regions (DETR). The latter's areas of responsibility included the environment, housing, transport services, rural affairs, planning, local government, regional development, regeneration, the construction industry, and health and safety.
- Following the general election of 2001 the DETR was dissolved and largely replaced by the new Department for Transport, Local Government and the Regions (DTLR). The latter retained most of the ex-DETR's areas of responsibility as above, except that environmental protection was lost to the new DEFRA (qv).
- In May 2002 the DTLR was dissolved and its responsibilities split in a reversion to the pre-1997 position. Transport was again given a department of its own, now called the Department for Transport or DfT (qv), and most of the DTLR's other responsibilities passed to the new Office of the Deputy Prime Minister (ODPM).
- In May 2006, following Mr John Prescott's resignation as Deputy Prime Minister, the ODPM was in turn dissolved and replaced by the newly created

Department for Communities and Local Government (DCLG) under Ms Rut
Kelly who had until then been Secretary of State for Education.
4 For more information visit:
• DCLG www.communities.gov.uk/
• Guide to Government www.number-10.gov.uk/output/Page30.

See also: *Government.*

Department for Constitutional Affairs (DCS)

1 In May 2007 the DCS, which had been in existence only since 2003, was abolished. Most of its duties were transferred to the newly created Ministry c Justice (qv). See also: *Electoral fraud.*
2 The now-superseded websitewww.dca.gov.uk/ has been retained to serve a purely archival function.

Department for Culture, Media and Sport (DCMS)

1 A government department created in 1997 with responsibility for museums galleries and libraries; the built heritage; the arts, sport and education; broadcasting, the media and tourism; and the creative industries. It is also responsible for the National Lottery (qv), which funds many building project each year and sets high standards of *accessibility* in those projects to which i contributes.
2 For more information visit:
• DCMS www.culture.gov.uk/
• Guide to Government www.number-10.gov.uk/output/Page30.a

See also: *Government.*

Department for Education and Skills

1 The DfES was superseded in June 2007 by the Department for Children, Schools and Families (qv).
2 During its lifetime the DfES fell very far short of fulfilling the mission statement given on its website, which boasted of 'creating opportunity, releasing potential, and achieving excellence for all'. Its actual performance may be judged by the following:
• As of 2008, around one in five adults in Britain is functionally illiterate. Fo the effect this has probably had on the lives of the children concerned see: *Literacy.*
• WORKING ON THE THREE R'S, a report commissioned by the DfES and published in August 2006 by the Confederation of British Industry (CBI), sta that a third of businesses were having to send staff for remedial lessons in ba literacy and numeracy that they should have learned at school. The report wa based on a survey of companies employing in total almost a million full-time staff.
• Also in August 2006, it was reported that a third of nurses with GCSE English and maths qualifications failed a basic English and maths test includ questions at the level of: 'How many minutes are there in half an hour?' and 8pm the same as 18.00, 19.00, 20.00 or 21.00 hours?'
Many more reports could be quoted, all pointing in the same direction – see: *Qualifications and Curriculum Authority.*
3 The difference in performance between those schools that were under the direct control of the DfES (comprehensives) and those with a substantial deg of freedom (independents) was extreme. As of 2006, of the 100 top-performi schools in the UK around five were comprehensives (representing perhaps 34 of all schools) and 65 were independents (representing perhaps 7% of all schools). While part of the explanation undoubtedly lies in different funding levels and in the different intakes of the two types of school, the disparities a too wide for that to be the whole story. For some comment on the damaging

consequences see: *Social mobility.*

4 Why did the DfES perform so catastrophically badly? It is worth recounting some of the reasons in the hope that the Department for Children, Schools and Families, which succeeds it, may do better. Independent commentators tend to dwell on four major things that went wrong with Britain's state education system, starting particularly in about the 1980s:

• The insufficiently tested ideas of educational theorists such as the Canadian Frank Green and Britain's own Professor Ted Wragg (see: *Literacy)* gained dominance in the educational and teacher-training establishment, who are bowled over by teaching methods that are 'creative' and 'child-centred', regardless of whether the latter actually work. In the fields of literacy and numeracy the new methods have demonstrably not worked.

• Most Ministers lacked the self-confidence to subject such modish theories to proper scrutiny, and tamely drifted along with the prevailing tide of educational fashion.

• Even if they were minded to get a grip on things, most Ministers have not stayed in their post for long enough to become effective. In the USA cabinet secretaries are appointed for their expertise and usually serve at least a four-year term. In Britain, recent terms of office of Secretaries of State for Education have been Kenneth Baker 1986–1989; John McGregor 1989–90; Kenneth Clarke 1990–92; John Patten 1992–94; Gillian Shephard 1994–97; David Blunkett 1997–2001; Estelle Morris 2002–02; Charles Clarke 2002–04; and Ruth Kelly 2004–06; Alan Johnson 2006–07.

• The DfES buried frontline staff under exhausting and demoralising piles of paperwork, thus making it very difficult for them to inspire and give full attention to their pupils. Precise figures are hard to come by, but according to Simon Jenkins, author of THATCHER AND SONS: A REVOLUTION IN THREE ACTS (see the bibliography), the DfES between 1997 and 2006 issued 500 regulations, 350 policy targets, 175 efficiency targets, 700 notes of guidance, 17 plans and 26 separate incentive grant streams. Sir Simon quotes Hansard as recording in 2001 that an annual average of 3 840 pages of instructions were being sent to schools in England. He also states that a survey in 2006 showed one in four head teachers considering quitting their jobs because of the volume of red tape. Instead of teaching, teachers read instructions and struggle with bureaucracy.

There must be a better way of doing things.

See also: *Office for Standards in Education; Qualifications and Curriculum Authority*; and *Social mobility*.

[De]partment for [Ed]ucation and Skills: [Sch]ools Capital [Ass]ets Design Team

1 The Schools Capital Assets Design Team is responsible for promoting best practice in design, use and management of school facilities. Its website, titled the School Buildings Information Centre (qv), is a prime information source for architects and others involved with schools design.

2 For more information visitwww.teachernet.gov.uk/

[De]partment for [Env]ironment, Food and [Ru]ral Affairs [[D]EFRA]

1 The objectives of DEFRA are to encourage the development of British farming, fishing, food, water, and related industries in ways that create a better quality of life for everyone, and are sustainable. The aspect of its work likely to be most relevant to users of this Guide is the protection and enhancement of the environment.

2 *Quangos* (qv) under DEFRA include the Countryside Agency; English Nature; and the Environment Agency.

3 For more information visit:

• DEFRA www.defra.gov.uk/

• Guide to Government www.number-10.gov.uk/output/Page30.as

See also: *Government*

Department for Innovation, Universities and Skills (DIUS)

1 In June 2007, immediately after Mr Gordon Brown became Prime Minister the Department for Education and Skills (qv) and the Department for Trade a Industry were abolished, and those of their functions that had to do with research, science, innovation, and skills were brought together in the newly created DIUS.
2 One effect of the reorganisation is that science policy will be brought closer education, which is a welcome development. For more information visit:
• DIUS www.dius.gov.uk/
• Guide to Government www.number-10.gov.uk/output/Page30.as

See also: *Education*; and *Government.*

Department for Justice

See: *Ministry of Justice.*

Department for Transport (DfT)

1 The responsibilities of the DfT include roads, vehicles and road safety; railways; aviation; shipping; integrated transport; the Blue Badge parking scheme (qv); and mobility and inclusion.
2 *Quangos* (qv) under the DfT include:
• The Disabled Persons Advisory Committee or DPTAC (qv). Visit www.dptac.gov.uk/
• The Mobility and Inclusion Unit (MIU) promotes *accessibility* (qv) and *inclusiveness* (qv) throughout the transport environment, particularly by publishing authoritative guides such as INCLUSIVE MOBILITY (qv). Visit www.dft.gov.uk/transportforyou/access/miu/
• The Mobility Advice and Vehicle Information Service for disabled people (MAVIS) offers advice on the most suitable adapted vehicle for clients' needs, and on other aspects of public and private transport and outdoor mobility. Vis www.dft.gov.uk/transportforyou/access/mavis/
• The Commission for Integrated Transport. Visitwww.cfit.gov.uk/
• The Rail Group (qv).
• Transport for London (qv).
3 For more information visit:
• Department for Transport www.dft.gov.uk/
• Guide to Government www.number-10.gov.uk/output/Page30.as

See also: *Government*; and *Transport.*

Department for Transport, Local Government and the Regions (DTLR)

See: *Department for Communities and Local Government.*

Department for Work and Pensions (DWP)

1 The DWP was created in 2001 with the aim of helping more people into wo while providing security for those who cannot work. The Disability Unit with the DWP is particularly relevant to users of the present work.
2 For more information visit:
• Department for Work and Pensions www.dwp.gov.uk/
• Guide to Government www.number-10.gov.uk/output/Page30.as

See also: *Government.*

partment of Health DH)	1 The Department of Health works to improve the health and well-being of people in England. The National Health Service is an altogether separate entity – see: NHS. 2 For more information visit: • Department of Health www.dh.gov.uk/ • Guide to Government www.number-10.gov.uk/output/Page30.asp See also: *Government; Health*; and *NHS.*
partment of Health tates and Facilities nagement	1 This Directorate was created in 2005 to replace the erstwhile NHS Estates, which used for many years to provide authoritative advice on the design and management of *hospitals* and other healthcare facilities. The publications of NHS Estates on hospital design (including critical matters such as *fire safety*) were essential references for anyone involved in the design of healthcare buildings. 2 For more information visitwww.dh.gov.uk/
partment of the vironment (DoE)	See: *Department for Communities and Local Government.*
partment of the vironment, nsport and the gions (DETR)	See: *Department for Communities and Local Government.*
partment of Trade d Industry (DTI)	See: *Department for Innovation, Universities and Skills.*
partment of nsport, Local vernment and the gions (DTLR)	See: *Department for Communities and Local Government.*
privation	See: *Neighbourhood renewal*; and *Poverty in the UK.*
prived areas	See: *Areas of deprivation.*
sign and Access atements	1 A Design and Access Statement is a document submitted to the planning authority by a developer when applying for planning permission, and its purpose is to demonstrate how the principles of *accessibility* (qv) and *inclusive design* (qv), including the specific needs of disabled persons, older people and children have been integrated into the proposed design. For more information on *planning permission* visitwww.planningportal.gov.uk/ 2 It is officially stated that the Statements that accompany most planning applications need only be short, and in view of the Government's pledge to simplify regulatory procedures (see: *Better regulation bodies*) planning authorities should be held to this assurance. 3 As of 2007, useful guides to the drafting of such Statements include the following. For publication details see the bibliography. • CIRCULAR 01/2006: GUIDANCE ON CHANGES TO THE DEVELOPMENT CONTROL SYSTEM is a 2-page circular from the Department for Communities and Local Government (qv).

• DESIGN AND ACCESS STATEMENTS: HOW TO WRITE, READ AND USE THEM, a free guide from CABE (qv), is clear and well-illustrated.
• DESIGN AND ACCESS STATEMENTS: REPORT FROM A 'LEARNING GROUP' COMPRISING 16 LOCAL PLANNING AUTHORITIES, is an evaluative study.
• WHEELCHAIR ACCESSIBLE HOUSING, a best-practice guide for housing in London, gives authoritative advice on the Statements that must accompany al planning applications for *housing* (qv) developments in London.
4 ACCESS STATEMENTS: ACHIEVING AN INCLUSIVE ENVIRONMENT BY ENSURING CONTINUITY THROUGHOUT THE PLANNING, DESIGN AND MANAGEMENT OF BUILDINGS AND SPACES is, in contrast to the above, a mind-numbing document that is impossible to reconcile with the Government's own principles of good regulation – see: *Consultants' charter*.

See also: *Inclusive design*.

Design codes — See: *Urban design*. Also refer to DESIGN CODES: THE ENGLISH PARTNERSHIPS EXPERIENCE (see the bibliography) and visit www.communities.gov.uk/publications/citiesandregions/preparingdesigncodes

Design Guides — See: *Authoritative practical guides*.

Design quality — See: *Architecture; CABE*; and *Design and access statements*.

Design Quality Indicators (DQI) — 1 DQIs were launched in 2002 and were intended to become a tool that everyone involved in a building project would be able to use to evaluate the quality of that building, starting with the first *briefings* and all the way through to the building-in-use. They are sponsored by CABE (qv) and the Construction Industry Council (qv).
2 For more information visitwww.dqi.org.uk/DQI/default.htm

Designated parking spaces — 1 These are parking spaces that are specifically provided for the use of *disabl people* (qv).
2 For more information see: *Car parking*.

Designing for Accessibility — 1 A core reference that should be on every architect's shelf. The simple, straightforward guidance given in its 46 pages will make it easy for building designers to get most of their early decisions broadly right. The more precise and detailed guidance given in references such as BS 8300 (qv) can then be incorporated relatively easily. While this guide does not have the force of law, conformity with its recommendations will make it easier to demonstrate that t requirements of the Disability Discrimination Act (qv) have been met.
2 For publication details see the bibliography.

See also: *Authoritative practical guides*.

Desks: adjustable — See: *Worktops: offices*.

DETR — See: *Department for Communities and Local Government*.

Development Trusts — 1 A network of independent, not-for-profit, community-based organisations which are engaged in the economic, environmental, and *social regeneration* (qv) of a defined area or community.
2 For more information visit www.dta.org.uk/

DfES — See: *Department for Education and Skills*

a Ride

1 A door-to-door service provided by Transport for London (qv) and some other authorities for *disabled people* (qv) who cannot use conventional public transport.
2 For more information visit www.tfl.gov.uk/dial-a-ride/

itised speech

See: *Speech aids*; and *Talking household appliances*.

legia

See: *Paralysis*.

ing room table ghts for wheelchair rs

See: *Tables*.

ect Democracy

See: *Localism; Post-democracy*; and *Think tanks*.

ect payments

1 *Disabled people* (qv), like most others, wish to run their own lives and take their own decisions instead of having things done for them and to them. A policy that would help them do so is for *social services* (qv) to make direct payments to beneficiaries, for them to spend as they wish on *carers* (qv), *aids and equipment* (qv), etc., instead of placing these decisions in the hands of officials. Such a policy would be unsuitable for people with severe mental problems, old people with Alzheimer's disease, and some others, but it ought to be the norm for the great majority of disabled people.
2 For a well-informed discussion of the issues involved refer to pp 30, 139–43, and 148–51 of DISABILITY RIGHTS AND WRONGS by Tom Shakespeare, who is himself disabled – see the bibliography.
3 As of 2008, useful websites include:

• Directgov: disabled people	www.disability.gov.uk/
• Independent Living Funds	www.ilf.org.uk/

See also: *Independent living*; and *Independent Living Funds*.

ect taxes

See: *Taxation*.

ability: definition in Disability crimination Act

1 The Disability Discrimination Act (qv), which effectively outlaws *discrimination* (qv) against *disabled people* (qv) in public life, defines 'disability' as a *physical* or *mental impairment* which has a *substantial* and *long-term* adverse effect on a person's ability to carry out *normal day-to-day activities*. In the Act the italicised terms have precise meanings, which may be summarised as follows:
2 *Physical impairments* may be congenital (ie present from birth) or caused by injury or illness. As of 2007 those recognised under the Act, if their effect upon the individual is sufficiently severe, include weakened limbs or organs which lead to *impaired* mobility, dexterity or physical co-ordination; or to impaired continence, speech, hearing, or eyesight. In particular:
• Conditions that are specifically recognised include *arthritis, blindness*, and *partial sightedness*. Since 2005 *multiple sclerosis, cancer,* and *HIV* have been deemed to be disabilities from the time of diagnosis. See these individual entries for more detail.
• Conditions that are specifically <u>not</u> recognised include addiction to alcohol, nicotine, or any other drug; seasonal hay fever; and defects in eye-sight that can be corrected by spectacles.
3 *Mental impairments* include clinically recognised mental illnesses (in 2005 the requirement in the 1995 Act for mental illnesses to be 'clinically well

recognised' was removed.). Also, if the effect is sufficiently severe, difficulty in concentrating, learning or understanding; and difficulty in perceiving risk danger.
In particular:
• Conditions that are specifically recognised under the DDA include *dyslexia* and *epilepsy*. See these individual entries for more detail, and see also: *Impaired understanding*; and *Mental illness*.
• Conditions that are specifically <u>not</u> recognised include a tendency to set fire, or to steal.
4 *Substantial* does not necessarily mean severe, only that the effect of the impairment is more than minor or trivial. As examples, the term is likely to include:
• An inability to see moving traffic clearly enough to safely cross a road.
• An inability to turn taps or knobs.
• An inability to remember and relay a simple message correctly.
The Act makes an exception to the common sense interpretation of *substanti*[al] in the case of severe disfigurement which, unless caused by tattooing or body piercing, or similar acts, is specifically recognised as a disability even if its adverse effect on the ability to carry out normal day-to-day activities is minor non-existent.
5 *Long-term* means that the effect must have lasted, or must be likely to last, at least 12 months in a detrimental manner. In the case of progressive conditions such as *cancer (qv), HIV infection (qv), multiple sclerosis,* and *muscular dystrophy,* the Act now covers people from the time of diagnosis (th is a stipulation introduced in 2005). The phrase *long-term* includes people wh have recovered from *mental illness* but are still experiencing prejudice.
6 *Normal day-to-day activities* are ones carried out by most people on a fairly regular and frequent basis. Examples are likely to include:
• Moving from place to place; catching a bus.
• Washing, eating, or turning on a TV set.
• Lifting, carrying, or moving ordinary objects.
• Recognising physical danger.
Activities that do not form a normal part of many peoples' lives, such as playing a musical instrument to professional standard, or carrying out every aspect of a particular job, are not likely to be included.
7 For comments on the numbers of people that could be defined as 'disabled' b the above definitions, see: *Disabled people: numbers.*
8 Readers are reminded that the present work only gives simplified summarie of original documents and does not purport to provide full, authoritative and up-to-date statements of the law. For latest information refer to DISABILITY DISCRIMINATION: LAW AND PRACTICE (see the bibliography) and the following websites:
• Directgov: disabled people www.disability.gov.uk/
• Equality & Human Rights Commission www.equalityhumanrights.com/
9 Other useful references include the following:
• CODE OF PRACTICE – RIGHTS OF ACCESS: SERVICES TO THE PUBLIC, PUBLIC AUTHORITY FUNCTIONS, PRIVATE CLUBS AND PREMISES. This *code of practice* (qv) covers the duty of *service providers* (qv) not to *discriminate* (qv) against *disabled* (qv) customers. It replaces the earlier CODE OF PRACTICE – RIGHTS OF ACCESS: GOODS, FACILITIES, SERVICES AND PREMISES, which was published in 2002 and is now obsolete.
10 Other useful websites include the following. They are given in alphabetica order:

• Directgov: disabled people	www.disability.gov.uk/
• Equality & Human Rights Commission	www.equalityhumanrights.co
• Office of Public Sector Information	www.opsi.gov.uk/acts.htm

sability: the edical and social odels of sabilities

1 As of 2008, the only definition of *disability* (qv) that matters in UK law is the one given in the Disability Discrimination Act, and briefly outlined above. But the issues involved may be better understood with benefit of the following notes on three mental models of disability which – whether consciously or not – underlie all legislation in this field:

• The 'medical' model of disability assumes that disability is almost entirely the result of a medical condition – for instance *blindness* (qv) or *paralysis* (qv). It was accepted as a matter of course until the 1980s and formed the basis of the World Health Organisation's INTERNATIONAL CLASSIFICATION OF IMPAIRMENTS DISABILITIES AND HANDICAPS, published in 1980 (see the bibliography).

• The 'social' model of disability was developed in Britain mainly by Vic Finkelstein, Paul Hunt and Mike Oliver in the 1970s and 1980s. It greatly downgrades the importance of medical condition and argues that people are disabled mainly by *mainstream* (qv) society's failure to cater for such conditions. There is truth in this, but some people will insist on taking things to extremes, and some influential voices now effectively dismiss medical condition as irrelevant, and place complete responsibility for disabled people's difficulties on a discriminatory society. For many conditions this clearly isn't true, but pressure from some academics and *pressure groups* (qv) has succeeded in creating a climate of opinion that sometimes ignores reality. As one example: educationists in thrall to the social model of disability are denying thousands of disabled children places in *special schools* (qv) where they could receive the care and protection they need, and forcing them against their own and their parents' wishes into *mainstream schools* (qv) where they suffer agonies of neglect and bullying – see: *Special educational needs.* Fortunately two recent events – publication of DISABILITY RIGHTS AND WRONGS (see below) by the disabled writer Tom Shakespeare in 2006, and an article in the November 2006 issue of DISABILITY NOW (qv) by the then Disability Rights Commissioner Agnes Fletcher – could be early signs of a shift away from these damaging and dogmatic positions. The medical and social models of disability both contain elements of the truth, and enlightened common sense is a more reliable guide to getting the balance right than *ideology* (qv).

• The 'architectural' model is a variation of the social model, focusing on the way in which the *built environment* (qv) can impede people from enjoying a safe and convenient life by creating barriers such as steps, narrow doors, unreachable switches, poor lighting, etc. The concept is sound, and quite rightly underpins many current regulatory documents such as APPROVED DOCUMENT M (qv) of the Building Regulations, BS 8300 (qv), and INCLUSIVE MOBILITY (qv) – see: *Authoritative practical guides.*

5 As of 2008, useful references include the following. They are given in alphabetical order. For publication details see the bibliography.

• DESIGNING FOR THE DISABLED: THE NEW PARADIGM by Selwyn Goldsmith, is a thorough and authoritative work by an author, himself disabled, who was one of the pioneers of the disability movement. For a discussion of the medical and social models of disability see pp 149–151. For a discussion of architectural disablement see pp 151–156.

• DISABILITY RIGHTS AND WRONGS by the disabled writer Tom Shakespeare rejects what Dr Shakespeare calls the 'dangerous polarisations' of the medical and social models. He argues instead for a reasonable approach based on common sense and goodwill. Refer in particular to Chapters 2 and 3.

See also: *Commission for Equality and Human Rights, Deformation professionelle;* and *Inclusive design.*

Disability Action	1 This is a near-equivalent in Northern Ireland of the Royal Association for Disability and Rehabilitation, also known as RADAR (qv). 2 For more information visit www.disabilityaction.org/
Disability Allowance	See: *Disability Living Allowance.*
Disability and Carer's Service (DCS)	1 The DCS in the Department for Work and Pensions (qv) helps *disabled people* (qv) and *carers* (qv) lead more *independent* (qv) lives by administering *benefits* (qv); providing advice, and supporting the Government's policy of increasing the *inclusion* (qv) of disabled people and carers in society. 2 For more information visit www.dwp.gov.uk/lifeevent/benefits/dcs/
Disability and impairment	See: *Impairment and disability.*
Disability Awareness In Action (DAA)	1 Disability Awareness in Action is an international human rights network, ru for and by disabled people. While its activities are universal it has a particular interest in, and focus upon, the needs of people in developing countries. 2 For more information visit: www.daa.org.uk/
Disability discrimination	1 *Discrimination* (qv) on the basis of *disability* (qv) is outlawed in the UK by anti-discrimination legislation – for an overview see: *Discrimination.* 2 For details refer to DISABILITY DISCRIMINATION: LAW AND PRACTICE (see the bibliography), and see *Office for Disability Issues* (qv) and the entry below.
Disability Discrimination Act	1 The Disability Discrimination Act was passed in 1995. Since then the original Act has been very substantially amended by: • The Special Educational Needs and Disability Act 2001 (qv), better known a SENDA. • The Disability Discrimination Act 1995 (Amendment) Regulations 2003. • The Disability Discrimination Act 2005. For complete texts of this legislation visitwww.opsi.gov.uk/acts.htm. 2 The following entries deal separately with the Disability Discrimination Act 1995 and the Disability Discrimination Act 2005, but the two will in due cour

become known simply as 'the Disability Discrimination Act'.
3 Throughout this text the italicised terms have particular meanings which are explained here or elsewhere in the present work.
4 The present work only gives simplified summaries of original documents and does not purport to provide full, authoritative and up-to-date statements of the law. For latest information refer to DISABILITY DISCRIMINATION: LAW AND PRACTICE (see the bibliography) and the following websites:
• Directgov: disabled people www.disability.gov.uk/
• Equality & Human Rights Commission www.equalityhumanrights.com/

sability scrimination Act 95

1 Stated shortly: the Disability Discrimination Act places a *civil duty* (qv) upon employers, service providers, and landlords not to treat *disabled people* (qv) less favourably than others, without *justification* (qv). The Act applies to employment and to most aspects of public life, but not inside households or inside small private clubs, which are held to be in the private realm.
2 The following entries give a broad-brush summary of the contents and provisions of the Act in the following sequence:

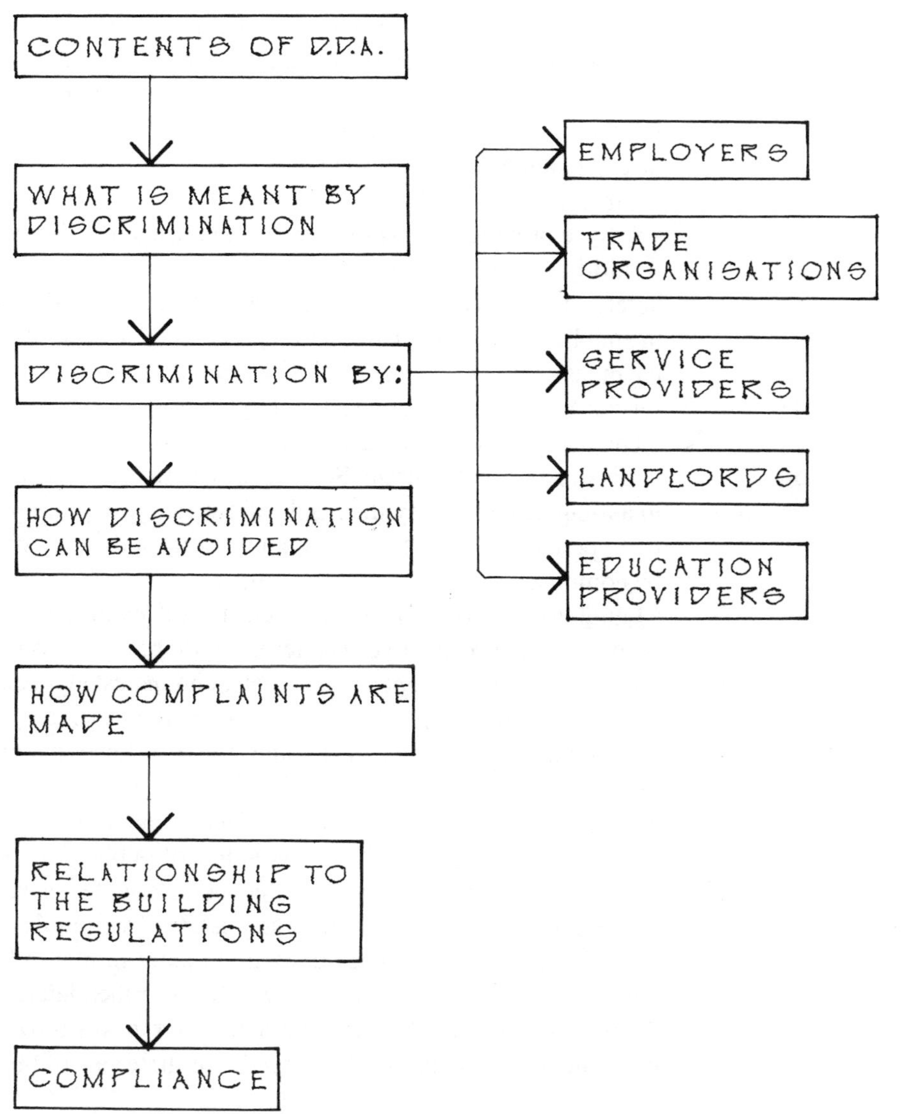

sability scrimination Act 95: contents

Leaving aside for the moment Parts 6 to 8, which give information of a supplemental nature, the major Parts of the Act are as follows:
1 Part 1, supplemented by further detail in Schedule 1, defines 'disabled

persons', who are stated to be people with a physical or mental impairment that has a substantial and long-term adverse effect on their ability to carry out normal day-to-day activities. This is a very wide-ranging formulation – for details see: *Disability: definition in the Disability Discrimination Act* (qv); an *Disabled people: numbers* (qv).

2 Part 2 creates a *civil duty* (qv) for:

• Employers (qv) not to discriminate against *disabled* employees or applicants for employment. The duty of non-discrimination is owed only to known disabled employees or job applicants, not to disabled people at large. This simplifies the employer's task, because the needs of these particular persons ca be readily established by a *workplace assessment* (qv) carried out by an *occupational therapist* (qv). The kinds of measures that may be needed to avoid or remove discrimination could for instance include some repositioning furniture; the provision of *aids and equipment* (qv) such as *textphones* (qv) or *inductionloops* (qv), or the installation of a *platform lift* (qv). For details see: *Adjustments under the Disability Discrimination Act* (qv).

• Trade organisations not to discriminate against *disabled* members or applicants for membership. The term trade organisations includes any trade o professional association such as a trade union, the RIBA, the BMA, the CBI, etc.; and as with employers the duty of non-discrimination is owed only to known members or applicants, not to disabled people in the population at larg For the kinds of measures that may be needed to avoid or remove discrimination see the notes above.

3 Part 3 creates a *civil duty* for:

• Service providers (qv) not to discriminate against *disabled* customers or would-be customers. The term 'service providers' includes not only shops an hotels, etc., but virtually every organisation that provides any kind of service t the public, regardless of size and regardless of whether the service is offered f payment or free of charge. Unlike employers and trade organisations above, service providers do not owe the duty of non-discrimination to particular identified persons; instead, it is an anticipatory and evolving duty that is owe to disabled people at large. Service providers therefore have a legal obligatio to anticipate and satisfy the needs of a varied group of unknown people, addi up to perhaps a fifth of the total population – a difficult problem that is best m by commissioning an *access audit* (qv).

• Landlords not to discriminate against *disabled* property buyers or tenants, both actual or prospective. The term 'landlord' includes property owners, housing providers, property managers, and the estate agents who arrange for selling or letting properties. As of 2007 there is an exemption for landlords letting out rooms in their own homes to six or fewer people, but the situation may change.

4 Part 4, inserted into the Disability Discrimination Act in 2001 by SENDA (qv), creates a *civil duty* for the providers of publicly funded education (privately funded education providers being counted as *service providers* und Part 3 above) not to discriminate against *disabled* pupils in relation to the provision of education. The duties are broadly similar to those under Part 3, with a few specific exceptions as will be described later.

5 Part 5 of the Disability Discrimination Act makes provision for anti-discrimination regulations to be made for three sets of public transport vehicl (qv):

• Taxis.

• Buses and coaches (termed *public service vehicles*).

• Rail vehicles.

Private vehicles are not covered by the Act, nor (as of 2007) are buses and coaches carrying fewer than nine passengers.

The scope and reach of Part 5 is in 2006 quite limited, but will be constantly developing over the next few years, and designers and managers must keep up with the latest information – see: *Transport: complying with the Disability Discrimination Act.*

6 As of 2008, useful websites include the following. They are listed in alphabetical order.

- Directgov: disabled people www.disability.gov.uk/
- Equality & Human Rights Commission www.equalityhumanrights.com/
- Office of Public Sector Information www.opsi.gov.uk/acts.htm

See also: *Adjustments under the Disability Discrimination Act.*

---

sability scrimination Act: 95: what is meant discrimination

1 Discrimination means (a) placing a *disabled person* (qv) at a substantial disadvantage in comparison with others (b) for a reason connected with his or her disability (c) without *justification* (qv). All three of these components should normally be present for a complaint of *discrimination* under the Act to succeed.

2 If for instance a *wheelchair user* (qv) is appointed to a job, finds that the entrance to the premises is obstructed by a doorstep that makes it difficult for him to enter, and the employer fails to remove or otherwise deal with the step, a complaint of discrimination will probably succeed because (a) the person is clearly being placed at a substantial disadvantage in comparison with non-disabled employees, and (b) this disadvantageous treatment is connected with his disability.

3 However, his complaint may fail if the employer can *justify* his refusal (a) on grounds of practicality (if for instance the matter could not realistically be dealt with within the period when the new employee is required); (b) on grounds of cost (if for instance the cost of the works could not be afforded by him); (c) on grounds of health and safety (if for instance giving equal treatment to the disabled person would endanger others); and so on.

4 For more detail on the above matters see: *Justification under the Disability Discrimination Act* (qv).

See also: *Discrimination.*

---

sability scrimination Act 95 – discrimination employers

1 Discrimination may occur if a *disabled person* (qv) is treated less favourably than others by (a) the manner in which a job is advertised; (b) the manner in which the job interview is conducted; (c) the terms on which the job is offered; (d) the conditions of work once the person is employed; or (e) the manner of a subsequent dismissal.

2 In all these contexts the purpose of the Act is that *disability* should not bar a person from employment unless (a) the disability would significantly impede that person from doing the work in question; and, importantly, (b) there is nothing the employer can reasonably be expected to do to overcome this. It is stated that the Act does not seek to stop employers from hiring the best person for the job, only to ensure that disabled applicants are given a level playing field on which to demonstrate their suitability for the job.

3 In October 2004 the provisions requiring disabled people not to be treated less favourably, and requiring the making of reasonable *adjustments*, were strengthened. As a consequence there are now three types of employment discrimination:

- Direct discrimination.
- Disability-related discrimination.
- Failure to make reasonable adjustments.

Under the strengthened provisions harassment and victimisation in employment

are also now specified.
4 The Disability Discrimination Act 2005 (qv) added new duties to those above, specifically for *public authorities* (qv). For more detail see these two entries.
5 As of 2008, useful references include the following. For publication details see the bibliography.
• CODE OF PRACTICE: EMPLOYMENT AND OCCUPATION. This *code of practice* (qv) explains how *disabled people* (qv) are protected from *discrimination* (qv) they are in employment or seeking employment. It also gives guidance on the law, which will help lawyers when representing disabled clients and assist courts and *tribunals* (qv).
6 Useful websites include:

• Department for Work and Pensions	www.dwp.gov.uk/employers/dd
• Equality & Human Rights Commission	www.equalityhumanrights.com
• Equality Direct	www.equalitydirect.org.uk/
• Employers' Forum on Disability	www.employers-forum.co.uk/
• Office for Disability Issues	www.officefordisability.gov.uk/

See also: *Disability Equality Duty*; and *Public Authorities*.

Disability Discrimination Act 1995 – discrimination by trade organisations

1 Discrimination may occur if a *disabled* (qv) applicant is refused membership or a disabled person who has been accepted for membership is treated less favourably than others, for reasons connected with their disability. If the treatment that is being complained of was for a reason that would also have applied to non-disabled people (for instance, lack of relevant qualifications or bad behaviour) then it would not be discriminatory under the Act.
2 As of 2008, useful references include the following. For publication details see the bibliography.
• CODE OF PRACTICE – TRADE ORGANISATIONS AND QUALIFICATIONS BODIES. This *code of practice* (qv) gives guidance on how *disabled people* (qv) are protected from *discrimination* (qv) by trade organisations and qualifications bodies.
3 Useful websites include:

• Commission for Equality and Human Rights	www.cehr.org.uk/
• Directgov: disabled people	www.disability.gov.uk/
• Equality & Human Rights Commission	www.equalityhumanrights.com

Disability Discrimination Act 1995 – discrimination by service providers

1 Discrimination may occur if a *disabled person* (qv) is refused service, is offered a worse service than others, or is served on worse terms than others, for reasons connected with their disability.

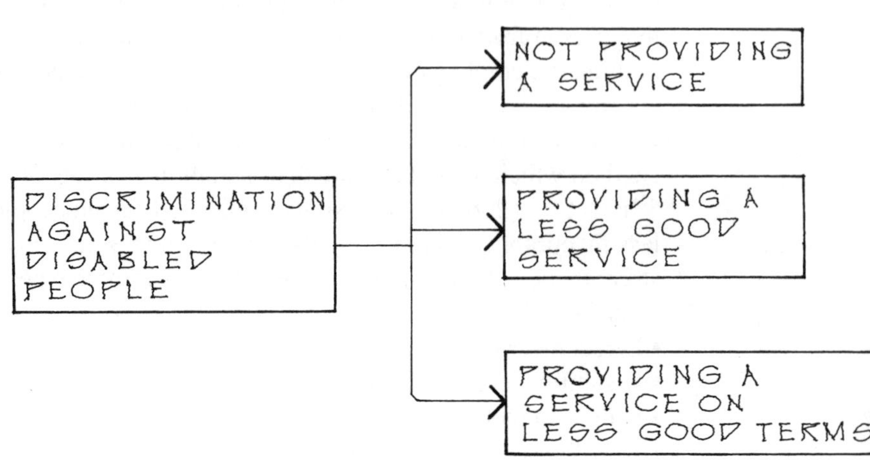

If the exclusion was for a reason that would also have applied to non-disabled people then it would not be discriminatory. As an example:
• If a group of people with cerebral palsy are refused entry to a fun-fair because their movements are conspicuously awkward and jerky, a complaint of discrimination will probably succeed because (a) they have received less favourable treatment than other people arriving at the gate, and (b) the less favourable treatment is connected with their disability. If on the other hand they are refused entry because of drunken or abusive behaviour, a complaint of discrimination will probably fail because their exclusion was for reasons that would also apply to others.
• A complaint will probably also fail if the above people were refused a ride on the carousel and the operator can *justify* (qv) his decision on grounds of health and safety – namely, that he had reason to believe that they would endanger the safety of themselves or others.
For more detail on the above matters see: *Service providers: complying with the Disability Discrimination Act* (qv).
2 Privately funded education providers are a special category of 'service provider' for whom the rules are somewhat different – eg, less favourable treatment could be additionally *justified* (qv) by a need to maintain prescribed academic standards. See: *Disability Discrimination Act 1995 – discrimination by education providers* (qv).
3 As of 2008, useful references include the following. They are given in alphabetical order. For publication details see the bibliography.
• CODE OF PRACTICE – RIGHTS OF ACCESS: SERVICES TO THE PUBLIC, PUBLIC AUTHORITY FUNCTIONS, PRIVATE CLUBS AND PREMISES. This *code of practice* (qv) covers the duty of *service providers* (qv) not to *discriminate* (qv) against *disabled* (qv) customers.
4 Useful websites include:
• Commission for Equality and Human Rights www.cehr.org.uk/
• Directgov: disabled people www.disability.gov.uk/
• Equality & Human Rights Commission www.equalityhumanrights.com/

See also: *Service providers: complying with the Disability Discrimination Act.*

---

sability scrimination Act 95 – discrimination landlords

1 Discrimination may occur if a *disabled person* (qv) is refused a sale or lease, or is disadvantaged by the manner in which a property is sold, let, or managed, for reasons connected with their disability. If the exclusion was for a reason that would also have applied to non-disabled people (for instance, evidence that they would be unable to pay) then it would not be discriminatory under the Disability Discrimination Act.
2 The present work does not deal further with these provisions. For more information refer to the following documents, publication details of which are given in the bibliography:
• CODE OF PRACTICE – RIGHTS OF ACCESS: SERVICES TO THE PUBLIC, PUBLIC AUTHORITY FUNCTIONS, PRIVATE CLUBS AND PREMISES.
• HOUSING AND THE DISABILITY EQUALITY DUTY.
3 As of 2008, useful websites include:
• Directgov: disabled people www.disability.gov.uk/
• Equality & Human Rights Commission www.equalityhumanrights.com/

---

sability scrimination Act 95 – discrimination education providers

1 Discrimination may occur if a *disabled* (qv) student is treated less favourably than others (a) in the admission or enrolment of students; (b) in the terms on which the admission is made; (c) by refusing to accept the application; (d) in the provision of services including educational courses, training, recreation, leisure and catering, and accommodation; or (e) in excluding a student. If the

action in question was for a reason that would also have applied to non-disabled people (for instance, a college student being asked to leave the hostel because of the number of noisy parties he has been giving) then it would not b discriminatory under the Act.

2 A key difference between educational institutions and other service provider is that educational institutions may offer an additional *justification* for providing less favourable treatment to disabled persons, namely the necessity maintaining academic or other prescribed standards.

3 The application of the Disability Discrimination Act to schools is amended b the provisions of the Special Educational Needs and Disability Act 2001 (qv), better known as SENDA. The result of such continual addition of new layers o legislation upon old is that laws which were intended to benefit children have become, instead, a veritable gold-mine for fee-earning specialist lawyers and consultants while children, parents, and teachers look in from the outside – se *Consultants' charter*; *Over-legislation*; and *Post-democracy.*

4 As of 2008, useful references include the following. They are given in alphabetical order. For publication details see the bibliography.

• CODE OF PRACTICE FOR SCHOOLS – DDA 1995: PART 4. Covers the duty of education providers not to *discriminate* (qv) against *disabled* (qv) pupils and prospective pupils in schools.

• CODE OF PRACTICE (REVISED) FOR PROVIDERS OF POST-16 EDUCATION AND RELATED SERVICES. Covers the duty of education providers not to discriminate against post-16 disabled pupils.

• CODE OF PRACTICE – RIGHTS OF ACCESS: SERVICES TO THE PUBLIC, PUBLIC AUTHORITY FUNCTIONS, PRIVATE CLUBS AND PREMISES. Covers the duty of service providers (qv) not to discriminate against disabled customers, which may include students.

• UNDERSTANDING THE DISABILITY DISCRIMINATION ACT: A GUIDE FOR COLLEGES, UNIVERSITIES, AND ADULT COMMUNITY LEARNING PROVIDERS IN GREAT BRITAIN. This guidance helps post-school education institutions improve disability equality and implement the latest provisions of the DDA.

5 Useful websites include:

• Directgov: disabled people www.disability.gov.uk/
• Equality & Human Rights Commission www.equalityhumanrights.com

See also: *Special educational needs.*

Disability Discrimination Act 1995: how discrimination can be avoided or put right

1 The diagram shows three ways in which discrimination can be avoided or pu right.

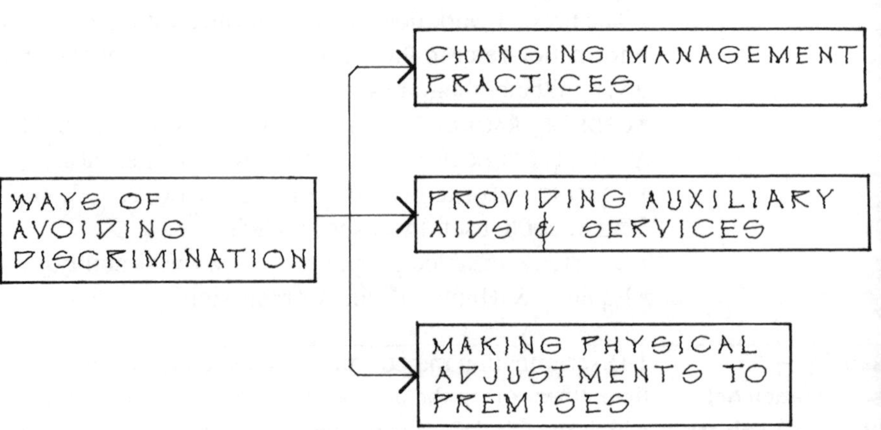

2 For a more detailed explanation see: *Adjustments under the Disability Discrimination Act*.

sability scrimination Act 95: how mplaints are made d dealt with

1 As of early 2008:
• In cases of alleged discrimination by an employer or trade organisation, complainants may sue for financial loss or for injury to feelings via an *employment tribunal* (qv).
• In cases of alleged discrimination by a service provider or landlord the complainant may sue via a county court.
2 Whatever its advantages, the above lawsuit-based approach is damaging, because lawyers have little innate interest in *conciliation* (qv) and finding common ground: their professional inclination is to fight as hard as possible for their client. Because of the adversial nature of the system it is quite likely that problems will be inflamed rather than soothed. To help avoid undue conflict there is an officially recognised independent conciliation service – see: *Mediation and the Disability Discrimination Act.*
3 Useful references include the following. They are given in alphabetical order. For publication details see the bibliography.
• CLAIMS ABOUT PHYSICAL BARRIERS TO ACCESS: A GUIDE TO THE DISABILITY DISCRIMINATION ACT.
4 Useful websites include:
• Directgov: disabled people www.disability.gov.uk/
• Equality & Human Rights Commission www.equalityhumanrights.com/

See also: *Tribunals.*

sability scrimination Act 95: relationship the Building gulations

1 The fact that a building, or part of a building, has been constructed in conformity with Part M of the Building Regulations (qv), which covers access to and use of buildings, does not in itself render the building owner immune to action under the Disability Discrimination Act. This is because Part M applies only to the design of certain physical elements of the building (eg stairs, ramps, doors, lobbies, lifts and lavatories), whereas the DDA applies to virtually every aspect of the design and management of the *public realm* (qv).
2 A degree of immunity exists in certain circumstances: if a physical feature of a building (for instance an entrance door, a corridor, a stair, or a lavatory) accorded with the provisions of APPROVED DOCUMENT M of the Building Regulations when constructed, then no *service provider* (qv) using that building can be required to make an adjustment to that feature within ten years of its construction or installation. For more on this refer to appendix B of CODE OF PRACTICE – RIGHTS OF ACCESS: SERVICES TO THE PUBLIC, PUBLIC AUTHORITY FUNCTIONS, PRIVATE CLUBS AND PREMISES – see below.
3 But the building containing that feature may still be judged discriminatory in a wider context. For example:
• A GP's surgery has an obstructive flight of steps at the front entrance, and the management decides to offer *disabled people* (qv) use of a side entrance, the design of which complies fully with APPROVED DOCUMENT M (qv) of the Building Regulations. If the side entrance is in an unpleasant alley where the rubbish bins are kept, this solution may not be acceptable under the DDA because it would offer disabled persons a substantially worse service than other people, who enter via a dignified main entrance.
4 In sum: while compliance with Part M of the Building Regulations will go some way towards making a building design safe against action under the DDA, it won't necessarily go all the way. For advice see the next entry.
5 The above example is suggestive only. For authoritative guidance refer to CODE OF PRACTICE – RIGHTS OF ACCESS: SERVICES TO THE PUBLIC, PUBLIC AUTHORITY FUNCTIONS, PRIVATE CLUBS AND PREMISES. Throughout this entry readers are reminded that the present work only gives simplified summaries of

original documents and does not purport to provide full, authoritative, and up-to-date statements of the law.
6 As of 2008, useful references include the following. For publication details see the bibliography.
• CODE OF PRACTICE – RIGHTS OF ACCESS: SERVICES TO THE PUBLIC, PUBLIC AUTHORITY FUNCTIONS, PRIVATE CLUBS AND PREMISES. This *code of practice* (qv) is a highly authoritative guide to the thinking of the Disability Rights Commission (qv) on these matters. Refer especially to Appendix B.
7 This text was written in early 2008, and useful websites for checking the lat situation include the following. They are listed in alphabetical order:
• Directgov: disabled people www.disability.gov.uk/
• Equality & Human Rights Commission www.equalityhumanrights.com/

Disability Discrimination Act 1995: compliance

1 The Disability Discrimination Act (qv) provides no procedure which, if followed, would categorically place the designers, owners, or managers of public premises on the right side of the law. As regards building design, pp 30 310 of CODE OF PRACTICE – RIGHTS OF ACCESS: SERVICES TO THE PUBLIC, PUBLIC AUTHORITY FUNCTIONS, PRIVATE CLUBS AND PREMISES (see the bibliography) refer vaguely to 'the wealth of published advice on the principl and practice of inclusive design' without identifying the documents concerne and other official sources advise designers to consult 'authoritative practical guides', again without identification.
2 Bearing in mind the severe penalties that may result from a breach of the A this situation completely contradicts the Government's own criteria for good regulation. The Better Regulation Task Force (qv), created in 1997 by the Cabinet Office (qv) to advise the authorities on good regulation making, identified two fundamental characteristics of good regulation:
• That those being regulated should be given a clear method of demonstrating compliance; and
• That the process of complying should not generate an unnecessary burden c paperwork.
The Disability Discrimination Act fails both of these tests.
2 This unsatisfactory situation could be remedied by two steps, both of which could readily be taken if the will existed:
• Publication of a clearly illustrated new Code of Practice (or a set of Codes different building types), very like the excellent APPROVED DOCUMENT M (qv) of the Building Regulations, but expanded to take in additional material from BS 8300 (qv), INCLUSIVE MOBILITY (qv), and one or two other sources.
• A change in the law to the effect that for most buildings, with certain specified exceptions, conformity with the new Code of Practice will guarant compliance with the Disability Discrimination Act.
Except for certain large, complex, or otherwise exceptional cases, planning applications would under this proposal be accompanied by:
• A simple certification that the design conforms in all respects with the new Code of Practice, in which case compliance with the Disability Discriminatic Act would be automatic.
or
• A Design and Access Statement (qv) stating that the design does not confor giving the reasons why, and justifying the alternative approach.
As such a new Code of Practice would consist mostly of existing material brought together in one place, compilation would be a straightforward task a is overdue.

See also: *Better regulation bodies; Consultants' charter; Design and access statement*; and *Over-regulation.*

sability scrimination Act 005

1 The DDA 2005 added new rights for *disabled people* to those contained in the DDA 1995. They include the following:

• First, the definition of *disability* given in the Disability Discrimination Act 1995 (qv) was widened in several respects, thus giving around a quarter of a million additional people protection under the DDA. All relevant entries in the present work have been updated accordingly – see for instance: *Cancer, HIV,* and *Multiple sclerosis.*

• Second, the application of the Disability Discrimination Act to public transport was extended to include the management of transport services. As an example: if a bus driver parks an accessible *bus* (qv) so far from the kerb that a *wheelchair user* finds it unreasonably difficult to enter, the latter now has redress under the DDA.

• Third, public authorities were given a new duty, known as the Disability Equality Duty or DED (qv), that requires them to actively promote disability equality in the way they carry out their business. See the next entry.

• Fourth, the letting of premises was brought more tightly within the ambit of the Disability Discrimination Act.

• Fifth, private clubs with 25 or more members were effectively given the same duties of non-discrimination to their members as those owed by *service providers* to their clients. And as with service providers, clubs owe the duty not to particular identified *disabled persons* (as is the case with *employers*) but to disabled people at large. For more detail on these matters see: *Service providers* (qv).

2 The present work only gives simplified summaries of original documents and does not purport to provide full, authoritative and up-to-date statements of the law. For latest information refer to DISABILITY DISCRIMINATION: LAW AND PRACTICE (see the bibliography) and the following websites:

• Directgov: disabled people www.disability.gov.uk/
• Equality & Human Rights Commission www.equalityhumanrights.com/

sability Equality Duty ED)

1 The Disability Discrimination Act 1995 (qv) was amended by the Disability Discrimination Act 2005 (qv) to place a duty on all *public sector authorities* (qv) to actively promote disability equality in all aspects of their work. It is known as the Disability Equality Duty, or more briefly as the 'duty to promote', which starts from the premise that *equality* (qv) of opportunity cannot be achieved simply by treating *disabled* (qv) and non-disabled people alike. It requires the entire public sector including the police, health services, schools, local authorities, NHS trusts, etc., to actively promote the equalisation of opportunities for *disabled people* by positive measures.

2 The 'duty to promote' is achieved by production of a Disability Equality Scheme stating precisely how the body concerned will go about eliminating unlawful discrimination and promoting equal opportunities. In producing the DES the organisation must (a) involve disabled people; (b) identify how it will gather and analyse evidence; (c) set out how it will assess the impact of its proposed measures; (d) produce an Action Plan; and (e) report annually on progress and conduct regular reviews of results achieved – all of this in great detail.

3 The process adds up to a staggering amount of bureaucracy and it must be asked whether the Better Regulation Commission (qv) reviewed the proposed measures for *proportionality* (qv) and the light regulatory touch to which the Government is formally committed, before imposing them upon already overstressed public authorities. The consultation paper A FRAMEWORK FOR FAIRNESS: PROPOSALS FOR A SINGLE EQUALITIES BILL FOR GREAT BRITAIN (qv)

has hinted that these unreasonable burdens might be greatly simplified. One must hope that this will happen. See in this connection: *Discrimination Law Review.*

4 As of 2008, useful references include the following. They are given in alphabetical order. For publication details see the bibliography.

- CODE OF PRACTICE. THE DUTY TO PROMOTE DISABILITY EQUALITY: STATUTORY CODE OF PRACTICE, ENGLAND AND WALES.
- MAKING THE DUTY WORK: A GUIDE TO THE DISABILITY EQUALITY DUTY FOR DISABLED PEOPLE AND THEIR ORGANISATIONS.
- PLANNING, BUILDINGS, STREETS AND DISABILITY EQUALITY – see in particular section 4.

5 Readers are reminded that the present work only gives simplified summaries of original documents and does not purport to provide full, authoritative and up-to-date statements of the law. This text was written in 2007, and useful websites for checking the latest situation include:

- Directgov: disabled people — www.disability.gov.uk/
- Employers' Forum on Disability — www.employers-forum.co.uk/
- Equality & Human Rights Commission — www.equalityhumanrights.com/
- Office for Disability Issues — www.officedisability.gov.uk/
- Office of Public Sector Information — www.opsi.gov.uk/acts.htm

See also: *Better regulation bodies; Over-regulation; Planning and the Disability Equality Duty*; and *Public authorities*.

Disability Equality Partnership

Visit www.heacademy.ac.uk/ourwork/learning/disability/dep

Disability Information Trust

Visit www.abilityonline.org.uk/ and see: *Disabled Living Foundation*; and *Ricability.*

Disability Law Service (DLS)

1 The DLS is a national registered charity that provides confidential and free legal advice for *disabled* adults, their families, and *carers.* It also offers disabled people a casework service.

2 For more information visitwww.abilityonline.org.uk/

See also: *Community Legal Service Partnerships*; and *Mediation.*

Disability Living Allowance (DLA)

1 An allowance for people who have needed help for at least 3 months because of a severe physical or mental illness or *disability*, and are likely to need it for another 6 months.

2 For more information visitwww.disability.gov.uk/

See also: *Social protection.*

Disability Now

1 A leading monthly periodical for and about *disabled people* (qv), and an essential information source for everyone wishing to keep up with events in this field. In addition to reporting and commenting on news and events, and featuring new products, it runs campaigns aimed at removing discrimination against disabled people in particular contexts – for instance, the misuse of car parking spaces designated for disabled people.

2 DISABILITY NOW is a wholly admirable publication, but being owned by the *pressure group* (qv) Scope it occasionally falls prey to the ideological arrogance that tends to warp the good sense of many pressure groups. As an example: when Lord Joffe's Assisted Dying for the Terminally Ill bill was re-introduced to Parliament in November 2005, the magazine polled the views of the disabled people it claimed to represent. The survey found that 93% of respondents

favoured the bill and 7% were against (see the issue of December 2005, p 17) but DISABILITY NOW nevertheless continued its vehement assaults upon the bill. See in this connection also: *Pressure groups;* and *Self-deliverance.*
3 The magazine has a well-produced and informative website – visit www.disabilitynow.org.uk/

See also: *Access by Design.*

---

sability Rights mmission (DRC) 1 This *quango* (qv) was responsible for working to eliminate *discrimination* (qv) against *disabled people* (qv) under the Disability Discrimination Act (qv) until October 2008, when its responsibilities were taken over by the Commission for Equality and Human Rights (qv).
2 The DRC's website has, as of October 2007, been incorporated into www.equalityhumanrights.com/

---

sability Sport ents See: *Recreation and sports activities for disabled people.*

---

sability Standard 07 1 A management tool created by the Employers' Forum on Disability (qv) to help businesses and public sector organisations comply with the Disability Discrimination Act (qv). It includes practice solutions, workshops, and other forms of assistance.
2 For more information visitwww.employers-forum.co.uk/

---

sability surveys 1 Of the numerous surveys carried out from time to time to establish the incidence of *disability* (qv) in the British population, the most important include the following:
• In 1985–1988 the Office of Population Censuses and Surveys, now absorbed into the Office of National Statistics (qv), undertook, on behalf of the then Department of Health and Social Security, an ambitious social survey on disability. The results were published in four volumes in 1988 and 1989 as the OPCS SURVEYS OF DISABILITY IN GREAT BRITAIN – see the bibliography. These findings formed the basis of benefit policies and policy changes from 1990 onward. For more detail visitwww.statistics.gov.uk/
• In 1996–97 the Centre for Population Studies, together with the London School of Hygiene and Tropical Medicine, carried out a new survey on behalf of the then Department of Social Security to update the earlier findings. Unlike the previous surveys, children were not included. The results were published in 1999 under the title DISABILITY IN GREAT BRITAIN: RESULTS FROM THE 1996/97 DISABILITY FOLLOW-UP TO THE FAMILY RESOURCES SURVEY. For current details see the bibliography, where it is listed under the short title DISABILITY IN GREAT BRITAIN. For more detail visitwww.dwp.gov.uk/
2 As of 2008, useful websites for more recent data include:
• Family Resources Survey
www.dwp.gov.uk/asd/frs/index/publications.asp
• Labour Force Survey
www.statistics.gov.uk/STATBASE/Source.asp?vlnk=358
• Social Trends
www.statistics.gov.uk/statbase/Product.asp?vlnk=5748&More=N

---

sability Unit See: *Department for Work and Pensions.*

---

sabled access See: *Access.*

---

sabled building ers See: *Disabled persons: numbers.*

Disabled facilities grant (DFG)	See: *Home improvement grants*; and visit www.communities.gov.uk/publications/housing/disabledfacilitiesgrant.
Disabled Living Centres (DLCs)	Visit www.direct.gov.uk/
Disabled Living Foundation (DLF)	1 The UK's most comprehensive source of information on *aids and equipment* (qv) for *disabled people* (qv). 2 For more information visit www.dlf.org.uk/ See also: *Disability Information Trust*; and *Ricability*.
Disabled parking	See: *Blue badge schemes*; and *Car parking*.
Disabled passengers	See the following entries: • Accessible cars. • Transport: complying with the Disability Discrimination Act. • Wheelchair users and public transport.
Disabled people	1 This term is avoided by some writers because it tends to imply that there are two distinct classes of citizens – those who are *disabled* (qv) and those who are not. They tend therefore to prefer softer phrases such as 'people with disabilities'. 2 Some supporters of the 'social model of disability' (see: *Disability: other definitions*) hold this usage to be inaccurate. In their view a paralysed leg (for instance) is an *impairment* (qv), and any disablement that might follow is inflicted upon that person by a discriminatory society, not by the leg. The paralysed leg can therefore not be called a disability. 3 These theoretical distinctions are of little consequence in everyday speech and here, as in most other aspects of life, it is best to use ones common sense and suit the language to the occasion. Refer in this connection to the discussion on pp10–53 of DISABILITY RIGHTS AND WRONGS by the disabled writer Tom Shakespeare – see the bibliography. See also: *Disability; Disability Discrimination Act*; and *Political correctness: speech codes*.
Disabled people and public transport	See: *Transport: public*; and *Wheelchair-users and public transport*.
Disabled people and risk assessments	See: *Risk assessments and disabled people*.
Disabled people in society	1 The principle that decisions on the needs of *disabled people* (qv) should be taken by disabled people and not by well-meaning non-disabled *carers* (qv), volunteers, professionals, and researchers, etc., has been at the heart of the disability rights movement for at least the past decade or two. It is based on the view that only disabled people are capable of knowing the needs and desires of themselves and of other disabled people. 2 This concept has had many sound and highly beneficial results – see for instance: *Independent living*. But it has also been taken to questionable lengths by some theorists and *pressure groups* (qv) who hold the state of being disabled to be so all-important that:

• Society may be divided into two mutually exclusive groups – disabled and non-disabled.
• Disabled people should be seen almost in ethnic terms, as a distinct group with their own culture, language, and ways of being.
3 Against this, the disabled author Tom Shakespeare argues that a complete writing-out of non-disabled people would be wrong and damaging to disabled people themselves. He gives three reasons:
• Unlike the division between, for instance, men and women, which is lifelong and mostly immutable, many disabled people have been so for only part of their lives, and/or may well cease to be disabled in the future.
• Disability is only part of a disabled person's identity. In some cases it may be overwhelmingly important, but in many cases it is not, with gender, ethnicity, or religion being just as, or quite possibly more important.
• Non-disabled people play a desirable and usually essential role in the lives of most disabled people. Indeed, many of the latter actively resist being corralled into exclusively disabled groups, and try to live as far as possible with non-disabled people.
In short, it is misleading to define disabled people by their disability alone; it is impossible for disabled people to be completely self-sufficient ('autarkic'); and most disabled people do not wish to live exclusively with other disabled people, preferring if possible to integrate with mainstream society.
4 These matters are explored at greater length on pp 185–97 of DISABILITY RIGHTS AND WRONGS by Tom Shakespeare, a thoughtful work of which Mr Bert Massie, chairman of the Disability Rights Commission, has said: 'If you read only one book on disability rights this year, make this the book'. For publication details see the bibliography.

See also: Inclusion; *Independent living*; and *Pressure groups*.

---

sabled people: mbers

1 Two distinct questions can be asked:
• What proportion of the general population are disabled ? This is something that – for instance – those who are responsible for health and/or social policies need to know.
• What proportion of the USERS OF PUBLIC BUILDINGS are disabled ? The answer to this question may differ somewhat from that above, and may be more useful to those who plan and manage the built environment.
2 The next two entries examine these questions in turn.

---

sabled persons as a oportion of the pulation

1 In 1996–97 a survey for the Department of Social Security found that there were about 8.5 million *disabled* (qv) adults in Great Britain, i.e. 20% of the adult population – see: *Disability surveys*.
2 In 2004 the government reported that the number of adults in the UK who were disabled in terms of the definition used in the Disability Discrimination Act (qv) was now estimated at 9.8 million instead of the previous figure of 8.6 million – a rise attributed in part to the growing number of old and very old people. This statistic does not include people who are in long-term care, or children. It is derived from the 2002–03 Family Resources Survey (FRS), an annual survey commissioned by the Department of Work and Pensions (qv), and is based on the FRS finding that 22% of respondents had a long-standing limiting illness or disability.
3 In 2006 the Disability Rights Commission's web page titled 'What is disability and who is disabled' produced two figures which suggest that the definitions of disability that are now in use cannot be valid:
• It stated that 'around one in five people of working age are considered by the Government and the DRC to be disabled'. As people in this age group are on

average much healthier than those beyond, it follows that a much larger proportion of the adult population as a whole come within the definition of disability – perhaps a quarter or more. By all reasonable standards this statisti is absurd.
• It also reported that 'a recent survey showed that 52% of people who qualified as "disabled" under the DDA, and had rights not to be treated unfair because of their disability or health condition, did not consider themselves disabled'.
4 Taken together these figures suggest that the disability rights movement has led legislators to quite unreal definitions and overstated statistics of incidence and that it is time for a fundamental review.

Disabled persons as a proportion of building users

1 No single definitive estimate is possible because so much depends on the place, the time, the building type, and the kind of people using that building. For instance, the ratio of children/adults/old people will be very different in a sports venue than in a church; and in any particular venue it may also vary wi the day of the week, or the time of day.
2 Mr Selwyn Goldsmith discusses these matters in some detail on pp 159–176 DESIGNING FOR THE DISABLED: THE NEW PARADIGM – see the bibliography. Broadly speaking he suggests that fewer than 1% of visitors to places such as shopping centres are likely to be obviously disabled. In some of the particular venues he examined in the 1990s, around one in 200 visitors might use *walkin sticks* (qv) or *crutches* (qv); one in 400 might be in *wheelchairs* (qv); and one i 10 000 might be *blind* (qv). In addition, more than 3% of visitors might be able-bodied people pushing baby buggies or other forms of pushchair – a user group very liable to be *architecturally disabled* (qv) by stiff or heavy doors, awkward steps and stairs, and narrow corridors.

See also: *Inclusive design*.

Disabled Persons (Independent Living) Bill

See: *Independent living*.

Disabled Persons' Transport Advisory Committee

See: *DPTAC*.

Disabled toilets

See: *British Toilet Association*; and *Lavatories*.

Discrimination

1 Discrimination means treating someone less favourably than others without *justification* (qv).
2 As of July 2007, the following UK laws and regulations prohibit such 'less favourable treatment' of specified groups of people:

• Employment Equality (Age) Regulations 2006.	(age).
• Disability Discrimination Act 1995 and 2005.	(disabilit
• Race Relations Act 1976.	(race).
• Race Relations (Amendment) Act 2000.	(race).
• Employment Equality (Religion or Belief) Regulations 2003.	(religion).
• Racial and Religious Hatred Act 2006.	(religion).
• Equality Act 2006 (Part 2).	(religion)
• Equal Pay Act 1970.	(sex).
• Sex Discrimination Act 1975.	(sex).
• Employment Equality (Sexual Orientation) Regulations 2003.	(sex).
• Equality Act (Sexual Orientation) Regulations 2007.	(sex).

3 In October 2008 the Single Equality Act 2007 (qv) is expected to replace the existing 9 major and 100 ancillary pieces of anti-discrimination legislation, but the details are unknown at time of writing. For latest information visit www.cehr.org.uk/

4 It must be hoped that the most unacceptable features of current legislation will not continue or resurface in the new Act. For instance:

• The Disability Discrimination Act (qv) uses a definition of *disability* (qv) that simply cannot be right – see: *Disabled persons as a proportion of the population* above.

• Under the Employment Equality (Age) Regulations 2006 (qv), recruitment advertisements may no longer request a date of birth or use phrases such as 'energetic' or 'bright young graduate required', for fear that older applicants may be deterred. In October 2006 the global recruitment consultancy Robert Walters, which employs 1 300 staff in 14 countries and places thousands of recruitment advertisements a year, felt it prudent to issue internal guidance banning words such as 'quick-learner', 'self-starter', 'high-flyer', 'dynamic', 'experienced', 'ambitious', or 'recent graduate' for fear of breaking the new law. Though specialising in the recruitment of high-calibre candidates for demanding positions it even advised against any reference to 'modern qualifications' such as GCSEs or to university degrees in subjects such as information technology.

• The effect of a variety of laws (including the Crime and Disorder Act 1998; the Powers of Criminal Courts (Sentencing) Act 2000 as amended; the Anti-Terrorism, Crime and Security Act 2001; and the Criminal Justice Act 2003) is that an assault upon someone who is black or homosexual could result in twice the length of sentence that would have followed had the victim been white or heterosexual. As an example, in June 2006 the murderers of a homosexual man, Mr Jody Dobrowksi, were sentenced to 28 years imprisonment instead of the 14 years that would have resulted had the victim not been homosexual. Can it be right that the killing of one person be regarded as only half as serious as the killing of another? And what has become of the hard-won principle that all citizens are equal under the law?

5 Some of these matters may be clearer by the time this is published.

See also: *Age discrimination; Equality; Equality and Human Rights Commission; Sex discrimination*; and *Victimhood.*

---

scrimination Law **view** — See: *Government Equalities Office; Framework for Fairness*; and *Single Equality Act.*

---

scrimination Law **view green paper** — See: *Framework for fairness.*

---

scrimination: **sitive** — 1 FRAMEWORK FOR FAIRNESS: PROPOSALS FOR A SINGLE EQUALITIES BILL FOR GREAT BRITAIN (qv) draws the following distinction between 'positive action' and 'positive discrimination'.

• Positive action means offering targeted assistance to certain people so that they can take full and equal advantage of particular opportunities.

• Positive discrimination means explicitly treating people more favourably on the grounds of race, sex, religion or belief, etc., for example by appointing someone to a job just because they are male or just because they are female, irrespective of merit (p61).

2 As of 2008, positive discrimination is prohibited under British and European law. For instance, the European Court of Justice has held that it is permissible for equally-qualified women to be given preference for promotion where there

are fewer women than men in the relevant post, but unlawful to appoint a candidate from an under-represented group over a better-qualified candidate simply to redress the under-representation (p 71).

3 In March 2008 Ms Harriet Harman, the Minister for Women (qv), launched a consultation on proposals to let employers take race or sex into consideratio when choosing between two job applicants with the same qualifications, but n to give preference to a candidate with worse credentials. Business leaders we critical, and even the Fawcett Society (qv) expressed scepticism.

4 More radically, when launching FAIRNESS AND FREEDOM: THE FINAL REPORT OF THE EQUALITIES REVIEW (qv) in February 2007 Mr Trevor Phillips, since appointed chairman of the Commission for Equality and Human Rights (qv), issued an outright call for positive discrimination and the introduction of numerical quotas in order to end 'immovable, persistent disadvantage'.

Disproportionate — See: *Proportionateness.*

District councils — See: *Local authorities*; and *Regions.*

District Health Authorities — See: *NHS structure.*

Dispute resolution — See: *Employment law: dispute resolution.*

Diversity

1 All other things being equal (which they mostly are not), diverse societies ar generally more interesting, more stimulating, and more dynamic and inventive than monocultures. They are also, according to Robert D Putnam (Malkin Professor of Public Policy at Harvard University) in his 2006 Johan Skytte Pri Lecture titled E PLURIBUS UNUM: DIVERSITY AND COMMUNITY IN THE TWENTY-FIRST CENTURY, more stressed and less happy. For details of the published lecture see the bibliography.

2 Britain is currently in a state of some bafflement on this contentious matter. Here are some milestones to where we are now:

• Until some decades ago there was wide agreement that a large degree of commonality is required for societies to work. In his 1931 book EQUALITY, the economic historian and socialist R H Tawney wrote: 'What a community requires…is a common culture, because, without it, it is not a community at all In his 1942 REPORT ON SOCIAL INSURANCE AND ALLIED SERVICES (usually referred to as 'The Beveridge Report'), which helped to form the future welfar state, the economist W H Beveridge recognised that an effective system of social insurance would require 'a sense of national unity overriding the interes of any class or nation'.

• More recently assumptions of this kind have weakened among the opinion-forming classes, and the Government's attitude to diversity for at least the past ten years, and possibly more, can be summed up as 'the more the merrier'. Ordinary people were less sure about this, but their opinions were not invited c listened to.

• In February 2004 the British journal PROSPECT published an essay titled 'Too Diverse?' in which the magazine's editor, Mr David Goodhart, tentatively explored what he called the 'progressive dilemma' — the potential conflict between social cohesion and the many kinds of diversity, including ethnic diversity, that have flourished in recent decades'. The reaction was spectacular. Mr Trevor Phillips, then chairman of the Commission for Racial Equality (qv), and now chairman of the Equality and Human Rights Commission (qv), asked 'Is this the wit and wisdom of Enoch Powell? Jottings from the BNP leaders' weblog?', and his was not a lone voice. The piece was reprinted in the

Guardian on 24 February 2004 as DISCOMFORT OF STRANGERS and may be read at www.guardian.co.uk/race/story/0,11374,1154684,00.html

• Since then the uproar has subsided, and in October 2005 Mr Phillips said: 'We are used to the idea of one nation. That is why the prospect of a Britain fragmented by race and religion is so alien to us. It simply is not in our nature'. That partial reversion to earlier ideas has now become not merely acceptable, but is widely quoted by officialdom.

3 A fair summary might be that a certain amount of cultural diversity is a stimulus, but if the diversity is so great that significant proportions of the population feel a greater sense of loyalty to their own particular sub-group than to society at large, then that society is in danger. Most of the things that make societies work (obeying the law, playing fair at election times, dutifully paying taxes for the benefit of others, etc.) depend on the willingness of citizens to subordinate their personal interests – sometimes to a painful degree – to the interests of perhaps 60 million other people whom they will never meet, but nevertheless feel to be 'part of the family'. If that innate sense of mutual obligation fades, no amount of law-making and policing will maintain an orderly society.

4 How then are we to handle the unprecedented degree of diversity that now characterises Britain in the best interests of everyone? The next entry offers some thoughts on this important and difficult issue.

Diversity: alternative models

1 There are broadly three models of diversity in current Western societies:

• The French ('assimilationist') model, which in effect does not recognise ethnic diversity. The French Government refuses to treat people of different races differently, or even to know the racial composition of France. So rigidly is this policy applied that it is illegal to collect data about the race or religion of French citizens. This refusal even to recognise the existence of France's communities of largely workless, poverty-stricken immigrants from North Africa, let alone give special attention to their problems, has led to extremes of alienation for which France is beginning to pay dearly.

• At the other end of the spectrum, the British ('multiculturalist') model, which goes so far in recognising ethnic diversity that a 4-page government leaflet may well contain three pages of vital information, and the fourth entirely taken up by tick-boxes offering helplines in a dozen languages (to take an actual example: Bengali, French, Gujarati, Hindi, Polish, Portuguese, Punjabi, Somali, Spanish, Tamil, and Urdu). The system enables – in some ways actually encourages – newcomers to live outside the host society most of their lives, maintaining their non-British customs and speaking their own language perhaps for generations on end. In short: not so much a melting pot as a mosaic.

• The American model, which falls somewhere between the two. It insists that all Americans are American first, and owe a total loyalty to America. But provided that people from minority groups honour that fundamental principle they are warmly encouraged to maintain, take pride in, and celebrate their particular cultures.

In sum, whereas (to take a hypothetical example) a Somali immigrant to France is expected to become French, and in Britain is permitted to effectively remain Somali, he or she would in the USA be permitted – even encouraged – to become Somali-American – hence the term 'hyphenated Americans', with the emphasis on 'American'.

2 Comparisons between the three countries are risky because current America, unlike France and Britain, is very largely a nation of immigrants with no ancient host culture to complicate the problem of absorption. But looking at the record dispassionately it does seem that the French model has worked worst, the American model best, and that Britain fulfills its familiar role of muddling.

The best way forward appears to the present author to be a continuation of Britain's traditional untidy, tolerant, non- ideological pragmatism, but steering closer to the American than the French model.

3 Here are two thought-provoking books, the first touching directly, the second more indirectly, on the above matters. For publication details see the bibliography.

• THE TWO FACES OF LIBERALISM by John Gray. Professor Gray argues that there are two quite different liberal traditions. First, the 'Rawlsian' model, which was taken for granted in monocultural Britain, and works towards the ideal of everyone being liberal and not disturbing the peace with public expressions of illiberal views. Against that, the 'modus vivendi' model, which he believes Britain must now adopt. It accepts that society contains people who completely reject liberalism, and will not refrain from expressing their views in public. Instead of vainly trying to convert them, this model seeks to find ways of keeping the peace between fundamentally different factions living side-by-side in a single society.

• LETTERS TO LILY ON HOW THE WORLD WORKS by Alan Macfarlane. Professor Macfarlane is an anthropologist and historian at Cambridge University, and in this book, written in the form of 30 informal letters to his granddaughter, he ranges widely through history and across the world's cultures.

See also: *Equality; Equality and Human Rights Commission*; and *Human rights*.

Diversity is (or should be) a two-way street

1 The foregoing entries focus on how best to recognise and honour the rights of ethnic, religious, or other minorities within a majority culture. But in Britain we have a new source of friction – the perverse use by officialdom of the doctrine of 'diversity' to cease recognising, and in some cases to start actively suppressing, traditional majority views and practices. The bureaucratic view sometimes comes close to interpreting 'diversity' as a one-way street. Here are some examples:

• In January 2006 the Birmingham University Christian Union, a Christian group which has existed for 76 years, had its bank account frozen by the university student authorities because (a) it wished to restrict its membership to Christians; and (b) its publicity material about itself and its activities referred to 'men and women', words which were held to discriminate against 'transsexual or trans-gendered people'.

• In November of the same year *The Times* reported that Christian students' unions at Exeter, Birmingham, Edinburgh and Heriot-Watt universities had been suspended by student guilds or had privileges removed, because they only permitted Christians to sit on committees or address meetings. In July 2007 the Christian Union at the University of Exeter announced that it would sue the University and the Students' Guild after an independent adjudicator ruled that non-Christians must be allowed to become members of the society.

• Also in 2006, attempts were made at Dartmoor to accredit the volunteer-run InnerChange programme, which sets out to turn prisoners from a life of crime by 'the transformation of lives through the love of God' – an initiative that has had some success in a field where success is very scarce indeed. The Area Psychologist of the Prison Service was asked to have a look at the scheme, and did not like it, expressing concern that the programme might proselytise, and that the people who ran it believed that their version of Christianity was 'right'. She also noted that the programme promoted the unique virtue of heterosexual marriage, which meant that it was discriminatory against homosexuality – an issue which would, she commented, 'prevent the Validation Panel approving the programme'. The Chaplain-General to the Prison Service, the Venerable William Noblett, commented that the programme 'did not sit well with multi-faith chaplaincy' and agreed that it should not be accredited.

• Britain has a dozen Roman Catholic adoption agencies which handle roughly a third of all adoptions of older and troubled children – ones who are generally very difficult to place. These bodies do excellent work, and their disruption rates are among the lowest in the field. In January 2007 the Government decided that from April of that year all adoption agencies must be willing to place children with homosexual couples. The Roman Catholic church stated that it accepted the legal entitlement of homosexual couples to adopt children, and would refer any such couples who approached it to a more suitable agency, but that it was unwilling to make such placements itself. In this way the church could continue to place hundreds of children in stable, loving homes, while other agencies filled their own particular niches – precisely the kind of policy that a genuine commitment to *diversity* would encourage. Instead the Government, led by spokespersons Harriet Harman MP and Angela Eagle MP, announced that after December 2008 there would be no exemptions to the Sexual Orientation Regulations, and that agencies such as those mentioned above would cease to operate.

• In February 2007 we read (a) of Ms Pamela Stevens, who had long and proven experience of looking after teenagers, applying to Kensington and Chelsea Council to become a foster mother for older children, and being turned down; (b) of Ms Cherie Colman, who had since the 1990s run a charity for the children of single mothers called Cheer, being refused a grant from the Department for Education; and (c) of the Spitalfields Crypt Trust, which has for 40 years been ministering to homeless alcoholics, part-funded by Tower Hamlets Council, being told by the Council that their grants may be taken away. The reasons given in each case was the same: that the organisation was Christian, which in the view of the funding bodies would prevent it from being 'fully diverse'.

Bearing in mind that the number of adoptions in Britain in 2006 was – at around 4 800 – lower than at any time since 1998, with dreadful consequences for the unplaced youngsters (see: *Children in care*); any further diminution in the pool of potential parents would be disastrous. That the probable shrinkages noted in the two paragraphs above should be caused quite knowingly by politicians more interested in *ideology* (qv) than in the welfare of children, can fairly be described as shocking.

2 Social consequences apart, policies such as those noted above conflict with the views of mainstream Britain, are inconsistent with any genuine commitment to tolerance and diversity, and need urgently to be rethought.

See also: *Equality; Institute of Community Cohesion*; and *Race Relations Act 1976*.

NA Database See: *National DNA Database.*

---

octors See: *General practitioners.*

---

octors' surgeries 1 For compliance with the Disability Discrimination Act (qv) see: *Health and welfare buildings.*

2 ENHANCING CARE PROVISION FOR BLIND AND PARTIALLY-SIGHTED PEOPLE IN GP SURGERIES: GUIDELINES FOR BEST PRACTICE is a document that gives advice on making surgeries more *accessible* (qv) to people with *impaired vision* (qv). For publication details see the bibliography.

Door bells

1 Standard door bells may not be clearly heard by people with *impaired hearin* (qv). The following publications give guidance. For publication details see the bibliography.

• SAFETY, SECURITY AND ENVIRONMENTAL CONTROL SYSTEMS, published by the Centre for Accessible Environments (qv).

• SOLUTIONS, the annual catalogue of the RNID (qv).

Door to Door

1 An official *transport* (qv) and *travel* (qv) website for people who are *disable* (qv) or have *impaired mobility* (qv). It is run by DPTAC (qv). Individual sections cover travel by road, rail, air, and sea; travel around London; going shopping, to work, to college, and to hospital; and going on holiday.

2 For more information visitwww.dptac.gov.uk/door-to-door/

3 As of 2008, other useful websites include:

• DirectGov www.direct.gov.uk/

• Transport Direct www.dft.gov.uk/transportforyou/transportdirect/

See also: *Transport: public*.

Doors in dwellings

1 The following references contain authoritative design guidance.

• Paras 6.22–5 of APPROVED DOCUMENT M (qv). The provisions in AD M must be heeded in order to comply with the Building Regulations (qv) in England and Wales.

• THE HABINTEG HOUSING ASSOCIATION DESIGN GUIDE (qv) applies most specifically to *housing association* (qv) schemes, but ought to be by the designer's side in the design of all housing developments and individual dwellings.

• THE WHEELCHAIR HOUSING DESIGN GUIDE (qv), also from the Habinteg Housing Association, is another illustrated manual whose usefulness extends f. beyond the special remit implied by the title.

• THE HOUSING DESIGN GUIDE, published at the behest of the Mayor of London, is a best-practice guide for housing in London.

For publication details of the above references see the bibliography.

2 The clear widths shown on the next page will in general satisfy the provision of APPROVED DOCUMENT M, though each particular case should be checked in detail before decisions are finalised.

Doors in public buildings: complying with the Disability Discrimination Act

1 Public buildings and spaces must in general comply with the Disability Discrimination Act (qv). In the design of doors and doorways, relevant *authoritative practical guides* (qv) include the following:

• APPROVED DOCUMENT M (qv): the provisions in section 2 (external doors) and section 3 (internal doors) must be heeded in order to comply with the Building Regulations for England and Wales.

• BS 8300 (qv): the recommendations in section 6 (external doors) and paras 7.2 and 7.3 (internal doors) apply to all buildings in the UK.

• INCLUSIVE MOBILITY (qv): the recommendations in section 8.2 apply to *transport-related buildings* such as bus and coach stations, railway stations, air terminals, and transport interchanges.

Except for AD M, the above guides do not have the force of law, but conformi with their recommendations will make it easier to demonstrate that the requirements of the Disability Discrimination Act have been met.

The clear widths shown on the next page will in general satisfy the provisions of AD M and BS 8300, though each particular case should be checked in detai before decisions are finalised.

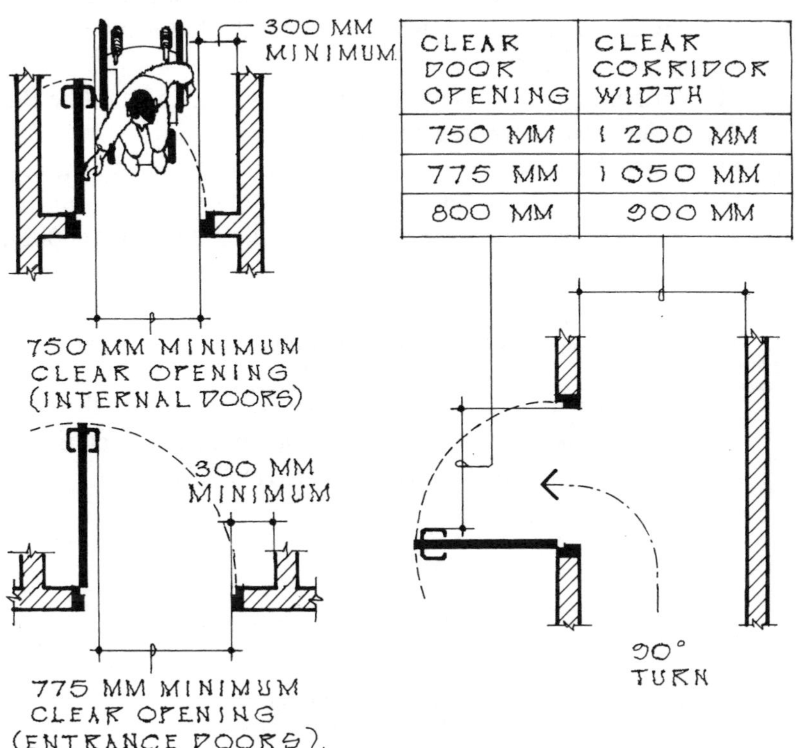

CLEAR DOOR OPENING	CLEAR CORRIDOR WIDTH
750 MM	1 200 MM
775 MM	1 050 MM
800 MM	900 MM

(A) DWELLINGS

FOR WHEELCHAIR HOUSING SEE TEXT

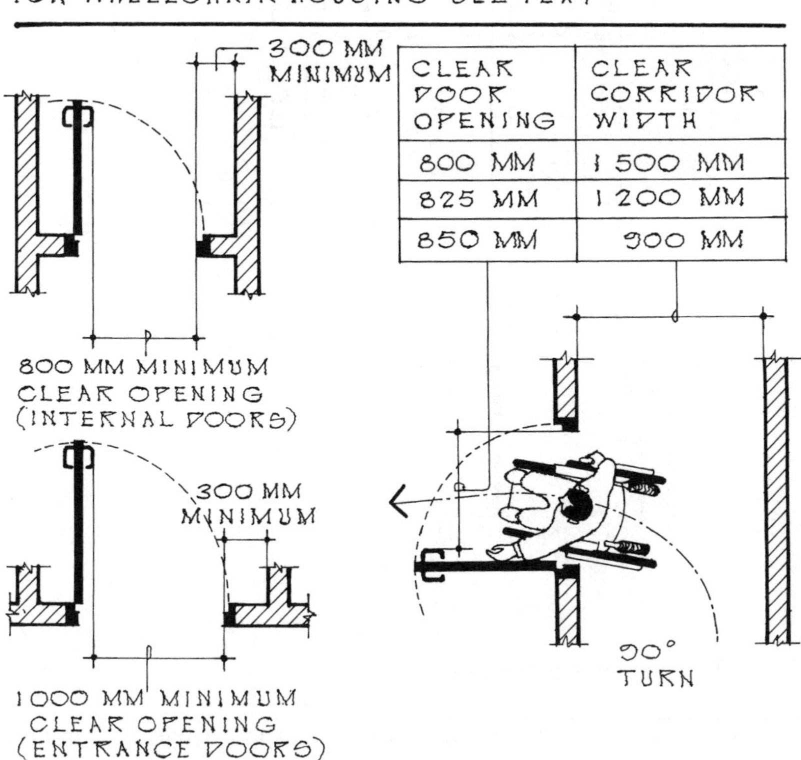

CLEAR DOOR OPENING	CLEAR CORRIDOR WIDTH
800 MM	1 500 MM
825 MM	1 200 MM
850 MM	900 MM

(B) NON-DOMESTIC BUILDINGS

2 Where doors form part of the *means of escape* (qv) in the event of *fire* (qv), then design and dimensions should be checked with the *fire authorities* (qv).
3 For places other than England and Wales the following regulatory documen should be consulted:

• In Northern Ireland: TECHNICAL BOOKLETS E, H, and R of the Building Regulations (Northern Ireland) (qv).
• In Scotland: the BUILDING (SCOTLAND) REGULATIONS and the associated TECHNICAL HANDBOOKS (qv).

4 In the USA, where the Americans with Disabilities Act applies, refer to the ADA AND ABA ACCESSIBILITY GUIDELINES (qv).

See also: *Corridors*; and *Means of escape*.

Doors in public buildings: design

1 The design details shown below will in general satisfy the provisions of APPROVED DOCUMENT M (qv) and BS 8300, though each particular case should be checked in detail before decisions are finalised.

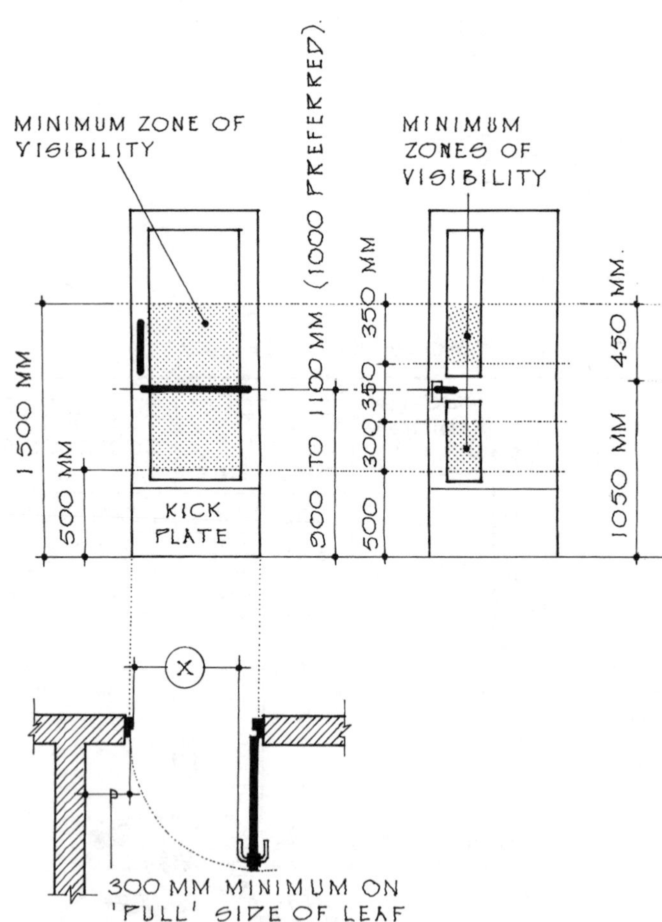

2 Dimension X is the clear width.

DPTAC

1 DPTAC, or the Disabled Persons' Transport Advisory Committee, is a *quango* (qv) under the Department for Transport (qv). Its main role is to advis the government and industry on the needs of *disabled persons* (qv) in *transpor* (qv). It publishes authoritative guides, the titles of which will be found on its website, and also provides public information and guidance websites includin the following:

• Door to Door (qv), a website that gives *disabled* (qv) and less mobile people information about travelling using all forms of transport.
• DPTAC Access Directory (qv) a website that aims to give designers a single gateway to the best information sources on inclusive design in transport and the built environment.
2 For more information visitwww.dptac.gov.uk/

TAC Access ectory

1 A continually updated database that aims to give designers a single gateway to the best information sources on *inclusive design* in transport and the built environment.
2 For a detailed list of the subject categories in the Directory's BUILT ENVIRONMENT database, which include the major building and transport types, consult www.dptac.gov.uk/thesaurus.htm
3 To use the DPTAC Directory visit:
www.dptac.gov.uk/adnotes.htm
www.dptac.gov.uk/pubs.htm

TAC Knowledge Map See: *DPTAC Access Directory.*

---

ving aids See: *Accessible cars.*

---

opped kerbs See: *Kerbs.*

---

al sensory pairment See: *Deafblindness.*

---

al-trained dogs See: *Assistance dogs.*

---

mbing down See: *Accessible.*

---

ties: general d specific See: *Disability Discrimination Act*; *Disability equality scheme*; and *Public authorities.*

---

ties under the sability scrimination Act DA)

1 The DDA imposes specific duties of non-discrimination against *disabled persons* upon the following parties:
• Employers.
• Trade organisations.
• Service providers.
• Landlords and their agents.
• Education providers.
• Public transport providers.
2 For definitions of these groups; an indication of the ways in which they might be guilty of discrimination; and notes on how they can avoid or eliminate such *discrimination*, see: *Disability Discrimination Act.*

---

ty to promote

1 For notes on what this duty amounts to see: *Age equality duty* (qv); *Disability equality duty* (qv); *Gender equality duty* (qv); and *Race equality duty* (qv).
2 The entries quoted above were written in May 2007, when the prescribed procedures for implementing the Duty involved authorities in staggering amounts of paperwork. Since then the consultation paper A FRAMEWORK FOR FAIRNESS: PROPOSALS FOR A SINGLE EQUALITIES BILL FOR GREAT BRITAIN (qv) contains the welcome suggestion that these burdensome provisions might be replaced with a simpler requirement for authorities to (a) set equality objectives and (b) take *proportionate* (qv) steps to achieve them.
3 As of 2007, useful references include the following. They are given in alphabetical order. For publication details see the bibliography.

• DISABILITY DISCRIMINATION: LAW AND PRACTICE by Brian Doyle.
• A FRAMEWORK FOR FAIRNESS: PROPOSALS FOR A SINGLE EQUALITIES BILL FOR GREAT BRITAIN.
• PLANNING, BUILDINGS, STREETS AND DISABILITY EQUALITY. This is a guide to the Disability Equality Duty (qv) and Disability Discrimination Act 2005 (qv) for *local authority* (qv) departments responsible for the planning, design and management of the *built environment* (qv) and *streets* (qv). It was published in 2006 by the Disability Rights Commission (qv) and contains useful procedural advice.
4 Useful websites include:

• Directgov: disabled people	www.disability.gov.uk/
• Equality & Human Rights Commission	www.equalityhumanrights.co

See also: *Over-regulation*; *and Single Equality Act.*

Dwellings: safety See: *Accidents.*

DWP See: *Department for Work and Pensions.*

Dyslexia 1 A condition broadly linked to *attention-deficit/hyperactivity disorder* (ADHD *autistic spectrum disorders* (ASD), and *dyspraxia*. Children affected by *dyslexia* have great difficulties in learning to read or spell, which are not attributable to lack of intelligence; and there may be associated problems wit arithmetic, with time or tense, with distinguishing between left and right, and with a sense of direction. Many children affected by dyslexia also show symptoms of dyspraxia.
2 This group of developmental problems as a whole are said to affect up to 20 of school-age children, and accounts for the vast majority of children with Special Educational Needs (qv). The associated difficulties usually persist int adulthood, and may in severe cases be very damaging to the individuals themselves, their families, and society.
3 Not everyone is convinced of the existence of dyslexia as a medical conditio and there is rising concern about deliberate misdiagnoses and sharp practice i schools and colleges:
• Writing in *The Times Education Supplement* of 2 September 2005, Professo Julian Elliott of Durham University argued that the term is largely an 'emotional construct', and asserted that the erroneous belief that dyslexics ar clever, but bad at reading for sound medical reasons, created an 'impassioned demand' to be labelled dyslexic. After 30 years in the field he had 'little confidence' in his ability to diagnose it, and considered that 'there is no sound, widely accepted body of scientific work that has shown there exists any particular teaching approach more appropriate for 'dyslexic' children than for other poor readers'. He repeated these remarks in May 2007, adding that dyslexia had to some degree become a social figleaf for parents who did not want their children labelled as having low intelligence. 'The disability lobby is so strong and the advantages, financial and otherwise, so great that they are diagnosing dyslexics all over the place', he said. 'At universities "dyslexic" students can get laptops, extra books and other equipment, sometimes to the value of almost £10 000 each'. Some students are milking the situation for what it's worth: 'They ask for different coloured exam papers, extra photocopying, anything they can get. And the numbers of people who do this are growing. If you are giving special needs provision without any particular criteria, it is obviously going to proliferate'. He said that the number of students receiving disability allowances at university had risen to 35 500.

• Schools are allowed to give pupils 25% extra time to complete GCSE or A-level exams if they have conditions such as dyslexia or dyspraxia. According to the Qualifications and Curriculum Authority (qv), quoted in the Sunday Times of 10 February 2008 under the heading 'Teachers bend exam time rules', the number of pupils given extra time jumped by 60% between 2005 and 2006, to 57 000. The report quotes Martin Turner, a consultant psychologist and former head of psychology at Dyslexia Action, as saying that he knew of a school where '40% of pupils got extra time for their GCSE papers for dyslexia, when the vast majority were not dyslexic at all'.

4 Dyslexia is specifically recognised as a *disability* under the Disability Discrimination Act (qv).

5 As of 2008, useful websites include:

• Directgov www.direct.gov.uk/
• Dyslexia Action www.dyslexiaaction.org.uk/
• Dyslexia Research Trust www.dyslexic.org.uk/

See also: *Attention-deficit/hyperactivity disorder*; and *Medicalisation*.

---

spraxia

1 As above, a condition broadly linked to ADHD, ASD, and *dyslexia*. Children affected by *dyspraxia* have great difficulties in carrying out purposeful activities to order. Symptoms include clumsiness, difficulty with catching a ball or balancing, tying shoelaces, or doing up buttons. The associated difficulties usually persist into adulthood, and may as above be very damaging to the individuals themselves, their families, and society.

2 Dyspraxia may qualify as a *disability* under the Disability Discrimination Act (qv).

3 As of 2008, useful websites include:

• Directgov www.direct.gov.uk/
• Dyslexia Action www.dyslexiaaction.org.uk/
• Dyslexia Research Trust www.dyslexic.org.uk/

E

ly excellence tres	See: *Sure Start.*
ly years: Sure rt	See: *Sure Start.*
t End, London	See: Human rights and social friction
nomic success	1 A truly *inclusive* (qv) society in which excellent health, education, housing, and the other good things in life are widely available, requires a high degree of economic prosperity. By all current evidence (see: *Happiness*) poor societies tend to be unhealthy and unhappy. It is therefore fortunate that economic success appears to depend more upon cultural attitudes that can be acquired by just about everyone, given the will and determination, than upon material circumstances. The Netherlands and Japan, to take two shining examples, are among the world's richest and most socially inclusive societies despite having virtually no natural resources with which to support their extremely dense (no pun intended) populations. 2 For readers who wish to pursue these matters, Professor David Landes stresses the primacy of cultural factors in his acclaimed study THE WEALTH AND POVERTY OF NATIONS (see for instance pp 253, 410, and 522–24), and so does Professor Jared Diamond in his masterly GUNS, GERMS AND STEEL: A SHORT HISTORY OF EVERYBODY FOR THE LAST 13 000 YEARS (see for instance pp 252 and 417). The following factors are among the ones that are fundamental to economic success: • An attachment to high academic standards. • A dedicated, disciplined, and purposeful attitude to work. • A love of technology. People and societies that cultivate these habits of mind can thrive almost anywhere. See also: *Culture*; *Happiness*; and *Inclusion*.
ucation Action es (EAZs)	Visit www.standards.dfee.gov.uk/eaz/
ucation: complying h the Disability crimination Act	See: *Disability Discrimination Act 1995 – discrimination by education providers.*
ucation: enditure and rformance	1 Expenditure on education in England grew in real terms from £28 billion in expenditure and 2001-02 to £41 billion in 2005-06. It is set to rise to £70 billion in 2006-07, £79 billion in 2007-08 – visit www.dfes.gov.uk/trends/index.cfm?fuseaction=home.showChart&cid=2&iid=5&chid=21 2 Public (ie taxpayer) funding per school pupil in England grew, again in real terms, from £3 200 in 1999-00 to £4 900 in 2005-06. It is set to increase to £5 000 in 2006-07, and £5 400 in 2007-08 – visit www.dfes.gov.uk/trends/index.cfm?fuseaction=home.showChart&cid=2&iid=5&chid=22. 3 In return for this deluge of money our education system is actually going backwards. EDUCATION AT A GLANCE 2004, published by the OECD in 2006, gives international comparisons of costs linked to outcomes. Britain comes top of 30 countries for increased spending, but has simultaneously slid down the performance table to the point that we are now the only country in the developed world to have dropped out of the top 15.

4 Useful references include the following. They are given in date order. For publication details see the bibliography.
• EDUCATION AT A GLANCE 2004, a regular report from the OECD, enables countries to compare their own performance with those of others.
• TEN YEARS OF BOLD EDUCATION BOASTS NOW LOOK SADLY HOLLOW. IT WILL BE HARD POLITICALLY BUT LABOUR MUST ACCEPT ITS VAUNTED POLICIES ON SCHOOL HAVEN'T WORKED is a study by Jenni Russell, and was published in the Guard on 14 November 2007.
• THE TRAJECTORY AND IMPACT OF NATIONAL REFORM: CURRICULUM AND ASSESSMENT IN ENGLISH PRIMARY SCHOOLS, one of the largest studies of its kin was carried out by the University of Cambridge Faculty of Education and wa published in February 2008. It reports 'a decrease in the overall quality of primary education' over the past 10 years.
• THE FAILED GENERATION, published in April 2008 by the Bow Group, shows that £70 billion has been wasted in failing to educate 4 million pupils.
5 Useful websites include the following. They are given in alphabetical order.

• CEM Centre	www.cemcentre.org/
• Dept for Children, Schools & Families	www.dfes.gov.uk/trends/index.cfm
• Dept for Children, Schools and Families	www.dcsf.gov.uk/
• DfES: Research and Statistics Gateway	www.dfes.gov.uk/rsgatew
• DfES: Trends in Education and Skills	www.dfes.gov.uk/trends
• Higher Education Statistics Agency	www.hesa.ac.uk
• Learning and Skills Council	www.lsc.gov.uk
• National Foundation for Educational Research	www.nfer.ac.uk
• OECD	www.oecd.org/
• Office for National Statistics (ONS)	www.statistics.gov.uk/
• Office for Standards in Education (Ofsted)	www.ofsted.gov.uk

See also: *Literacy; Qualifications and Curriculum Authority;* and *Social mobil*

Education: national curriculum

1 The National Curriculum was introduced in 1988 with the intention of raising educational standards. It now underpins one of the most centralised educational systems in the world. But instead of rising, as intended, standard in the state school system are falling (see: *Qualifications and Curriculum Authority*), and the major study THE TRAJECTORY AND IMPACT OF NATIONAL REFORM: CURRICULUM AND ASSESSMENT IN ENGLISH PRIMARY SCHOOLS, quoted in the previous entry, pins the blame for a 'decrease in the overall quality of primary education' over the past 10 years on 'the narrowing of the curriculun and the intensity of testing'. Professor Chris Woodhead charges that the NC h been captured by the very people who were debasing standards in individual classrooms, and are now debasing standards from the centre. What is necessa he argues, is to shift power from small groups of Ministers and civil servants to millions of individual parents, who overwhelmingly want their children to be able to read, write and do sums, and acquire an education that will open th door to good jobs. See: *Education: voucher systems*; and *Social mobility*.
2 READY TO READ? by Anastasia de Waal and Nicholas Cowen was published in June 2007 by the *think tank* (qv) Civitas, which has a deep and genuine interest in matters educational.

See also: *Literacy*; *Qualifications and Curriculum Authority*; and *Social mobili*

Education: pre-school

1 Education in Britain is not compulsory below the age of five. However, the post-1997 Labour Government gave pre-school education a high priority, and partly as a result the proportion of three- and four-year- olds enrolled in all infant schools in the UK rose from 21% in 1970–71 to 65% in 2004–05 – see also: *Childcare*.

2 As of 2008, useful references include the following. For publication details see the bibliography.
• EDUCATION AT A GLANCE 2004, an international comparison published by the Organisation for Economic Co-operation and Development (OECD) in 2006, found that funding per child was highest in the USA; the UK was second at $7 153 (£3 817) per head; and the OECD average was around $4 200.
• SOCIAL TRENDS (qv), published annually by the Office for National Statistics (qv). Refer to Chapter 4.
3 Useful websites include the following. They are given in alphabetical order.
• CEM Centre www.cemcentre.org/
• Directgov www.direct.gov.uk/EducationAndLearning/fs/en
• Directgov schoolsfinder http://www.schoolsfinder.direct.gov.uk//
• OECD www.pisa.oecd.org/

See also: *Childcare*; and *Neighbourhood Nursery Initiative*.

---

ducation: primary nd secondary

1 Since June 2007 primary and secondary education have come under the Department for Children, Schools and Families (qv), which looks after the learning of all 3–19-year-olds outside higher education.
2 Education in Britain is compulsory between the ages of 5 to 16. In 2004–05 there were just over 5 million pupils in primary schools in the UK, and somewhat over 4 million pupils in secondary schools. These schools are basically of two types:
• 'Maintained' schools, also called public sector schools, are funded by *local authorities* (qv) and are very closely controlled by state legislation. Maintained secondary schools include non-selective comprehensive schools with around 3.4 million pupils; and selective grammar schools with just over 200 000 pupils.
• 'Non-maintained' or independent schools are not funded by the state. Their income derives from pupil fees and perhaps endowments, and while they must meet certain minimum standards set by the state they are largely self-governing. Around 650 000 pupils attend mainstream (qv) independent schools.
3 The difference in performance between the state and non-state sectors is a national scandal:
• In 2007 not one of the Britain's 100 top-performing schools was from the huge comprehensive sector.
• Between 2000 and 2003 the OECD's Programme for International Student Assessment (Pisa) showed the UK slipping from 4th place to 11th in science and from 8th place to 18th in mathematics. But when Pisa looked separately at the state and private sectors in 31 developed countries, England's independent schools came top in the world – visit www.pisa.oecd.org/
4 It cannot be right that Britain is increasingly having to rely on a tiny pool of independent schools, educating around 7% of pupils, to supply most of its top-level thinkers and managers, and that *social mobility* (qv) in the UK is actually declining owing to the poor education offered by much of the comprehensive sector. Fortunately improvement is possible. In 2007 the Conservative Party's 'Comprehensively Excellent' campaign identified the best 100 comprehensives in England, and also the factors that made them so successful. It found that the overwhelming majority set classes by ability, have strict discipline policies, and traditional uniforms. The very best have longer school days with committed teachers providing extra-curricular activities. These policies could be widely implemented within the current education budget, which grew from £27.8 billion in 2001-02 in England, to £40.8 billion in 2005-06 – a real-terms rise of 47% – visit www.dfes.gov.uk/trends/index.cfm?fuseaction=home.showChart&cid=2&iid=5&chid=21.
5 As of 2008, useful references include the following. They are given in alphabetical order. For publication details see the bibliography.

• PISA – SHOW'S OVER: INTERNATIONAL STUDY EXPOSES GOVERNMENT STANDARDS CHARADE. PISA makes regular international comparisons of school performance. Between 2000 and 2006 UK 15-year-olds fell from 7th to 17th place in reading skills, and from 8th to 24th place in maths.
• PROGRESS IN INTERNATIONAL READING LITERACY (PIRLS) makes regular international comparisons of reading, writing, and comprehension skills. Between 2001 and 2006 English pupils fell from 3rd place to 19th.
• SOCIAL TRENDS (qv), published annually by the Office for National Statistics (qv), gives latest statistics plus a well-informed discussion on some of the above matters. Refer to Chapter 3.
6 Useful websites include the following. They are given in alphabetical order.

• CEM Centre	www.cemcentre.org/
• Dept for Children, Schools & Families	www.dfes.gov.uk/trends/index.cfm
• Directgov	www.direct.gov.uk/EducationAndLearning/fs/e
• Centre for Policy Studies	www.cps.org.uk/
• Kent County Council	www.kent.gov.uk/education-and-learning/

See also: *Literacy; Public services: quality*; and *Social mobility*.

Education: reform

See: *Social mobility*.

Education: tertiary

1 Post-16 education, which comes under the Department for Innovation, Universities and Skills (qv), is optional and comprises two sectors:
• Further education and training, which in 2003-04 included 4.9 million students (59% women and 41% men).
• Higher education, which in the same year included 2.4 million students (57% women and 43% men)
2 These figures represent a huge and welcome growth. Thirty years ago only around 6% of young adults entered university, while the figure in 2006 was 42% of 18 to 30 year-olds. But the official *target* (qv) of pushing up this figure to 50% of school-leavers by 2010 is misconceived and unrealistic. It was reported in March 2008 that the growth since 1999 had only been from 39.2% to 39.8% despite a massive input of public funds, and we know from other sources that the annual drop-out rate has become alarming.
3 Some sources suggest that the UK's university education system compares very well with those elsewhere. According to *The Times Higher Education Supplement*, 26 of the world's top 150 universities are in Britain, 5 in France, and 5 in Germany. It also states that 6 of the world's top 10 universities are in the USA and 4 in the UK (Oxford, Cambridge, Imperial, and UCL) – visit the TES 'World University Rankings' at www.thes.co.uk/worldrankings/ For a less flattering view of our universities refer to THE ACADEMIC EXPERIENCE OF STUDENTS IN ENGLISH UNIVERSITIES – see below.
4 To maintain or improve upon this ranking in a competitive world, British governments might do well to take a few lessons from the world's front-runner which is the USA. For instance:
• Increase state funding. In addition to the huge private donations and endowments enjoyed by American universities, government funding in the US is 2.6% of GDP compared with 1.1% in the UK.
• Give generous tax breaks to people, especially the rich and super-rich, who make donations to universities.
• Ease the cap on places so that successful universities can expand.
5 As of 2008, useful references include the following. They are given in alphabetical order. For publication details see the bibliography.
• THE ACADEMIC EXPERIENCE OF STUDENTS IN ENGLISH UNIVERSITIES by Tom Sastry and Bahram Bekhradnia is from the Higher Education Policy Institute.

• THE NON-COURSES REPORT 2007 lists around 400 'mickey mouse' higher education courses offered by 90 UK colleges and universities, giving students and taxpayers the illusion rather than the reality of education.

6 Useful websites include the following. They are given in alphabetical order.

• Aimhigher www.aimhigher.ac.uk/
• CEM Centre www.cemcentre.org/
• Directgov www.direct.gov.uk/EducationAndLearning/fs/en
• World University Rankings www.thes.co.uk/worldrankings/

See also: *Aimhigher; Office for Standards in Education; and Qualifications and Curriculum Authority*;

---

ucation: voucher stems

1 Sweden has had a voucher system since 1992. Any group of qualified teachers can set up a school if they can meet certain minimum standards and can show that there is a demand. Unlike Britain's approximately 50 City Academies these schools, of which there are now around 900, (a) did not first have to find a sponsor willing to put up £2 million; (b) are not restricted to certain 'deprived' areas; and (c) do not mainly replace existing schools. The state pays such schools the average cost of educating a child in the local state school (around £5 000) per enrolled pupil, and schools must accept applicants on a first-come first-served basis, with no power to select pupils. The voucher cannot be 'topped up', i.e. schools cannot charge any additional fees.

2 Roughly similar systems now also operate in Denmark and the Netherlands, are well-established in Milwaukee in the USA, and are being experimented with elsewhere in the USA (the so-called 'charter schools') and in Canada.

3 The magic ingredient in all cases is parents' right to take their financial entitlement to any school they wish. That right is guaranteed in the Netherlands (where two thirds of schools are funded by the state but run privately), Sweden, and Denmark. It is not only a democratic entitlement, but has proven to be a powerful weapon for encouraging good schools and eliminating bad ones.

4 As of 2007, useful websites include the following.

• Reform www.reform.co.uk/website/education.aspx
• Specialist Schools Trust www.specialistschoolstrust.org.uk/
• Sutton Trust www.suttontrust.com/

See also: *Social mobility*.

---

lucational tainment and culture

See: *Culture*.

---

lucational, cultural d scientific ildings: basic anning data

1 Distilled planning data for most of the major building types is given in the METRIC HANDBOOK: PLANNING AND DESIGN DATA – see the bibliography.

2 For general design guidance, commentary, and case examples visit:

• CABE www.cabe.org.uk/default.aspx?contentitemid=37

---

lucational, cultural d scientific ildings: design: mplying with the sability scrimination Act

1 Buildings used by the public must in general comply with the Disability Discrimination Act (qv). In the case of universities and colleges (qv); schools (qv); libraries (qv); laboratories; (qv) museums (qv); and art galleries (qv), *authoritative practical guides* (qv) include DESIGNING FOR ACCESSIBILITY (qv) plus the following:

• APPROVED DOCUMENT M (qv), the provisions of which must be heeded in order to comply with the Building Regulations (qv) for England and Wales.

• BS 8300, the provisions of which apply to all buildings in the UK. Para 13.9 gives brief design recommendations for all the building types listed above. BS 8300 does not have the force of law, but conformity with its recommendations will make it easier to demonstrate that the requirements of the Disability Discrimination Act have been met.
In places other than England and Wales the following regulatory documents should be consulted:
• In Northern Ireland: TECHNICAL BOOKLET R of the Building Regulations (Northern Ireland) (qv).
• In Scotland: the TECHNICAL HANDBOOKS of the Building (Scotland) Regulations (qv).
2 For further publications, case studies, and commentary on design quality vis www.cabe.org.uk/default.aspx?contentitemid=40

See also: *Exhibition buildings and spaces*.

Educational, cultural and scientific buildings: management: complying with the Disability Discrimination Act

See: *Service providers*.

Egress

See: *Means of escape*.

Electoral Commission

1 The Electoral Commission is a *quango* (qv) with the task of fostering public confidence in British democracy by 'promoting integrity, involvement, and effectiveness in the democratic process'.
2 As with so many other bodies which are appointed by, funded by, or are answerable to the people whose behaviour they are supposed to be monitoring (see for instance *Qualifications and Curriculum Authority*), the Commission has shown itself to be a toothless watchdog. In January 2007 the Committee o Standards in Public Life (qv) published a REVIEW OF THE ELECTORAL COMMISSION (see the bibliography) which accused the Commission of 'lacking the leadership, knowledge, or courage to enforce its regulatory duties'. The impression of ineffectiveness was confirmed in March 2007 when the Department for Constitutional Affairs (qv) ignored the Commission's express doubts about electronic vote-counting in the forthcoming May elections, and t Commission meekly accepted this disparaging rebuff.
3 For more information visit www.electoralcommission.org.uk/

See also: *Electoral fraud*.

Electoral fraud

1 The incorruptability of the voting process is a cornerstone of democracy, an a matter in which the UK until fairly recently set an example to the world. Bu starting in June 2004, when the Government defied advice from the Electoral Commission (qv) and others, and insisted that it would press ahead with all-postal ballots for three referenda on regional assemblies (see: *Regions*), a seri of thoroughly unsavoury events have sullied that record.
• In April 2005 Mr Richard Mawrey, QC, a senior judge who had presided ove a special election court in Birmingham to look into allegations of fraud in tha city's council elections in June 2004, wrote a 192 page report that expressed outrage at the scale of criminality that had been uncovered. He savagely criticised virtually everyone involved in the running of that election – the loca Labour party, the candidates, the returning officer, the police, and the

Government itself. For details refer to the report by Sandra Laville titled JUDGE SLATES 'BANANA REPUBLIC' POSTAL VOTING SYSTEM in the *Guardian* newspaper on 5 April 2005 – see the bibliography.
• On 4 February 2007 Sir Alistair Graham, chairman of the Committee on Standards in Public Life (qv), published an article in the *Sunday Times* titled 'Prepare for fraud', in which he warned that electoral fraud had become a real and potent threat to our democracy, described the Government as being in denial about the situation, and called for urgent and decisive action. He directed particular blame at the Department for Constitutional Affairs (qv) and the Electoral Commission (qv).
• In June 2007 the Open Rights Group (qv) published a report titled MAY 2007 ELECTION REPORT: FINDINGS OF THE OPEN RIGHTS GROUP ELECTION OBSERVATION MISSION IN SCOTLAND AND ENGLAND (see the bibliography) which recorded 'chaotic scenes' at vote counts with malfunctioning scanners, software errors, and unreliable computers. In Dereham Humbleford, Norfolk, the only *ward* (qv) where votes were counted both manually and electronically, manual counting produced 56% more votes than electronic counting (page 2). The report concludes that 'ORG cannot express confidence in the results for areas observed'.
• In March 2008, in the run-up to nationwide local elections in May, Mr Richard Mawrey QC repeated his earlier warnings about postal voting. He charged that the current system made 'wholesale electoral fraud both easy and profitable'; accused politicians of failing to act after past scandals; and urged sweeping reforms to electoral law dealing with corruption. As before his warnings elicited no meaningful government response.
2 All in all, a sad and shaming story.

See also: *Polling stations*; and *Privacy*.

---

ectrical outlets and ntrols
See: *Controls and switches*.

---

ectrical switches for sabled people
See: *Plate or touch switches*.

---

ectronic assistive chnology (EAT)
See: *Assistive technology: high-tech*.

---

nergency egress
See: *Means of escape*; and *Signage: safety*.

---

nigration
See: *Population*.

---

nployers and the sability scrimination Act
See: *Disability Discrimination Act 1995 – discrimination by employers*; and *Disability Equality Scheme*.

---

nployers' Forum on sability
1 A charity that provides UK businesses with information on *disability* (qv) issues, and on the application of the *Disability Discrimination Act* (qv) to employment. Its guidance is highly authoritative.
2 For more information visit www.employers-forum.co.uk/

---

nployment
1 Starting with some basic definitions:
• 'Economically active' people are those UK citizens aged 16 or over who are either (a) in work ('employed'), or (b) not in work but actively seeking work ('unemployed'). This group is also called the *labour force*.

• 'Economically inactive' people are those UK citizens aged 16 or over who are neither in work ('employed'), nor seeking work ('unemployed'). Over half are of retirement age. The rest include 'housewives' and 'house husbands' who are looking after the home; people who are unable to work owing to *disability* (qv) or long-term illness (see: *Incapacity benefit*); people who have given up trying to get a job because of educational deficiencies (see: *Literacy*) or other difficulties – see: *Neets*.

2 The graph below, from figure 4.4 in SOCIAL TRENDS 37 (qv), shows that while the proportion of working-age people in the UK who were in work in 2005 ended up (after a dip in the 1980s) close to where it had been in 1971, at around 75%, the employment rate for working-age men had fallen steeply from 92% in 1971 to 79% in 2005, while that for working-age women had risen from 56% to 70%. Clearly the workforce is becoming steadily more female.

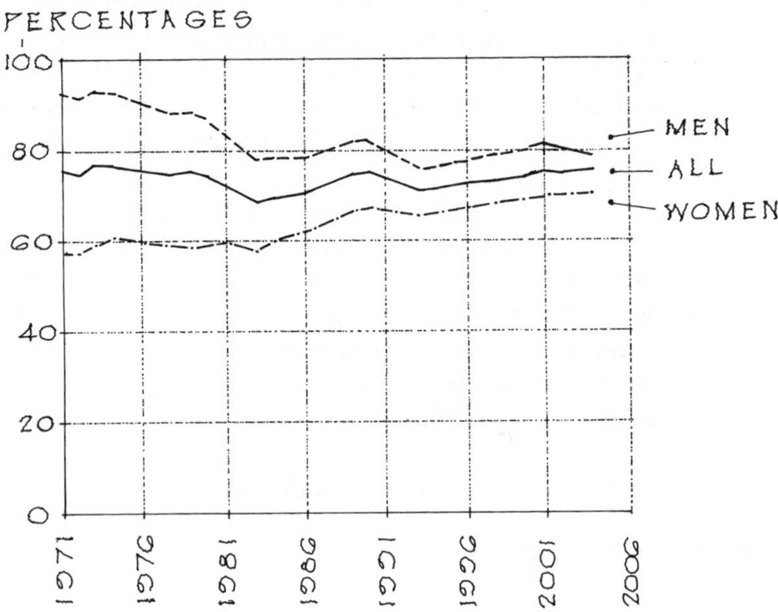

3 The next set of graphs, from Figure 4.4 in SOCIAL TRENDS 37 (which has a slightly more up-to-date version than that shown below) look at unemployment, the converse of employment, and do so in absolute figures rather than percentages. They show the number of people aged 16 or over who were out of work between 1971 and 2005. Converted to percentages, these figures equate to an overall unemployment rate in the UK in 2005 of 4.7%, which was the 3rd lowest in the EU after Ireland and the Netherlands.

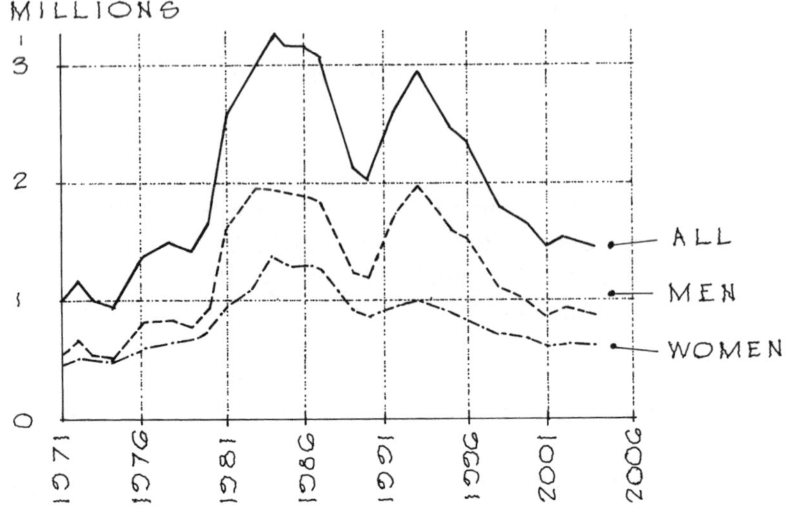

4 The above are simplified summaries only. For more detail – for instance the number of workless households, and the different employment/unemployment rates for the UK's *ethnic groups* (qv) – refer to Chapter 4 of the latest edition of SOCIAL TRENDS (qv).
5 As of 2008, useful references include the following. They are given in alphabetical order. For publication details see the bibliography.
• THE REAL LEVEL OF UNEMPLOYMENT 2002 by Professor Steve Fothergill and others of Sheffield Hallam University, published in June 2007, claims that official unemployment figures are a 'work of fiction' and that there are three times the official number of jobless in Britain. The report states that on top of the 880 400 people who are out of work and claiming benefits there are a further 1.7 million hidden jobless people. Almost a million are claiming *incapacity benefit* (qv) even though they are fit for work, and another 650 000 are out of work but do not claim benefit for various reasons. The study claims that the 'real' rate of unemployment is therefore around 2.6 million, rather than the official rate of 880 400.
• SOCIAL TRENDS (qv), published annually by the Office for National Statistics (qv), gives latest statistics plus a well-informed discussion on some of the above matters. Refer to Chapter 4.

See also: *Incapacity benefit; Jobseeker's allowance; Literacy; New deal*; and *Social exclusion*.

[Em]ployment Equality [(A]ge) Regulations [2]006

1 These Regulations outlaw discrimination on grounds of age – for details visit: www.agepositive.gov.uk/agepartnershipgroup/
2 Under the Regulations, recruiting advertisements may no longer request a date of birth or use phrases such as 'bright young graduate required', or 'energetic', as these might deter older applicants. As one example of the manner in which prudent recruiters are interpreting the law: in October 2006 the Robert Walters agency, which employs 1 300 staff in 14 countries and places thousands of recruitment advertisements a year, issued internal guidance banning words such as 'quick-learner', 'self-starter', 'high-flyer', 'dynamic', 'experienced', 'ambitious', or 'recent graduate' for fear of breaking the new law. The agency even felt obliged to advise employers against any reference to 'modern qualifications' such as GCSEs or to university degrees in subjects such as information technology – matters which are, one would assume, of central importance in many jobs. For the similarly bizarre implications of another age-related law, see: *Equality Act*.
3 As of 2008 these questions are untested in the courts, and it is unclear whether the above examples represent the intended outcomes of the Regulations. As the full implications of these regulations will only become clear over time, even worse may be in store for employers. How, it may be asked, did such a thoroughly flawed set of regulations get through Parliament? The answer is given by the Labour MP Gisela Stuart on pp 83–84 of TOWARDS A NEW CONSTITUTIONAL SETTLEMENT (see the bibliography). The regulations arrived from the European Commission (qv) and were virtually nodded through Parliament under Section 2(2) of the European Communities Act 1972 (qv), enacting EC directive 2000/78. The House of Commons as a whole never debated them; the Commons standing committee on delegated legislation considered them for 40 minutes; and the House of Lords considered them for 32 minutes. Longer consideration, as Ms Stuart points out, would have been quite pointless as no changes would have been permitted. This is the destination to which British politicians, negotiating with the EU on behalf of UK citizens who look in, powerless, from the outside, have brought us. See also: *Constitution; Parliament*; and *Post-democracy*.

4 As of 2008, useful websites include the following. They are given in alphabetical order.
- Age Positive www.agepositive.gov.uk/
- Age Partnership Group www.agepositive.gov.uk/agepartnershipgroup
- Directgov www.direct.gov.uk/
- Equality & Human Rights Commission www.equalityhumanrights.com

See also: *Ageing; Commission for Equality and Human Rights; Discriminatio* and *Single Equality Act.*

Employment law

1 In the matter of combating discrimination there are, in 2007, four major pieces of legislation governing the workplace.
- Age discrimination is outlawed by the Employment Equality (Age) Regulations (qv).
- Disability discrimination is outlawed by the Disability Discrimination Act (qv) – refer to Chapters 3 and 4 of DISABILITY DISCRIMINATION: LAW AND PRACTICE BY Brian Doyle – see the bibliography.
- Race discrimination is outlawed by the Race Discrimination Act (qv).
- Sex discrimination is outlawed by the Sex Discrimination Act (qv).

There are in addition many other laws and regulations which have a bearing on employer/employee relations. For a portal to these visit in the first instance www.direct.gov.uk/

2 As of April 2008 it is not known what effect the forthcoming Single Equalit Act (qv) will have upon the situation outlined above, and readers should chec the latest situation. Useful websites include the following. They are given in alphabetical order.
- Age Partnership Group www.agepositive.gov.uk/
- Equality & Human Rights Commission www.equalityhumanrights.com
- Department for Work and Pensions www.dwp.gov.uk/employers/dda/
- Directgov www.direct.gov.uk/
- Employers' Forum on Disability www.employers-forum.co.uk/
- Equality Direct www.equalitydirect.org.uk/

See also: *Equality and Human Rights Commission; Discrimination*; and *Singl Equality Act.*

Employment Rights Act 1996

1 The Employment Rights Act 1996 requires that certain terms and conditions of employment must be set out in a single document. The latter can be a writt 'contract of employment' or a 'statement of the main terms and conditions of employment'. The written terms and conditions will contain both contractual and statutory rights – that is, both those protected by law and those negotiated directly between the employer and the employee or representative.

2 As of 2008, useful websites include the following. They are given in alphabetical order.
- Directgov www.direct.gov.uk/
- Employers' Forum on Disability www.employers-forum.co.uk/
- Equality Direct www.equalitydirect.org.uk/

Employment Tribunals See: *Tribunals.*

Employment Zones (EZs) Visit www.jobcentreplus.gov.uk/JCP/Customers

End Child Poverty A coalition of 90 children's charities. Visit www.endchildpoverty.org.uk/

End of life autonomy See: *Self-deliverance.*

glish Community re Association CCA) Visit www.ecca.org.uk/ and see: *Residential care homes*.

glish Federation of sability Sport See: *Recreation and sports activities for disabled people*.

glish Heritage (EH) Visit www.english-heritage.org.uk/

glish Partnerships P) Visit www.englishpartnerships.co.uk/

tertainment-related ildings: basic anning data

1 Distilled planning data for most of the major building types is given in the METRIC HANDBOOK: PLANNING AND DESIGN DATA. For publication details see the bibliography.

2 For general design guidance, commentary, and case examples visit:

- CABE www.cabe.org.uk/default.aspx?contentitemid=42

tertainment-related ildings: design: mplying with the sability scrimination Act

1 Buildings used by the public must in general comply with the Disability Discrimination Act (qv). In the case of theatres, cinemas, and concert halls *authoritative practical guides* (qv) include DESIGNING FOR ACCESSIBILITY (qv) plus the following:

2 APPROVED DOCUMENT M (qv), the provisions of which must be heeded in order to comply with the Building Regulations (qv) for England and Wales. Paras 4.10–4.12 and 4.35–4.36 contain important provisions on *wheelchair* (qv) access.

3 BS 8300, the provisions of which apply to all buildings in the UK. Para 13.6 gives brief design recommendations for theatres, cinemas, and concert halls. BS 8300 does not have the force of law, but conformity with its recommendations will make it easier to demonstrate that the requirements of the Disability Discrimination Act have been met.

4 In places other than England and Wales the following regulatory documents should be consulted:

- In Northern Ireland: TECHNICAL BOOKLET R of the Building Regulations (Northern Ireland) (qv).
- In Scotland: the TECHNICAL HANDBOOKS of the Building (Scotland) Regulations (qv).
- In the USA, where the Americans with Disabilities Act applies: the ADA AND ABA ACCESSIBILITY GUIDELINES (qv).

5 For further publications, case studies, and commentary on design quality visit www.cabe.org.uk/default.aspx?contentitemid=42

See also: *Auditorium design, cinemas; Concert halls*; and *Theatres*.

tertainment-related ildings: anagement: mplying with the sability scrimination Act See: *Service providers*.

tertainment-related ildings: emergency cape See: *Auditorium design: emergency escape*.

Entrance halls and reception areas

1 The entrance hall is a visitor's first point of contact with an unfamiliar building, and should enable people to see immediately and clearly how they ca get to where they want to be. To this end:

• The routes to all parts of the building should start from the entrance lobby, as shown below, and not from hidden locations elsewhere as, to take a familiar an egregious example, the Barbican Centre in London.

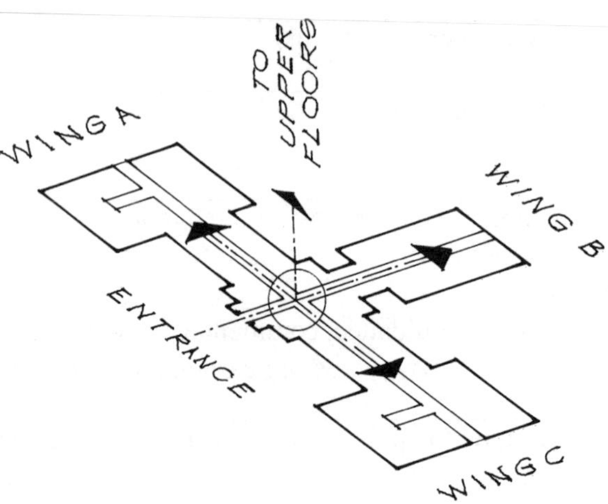

• All routes should be clearly signposted – see: *Signage*.

• Visitors will be greatly helped by an information desk near the entrance, manned or womanned by a knowledgeable member of staff. For the design of such a desk see: *Counters and reception desks*.

2 The plan below sets out some of the above principles in schematic fashion.

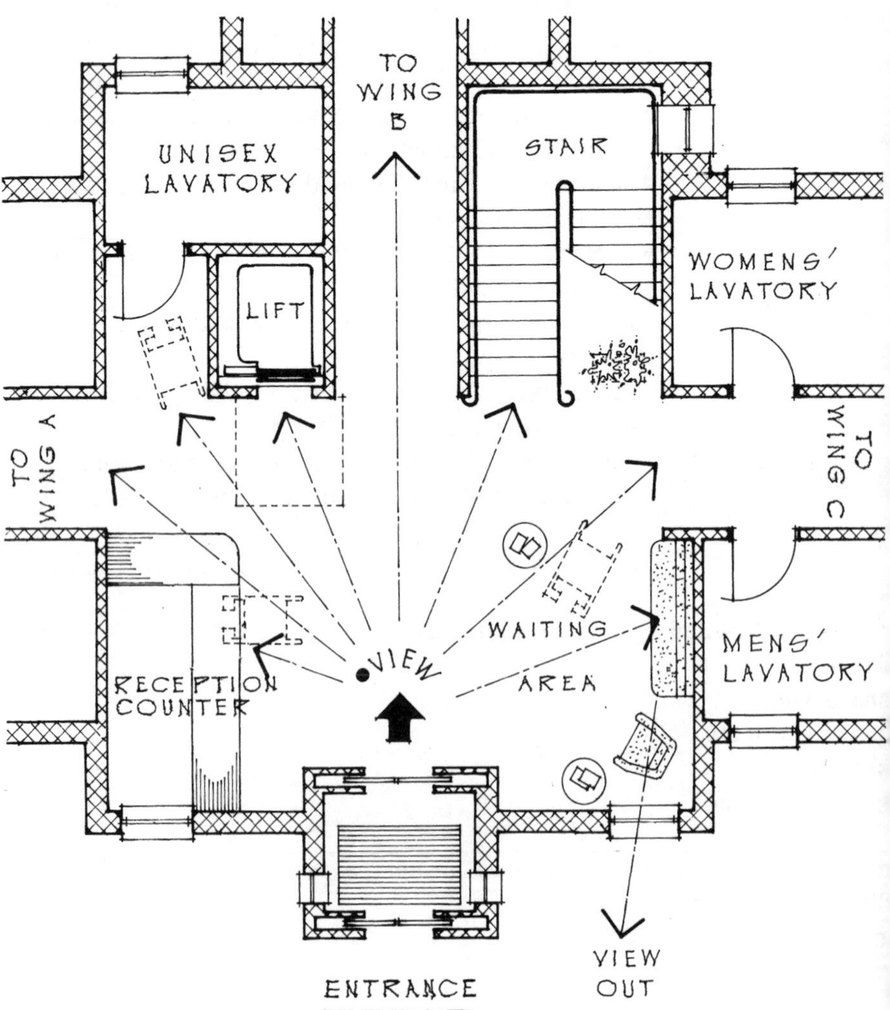

trance halls and ception areas: mplying with the sability scrimination Act

1 Public buildings and spaces must in general comply with the Disability Discrimination Act (qv). In the design of entrance halls and reception areas, *authoritative practical guides* (qv) include DESIGNING FOR ACCESSIBILITY (qv) plus the following:

- APPROVED DOCUMENT M (qv), the provisions of which must be heeded in order to comply with the Building Regulations for England and Wales. The provisions in paras 3.2–3.6 apply.
- BS 8300, the provisions of which apply to all buildings in the UK. The recommendations in para 7.1 apply to entrance halls and reception areas.
- INCLUSIVE MOBILITY (qv) the recommendations of which apply to *transport-related buildings* (qv) such as bus and coach stations, railway stations, air terminals, and transport interchanges. The recommendations in sections 9.3–9.4 apply to seating and waiting areas.

The latter two references do not have the force of law, but conformity with their recommendations will make it easier to demonstrate that the requirements of the Disability Discrimination Act have been met.

2 In places other than England and Wales the following regulatory documents should be consulted:

- In Northern Ireland: TECHNICAL BOOKLET R of the Building Regulations (Northern Ireland) (qv).
- In Scotland: the TECHNICAL HANDBOOKS of the Building (Scotland) Regulations (qv).
- In the USA, where the Americans with Disabilities Act applies: the ADA AND ABA ACCESSIBILITY GUIDELINES (qv).

See also: *Counters; Lobbies: entrance*; and *Waiting areas*.

trance lobbies

See: *Lobbies*; and *Lighting: transitional.*

trance thresholds

See: *Thresholds: level.*

trances to historic ildings

1 The entrances to existing – and especially historic-buildings frequently have imposing flights of steps and heavy doors which create problems not only for *disabled people* (qv), and especially *wheelchair-users* (qv), but also for many frail and elderly people and for anyone pushing a *pushchair* (qv).

2 Para 13.10 of BS 8300 (which is recognised as an *authoritative practical guide* (qv) under the Disability Discrimination Act) recommends that 'historic buildings should be made accessible to disabled people, wherever possible, without compromising conservation and heritage issues'. The final words in that phrase are crucial and should be noted.

3 Under the Disability Discrimination Act (qv) a physical feature that creates a barrier for disabled people can be dealt with by three types of *adjustment* (qv): the feature can be removed, altered, or avoided.

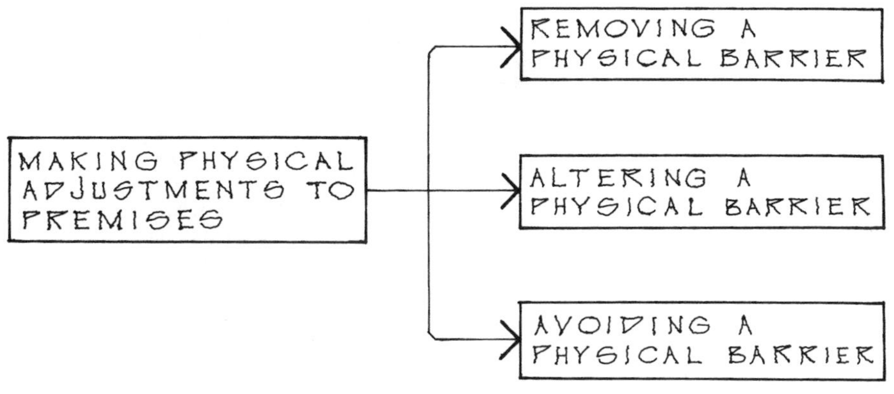

If physical adjustments are found not to be feasible then the provision of an *auxiliary aid* (qv) such as a *steplift* (qv), or even in extremis a *stairclimber* (qv) may resolve the problem.

4 The sketch below shows a *principal entrance* (qv) to a public building that is rendered inaccessible by a steep flight of stairs.

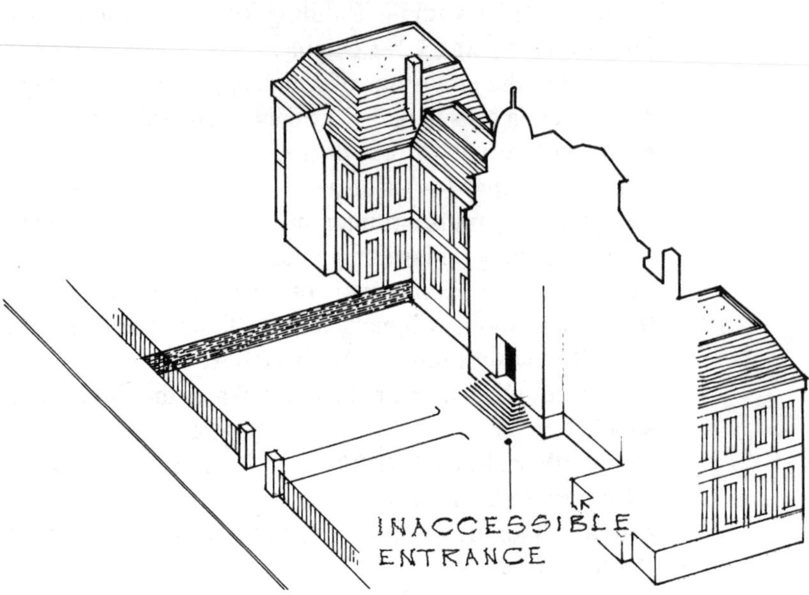

5 In figure A below, the existing steps have been removed (in terms of the diagram above) by gently regrading the ground to form a *level threshold* (qv). When dealing with historic properties this is likely to be the most unobtrusive and generally satisfactory solution, both aesthetically and functionally. Case examples include:

- The National Maritime Museum in Greenwich, London SE10.
- The Queen's House at the same site. Refer to pp 11–15 of issue 96 of ACCESS BY DESIGN (qv), the journal of the Centre for Accessible Environments (see the bibliography
- The west entrance to Arundel Roman Catholic Cathedral, West Sussex. Refer to pp 10–11 of issue 84 of ACCESS BY DESIGN (qv).

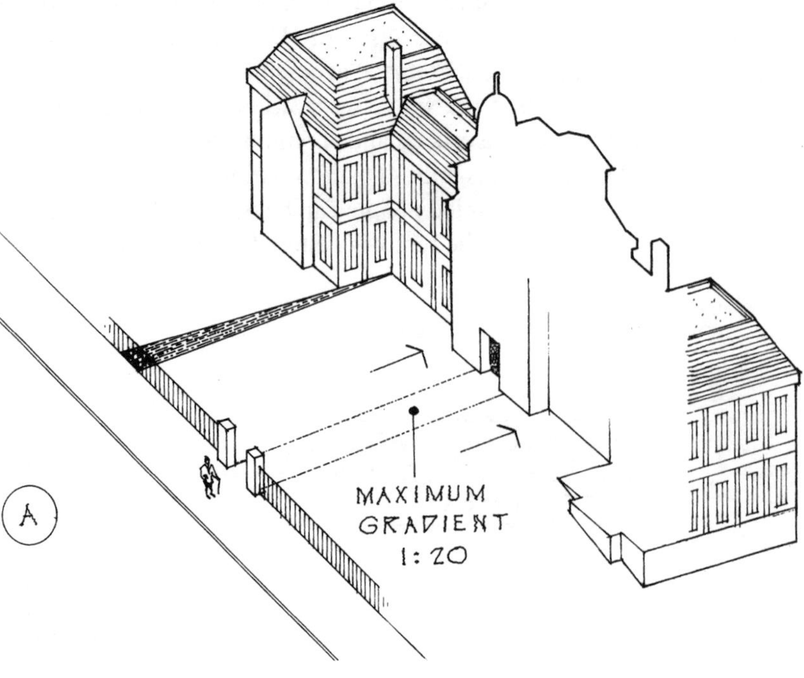

6 In figure B below, the existing steps have been altered by the formation of a pair of *ramps* (qv) and a new, accessible *stair* (qv). This sketch is purely schematic, and does not purport to show a real design. The best examples should be studied. They include:

- The main entrance to the Victoria & Albert Museum in Cromwell Road, London SW7.
- Civic buildings with grand entrances with a portico or porte-cochere within which ramps can be concealed (see the Royal Academy in Piccadilly, London SW1), or substantial stone balustrades behind which ramps can be semi-concealed (see the Municipal Buildings in Liverpool).

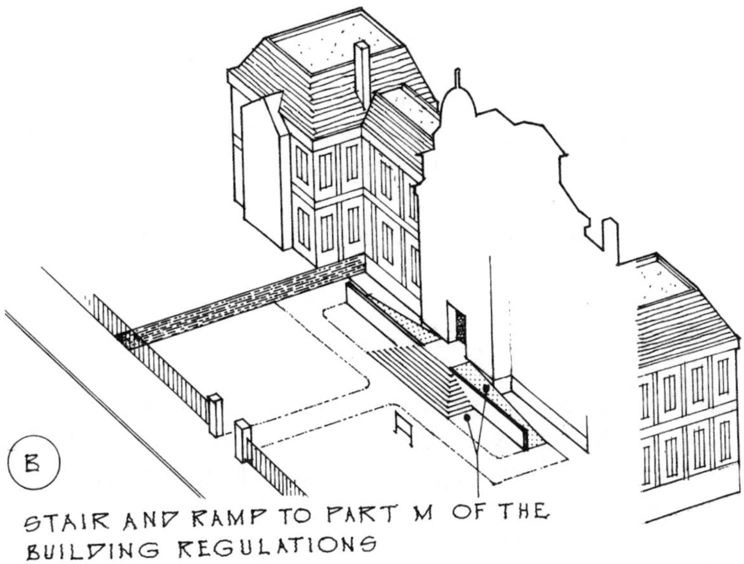

7 In figure C below the existing steps have been avoided by forming a new *principal entrance* (qv) nearby, the latter incorporating a *level threshold* (qv). Such a solution may require considerable internal replanning of the building, and is probably most appropriate if a building is being extended.

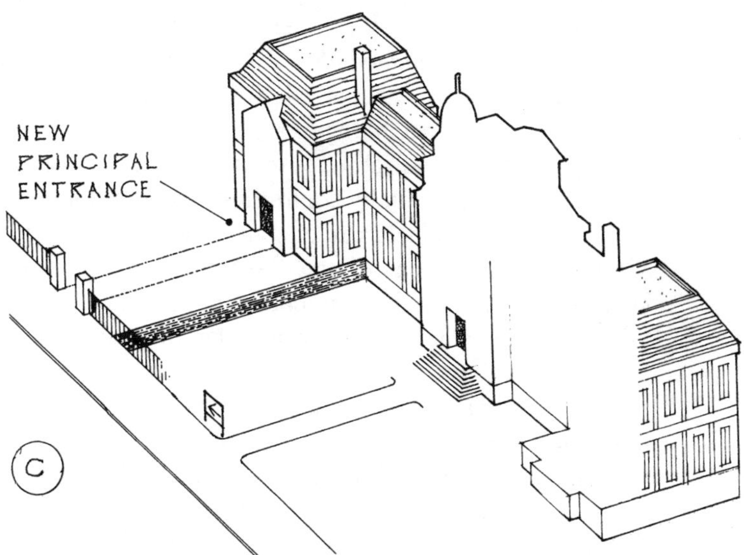

Examples of the above approach in London include:

- The Tate Britain Museum in Pimlico, SW1, where the existing Millbank entrance with its imposing flight of steps very properly remains in use as the grand entrance, which well over 90% of all users love and enjoy, and the new entrance in Atterbury Street provides an *accessible* (qv) alternative.

• The Horniman Museum in Dulwich, where the existing entrance was closed and replaced by the new entrance. This avoids the undesirable situation of having a 'first class' and a 'second class' entrance; but against that it is sad to see C H Townsend's charming turn-of-century Arts and Crafts entrance façade reduced to a functionless wall.

The message that emerges from these contrasting case examples is that there is no 'correct' or 'incorrect' way of solving this problem. Each individual case must be considered on its merits, and a balanced judgment striven for. See in this connection the concluding comments under: *Local access groups* (qv).

8 If an entrance cannot be made *accessible* by means of physical adjustments, or if the required adjustments would have an unacceptable effect on the architectural character of the property, then the provision of an *auxiliary aid* (qv) such as a *platform lift* (qv) or *stairclimber* (qv) may be the solution. Examples include *vertical platform lifts* (qv) at:

• The Great Russell Street entrance to the British Museum in Bloomsbury, London WC1. The lift is to the left of the flight of stairs leading up to the main entrance.

• Southwark Cathedral, Montague Close, London SE5. The lift is at the west end of the cathedral.

9 For more detailed notes on the above matters see: *Adjustments under the Disability Discrimination Act.*

Environment Agency (EA)

1 A *quango* (qv) under the Department for Environment, Food and Rural Affairs (qv), with responsibility for protecting and enhancing the natural environment in England and Wales. Its responsibilities include air quality; conservation and ecology, especially along rivers and in wetlands; flood control; land quality; waste management; water quality; water resources, and recreation around inland and coastal waters. It operates a *flood* (qv) warning system to which householders in low-lying areas can subscribe.

2 For more information visit www.environment-agency.gov.uk/

Environmental control systems

1 Automatic or remote methods of controlling such aspects of the environment as heating, lighting, and security can greatly improve the quality of life of *disabled* (qv), frail and elderly people, enabling them to enjoy degrees of comfort, safety, and *independence* (qv) that were previously unthinkable. Highly sophisticated systems, all of which are examples of *electronic aids and equipment* (qv) are beginning to appear on the market, enabling people with even the most severe disabilities to partly manage their home from their bed, armchair, or wheelchair, by controlling a host of domestic appliances and services from a handheld pad or some other suitable device. At present such systems are very expensive, but costs may be expected to fall over time. All these matters are discussed in more detail below.

2 *Disabled people* (qv) who consider that they need an environmental control system of the above kind should contact an *occupational therapist* (qv) in the local *social services* (qv) department. Alternatively their GP (qv) or community nurse could refer them to a medical consultant assessor for that area. Unfortunately delivery falls far short of theoretical entitlement (see: *Aids and equipment: entitlement*) but all claims should be pursued.

3 For the ultimate development of technologies such as the above see: *Smart homes*.

4 As of 2008, useful references include the following. They are given in alphabetical order. For publication details see the bibliography.

• SAFETY, SECURITY AND ENVIRONMENTAL CONTROLS. A SPECIFIERS' GUIDE ON ELECTRONIC ASSISTIVE TECHNOLOGY IN THE HOME, published by the Centre

for Accessible Environments (qv). Written in simple, clear language, and contains several enlightening case examples.
5 Useful websites include the following. They are given in alphabetical order.

• Automated Door Systems	www.automateddoorsystems.co.uk/
• Disabled Living Foundation	www.dlf.org.uk/
• EIB	www.eibshop.co.uk/
• Let's automate	www.letsautomate.com/
• Novomed	www.novomed.net/
• Possum/Gewa	www.possum.co.uk/
• QED	www.qedltd.com/wolfson
• Rehab Teq	www.justmobility.com/
• Ricability	www.ricability.org.uk/
• RSL Steeper	www.rslsteeper.co.uk/
• Smarta	www.smartasystems.co.uk/
• Toby Churchill	www.toby-churchill.com/

See also: *Aids and equipment; Assistive technology*; and *Smart homes*.

---

nvironmental control stems: examples

1 Examples of the kinds of operation that can be carried out by a *disabled person* (qv) from bed, armchair, or wheelchair include:

ACCESS CONTROLS:
• Locking/unlocking doors.

ENVIRONMENTAL CONTROLS:
• Opening/closing doors.
• Opening/closing windows.
• Drawing curtains and raising/lowering blinds.
• Raising/lowering powered beds and chairs.
• Switching lights on/off.
• Operating a page-turner or talking book.

COMMUNICATIONS:
• Operating radio, TV, video, and hi-fi.
• Operating a loudspeaking telephone.
• Operating an intercom, alarm, or warden calling service.
• Operating a personal computer.
• Operating an intercom, alarm, or warden calling service.

2 The diagrams on the next page, based on figures in the CAE publication SAFETY, SECURITY AND ENVIRONMENTAL CONTROLS, illustrate a system for a living room (above) and bedroom (below). Both are intended only as schematic examples.
3 Examples of the types of devices that may be used to direct the above operations include:
• For people with good *dexterity* (qv): keypads with pushbuttons. The keypad can be handheld, placed on a table, or mounted on a bed, armchair, or wheelchair.
• For people with *impaired dexterity* (qv): joysticks, click-switches, squeeze-switches, push-plates, foot-switches, or chin-switches.
• For more severely disabled people: suck/puff switches, or *voice-activated controls* (qv).
4 For a government-approved system that is available to severely disabled people via the Health Authority, visit www.possum.co.uk/

See also: *Aids and equipment; Assistive technology; Automatic control*; and *Communication aids*.

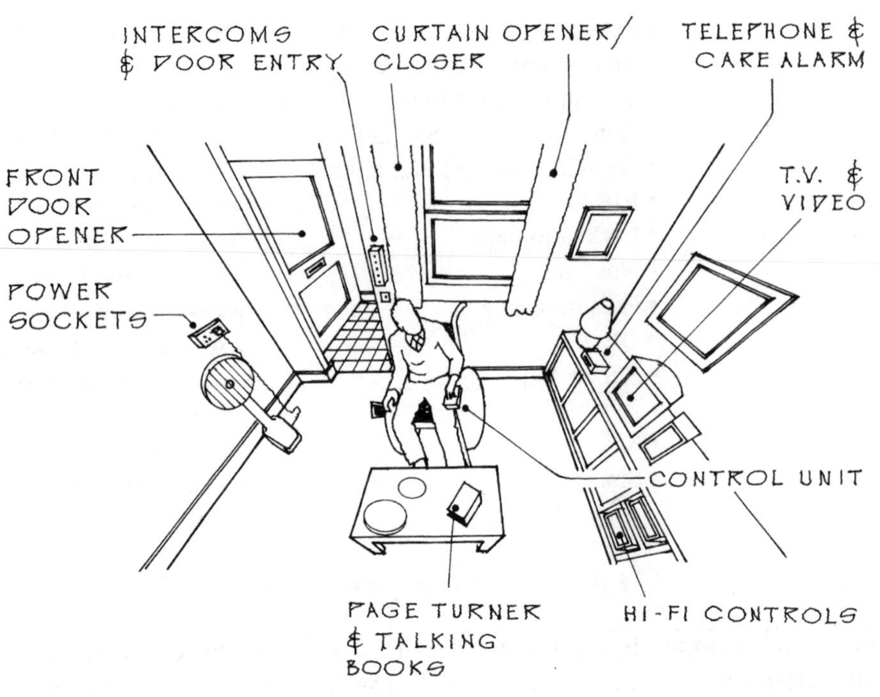
INTERCOMS
& DOOR ENTRY
CURTAIN OPENER/
CLOSER
TELEPHONE &
CARE ALARM
FRONT
DOOR
OPENER
T.V. &
VIDEO
POWER
SOCKETS
CONTROL UNIT
PAGE TURNER
& TALKING
BOOKS
HI-FI CONTROLS

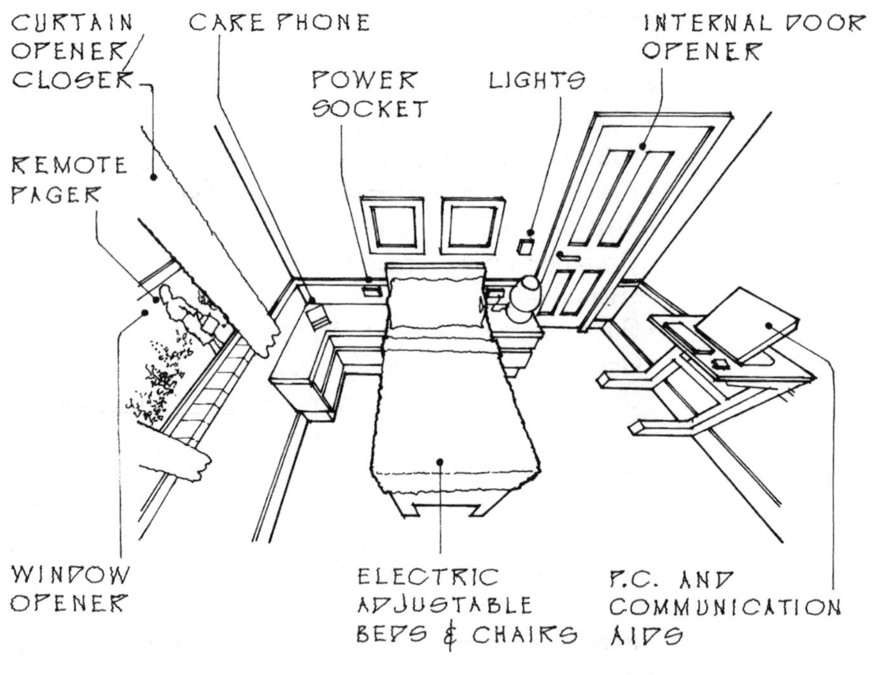
CURTAIN
OPENER/
CLOSER
CARE PHONE
INTERNAL DOOR
OPENER
POWER
SOCKET
LIGHTS
REMOTE
PAGER
WINDOW
OPENER
ELECTRIC
ADJUSTABLE
BEDS & CHAIRS
P.C. AND
COMMUNICATION
AIDS

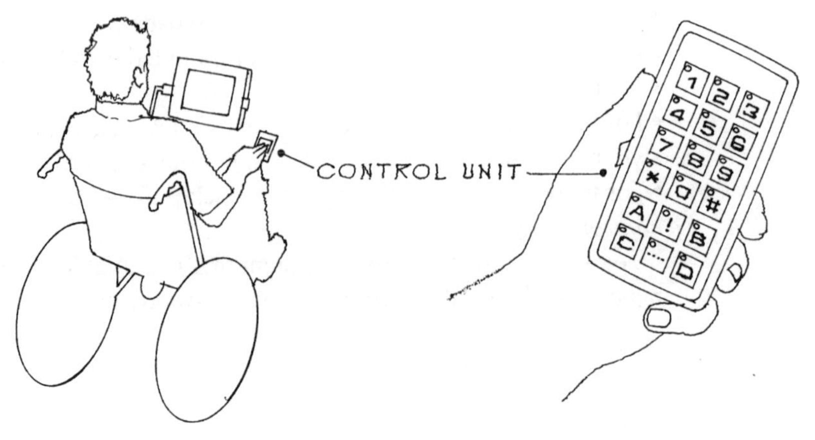
CONTROL UNIT
1
2
3
4
5
6
7
8
9
*
0
#
A
!
B
C
...
D

ilepsy

1 A chronic functional disease of the nervous system, characterised by recurring attacks of sudden insensibility or *impairment* (qv) of consciousness, commonly accompanied by peculiar convulsive seizures.
2 Epilepsy is recognised as a *disability* under the Disability Discrimination Act (qv).
3 A of 2008, useful websites include the following. They are given in alphabetical order.

- Directgov www.direct.gov.uk/
- Directgov: disabled people www.disability.gov.uk/
- Disability Rights Commission www.drc-gb.org/
- NHS Direct www.nhsdirect.nhs.uk/

ual Opportunities mmission (EOC)

See: *Equality and Human Rights Commission.*

ual Pay Act 1970

See: *Sex discrimination.*

ualities Office

See: *Government Equalities Office.*

ualities Review

See: *Government Equalities Office*; and *Single Equality Act.*

uality

1 As of July 2007, the following three documents usefully illuminate current thinking on this important but difficult subject. The first document reports on the attitudes of the general public to these matters; and the second and third embody the attitudes of the legislators who must convert that public will into law.
2 TALKING EQUALITY: WHAT MEN AND WOMEN THINK ABOUT EQUALITY IN BRITAIN TODAY by Melanie Howard and Sue Tibballs, published by the Equal Opportunities Commission (qv) in 2003, presents the findings of new research into the attitudes to equality and diversity held by the general public. Key findings include the following:

- People generally are aware that social inequality and discrimination remain widespread in Britain, and are concerned about this. They believe that society should be fairer and more tolerant but their thoughts, language, and preferred actions with respect to these issues are markedly different from the ones held by 'equality professionals' (p 7).
- In thinking about how to tackle these problems most people are comfortable with the idea of equal opportunities, or protection from discrimination. But in talking about these matters they use phrases such as 'fairness', 'tolerance' and 'having the same chance in life' rather than 'equality', which they find 'woolly' and 'abstract' (p 7). They are sceptical about the idea of equal outcomes, and strongly dislike measures such as *positive discrimination* (p 8).
- On equality issues generally, 'many people felt that … people are not born with the same skills and motivations, so there will always be inequalities' (p 23). As a Glaswegian woman put it to the researchers: 'Everybody can start equal but they aren't going to end up in the same place because you make different choices'.
- Coming more specifically to the issue of *sex discrimination* (qv) which was the main focus of this particular study, there is 'little support for the idea that women, as a group, are unequal in society today' (p 8) – a view that was expressed by women as well as men (p 29). There is 'almost universal agreement that women and men are different by design, and that this largely accounts for the differences in women's and men's lives' (p 34). Ominously for policy-makers and legislators, 'one of the central themes emerging from our

analysis of the research is the difference between professional and public view of sex inequality'. The researchers were struck by the extent of the differences in the views and beliefs expressed by respondents compared to those of policy analysts, policy makers, academic sociologists, *think tanks* (qv), and women's representative groups (p 44). 'The evidence from this research is that the professional equality community is in danger of losing touch with the audience that they aim to reach and serve. People ... recognised inequality and prejudic but do not share some equality concepts, such as the idea of equal outcomes; not share some of the terminology; and do not share the view that strategies should include group or collective action' (p 10).

• That last point leads straight to the thorny question of 'individual rights' vs 'group rights' as the preferred basis for legislation (see: *Equality and Human Rights Commission*). The researchers found that 'people see inequality as specific to individual circumstances, so there is resistance to group claims – f example, *all* women, or *all* ethnic minority people experience inequality' (p 2 The main areas in which further action was favoured were *disability* (qv) and *age* (qv), but these were seen as issues of health rather than prejudice (p 26 an p 28). In sum: '70% of people say that it is very important to be in control of one's own life'; 'people increasingly expect to take responsibility for their ow lives and outcomes'; and 'this tends to result in people acting independently an not seeking group responses' (p 47).

3 A FRAMEWORK FOR FAIRNESS: PROPOSALS FOR A SINGLE EQUALITIES BILL FOR GREAT BRITAIN (qv) is a consultation paper published in June 2007 by the Department for Communities and Local Government (qv). It sets out the Government's proposals for a Single Equality Bill for Great Britain. The Singl Equality Act (qv) that is expected to result from this consultation was envisaged as not merely consolidating the existing anti-discrimination legislation (nine pieces of major legislation plus around 100 pieces of ancillar legislation) but entirely rethinking the subject and 'taking a *proportionate* (qv approach to the regulatory burdens this necessarily imposes on private, public and voluntary sector businesses, authorities and organisations'. The opening pages of the report usefully remind us that there are only 15 ethnic minority MPs in the House of Commons (2.3% compared to 7.9% in the population); only 121 women directors in FTSE100 companies (10.5%), of whom only fou are from ethnic minority communities; and that only 13% of editors of nationa newspapers are women. But they give no hint as to whether such statistical disparities will in future be taken ipso facto as evidence of unfair *discriminatio* (qv), or considered more cautiously with the healthy sense of realism that infuses TALKING EQUALITY above. For instance, does the statistic that 'only 13% of editors of national newspapers are women' mean that newspaper proprietors discriminate unfairly against women? Perhaps, but here is an alternative view from the journalist Amanda Platell in an article titled 'Wome deserve the same pay as men as long as they work for it', published in the *Dai Mail* on 31 July 2007: 'Having worked as an executive on national newspaper for a couple of decades and at the heart of Parliament for some years, I can assure you that when a talented woman was prepared to put in the hours – insid and outside the office – not only did she deserve to earn exactly the same as t men, that's exactly what she got. So why, then, were a lot of the women I employed paid less than a man in a similar post? Quite simply because they were either less talented or, far more often, less committed. The women simpl weren't putting in the same hard graft as the men. And frankly, I don't blame them one jot. We women can't on the one hand celebrate our skills as multi-taskers, as we rightly do – running homes, wiping noses, holding down career nursing egos (and that's just our husbands) – and then lament the fact we have chosen – let me repeat that word *chosen* – to prioritise our families over our

careers. There is a price to be paid, and it's in our pay packets'. The financial commentator Ruth Lea has repeatedly written on this topic, for instance in an article titled 'Employers tread warily in this minefield of sensitivities', published in the *Daily Telegraph* on 19 October 2006, and in papers for the *think tank* (qv) the Centre for Policy Studies.

4 FAIRNESS AND FREEDOM: THE FINAL REPORT OF THE EQUALITIES REVIEW (qv), a Government paper published in February 2007, was written in parallel with A FRAMEWORK FOR FAIRNESS, above, to prepare the ground for the Single Equality Act (qv). Together with that document it reflects latest government thinking on 'equality'. The drafting committee was chaired by Mr Trevor Phillips, chairman-designate of the Equality and Human Rights Commission (qv), and their task was to look at the broader issues leading to an unequal society. The Report starts well, with a mea culpa from Mr Phillips that 'in the public mind, recent history has associated the idea of equality with bureaucratic finger-wagging and legal restriction' (p 2). But it then goes on to propose a set of implementation tools so stupefyingly bureaucratic (it includes, for instance, as part of a 'practical toolkit' for managers, a 'measurement framework' that comprises 73 'central and valuable capabilities for adults', with a separate framework for children still to come) that one must ask whether the authors were aware of the Government's pledge to cut the administrative burdens on businesses, voluntary societies, and government departments by at least one quarter by 2010 see: *Better regulation bodies*. Such misgivings were deepened by a speech made by Mr Phillips when launching the Report, in which he called for *positive discrimination* (qv) and the introduction of numerical *quotas* (qv) in order to end 'immovable, persistent disadvantage'. If these omens are correct, then far from healing the breach between legislators and the general public that was revealed in TALKING EQUALITY above, the Government is drifting even further away from the opinions of mainstream Britain. See in this connection: *Post-democracy*.

5 As of 2008, useful references include the following. They are given in alphabetical order. For publication details see the bibliography.

- FAIRNESS AND FREEDOM: THE FINAL REPORT OF THE EQUALITIES REVIEW.
- A FRAMEWORK FOR FAIRNESS: PROPOSALS FOR A SINGLE EQUALITIES BILL FOR GREAT BRITAIN.
- TALKING EQUALITY: WHAT MEN AND WOMEN THINK ABOUT EQUALITY IN BRITAIN TODAY by Melanie Howard and Sue Tibballs.

6 Useful websites include the following, given in alphabetical order.

• Directgov	www.direct.gov.uk/
• Equalities Review	www.theequalitiesreview.org.uk/
• Equality & Human Rights Commission	www.equalityhumanrights.com/
• Government Equalities Office	www.dwp.gov.uk/aboutus/geo.asp

See also: *Discrimination; Human rights*; and *Post-democracy*.

uality 2025 See: *Office for Disability Issues*.

uality Act 2006 1 This Act was passed with three purposes:

- To provide for the creation of a single anti-discrimination Commission to replace the previous three – the Commission for Racial Equality (qv); the Disability Rights Commission (qv); and the Equal Opportunities Commission (qv). The new body is called the Equality and Human Rights Commission – for more detail see that entry.
- To outlaw (apart from certain specific exemptions) all discrimination in the provision of goods, facilities and services; in the management of premises; in education; and in the exercise of public functions.

• To place upon public authorities a *duty to promote* (qv) equality of opportunity in the workplace.
2 In May 2007 it emerged that the Act was likely to prohibit Saga Holidays from advertising its holidays for 'over-fifties'. These holidays are highly valued by older people who prefer the restful companionship of others of mature years, and the Government has responded to widespread outrage by promising to take a 'common-sense' approach to the application of the law. W await evidence on what this pledge will amount to in practice, but our experience of governmental promises noted under *Civil liberties* (qv), and th iniquities noted under *Diversity is a two-way street* (qv), give no grounds for optimism. For the similarly unjust workings of another presumably well-intended Act see: *Employment Equality (Age) Regulations 2006.*
3 For more detailed information visit:

• Directgov	www.direct.gov.uk/
• Equalities Review	www.theequalitiesreview.org.u
• Equality & Human Rights Commission	www.equalityhumanrights.con
• Office of Public Sector Information	www.opsi.gov.uk/acts.htm

See also: *Commission for Equality and Human Rights; and Single Equality A*

Equality and Diversity Forum 1 The Forum promotes dialogue and understanding across the separate equali (qv) strands (age, disability, gender, race, religion, sexual orientation, and broader equality and human rights issues) and to build consensus on these matters. It works at developing best practice and tries to advance the policy debate in a constructive manner.
2 For more information visit www.edf.org.uk/

See also: *Equality; Commission for Equality and Human Rights*; and *Single Equality Act.*

Equality and human rights See: *Commission for Equality and Human Rights; Equality Act; and Human rights.*

Equality and Human Rights Commission (EHRC) 1 The EHRC was established by the Equality Act 2006 (qv) and is a *quango* (qv). It is sponsored by the Government Equalities Office (qv), but according to early reports the EHRC must account for its actions to six Ministers in thre departments – the Government Equalities Office (qv) for women's issues; the Department for Communities and Local Government (qv) for race and faith issues; and the Department for Business, Enterprise and Regulatory Reform (qv). The opportunities for generating interdepartmental committees, meeting circulars, and other forms of *red tape* (qv) promise to be hyperbolic.
2 The EHRC replaces the Commission for Racial Equality (qv), the Disability Rights Commission (qv), and the Equal Opportunities Commission (qv) – no called the 'legacy commissions' – and amalgamates their responsibilities for eliminating *discrimination* (qv) on grounds of race, disability, and gender, wi the additional strands of sexuality, religion, and age. For more detail, and notes on some of the difficult paradoxes the EHRC will have to resolve, see: *Diversity; Equality*; and *Victimhood.*
3 Academics involved in the development of the EHRC discussed a number o different equality models that might underpin its workings, including the following:
• The 'individual justice' model, with its emphasis on redress for individual victims of discrimination. Mainstream public opinion appears to lean fairly strongly in this direction – see: *Equality.*

• The 'group justice' model, which holds that entire groups of people share a common fate of social disadvantage and must therefore be protected as groups. This model has its points, but as noted under *Equality* (qv) public opinion is sceptical about the concept, except perhaps in the case of *disability* and *age*, and its application gives rise to serious anomalies – see *Victimhood.* In early 2008 this appears to be the basis upon which the EHRC will operate, but matters may be clearer by the time these words are published.
• The 'identity' model, which is criticised by the Nobel laureate Amartya Sen in his book THE ILLUSION OF DESTINY – see the bibliography. .
• The 'participatory' model, which holds that that it is the 'taking part' that really counts.
The book DISABILITY RIGHTS AND WRONGS (see below) usefully discusses the various theoretical models of equality.
4 As of April 2008 we do not know how these matters are going to work out. The present author would hope that, whatever course is chosen:
• It will be one that enjoys the support of the broad mass of the population (see: *Equality*) and tries to bring society together rather than divide it (see: *Victimhood*).
• It will be one that reduces, not expands, the mountain of *red tape* (qv) now burdening businesses and society. Already equality regulations are so complex that the EHRC found it impossible to comply with the rules it sets for the rest of us, and failed to have an *Equality policy and action plan* (qv) in place by 1 January 2008. Ministers dug the EHRC out of this hole by getting Parliament to pass a special *statutory instrument* (qv) extending the deadline to 1 April for this body, but for no-one else. This scandalous episode proves beyond doubt the urgent need to scrap *Equality policy and action plans* and substitute simple procedures that accord with *Better regulation* (qv) principles.
5 As of 2008, useful websites include the following. They are given in alphabetical order.

• Directgov	www.direct.gov.uk/
• Equalities Review	www.theequalitiesreview.org.uk/
• Equality & Human Rights Commission	www.equalityhumanrights.com/
• Government Equalities Office	www.dwp.gov.uk/aboutus/geo.asp

See also: *Diversity; Equality; Government Equalities Office; Institute of Community Cohesion*; and *Single Equality Act.*

[Eq]uality policy [an]d action plan

See: *Duty to promote.*

[Er]gonomics

1 Ergonomics is the science of achieving a good fit between people and the things they do, the objects they use, and the environments in which they live, work, play or travel.
2 The required 'fit' is not only physical (for instance a comfortable position, size, and brightness for a computer screen) but also psychological (for instance easily understood instructions and software). The aim throughout is to make life safe, easy and comfortable for people – and that means all people, not just the young, the fit, and the clever – see: *Inclusive design.*
3 The science of ergonomics has a number of sub-branches. These include:
• *Anthropometrics* (qv), which deal with body measurements.
• Cognitive ergonomics, which deal with information processing.
• Environmental ergonomics.
4 As of 2008, useful websites include:
• The Ergonomics Society www.ergonomics.org.uk/

See also: *Anthropometrics.*

Escalators: public buildings: complying with the Disability Discrimination Act

1 Escalators can move large numbers of people very comfortably and efficiently, provided the vertical distance is not too great – say 6 m, or with certain provisos 13 m, though there is in theory no limit to the vertical rise. They do not suit all *disabled* (qv), frail, or elderly people, or people with *guide dogs* (qv) or *pushchairs* (qv), who would be better served by a *passenger lift* (qv). But if provided near a lift, an escalator will take much of the load off the latter and increase its availability to other users.

2 Escalators are not suitable as a *means of escape* (qv) in the event of fire, and should not be relied upon for that purpose, though APPROVED DOCUMENT B (qv) of the Building Regulations (qv) recognises that they will in practice probably be used as such.

3 Public buildings and spaces must in general comply with the Disability Discrimination Act (qv). In the provision and design of escalators, relevant *authoritative practical guides* (qv) include the following:

• BS 8300 (qv): the recommendations in para 8.5 apply to all buildings in general.

• INCLUSIVE MOBILITY (qv): the recommendations in para 8.4.4 apply to *transport-related buildings* such as bus and coach stations, railway stations, air terminals, and transport interchanges. They ought in fact to be considered in all heavily trafficked public buildings and spaces.

The above guides do not have the force of law, but conformity with their recommendations will make it easier to demonstrate that the requirements of the Disability Discrimination Act have been met.

4 In the USA, where the Americans with Disabilities Act applies, refer to the ADA AND ABA ACCESSIBILITY GUIDELINES (qv).

5 As of 2008 useful websites include:

• Lift and Escalator Industry Association www.leia.co.uk/
• Kone www.kone.com/
• Otis www.otis.com/
• Schindler www.schindlerlifts.co.uk/

See also: *Passenger conveyors; and Travelators.*

Escape plans See: *Evacuation plans.*

Escape routes See: *Means of escape.*

Ethnic minorities See: *Population.*

Ethnic minorities and educational attainment See: *Culture.*

Ethnic minorities and poverty See: *Poverty.*

Ethnic Minority Advisory Group (EMAG)

1 The Ethnic Minority Advisory Group was launched in July 2006 by the Department for Work and Pensions (qv) to help boost employment opportunities for people from *ethnic minorities* (qv).

2 For more information visit

• Department for Work & Pensions www.dwp.gov.uk/
• EMETF www.emetaskforce.gov.uk/

See also: *Employment.*

ropean Commission C)

1 The Commission, which is appointed and not elected, is responsible for initiating all EU legislation, for planning and implementing EU policies, for managing EU programmes, and for executing the £80 billion EU budget.
2 The Commission has some peculiar practices.
• When initiating new EU legislation (which in due course becomes binding on all member states, including the UK) the Commission has since its inception followed the practice of debating and agreeing proposed legislation behind closed doors, with members sworn to secrecy – reputedly the only legislature in the world to do so except those of Cuba and North Korea. After years of popular protest against this anti-democratic custom the Commission recently started taking the final Council vote on some EU laws – those needing both Council and Parliament approval – in public. But debates remain secret and confidential, and no minutes are published.
• With respect to the £80 billion EU budget, the European Court of Auditors has, as of 2008, refused to sign off this document for 13 successive years owing to the irregularities contained therein. Whistle-blowers such as the EC's internal auditor Paul van Buitenen (who repeatedly raised concerns about fraud and in 1998 sent a dossier to the European Parliament) and its chief accounting officer Marta Andreasen (who refused to sign off the 2001 accounts) were suspended, moved to different jobs, or dismissed.
3 The time is overdue for a thorough public debate on the UK's relationship with this body and hence with the EU – see: *Constitution*.

See also: *Council of Europe*; and *European Convention on Human Rights*.

ropean mmunities Act 72

1 This Act gave EU decisions automatic primacy over British Acts of Parliament. When the law was passed, it was generally assumed that such precedence would be confined to certain cross-border questions such as trade, competition, and pollution. But 34 years on Brussels has become the primary source of UK legislation, accounting for 50% to 80% of our laws. The UK Government has steadfastly refused to name a precise figure, claiming that it would be too expensive to compile the data, but the German Government has been more forthcoming. In reply to a question by Mr Johannes Singhammer, MP for the CSU, the Federal Justice Ministry replied that 19 000 out of 23 167 legislative Acts passed since 1998 originated in the EU – i.e. around 80%. Studies in the UK and other EU states suggest that this percentage may also be approximately correct for the UK.
2 Parliamentary debates over issues such as Britain's border controls, our methods of refuse disposal, and a host of other matters that citizens care deeply about, are therefore largely meaningless. Once the unelected European Commission, sitting in secret in Brussels, has decided upon a new law, the latter will make its way irresistibly onto the UK lawbook under the European Communities Act and there is nothing that we or our MPs can do about it.
3 This not what is meant by an *inclusive* (qv) democracy. The time is overdue for a new constitutional settlement to restore Parliament (qv) to its pre-1972 status as the sole maker of all UK laws – see: *Constitution*.
4 Christopher Booker's Notebook, published each week in the *Sunday Telegraph*, provides a well-informed commentary on these matters.
5 As of 2008, useful websites include:

• Ministry of Justice	www.dca.gov.uk/constitution.htm
• Office of Public Sector Information	www.opsi.gov.uk/acts.htm

See also: *Constitution*; and *Parliament*

European Convention on Human Rights

1 The Convention for the Protection of Human Rights and Fundamental Freedoms was drawn up by the Council of Europe (qv) in 1950, taking as its starting point the 1948 Universal Declaration of Human Rights (qv). It came into force in September 1953. In addition to laying down a set of civil and political rights and freedoms, the Convention also set up a mechanism for the enforcement of those rights and freedoms. This responsibility was entrusted t three institutions:

• The European Commission of Human Rights, created in 1954.

• The Committee of Ministers of the Council of Europe. The Committee is composed of the Ministers of Foreign Affairs of the member States or their representatives.

• The European Court of Human Rights (ECHR), which was created in 1959. The Court is based in Strasbourg and has the duty of interpreting and applyin the Convention. It is made up of judges (many of whom are in fact retired politicians with no legal training) nominated by each of the member countries of the Council of Europe. Since 1959 the UK, which signed up to the Convention in 1953, has been frequently found by the Court to be in breach thereof.

2 For more information on the European Court of Human Rights visit www.echr.coe.int/echr

See also: *Constitution; Human rights; Human Rights Act*; and *Universal Declaration of Human rights*.

European Council of Ministers

1 The Council of Ministers (also known as the Council of the European Unio is made up of relevant government ministers from member states.

2 One of the Council's responsibilities is to give final approval to EU legislation on behalf of the member states. When coming to these decisions th Council meets in secret, and minutes are not published.

See also: *European Commission*.

European Court of Human Rights

See: *European Convention on Human Rights*.

Euthanasia

See: *Self-deliverance*.

Evacuation

1 There are two complementary aspects to the subject of safe escape from fire in buildings:

• Design precautions, relating to the building layout, construction, and service (lighting, alarms, and sprinkler systems) etc. These are known as *means of escape* (qv).

• Management precautions, known as *evacuation*. Except for private *dwellings*, in which occupants are expected to organise themselves, it is the responsibility of premises managers to ensure that there are procedures in pla to ensure that all occupants of a building know what to do in case of emergen and receive reasonable assistance.

2 These are specialised subjects and it is essential to get expert advice, startin in the first instance with the local *Building Control Body* (qv) and *Fire Authority* (qv).

3 As of 2008, useful references include the following. They are given in alphabetical order. For publication details see the bibliography.

• APPROVED DOCUMENT B (qv) of the Building Regulations. The provisions of AD B must be heeded in order to comply with the Building Regulations for

England and Wales. In Northern Ireland refer to TECHNICAL BOOKLET E of the Building Regulations (Northern Ireland); and in Scotland to the TECHNICAL HANDBOOKS of the Building (Scotland) Regulations (qv).
• BS 5588–8: FIRE PRECAUTIONS IN THE DESIGN, CONSTRUCTION AND USE OF BUILDINGS – CODE OF PRACTICE FOR MEANS OF ESCAPE FOR DISABLED PEOPLE (qv) deals particularly with means of escape for *disabled people* (qv), especially *wheelchair-users* (qv).
• FIRE SAFETY RISK ASSESSMENT GUIDES. These are individual guides from the Department for Communities and Local Government (qv) on specific building types including healthcare premises, residential care premises, educational premises, theatres and cinemas, factories and warehouses, offices and shops, sleeping accommodation, places of assembly, outdoor spaces, and others. They are reasonably priced and aimed at non-specialists.
• PERSONAL EMERGENCY EGRESS PLANS is a practical manual on how to plan and implement the management elements of a fire safety strategy for the safe evacuation of disabled people.
4 Useful websites include:

• Fire gateway	www.fire.gov.uk/
• Fire Protection Association	www.thefpa.co.uk/

See *also: BS 5588: Part 8; Hazard; Means of escape*; and *Risk*.

Evacuation chairs

1 Manually carrying elderly or *disabled* (qv) people up or down stairs in the event of a fire or other emergency is extremely dangerous. Evacuation chairs are specially designed chairs in which such people can be safely moved by designated staff – see: BS 5588.

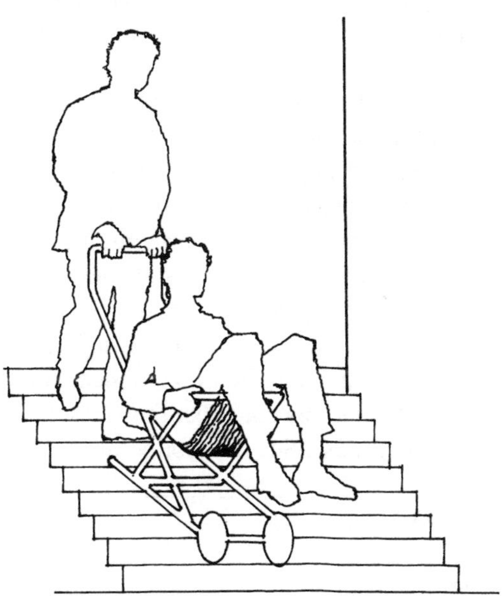

2 As of 2008, useful websites include the following. They are given in alphabetical order.

• Disabled Living Foundation	www.dlf.org.uk/
• Evac+Chair	www.evac-chair.com/
• KTP	www.loops.connectfree.co.uk/page16.html
• Mobility-MHS	www.mobility-mhs.co.uk/
• OTDirect	www.otdirect.co.uk/link-equ.html#Platform
• Ricability	www.ricability.org.uk/

See also: *Stairclimbers*.

Evacuation lifts See: *Lifts: evacuation.*

Every child matters 1 Following Lord Laming's report into the death of Victoria Climbié (qv), the Government published in 2003 a *Green paper* (qv) titled EVERY CHILD MATTERS, to accompany its formal response to Lord Laming's report. After a formal consultation process it then published EVERY CHILD MATTERS: THE NEXT STEPS and passed the Children Act 2004 (qv), which set out to provide the legislative spine for services that would better serve the needs of *children* (qv) and *families* (qv).

2 Central to this programme is a colossal computerised database first called th 'Information Sharing Index'; then renamed the 'Children's Index'; and now known as 'ContactPoint'. For comments on this thoroughly Orwellian project see: *ContactPoint* (qv).

3 As of 2008, useful websites include:

- Connexions www.connexions.gov.uk/
- Every child matters www.everychildmatters.gov.uk/

See also: *Children: looked-after*; and *ContactPoint.*

Excellence Challenge See: *Aimhigher.*

Excellence in Cities 1 This programme, which aimed to raise standards in urban schools, ran from 1999 until 2006 in 47 areas of England – visit www.literacytrust.org.uk/Database/secondary/excellence.html#Backgroundto. Its website subsequently claimed that it had 'made a unique contribution to the raising of attainment of disadvantaged pupils in our most deprived cities, towns, and rural areas', and that 'its work, through more than 1 300 secondary and 3 600 primary schools nationwide, has transformed the life chances of many thousands of young people'.

2 AN ANALYSIS OF THE VALUE ADDED BY SECONDARY SCHOOLS IN ENGLAND: IS THE VALUE ADDED INDICATOR OF ANY VALUE?, written in 2007 by Professor Jim Taylor and Anh Ngoc Nguyen of the Department of Economics in the Management School at Lancaster University, questioned these claims. It foun that the Government's specialist schools programme, and the Excellence in Cities initiative, had produced 'meagre' results despite spending billions of pounds of taxpayers' money. Educational resources had been allocated 'inefficiently and inequitably', and while there had been an improvement in exam results, 'only one third of the improvement could be attributed to government policy'. The authors state that these results 'suggest a substantial misallocation of public funds'. See the bibliography.

Excluded pupils See: *Schools: exclusions.*

Exclusion See: *Social exclusion.*

Executive Agencies
Executive bodies
Executive public bodies *See: Quangos.*

Exemplar schools See: *School buildings*; and *Building Schools for the Future.*

Exercise and rehabilitation aids 1 For examples of therapeutic *aids and equipment* (qv) that could assist in the rehabilitation of people with *disabilities* (qv), including children, visit www.dlf.org.uk/

2 As of 2008, other useful websites include the following. They are given in alphabetical order.

- Cyclone www.cyclonemobility.com/
- Medimotion www.medimotion.co.uk/
- OTDirect www.otdirect.co.uk/link-equ.html#Platform

See also: *Aids and equipment*; and *Recreation and sports activities for disabled people*.

ibition spaces: sic planning data

1 Distilled planning data on all the major building types is given in the METRIC HANDBOOK: PLANNING AND DESIGN DATA. For publication details see the bibliography.

2 For general design guidance, commentary, and case examples visit:

- CABE www.cabe.org.uk/default.aspx?contentitemid=42

ibition spaces: nplying with the ability crimination Act

1 For general guidance see: *Educational, cultural and scientific buildings*. As of 2008, sources of more specific information on the design and management of exhibition spaces include the following. They are given in alphabetical order. For publication details see the bibliography.

- ACCESS FOR DISABLED PEOPLE TO ARTS PREMISES TODAY: THE JOURNEY SEQUENCE by C Wycliffe Noble and Geoffrey Lord.
- DESIGNING GALLERIES: THE COMPLETE GUIDE TO DEVELOPING AND DESIGNING SPACES AND SERVICES FOR TEMPORARY EXHIBITIONS by Mike Sixsmith. A practical guide from Arts Council England (qv) for architects, local authorities, galleries, funding bodies, and curators.
- MUSEUMS AND ART GALLERIES: MAKING EXISTING BUILDINGS ACCESSIBLE by Adrian Cave. Technical guidance and 14 case studies from the Centre for Accessible Environments (qv).

2 Useful websites include the following. They are given in alphabetical order:

- Art Fund www.artfund.org/
- Arts Access UK www.lonsas.org.uk/
- Arts Council England www.artscouncil.org.uk/
- Artsline www.artsline.org.uk/
- Disability Arts Forum www.ndaf.org/
- DPTAC Access Directory www.dptac.gov.uk/adnotes.htm
- DPTAC Publications www.dptac.gov.uk/pubs.htm
- Good Gallery Guide www.goodgalleryguide.com/
- Gulbenkian prize www.thegulbenkianprize.org.uk/
- Magic www.magicdeaf.org.uk/
- Magda www.magda.org.uk/
- Museums Association www.museumsassociation.org/
- Open the Door www.shapearts.org.uk/
- RNIB www.rnib.org.uk/
- Shape www.shapearts.org.uk/

See also: *Audioguides; Museums: case examples; Touch tours*; and *Virtual tours*.

its

See: *BS 5588*; and *Escape routes*.

perts

1 Experts have an essential role to play in policy formation and other public matters. But should their views always outweigh those of the general public, as they currently tend to do? In pondering this question the following books are worth reading:

• EXPERT POLITICAL JUDGMENT: HOW GOOD IS IT? HOW CAN WE KNOW? by Philip Tetlock, is based on a long-running study of over 80 000 predictions by 280 specialist commentators on political trends. The author came to the conclusion that the experts would have obtained more accurate results by blindly sticking a pin in a sheet of paper. Drawing on Isaiah Berlin's classification of thinkers into 'foxes' (who know many little things) and 'hedgehogs' (who know one big thing, and know it very well indeed), Profess Tetlock deduced that when dealing with ill-defined problems of a changeable nature, people with wide and multifarious experience are better at predicting t future than dedicated specialists devotedly concentrating their attention on th particular specialised field.

• THE WISDOM OF CROWDS by James Surowiecki similarly argues, on the basis of many quoted examples, that in matters such as economics and psychology the judgment of large numbers of people without specialised knowledge in those fields usually proves to be more reliable than that of highly trained experts.

2 Experts undoubtedly possess a superior judgment in areas such as mathematics, physics, chemistry, and genetics, where lay people have little relevant knowledge to offer. But this is not necessarily true of *architecture* (q in the sense of judging the quality of buildings, streets, and cities; of *childcar* (qv), or of policies in the areas of *equality* (qv), *discrimination* (qv), or *crime* (qv). In these matters ordinary people have direct experience that may be not merely equal in value to the views of specialists, but in many cases of greater value. If, for instance, successive Home Secretaries after the mid-1950s had heeded widely-held public views on crime and imprisonment (as, to his credi Mr Michael Howard did from 1993 to 1997, with a resultant fall in the numbe of recorded crimes from 5.52 million to 4.59 million), instead of despising them, millions of crimes could have been avoided, and millions of victims spared the trauma of being burgled, mugged, or otherwise abused. For other examples see: *Déformation professionelle*.

External surfaces See: *Surfaces: outdoor*.

Extradition Act 2003 See: *Law: rule of*.

Eyesight See: *Impaired vision*.

F

…ness and Freedom: …final report of the …alities Review See: *Single Equality Act.*

…ilies 1 A family is officially defined as two or more related people living together, whereas the term *household* (qv) refers to one or more people, who may or may not be related, occupying a house. Therefore a person living alone is a one-person household, not a family; and two or more unrelated adults who live in the same house are again a household but not a family. These distinctions need to be understood when using official statistics.

2 In 2005 there were 17.1 million families in the UK, of which around 70% (a steadily declining proportion) were headed by a married couple. Two trends are worth noting:

• The graph below, from Figure 2.9 in SOCIAL TRENDS 37 (qv), shows the number of marriages and divorces in the UK between 1950 and 2005.

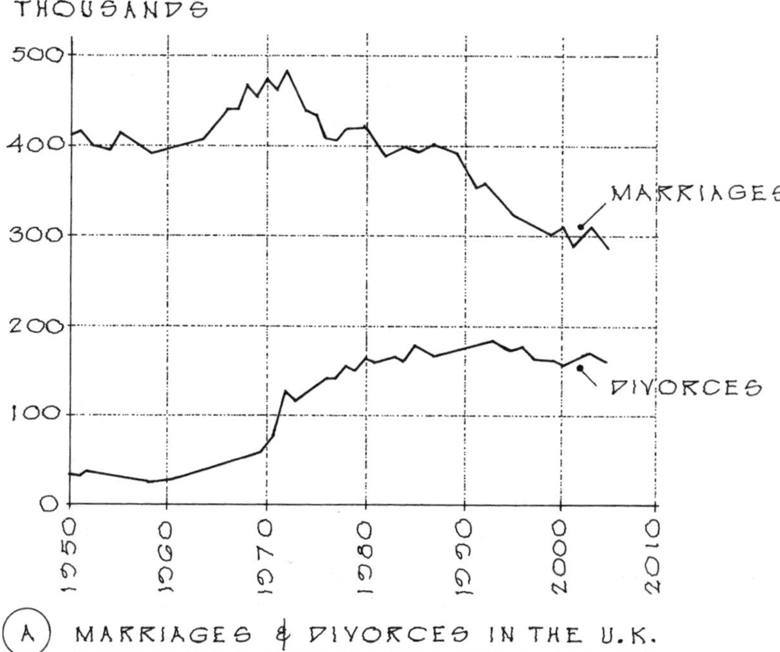

• The next graph, from Figure 2.16 in SOCIAL TRENDS 31 and Figure 2.18 in SOCIAL TRENDS 37, shows that in 1980 only 12% of all births in the UK were outside marriage, but by 2005 this had risen to 43% – see in this context also: *Children born outside marriage.*

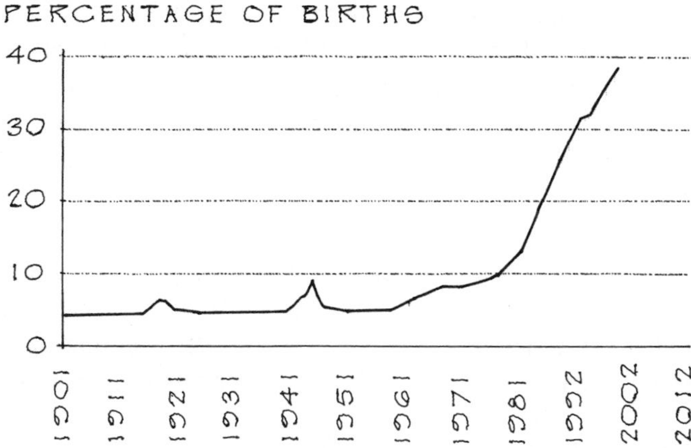

See also: *Children born outside marriage*, and refer to the sources quoted ab for more detail.
3 From children's point of view the available evidence indicates that, all othe things being equal, children do best when brought up by married couples; somewhat less well if brought up by unmarried couples in a stable relationsh and are at great risk if brought up by unmarried couples in unstable relationships. Yet as of 2007 *tax credit* (qv) and benefit policies heavily discriminate against two-parent families and actively discourage single parer from forming stable relationships and a study published in January 2008 found that the financial penalties for parents who live together are actu getting worse – see: *Children: single-parent.*
4 As of 2008, useful references include the following. They are given in alphabetical order. For publication details see the bibliography.
• DOES MARRIAGE MATTER? by Rebecca O'Neill. This 52-page report from the *think tank* (qv) Civitas, published in 2003, summarises in easy language a la body of research on marriage.
• FAMILIES WITHOUT FATHERHOOD by Norman Dennis and George Erdos; with a foreword by Professor A H Halsey. This 152-page report, also from Civitas, was first published in 1992 and is now in its 3rd edition.
• FAMILY POLICY, FAMILY CHANGES by Patricia Morgan compares the state of the family in secular Sweden, Catholic Italy, and Britain. The differing degre to which the state interferes in family life, especially the rearing of children, these countries, are instructive.
• FOCUS ON FAMILIES 2007, from the Office for National Statistics, is a fact-filled annual compendium. It shows (a) that the number of co-habiting but unmarried couples is rapidly growing, and (b) that the children of such famil get worse results at school, leave education earlier, and have a higher risk of developing a serious illness than the children of married parents.
• FREEDOM'S ORPHANS: RAISING YOUTH IN A CHANGING WORLD by Julia Margo. This report from the *think tank* (qv) the Institute for Public Policy Research draws on evidence from across the world to argue that the frequent condemnation of young people, and converse attempts to absolve them from blame, are equally misplaced.
• THE NATIONALISATION OF CHILDHOOD by Jill Kirby draws on a wealth of factual evidence to argue that the Government is systematically undermining institution of the family, and working to replace parental authority over their offspring with that of the state.
• THE PRICE OF PARENTHOOD by Jill Kirby looks particularly at the impact of the tax and benefit systems upon two-parent families.
• SOCIAL TRENDS (qv), an annual publication of the Office of National Statistics (qv), gives latest statistical data plus background explanation.
• THE SUBVERSIVE FAMILY: AN ALTERNATIVE HISTORY OF LOVE AND MARRIAGE by Ferdinand Mount argues that the nuclear family is a way of livi so deep-rooted and durable that it must come naturally to humankind.
5 Useful websites include the following. They are given in alphabetical orde
• Civitas www.civitas.org.uk/pubs/familyMain.php
• Directgov www.direct.gov.uk/
• Families Link International www.familieslink.co.uk/
• Family and Youth Concern www.famyouth.org.uk/publications.php
• Frank Field www.frankfield.co.uk/type3.asp?id=20&type=3
• One Parent Families www.oneparentfamilies.org.uk/
• Social Trends www.statistics.gov.uk/socialtrends

See also: *Childcare; Children born outside marriage*; and *Children: single parent.*

nily centres	See: *Residential family centres.*
nily courts	Visit www.direct.gov.uk/Parents/FamilyIssu
vcett Society	1 The Fawcett Society campaigns for *equality* (qv) for *women* (qv). 2 Visit www.fawcettsociety.org.uk/
ries	See: *Ships and ferries: complying with the Disability Discrimination Act.*
ld, Frank	See: *Frank Field.*
e	Visit www.fire.gov.uk/
e alarms	See: *Alarms.*
e authorities	Visit www.fire.gov.uk/
e escape	See: *BS 5588 Part 8; Evacuation*; and *Means of escape.*
e-fighting lifts	See: *Lifts: fire-fighting.*
e safety	1 The most useful single reference as of April 2008 is IP 2111:2008 – A COMPREHENSIVE GUIDE TO FIRE SAFETY – see the bibliography. 2 Useful websites include www.fire.gov.uk/ See also: *BS 5588 Part 8; Evacuation; Means of escape;* and *Part B.*
e statistics	Visit: www.communities.gov.uk/index.asp?id=1124891
ght Rights	1 A campaign to tackle the problem of airlines damaging the *wheelchairs* (qv) of their *disabled* (qv) passengers. Visit www.disabilitynow.org.uk/ See also: *Air travel*; and *Wheelchair-users* and *public transport.*
od risk and level esholds	See: *Thresholds: level.*
ors: dwellings	1 The following references contain authoritative recommendations on floor finishes, which have *risk* (qv) implications. They are given in alphabetical order. For publication details see the bibliography. • BS 8300 (qv): the recommendations in para 9.1.3 and the data on slip resistance in annex C apply to all buildings in the UK. • THE HABINTEG HOUSING ASSOCIATION DESIGN GUIDE (qv): the recommendations in para 3.15 apply most specifically to *housing association* (qv) schemes, but designers would do well to consider them in the design of all housing developments and individual dwellings. • THE HOUSING DESIGN GUIDE is a best-practice guide for housing in London. • WHEELCHAIR ACCESSIBLE HOUSING: DESIGNING HOMES THAT CAN BE EASILY ADAPTED FOR RESIDENTS WHO ARE WHEELCHAIR USERS is a best-practice guide for housing in London, and is meant to be used in conjunction with the next title. • THE WHEELCHAIR HOUSING DESIGN GUIDE (qv): a number of brief recommendations on pp 30 (garages), 43 (entrance matting), and 74–86 (flooring in general) apply specifically to *wheelchair housing* (qv), but again

designers would do well to consider them in the design of all housing developments and individual dwellings.

Floors: public spaces: complying with the Disability Discrimination Act

1 Public buildings and spaces must in general comply with the Disability Discrimination Act (qv). In the case of floor finishes, relevant *authoritative practical guides* (qv) include DESIGNING FOR ACCESSIBILITY (qv), plus the following:

- APPROVED DOCUMENT M (qv), the provisions of which must be heeded in order to comply with the Building Regulations (qv) for England and Wales. The provisions in paras 2.7, 2.29, 3.6, 3.14–3.16, 5.18, and 5.21 apply.
- BS 8300, the provisions of which apply to all buildings in the UK. The recommendations in para 9.1.3 and the data on slip resistance in annex C appl BS 8300 does not have the force of law, but conformity with its recommendations will make it easier to demonstrate that the requirements of t Disability Discrimination Act have been met.

In places other than England and Wales the following regulatory documents should be consulted:

- In Northern Ireland: TECHNICAL BOOKLET R of the Building Regulations (Northern Ireland) (qv).
- In Scotland: the TECHNICAL HANDBOOKS of the Building (Scotland) Regulations (qv).

2 In the USA, where the Americans with Disabilities Act applies, refer to the ADA AND ABA ACCESSIBILITY GUIDELINES (qv).

See also: *Surfaces*.

Floors: slip resistance

1 Annex C of BS 8300 (qv) lists the slip resistance of around two dozen floor finishes, both when dry and wet. Carpet, clay pavers and tiles, concrete, granolithic, linoleum, mastic asphalt, treated pvc, and resin surfaces tend to have a good to moderate slip resistance. At the other end of the scale ceramic tiles, glass, rubber, terrazzo, timber, and untreated pvc tend to perform badly very badly when wet.
2 For publication details of BS 8300 see the bibliography.

Flush kerbs

See: *Kerbs*.

Flush thresholds

See: *Thresholds*.

Food preparation aids

See: *Products for easier living*.

Footbridges: complying with the Disability Discrimination Act

1 Facilities used by the public must in general comply with the Disability Discrimination Act (qv).
2 Many older and *disabled people* (qv), including all *wheelchair-users* (qv), cannot use a stepped footbridge across a road or railway line. This in turn makes it impossible for many of them to undertake a journey by rail, as they dare not set out on a journey if there is a significant risk of them being strand somewhere along the way. The ideal solution is to provide a *vertical platform lift* (qv), as for instance at the St Paul's end of the London's Millennium bridg across the Thames. If such a lift is provided, it must by law be kept in constan working order.
3 Unfortunately, equipping footbridges with reliable, well-maintained lifts is very expensive. For the particular case of footbridges across railway tracks, see: *Rail travel: accessibility*.

4 Turning to the design of footbridges in general, *authoritative practical practical guides* (qv) include the following. For publication details see the bibliography.
• INCLUSIVE MOBILITY (qv), which is the most authoritative reference for the *inclusive* (qv) design of *pedestrian infrastructure* (qv) and of *transport-related buildings* (qv) such as bus and coach stations, railway stations, air terminals, and transport interchanges. It is an advisory document and does not have the force of law, but conformity with the recommendations in section 8.4.6 will help to demonstrate that the requirements of the Act have been met.

See also: *Pedestrian environments; Rail travel: accessibility; Transport-related buildings*; and *Underpasses*.

---

otpaths and otways: complying h the Disability scrimination Act

1 Public routes must in general comply with the Disability Discrimination Act (qv). Strictly speaking *footways* are pedestrian walkways adjacent to streets, while *footpaths* are not contiguous with streets; but in everyday language the terms are fairly interchangeable and in most cases the design rules for one also apply to the other. Relevant design guidance is given in the following entries:
• Where footpaths and footways form part of the public environment, for instance the pavements alongside streets or the walkways through public squares and parks, see: *Pedestrian environments*; and *Street furniture*.
• Where footpaths lead from a site boundary or car park to a building entrance, see: *Access routes*.
• For footpaths in the countryside, see: *Countryside facilities*.
2 See also: *Authoritative practical guides*.

See also: *Approach routes; Pedestrian environment; and Street furniture*.

---

undation for ormation Policy search

1 The FIPR is an independent body that studies the interaction between information technology (see: *IT*) and society with the aim of identifying technical developments that threaten *privacy* (qv) or have other significant social impacts. What its website calls 'hot topics' include surveillance, copyright, e-democracy, and health privacy.
2 Visit www.fipr.org/

See also: *ContactPoint; IT systems*; and *Privacy*.

---

mework for rness

See: *Equality; Government Equalities Office*; and *Single Equality Act*.

---

nk Field

See: *Anti-social behaviour; Children: single-parent; Human rights and social friction; New Deal: employment; Think tanks: Reform*; and *Welfare dependency*.

---

edom

See: *Civil rights; Human rights*; and *Speech: freedom of*.

---

edom of ormation Act 2000

1 The Freedom of Information Act gives everyone the right to gain access to information held by the public sector. This right applies to information held by *Parliament* (qv), *government departments* (qv), *local authorities* (qv), the *NHS* (qv), *GPs* (qv), and other organisations. The Act is administered by the Information Commissioner (qv).
2 While the Act has achieved some good things, it has been partly neutered by Ministers and Whitehall officials. As one example: in 2005 Miss Heather Brooke, a Director of the 'Your Right to Know' campaign, requested details of

the names and salaries of the staff employed by MPs and funded by the taxpayer. Eighteen months later, in September 2006, when the Information Commissioner seemed about to rule in her favour, the Speaker of the House o Commons issued a 'certificate of exemption' which effectively exempted Parliament from the provisions of the Act. Under the terms of the certificate t Commissioner is not permitted even to rule whether or not it is in the public interest for the names and salaries of MPs' staff to be released.

3 Other examples include the following:

• As of early 2007, 26 083 out of the 62 852 requests for information made to central government since 1 January 2005 were refused.

• A notional cost limit of £600 has been set, beyond which officials can refus to co-operate any further. As this cost includes the time spent by officials on considering requests and consulting with Ministers, questions can be quashed simply by spending enough time on the preliminaries to push the cost above t limit. In March 2007 Baroness Ashton of Upholland, the Information Rights Minister, announced that there would be a second consultation into plans to change the way in which the cost of freedom of information requests is calculated – a move that raised hopes that these restrictions might be abandoned.

4 As of 2007, useful references include the following. They are given in alphabetical order. For publication details see the bibliography.

• YOUR RIGHT TO KNOW: A CITIZEN'S GUIDE TO THE FREEDOM OF INFORMATION ACT by Heather Brooke.

5 Useful websites include the following. They are given in alphabetical order.

• Directgov www.direct.gov.uk/
• Foundation for Information Policy Research www.fipr.org/
• Freedominfo.org www.freedominfo.org/countries/united_kingdom.ht
• Information Commissioner's Office www.ico.gov.uk/
• Ministry of Justice www.justice.gov.uk/
• Office of Public Sector Information www.opsi.gov.uk/acts.htm
• Your right to know www.yrtk.org/

See also: *Data Protection Act; Information Commissioner's Office; Ministry Justice; Privacy; and Speech: freedom of.*

Freedom of speech See: *Speech: freedom of.*

FSMs 1 FSM stands for 'free school meals' and provides a rough guide to the degre of deprivation in particular school intake or catchment areas. The national average for FSMs may be around 14%, but in very *deprived* (qv) areas it coul be 40% or more.

2 Useful websites include:

• DfES www.dfes.gov.uk/foischeme/_documents/DfES_FoI_116.xls
• DfES www.dfes.gov.uk/foischeme/_documents/DfES_FoI_117.xls

Furniture: adjustable See: *Adjustable furniture.*

Further education and Training See: *Education: tertiary.*

G

lleries: art	See: *Exhibition design*; and *Exhibition management*.
ambling Act 2005	1 The pernicious outcomes of this Act include the following: • Modest community events run by *charities* (qv) and *voluntary and community sector organisations* (qv) – the very activities which governments of all stripes constantly assure us are the lifeblood of a healthy society – are being closed down. These bodies cannot cope with the onerous record-keeping, or afford the substantial application and annual licence fees (£1 900 and £1 500 respectively, in a case known to the author), that are now required under the Act. • Eight large and eight medium-sized commercial casinos, plus perhaps in due course 88 smaller ones, will be enabled to open (if local authorities so wish) in locations including some of Britain's poorest boroughs. This widely criticised outcome is a scaled-down version of what the Act was intended to bring about. Initially Secretary of State Tessa Jowell talked approvingly of around 40 supercasinos, each with up to 1 250 unlimited-jackpot slot machines. This grotesque proposal generated intense opposition, and warnings of grave social consequences by government advisors such as Professor James Orford (who teaches clinical and community psychology at the University of Birmingham) and Professor Mark Griffiths of Nottingham Trent University (who has been studying gambling addiction for 20 years). Her first plan was then replaced by the one above, announced in February 2008. Dr Emanuel Moran, a specialist advisor on pathological gambling to the Royal College of Psychiatrists, called Ms Jowell's latest decision 'totally reckless' and 'shamefully irresponsible' and predicted that these casinos would bring misery to poor people. 2 As of 2008, useful websites include: • Directgov www.direct.gov.uk/ • Office of Public Sector Information www.opsi.gov.uk/ See also: *Over-regulation*; and *Voluntary and community sector*.
ang culture	See: *Crime: youth*.
ender discrimination	See: *Sex discrimination*.
ender Equality Duty	1 For details of this Duty (qv) visit www.eoc.org.uk/Default.aspx?page=17686 2 See also: *Discrimination; Duty to promote; Equality; Sex Discrimination;* and *Single Equality Act*.
ender pay gap	See: *Incomes: gender pay gap*; and *Sex Discrimination*.
eneral needs ousing	See: *Housing: accessibility*.
eneral Practitioners Ps)	1 GPs are among the health professionals who offer a first line of contact between patients and the NHS. Usually a small group of GPs work together in a practice that might be called a surgery, a clinic, or a health centre. Very few now work by themselves as one-person practices. 2 The work and remuneration of GPs was radically affected by the Quality and Outcomes Framework in 2004. Under this system GPs' pay is based upon points, which are earned by demonstrating that various centrally-imposed targets have been met. Thus, as of 2007, a GP can earn an annual bonus of £8 472 by recording whether patients suffering from heart disease are smokers and offering them advice on giving up smoking; a further bonus of up

to £11 587 by compiling a register of patients with diabetes, taking their blood pressure and treating their symptoms; and so on.
3 The system was launched on the basis of a calculation that GPs would generally earn 700 or 800 points in the course of a year and that their pay would rise moderately. Instead, according to a National Audit Office (qv) report published in February 2008, their earnings rose by 56% over the two years to March 2006 to an average £113 614 a year per partner in England. The report states that the new system has cost taxpayers £1.76 billion more than the Government had anticipated, without commensurate improvements in healthcare.
4 Despite this pay rise at a time that their workload was reduced to weekdays only, a poll of GPs and hospital doctors in April 2007, commissioned by the magazine HOSPITAL DOCTOR, found that only 2% described their morale as 'excellent' while 54% said it was 'poor' or 'terrible. Most blamed their growing pessimism on the extra workload imposed upon them by government targets and the NHS (qv) reform programme.
5 In September 2007 it was revealed that the Government has plans to progressively cajole GPs out of their traditional surgeries into large polyclinics serving around 50 000 patients each. These would be less personal, would oblige patients to travel further from home for their regular needs, but fits in neatly with current governmental policies of taking services (from Accident an Emergency departments to police stations) further and further away from the people they serve in the name of rationalisation.
6 As of 2008, useful websites include:

• Department of Health	www.dh.gov.uk/
• Quality & Outcomes Framework information	www.ic.nhs.uk/services/qo

See also: *Doctors for Reform; Doctors' surgeries*; and NHS.

General Social Care Council (GSCC)

1 This *quango* (qv) is the social care workforce regulator and guardian of standards. It has published the CODES OF PRACTICE: EMPLOYERS AND SOCIAL CARE WORKERS – see the bibliography.
2 As of 2008, useful websites include the following. They are given in alphabetical order.

• Commission for Social Care Inspection	www.csci.org.uk/
• Directgov	www.direct.gov.uk/
• English Community Care Association	www.ecca.org.uk/
• National Care Homes Association	www.directions-plus.org.uk
• National Care Standards Commission	www.dh.gov.uk/

See also: *Care homes; Care Standards Act; Commission for Social Care Inspection; National Care Standards Commission*; and *Social Care*.

Geniuses: shortage of See: *Architecture*.

Gini coefficient

1 This is an internationally used measure of income inequality, developed by the Italian statistician Corrado Gini in 1912. The coefficient is a number between 0 and 1, where 0 corresponds with perfect equality (i.e. everyone has exactly the same income) and 1 corresponds with perfect inequality (i.e. one person has all the income and everyone else has none).
2 Figures close to 0 or 1 are of course purely theoretical, and real-world coefficients tend to fall between about 0.24 (unusually equal) and 0.51 (very unequal). Those for western Europe are seldom higher than 0.36. A typical Scandinavian figure is 0.24 or 0.25, indicating a markedly equal society, which probably explains in part why the Scandinavian countries score well on *happiness* (qv) surveys.

3 The figure below, from Figure 2.1 of FAIRNESS AND FREEDOM: THE FINAL REPORT OF THE EQUALITIES REVIEW (see the bibliography), shows that Britain's coefficient was stable at around 0.25 in the 1960s and 1970s, but started to rise under Mrs Margaret Thatcher and seems to have stabilised at around 0.34. This graph shows the 'BHC Gini coefficient', ie before housing costs were taken into account.

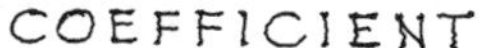

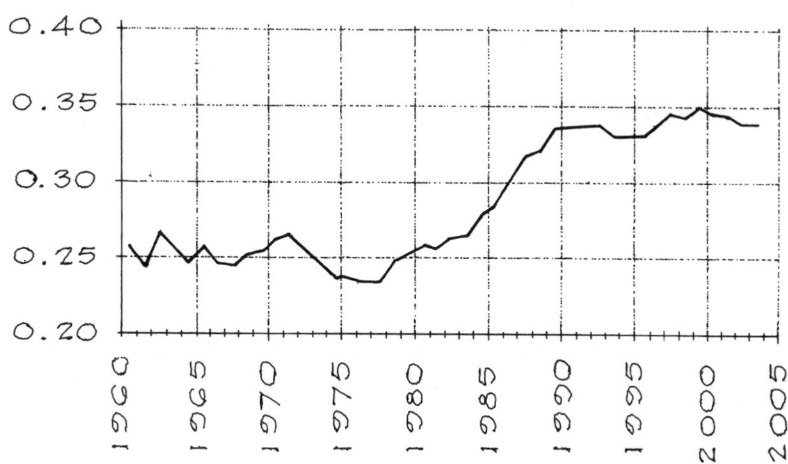

4 According to 2004 World Bank figures, which are slightly different from those above from the Institute for Fiscal Studies, the UK's Gini coefficient then stood at 0.36; that of the USA at 0.41; that of China at 4.5; and that of Nigeria at 0.51 (rounded figures).
5 For comparable data from the Office for National Statistics visit www.statistics.gov.uk/CCI/nugget.asp?ID=332&Pos=&ColRank=1&Rank=176

See also: *Incomes: inequality*; and *Social mobility*.

lass: safety

1 When standard (i.e. annealed) glass shatters it breaks into razor-sharp shards which can cause horrific injuries or death. People can suffer terrible cuts falling through – for instance – an annealed glass door, and following an explosion in or near a building over 90% of injuries and fatalities are commonly caused by flying glass. It is essential that one of three methods be used to make large sheets of glass in public places safe:
• Such panes could consist of *laminated* safety glass, which is the most expensive option. This product consists of layers of glass (which for added safety may be toughened, see below) with interlayers of extremely tough plastic, producing a sandwich of material that is able to offer a high degree of resistance to explosions or bullets. If the glass does break it remains in place, swinging outwards, in the case of an explosion, rather like a curtain.
• Panes could consist of *toughened* (i.e. heat-treated) safety glass. This is up to five times stronger than annealed glass, and if the glass does break it disintegrates into crumb-like fragments which are unlikely to cause serious injury. Toughened safety glass might cost around 80% as much as an equivalent area of laminated glass.
• Large panes could have a transparent *safety film* glued to one face. Though they are paper-thin, such films will hold annealed glass in place if broken, acting very like laminated glass. The application of a safety film will also make it more difficult for burglars to break in, and cost is typically less than a third of that of an equivalent area of laminated glass.
2 For more information visit www.ggf.co.uk/

Glass: safety markings 1 Glass doors and screens should have permanent, clearly visible marks at eye level to prevent people colliding with the glass. Relevant *authoritative practical guides* (qv) include the following:

• APPROVED DOCUMENT M (qv), whose provisions must be heeded in order to comply with the Building Regulations for England and Wales.

• BS 8300 (qv), which does not have the force of law, but conformity with its recommendations will make it easier to demonstrate that the requirements of the Disability Discrimination Act have been met.

2 Relevant paragraph of the above references are tabulated below.

ASPECT	DOCUMENT: RELEVANT PARAGRAPHS	
	APPROVED DOCUMENT M	BS 8300
GLASS DOORS	2.22–2.24; 3.10	6.4.4
GLASS WALLS & SCREENS	3.14	9.1.5

3 The markings below will in general satisfy the provisions of AD M and BS 8300, though each particular case should be checked in detail before decisions are finalised.

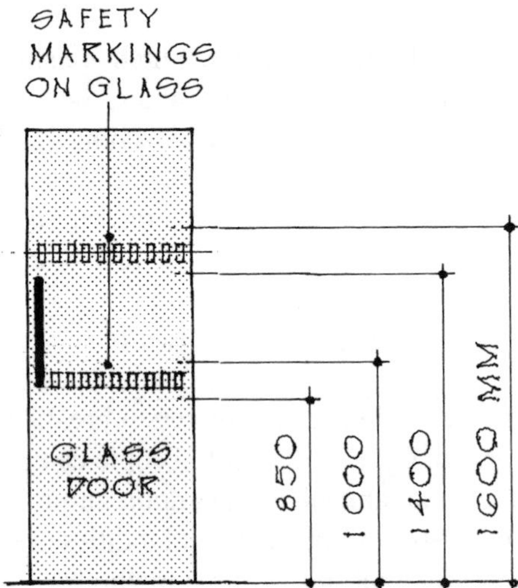

For places other than England and Wales the following regulatory documents should be consulted:

• In Northern Ireland: TECHNICAL BOOKLET R of the Building Regulations (Northern Ireland) (qv).

• In Scotland: the TECHNICAL HANDBOOKS of the Building (Scotland) Regulations (qv).

4 In the USA, where the Americans with Disabilities Act applies, refer to the ADA AND ABA ACCESSIBILITY GUIDELINES (qv).

5 As of 2008, useful references include the following. For publication details see the bibliography:

• GLASS IN BUILDINGS by Ann Alderson, published by the Centre for Accessible Environments (qv).

See also: *Hazards: protection*; and *Street works: safety markings*.

Global warming See: *Kyoto protocol*.

Gobbledygook See: *Architecture*; and *Jargon*.

od Loo Design Guide 1 This is a core reference that should be on every architect's shelf. The clear guidance given in its 50 pages will help building designers get an important design element right. While this document does not have the force of law, conformity with its contents will make it easier to demonstrate that the requirements of the Disability Discrimination Act (qv) have been met.
2 For publication details see the bibliography.

See also: *Authoritative practical guides*; and *Lavatories*.

vernment 1 Traditionally the UK has had two levels of government:
• *Central* government, run from Westminster and commonly known by that name. For guidance visit www.number-10.gov.uk/output/Page30.asp
• *Local* government, run from town halls. For guidance visit www.info4local.gov.uk/
2 After 1997 an intermediate level of *regional* government was progressively introduced between these layers – an elected National Assembly for Wales; Parliament for Scotland; an elected London Assembly, and – iniquitously – eight unelected bodies in eight newly-created English regions about which citizens had never been consulted (see: *Regions*). In England these bodies are taking power away from local rather than central government, which is the reverse of 'devolution', as the Government likes to call its regional plans, and the reverse of true democracy – see: *Localism*.
3 As of 2008, useful references include the following. For publication details see the bibliography.
• BIG BANG LOCALISM by Simon Jenkins. The author argues that Britain has one of the most centralised governments in the world and proposes a radical devolution of power to local level.
• WHITTAKERS ALMANAC is an invaluable information source on all matters governmental, and is held by most public libraries.
4 Useful websites include:
• Directgov www.direct.gov.uk/
• Info4local www.info4local.gov.uk/

See also: *Localism; Post-democracy*; and *Regions*.

vernment Equalities fice 1 In July 2007 the newly-installed Prime Minister Gordon Brown announced the establishment of a new body, the GEO, within the Department for Work and Pensions (qv). The Office would take responsibility for the Government's overall strategy on *equality* (qv) issues. Its work includes:
• Creating a more integrated approach on equality issues across government departments.
• Taking forward the priorities of the *Minister for Women* (qv), as announced on 17 July 2007.
• Taking forward the *Discrimination Law Review* (qv) and looking at the whole of *discrimination* (qv) law, leading to a *Single Equality Act* (qv).
• Responding to the *Equalities Review* (qv).
2 The GEO sponsors the following *quangos* (qv):
• Commission for Equality and Human Rights (qv).
• Women's National Commission (qv).
3 For more information visit www.dwp.gov.uk/aboutus/geo.asp

See also: *Commission for Equality and Human Rights*; *Equality*; and *Single Equality Act*.

Government: central, regional, and local

1 Central government in England and the rest of the UK (commonly referred t as Westminster) is run by the following departments. They are given in alphabetical order:

Cabinet Office	www.cabinetoffice.gov.uk
Department for Business, Enterprise & Regulatory Reform	www.berr.gov.uk/
Department for Children, Schools & Families	www.dfes.gov.uk
Department for Communities & Local Government	www.communities.gov.uk
Department for Culture, Media & Sport	www.culture.gov.uk/
Department for Environment, Food and Rural Affairs	www.defra.gov.uk/
Department for Innovation, Universities & Skills	www.dius.gov.uk/
Department for Transport	www.dft.gov.uk/
Department for Work & Pensions	www.dwp.gov.uk/
Department of Health	www.dh.gov.uk/
Home Office	www.homeoffice.gov.uk/
HM Treasury	www.hm-treasury.gov.uk/
Ministry of Justice	www.justice.gov.uk/

Most of these departments have entries in the present work.

2 Regional government in England consists of eight unelected regional bodies plus one that is elected (London). The nine regions are listed below together with the websites of their respective Government Offices (qv):

North East	www.go-ne.gov.uk
Yorkshire and the Humber	www.goyh.gov.uk
East Midlands	www.go-em.gov.uk
East of England	www.go-east.gov.uk
South East	www.go-se.gov.uk
London	www.go-london.gov.uk
South West	www.gosw.gov.uk
West Midlands	www.go-wm.gov.uk
North West	www.go-nw.gov.uk

For a map of the regions, and more detail on their governmental structures, se *Regions*.

3 Local government (qv) in England has traditionally been based upon the following pattern:

- In non-metropolitan areas: a two tier system of *county councils* at the upper level, and below that a number of *district councils*. County councils typically provided education and social services, leaving housing, environmental health etc., to the district councils. For a map of England's forty six counties see: *Regions*.
- In the six metropolitan areas: a single tier system of *metropolitan councils* (called *borough councils* in London).

But in some non-metropolitan areas the two-tier system was replaced between 1995 and 1998 by a single tier system of *unitary authorities* – ie ones which do not split up the various services between them, but provide the full range o local authority services themselves. The term usually applies to urban or other built up areas.

4 For more up-to-date information refer to WHITTAKERS ALMANAC or visit the following websites:

- Directgov www.direct.gov.uk/
- Info4local www.info4local.gov.uk/

See also: *Local authorities; Regions*; and *Wards*.

Government: London 1 As noted above London is one of the nine *regions* of England, and the only one with an elected assembly. It is governed by the Greater London Authority (GLA), which has responsibility for the 32 London boroughs and the Corporation of London. At its head is a directly elected mayor (as of 2007, Mr Ken Livingstone) whose activities are scrutinised by a London Assembly consisting of 25 elected members.

2 The GLA has eight areas of responsibility – transport; planning; economic development and regeneration; the environment; police; fire and emergency planning; culture; and health. The bodies that co-ordinate these functions and report to the GLA are:

Transport for London (TfL) www.tfl.gov.uk/
London Development Agency (LDA) www.lda.gov.uk/
Metropolitan Police Authority (MPA) www.mpa.gov.uk/
London Fire and Emergency Planning Authority (LFEPA) www.london-fire.gov.uk/

3 The Greater London Authority Act 1999 places responsibility for strategic planning in London on the Mayor, and requires him to produce a Spatial Development Strategy for London, and to keep it under review. For details see: *London plan*.

4 As of 2008, useful websites include:

• London www.london.gov.uk/

Government: Northern Ireland Visit www.nics.gov.uk/

Government: Scotland 1 In September 1997 the people of Scotland voted for a devolved Scottish Parliament (see: *Government* above). Roughly 60% of the electorate participated, and of those 74% voted Yes. The Parliament was duly established and started work in 1999.

2 Devolution was intended to boost Scotland's ability to solve its own problems, but its dependency on England appears instead to have increased. Figures published in December 2006 in 'Government Expenditure and Revenue in Scotland 2004–05' showed that government expenditure per head of population was £6 361 a year in England and £7 597 in Scotland. But GDP per capita in Scotland is 5% below that of England; and the more generous expenditure on Scottish citizens depends on an annual subsidy from English taxpayers of around £11 billion.

3 This financial anomaly is aggravated by a political one. Scottish devolution was framed in such a way that Westminster MPs with Scottish seats are able to vote on matters affecting English constituencies, but MPs with English constituencies have no corresponding influence on very many Scottish matters. Thus were tuition fees imposed upon English students with the help of the votes of Scottish MPs at Westminster, but not imposed upon Scottish students because the Scottish Parliament preferred not to – for more on this see: *West Lothian question*.

4 The above anomalies cannot endure, and will have to be remedied by a new *constitutional* (qv) settlement. The matter is discussed, and a way forward proposed, in the pamphlet ABC: A BALANCED CONSTITUTION FOR THE 21ST CENTURY by Martin Howe – see the bibliography.

5 For details of current Scottish government structures refer to WHITTAKERS ALMANAC or visit www.scottish.parliament.uk/

Government: Wales Visit www.assemblywales.org/

Government websites

1 As of 2008, government gateway websites include the following:
• www.direct.gov.uk/ gives consumers access to information about all the government services that are available to them.
• www.info4local.gov.uk/ gives access to local government (qv) services, central government departments (qv), and quangos (qv).
• www.opsi.gov.uk/acts.htm contains the full texts of all *Acts* (qv) of Parliament since 1987.
• www.planningportal.gov.uk/ gives access to information on all *planning* (qv) matters in the UK, and to the *Building Regulations* (qv) and *Approved Documents* (qv), or England and Wales, Scotland, and Northern Ireland.

Governmental bodies

See: *Quangos*.

GPs

See: *General Practitioners*.

GP's surgeries

See: *Doctors' surgeries*.

Grab rails and grab handles

See: *Support rails*.

Gradients

See: *Access routes*; and *Pedestrian movement: gradients*.

Gravel surfaces

1 Loose gravel surfaces are very difficult to negotiate by *wheelchair-users* (qv), people pushing *pushchairs* (qv), and people using *walking and standing aids* (qv) such as walking frames.
2 While homeowners may do as they please on the *access routes* (qv) to, or *footpaths* (qv) around private dwellings, loose gravel paths should not be used on premises that are open to the public. Useful alternatives in these cases include the following:
3 Bonded gravel surfaces have virtually the same appearance as loose gravel, but without the disadvantages. They consist of a layer of gravel or other stone aggregate, resin-bonded to an asphalt or other substrate to give a firm and durable surface. Aggregates are available in many colours. Case examples may be seen at:
• London's Green Park, Hyde Park, and Kensington Gardens. The extensive pathway surfaces are in varying states of repair, and offer realistic examples of how the finish stands up to heavy wear in all weathers, with less than perfect maintenance.
• The Horniman Museum in Dulwich, South London. A new *access route* (qv) leads through lush greensward, and is surfaced in golden-brown Addastone 'amber flint' bonded gravel, forming an attractive approach to the new museum entrance.
4 If a loose gravel surface is insisted upon, for instance to maintain the character of an *historic site* (qv), a smooth-surfaced pathway can be formed within the larger gravel surface, leading for instance from the site entrance to the front door. Materials could include York stone pavings, asphalt, or bonded gravel as above.
5 In the case of parks or rural areas where a completely natural character is important, and where users are likely to be able-bodied walkers, cyclists, and horse-riders, self-binding gravel may be an appropriate surface. This is a very cohesive type of fine gravel that looks like ordinary loose sand, but which packs together tightly and – unless severely disturbed – maintains a firm surface in all weathers. Examples in or near London include the pathways across Richmond Park.

6 As of 2008, useful references include the following. For publication details see the bibliography.

- EASY ACCESS TO HISTORIC BUILDINGS.
- EASY ACCESS TO HISTORIC LANDSCAPES.

These informative and beautifully illustrated twin publications from English Heritage give excellent guidance on how to reconcile the requirement for *accessibility* (qv) with the special nature of historic buildings

7 Useful websites include:

- Addastone www.addagrip.co.uk/
- Grundon www.grundon.com/

See also: *Access routes; Cobbled surfaces*; and *Historic buildings and sites*.

---

eater London ıthority (GLA) See: *Government: London*.

---

een spaces: See: *Countryside facilities*; and *Parks*.

---

een paper 1 A consultation document that sets out the government's views on a policy area (for instance, age discrimination), and invites views and discussion. It is the first step in a policy-making process that usually leads in due course to the enactments of laws.

2 See also W*hite paper*.

---

eenspace 1 Greenspace is a *charity* (qv), and the nation's leading network of information on the improvement of all parks and green spaces. Its membership includes more than half of the local authorities in the country.

2 For more information visit www.green-space.org.uk/

---

ipping and reaching ds See: *Products for easier living*.

ıardings 1 In balustrades the guarding, as distinct from the *handrail* (qv), is that element that stops people falling over the edge.

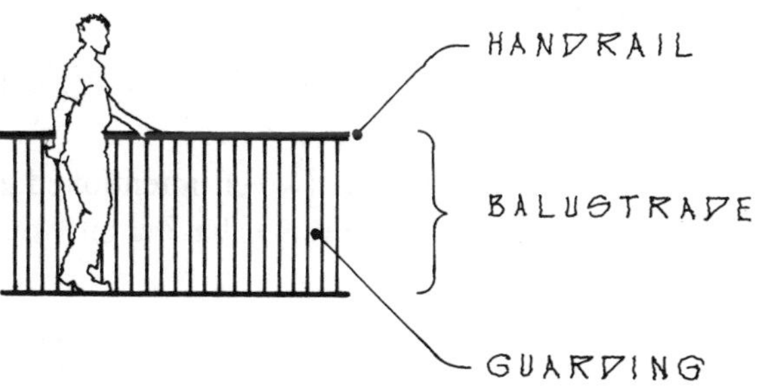

2 In England and Wales the provisions of section 3 of APPROVED DOCUMENT K (qv) of the Building Regulations must be heeded in order to comply with the law. Though checks should be carried out before taking any final decisions, AD K generally requires guardings to have a minimum height of 900 mm alongside the flights of *ramps* (qv) and *stairs* (qv), and a minimum height of 1 100 mm alongside landings. A 100 mm sphere should not be able to pass through any opening in the guarding, and the guarding should not be readily climbable by children.

For places other than England and Wales the following regulatory documents should be consulted:

- In Northern Ireland: TECHNICAL BOOKLET R of the Building Regulations (Northern Ireland) (qv).

• In Scotland: the TECHNICAL HANDBOOKS of the Building (Scotland) Regulations (qv).
• In the USA, where the Americans with Disabilities Act applies, refer to the ADA AND ABA ACCESSIBILITY GUIDELINES (qv).
3 In sports grounds, where safety is a particular concern, para 7.7 of GUIDE TO SAFETY AT SPORTS GROUNDS (qv) recommends a minimum height of 1 100 mm in all situations, including the flights of ramps and stairs.

See also: *Handrails; Ramps*; and *Stairs*.

Guidance path surface 1 Guidance path surfaces, consisting of a series of raised, flat-topped bars running in the direction of pedestrian travel, are one of six types of *tactile pavings* (qv). Their particular function is to guide people with *impaired vision* (qv) along a route where traditional cues such as a property line or *kerb* (qv) edge do not exist. They can also be used to guide people round obstacles, for example *street furniture* (qv) in a pedestrianised area. The message is: 'Walk this way'.
2 The rules for correct use in particular situations are very precise and designers must in all cases consult Chapter 6 of GUIDANCE ON THE USE OF TACTILE PAVING SURFACES (qv) and para 4.6 of INCLUSIVE MOBILITY (qv) – see the bibliography.
3 For details of the six types of tactile paving, including corduroy surfaces, see *Tactile pavings*.

Guide dogs See: *Assistance dogs*.

Guide to safety at sports grounds 1 Popularly known as the GREEN GUIDE, this is an essential reference for anyone involved in the design of sports facilities.
2 This reference should be used in conjunction with ACCESSIBLE STADIA: A GOOD PRACTICE GUIDE TO THE DESIGN OF FACILITIES TO MEET THE NEEDS OF DISABLED SPECTATORS AND OTHER USERS (qv), which updates it in some respects.
3 For publication details of both references see the bibliography.

See also: *Recreation and sports facilities*.

Guillotine 1 This is the colloquial name for a procedure that is used to cut short Parliamentary debates.
2 For a note on the way this originally well-intentioned device is now being misused, and the destructive effect this is having on Britain's democracy, see: *Parliament*.

H

abinteg

1 Habinteg is a *housing association* (qv) that specialises in wheelchair-accessible housing.
2 It is also an authoritative publisher of practical design guides in the field of *social housing* (qv) in general. Conformity with the following references, in situations where they are applicable, will help to demonstrate that the requirements of the Disability Discrimination Act (qv) have been met. For publication details see the bibliography.

- The HABINTEG HOUSING ASSOCIATION DESIGN GUIDE is a leading design manual for *general needs housing* (qv). It incorporates *lifetime home* (qv) standards, which are increasingly being required for all new social housing in the UK.
- The WHEELCHAIR HOUSING DESIGN GUIDE is a leading design manual for *special needs housing* (qv) for *wheelchair-users* (qv).

Though applying most specifically to housing association schemes, these guides ought to be by the designer's side in the design of all housing developments and individual dwellings.
3 For more information visit www.habinteg.org.uk/

See also: *Disability Discrimination Act: compliance by designers; Housing Associations; Lifetime Homes*; and *Wheelchair housing.*

andrails: complying ith the Disability iscrimination Act

1 Handrails (also in some situations called *grab rails* or *support rails*) are rails that people may grip for guidance or support and to prevent accidents. Situations in which they are recommended include the following:

- In *bathrooms* and *showers* (qv) to help people steady themselves while getting in or out of the bath; to assist them in raising or lowering themselves; and to help them avoid slipping and falling. See: *Support rails.*
- In buses, coaches, trains, and other public service vehicles. See: *Support rails*; and *Vehicles: public.*
- At *stairs* (qv) and *ramps* (qv) – see below.

2 Private dwellings are not subject to the Disability Discrimination Act (qv), but buildings and spaces that may be used by the public are. Relevant *authoritative practical guides* (qv) for designing the latter include DESIGNING FOR ACCESSIBILITY (qv) plus the following:

- APPROVED DOCUMENT M (qv), the provisions of which must be heeded in order to comply with the Building Regulations for England and Wales.
- BS 8300 (qv) which applies to all buildings in the UK.
- INCLUSIVE MOBILITY (qv), which is the most authoritative reference for the *inclusive* (qv) design of *pedestrian infrastructure* (qv) and of *transport-related buildings* (qv) such as bus and coach stations, railway stations, air terminals, and transport interchanges.
- GUIDE TO SAFETY AT SPORTS GROUNDS (qv) which applies to sports grounds in general.

Except for AD M, the above guides do not have the force of law, but conformity with their recommendations will help to demonstrate that the requirements of the DDA have been met. The relevant clauses in each are set out below:

DOCUMENT	RELEVANT PARAGRAPHS
APPROVED DOCUMENT M	1.26; 1.32; 1.34–1.37; 3.54–3.55
BS 8300	5.8.5; 5.10; 8.1.6; 8.2.5; 8.3
INCLUSIVE MOBILITY	8.4.3
GUIDE TO SAFETY AT SPORTS GROUNDS	7.8

For places other than England and Wales the following regulatory documents should be consulted:

• In Northern Ireland: TECHNICAL BOOKLET R of the Building Regulations (Northern Ireland) (qv).

• In Scotland: the TECHNICAL HANDBOOKS of the Building (Scotland) Regulations (qv).

3 Details that will in general satisfy the above requirements and recommendations are given below.

See also: *Support rails*.

Handrails: suitable details

1 All of the above documents draw a clear distinction between the 'handrail' (a rail that people may grip for guidance or support) and the 'guarding' (a screen to stop people falling over the edge). The two may be of the same height, but do not have to be – see: *Ramps* (qv) and *Stairs* (qv).

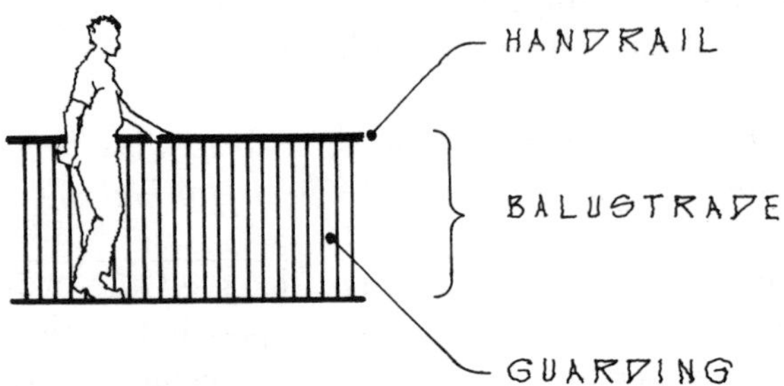

2 To suit the whole population, including people with weak or arthritic hands, handrails should have an easily grippable profile and a smooth surface.

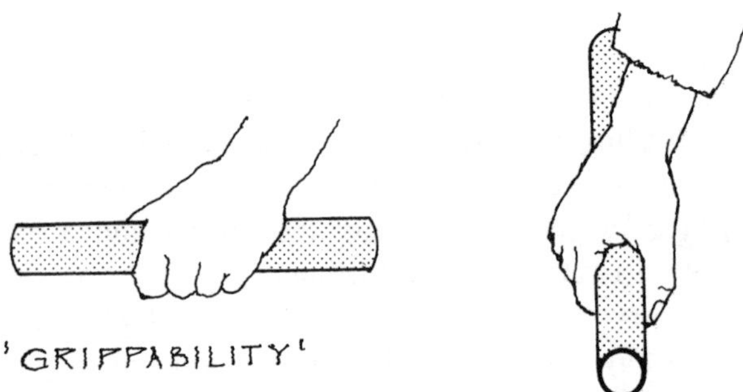

3 Support brackets should be slim, and fixed to the bottom of the handrail (no for instance the side) so that the hand can slide easily along the rail without encountering obstructions.

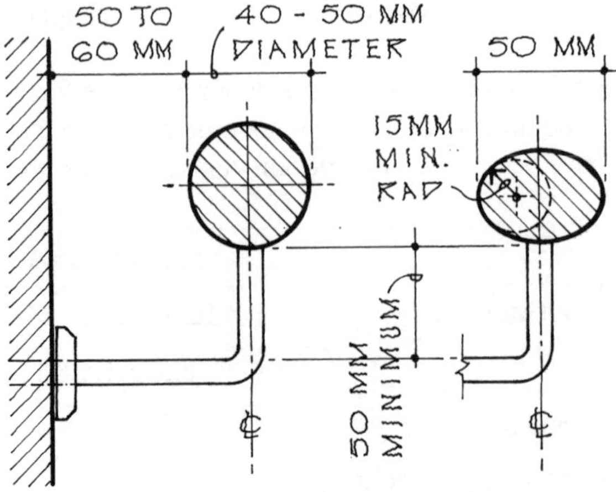

4 Guardings and handrails may be of the same height, but do not have to be – see: *Ramps* (qv) and *Stairs* (qv). Detail A below shows an example of a handrail that is at the same level as the top of the guarding, the latter in this case being a sheet of safety glass. Details B and D show situations where the guarding is higher than convenient handrail height. The detail at the left aims for transparency, that at the right for an appearance of solidity. Litter is liable to accumulate in the channel of the latter detail. It should only be used in clean indoor locations with frequent and regular cleaning, and the ends of the channel must be left open for easy sweeping.

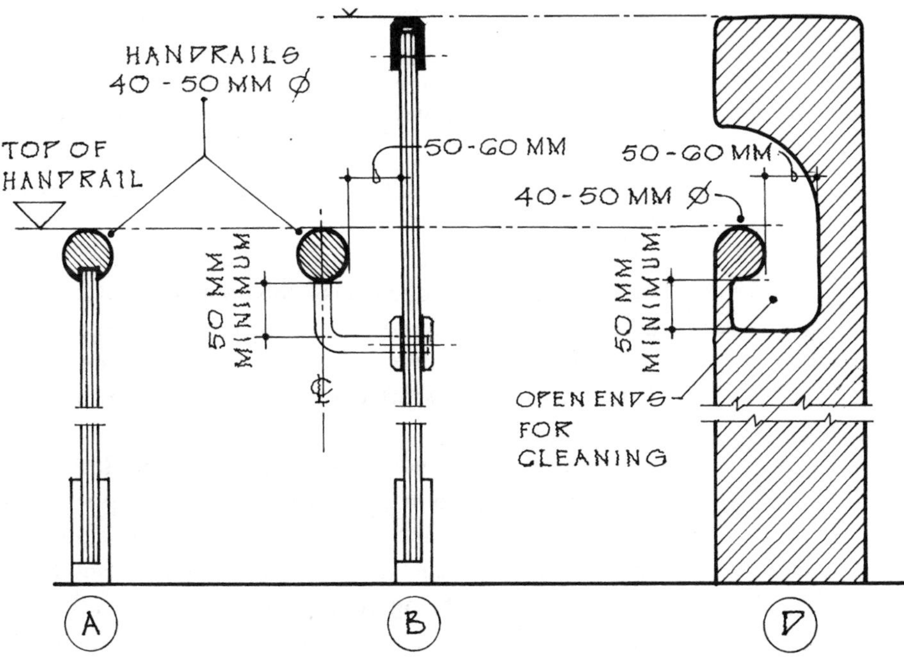

5 Though checks should be made before any decisions are finalised, the details shown above will in general conform with *authoritative practical guides* (qv).

See also: *Guardings*.

ppiness

1 Happiness is difficult to define or measure, but trying to do both probably makes researchers happier and could even yield a few insights. As of 2008, useful studies include the following:

2 THE WORLD MAP OF HAPPINESS, published in July 2006 by Professor Adrian White, an analytical social psychologist at the University of Leicester, summarises the results of over 100 studies by Unesco and other organisations. National rankings, from happiest to unhappiest, include Denmark 1st, Switzerland 2nd, Austria 3rd, Iceland 4th, Bahamas 5th, Finland 6th, Sweden 7th, Bhutan 8th, Ireland 11th, USA 23rd, Germany 35th, UK 41st, Spain 46th, Italy 50th, China 82nd, Japan 90th, India 125th. The unhappiest countries are Ukraine at 174th, Moldova 175th, Democratic Republic of Congo 176th, Zimbabwe 177th, and Burundi 178th. One analysis of these findings looked at the links between health (happiness and life expectancy), wealth (GDP per capita), and education (access to secondary level education), and concluded that happiness correlated most closely with good health, followed by wealth, and then education.

3 The January 2006 issue of the JOURNAL OF HAPPINESS STUDIES featured a survey conducted by Professor Ruut Veenhoven, a specialist in happiness

studies. Around 100 000 people in 90 countries were questioned. Rankings, from happiest to unhappiest, include Malta 1st, Denmark/Switzerland/ Colom 2nd; Iceland 3rd, Ireland/Netherlands 4th, Canada/Finland/Ghana/Sweden 5th, Luxembourg 6th, Guatemala/Norway/ Uruguay/USA 7th, New Zealand/ Australia/ Mexico 8th, Belgium 9th, and UK/Germany 10th.
4 In April 2007 the Faculty of Economics at the University of Cambridge published its latest analysis of the regular European Social Survey. Denmark come first, followed by the rest of Scandinavia and the Netherlands. Britain was 9th out of 15 countries (with East Anglia, London, and the East Midlands the happiest regions), just ahead of France and Germany. Greece, Portugal and Italy came last. The authors commented that one major key to happiness appears to be trust – trust in one's government, trust in the police and the justi system, and trust in the people one knows personally.
5 The consistently high ranking of Scandinavia and Switzerland in the above surveys suggests to the present author that people tend to be happy in societie that:
• Are rich, but with narrow differentials between the richest and poorest. See: *Gini coefficient*; and *Incomes: inequality*.
• Have honest government and excellent social services.
• Are culturally homogeneous, with populations that have over time settled int accepted ways of life not greatly disturbed by large inflows of outsiders with markedly different values and customs. See: *Diversity*.
6 As of 2008, useful references include the following. For publication details see the bibliography.
• HAPPINESS, ECONOMICS AND PUBLIC POLICY by Helen Johns and Paul Ormerod, published in 2007 by the *think tank* (qv) the Institute of Economic Affairs, is sceptical about links between economic growth and happiness, and t ability of governments to make people happier. But the authors do suggest th stable family life, being married, having financial security and good health, having religious faith, and living in a cohesive community where people can t trusted all contribute to happiness.
• HAPPINESS: LESSONS FROM A NEW SCIENCE by Richard Layard.
7 Useful websites include the following. They are given in alphabetical order.
• European Social Survey www.europeansocialsurvey.org/
• Journal of Happiness Studies www.springerlink.com/content/104910/
• Ruut Veenhoven www2.eur.nl/fsw/research/veenhoven/
• University of Cambridge www.admin.cam.ac.uk/news/dp/20070417(
• University of Leicester www2.le.ac.uk/ebulletin/news/press releases/2000-2009/2006/07/nparticle.2006-07-28.244832382
• World Database of Happiness www1.eur.nl/fsw/happiness/

See also: *Economic success; Incomes: inequality*; and *Inclusion*.

Happiness and homes 1 The HALIFAX HAPPIEST HOME REPORT (see the bibliography), prepared in 2005 for Halifax General Insurance by the Social Issues Research Centre in Oxford, found that most people rated their homes as happy, 80% of them giving their dwelling a score of 7/10 or more on the 'happiness' scale. Other interesting findings are that:
• Home happiness was directly linked to the age of the occupants. The over 50 s had significantly happier homes than younger people.
• The happiest house types in Britain were found to be *bungalows* (qv), the occupants of which rated them 8.15 out of 10 on a 'happiness' scale.
• Happiness was found to increase linearly with the number of bedrooms, and the happiest homes had a ratio of more than 1.5 bedrooms per occupant.
• Home-owners without a mortgage are happier than those with.

• The happiness of a home was, rather surprisingly, found to be independent of socio-economic group.
For more information visit www.sirc.org/publik/happy_homes.pdf
2 The preference, above, for bungalows above other house types confirms the findings of a public opinion survey conducted by Mori and announced on 25 June 2002 by CABE (qv) at the annual Housing Design Awards.
3 Two other surveys have found detached houses to be most popular, with bungalows second. These are:
• A survey conducted by NOP in spring 2004 for Halifax Estate Agents.
• WHAT HOME BUYERS WANT (see the bibliography), a report published by CABE (qv) in March 2005.
• Virtually all surveys find homes in medium-rise blocks to be unpopular, and those in high-rise blocks very unpopular. Most people want to live on or near the ground – a point that *local authorities* and architects might note. See in this connection also: *Architecture* (qv) and *le Corbusier* (qv).

---

te crime

1 Hate crime is described by the Home Office (qv) as any criminal offence committed against a person or property that is motivated by an offender's hatred of someone on grounds of gender, gender identity, or sexual orientation; disability; race, colour, or ethnic origin; nationality or national origin; or religion. The Home Office website further defines hate crime as 'any incident, which constitutes a criminal offence, which is perceived by the victim or by any other person as being motivated by prejudice or hate'.
2 This legislation overturns two fundamental principles of British law, both of which were hard-won, have stood the test of time, and are central to our sense of fairness and democracy:
• First, that all citizens should be treated equally under the law. Why should a crime motivated by hatred of (say) Jews or homosexuals count as a hate crime, but not one that is motivated by hatred of tramps or beggars? Why should such a crime be punished twice as severely as the identical crime inflicted on a victim not coming within one of the defined groups? Is the life of a Christian worth half that of Jew or Muslim, or the life of a heterosexual worth half that of a homosexual? For more on this see: *Victimhood*.
• Second, that crimes should be defined as factually and objectively as possible, with minimal recourse to variable subjective opinion. In March 1999 the Macpherson Report turned this precept on its head by introducing the wholly subjective criterion, noted above, that 'a racist incident is any incident which is perceived (sic) to be racist by the victim or any other person'.
3 Laws that flout the sense of justice held by the broad mainstream of society risk bringing the entire law-making process into disrepute, and generating damaging backlashes. Our legal masters would be wise, before proceeding any further along their present course, to engage in a true national debate on where they are taking us – that is to say, one that includes all of society and not just ministers, their civil servants, and a closed circle of *experts* (qv), *think tanks* (qv) and single-issue *pressure groups* (qv).
4 As of 2008, useful references include the following. For publication details see the bibliography.
• THE STEPHEN LAWRENCE INQUIRY: REPORT OF AN INQUIRY BY SIR WILLIAM MACPHERSON OF CLUNY. For the formulation that 'a racist incident is any incident which is perceived to be racist by the victim or any other person' see pp 313–14, volume 1.
• WE'RE (NEARLY) ALL VICTIMS NOW! HOW POLITICAL CORRECTNESS IS UNDERMINING OUR LIBERAL CULTURE. This polemic by David G Green, Director of the *think tank* (qv) Civitas, was published in 2006.
5 Useful websites include the following. They are given in alphabetical order:
• Civitas www.civitas.org.uk/

- Equality & Human Rights Commission www.equalityhumanrights.com
- Home Office
  www.homeoffice.gov.uk/crime-victims/reducing-crime/hate-crim
- Office of Public Sector Information www.opsi.gov.uk/acts.htm

See also: *Discrimination; Equality*; and *Victimhood.*

Hazard markings

See: *Glass: safety markings; Hazards: protection*; and *Street works: safety markings*.

Hazard warning surfaces

See: *Tactile pavings*.

Hazards

See: *Risk*.

Hazards: protection

1 On *access routes* (qv) and other *pedestrian* (qv) areas, hazards (qv) such as posts and columns, outward-opening windows, and trees and shrubs can cause injury by collision, particularly to people with *impaired vision* (qv). Para 5.7 of BS 8300 (qv), which is an *authoritative practical guide* (qv), recommends a number of precautionary measures, including the following.
2 Litter bins, seats, bollards, flower beds, lighting columns, and other *street furniture* (qv) should be located outside the walkway area, so that the latter is completely clear of obstacles. The figure shows some of the recommended safety precautions at litter bins, bollards, and planting boxes etc.

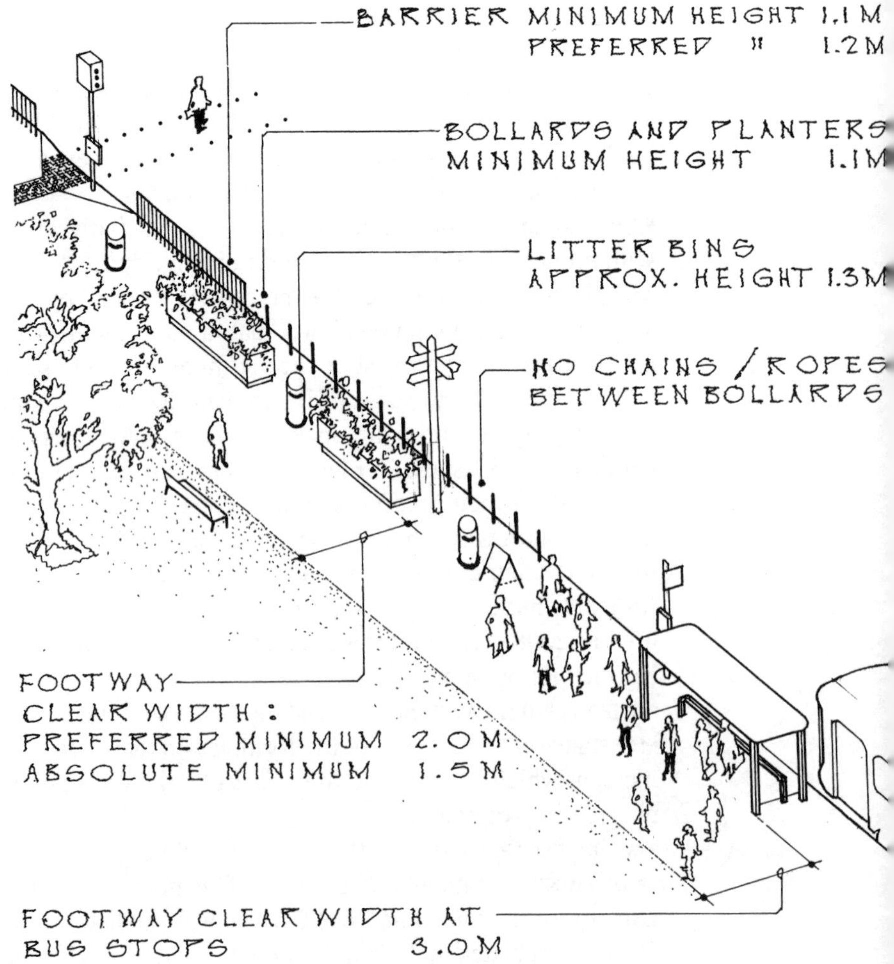

3 If the presence of a column or post within the walkway cannot be avoided, then the obstacle should be marked with a warning band 150 mm deep, with its bottom edge 1 500 mm above paving level – see: *Street works*.

4 If an open window or door, a telephone booth, a shop display case, or any other part of a building projects more than 100 mm into a pedestrian walkway then one of the following precautions should be applied:

• A cane-detectable *kerb* (qv) or other solid barrier at ground level, plus a rail or some other form of *guarding* (qv) between 900 mm and 1 100 mm high, to guide people round the hazard.

Or

• A cane-detectable deterrent surface that will guide people round the hazard.

Or

• A flowerbed or some similar feature, that is at least 900 mm high, to prevent people walking into the hazard.

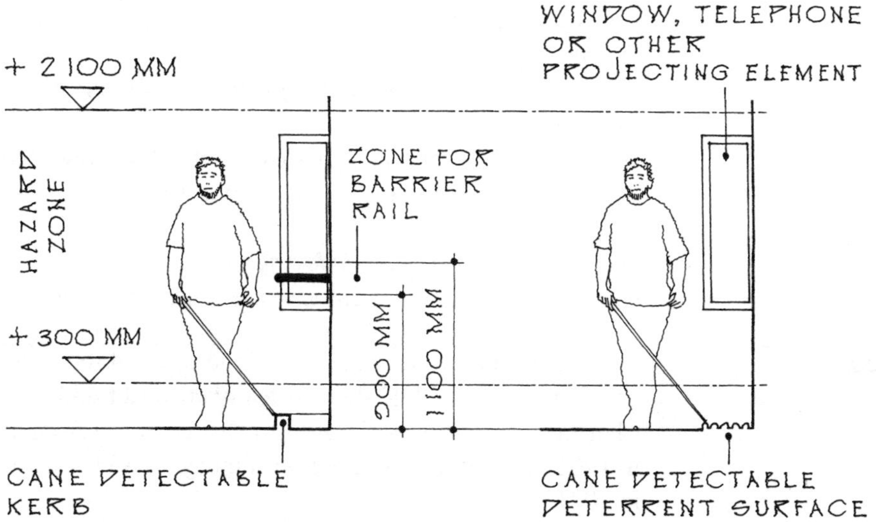

5 Similar measures must be taken if any part of a free-standing *stair* (qv) or *ramp* (qv) has a soffit lower than 2 100 mm above paving level.

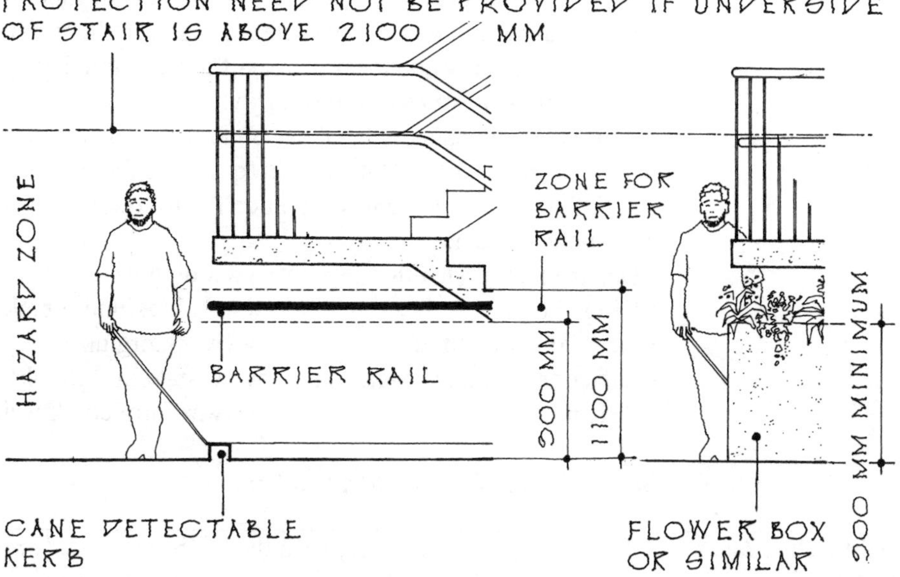

6 Though checks should be carried out before decisions are finalised, the details shown above will in general conform with the following *authoritative practical guides* (qv).

• APPROVED DOCUMENT M (qv): the provisions in paras 1.38–1.39 must be heeded in order to comply with the Building Regulations for England and Wales.
• BS 8300 (qv): the recommendations in para 5.7 apply to all buildings in the UK.
• INCLUSIVE MOBILITY (qv): the recommendations in paras 3.7–3.8 apply to *pedestrian environments* (qv) in general – i.e. the public footways and footpat pedestrian precincts, parks, squares, and other walkable surfaces that make u our cities, towns, and countryside.
Except for APPROVED DOCUMENT M, the above guides do not have the force of law, but conformity with their recommendations will make it easier to demonstrate that the requirements of the Disability Discrimination Act have been met.
For places other than England and Wales the following regulatory documents should be consulted:
• In Northern Ireland: TECHNICAL BOOKLET R of the Building Regulations (Northern Ireland) (qv).
• In Scotland: the TECHNICAL HANDBOOKS of the Building (Scotland) Regulations (qv).
7 In the USA, where the Americans with Disabilities Act applies, refer to the ADA AND ABA ACCESSIBILITY GUIDELINES (qv),

See also: *Access routes; Pedestrian environments; Canes for blind people; Glass: safety markings; Street works: safety markings*; and *Tactile pavings*.

Health

1 Life expectancy is a widely used indicator of the state of the nation's health and figures for the UK show a steady upward trend:
• For men, life expectancy at birth has increased from 45 years for babies bor in 1901, to 77 years for those born today, to 80 years (projected) for those bo in 2021.
• For women, life expectancy at birth has increased from 49 years for babies born in 1901, to 81 years for those born today, to 84 years (projected) for tho born in 2021.
These are *average* (qv) figures, and while people of all backgrounds are livin longer, poor people are benefiting less than the affluent. According to SOCIAL TRENDS (qv) the gap in life expectancy between the worst-off and the best-off people is currently about seven years for men, and three years for women. Se in this context also: *Health inequality*.
2 More detailed data on the above and related matters is given in Chapter 7 o the latest edition of SOCIAL TRENDS (qv) – see the bibliography.
3 As of 2008, useful websites include the following. They are given in alphabetical order.

• Department of Health	www.dh.gov.uk/
• Kings Fund	www.kingsfund.org.uk/
• New Policy Institute	www.npi.org.uk/
• Office for National Statistics	www.statistics.gov.uk/
• Reform	www.reform.co.uk/website/health.aspx

See also: *Health inequality*; and *NHS*.

Health Action Zones (HAZs)

1 These are partnerships between the NHS (qv), *local authorities* (qv), community groups, and the *voluntary* (qv) and business sectors in areas of hi *deprivation* (qv); and they are aimed at tackling *health inequality* (qv) and pc health.
2 For more information visit www.investingforhealthni.gov.uk/zones.asp

lth and safety

See: *Health and Safety Commission*; and *Royal Society for the Prevention of Accidents*.

lth and Safety mission (HSC)

1 The Health and Safety Commission was created under the Health and Safety at Work etc. Act 1974 (qv) with the purpose of reforming health and safety law; proposing new regulations; and generally to promote the protection of the public and people at work from the *hazards* (qv) that arise from industrial and commercial activity, including major industrial accidents and the transportation of hazardous materials.
2 Visit www.hse.gov.uk/

lth and Safety cutive (HSE)

1 The HSE is the executive instrument of the Health and Safety Commission (qv). It advises the Commission on the creation of safety standards through regulations and practical guidance. Its aims are worthy, but much of the *over-regulation* (qv) that now afflicts the UK originates with this nannying and all-intrusive body.
2 Visit www.hse.gov.uk/

See also: *Risk assessments*.

lth and safety in sing

See: *Housing Fitness Standard*.

lth and welfare dings: basic ning data

1 Distilled planning data for most of the major building types is given in the METRIC HANDBOOK: PLANNING AND DESIGN DATA. For publication details see the bibliography.
2 For further information, case studies, and commentary on design quality visit the following websites:

- CABE www.cabe.org.uk/default.aspx?contentitemid=38
- Estates and Facilities Management www.dh.gov.uk/

lth and welfare dings: design: plying with the bility crimination Act

1 Buildings used by the public must in general comply with the Disability Discrimination Act (qv). In the case of hospitals; health centres; doctors' and dentists' surgeries; opticians' facilities; and older people's day centres, *authoritative practical guides* (qv) include DESIGNING FOR ACCESSIBILITY (qv) plus the following:
2 BS 8300, the provisions of which apply to all buildings in the UK. Para 13.4 gives brief design recommendations for all the building types listed above. BS 8300 does not have the force of law, but conformity with its recommendations will make it easier to demonstrate that the requirements of the Disability Discrimination Act have been met.
3 As of 2008, useful references include the following. They are given in alphabetical order. For publication details see the bibliography.

- Estates and Facilities Management www.dh.gov.uk/
- Kings Fund www.kingsfund.org.uk/

See also: *Hospital design: quality*.

lth authorities

See: *NHS*.

lth Improvement grammes (HImPs)

1 These local plans to improve health and healthcare are drawn up by *primary care* professionals working in conjunction with other agencies, such as *local authorities* (qv) and the *voluntary sector* (qv).
2 For more information visit www.nhs.uk/

Health Protection Agency (HPA)	1 The Health Protection Agency is a *quango* (qv) whose task is to protect UK public health by providing support and advice to the NHS (qv), *local authori* (qv), the Department of Health (qv), and other relevant bodies. 2 For more information visit www.hpa.org.uk/
Health inequality	1 This term refers to the gap in health, and in access to health services, betw different social classes and ethnic groups and between populations in differe geographical areas. According to TACKLING HEALTH INEQUALITIES: 2007 STATU REPORT ON THE PROGRAMME FOR ACTION, published in March 2008 by the Department of Health (see the bibliography), most people are getting health but the gap between rich and poor is widening. 2 As of 2008, useful websites include the following. They are given in alphabetical order. • Department of Health www.dh.gov.uk/ • Focus on Social Inequalities www.statistics.gov.uk/focuson/socialinequaliti • Kings Fund www.kingsfund.org.uk/ • Office for National Statistics www.statistics.gov.uk/ • Reform www.reform.co.uk/website/health.aspx See also: *Health; Poverty*; and *Social exclusion*.
Health scares	1 Various busybodies and vested interests strive constantly to make Britain's unprecedentedly healthy population fearful about its food (the best they have ever enjoyed), its water (the cleanest they have ever had), going out in the sun, the environment in general, and just about every aspect of normal life. The culprits include pharmaceutical companies trying to sell products (see: *Medicalisation*); academics trying to win attention, publishers trying to sell books, and single-issue *pressure groups* (qv) trying to frighten population at large into sharing their skewed views of the world. A particula sad consequence of their scare-mongering is the extent to which guilt-ridden adults are robbing youngsters of the fun to which children are entitled by not letting them play freely in the mild British sun, as children through the ages have safely done, without being made miserable by unpleasant layers o protective clothing and/or sticky pharmaceutical cream – see: *Children's pla* 2 As of 2008, the following books offer interesting background reading. For publication details see the bibliography. • HUBBUB: FILTH, NOISE AND STENCH IN ENGLAND, 1600–1717 by Emily Cockayne. outlines the grim realities of life in the past under chapter headin such as Ugly, Itchy, Mouldy, Noisy, Grotty, Busy, Dirty, and Gloomy. Not fo the faint-hearted. • THE IMPROVING STATE OF THE WORLD: WHY WE'RE LIVING LONGER, HEALTHIER, MORE COMFORTABLE LIVES ON A CLEANER PLANET by Indur Goklany. • PANIC NATION: UNPICKING THE MYTHS WE'RE TOLD ABOUT FOOD AND HEALTH by Stanley Feldman and Vincent Marks. • SCARED TO DEATH: FROM BSE TO GLOBAL WARMING – HOW SCARES ARE COSTING US THE EARTH by Christopher Booker and Richard North. See also: *Medicalisation*.
Health service ombudsman	Visit www.ombudsman.org.uk/
Health Trusts	See: *NHS structure*.

althcare nmission

1 The Healthcare Commission (more properly known as the Commission for Healthcare Inspection and Audit) is a *quango* (qv) that was launched in April 2004 to replace the Commission for Health Improvement. It carries out independent assessments of the performance of those who provide services.
2 The HC will be absorbed in a new Care Quality Commission in 2008. Meanwhile, visit www.healthcarecommission.org.uk/

althcare services

Visit www.nhs.uk/England/AuthoritiesTrusts/Default.cmsx

althy Hospitals npaign

1 A campaign launched by the CABE (qv), in November 2003, in partnership with the Royal College of Nursing, to call for radical improvements in the design of new hospitals.
2 For more information visit www.mascot-creative.co.uk/mascot2/content/w_hh.html

See also: *Hospital design: quality*.

althy Living Centres .Cs)

Visit www.dh.gov.uk/

aring

See: *Impaired hearing*.

aring-aids and iipment

1 Hearing aids and equipment comprise two groups of devices:
• Hearing aids, which are personal aids.
• Hearing enhancement systems, which are installed in buildings.

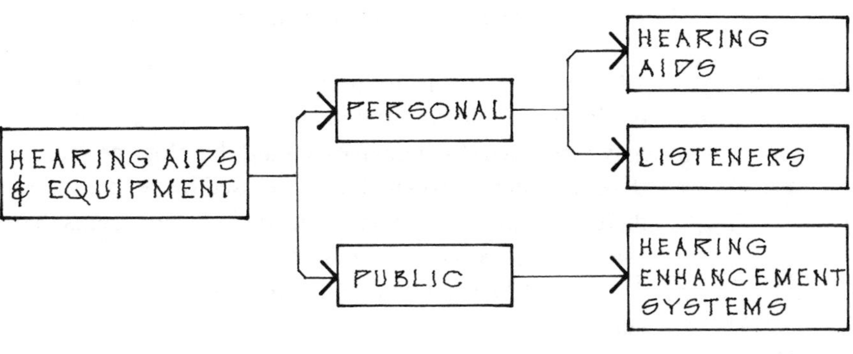

2 Both are forms of *communication aids and systems* (qv). More details on both are given below.

See also: *Aids and equipment*; and *Impaired hearing*.

aring aids and teners

1 *Hearing aids* are devices worn in or behind the user's ear to amplify received sound; *listeners* are handheld or neck-worn devices used by deaf or near-deaf people either on their own, or in conjunction with hearing aids, to enhance received sound.
2 SOLUTIONS, the annual catalogue of the RNID (qv) gives details of these and other products for people with *impaired hearing* (qv). For publication details see the bibliography.

aring enhancement stems: complying h the Disability scrimination Act

1 Buildings used by the public must in general comply with the Disability Discrimination Act (qv), which means that such buildings should be easily usable by people with *impaired hearing* (qv).
2 Relevant *authoritative practical guides* (qv) include DESIGNING FOR ACCESSIBILITY (qv) plus the following:

• APPROVED DOCUMENT M (qv) of the Building Regulations. The provisions in paras 0.14, 3.6, and 4.31–4.36 must be heeded in order to comply with the Building Regulations for England and Wales.
• BS 8300 (qv). The recommendations in section 9.3 apply to all buildings in the UK.
• INCLUSIVE MOBILITY (qv). The recommendations in sections 9.1–9.4 apply to *transport-related buildings* (qv) such as bus and coach stations, railway station air terminals, and transport interchanges.
Except for AD M, the above guides do not have the force of law, but conformi with their recommendations will make it easier to demonstrate that the requirements of the Disability Discrimination Act have been met.
3 Put simply, the requirement is that 'a hearing enhancement system, using induction loop, infrared, or radio transmission, should be installed in rooms an spaces used for meetings, lectures, classes, performances, spectator sport or films, and at service and reception counters where the noise level is high or where glazed screens are used'.
In places other than England and Wales the following regulatory documents should be consulted:
• In Northern Ireland: TECHNICAL BOOKLET R of the Building Regulations (Northern Ireland) (qv).
• In Scotland: the TECHNICAL HANDBOOKS of the Building (Scotland) Regulations (qv).
4 In the USA, where the Americans with Disabilities Act applies, refer to the ADA AND ABA ACCESSIBILITY GUIDELINES (qv).

See also: *Acoustics; Impaired hearing; Induction communication systems; Infrared communication systems*; and *Public address systems.*

Heat emitters

1 Radiators with surface temperatures of 41°C or higher can burn *disabled* (q or elderly people who fall against them. If they are in a position where harm may be caused, such devices should therefore be screened. This is particularly important in rooms which may be used by people with little or no feeling in their limbs or torso. Such rooms include bathrooms, bedrooms, changing and shower areas, kitchens, and lavatories.
2 The most relevant clauses in *authoritative practical guides* (qv) are tabulate below:

ROOM	DOCUMENT: RELEVANT PARAGRAPHS	
	APPROVED DOCUMENT M	BS83
TOILETS & SANITARY ACCOMMODATION	5.4 AND 5.10	
KITCHENS		12.1.
BATHROOMS		12.2.
CHANGING & SHOWER AREAS		12.3.
LAVATORIES		12.4.
BEDROOMS		12.5.

Height above floor level for fittings and work surfaces

1 For compliance with the Disability Discrimination Act (qv), the guidance i the following paragraphs of BS 8300 (qv) should be heeded:

• Alarm pull cords	12.4.10
• Automatic teller machines	10.2
• Bathroom fittings	12.2

• Changing cubicle fittings	12.3.3–12.3.6
• Clothes hooks	2.2.11
• Coin and card operated devices	10.2
• Controls and switches	10.5.2
• Counters and reception desks	10.5.2
• Door entry control systems	10.2.5
• Door handles	6.5.1
• Electrical outlets & switches	10.5.2
• Guardings	5.7.2–5.7.3
• Hand rails	5.9.7, 5.10, 8.1.6, 8.2.5, 8.3
• Kitchen fittings	12.1
• Lavatory fittings	12.4.3
• Mirrors	12.4.6.2
• Public telephone controls	10.4.2
• Seats in waiting areas etc.	11.2.4
• Shower fittings	12.3.6
• Storage shelves	10.1.3
• Support rails	12.3.3–12.3.6
• Tables in restaurants and cafes	13.5.3
• Wash basins	12.4.5
• Window controls	10.3.2
• Window sills and transoms	10.3.1
• Work surfaces in kitchens	12.1.3–12.1.4

2 BS 8300 (qv) is an advisory document and does not have the force of law, but conformity with its recommendations will make it easier to demonstrate that the requirements of the Disability Discrimination Act have been met.

See also: *Authoritative practical guides*; and *BS 8300: contents.*

---

ght-adjustable iture See: *Adjustable furniture.*

Majesty's See: *HM*

---

itage buildings sites See: *Historic buildings and sites.*

---

itage Lottery Fund See: *National Lottery.*

---

h-tech aids and ipment See: *Aids and equipment*; and *Assistive technology.*

---

her Education ulation Review up (HERRG)

1 This body was created to cut *red tape* (qv) - see: *Cutting Burdens on Teachers.* Instead, it has generated a truly impressive amount of paperwork and has not reduced the bureaucratic load on frontline staff by one iota.
2 For more information on the HERRG visit www.dfes.gov.uk/hegateway/hereform/improvingregulation/index.cfm

See also: *Better regulation bodies.*

Ps See: *Health Improvement Programmes.*

---

s See: *Home Information Packs.*

Historic buildings and sites: design: complying with the Disability Discrimination Act

1 Buildings and spaces used by the public must in general comply with the Disability Discrimination Act (qv). In the case of historic buildings *authoritative practical guides* (qv) include DESIGNING FOR ACCESSIBILITY (qv) plus the following:

• APPROVED DOCUMENT M (qv), the provisions of which must be heeded in order to comply with the Building Regulations (qv) for England and Wales. Paras 0.17–0.19 and 3.22 deal specifically with historic buildings.

• BS 8300, the provisions of which apply to all buildings in the UK. Para 13. gives brief design recommendations with respect to historic premises. BS 830(does not have the force of law, but conformity with its recommendations will make it easier to demonstrate that the requirements of the Disability Discrimination Act have been met.

In places other than England and Wales the following regulatory documents should be consulted:

• In Northern Ireland: TECHNICAL BOOKLET R of the Building Regulations (Northern Ireland) (qv).

• In Scotland: the TECHNICAL HANDBOOKS of the Building (Scotland) Regulations (qv).

2 It should be noted that:

• Para 20(2) of the Disability Discrimination Act (qv) only requires *service providers* (qv) to take such steps as are 'reasonable, in all the circumstances c the case', to ensure that access is not 'impossible' or 'unreasonably difficult' fo disabled persons. The standards of accessibility that are demanded of new buildings cannot necessarily be applied to existing premises, and the term 'reasonable' must be given full weight.

• Para 0.18 of APPROVED DOCUMENT M (qv) of the Building Regulations states that 'the aim should be to improve accessibility where and to the extent that it practically possible, always provided that the work does not prejudice the character of the historic building, or increase the risk of long-term deteriorati to the building fabric or fittings'.

• Para 13.10 of BS 8300 (qv) states that 'historic buildings should be made accessible to disabled people, wherever possible, without compromising conservation and heritage issues'.

These authoritative references state unanimously that only 'reasonable' adjustments need be made to historic properties, yet we read reports of much loved historic premises being shut down or refused the grants that are needed for their continued existence, simply because they could not be brought fully u to the standards of physical access that are required of new buildings. Such breaches of the law should be vigorously resisted.

3 Examples in London of adapted historic properties where *accessibility* (qv) has been achieved while protecting (or even enhancing) the architectural qua and historic integrity of the site include:

• The Queen's House at Greenwich. Note particularly the main entrance, where the original entrance steps have been eliminated by gently regrading th site – see: *Entrances to historic buildings*.

• Southwark Cathedral, SE 1. A major refurbishment programme was carried out starting in 1997, a key feature of which is Lancelot's Link, an internal str linking the new buildings with the cathedral.

4 As of 2008, useful references include the following. They are given in alphabetical order. For publication details see the bibliography.

• BUILDING REGULATIONS AND HISTORIC BUILDINGS, a free booklet from English Heritage which gives guidance on how to reconcile the needs of ener conservation, as embodied in Part L of the Building Regulations (qv), with th special nature of historic buildings.

• CABE AND THE HISTORIC ENVIRONMENT sets out the Commission for the Built Environment's approach to historic conservation.
• EASY ACCESS TO HISTORIC BUILDINGS.
• EASY ACCESS TO HISTORIC LANDSCAPES.
These informative and beautifully illustrated twin publications from English Heritage give excellent guidance.
5 Useful websites include:
• CABE www.cabe.org.uk/
• English Heritage www.english-heritage.org.uk/

See also: *Cobbled surfaces; Gravel surfaces; Local access groups*; and *Reversibility*.

V infection

1 The human immunodeficiency virus (HIV) attacks the body's immune system, making it hard to fight off infections. HIV infection is specifically recognised as a *disability* under the Disability Discrimination Act (qv) and this recognition is effective from the time of diagnosis, and not – as is the case with most other conditions – from the point at which the effects of the condition have become 'substantial'.
2 According to the Health Protection Agency (qv), there were approximately 63 000 adults in the UK with HIV in 2006, around a third of them unaware of their condition. Thanks to new treatments the number of deaths attributed to HIV fell from a peak of 1 500 in 1996 to around 500 a year in 2006.
3 For more information visit www.hpa.org.uk/

See also: *Aids:* and *Disability Discrimination Act.*

Courts Service

1 This Executive Agency (qv) within the Ministry of Justice (qv) is responsible for managing the magistrates' courts, the Crown Court, county courts, the High Court and the Court of Appeal in England and Wales. Its tasks are to reduce crime and anti-social behaviour, protect the rights of the law abiding citizen, and make our communities safer. Its success in these matters, over the past half century, may be judged by the graph of Recorded Crime Statistics under *Crime* (qv).
2 For more information visit www.hmcourts-service.gov.uk/

Inspectorate MI)

See: *Office for Standards in Education.*

Inspectorate of nstabulary

1 HM Inspectors of Constabulary are appointed by the Crown on the recommendation of the Home Secretary, and their task is to make policing in England, Wales, and Northern Ireland more efficient and effective. Their effectiveness in these matters, over the past half century, may be judged by the graph of Recorded Crime Statistics under *Crime* (qv).
2 For more information visit http://inspectorates.homeoffice.gov.uk/hmic/

Inspectorate of ons

1 The HMIP's task is to report to the Home Secretary on the treatment and conditions of *prisoners* (qv) in England and Wales. It claims to be an independent inspectorate, but as the Chief Inspector of Prisons is appointed by the person whose Department is being inspected – the Home Secretary – is open to doubt. Sir David (now Lord) Ramsbotham, who was appointed in 1995 by the then Home Secretary Mr Michael Howard, was admirably robust, but few of his successors are likely to have his unusual background and vigorous independence of mind.
2 For more information visit www.inspectorates.homeoffice.gov.uk/hmiprisons/

HM Inspectorate of Probation

1 The Inspectorate has responsibility for inspecting the work and performance of the National Probation Service (qv) and of Youth Offending teams. Its aim is t make the public safer by supervising offenders and reducing their rate of re-offending. Its website states that it is 'completely independent', but as it is funded by the Ministry of Justice, and reports to the Secretary of State for that department rather than Parliament, it clearly is not.
2 The failure by successive governments to build enough *prisons* (qv) to hold Britain's convicted criminals, resulting in the constant early release of people who in the interests of public safety should be in custody, and then expecting a grossly understaffed probation service to stop them re-offending, has in recent years made the work of this service virtually impossible.
3 For more information visit:
http://inspectorates.homeoffice.gov.uk/hmiprobation/about-us.html/

See also: *Crime; National Probation Service*; and *Prisons*.

HM Land Registry

1 HMLR registers the ownership of land in England and Wales. Its services include a House Price Index that gives average *house prices* (qv) for England and Wales both overall, and at regional or local levels.
2 For more information visit www.landreg.gov.uk/

See also: *House prices*.

HMOs

See: *Houses in multiple occupation*.

HMO grant

1 A type of *home improvement grant* (qv). For more information visit www.direct.gov.uk/
2 See also: *Houses in multiple occupation*.

HM Prison Service

1 Her Majesty's Prison Service aims to serve the public by keeping in custody those committed by the courts; looking after them with humanity; and helping help them lead law-abiding and useful lives in custody and after release.
2 The Service has for many years been thwarted in the latter task by successiv governments' failure to build enough *prisons* (qv) to hold Britain's convicted criminals. Not only are tens of thousands of persistent criminals who ought to be locked up out on the streets, making other people's lives a misery (across th EU there are on average 17.7 prisoners for every 1 000 recorded crimes; in the UK only 12.7), but the minority who are in prison cannot be effectively educated, trained, and rehabilitated owing to chronic overcrowding and the constant moving of prisoners from one prison to another. Here lies one of the explanations for the vast rise in the Recorded Crime Statistics shown under *Crime* (qv).
3 For more information visit www.hmprisonservice.gov.uk/

See also: *Crime*; and *Prisons*.

Hoists

1 The manual lifting of *disabled* (qv) or sick people out of beds or baths, and moving them from one position or place to another, places a great strain on bot *carer* (qv) and patient, with a *risk* of injury to both.
2 Hoists are a category of *aids and equipment* (qv) for use in hospitals, care homes, dwellings, etc., to reduce this risk. There are many types, as shown in the references below. The diagram on the next page shows a mobile hoist (lef and a ceiling-mounted model (right). In the latter the components are (1) trac (2) trolley; (3) spreader bar; and (4) sling.

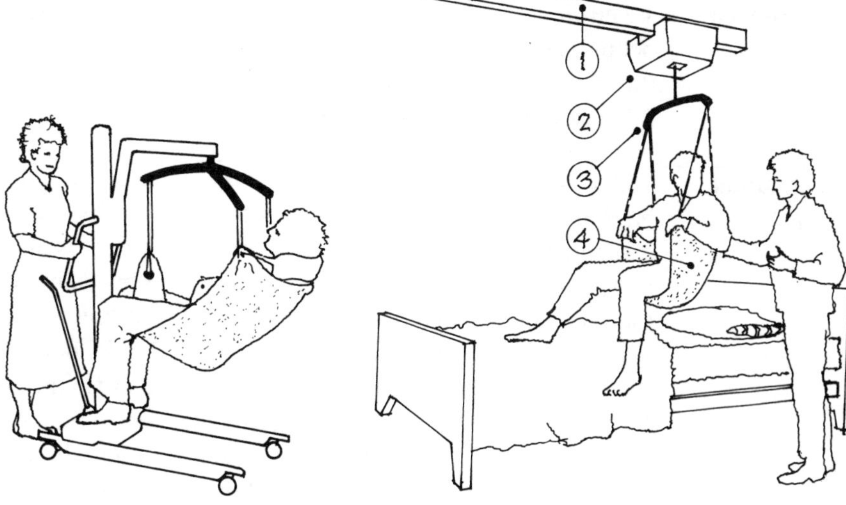

3 As of 2008, useful references include the following. For publication details see the bibliography.

- CHOOSING A MOBILE HOIST
- CHOOSING AN OVERHEAD HOIST

4 Useful websites include the following. They are given in alphabetical order.

- Disabled Living Foundation www.dlf.org.uk/public/factsheets.html
- Disabled Living Online www.disabledliving.co.uk/
- OTDirect www.otdirect.co.uk/link-equ.html#Platform
- Patient UK www.patient.co.uk/showdoc/12/

---

ome adaptations

1 Most *disabled* (qv) or older people prefer to 'stay put' in their own homes for as long as possible before being forced to 'move on'. Advances in *assistive technology* (qv) are continually extending the range of adaptations that can be made to an existing dwelling to help them do so. For a selection of helpful *aids and equipment* (qv) see that entry.

2 As of 2008, useful references include the following. For publication details see the bibliography.

- HOUSE ADAPTATIONS: A WORKING FILE FOR OCCUPATIONAL THERAPISTS by Stephen Thorpe, published by the Centre for Accessible Environments (qv).
- MINOR ADAPTATIONS WITHOUT DELAY: A PRACTICAL GUIDE AND TECHNICAL SPECIFICATIONS FOR HOUSING ASSOCIATIONS, published by the Housing Corporation (qv).

3 As of 2008, useful websites include the following. They are given in alphabetical order.

- Citizens Advice Bureau www.adviceguide.org.uk/index/family_parent/housing/help_with_home_improvements.htm
- Directgov www.direct.gov.uk/
- Directgov: disabled people www.disability.gov.uk/
- Disabled Living Foundation www.dlf.org.uk/
- Ricability www.ricability.org.uk/

See also: *Ageing; Aids and equipment; Environmental control systems; Home improvement grants*; and *Smart homes*.

---

me adaptations: nt aid

See: *Home improvement grants*.

---

me Improvement ncies (HIA)

Visit www.communities.gov.uk/index.asp?id=1152817

Home improvement grants

1 The Government offers financial help with home improvements to many people, especially those who are elderly, in bad health, and/or *disabled* (qv), and therefore in greatest need. There are many types of grant available.
2 For details visit www.direct.gov.uk/

Home Information Packs (HIPs)

1 Under this legislation every home that is sold in England and Wales must have an HIP, which generally costs the seller £250 to £400. Ministers assured a sceptical public that HIPs would make house sales quicker and cheaper. But in March 2008 the consumer product-testing magazine WHICH? savaged HIPs; the Law Society charged that they represented 'the worst piece of consumer legislation in 50 years'; and according to a MORI poll 41% of buyers found that they actually made buying a house more difficult. The Minister responsible for this money-wasting fiasco, Ms Yvette Cooper, had meanwhile been promoted to the second most powerful position in the Treasury.
2 For more information visit www.homeinformationpacks.gov.uk/

Home Office

1 This is one of the major departments of central *government* (qv). Until May 2007 it was responsible for internal affairs in England and Wales including community policy; crime reduction; criminal justice; drugs prevention; immigration and nationality; passports; race equality and diversity; and research and statistics.
2 In that month the new Ministry of Justice (qv) came into being, and took over several responsibilities of the Home Office including criminal law and sentencing; reducing re-offending; and prisons and probation. The new slimmed-down Home Office is now stated to concentrate all its attention on the core issue of protecting the public.
3 For more information visit www.homeoffice.gov.uk/

See also: *Crime*; and *Government*.

Home repair assistance

See: *Home improvement grants*.

Home zones

1 Home zones are streets or groups of streets that have been modified to meet the interests of pedestrians and cyclists rather than motorists, thus opening up the streets for social use and children's play. Vehicles are typically allowed to travel only slightly faster than walking pace by means of speed limits and chicanes; which then allows the street to be redesigned to include children's play areas, larger gardens, trees, seats for chatting, and cycle paths.
2 As of 2008, useful references include the following. For publication details see the bibliography.
• LIVING WITH RISK: PROMOTING BETTER PUBLIC SPACE DESIGN, published by CABE (qv), is a study of 10 public spaces and streets across England.
3 Useful websites include:

• Home Zone News	www.homezonenews.org.uk/
• Home Zones	www.homezones.org/

See also: *Traffic calming; Streets for people*; and *Traffic advisory leaflets*.

Homebuy

See: *Housing: affordable*.

Homeless Link

1 Homeless Link is an umbrella body for agencies dealing with homeless people (see next entry).
2 For more information visit www.homeless.org.uk/

omelessness

1 Homeless people are among the poorest and most disadvantaged members of society. They may get into that situation for a variety of reasons. Risk factors include a loss of income through redundancy; relationship breakdown; eviction; and drug, alcohol and mental health problems. Local housing authorities have a statutory obligation known as the 'main homeless duty' to ensure that suitable accommodation is available for applicants who are eligible for assistance, have become homeless through no fault of their own, and come within a priority need group. Such groups include families with children and households that include someone who is vulnerable, for example because of pregnancy, domestic violence, old age, or physical or mental *disability* (qv).

2 As of 2007 two factors are helping to make the situation worse:

• The average house price paid by first-time buyers has doubled in five years, and average house prices are now beyond the reach of first-time buyers in 93% of towns, compared with 37% of towns in 2001– for more on this see: *Houses: affordability*.

• Large numbers of failed asylum seekers are ending up completely destitute, not looked after by the welfare services, and not merely sleeping rough but probably ill and malnourished if not half-starving. This situation may have deteriorated by the time these words are published.

3 Useful references include the following. For publication details see the bibliography.

• SOCIAL TRENDS (qv), an annual publication of the Office of National Statistics (qv), gives latest statistical data plus background explanation.

4 Useful websites include the following. They are given in alphabetical order.

• Centrepoint www.centrepoint.org.uk/
• Crash www.crash.org.uk/homelessness/intro.lml
• Directgov www.direct.gov.uk/
• Guardian http://society.guardian.co.uk/homelessness/page/0,,527295,00.html
• Homeless Link www.homeless.org.uk/
• Homelessness Code of Guidance for Local Authorities www.communities.gov.uk/index.asp?id=1149848
• Neighbourhood Renewal Unit www.neighbourhood.gov.uk/page.asp?id=634
• New Policy Institute www.npi.org.uk/projects/homelessness.htm
• Shelter http//england.shelter.org.uk/home/index.cfm

See also: *Child poverty; Day centres; Housing: affordability; Move-on accommodation*; and *Social exclusion*.

omelifts

See: *Through-floor lifts*.

omes and happiness

See: *Happiness and homes*.

omes: safety

See: *Accidents at home*.

omosexuals

1 In the UK, unfair *discrimination* (qv) in public life against homosexual people is outlawed by the Equality Act 2006 (qv) and various other measures – see: *Discrimination* (qv); and *Equality* (qv).

2 The number of people affected by the above legislation is most probably around 5% of the population – slightly more for men, somewhat less for women. These figures are from the 1990–91 National Survey of Sexual Attitudes and Lifestyles (Natsal), carried out by the UCL Centre for Sexual Health & HIV Research with financial support from the Wellcome Trust. This survey remains, as of 2007, the largest and most authoritative such study ever done in the UK. Its findings were updated by a somewhat smaller follow-up survey, carried out

by the same institution in 1999–2001 the same institution in 1999–2001, which suggested that the figure might be more like 6%. For information on the latter visit www.ucl.ac.uk/sexual-health/research/natsal-publications.html.
3 The main study quoted above was published in 1994 under the title SEXUAL BEHAVIOUR IN BRITAIN: THE NATIONAL SURVEY OF SEXUAL ATTITUDES AND LIFESTYLES by K Wellings, J Field, A M Johnson, and J Wadsworth – see the bibliography.
4 As of 2008, useful websites include:
• Professor Kay Wellings www.lshtm.ac.uk/pehru/staff/kwellings.html
• UCL Centre for Sexual Health & HIV Research www.ucl.ac.uk/sexual-health/research/sex-attitudes.htm

See also: *Discrimination; Equality*; and *Equality Act 2006*.

Hospital design

See: *Health and welfare buildings*.

Hospital design: quality

1 The NHS Plan of June 2001 created ambitious targets for a huge programme of healthcare building over the next decade, including:
• 300 new *GP's surgeries* by the end of 2004.
• 20 diagnostic and treatment centres for day surgery by 2005.
• 750 one-stop *primary care centres* by 2006.
• Over 100 new state of the art *hospitals* by 2010, at a projected total cost of £11.4 billion.
2 Most of these projects are being realised via the Private Finance Initiative or PFI (qv), whereby private consortia fund the design, construction, and operating costs of the buildings and the NHS leases them back over a period of 25 to 30 years – a method which according to many critics has led to ugly, user-unfriendly and sub-standard hospitals.
3 As of 2008, useful websites include the following. They are given in alphabetical order.
• CABE www.cabe.org.uk/default.aspx?contentitemid=38
• King's Fund www.kingsfund.org.uk/

See also: *Health and welfare buildings*.

Hospital services

Visit www.nhs.uk/England/AuthoritiesTrusts/Default.cmsx

Hostels

See: *College halls of residence*.

Hotel and motel design: complying with the Disability Discrimination Act

1 Buildings used by the public must in general comply with the Disability Discrimination Act (qv). In the design of hotels and motels, *authoritative practical guides* (qv) include the following:
• Para 13.11 of BS 8300 refers briefly to hotels, motels, hostels, and residential clubs; bed and breakfast guest accommodation; self-catering holiday accommodation; and accommodation providing holiday care.
BS 8300 does not have the force of law, but conformity with its recommendations will make it easier to demonstrate that the requirements of the Disability Discrimination Act have been met.
2 As of 2008, useful websites include:
• NQAS www.qualityintourism.com/asp/letsgetassessed.asp

See also: *Travel and tourism-related buildings*; and *Tourism*.

Hotels and motels: access awards

See: *Access awards*.

ouse adaptations See*: Home adaptations*.

ouse Adaptations dvisory Service AAS) Visit www.cae.org.uk/haas.html

ouse of Commons
ouse of Lords See: *Parliament*.

ouse prices 1 According to SOCIAL TRENDS 37 (qv) the average price paid by first-time buyers rose by 204% between 1995 and 2005, while their average incomes increased by 92%. For an illustration of what this means in practice: in March 2007 the *Daily Telegraph* reported that average house prices had jumped by more than £3 400 during the single month of February, thus earning more on average than their owners did.
2 Such increases in the value of properties give the better-off classes a feeling of well-being, but represent a massive transfer of wealth to those who own property from those who do not. For the latter, getting a foot on the home ownership ladder is a rapidly receding prospect: in 2007 average house prices were unaffordable by first-time buyers in 93% of towns, compared with 37% of towns in 2001.
3 For some of the damaging side-effects of exorbitant house prices see: *Childcare*; and *Homelessness*.
4 For detailed average prices by date and geographic location, visit www.landreg.gov.uk/houseprices/

See also: *Housing: affordability*.

ousehold income See: *Incomes*.

ouseholds 1 A 'household' is a number of people occupying a house. Though most households contain one family, some contain multiple families, while others do not contain a family at all (for example, where the *household* consists of only one person or of non-related adults). The words household and *family* (qv) therefore mean different things and should not be used indiscriminately.
2 The number of households in the UK is growing much faster than the rate of population growth. This is because:
• Families are becoming on average smaller.
• There are more lone-parent families.
• There are more people living on their own.
Very recently an unprecedented surge in the number of immigrants has also contributed to the rapid increase in number of households, but no-one knows by how much. The Office for National Statistics (qv) has promised an updated projection.
3 Three matters connected with of the above trends are having important social consequences:
• Because the rate of house-building has not kept up with the rapid rise in new household-formation, house prices have skyrocketed, especially in London and the South East – see: *Housing: affordability*.
• For the same reason the problem of *homelessness* (qv) has recently worsened greatly.
• A sharp rise in the number of one-person households is having important social consequences – greater individual freedom on the positive side; but more loneliness and social isolation on the negative.

4 The bar chart, from Figure 2.4 in SOCIAL TRENDS 36 (qv), shows the number of people living alone, by age and sex, in Great Britain; and how the numbers have changed for men and women since 1986.

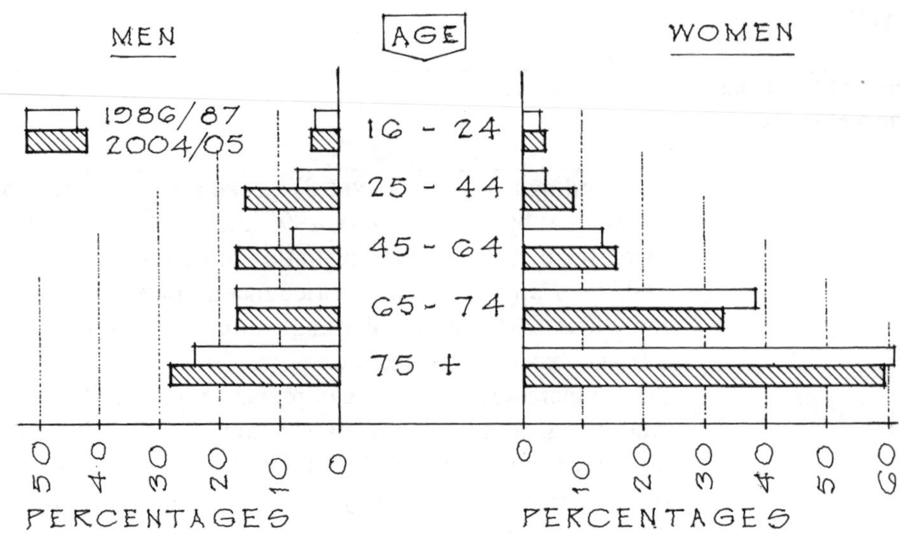

5 The next pair of graphs, from Figure 2.5 in SOCIAL TRENDS 31, project the number of one-person households in England and Wales to the year 2021. It is immediately noticeable that by 2021 the male households will be comprised mostly of under-65 s, with few over-65 s, while the proportion of over-65 s in female households is high. These are trends for which social services and local authority housing departments will have to plan.

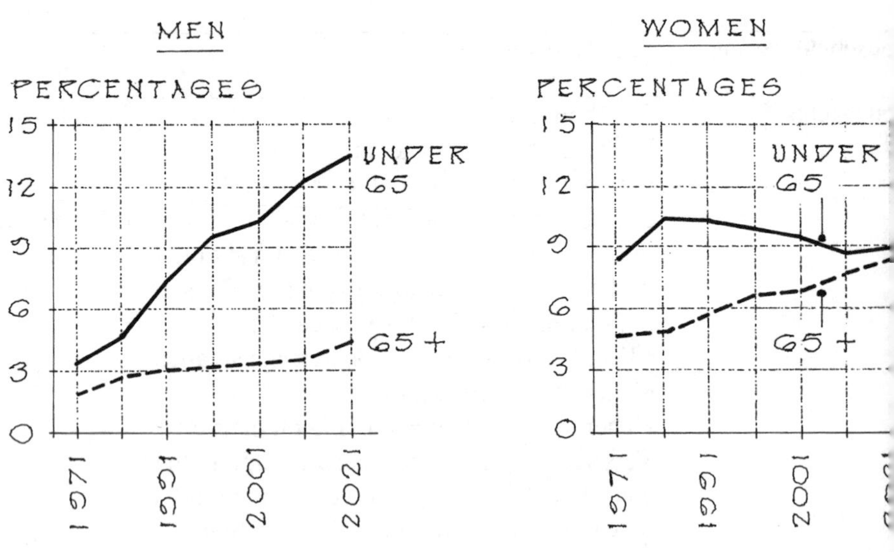

6 For more detailed information plus commentary on the above and related matters, refer to Chapter 2 of the latest edition of SOCIAL TRENDS (qv).

House renovation grants — See: *Home improvement grants.*

Houses in Multiple Occupation (HMO)

1 HMOs provide affordable housing for many young and vulnerable tenants, but sometimes have unscrupulous landlords, suffer from bad management, and may have inadequate facilities and poor fire safety standards.
2 In an attempt to overcome these and other problems, the 2004 Housing Act, which came into effect in July 2006, introduced the licensing of HMOs. Briefly, anyone living in a property in which three or more people live, at least

one of whom is unrelated to the others and where all of them share the kitchen and toilet, etc., is probably living in an HMO. Houses converted into bedsits, for instance for students, would be a common example. But the matter can be extremely complicated – for more particulars visit www.direct.gov.uk/or www.communities.gov.uk/index.asp?id=1151996

3 If a property does come within the definition of an HMO, a stupefying mass of red tape, too complex to describe here, immediately descends upon the owner – not to mention having to pay £1 000 for a licence. As with the Gambling Act qv) and the Licensing Act (qv) one can legitimately ask whether these measures were approved by the Better Regulation Commission (qv) as being *proportionate* (qv) to the ends being served. Common sense suggests that they are not. At a guess:

• Some landlords will reduce the number of people to whom they let rooms, to avoid coming within the HMO regulations, thus shrinking the stock of housing accommodation at a time of acute housing shortage.

• Many landlords will decide that complying with the regulations is simply too bothersome; will ignore them; and probably get away with it. Local authorities are already so overburdened with red tape that this kind of policing is becoming almost impossible.

As Chancellor, our current Prime Minister Gordon Brown committed himself to cutting the burden of *red tape* (qv) by a quarter by the year 2010. A radical redrafting of the 2004 Housing Act would be a good start.

4 As of 2008, useful websites include the following. They are given in alphabetical order.

• Better Regulation Commission www.brc.gov.uk/
• Department of Communities & Local Government www.communities.gov.uk/index.asp?id=1151996
• Directgov www.direct.gov.uk/
• Office of Public Sector Information www.opsi.gov.uk/acts.htm

---

Housing: accessibility 1 In terms of *accessibility* there are two broad categories of *social housing* – 'general needs' and 'special needs' as indicated below:

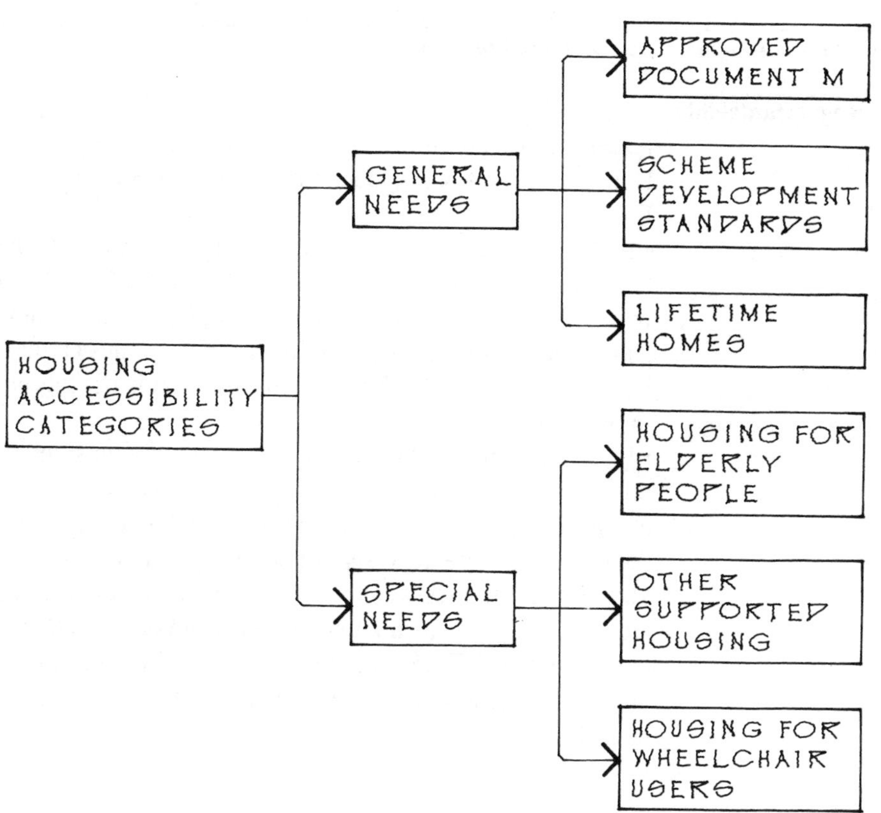

• General needs housing, also called *mainstream housing*, is designed for the majority of the population (perhaps around four fifths) who are not *disabled* (qv) or otherwise in need of a specially-designed dwelling. The provisions of APPROVED DOCUMENT M, which must be heeded in order to comply with the Building Regulations in England and Wales, set the minimum design standard. *Scheme development standards* (qv), which set some additional requirements, apply to all schemes that receive *social housing grant* (qv). *Lifetime homes* (qv) represent the highest of these standards, and it is strongly recommended that all new housing be designed to this level. The HABINTEG HOUSING ASSOCIATION DESIGN GUIDE (qv) gives guidance.
• Special needs dwellings are designed for people who, because they are *ageing* (qv), *disabled* (qv), or for other reasons, cannot live safely and comfortably in general needs accommodation. For more detail on the categories shown on the diagram see: *Housing: special needs*.
2 Local authorities may have recommendations or requirements on the ratio of 'general needs' to 'special needs' housing to be provided in any particular development, and should be consulted. As of 2005 mayor Ken Livingstone's London Plan (qv), for instance, required all new housing in London to be built to *Lifetime Home* (qv) standards, and 10% either to *wheelchair housing* (qv) standard, or for easy adaptation thereto. For more recent information refer to the HOUSING DESIGN GUIDE for London (see the bibliography), which is not yet published as this is written.

See also: *Housing: special needs*.

Housing Action Trusts (HACs)

1 Housing Action Trusts are *quangos* (qv) with five specific objectives as set out in section 63(1) of the Housing Act 1988:
• To repair and improve the housing in their areas.
• To manage their housing effectively.
• To encourage diversity of tenure.
• To improve the social, environmental and living conditions of their areas.
2 For more information visit www.communities.gov.uk/index.asp?id=115288

Housing adaptations

See: *Home improvements*.

Housing: affordability

1 HOME TRUTHS: THE CASE FOR 70 000 SOCIAL HOMES A YEAR, a study researched by Oxford Economics and published in August 2007 by the National Housing Federation, made the following statements about the housing market in England:
• In 2006 house prices rose to an average of over £206 000, taking average house prices in England to a massively unaffordable 11 times average incomes. In London the figure is 13 times average incomes. (Author's interpolation: taking on a mortgage for more than about 4 times annual income is generally regarded as very risky. Much more than that is not merely risky but impossible.)
• We know that people in difficult housing circumstances are very likely to lead blighted lives. But there is no improvement in sight, because hyperbolic house prices have their roots in a mismatch between housing supply and demand, and there is no sign of supply getting anywhere near demand. In 2006 165 000 new homes were built in England, but on average 223 000 households are expected to be formed every year from now to 2026. Even if Prime Minister Gordon Brown succeeds in his plans to build 3 million homes by 2020, that will not be sufficient to meet the demand created by a projected 3.9million extra households by 2021.

• Lack of sufficient housing supply is storing up huge social problems. Therefore the NHF called on the Government to help fund a social house-building programme averaging 70 000 new homes a year for the years 2008–11. That equals a 75% increase in annual supply.

2 In the following graph of house-building completions in the UK, from Figure 10.3 in SOCIAL TRENDS 36, the white area above the baseline shows private sector completions; the rapidly shrinking shaded area above that shows public sector completions; and the top line is the sum of the two – total completions. It is seen that these fell from a peak of 426 000 in 1968 (53% by the private sector; 47% by the social sector) to 206 000 in 2004–05.

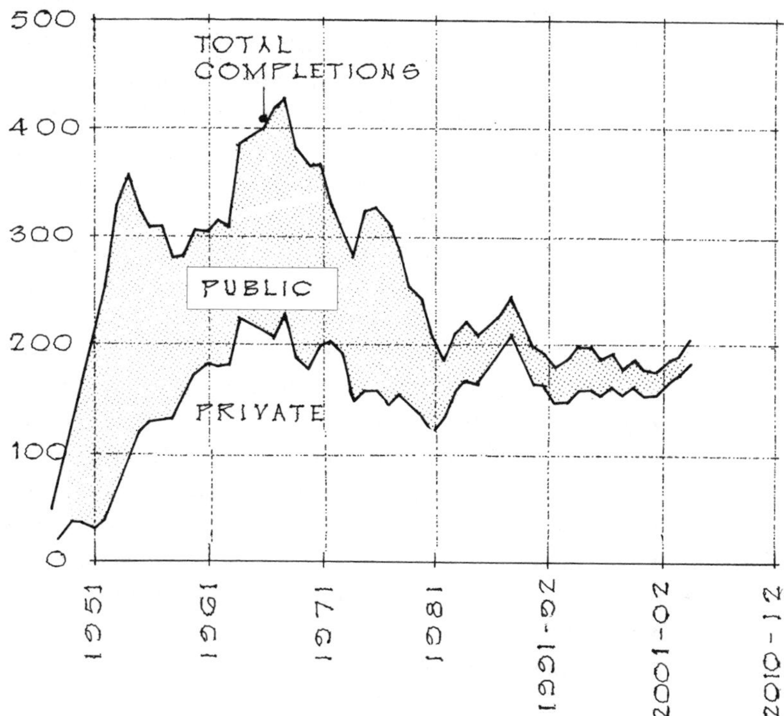

3 As of 2008, useful references include the following. They are given in date order. For publication details see the bibliography.

• In March 2004 a study commissioned by the Treasury, THE BARKER REVIEW: DELIVERING STABILITY – SECURING OUR FUTURE HOUSING NEEDS, concluded that the number of new houses and flats built each year should increase by at least 70 000 and perhaps 120 000 above the current rate.

• In December 2005 THE GOVERNMENT'S RESPONSE TO KATE BARKER'S REVIEW OF HOUSING SUPPLY accepted this need and (a) proposed a reformation of the planning system; (b) foreshadowed increased investment in infrastructure to support sustainable housing growth; and (c) set out a plan to increase the housing supply to 200 000 dwellings per annum in England by 2015 – a figure that Ms Ruth Kelly, Secretary of State for Communities and Local Government (qv), said in May 2007 might have to be raised to 225 000 houses and flats per annum to meet the latest forecasts of demand.

• Also in 2005 the report UNAFFORDABLE HOUSING: FABLES AND MYTHS by Alan W Evans and Oliver Marc Hartwich, which was supported by the Joseph Rowntree Foundation (qv), called for a change in the British planning system to encourage developers to build much more housing, and vastly better housing, than at present. This document, which was named 'Pamphlet of the Year' in

2005 by *Prospect* magazine, was followed in 2006 by a second report BETTER HOMES, GREENER CITIES.

4 Following on from the above:

• In 2006 Mr Adam Sampson, the Director of Shelter, stated that 216 000 new households are formed each year in England, but only 160 000 new homes are built, with the result that 'everyone is being forced to step down a rung of the housing ladder, and the people on the last rung step off'. Shelter has published a report titled BUILDING FOR THE FUTURE, produced by the Cambridge Centre for Housing and Planning Research, that calls for over £1 billion of extra government funding to be invested each year between 2008 and 2011 to build 20 000 additional socially rented homes per annum. The alternative, it asserts, will be a true housing crisis.

• In May 2007 Mr David Orr, chief executive of the National Housing Federation (qv), wrote in the *Guardian* newspaper: 'The Government must … provide the resources for 210 000 new affordable homes over the next three years. This means increasing spending on housing by £1.5 billion a year. Housing associations will match this, pound for pound, with private borrowing and their own reserves'.

• In August 2007 the National Housing Federation published HOME TRUTHS: TH CASE FOR 70 000 SOCIAL HOMES A YEAR (see summary above).

• In October 2007 the National Housing and Planning Advice Unit (qv) called for official plans to build 3 million new homes in England by 2020 to be increased by 250 000, to avoid average house prices rising to at least 11, and perhaps 13 times average earnings.

5 Useful websites include the following. They are given in alphabetical order.

• Defend Council Housing www.defendcouncilhousing.org.uk
• Department for Communities and Local Government www.communities.gov.uk/index.asp?id=1503252
• House Builders Federation www.hbf.co.uk/
• Housing Corporation www.housingcorp.gov.uk/
• Joseph Rowntree Foundation www.jrf.org.uk/
• National Housing Federation www.housing.org.uk/
• Rural Communities www.ruralcommunities.gov.uk/files/LA%20aff%20map.pd
• Shelter http://england.shelter.org.uk/
• Social Trends www.statistics.gov.uk/

See also: *Child poverty; House prices*; and *Housing: affordable*.

Housing: affordable

1 There are two categories of *affordable housing*, both of which are distinct from the altogether different category of *market housing* (qv):

• Low-cost *social housing* is heavily subsidised rented accommodation for people on low incomes. It is provided either by *local authorities* (qv); or by *registered social landlords* (qv) such as *housing associations* (qv). The quanti of social housing being built in the UK has fallen steeply over the past 25 years see the figure under *Housing: affordability*.

• Mid-cost *intermediate housing* is housing for people who cannot afford market housing but earn too much to qualify for *social housing*. It may be targeted at essential public sector workers such as *key workers* (qv). Beneficiaries are either given help to buy their homes, or charged rents that ar set above social housing target rents but below open market rents. As of 2007, such schemes have included the Starter home initiative; Key worker living programme; and Homebuy.

All of the above schemes are subject to constant change. For latest details visi www.direct.gov.uk/

2 Government funding schemes are well-intended, but cannot be effective unless more dwellings are built overall. As always, when there is a gross mismatch between the demand and supply of an essential commodity, every attempt to circumvent or camouflage the underlying problem throws up a new dilemma. Thus:

• It emerged in June 2006 that flats built as 'affordable housing' in London's Chelsea Bridge Wharf development by Berkeley Homes were being sold perfectly legally at prices ranging from £200 000 to £385 000, whereas *key workers* could probably not afford more than £172 000. Quite separately, but at about the same time, the London Assembly (qv) published a report that accused Mayor Ken Livingstone of failing to define what 'affordable' means, and recommended that affordable housing should have a price cap to ensure that it would actually be available to the people who need it. The report failed to estimate the subsidies that would be needed over the next five or ten years to make such ideas work, and failed to suggest who would provide the colossal sums that would be required.

• If such funding could be found there is a more fundamental problem, identified by the chief economist at the Royal Institute of Chartered Surveyors when the Government launched the Open Market HomeBuy scheme for key workers in October 2006. He warned that the scheme would not merely be ineffective, but could actually make the problem of housing affordability worse by increasing demand at the lower end of the market when there was no effective increase in supply. 'Building more houses is the only way to solve this problem', he said; and a statement issued by the Department for Communities and Local Government (qv) concurred.

3 For more detailed information on these and related matters, refer to Chapter 10 of the latest edition of SOCIAL TRENDS (qv) – see the bibliography.

4 As of 2008, useful websites include the following. They are given in alphabetical order.

• Directgov www.direct.gov.uk/
• *Guardian* http://society.guardian.co.uk/keyworkers/page/0,,546855,00.html
• Housing Options www.housingoptions.co.uk/ho2/#
• IDeA Knowledge www.idea.gov.uk/
• Key Worker Living www.housingcorp.gov.uk/server/show/nav.547
• National Affordable Housing Programme www.housingcorp.gov.uk/server/show/nav.446
• Rural Communities www.ruralcommunities.gov.uk/files/LA%20aff%20map.pdf

ousing Association rants

See: *Home improvement grants.*

ousing Associations As)

1 Housing Associations are not-for-profit organisations run by *voluntary* (qv) committees for the provision of low-cost housing. They are the predominant type of *Registered Social Landlord* (qv), and are supervised by the Housing Corporation (qv).

2 For contact details of the HAs that operate in each particular area visit www.housingcorp.gov.uk/server/show/conWebDoc.1153

3 As of 2008, other useful websites include:

• Directgov www.direct.gov.uk/
• Habinteg www.habinteg.org.uk/
• Housing Corporation www.housingcorp.gov.uk/

See also: *Habinteg; Housing Corporation; Registered Social Landlords*; and *Scheme Development Standards.*

Housing Benefit

1 This is a means-tested benefit to help tenants of *Housing Associations* (qv), *local authorities* (qv), and private landlords pay their rent if they get into financial difficulties. Real spending on housing benefit leapt from £4.5 billion in 1980–81 to £9.5 billion in 1985–86, and then to £17 billion in 2005–06 (all at 2005–06 prices).
2 It is a much misunderstood benefit. For instance, the charity Age Concern found in April 2007 that 2.2 million pensioner households were unaware that of their entitlement to HB, and failing to claim £1.4 billion to which they were entitled and which would make a real difference to their living standards. It launched the 'Your Rights' campaign to help rectify the situation.
3 Housing Benefit was amended in certain respects by the Welfare Reform Act 2007 (qv), but a need remains for more radical reforms
4 As of 2008, useful references include the following. For publication details see the bibliography:
• REFORMING WELFARE by Nicholas Boys Smith. This paper was written before the Welfare Reform Act came into force.
5 Useful websites include the following. They are given in alphabetical order.
• Department for Work and Pensions www.dwp.gov.uk/
• Directgov www.direct.gov.uk/
• Frank Field www.frankfield.co.uk/type3.asp?id=20&type=3

See also: *Welfare reform.*

Housing: best value

1 The widely-applied government policy of *best value* (qv) requires local government departments to continuously assess the quality of their service delivery by (a) comparing their performance with that of other authorities; (b) consulting the people who use those services, and (c) using competition to get the best service available. It is an extremely cumbersome process, and not all commentators are convinced that it is the most effective means for achieving the desired ends.
2 As of 2008, useful websites include the following. They are given in alphabetical order:
• *Guardian* http://society.guardian.co.uk/bestvalue/page/0,,531381,00.html
• Local Government Performance www.bvpi.gov.uk/pages/Index.asp
• Department for Communities and Local Government www.communities.gov.uk/index.asp?id=1153107

See also: *Best value.*

Housing Corporation

1 A *quango* (qv) sponsored by the Department for Communities and Local Government (qv) with the task of funding and regulating *Registered Social Landlords* (qv) in England.
2 Since 2008 the Housing Corporation has also been referred to as the National Affordable Homes Agency.
3 For more information visit www.housingcorp.gov.uk/

Housing: decent homes standard

1 In July 2000 the Government set a 10-year target with the aim of bringing all *social housing* (qv) up to a 'decent standard' by 2010. A 'decent' home is officially defined as one that is wind and weather tight, warm, and has modern facilities.
2 As of 2008, useful websites include the following:
• Decent Homes www.communities.gov.uk/index.asp?id=1152137
• Directgov www.direct.gov.uk/

ousing: design

1 As of 2008, key *authoritative practical guides* (qv) for housing design include the following. They are given in alphabetical order. For publication details see the bibliography.

• APPROVED DOCUMENT M (qv), the provisions of which must be heeded in order to comply with the Building Regulations for England and Wales. In Northern Ireland refer to TECHNICAL BOOKLET R of the Building Regulations (Northern Ireland); and in Scotland to the TECHNICAL HANDBOOKS of the Building (Scotland) Regulations (qv).

• THE HABINTEG HOUSING ASSOCIATION DESIGN GUIDE (qv) applies most specifically to *housing association* (qv) schemes, but relevant to all housing.

• THE HOUSING DESIGN GUIDE is a best-practice guide for housing in London.

• WHEELCHAIR ACCESSIBLE HOUSING: DESIGNING HOMES THAT CAN BE EASILY ADAPTED FOR RESIDENTS WHO ARE WHEELCHAIR USERS is a best-practice guide for housing in London, and is meant to be used in conjunction with the next title.

• THE WHEELCHAIR HOUSING DESIGN GUIDE (qv) applies most specifically to *housing association* (qv) schemes, but relevant to all housing.

2 Other references include:

• ACCESSIBLE HOUSING: QUALITY, DISABILITY AND DESIGN by Rob Imrie.

• WHAT HOME BUYERS WANT: ATTITUDES AND DECISION MAKING AMONG CONSUMERS, published by Cabe (qv).

• EVALUATING HOUSING PRINCIPLES STEP BY STEP, also from Cabe.

3 Useful websites include:

• CABE www.cabe.org.uk/default.aspx?contentitemid=39

See also: *Housing: accessibility*.

ousing estates

1 British housing estates represent a noble idea, the development of which can be traced from Victorian philanthropists such as Titus Salt, who provided subsidised housing for his workers, through visionaries like Ebenezer Howard, who created the first garden city at Letchworth, to Lloyd George, who declared that troops returning from World War 1 must have 'homes fit for heroes', to the Labour politician Aneurin Bevan, who dreamed in the 1940s of creating towns where 'the doctor, the grocer, the butcher and the farm labourer' could live as neighbours in houses of equal quality.

2 Since then that noble dream has soured. Here are two particularly interesting recent references.

• ESTATES: AN INTIMATE HISTORY, a book that was thought by the *Observer* newspaper's commentator Will Hutton to be as important as the 1966 BBC drama 'Cathy Come Home', states that Britain's estates, which were once the home of an upwardly mobile working class, have become dense concentrations of poverty, crime and disaffection, and a symbol of failure for everyone, especially those who live in them. The author, Lynsey Hanley, grew up on a gigantic Birmingham estate, argues that it is not just bricks and mortar that now create an almost impermeable barrier between the council-house dweller and the rest of the community, but a wall in the mind – the conviction that 'the aspirations and expectations of the rest of society are not for you'.

• ENDS AND MEANS: THE FUTURE ROLES OF SOCIAL HOUSING IN ENGLAND, a report by Professor John Hills of the London School of Economics, commissioned by Ms Ruth Kelly, Secretary of State for Communities and Local Government and published in February 2007, raises the question: 'What is council housing trying to do?'. If the purpose is to provide poor tenants with the first rung of a ladder of opportunity for either work or housing, then it has

failed utterly: three-quarters of tenants claim benefit and never leave the estate. Council housing is in fact a living tomb – tenants dare not give the house up because they might never get another, but staying means being trapped in a ghetto of both place and mind. Again in the words of Will Hutton: the answer to the oft-asked question 'What ails Britain?' is 'British council estates. We made them; now we need to unmake them, doing whatever it takes'.
• THE SOCIAL ENTREPRENEUR: MAKING COMMUNITIES WORK by Andrew Mawson offers sensible ideas for improvement. For publication details of these references see the bibliography.

See also: *Anti-social behaviour*; and *Underclass*.

Housing Fitness Standard Visit www.communities.gov.uk/

Housing Forum Visit www.constructingexcellence.org.uk/

Housing: general needs See: *Housing: Accessibility*.

Housing health and safety rating system See: *Housing Fitness Standard*.

Housing in multiple occupation See: *Houses in multiple occupation*.

Housing Inspectorate Visit www.audit-commission.gov.uk/

Housing: intermediate See: *Housing: affordable*.

Housing: key worker See: *Housing: affordable*.

Housing: large scale voluntary transfer (LSVT) 1 This term refers to the selling-off of 500 *council homes* or more to a *Housing Association* (qv) following a successful ballot of the tenants involved.
2 See also: *Housing: right to buy*.

Housing: London 1 The graph shows the number of housing completions in London, by category, from 1971 to 2005–06.

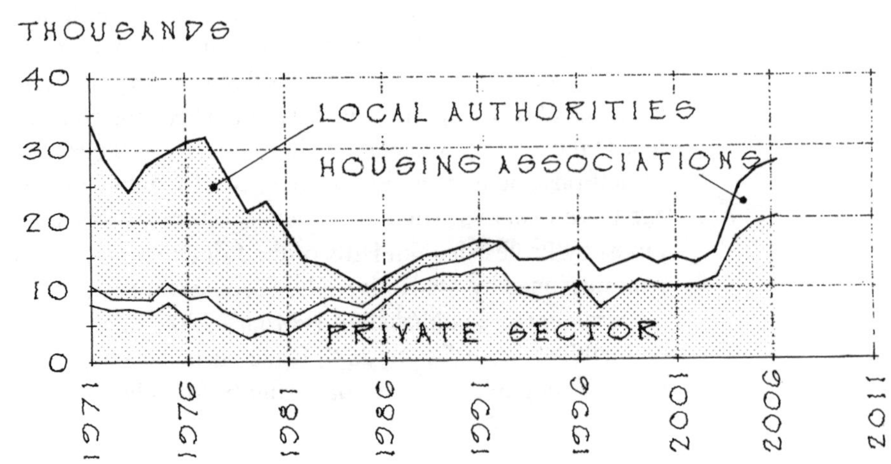

2 In September 2007 it was announced that the Mayor of London would publish a statutory Housing Strategy for London in January 2008, and at the same time a HOUSING DESIGN GUIDE (see the bibliography), written by Design for London, that would promote *planning* (qv) reform and act as a best-practice guide for housing in London. The GUIDE was expected to look at space standards, floor-to-ceiling heights, and the quality of internal layout with the aim of 'achieving high-quality housing designs at high density, that will endure'. These initiatives form part of Mayor Ken Livingstone's plan to deliver 50 000 more *affordable homes* (qv) in London over the next 3 years, which in turn fits in with Prime Minister Gordon Brown's target of 3 million new homes in Britain by 2020 (see: *Housing: affordability*). At time of writing it is anticipated that the Mayor will have direct control of £1 billion a year to expand the housing supply, but that control of design quality will remain with *local authorities* (qv).
3 By the time this is published the above plans will no doubt be much clearer. For latest information visit www.designforlondon.gov.uk/

See also: *Housing: affordability*; and *Housing: design*.

·using: low cost See: *Housing: affordable*.

·using: mainstream See: *Housing: accessibility*.

·using Management ·newal Areas ·MRAs) 1 HMRAs bring together *local authorities* (qv) and other agencies in areas where the housing market is thought to be failing.
2 For more information contact the relevant local authority.

·using: market 1 *Market housing* is owner-occupied and private rented housing that does not meet the affordability and access criteria for *social housing* (qv) or *intermediate housing* (qv).
2 See: *Housing: affordable*.

·using market ·newal programme See: *Pathfinders*.

·using Ombudsman ·rvice See: *Independent Housing Ombudsman*.

·using: pathfinder ·rtnerships See: *Pathfinders*.

using policy and ·cial friction See: *Human rights and social friction*.

·using: right to buy 1 The Right to Buy scheme has helped almost 1.6 million council tenants in England buy their own homes, and it has – on the positive side – encouraged some of the more affluent tenants to remain in the neighbourhoods they have lived in for many years, thus helping to create stable, mixed-income communities.
2 But selling off *social housing* (qv) without simultaneously building a new unit for every one that is sold, which is what has been happening under all governments since that of Mrs Margaret Thatcher, has led to an agonising shortage of affordable rented homes for those who will never be able to buy – even with government help – and who rely on this safety net for a decent life. The results of this short-sighted policy have been disastrous – see: *Housing: affordability*.

3 Figure 10.6 in SOCIAL TRENDS 37 (qv) shows the changing numbers of right to buy sales, large scale voluntary transfers, and other sales and transfers in Great Britain from 1981 to 2005.
4 As of 2008, useful websites include the following. They are given in alphabetical order.
- Directgov www.direct.gov.uk/
- Housing Corporation www.housingcorp.gov.uk/

See also: *Housing: affordability; Housing: affordable*; and *Housing: large sc voluntary transfers*.

Housing: rural

See: *Affordable Rural Housing Commission.*

Housing: sheltered

See: *Housing: special needs.*

Housing shortfall

See: *Housing: affordability.*

Housing: social

1 Social housing is defined as low-cost housing, let at rents no higher than Housing Corporation (qv) target rents, provided in most cases by a *registered social landlord* (qv) or a *local authority* (qv) housing authority using public subsidy.
2 As of 2008, useful websites include the following. They are given in alphabetical order.
- Defend Council Housing www.defendcouncilhousing.org.uk/
- *Guardian* http://www.society.guardian.co.uk/internet/page/0,,526807,00.ht
- Housing Corporation www.housingcorp.gov.uk/
- Inside Housing www.insidehousing.co.uk/
- New Policy Institute www.npi.org.uk/projects/housing%20corporation.h

See also: *Housing: affordable.*

Housing: special needs

1 In terms of *accessibility* there are two broad categories of housing – 'gener needs' (qv) for the great majority of people, and 'special needs' for that minority who for reasons of age, infirmity, disability, or some other reason w not be safe or comfortable in a conventional dwelling.
2 The figure below summarises the main types of special needs housing. Definitions overlap somewhat, but broadly speaking there are three categorie

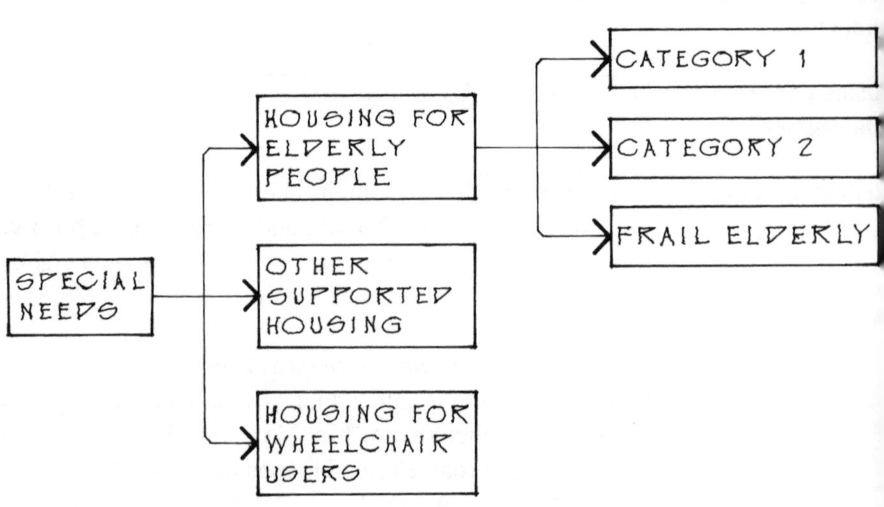

3 Housing for elderly people. These dwellings tend to be in the form of flats or *bungalows* (qv) served by a warden and containing shared facilities such as a common room. Such warden-supported accommodation is commonly referred to as 'sheltered housing', and as of 2007 there are three classes:

- Category 1 sheltered housing for people who are fairly active and independent.
- Category 2 sheltered housing for people who are less so.
- Frail elderly (also called extra care) sheltered housing for people needing intensive warden support.

For more detail refer to p 7 of the Housing Corporation (qv) publication SCHEME DEVELOPMENT STANDARDS – see the bibliography.

4 Other supported housing, a term that refers more to management than to design matters, is aimed at helping a variety of people who need intensive help and support. Examples include *disabled people* (qv); people with drug or alcohol related problems; people leaving prison; young people leaving care; women at risk of domestic violence; and single *homeless* (qv) people. For more information visit www.direct.gov.uk/

5 Housing for *wheelchair-users* (qv), more commonly known as 'wheelchair housing', is intended for people who use *wheelchairs* (qv) most or all of the time. The London Plan (qv) requires 10% of all new dwellings to be wheelchair housing, or to be easily adaptable to wheelchair standards. As of 2007, the WHEELCHAIR HOUSING DESIGN GUIDE, published by Habinteg (qv), is the most authoritative design manual for such dwellings – see the bibliography.

6 Useful websites include the following. They are given in alphabetical order.

- Directgov www.direct.gov.uk/
- Housing Corporation www.housingcorp.gov.uk/
- Sheltered and Supported Housing Service www.cshs.co.uk/

‹sing supply See: *Housing: affordability*.

‹sing: supported See: *Housing: special needs*.

‹sing types: ‹ularity See: *Happiness and housing*.

‹nan rights 1 Human rights are usually stated to be those basic rights to which all people (not just a favoured few) are entitled simply by virtue of being human beings: the right to life and liberty; freedom of thought and expression; and equality before the law. Such an idea was unknown in the world until about the 17th century, when it began to arise in the minds of thinkers such as John Locke in Britain and western Europe and their cousins in northern America. Emerging gradually via the English Bill of Rights of 1689 and the American Bill of Rights of 1791, the principle was most famously stated in the French Declaration of the Rights and Duties of Man and the Citizen of 1795, which defined a set of individual and collective rights that apply not merely to French citizens but to all men and women without exception.

2 Slowly, haphazardly, and imperfectly some approximation of this ideal state of affairs has been achieved in western Europe, northern America, Australasia, and Japan. But despite welcome progress in Eastern Europe, India, and South Africa, most people still arrive in the world with no effective entitlement to the rights noted in the opening sentence above.

3 As of 2008, useful references include the following. For publication details see the bibliography.

- TOWARDS THE LIGHT: THE STORY OF THE STRUGGLES FOR LIBERTY & RIGHTS THAT MADE THE MODERN WEST by A C Grayling.

4 Useful websites include the following. They are given in alphabetical order

- British Institute of Human Rights www.bihr.org.uk/
- Equality & Human Rights Commission www.equalityhumanrights.com/
- Human Rights Watch http://hrw.org/
- Justice www.justice.org.uk/
- Liberty www.liberty-human-rights.org.uk/
- Ministry of Justice www.justice.gov.uk/
- openDemocracy www.opendemocracy.net/
- Wikipedia http://en.wikipedia.org/wiki/Human_rights#Human_rights_history
- Yahoo Directory of Human Rights Organisations http://dir.yahoo.com/Society_and_Culture/Issues_and_Causes/Human_Rights/Organizations/
- Your Rights www.yourrights.org.uk/

See also: *Civil liberties; Commission for Human Rights and Equality; Huma Rights Act 1998*; and *Universal Declaration of Human Rights.*

Human Rights Act 1998

1 The Human Rights Act (1998) came into force in the UK on 2 October 200 and incorporates into UK law the1950 European Convention on Human Righ (qv), thus making this convention justiciable in British law. It lays down 18 personal rights and freedoms that can be defended in court. These include people's right not to be treated differently because of their race, religion, sex, political views or any other status, unless such treatment can be objectively *justified* (qv). For the full text visit www.opsi.gov.uk/acts.htm

2 When adopted the Act was widely welcomed, but it has since lost much of that popular support. There are two principal reasons for this:

- Use of the Act to protect and indeed expand the rights of wrongdoers. The public are bemused to learn (a) that four successive Home Secretaries have found it impossible to deport a group of asylum seekers who hijacked an aircr to enter the UK in February 2000; (b) that prison governors are unable to stop inmates (including at least one sex murderer) from watching pornographic fil in their cells; and (c) that probation officers have been instructed not to moni early-released serious offenders (one of whom went on to kill Naomi Bryant) too closely for fear of infringing their human rights. Not all reported cases of this kind can be properly attributed to the Human Rights Act, but some can, a they affront people's sense of justice.
- The Act arrived on the UK statute book by processes to which the British electorate were onlookers. The people were not asked to comment on consultative drafts, or invited to vote on any proposal, and are gradually discovering – with a mixture of surprise and resentment – what their political masters have signed up to on their behalf.

3 The UK does need a Human Rights Act, and probably one that has much in common with the present HRA – though with a few key differences. What is beyond dispute is that British citizens are entitled to an Act that has been draft in Britain, and passed through the UK parliament, with their participation anc consent. The time is ripe for such a democratising advance.

4 As of 2008, useful references include the following. They are given in alphabetical order. For publication details see the bibliography.

- ABC: A BALANCED CONSTITUTION FOR THE 21ST CENTURY by the constitutional lawyer Martin Howe was published in 2006 by the *think tank* (qv) politeia.
- A GUIDE TO THE HUMAN RIGHTS ACT 1998 is a free 47-page booklet from the Ministry of Justice (qv).

See also: *Constitution*; and *European Convention on Human Rights.*

[Hu]man rights and [soc]ial friction

1 In researching THE NEW EAST END: KINSHIP, RACE AND CONFLICT (see CONFLICT (see the bibliography), which should be required reading for all housing ministers and their advisors, Geoff Dench and Kate Gavron explored the root causes of white working class resentment against immigrants (for instance Bangladeshis) in London's East End. Asked what they objected to about immigrants, local people repeatedly identified housing as the key issue. Their sense of fairness was affronted by the way in which recently-arrived immigrants could jump to the top of the queue and take the best homes over the heads of local applicants who had patiently waited their turn for years or decades. In their view, people who have waited longest should rise to the top of the queue and take the best housing, and newcomers should be offered the accommodation they have vacated.

2 Elsewhere in Britain similar resentments are widely sparked – and not only among whites – by the manner in which single parents or homeless people scoop the housing jackpot.

3 What emerges from studies such as this is that ordinary people have always understood 'their' welfare state (the one founded by the Labour government of 1945, and built up ever since by governments of all political hues) as a kind of insurance scheme into which citizens dutifully pay their contributions in the form of taxes, thereby insuring their families against hard times, and from which there can, as a general rule, be no payout before some contribution has been made.

4 Mr Frank Field (qv), the Labour MP for Birkenhead and one of Britain's most respected commentators on these matters, has suggested three steps towards a fair, workable, and popular system of welfare support.

- Instruct all housing authorities that length of service as good tenants should be the crucial determinant of housing allocation. Newer arrivals can then be offered the accommodation thus vacated – but again on a track record of good citizenship rather than human rights.
- Partially freeze the benefit levels for single people until the rate for a couple equals twice that of the single person. The welfare system would then no longer be seen as discriminating financially against couples who live together, particularly so if they have children.
- Impose a contributory period before welfare can be drawn, and focus the debate on how long this period should be.

Mr Field recognises that such policies may well conflict with *human rights* (qv) legislation, whereby state assistance must be directed at those individuals who have the greatest need, regardless of other considerations. But he holds that the issue of social fracturing is now so ominous that exemptions from current human rights legislation should, if such a step proves to be necessary, be insisted upon.

5 As of 2008, useful websites include:

- Frank Field www.frankfield.co.uk/type3.asp?id=20&type=3

[Hu]man rights: [pro]tection of

See: *Commission for Equality and Human Rights.*

I

ards	See: *National Identity Scheme.*
s: the power of	See: *Think tanks.*
logy	See: *Déformation professionelle*; *Pressure groups*; and *Speech: Freedom of.*
tity and Passport vice (IPS)	1 The Identity and Passport Service is an Executive Agency (see: *Quangos*) and has responsibility for passports and the *National Identity Scheme* (qv) 2 For more information visit www.ukpa.gov.uk/
tity cards	See: *National Identity Scheme.*
tity Cards Act 6	1 The Identity Cards Act, which was passed in March 2006, paves the way for *ID cards (qv),* the *National Identity Register* (qv) and the *National Identity Scheme* (qv). 2 The Act: • Establishes the National Identity Register. • Provides powers to issue identity cards. • Ensures checks can be made against other databases to confirm an applicant's identity and guard against fraud. • Sets out what information would be held and what safeguards would be in place. • Enables public and private sector organisations to verify a person's identity by checking against the National Identity Register, with that person's consent, to validate their identity before providing services. • Includes enabling powers so that in the future access to specified public services could be linked to the production of a valid identity card. • Provides for it to become compulsory to register and be issued with a card, including penalties against failure to register. 3 Many people seem to believe that the proposed ID cards are simple plastic identifiers which will make little difference to their everyday freedoms. They are wrong. Every adult citizen of the UK (but not foreigners who have been in Britain for less than six months, thus giving would-be terrorists an opening wider than a barn door) (the *Observer's* columnist Mr Henry Porter) will be required by law to give the state 49 pieces of information about themselves, including fingerprints and a facial image, for their compulsory ID card. Further, every citizen (read this sentence slowly) will be obliged to inform the authorities each time he or she changes their address, or be fined up to £1 000 per offence. 4 For a comment by Dr Ian Angell, Professor of Information Systems at the London School of Economics, refer to ID CARDS ARE THE ULTIMATE IDENTITY THEFT. COMPUTER SYSTEMS ALWAYS FAIL – AND THE NATIONAL DATABASE WILL DO SO BIG TIME – see the bibliography. 5 As of 2008, useful webites include: • Directgov www.direct.gov.uk/ • Information Commissioner's Office www.ico.gov.uk/ • Liberty www.liberty-human-rights.org.uk/ • National Identity Scheme www.identitycards.gov.uk/scheme.asp • No2ID www.no2id.net/ • Office of Public Sector Information www.opsi.gov.uk/acts.htm See also: *Constitution; Civil liberties; IT systems*; and *Post-democracy.*
eracy	See: *Literacy.*
minance	See: *Lighting design.*

Immigration See: *Population*.

Impact Assessments See: *Better regulation bodies*; and *Regulatory impact assessments*.

Impaired communication

1 Impaired communication can arise from several causes, of which the most common is *stroke* (qv). For *aids and equipment* (qv) that may help people wi impaired speech see: *Speech aids* (qv). For sign languages see: *Sign and symbol systems* (qv).
2 As of 2008, useful websites include:

- Afasic www.afasic.org.uk/
- Directgov www.direct.gov.uk/
- Directgov: disabled people www.disability.gov.uk/
- Joseph Rowntree Foundation www.jrf.org.uk/
- Royal College of Speech and Language Therapists www.rcslt.org.uk
- Signalong www.signalong.org.uk/
- Speakability www.speakability.org.uk/

See also: *Aphasia; Sign and symbol systems*; and *Speech aids*.

Impaired dexterity: complying with the Disability Discrimination Act

1 The most common causes of weak, stiff, or painful hands are *arthritis* (qv) and/or *ageing* (qv). Whatever the cause, if the effect is severe enough then impaired dexterity is recognised as a *disability* (qv) under the Disability Discrimination Act (qv).
2 For compliance with the Disability Discrimination Act (qv), which applies to all buildings used by the public, but not to private dwellings, relevant *authoritative practical guides* (qv) include DESIGNING FOR ACCESSIBILITY (qv) plus the following:

- APPROVED DOCUMENT M (qv), the provisions of which must be heeded in ord to comply with the Building Regulations for England and Wales.
- BS 8300 (qv) which applies to all buildings in the UK.
- INCLUSIVE MOBILITY (qv), which is the most authoritative reference for the *inclusive* (qv) design of *pedestrian infrastructure* (qv) and of *transport-relate buildings* (qv) such as bus and coach stations, railway stations, air terminals, and transport interchanges.

While the latter two documents are advisory and do not have the force of law conformity with their recommendations will make it easier to demonstrate th the requirements of the DDA have been met.
Relevant clauses in the above documents include the following:

FEATURE	DOCUMENT: RELEVANT PARAGRAPHS	
	APPROVED DOCUMENT M	BS 8300
DOOR FURNITURE	2.17	6.5
CONTROLS AND SWITCHES	4.25–4.30; 8.1–8.3	10.3–10.5
TAPS	5.4, 5.18	12.1.6, 12.2 12.3.6, 12.4

In places other than England and Wales the following regulatory documents should be consulted:

- In Northern Ireland: TECHNICAL BOOKLET R of the Building Regulations (Northern Ireland) (qv).

• In Scotland: the TECHNICAL HANDBOOKS of the Building (Scotland) Regulations (qv).
3 In the USA, where the Americans with Disabilities Act applies, refer to the ADA AND ABA ACCESSIBILITY GUIDELINES (qv).
4 As of 2008, useful websites include:

• Directgov: disabled people	www.disability.gov.uk/
• Disabled Living Foundation	www.dlf.org.uk/
• Disabled Living Online	www.disabledliving.co.uk/
• Ricability	www.ricability.org.uk/

See also: *Aids and equipment; Inclusive design*; and *Products for easier living*.

aired hearing

1 Impaired hearing is recognised as a *disability* (qv) under the Disability Discrimination Act (qv) if the effect is severe enough.
2 Four degrees of hearing loss are generally recognised:
• People with mild hearing loss have some difficulty in following speech, particularly in noisy situations, and may possibly wear *hearing aids* (qv).
• Those with moderate hearing loss will probably wear hearing aids, and may be able to use an ordinary voice *telephone* (qv) if it has adjustable volume or is designed to work with hearing aids.
• Those with severe hearing loss get little help from a hearing aid, find it hard to use a telephone even if the sound is amplified, and may prefer to communicate by non-auditory devices such as a *textphone* (qv).
• Those who are profoundly deaf are totally dependent on *lip-reading* (qv), *sign language* (qv), and devices such as a textphone.
3 As shown in the diagram below, helpful measures in buildings are of two kinds – (a) 'passive measures', relying on sensible design of the building fabric; and (b) 'active measures', using energy-driven aids and equipment.

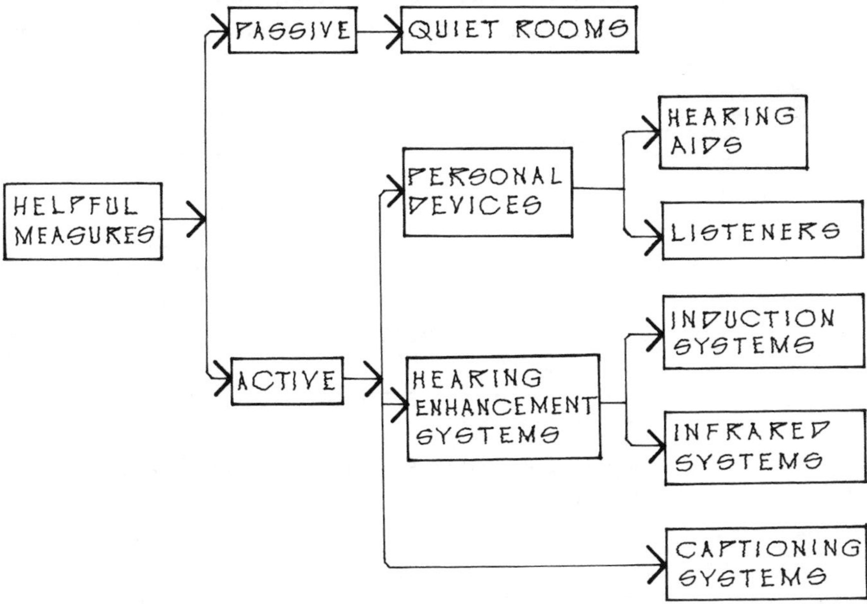

4 With respect to passive measures, people with impaired hearing (and, it may be added, most other people too) are greatly helped by rooms and spaces with quiet acoustics. This is achieved by having a minimum of hard, sound-reflective surfaces, and a generous provision of carpeting, sound-absorbent ceilings, and other acoustically absorbent surfaces.

5 With respect to active measures, people with impaired hearing can be furt assisted by, in addition to their own personal *hearing aids* (qv) and *listeners* (qv); the following installed systems:

• At ticket offices and inquiry desks, and in conference and lecture halls and meeting rooms: the installation of *hearing enhancement systems* (qv) such a *induction loops* (qv).

• In cinemas, theatres and concert halls where sound quality must be high, a provided people do not mind having to wear a neck-hung special receiver, *infra-red communication systems* (qv) may be more appropriate than inducti loops.

• In entertainment, cultural, educational, and training venues the use of captioning and sub-titling systems can help the hard of hearing to follow the proceedings on a more equal footing with others – see: *Captioning* (qv).

For other useful products visit www.rnid.org.uk/

Impaired hearing: complying with the Disability Discrimination Act

1 For compliance with the Disability Discrimination Act (qv), which applies all buildings used by the public, but not to private dwellings, relevant *authoritative practical guides* (qv) include DESIGNING FOR ACCESSIBILITY (qv plus the following:

• APPROVED DOCUMENT M (qv), the provisions of which must be heeded in or to comply with the Building Regulations for England and Wales.

• BS 8300 (qv) which applies to all buildings in the UK.

• INCLUSIVE MOBILITY (qv), which is the most authoritative reference for the *inclusive* (qv) design of *pedestrian infrastructure* (qv) and of *transport-relat buildings* (qv) such as bus and coach stations, railway stations, air terminals, and transport interchanges.

While the latter two documents are advisory and do not have the force of la conformity with their recommendations will make it easier to demonstrate tl the requirements of the DDA have been met.

Relevant clauses in the above documents include the following:

FEATURE	DOCUMENT: RELEVANT PARAGRAPHS		
	APPROVED DOCUMENT M	BS 8300	INCLUS MOBILIT
QUIET ROOMS AND SPACES	4.33	9.1.2	–
AUDIBLE SIGNS	–	9.2.5	–
AUDIBLE INFORMATION SYSTEMS	4.31–4.36	9.3	10.1.8

In places other than England and Wales the following regulatory documents should be consulted:

• In Northern Ireland: TECHNICAL BOOKLET R of the Building Regulations (Northern Ireland) (qv).

• In Scotland: the TECHNICAL HANDBOOKS of the Building (Scotland) Regulations (qv).

2 As of 2008, useful references in addition to those already mentioned inclu the following. For publication details see the bibliography.

• SOLUTIONS, the annual catalogue of the RNID.

3 Useful websites include:

• Directgov: disabled people www.disability.gov.uk/
• RNID www.rnid.org.uk/

See also: *Aids and equipment; Deafness; Signs*; and *Telephone services for d people.*

aired mobility

1 Impaired mobility is recognised as a *disability* (qv) under the Disability Discrimination Act (qv) if the effect is severe enough.

2 It is estimated that nearly 70% of all *disabled people* suffer to some degree from impaired mobility.

- There are two broad categories of impairment:
- Less seriously impaired people are *ambulant*, able to walk but probably needing some form of *walking or standing aid* (qv) to help them get about.
- More seriously impaired people will, assuming they are not actually bed-bound, need *wheelchairs* (qv) and/or other *small personal vehicles* (qv) to get about. These people, many of whom are over-65 s, may in fact not be totally dependent on their wheelchairs and could move short distances, for instance in their homes, provided there are plenty of easily reachable *handrails* (qv) and *support rails* (qv) at strategic positions.

3 The diagram suggests a taxonomy of mobility aids for both ambulant and non-ambulant disabled people.

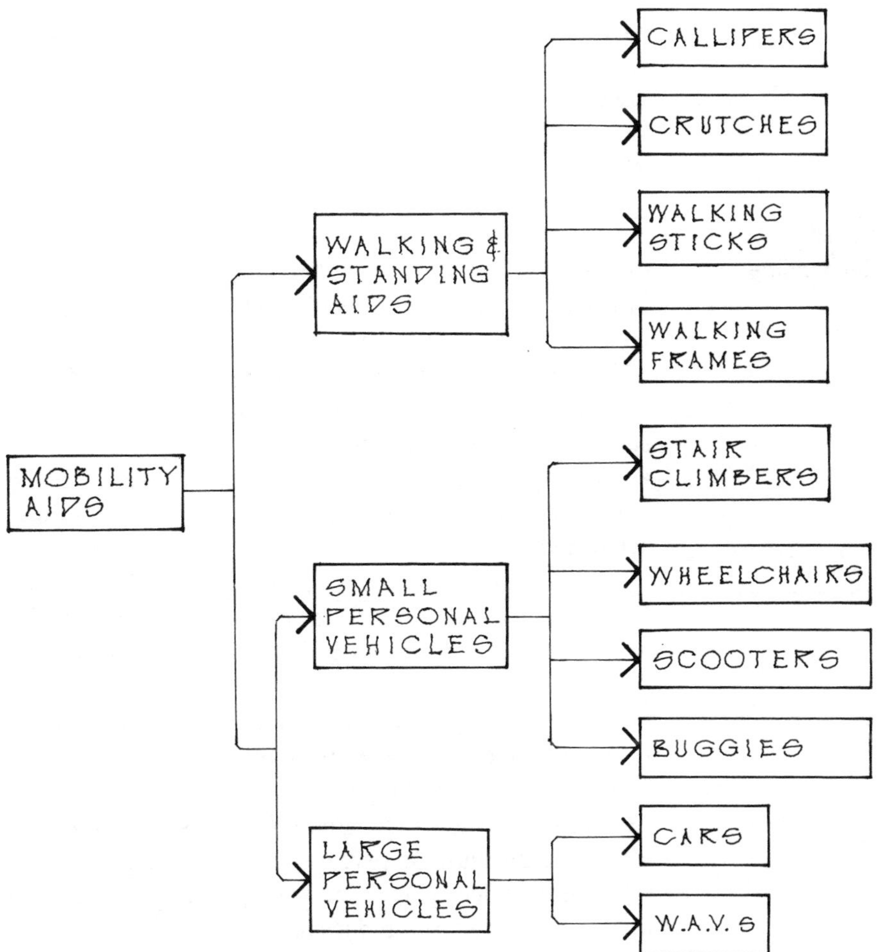

4 In designing private dwellings which may be used by people with impaired mobility, recommended references include the following:

- THE HABINTEG HOUSING ASSOCIATION DESIGN GUIDE (qv) applies most specifically to *housing association* (qv) schemes, but ought to be by the designer's side in the design of all housing developments and individual dwellings.
- THE WHEELCHAIR HOUSING DESIGN GUIDE (qv), also from the Habinteg Housing Association, is another illustrated manual whose usefulness extends far beyond the special remit implied by the title.

5 In designing buildings used by the public, where the Disability Discriminat Act (qv) applies, relevant *authoritative practical guides* (qv) include DESIGNI FOR ACCESSIBILITY (qv) plus the following:
• APPROVED DOCUMENT M (qv), the provisions of which must be heeded in order to comply with the Building Regulations for England and Wales.
• BS 8300 (qv) which applies to all buildings in the UK.
• INCLUSIVE MOBILITY (qv), which is the most authoritative reference for the *inclusive* (qv) design of *pedestrian infrastructure* (qv) and of *transport-relate buildings* (qv) such as bus and coach stations, railway stations, air terminals, and transport interchanges.
While the latter two documents are advisory and do not have the force of law conformity with their recommendations will make it easier to demonstrate th the requirements of the DDA have been met.
6 As of 2008, useful websites include:
• Directgov: disabled people www.disability.gov.uk/
• Disabled Living Foundation www.dlf.org.uk/
• Disabled Living Online www.disabledliving.co.uk/
• Ricability
www.ricability.org.uk/ www.ricability.org.uk/

See also: *Mobility aids and equipment; Wheelchair-users*; and *Wheelchairs*.

Impaired understanding

1 It is less easy to be precise about *mental impairments* (qv) than about physi ones such as *deafness* (qv) *blindness* (qv), or *paralysis* (qv), but there are broadly two categories of people who have greater than normal difficulty in concentrating, learning and understanding:
• Those with *learning difficulties* (qv) – i.e. moderate difficulties that can be largely overcome with good care and good schooling, enabling such people tc successfully blend into society.
• Those with a *learning disability* (qv) – i.e. severe and complex difficulties t amount to a lifelong condition, the most common example being Down's syndrome.
2 Impaired understanding is recognised as a *disability* (qv) under the Disabili Discrimination Act (qv) if the effect is severe enough – see: *Disability: definition in the Disability Discrimination Act*. Conditions covered under the Act include:
• Clinically recognised mental illnesses. The protection of the Act extends to people who have recovered from mental illness but are still experiencing prejudice as a result thereof.
• Impairments such as severe *dyslexia* (qv) and *epilepsy* (qv).
• A severe difficulty in perceiving risk or danger.
• A severe difficulty in concentrating, learning, and understanding – common referred to as a *learning difficulty*.
3 As of 2008, useful references include the following.
• DESIGNING FOR SPECIAL NEEDS: an architect's guide to briefing and designing for people with learning disabilities. See the bibliography.
4 Useful websites include the following. They are given in alphabetical order.
• Directgov: disabled people www.disability.gov.uk/
• Mencap www.mencap.org.uk/
• Mind www.mind.org.uk/
• National Institute for Mental Health www.nimhe.csip.org.uk/
5 Suppliers of learning aids and play equipment for children with impaired understanding include:
• Possum www.possum.co.uk/
• Rompa www.rompa.com/

- Wicksteed www.wicksteed.co.uk/
- Winslow www.winslow-cat.com/

See also: *Communication aids; Computers; Learning aids; Learning disability*; and *Speech aids*.

[Imp]aired vision: [com]plying with the [Disa]bility [Disc]rimination Act

1 There are many different types and degrees of visual impairment. Very few people – a minority even of those formally categorised as *blind* (qv) – cannot see anything at all. More typical problems include a lack of central vision; a lack of side vision; seeing everything as a blur; or seeing the world as a patchwork of blanks and defined areas. For more detail visit www.rnib.org.uk/

2 Whatever the cause, impaired vision is recognised as a *disability* (qv) under the Disability Discrimination Act (qv) if the effect is severe enough, but not if the impairment can be corrected by wearing spectacles.

3 For compliance with the Disability Discrimination Act (qv), which applies to all buildings used by the public, but not to private dwellings, relevant *authoritative practical guides* (qv) include DESIGNING FOR ACCESSIBILITY (qv) plus the following:

- APPROVED DOCUMENT M (qv), whose provisions must be heeded in order to comply with the Building Regulations in England and Wales.
- BS 8300 (qv) which is the most authoritative general reference for *inclusive* building design, and is an advisory document.
- INCLUSIVE MOBILITY (qv), which is the most authoritative reference for the *inclusive* (qv) design of *pedestrian infrastructure* (qv) and of *transport-related buildings* (qv) such as bus and coach stations, railway stations, air terminals, and transport interchanges.

While the latter two documents are advisory and do not have the force of law, conformity with their recommendations will make it easier to demonstrate that the requirements of the DDA have been met.

4 Relevant clauses in the above documents include the following:

FEATURE	DOCUMENT: RELEVANT PARAGRAPHS		
	APPROVED DOCUMENT M	BS 8300	INCLUSIVE MOBILITY
HAZARD PROTECTION	1.38–1.39	5.7.2–5.7.3	3.7–3.11
STAIR NOSINGS	1.33	5.9.4	8.4.1
STAIR & RAMP HANDRAILS	1.37	5.10	8.4.3
TACTILE WARNING SURFACES	1.33	5.9.6	4.1–4.7
SURFACE FINISHES & COLOURS	4.32–4.34	9.1	–
SIGNS & LETTERING	–	9.2, 11.4.8	10.1
LIGHTING	3.12, 4.3	49.4	11

For places other than England and Wales the following regulatory documents should be consulted:

- In Northern Ireland: TECHNICAL BOOKLET R of the Building Regulations (Northern Ireland) (qv).
- In Scotland: the TECHNICAL HANDBOOKS of the Building (Scotland) Regulations (qv).

5 In the USA, where the Americans with Disabilities Act applies, refer to the ADA AND ABA ACCESSIBILITY GUIDELINES (qv).

6 As of 2008, other useful references include the following. For publication details see the bibliography.

• HOUSTING FOR PEOPLE WITH SIGHT LOSS: A THOMAS POCKLINGTON TRUST DES GUIDE.
• SEE IT RIGHT: MAKING INFORMATION ACCESSIBLE FOR PEOPLE WITH SIGHT PROBLEMS.
7 Useful websites include:

• Directgov	www.direct.gov.uk/
• Directgov: disabled people	www.disability.gov.uk/
• RNIB	www.rnib.org.uk/
• Specs	www.eyeconditions.org.uk/

See also: *Assistance dogs; Canes for blind people*; and *Colour and tonal contrast*.

Impairment and disability

1 Some theorists, especially in Britain, draw a hard distinction between the terms *impairment* and *disability*, and hold that it is wrong to speak of 'a per with disabilities'. They reason that:
• An 'impairment' is a reduced functioning in some part of a person's body mind.
• 'Disability' is the negative effect suffered by such people if society fails to provide for that impairment.
From this point of view someone is *disabled* not by having a paralysed lowe body, but by management systems and buildings or vehicles that do not suit people with paralysed lower bodies.
2 The disabled writer Tom Shakespeare suggests that the British preoccupati with terminological wranglings of this kind is unhelpful and ought to be abandoned. In America highly effective anti-discrimination policies are implemented without bothering overmuch about precise definitions. For a w informed discussion of these issues refer to Part 1 of his book DISABILITY RIG AND WRONGS, of which Mr Bert Massie, chairman of the Disability Rights Commission (qv) has said: 'If you read only one book on disability rights thi year, make this the book'. For publication details see the bibliography.

Implementation Review Unit (IRU)

1 For the function of the Implementation Review Unit see: *Cutting Burdens Teachers* and visit www.dfes.gov.uk/reducingbureaucracy/
2 For more information on the IRU visit www.dfes.gov.uk/iru/

See also: *Better regulation bodies*

Incapacity Benefit (IB)

1 Incapacity Benefit replaced Sickness Benefit and Invalidity Benefit in Apr 2005 and is paid to people of working age who have been assessed as being unable to work owing to illness or *disability* (qv). IB is currently claimed by 2.6 million people compared with 700 000 in the 1980s. It seems certain that very many of the 2.6 million are in fact capable of work. As worklessness impoverishes and demoralises the jobless and their families, and IB costs taxpayers £12.5 billion every year, helping 1.5 million claimants into work (as the Government aims to do by 2015) would benefit everyone. Mr David Freud, who advises both the Treasury and the Conservative Party, estimated i February 2008 that it would make sense to pay jobfinding firms up to £62 00 to get an average IB claimant into stable employment.
2 In January 2008 the Leader of the Opposition, Mr David Cameron, announced that a future Conservative government would (in a dramatic break with recent governments of all political stripes) test every IB claimant in the country to judge their ability to work, and transfer those deemed capable of working to Jobseeker's Allowance (qv).

3 As of 2008, useful references include the following. They are given in alphabetical order. For publication details see the bibliography.

• THE REAL LEVEL OF UNEMPLOYMENT 2002 by Professor Steve Fothergill and others of Sheffield Hallam University claims that almost a million people claiming *incapacity benefit* (qv) are in fact fit for work.

4 Useful websites include the following. They are given in alphabetical order:

• Directgov www.direct.gov.uk/
• Directgov: disabled people www.disability.gov.uk/
• Frank Field www.frankfield.co.uk/type3.asp?id=20&type=3
• Jobcentre Plus www.jobcentreplus.gov.uk/
• Office for National Statistics www.statistics.gov.uk/

See also: *Employment; Jobseeker's allowance; Neets; Underclass*; and *Welfare dependency*.

Incline platform lifts

1 Incline platform lifts are one of two categories of stairlift as shown below.

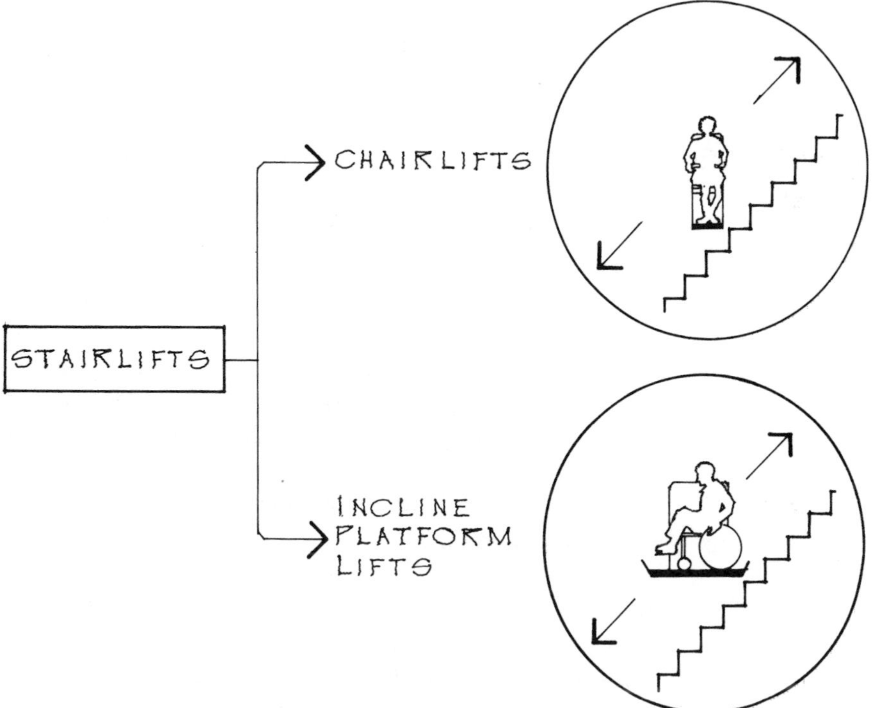

Unlike *chairlifts* (qv), which take only a passenger, incline platform lifts can carry a passenger sitting in his or her *wheelchair* (qv). This is a great advantage to the many people who find the process of transferring from wheelchair to chairlift dangerous if not impossible.

2 A typical example is shown below.

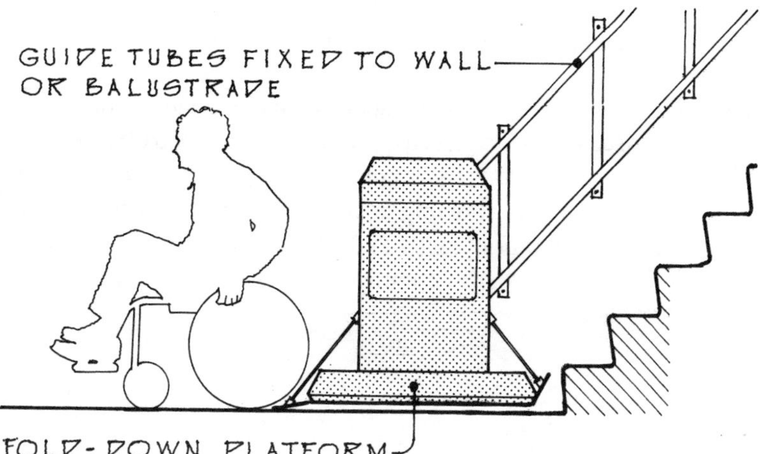

While these lifts are appropriate for certain situations, the support rails on which they run are obtrusive and usually very unattractive, which makes then an inappropriate choice in elegant interiors – see: *Historic buildings*.

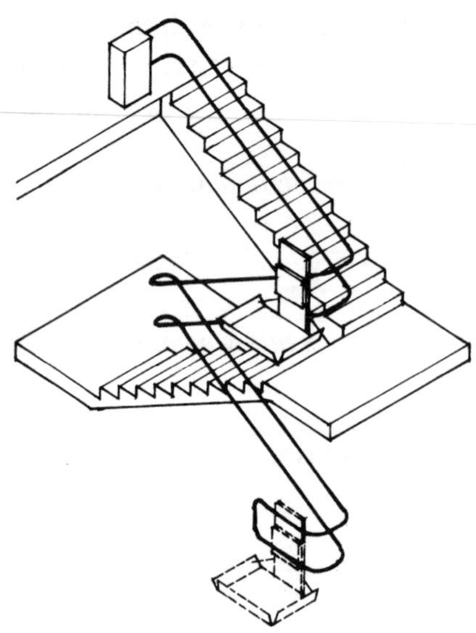

Incline platform lifts in dwellings

1 In private dwellings: there are no statutory regulations covering the installation of lifts, and homeowners may do as they please. They will usually want a lift from ground floor to first floor, to help a disabled occupant get up bedroom level, and in such situations a *through-floor lift* (qv) is an alternative answer. For possible grant aid see: *Aids and equipment: entitlement* (qv) and *Home improvement grants* (qv).

2 In blocks of flats: APPROVED DOCUMENT M (qv) contains provisions on *passenger lifts* (qv) that must be heeded in order to comply with the Building Regulations – see section 9. As of 2007 this makes no reference to *stairlifts* (c of any kind.

3 As of 2008, useful references include the following. For publication details see the bibliography.

- CHOOSING EQUIPMENT TO GET UP AND DOWN STAIRS, published by the Disabled Living Foundation (qv).
- PLATFORM LIFTS, published by the Centre for Accessible Environments (qv), gives well-informed advice and shows case examples of excellence. It deals mainly with public buildings, but contains much useful information and shou be consulted.

4 Useful websites include the following. They are given in alphabetical order.

- Disabled Living Foundation www.dlf.org.uk/public/factsheets.html
- Disabled Living Online www.disabledliving.co.uk/
- OTDirect www.otdirect.co.uk/link-equ.html#Platfo
- Ricability www.ricability.org.uk/

See also: *Chairlifts; Lifts*; and *Platform lifts*.

Incline platform lifts in public buildings: complying with the Disability Discrimination Act

1 Public buildings and spaces must in general comply with the Disability Discrimination Act (qv). In the case of chairlifts, relevant *authoritative practical guides* (qv) include DESIGNING FOR ACCESSIBILITY (qv) and in addition the following documents. For publication details see the bibliograph

- APPROVED DOCUMENT M (qv): the provisions in paras 3.17 to 3.49 must be heeded in order to comply with the Building Regulations for England and Wales.

• BS 8300 (qv): the recommendations in section 8.4 apply to all buildings in the UK. BS 8300 does not have the force of law, but conformity with its recommendations will make it easier to demonstrate that the requirements of the Disability Discrimination Act have been met.
For places other than England and Wales the following regulatory documents should be consulted:
• In Northern Ireland: TECHNICAL BOOKLET R of the Building Regulations (Northern Ireland) (qv).
• In Scotland: the TECHNICAL HANDBOOKS of the Building (Scotland) Regulations (qv).
The general import of the above recommendations is that:
• The preferred option in all public buildings is a *passenger lift* (qv) that will accommodate a *wheelchair-user* (qv). Section 8.4 of BS 8300 gives detailed recommendations on such lifts.
• If a passenger lift is not feasible, and travel is only between two levels, a *vertical platform lift* (qv).
• If that too is not feasible, and travel is only between two levels, an *incline platform lift* – provided its installation does not conflict with *means of escape* (qv) on staircases.
• If even that is not feasible, a *chairlift* (qv) is the last and least desirable option.
2 In the USA, where the Americans with Disabilities Act applies, refer to the ADA AND ABA ACCESSIBILITY GUIDELINES (qv).

See also: *Lifts*.

---

lusion

1 An inclusive society is one in which everyone (or very nearly everyone) has dignified and easy access to the full range of opportunities that society has to offer. Such a society will generally be characterised by a high degree of safety, comfort, and freedom for all citizens.
2 All the entries in the present work were motivated by that ideal, and are written in that spirit.

---

lusive design

1 'Inclusive design' has in the UK become the preferred term for a design ethos that is also sometimes called (in alphabetical order) 'barrier-free design'; 'design for all'; 'non-handicapping design'; 'trans-generational design'; or 'universal design'.
2 The Centre for Universal Design has defined the concept thus: 'The design of products and environments to be usable by all people, to the greatest extent possible, without the need for adaptation or specialised design'. In practice this comes down to doing three things:
• Designing all products and environments for the broadest possible 'average' – see: PERCENTILES: USE IN DESIGN.
• Providing for 'adjustability' by individual users, wherever this is possible. For instance: as no single *seat* (qv) height can ever suit everyone, people should be able to easily adjust the height of their office chair. Again, as no particular brightness of task lighting will ever suit everyone, and at all times, general background lighting should be augmented by desk lights that are under individual control.
• Making 'special provision' for those people who have characteristics and needs that differ so much from the average that they cannot be satisfied even by the most conscientiously designed 'normal provision'.
3 Useful discussions will be found in the following references, given here in alphabetical order. For publication details see the bibliography.
• BODYSPACE by Stephen Pheasant. See especially pp 3–14 et seq.

- DESIGNING FOR THE DISABLED: THE NEW PARADIGM by Selwyn Goldsmith. See especially pp 118, 121, and 384.
- DISABILITY RIGHTS AND WRONGS by Tom Shakespeare. See pp 44–50.
- ENGLISH PARTNERSHIPS GUIDANCE NOTE: INCLUSIVE DESIGN by Rita Newton and Marcus Ormerod lists virtually all official standards and guidance documents as at 2007.
- UNIVERSAL DESIGN by Selwyn Goldsmith. See especially pp 4 and 11–12.

See also: *Authoritative practical guides*.

Inclusive design awards

See: *Access awards*.

Inclusive mobility

1 INCLUSIVE MOBILITY: A GUIDE TO BEST PRACTICE ON ACCESS TO PEDESTRIAN AND TRANSPORT INFRASTRUCTURE is, as of 2007, the most authoritative reference for the *inclusive* design of *pedestrian environments* (qv) and of *transport-related buildings* (qv). It does not have the force of law, but conformity with its recommendations will help to demonstrate that the requirements of the Disability Discrimination Act (qv) have been met. It is a core reference for planners and designers of the built environment.
2 For publication details see the bibliography.

See also: *Authoritative practical guides*.

Income: median

1 The commonly-used phrase 'average income' can refer to either the 'media income' or to 'mean income'. These terms mean slightly different things, an each has its appropriate uses. To see the difference, consider this set of five annual incomes:

£3 000, £9 000, £15 000, £17 000, £44 000.

The median income is the middle value, with equal numbers of incomes bel and above. Therefore it is £15 000.
The mean income is the sum of all five incomes, divided by five. Therefore i £88 000 ÷ 5 = £17 600.
It will be seen that the 'mean' average was markedly influenced by the comparatively high value of £44 000, whereas the 'median' average is not.
2 The median is conventionally accepted as the most useful kind of average t use when discussing incomes. The graph below, from Figure 5.12 in SOCIAL TRENDS 36 (qv), shows the distribution of weekly household disposable incon in Great Britain in 2003–04, with both the median and the mean marked.

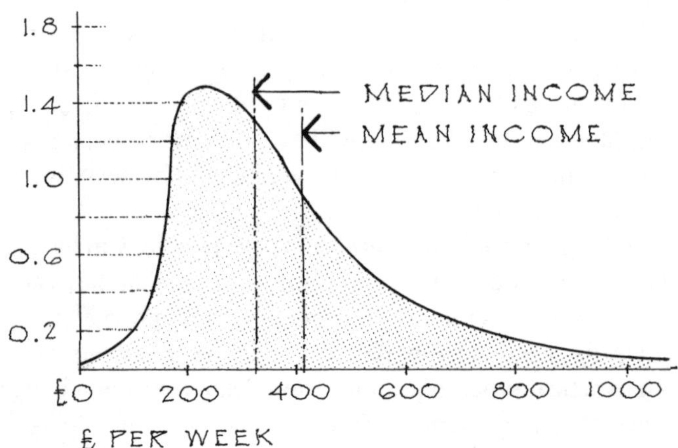

4 In 2005–06 the median household income of British families was approximately £24 700 – see: *Taxation*.

See also: *Norm*; and *Poverty*.

come support

1 Income support is a state benefit that is available, with certain qualifications, to people who cannot be available for full-time work and do not have enough money to live on.
2 Like many other Government programmes this one has become complex beyond comprehension, is littered with traps, and the safest thing for many beneficiaries is not to go back to work. REFORMING WELFARE by Nicholas Boys Smith, a study from the cross-party *think tank* Reform (qv), analyses the problem and proposes a better approach – see the bibliography.
3 As of 2008, useful websites include:

- Directgov www.direct.gov.uk/
- Jobcentre Plus www.jobcentreplus.gov.uk/

See also: *Jobseeker's allowance; Poverty; Social protection; Unemployment*; and *Welfare reform*.

comes: inequality

1 Twenty years ago FTSE chief executives earned 17 times the pay of their workers; today they earn 75 times more and the gap is growing. Those simple facts summarise what is happening to British society today, and the following paragraphs tell the story in greater detail.
2 First, the entire British population is on *average* (qv) becoming richer. The graph below, from Figure 5.1 in SOCIAL TRENDS 37, shows that real household disposable income per head of the UK population has increased by one and a third times since 1971.

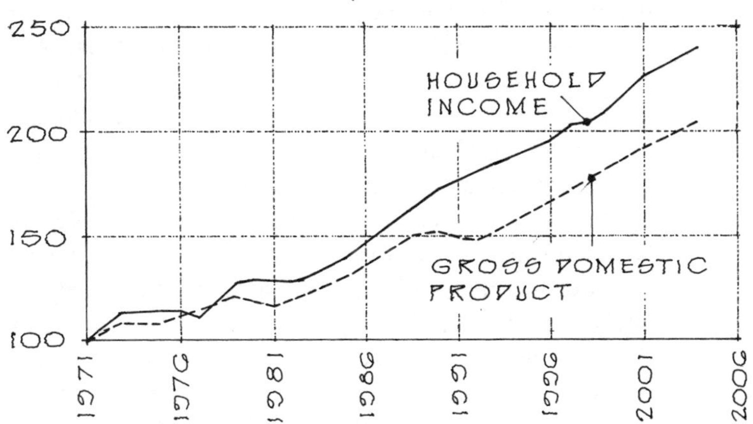

(A) REAL HOUSEHOLD DISPOSABLE INCOME

3 But second, as shown by the graph at the top of the next page, which is from Figure 5.11 in SOCIAL TRENDS 37, the top 10% of earners have benefited most and the bottom 10% least and the gap is widening. For a simple depiction of this rising degree of economic inequality see: *Gini coefficient* (qv), and for latest data plus commentary consult Chapter 5 of the latest edition of the annual publication SOCIAL TRENDS (qv).
4 As of 2008 other useful references on these trends, and on the social consequences that may follow, include the following. They are given in alphabetical order. For publication details see the bibliography.

- FALLING BEHIND: HOW RISING INEQUALITY HARMS THE MIDDLE CLASS by Robert Frank. In this short book (160 pp) Professor Frank, an American, argues

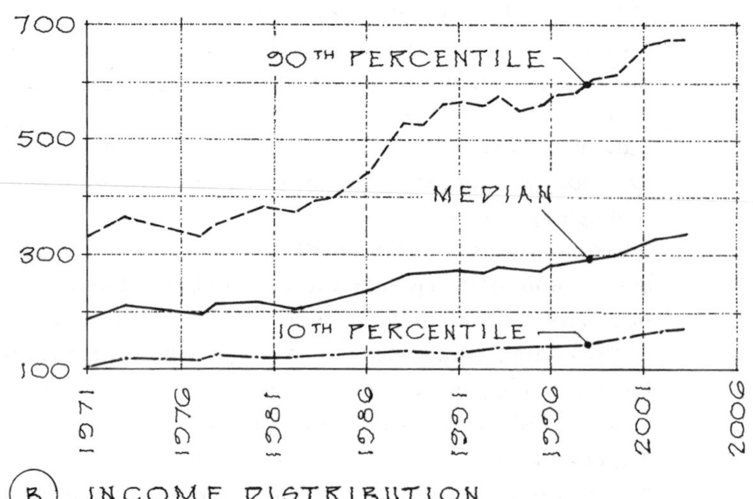

(B) INCOME DISTRIBUTION

that the rise of an overclass is indirectly affecting the quality of life of the rest of the population, and not in a beneficial way.

• THE IMPACT OF INEQUALITY: HOW TO MAKE SICK SOCIETIES HEALTHIER by Richard R Wilkinson. The author, who is Professor of Social Epidemiology at the University of Nottingham and Visiting Professor at the International Centr for Health and Society at University College London, believes that average lif expectancy in the UK would increase, homicide rates and levels of violence would fall, people would trust each other more, and all sectors of society woul benefit if we changed from being one of the most unequal of European countries to being among the most equal.

• PERMANENT DIFFERENCES? INCOME AND EXPENDITURE INEQUALITY IN THE 1990s AND 2000s by Alissa Goodman and Zoë Oldfield, and published by the Institute for Fiscal Studies, compares income and expenditure inequalities in th 1990s and early 2000s with those of previous decades.

• POVERTY, WEALTH, AND PLACE IN BRITAIN, 1968–2005, published in 2007 by the Joseph Rowntree Foundation (qv), is a groundbreaking study. The authors use a wealth of figures to show that the UK has the highest inequality levels fo 40 years, and that the gap is increasing.

• REDUCING INEQUALITIES, a report published in September 2007 by the National Children's Bureau, the Institute of Education at the University of London, and the Family and Parenting Institute, analysed data from a study th tracked the lives of 17 000 people born in 1970. It found that inequalities in British society are entrenched; social mobility is a myth; and despite billions c pounds of government funding to cut child poverty, the gap between the poore and richest children is probably wider now than it was 30 years ago.

• REFORMING WELFARE by Nicholas Boys Smith, a study from the cross-party *think tank* (qv) Reform, points out that the current *benefits* (qv) and *tax credit* (qv) systems have not reduced poverty and proposes a different approach.

• RICHISTAN: A JOURNEY THROUGH THE 21st CENTURY WEALTH BOOM AND THE LIVES OF THE RICH by Robert Frank. Professor Frank argues that America's super-rich are creating for themselves a state within a state, visits it, and sends out dispatches like a correspondent in a foreign land.

5 Useful websites include the following. They are given in alphabetical order.

• Dept for Work and Pensions www.dwp.gov.uk/
• Focus on Social Inequalities www.statistics.gov.uk/focuson/socialinequalities/
• New Policy Institute www.npi.org.uk/projects/low%20pay.htm

- Office for National Statistics www.statistics.gov.uk/cci/nugget.asp?id=332
- Office for National Statistics www.statistics.gov.uk/cci/nugget.asp?id=1005

See also: *Poverty; Exclusion*; and *Taxation*.

Incomes: inequality: what can be done?

1 The following shaming details fill out the picture sketched above:

- At the top of the pile, the remuneration specialists Incomes Data Services reported in October 2007 that average salary-plus-bonuses earned by Chief Executives of Britain's top 100 firms was £3 million per person per annum (a rise of 100% in 5 years), while the average annual earnings of British full-time workers were £30 000 per person (a rise of 20% in 5 years). According to earlier press reports the total amount given in bonuses to highly paid executives in 2006 is said to have been £21 billion; individual packages of £10 to £50 million were not uncommon, and 4 500 people are said to have received bonuses of more than £1 million.
- At the bottom, the workers who do the dirty work of cleaning these people's premises struggled to earn much more than £5 an hour.

To cap it all, very poor people who try to better themselves through work may find themselves paying a marginal taxation (qv) as high as 90%, which is over twice the rate paid by top earners.

2 In addition to being morally repugnant such inequalities are becoming positively damaging to society in four ways:

- *House prices* (qv) are rocketing out of sight, partly because a slow-growing supply of desirable houses is now being chased by hundreds of thousands of buyers who barely notice the number of zeros on the price tag. It is difficult to imagine a scale of house-building that can bring an end to this mismatch between supply and demand with top incomes at current levels.
- Good workplace relations, which are essential to efficient businesses and a happy society, cannot survive a situation in which top earners enjoy annual increases of 20% to 30% on income of millions, while those at the bottom fight for 2% or 3% annual increases on incomes of £10 an hour.
- A law-abiding society depends ultimately on people feeling themselves to be part of the same family, with mutual obligations towards each other. If that goodwill disappears (and it is beginning to do so) then no amount of surveillance and policing can enforce good behaviour.
- Finally, if governing and decision-making elites are able to isolate themselves from major social and environmental problems, as the rich and super-rich are increasingly doing in their gated communities, special holiday resorts, and private planes, it becomes less likely that they will address themselves to solving those problems. HOW SOCIETIES FAIL – AND SOMETIMES SUCCEED by Professor Jared Diamond touches upon this matter.

3 What can be done?

4 An area that falls squarely within the powers of government is the tax system, and this should be made fairer in three ways.

- First, taking the low-paid out of income tax altogether has become unanswerable. The graph on the next page, from Figure 5.12 in SOCIAL TRENDS 36 (see the bibliography), shows weekly household disposable incomes in Great Britain in 2003–04. For reasons of space the graph must stop at £1 000 a week: to show the average annual pay of £2.8 million for the chief executives mentioned above, the diagram would have to be extended 10 or 15 feet to the right. What moral or practical justification can there be for trying to extract income tax from the meagre earnings of people to the left of the 60% poverty line (who have already paid out 30% of their disposable household income on *indirect taxes*) when the same amount could be contributed by those along the 15-foot long tail to the right of that line with barely a twinge?

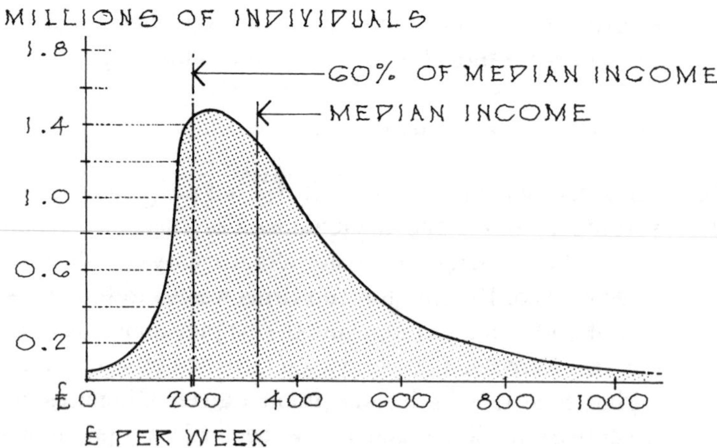

• Second, reconsidering the very principle of indirect taxes such as VAT (whic applies in Britain only because of our EU membership). These taxes absorb almost a third of the disposable household income of the poorest people in Britain, are impossible to avoid, and are morally indefensible in a society as economically unequal as the UK today.

• Third, introducing carefully-designed tax increases on high earners, not to punish them (driving away top-performing executives and companies would d no-one any good) but to raise money for tackling problems that are inflicting great damage on society. Speaking at the annual Allen Lane lecture in Februa 2008 Mr Frank Field MP (qv) suggested a 10% tax increase on those earning more than £150 000 a year (probably yielding the Treasury £3.6 billion a year which could be completely offset by charitable donations.

5 For an interesting development on tax evasion by the super-rich refer to the article AT LEAST GERMANY STAMPS ON TAX HAVENS by Nick Cohen, published in the *Observer* on 24 February 2008 – visit www.guardian.co.uk/commentisfree/2008/feb/24/

Incomes: gender pay gap

1 Women tend to be paid less than their male counterparts, though the gap for full time workers in Great Britain is slowly narrowing. According to SOCIAL TRENDS 36 (qv) the median net income of men rose by 13% between 1996–67 and 2003–04, but that of women by 29%. The graph below, from Figure 5.6 from SOCIAL TRENDS 36, illustrates the same trend for full time male and femal workers over a longer period, from 1986.

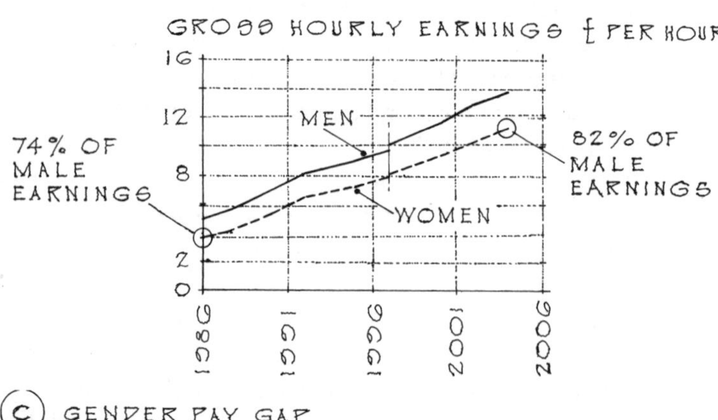

(C) GENDER PAY GAP

2 Looking at employment of all kinds, there are four main reasons for disparities between men's and women's pay:

• The discriminatory practice of paying women less than men for doing the same or equivalent jobs.

• The fact that women's employment tends to be concentrated in the lower-paid, lower-skilled jobs.
• The fact that women tend to take time out of the labour market around the time of the birth of their children.
• The fact that women with children are more likely than men to give their family a higher priority than their career, and more likely to want to work part-time.
3 The first of the above is a clear example of *discrimination* (qv). The other three may be, but are also closely bound up with life and career choices.
• A study published in September 2007 by Professor Alison Booth of Essex University and Jan van Ours of Tilburg University found that women are happiest with part-time jobs that allow them to combine work and family life, while men are happiest working full time. The survey, claimed to be the first detailed academic study of British couples and their work-life balance, was based on interviews with 3 800 couples who were questioned repeatedly over an eight-year period in the British Household Panel Survey. Such gender preferences are bound to have an influence on career choice, career success, and incomes.
• More informally, writing in the *Daily Mail* on 31 July 2007 under the title 'Women deserve the same pay as men as long as they work for it', the journalist Amanda Platell observed: 'Having worked as an executive on national newspapers for a couple of decades and at the heart of Parliament for some years, I can assure you that when a talented woman was prepared to put in the hours – inside and outside the office – not only did she deserve to earn exactly the same as the men, that's exactly what she got. So why, then, were a lot of the women I employed paid less than a man in a similar post? Quite simply because they were either less talented or, far more often, less committed. The women simply weren't putting in the same hard graft as the men. And frankly, I don't blame them one jot. We women can't on the one hand celebrate our skills as multi-taskers, as we rightly do – running homes, wiping noses, holding down careers, nursing egos (and that's just our husbands) – and then lament the fact we have chosen – let me repeat that word *chosen* – to prioritise our families over our careers. There is a price to be paid, and it's in our pay packets'.
• Rosie Boycott, co-founder of the feminist magazine 'Spare Rib' (1971) and Virago Press (1973), made the same point in her review of the THE SEXUAL PARADOX by Susan Pinker, published in the Daily Mail of 22 April 2008 under the title 'As a feminist I thought it was men keeping women from top jobs'.
4 As of 2007, useful websites include the following:

• Directgov	www.direct.gov.uk/
• Equality & Human Rights Commission	www.equalityhumanrights.com/
• Fawcett Society	www.fawcettsociety.org.uk/
• Office for National Statistics	www.statistics.gov.uk/cci/nugget.asp?id=167
• Social Trends	www.statistics.gov.uk/socialtrends
• Women & Work Commission	www.womenandequalityunit.gov.uk/

See also: *Discrimination; Equality*; and *Women*.

[In]dependent [Bu]reaucracy [R]eduction Group [(B]RG)

1 The BRG was launched in 2005 with a remit to cut red tape and reduce bureaucracy across the whole learning and skills sector. Since then it has generated an impressive amount of paperwork and the exhausting bureaucratic load on frontline staff has increased, not shrunk.
2 For more detail visit www.dfes.gov.uk/furthereducation/index.cfm?fuseaction=content.view&CategoryID=25

See also: *Better Regulation Bodies*.

Independent Housing Ombudsman	Visit www.ihos.org.uk/
Independent living	1 Independence does not mean living without help, which is impossible. It means assisting frail, elderly, or disabled people in making their own decision instead having decisions taken for them. This approach is undoubtedly right for most physically disabled people, but unfortunately cannot apply to the many people who are mentally retarded, suffer from Alzheimer's disease, or have other forms of *impaired understanding* (qv). However, in general it is a fundamental principle that should apply in all areas of life, throughout life, and finally in the manner in which people depart from this life – see: *Self-deliverance*. 2 In 2008 a Disabled Persons (Independent Living) Bill was brought before th House of Commons, with the aim of giving disabled people a legal right to services such as housing adaptations and alterations. Such an Act would undoubtedly be extremely helpful to disabled people and also make long-term savings to social care budgets; but at time of writing the fate of the Bill is not yet clear. 3 For a balanced discussion refer to pp 24, 30, 136–148, 151, 189–90, 194, an 199 of DISABILITY RIGHTS AND WRONGS by Tom Shakespeare, a work of which Mr Bert Massie, chairman of the Disability Rights Commission (qv), wrote: ' you read only one book on disability rights this year, make this the book'. 4 As of 2008, useful references include the following. They are given in alphabetical order. For publication details see the bibliography. • IMPROVING THE LIFE CHANCES OF DISABLED PEOPLE, published in 2005 by the Cabinet office (qv), called for practical measures in four key areas, which included 'helping disabled people to achieve independent living'. • INDEPENDENT LIVING: THE RIGHT TO BE EQUAL CITIZENS by Sarah Gillinson, Hannah Green, and Paul Miller. 5 Useful websites include the following. They are given in alphabetical order. • Directgov: disabled people www.disability.gov.uk/ • Independent Living Funds www.ilf.org.uk/ • National Centre for Independent Living www.ncil.org.uk/ • Office for Disability Issues www.officefordisability.gov.uk/ See also: *Aids and equipment; Assistive technology; Direct payments; Environmental control systems*; and Products for easier living.
Independent Police Complaints Commission (IPCC)	1 The IPCC was set up in 2004 to provide a fair and effective system for dealing with complaints against police officers – visit www.ipcc.gov.uk/ 2 After years of truly scandalous dysfunctionality it came close to collapse in February 2008 when 100 lawyers who specialise in handling police complaint resigned in despair from its advisory board – visit www.guardian.co.uk/politics/2008/feb/25/police.law1
Index of Deprivation	1 The Indices of Deprivation 2000 are measures of deprivation for every *ward* (qv) and *local authority* (qv) area in England. They combine a number of indicators which cover a range of domains (income, employment, health deprivation and disability, education skills and training, housing and geographical access to services) into a single deprivation score for each specific area. 2 For more information visit www.communities.gov.uk/archived/publications/citiesandregions/indicesdeprivation
Indirect taxes	See: *Taxation*.

[Ind]uction and infrared [co]mmunication [sys]tems 1 These are common types of *hearing enhancement system* (qv) that may be installed (a) in rooms and spaces used for meetings, lectures, classes, performances, spectator sports, or films; and (b) at *counters* (qv) or reception desks where the background noise level is high or there is a glass screen between the assistant and the person being served. They enable speech or music to be picked up by a microphone, transmitted electronically to auditors wearing suitable reception devices (which could be a *hearing aid* (qv) set to the 'T' position), and then heard clearly, without background noise.

2 Induction systems can take two forms – (a) a wire loop laid around a small or large space, enabling people sitting within the loop to pick up the signal and hear what is being said; or (b) a coupler built into a telephone handset, offering telephone users with hearings aids an enhanced sound quality.

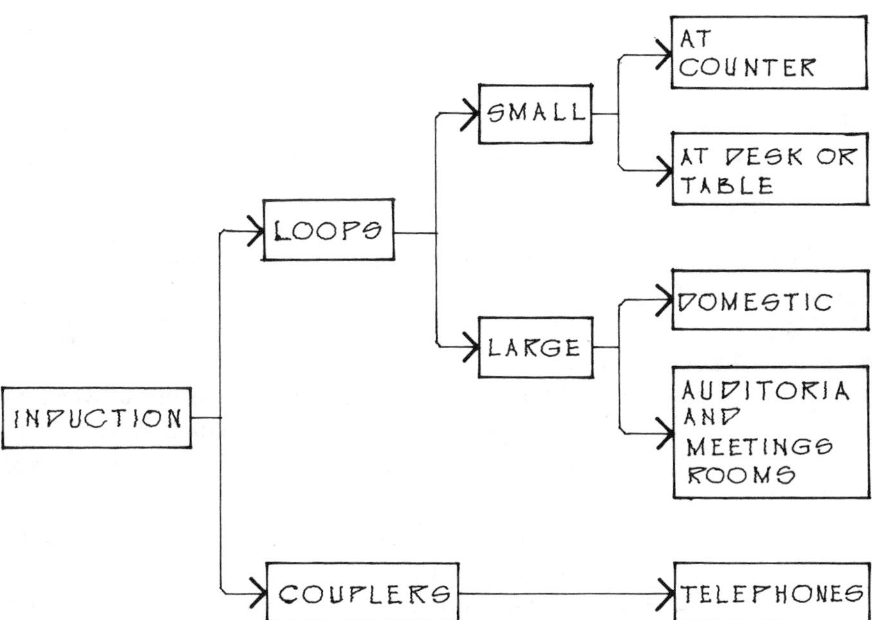

Induction systems are well suited to situations where:

- The listener is trying to hear information from a particular source, eg an informant at a ticket office or information point, or a speaker on a platform.
- Speakers can be relied on to speak into a microphone.
- The risk of the sound being picked up by people with receivers or *hearing-aids* in adjoining rooms is acceptable.
- Sound quality need not be high.

3 Infrared systems are used particularly in lecture rooms, concert halls, cinemas, and theatres. As with induction systems there is a microphone to pick up the speech or music, but instead of being fed into a wire loop the electronic signal is radiated (rather like a beam of invisible light) across the listening area, and then received and converted back to sound by receiving devices incorporated either in headphones, or neck-hung receivers, worn by listeners in the audience. Infrared systems are well suited to situations where:

- Sound quality must be high, eg cinemas, theatres, and concert halls.
- There is a direct sight-line between transmitter and receiver.
- More than one channel is required, eg different information streams must be transmitted to different people.
- Speakers or entertainers may want to move about.
- Confidentiality is important.

Infrared systems are however disliked by some people because headphones or neck-hung receiver must be worn, which are cumbersome and make the weare[r] feel conspicuous.
4 As of 2008, useful websites include the following. They are given in alphabetical order.

• CIE Audio	www.cie-audio.com/
• C-Tec	www.c-tec.co.uk/
• Eastern Electronics	www.easternelectronics.co.uk/
• Force 10	www.forcetenco.co.uk/
• Gordon Morris	www.gordonmorris.co.uk/
• Northern Acoustics	www.northern-acoustics.co.uk/
• RNID	www.rnid.org.uk/www.rnid.org.uk/

See also: *Assistive technology; Communication; Hearing aids and equipment*; and *Impaired hearing*.

Industrial buildings: basic planning data

1 Distilled planning data for most of the major building types is given in the METRIC HANDBOOK: PLANNING AND DESIGN DATA.
2 For publication details see the bibliography.

Industrial buildings: design: complying with the Disability Discrimination Act

1 Buildings used by the public must in general comply with the Disability Discrimination Act (qv). In the case of factories and warehouses, *authoritative practical guides* (qv) include DESIGNING FOR ACCESSIBILITY (qv) plus the following:
2 APPROVED DOCUMENT M (qv), the provisions of which must be heeded in order to comply with the Building Regulations (qv) for England and Wales. For places other than England and Wales the following regulatory documents should be consulted:
• In Northern Ireland: TECHNICAL BOOKLET R of the Building Regulations (Northern Ireland) (qv).
• In Scotland: the TECHNICAL HANDBOOKS of the Building (Scotland) Regulations (qv).
3 BS 8300, the provisions of which apply to all buildings in the UK. Para 13.2 gives brief design recommendations for both of the above building types. BS 8300 does not have the force of law, but conformity with its recommendations will make it easier to demonstrate that the requirements of the Disability Discrimination Act have been met.
4 In the USA, where the Americans with Disabilities Act applies, refer to the ADA AND ABA ACCESSIBILITY GUIDELINES (qv).

See also: *Workplaces*.

Industrial injuries

See: *Accidents*; and visit www.hse.gov.uk/

Inequality

See: *Health inequality; Incomes: inequality*; and *Social inequality*.

Infant care

See: *Childcare*.

Info4local

1 This is a website that provides quick access to information from more than 6[…] *government departments* (qv) and *quangos* (qv).
2 Visit www.info4local.gov.uk/

See also: *Government websites*.

Information centres

See: *Libraries*.

formation Commissioner's Office 1 The Information Commissioner's Office describes itself as an independent official body (see: *Quangos*). Its head, the Information Commissioner, is responsible for administering the provisions of the Data Protection Act 1998 (qv) and the Freedom of Information Act 2000 (qv).
2 As of 2008 the Commissioner has issued some robust reports and statements – see for instance: *ContactPoint* (qv) and *Privacy* (qv) – but the fact that the IOC's budget and staffing are controlled by the Ministry of Justice (qv) must cast some doubt on the degree and long-term reliability of its independence. As with various HM Inspectorates (qv) and other watchdogs, complete financial and organisational independence would be far better.
3 For more information visit www.informationcommissioner.gov.uk/

See also: *Data Protection Act; Freedom of Information Act 2000*; and *Privacy*.

formation: freedom See: *Freedom of Information*

formation Sharing ndex See: *ContactPoint*.

formation surfaces 1 Information surfaces are one of six types of *tactile paving*, and its particular use is to help visually impaired people locate amenities such as a telephone box or ticket office.
2 For more information see: *Tactile paving surfaces*.

quiries Act 2005 See: *Civil liberties*.

stitute of Community ohesion Visit www.coventry.ac.uk/icoco/a/264

tegrated transport trategy 1 The integration of land-use and transportation planning to allow transport provision and the demand for travel to be planned and managed together, balancing the use of different modes of transport to encourage easy transfer between them and reduced reliance on the private car.
2 For more information visit www.cfit.gov.uk/

tegration: racial See: *Population: racial integration*.

termediate housing See: *Housing: affordable*.

ternal lobbies See: *Lobbies*.

validity benefit See: *Incapacity Benefit*.

vestigation of egulatory Powers Act 000 See: *Civil liberties*.

visible children See: *Neets*.

systems 1 IT (for 'information technology') systems are ones that use computers, digital electronics, telecommunications, and related technologies to store, process, and transmit information. Badly designed, inadequately piloted, and hastily-launched IT systems lie at the heart of so many failed Government programmes that a brief résumé is in order.

2 NOT FIT FOR PURPOSE by Bobbie Johnson and David Hencke, a report published in the *Guardian* on 5 January 2008, states that the Government has wasted £2 billion of taxpayers' money on faulty IT schemes. It quotes Mr Joe Harley, programme and systems delivery officer at the Department for Work and Pensions (qv), as saying that the Government's £14 billion annual spend on IT could be used to build thousands of schools every year, or to employ hundreds of thousands of nurses in the NHS. 'Today only 30%, we estimate, of our projects and programmes are successful', he told a conference. 'It is not sustainable for us as a Government to continue to spend at these levels. We need to up the quality of what we do at a reduced cost of doing so'.

3 Worse than such waste of taxpayers' money is the fact that many of these expensive Government IT systems are still not working, and some (most scandalously the £15 billion or perhaps £20 billion NHS Care Records Service) almost certainly never will work. The following cases are given in alphabetical order:

• The Child Support Agency (qv) was launched in 1993 with the aim of making absent parents pay a fair share of the costs of raising the children they had left behind. It turned out to be an expensive disaster. In the 13 years of its existence millions of children received little or no maintenance money even though taxpayers paid out almost £50 billion. When in July 2006 Ministers finally despaired of the CSA and dissolved it, the computerised system still had 500 identified faults despite expenditure of around £91 million on private consultants.

• C-NOMIS is a National Offender Management Service (NOMIS) database that was intended to hold a single accurate profile for each offender, thus enabling officials to track criminals all the way through the court and *prison* (qv) system, and then after their release. In March 2007 the Home Secretary announced that the programme would be implemented by July 2008. In June 2007 the project head, Mr Mike Manisky, candidly admitted: 'We can't even ge the basics right. The whole thing is actually very badly thought through'. In August 2007 the Justice Minister, Mr David Hanson, announced that programme costs were out of control, and as this was written in September 200 it seemed a safe bet that C-NOMIS would be abandoned by early 2008 as an irretrievable failure.

• The Courts Service computer, it was revealed in September 2007 by the Law Society's magazine *Gazette*, is unable to operate a scheme launched in April 2007, requiring all offenders who are fined to pay a £15 surcharge to their victims.

• The Digitised of Vital Events (DOVE) project to place the births, marriages and deaths records of the English and Welsh populations on a searchable electronic database was intended to be operational by 2008. In September 2007 it is not expected to be completed before mid-2009 at the earliest, and that date looks optimistic.

• The Joint Personnel Administration (JPA) computerised pay system for Britain's armed forces was introduced in March 2006. In December 2007 it wa admitted to be in chaos, with at least 16 000 service men and women being pai the wrong salaries and many suffering gross hardship as a result.

• Modernising Medical Careers (MMC) is a Government programme using an IT system called the Medical Training and Application Service (qv) to handle the applications of junior doctors for specialist registrar jobs. The MTAS was introduced by the Department of Health in early 2007 but had to be abandoned after twice being found to be insecure, and after its other shortcomings had caused untold damage to the future careers of young doctors. For a postmorten refer to ASPIRING TO EXCELLENCE – see the bibliography.

• The National Probation Service (qv) launched a National Probation Service Information Systems Strategy (NPSISS) in 1993, with the aim of having in due course an effective computer-based Case Recording and Management System (CRAMS). For an account of the problems that subsequently plagued this system refer to the National Audit Office report THE IMPLEMENTATION OF THE NATIONAL PROBATION SERVICE INFORMATION SYSTEMS STRATEGY – see the bibliography.
• The NHS Care Records Service (qv) is a central computerised database that is intended ultimately to hold the medical records of every NHS patient in England. Since its launch the estimated total cost has escalated from an initial £12 billion to £15 billion, with no upper limit in sight. In October 2006, a group of 23 leading British computer scientists publicly questioned whether the programme could ever work, and called urgently for an independent inquiry before proceeding any further. In February 2007 Mr Andrew Rollerson, a top executive with one of the system's main suppliers, Fujitsu, told a conference of IT executives that the programme had not been properly thought out; that it 'isn't working, and it isn't going to work'. In November 2007 a poll of 1 000 GPs found that two-thirds of doctors – the very people who were supposed to be using the NHS CRS – were so worried about confidentiality that they proposed to boycott the database.
• The Police National Computer (qv) is intended to help police officers, the Criminal Records Bureau (qv), the Special Branch, and M15 vet people who have been stopped for questioning, or have applied for sensitive posts. Its reliability has several times been questioned, for instance in the House of Commons in January 2007 when the Home Secretary conceded that over half of 540 serious criminals (including rapists and murderers) had still not been put on the database, and admitted to a failure to log on 27 000 crimes committed by Britons abroad. Some commentators believe that the PNC is not fit for purpose now, and may never be.
• The Registration Online (RON) system that is intended to register births and deaths electronically, instead of by hand, failed when introduced in spring 2007, and kept failing. As of September 2007, Registrars in many parts of the country have been told to abandon the system until the problems are resolved.
• The Tax Credits (qv) scheme was launched in April 2003 with the aim of alleviating hardship among vulnerable poor people. Its computer systems were rushed into operation without adequate preparation, and collapsed within a few months after being launched. Helplines went into meltdown, leaving callers, many of whom were completely dependent on bureaucratic advice to get their money out of the system, distraught. Despite massive funding from taxpayers the systems have malfunctioned ever since.

4 In addition to the above, the following cases were briefly noted in the *Times* on 9 August 2007 under the title 'Systems down':

• In 1999 a £1 billion IT project involving the Post Office, the Benefits Agency, and ICL with the intention of creating a swipe-card for benefits collection, collapsed.
• An IT system aimed at speeding up asylum applications was abandoned in 2001 after a £77 million contract with Siemens failed to deliver.
• The £118 million Integris IT system for the National Probation Service (qv) was completed late and £22 million over budget.
• As of September 2007 the cost of the Libra IT system for magistrates' courts had risen from £146 million in 1998 for a 10-year contract to £232 million for an 8-year contract.

5 In short, British governments have for 20 years been demonstrating a persistent and deep-seated inability to make large IT systems work. What is the explanation? Here are two views:

• In June 2007 the Select Committee on Public Accounts of the House of Commons published a report titled DELIVERING SUCCESSFUL IT-ENABLED BUSINESS CHANGE (see the bibliography) which placed the blame squarely on amateurishness and irresponsibility at the highest levels. Projects had gone billions of pounds over budget and run years behind schedule, yet in a fifth of 'mission critical and high risk' cases Ministers had never called a single meeting with their senior Whitehall officials to discuss progress. In a further quarter of projects Ministers had called meetings fewer than four times a year They had neglected to ensure that they were fully briefed on hugely expensive projects on which the wellbeing or even the lives of millions of citizens depended.

• Addressing the *Guardian* newspaper's Public Services Summit in January 2007, Mr Nick Manning, head of public sector management and performance the Organisation for Economic Cooperation and Development (OECD), accuse British governments of constant and 'mindless promotion of change' despite little proof that the proposed reforms would actually work. He characterised th attitude as 'We know that plan A isn't working, so plan B must be better', whereupon Plan B would be implemented with as little preparation and testing as its failed predecessor.

6 It is a shocking story, which makes nonsense of all Government assurances that future IT-based systems such as ContactPoint (qv) and the National Identi Register (qv) launched under Prime Minister Tony Blair's *Transformational government* (qv) initiative can be trusted. Underlying the particular manageria failings noted above, the present author would suggest the following more fundamental problems, which persist from Prime Minister to Prime Minister and are intrinsic to the British way of conducting government:

• Ministerial inexperience. Whereas cabinet secretaries in the USA are appointed for their expertise and usually serve at least a four-year term in orde to develop a commanding knowledge of their portfolio, their counterparts in t UK generally take responsibility for matters of which they have no prior knowledge whatsoever, and are then moved on at three, two, or even one-year intervals to suit prime ministerial whim.

• Ministerial impatience. Successful businesses never introduce an IT system without a full pilot run, followed by a rigorous assessment. Most of the schemes noted above were rushed into operation without piloting and without back-up systems in case of failure.

• Ministerial arrogance. It is a recurrent feature of the above schemes that warnings of problems were contemptuously brushed aside and ignored. Ministers with little or no experience of life outside politics appear to believe that if they announce firmly enough that something will happen by a certain date, then it will magically do so.

7 As of 2008, useful websites include the following. They are given in alphabetical order.

• Foundation for Information Policy Research www.fipr.org/
• Michael Cross at Technology Guardian
http://browse.guardian.co.uk/search?search=michael+cross&N=4294962709
• NHS Confidentiality Campaign www.nhsconfidentiality.org/
• No2ID www.no2id.net/
• Open Rights Group www.openrightsgroup.org/

See also: *Jargon; Privacy*; and *Transformational government*.

J

gon

1 Jargon is impressive-sounding language used to mask a speaker's ignorance, or to give trivial thoughts an air of profundity.

2 Compare for instance the following pronouncements on education, reported in the *Daily Telegraph* on 4 January 2007. Both refer to Ofsted's (qv) 60-page report, titled THE 2020 REVIEW, on how schools in England should develop over the next 15 years.

• Teachers will be 'inviting their pupils to work in curriculum teams to review schemes of work and develop plans to improve teaching and learning'. They will be participating in a 'system-wide strategy to help the profession innovate in a disciplined way'. The aim is 'personalised learning', which 'does not mean that teachers are supposed to teach every child in a completely different way from the rest', but is 'an approach to learning that enables teachers to achieve exactly what they want … using their expertise and flexibilities that already exist in the system'. (From the report itself by Ofsted's Chief Inspector Christine Gilbert, and the official launch of the report by Secretary of State Alan Johnson.)

• 'Teachers need to know and be enthusiastic about their subjects. They should care for and have high expectations of their pupils. They should be able to keep control, explain things clearly, and help children who have difficulty understanding'. (From a comment on the Report by Professor Chris Woodhead, former Chief Inspector of Schools.)

The second statement is written in English. The reader can readily form a mental image of how these words might be acted upon, and how one would know whether the results are turning out as promised.

The preceding statements are jargon. They seem to mean something, but what? A 'system-wide strategy to help the profession innovate … in a disciplined way' could be an effective plan of action or a useless pile of paper. It might produce school-leavers who are fluent at reading, writing, and arithmetic, or ones who cannot write a coherent page of text or do everyday sums. It is impossible to know, because these gaseous words convey no tangible image of either the process, or its intended outcomes. Instead of expressing thoughts, the speakers bolted together a series of prefabricated technocratic phrases and hoped we would be impressed.

3 A second type of jargon uses words we don't understand to baffle us and effectively exclude us from the discussion. In subjects such as advanced mathematics, physics or chemistry such specialised vocabularies may be unavoidable. In everyday matters such as education, social work, politics, crime, equality, discrimination, or architecture they generally are not, and in these cases it is a sound rule to disbelieve (or at any rate be extremely wary of) anything stated in language one cannot understand.

4 These are matters of practical importance. It is depressing enough that the people running our education system should speak as noted above, but they were only following what is now standard practice in mission statements, funding applications, and other formal documents. A bid phrased in the words of Professor Woodhead wouldn't stand a chance. The language that wins public service contracts resolutely avoids clarity and concreteness and speaks instead of system-wide strategies, improvement levers, trending, benchmarking, base-lining, etc – for an amusing but devastating account read GOVERNMENT'S WAY WITH WORDS IMPACTS THE LIFE-CHANCES OF ALL (see the bibliography). Thus do consultants bamboozle non-specialist clients, and as the latter don't speak the language they find it difficult to ask sharp questions, or to prove that what they expected isn't being delivered.

5 Which probably helps to explain, at least in part, the confusions and failures noted under *IT systems* (qv). And citizens are the losers.

Jobcentre Plus	Visit www.jobcentreplus.gov.uk/
Jobseeker's allowance (JSA)	1 JSA is available for people who are not working (or working less than 16 hours a week); are aged under 65 for men and under 60 for women; are capab[le] of working; and are actively seeking work. 2 In January 2008 the Leader of the Opposition, Mr David Cameron, announced that under a future Conservative government JSA would last for 2 years, after which claimants would move to an Australian-type *workfare* (qv) scheme. 3 Useful websites include the following. They are given in alphabetical order. • Directgov www.direct.gov.uk/ • Frank Field www.frankfield.co.uk/type3.asp?id=20&type=3 • Jobcentre Plus www.jobcentreplus.gov.uk/ See also: *Incapacity benefit*; and *Unemployment*.
Joseph Rowntree Foundation (JRF)	1 The JRF was established in 1904 and is now one of the largest independent social policy research and development charities in the UK. It supports a broa[d] programme of projects in *housing* (qv), *social care* (qv), and social policy. It is widely respected as a source of high-quality information, but the amount of work it is doing for the Government is in some ways becoming a cause for concern – see: *Charities and the state* (qv). 2 For more information visit www.jrf.org.uk/ See also: *Think tanks*.
Joseph Rowntree Housing Trust (JRHT)	1 The Joseph Rowntree Housing Trust is part of the Joseph Rowntree Foundation. It carries out practical innovative projects in *housing* (qv) and *car[e]* (qv), and its work is highly regarded. 2 For more information visit www.jrht.org.uk/ See also: *Environmental controls; Housing*; and *Smart homes*.
Journey sequence	1 This is a recommended mental model to follow when designing buildings fo[r] *accessibility* (qv); when auditing them; and when writing technical guides for designers. It was pioneered by the architects C Wycliffe Noble and Geoffrey Lord, and is more fully explained on pp 2–5 of their book ACCESS FOR DISABLE[D] PEOPLE TO ARTS PREMISES TODAY: THE JOURNEY SEQUENCE – see the bibliography. The bulk of the 162-page book comprises case examples of arts premises. 2 The term means simply that the planner, designer, assessor, or technical wri[ter] has in his or her mind the same sequence that a building user would follow. Such a person would normally: • Approach the site by public transport, by car, or on foot. • Park the car, if that was the mode of travel. • Approach the building. • Locate the principal entrance. • Enter the building. • Move along the internal circulation routes, both horizontal and vertical. • Use the facilities, including lavatories. • Leave the building. This includes *emergency egress* (qv). 3 *Authoritative practical guides* (qv) such as DESIGNING FOR ACCESSIBILITY (qv); APPROVED DOCUMENT M (qv) of the Building Regulations; BS 8300 (qv) and others are all structured in this way; *access auditing* (qv) is done in this

sequence; and designers should have this model in mind when checking that their building designs conform with people's needs. Such a methodical approach helps to ensure that nothing is overlooked, and that everything fits together.

licial system

1 JUSTICE – A NEW APPROACH by Lord Falconer, Lord Chancellor and Secretary of State for Justice, sets out in some detail the objectives of the Ministry of Justice (qv), a new Government department that was formed in May 2007 to take over some of the functions previously carried out by the Home Office (qv). The objective is summarised as being 'to improve the justice system for the public'.
2 This is good news, but only if the judgement of what is wanted by the public rests with the public itself, and not with a closed circle of politicians, civil servants, *experts* (qv), *pressure groups* (qv) and *think tanks* (qv) who claim to speak for them. See in this connection: *Post-democracy*.

See also: *Civil liberties; Crime*; and *Privacy*.

tification under the ability crimination Act)A)

1 The Disability Discrimination Act (qv), which effectively outlaws *discrimination* (qv) against *disabled people* (qv) in public life, recognises certain circumstances in which less favourable treatment for disabled people could be justified.
2 A selection of hypothetical examples is given below.
3 Readers are reminded that the present work only gives simplified summaries of original documents and does not purport to provide full, authoritative and up-to-date statements of the law. For latest information refer to DISABILITY DISCRIMINATION: LAW AND PRACTICE (see the bibliography) and the following websites:

- Directgov: disabled people — www.disability.gov.uk/
- Equality & Human Rights Commission — www.equalityhumanrights.com/

See also: *Disability Discrimination Act*.

tification under Disability crimination Act relation to ployers

1 Here the DDA lays down that justification for less favourable treatment must be based on a reason that is (a) genuinely relevant to the particular case, and (b) substantial, not trivial or minor. For instance:
2 An employee with *mental impairments* (qv) is moved to an isolated position because he mumbles to himself.

- If it can be convincingly demonstrated that the mumbling was having a substantial adverse effect on other peoples' work, then the less favourable treatment could probably be justified on the basis of being both relevant and substantial.
- If it cannot, then a plea of justification would probably fail.

3 This example is suggestive only; for authoritative guidance refer to pp 83–91 of CODE OF PRACTICE: EMPLOYMENT AND OCCUPATION – see the bibliography.

stification under Disability crimination Act relation to rvice providers and dlords

1 Here the DDA recognises a number of grounds on which a failure to provide disabled people with the same level of service as others could be justified:
2 First, if providing the same service to disabled people would endanger either the disabled person or other people. This is because no law can ever require anyone to do anything that would endanger anyone's health or safety. For instance:

- A fairground operator could be justified in refusing to allow a *wheelchair-user* (qv) onto a ride.

• A swimming pool operator could be justified in refusing a child with cerebr palsy from using the pool if he has good reason to believe that the child woul be in danger of drowning.

3 Second, if the disabled person is incapable of giving informed consent or entering into a legally enforceable agreement. For instance:

• A service provider (eg shop) or landlord could be justified in refusing to sign a contract with a person suffering from a *mental impairment* (qv) that prevents them understanding, for instance, the payments to which they are committing themselves. But a refusal would not be justified if the customer was accompanied by a competent person with power of attorney.

4 Third, if providing the service to a disabled person would ruin the service f others. For instance:

• A theatre manager could be justified in refusing entry to someone suffering from a behavioural disorder that causes them to shout or otherwise ruin the occasion for everyone around.

• A landlord could be justified in refusing to let a disabled person use a share kitchen, if the nature of their disability would prevent other tenants from usin the kitchen.

5 Fourth, if providing the same service to disabled people would jeopardise th service for everyone. For instance:

• A cinema owner could be justified in setting aside a limited area for wheelchair-users, if making the whole auditorium wheelchair-accessible wou cost so much as to put the cinema out of business.

• A tour guide could be justified in refusing to allow a severely disabled perso on a tour of city walls, if he has good reason to believe that the extra help the guide would have to give that person would prevent the party from completin the tour.

6 Fifth, if serving a disabled person would cost the provider more in labour or materials than serving other customers. For instance:

• A tailor could be justified in refusing to make a suit for a disabled person at the same price as other customers, if it would cost him more to make a suit tha departs from his usual range. In such a case the service provider would be justified in charging a disabled customer more than other people – something that would normally be unlawful.

7 These examples are suggestive only; for authoritative guidance refer to CODE OF PRACTICE – RIGHTS OF ACCESS: SERVICES TO THE PUBLIC, PUBLIC AUTHORITY FUNCTIONS, PRIVATE CLUBS AND PREMISES – see the bibliography.

Justification under the Disability Discrimination Act in relation to schools

1 Here the DDA recognises two grounds on which less favourable treatment o *disabled* (qv) pupils could be justified.

2 First, if it is the result of a 'permitted form of selection'. Thus: (a) grammar schools may select their intake; and (b) specialist schools may take into account the aptitude of applicants for their particular subject areas. For instance:

• A girl with *learning difficulties* (qv) fails the entrance test of a school that selects its intake on the basis of academic ability. Provided the school operate the selection objectively, the girl's less favourable treatment is likely to be justified because it was the result of a 'permitted form of selection.'

3 Second, if it is for a reason that is (a) genuinely relevant to the circumstance of the particular case and (b) substantial, not trivial or minor. For instance:

• A pupil with a *mental impairment* (qv) that causes him to use abusive language is banned from a school visit, on the basis that including him would make it impossible to maintain discipline. Provided the school has seriously considered ways in which the pupil's behaviour might have been managed, or

reasonable *adjustments* (qv) made to its policies and procedures, this may well constitute a material and substantial reason and be justified.
4 In all cases of this kind the education provider must look into the particular case in detail, and seriously investigate possible solutions. Making general assumptions about an individual on the basis of his or her presumed disability – say *autism* (qv) or *dyslexia* (qv) – and about the difficulties that might ensue, is not acceptable.
5 These examples are suggestive only; for authoritative guidance refer to CODE OF PRACTICE FOR SCHOOLS – DDA 1995: PART 4 – see the bibliography.

tification under Disability crimination Act relation to st-16 colleges

1 Here the DDA recognises a number of grounds on which less favourable treatment of disabled students could be justified.
2 First, if the less favourable treatment of that person is necessary to maintain academic or other prescribed standards. The DDA does not require a responsible body to do anything that would undermine the academic standards of a particular course, as illustrated by the following example:
• A young man with *learning difficulties* (qv) applies to do a biology degree. The university investigates his situation and concludes that even if it were to take all reasonable steps to eliminate the disadvantages caused by his disability, there would be no prospect of his successfully completing the course. To accept him would undermine the academic standards of that course, therefore a rejection of his application would probably be justified.
3 Second, if it is for a reason that is (a) genuinely relevant to the circumstances of the particular case and (b) substantial, not trivial or minor. As an example:
• A student with emotional and behavioural difficulties applies for a college course. Staff know that he is so disruptive as to prevent other students from learning. The college investigates his situation and concludes that even if it were to take all reasonable steps to eliminate the disadvantages caused by his behaviour, which is related to a mental disability (qv), there would be no prospect of solving the problem. Therefore a rejection of his application would probably be justified.
4 In all cases of this kind the education provider must look into the particular case in detail, and seriously investigate possible solutions. Making general assumptions about an individual on the basis of his or her presumed disability – say *autism* (qv) or *dyslexia* (qv) – and about the difficulties that might ensue, is not acceptable.
5 These examples are suggestive only; for authoritative guidance refer to CODE OF PRACTICE (REVISED) FOR PROVIDERS OF POST-16 EDUCATION AND RELATED SERVICES – see the bibliography.

stification under Disability crimination Act relation to transport oviders

1 Here the DDA recognises a number of grounds on which less favourable treatment of *disabled people* (qv) could be justified. They relate to:
• Health or safety.
• The disabled person being incapable of entering into a contract.
• The transport provider being otherwise unable to provide the service to the public.
• The greater cost of providing a tailor-made service.
2 These matters are explained more fully on pp 73–83 of PROVISION AND USE OF TRANSPORT VEHICLES, STATUTORY CODE OF PRACTICE, SUPPLEMENT TO PART 3 CODE OF PRACTICE – see the bibliography

K

rbs

1 <u>Raised</u> or upstand kerbs, defined as kerbs with a vertical upstand of 25 mm or more, are the normal form of edging between pedestrian *footways* (pavements) and vehicular *carriageways* (streets). People who are *blind* (qv) or have severely *impaired vision* (qv) use them as an essential warning device that they are about to step onto a carriageway. Raised kerbs have the disadvantage that they obstruct *wheelchair users* (qv).

2 <u>Dropped</u> (also called flush or level) kerbs are set flush with the carriageway, or have a maximum upstand of 6 mm with a rounded bullnose edge in order not to obstruct wheelchair users. A length of dropped kerbing should be provided wherever wheelchair users may wish to cross a carriageway. The presence of a dropped kerb must always be indicated by *blister-surfaced* (qv) pavings to warn blind people that they are about step into a carriageway.

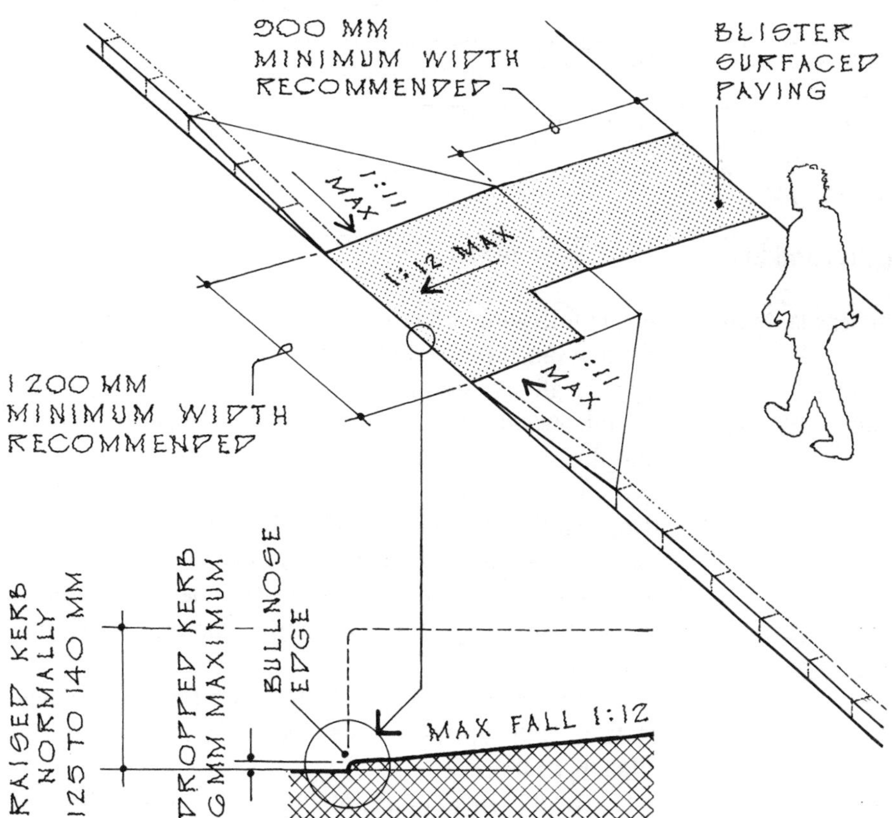

3 All public spaces must comply with the Disability Discrimination Act (qv). Para 3.13 of INCLUSIVE MOBILITY (qv), which is an *authoritative practical guide* (qv) for the application of the Act to the design of *pedestrian infrastructure* (qv), recommends that dropped kerbs be provided at the following locations:

- At all *pedestrian street crossings*, and at other crossings used by pedestrians – eg access points to *car parking* areas.
- At approximately 100 metre intervals on longer side roads and residential roads, so that wheelchair users need not travel too far in order to find a crossing point.

The minimum width of the flush area should be 1.2 m, or up to 3.0 m where there are heavy pedestrian flows; but 1.0 m is acceptable adjacent to *car parking* (qv) reserved for *disabled* (qv) users.

See also: *Blister surfaces; Pedestrian crossings*; and *Tactile pavings*.

Key Performance Indicators (KPIs) — See: *Design Quality Indicators*.

Key scheme — See: *RADAR*.

Key workers — 1 This term usually refers to public sector staff such as nurses, police officers, and teachers who are crucial to the economy and vital for better public service but in many cases are relatively poorly paid. It is often used (particularly in London) in relation to the lack of *affordable housing* (qv) for such people in areas of high house prices. They are also called 'named workers'.
2 To check whether a particular person could qualify as a key worker visit www.housingoptions.co.uk/ho2/#
3 For key worker housing schemes see: *Housing: affordable*.

King's Fund — See: *Think tanks*.

Kitchen tables — See: *Tables*.

Kitchen taps — See: *Taps*.

Kitchen worktops — See: *Worktops: kitchen*.

Kitchens in dwellings — 1 The following references contain authoritative guidance:

- THE HABINTEG HOUSING ASSOCIATION DESIGN GUIDE (qv) applies most specifically to *housing association* (qv) schemes, but ought to be by the designer's side in the design of all housing developments and individual dwellings.
- THE WHEELCHAIR HOUSING DESIGN GUIDE (qv), also from the Habinteg Housing Association, is another illustrated manual whose usefulness extends f beyond the special remit implied by the title.
- HOUSING DESIGN GUIDE, published by Design for London, is an official guide issued by the Mayor of London, that applies specifically to housing in London see: *London: housing*.

For publication details of the above see the bibliography.
2 As of 2008, useful websites include the following. They are given in alphabetical order.

• Disabled Living Foundation.	www.dlf.org.uk/
• Disabled Living Online	www.disabledliving.co.uk/
• Ricability	www.ricability.org.uk/

Kitchens in public buildings: complying with the Disability Discrimination Act — 1 Buildings used by the public must in general comply with the Disability Discrimination Act (qv). In the case of kitchens in public buildings, *authoritative practical guides* (qv) include DESIGNING FOR ACCESSIBILITY (qv) and in addition the following documents. For publication details see the bibliography.

- BS 8300 (qv). Refer to section 12.1, which makes recommendations for kitchen facilities generally. BS 8300 does not have the force of law, but conformity with its recommendations will make it easier to demonstrate that the requirements of the Disability Discrimination Act have been met.

See also: *Refreshment facilities*; and *Worktops: kitchens*.

Knowledge Map — See: *DPTAC Access Directory*.

Kyoto Protocol

1 The 2005 'Kyoto Protocol to the United Nations Framework Convention on Climate Change', to give the document its full name, is an agreement by participating countries to reduce their emission of carbon dioxide and five other greenhouse gases in order to combat global warming.

2 The cost of implementation has been estimated at between US $150 billion and $350 billion per annum, starting in 2010; and the predicted result is that average global temperatures would be less than one half of one degree Celsius lower, after a century of full implementation, than they would otherwise have been.

3 This is a temperature difference too small to be detected by the human body if it happened from one hour to the next, and too small to be registered by most thermometers. Taxpayers and consumers are entitled to ask whether such a trivial result after a century of spending several billion pounds per annum is a sensible use of money, or whether these colossal sums could be better spent on preparing for rising sea levels and other precautions.

4 As of 2008, useful references include the following. They are given in alphabetical order. For publication details see the bibliography.

- THE ECONOMICS AND POLITICS OF CLIMATE CHANGE: AN APPEAL TO REASON, published by the Centre for Policy Studies is a simple, clear lecture that may be read or heard at www.cps.org.uk/historiccatalogue/default.asp
- UNSTOPPABLE GLOBAL WARMING EVERY 1,500 YEARS by S Fred Singer and Dennis T Avery sets out the evidence for the hypothesis that current global temperature is part of a regular and unstoppable natural cycle, and that the expenditure of public money would be far better directed to other ends.

5 Useful websites include http://scienceandpublicpolicy.org/monckton/goreerrors.html

- The daily and weekly lay media are depressingly short of accurate commentary on these matters. Christopher Booker's Notebook, published each week in the *Sunday Telegraph*, is one of the few that can be recommended.

L

aboratory design: asic planning ata

1 Distilled planning data for most of the major building types is given in the METRIC HANDBOOK: PLANNING AND DESIGN DATA.
2 For publication details see the bibliography.

aboratory design: omplying with the isability iscrimination Act

See: *Educational, cultural and scientific buildings.*

abour force

See: *Employment.*

amps: touch

See: *Touch lamps.*

andings

See: *Ramps*; and *Stairs.*

andlords: complying ith the Disability iscrimination Act

See: *Disability Discrimination Act.*

anguage and sability

See: *Political correctness: speech codes.*

rge personal hicles for sabled people

See: *Mobility aids.*

vatories

1 BS 8300 (qv) employs the somewhat old-fashioned term 'lavatories' rather than the more widely-used word 'toilets'. The present work mostly follows BS terminology.
2 Unisex WCs on the other hand are sometimes referred to as 'unisex toilets'.

vatories in dwellings

1 The following references contain authoritative design guidance. For publication details see the bibliography.

- Section 10 of APPROVED DOCUMENT M (qv). The provisions in AD M must be heeded in order to comply with the Building Regulations for England and Wales. In Northern Ireland refer to TECHNICAL BOOKLET R of the Building Regulations (Northern Ireland); and in Scotland to the TECHNICAL HANDBOOKS of the Building (Scotland) Regulations (qv).
- THE HABINTEG HOUSING ASSOCIATION DESIGN GUIDE (qv) applies most specifically to *housing association* (qv) schemes, but ought to be by the designer's side in the design of all housing developments and individual dwellings.
- THE WHEELCHAIR HOUSING DESIGN GUIDE (qv), also from the Habinteg Housing Association (qv), is another housing design reference whose usefulness extends well beyond the special remit implied by the title.
- HOUSING DESIGN GUIDE, published by Design for London, is an official guide that applies specifically to housing in London – see: *London: housing.*

2 For *general needs housing* (qv) a downstairs lavatory for the use of *disabled* (qv) visitors, with internal clear dimensions of 2 m × 1 m (or as an absolute minimum 1.9 m × 0.9 m) as shown on the diagram on the next page, will in general satisfy the provisions in the above references, though each particular case should be checked in detail before decisions are finalised.
3 For *wheelchair housing* (qv) see section 11 of THE WHEELCHAIR HOUSING DESIGN GUIDE.

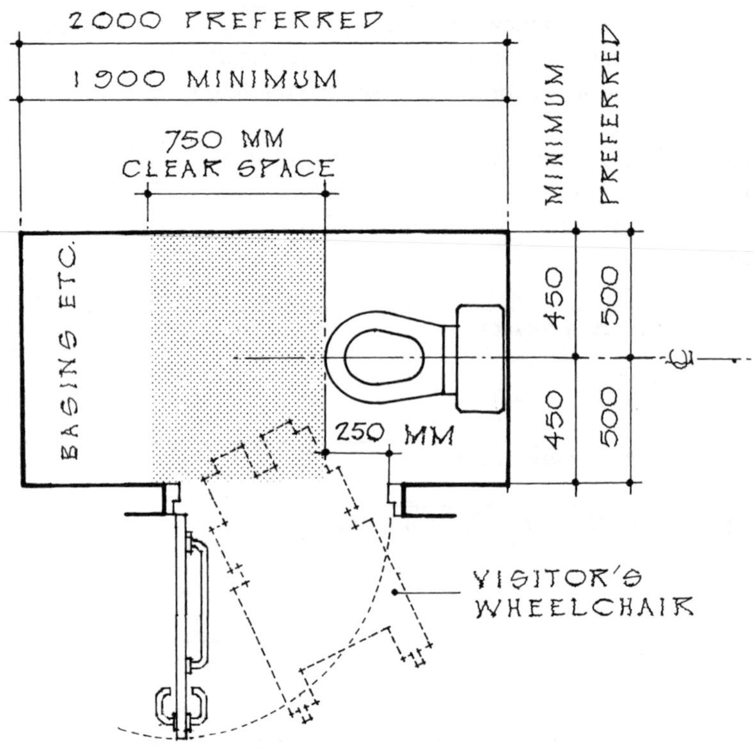

Lavatories in public buildings: complying with the Disability Discrimination Act

1 Buildings used by the public must in general comply with the Disability Discrimination Act (qv). In the case of lavatories in public buildings, *authoritative practical guides* (qv) include DESIGNING FOR ACCESSIBILITY (qv) and in addition the following documents. For publication details see the bibliography.

- APPROVED DOCUMENT M (qv), the provisions of which must be heeded in order to comply with the Building Regulations (qv) for England and Wales. Fo more detail see the table below.
- BS 8300 (qv), which applies to all buildings in the UK. For more detail see the table below.
- THE GOOD LOO DESIGN GUIDE (qv), a core reference that should be on every architect's shelf.
- INCLUSIVE MOBILITY (qv), which applies to *pedestrian infrastructure* and to transport-related facilities such as bus and coach stations, railway stations, air terminals, and transport interchanges; and is an advisory document. Refer in particular to para 9.6.
- ACCESSIBLE STADIA (qv), which is as of 2008 the most authoritative general reference for the *inclusive* design of sports and recreation facilities. Refer in particular to paras 2.26–2.29, and pp 74–75.
- GUIDE TO SAFETY AT SPORTS GROUNDS (qv), which is an essential reference in the design of sports facilities. Refer in particular to paras 2.22, 8.5, 13.19, and 20.7.

Except for AD M the above references do not have the force of law, but conformity with their recommendations will make it easier to demonstrate tha the requirements of the Disability Discrimination Act have been met.

For places other than England and Wales the following regulatory documents should be consulted:

- In Northern Ireland: TECHNICAL BOOKLET R of the Building Regulations (Northern Ireland) (qv).
- In Scotland: the TECHNICAL HANDBOOKS of the Building (Scotland) Regulations (qv).

2 The relevant clauses in AD M and BS 8300 are set out below.

ASPECT	DOCUMENT: RELEVANT PARAGRAPHS	
	AD M	BS 8300
GENERAL NOTES, SCALE OF PROVISION, ETC.	5.1–5.14	12.4–4.2
UNISEX ACCESSIBLE CORNER WC	5.8–5.10	12.4.3.1
UNISEX ACCESSIBLE PENINSULAR WC (ASSISTED USE)	–	12.4.3.2
WC COMPARTMENT FOR AMBULANT DISABLED PERSON	5.6–5.14	12.4.3.3
BABY-CHANGING FACILITIES	5.12	–
WC PAN AND CISTERN	5.10	12.4.3.4
GRAB RAILS	5.4–5.14	12.4.
URINALS	5.13	12.4.4
WASH BASINS	5.3–5.4	12.4.5
ACCESSORIES (PAPER & SOAP DISPENSERS ETC.)	–	12.4.6
DOORS & DOOR FURNITURE	5.3–5.4	12.4.7
LOBBIES	3.15–3.16	12.4.8
FINISHES, COLOURS & LIGHTING	5.4	12.4.9–12
ALARMS	5.4	12.4.10
RADIATORS AND OTHER HEAT EMITTERS	5.4	12.4 11

3 For ambulant people (i.e. non-wheelchair-users) a cubicle with internal clear dimensions of 1.5 m × 0.8 m as shown below will in general satisfy the provisions in the above references, though each particular case should be checked in detail before decisions are finalised.

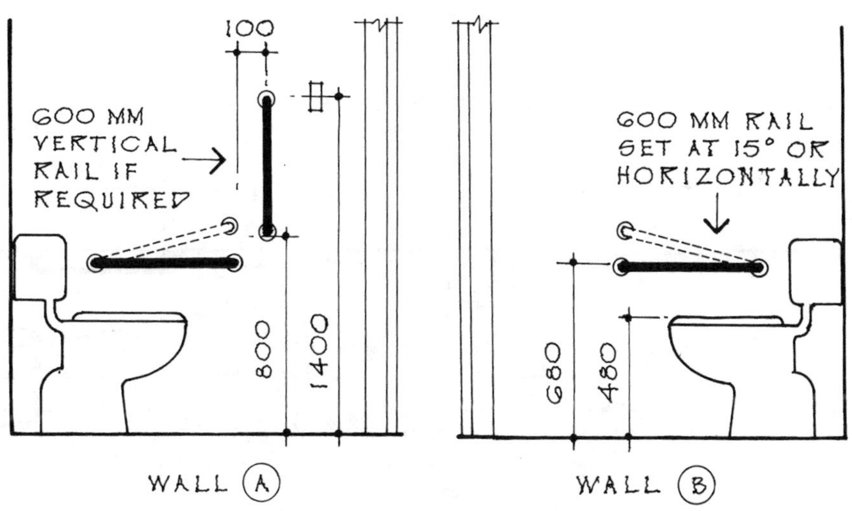

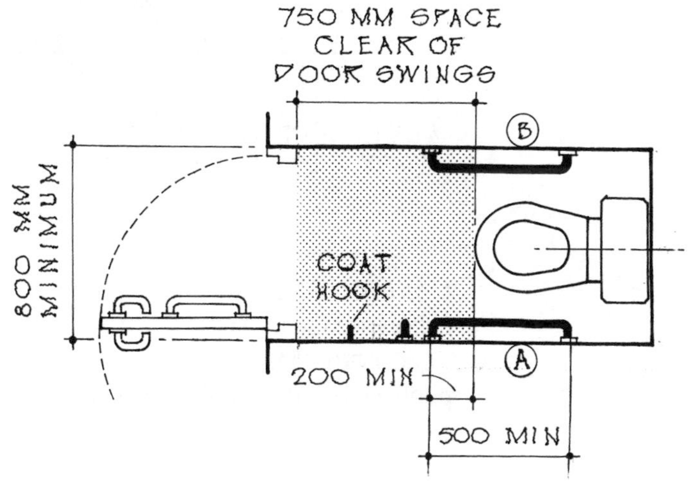

4 For *wheelchair-users* (qv), a cubicle with internal clear dimensions of 2.2 m 1.5 m as shown below will in general satisfy the provisions in the aforementioned references, though each particular case should be checked in detail before decisions are finalised. Alternative layouts requiring cubicle dimensions of, for instance, 2.2 m × 2.4 m will in some situations be preferal and user needs should be carefully considered before coming to a decision.

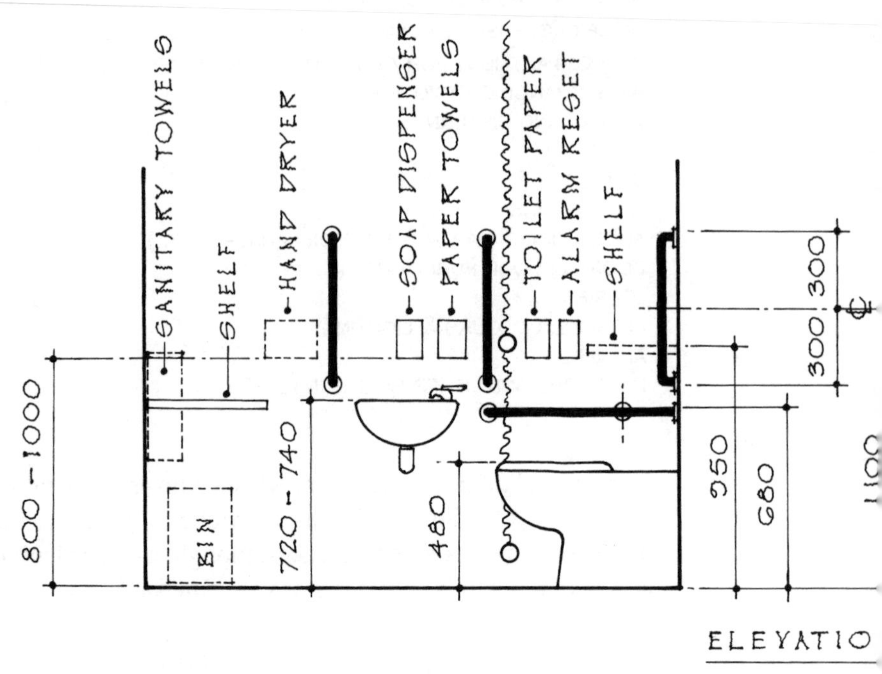

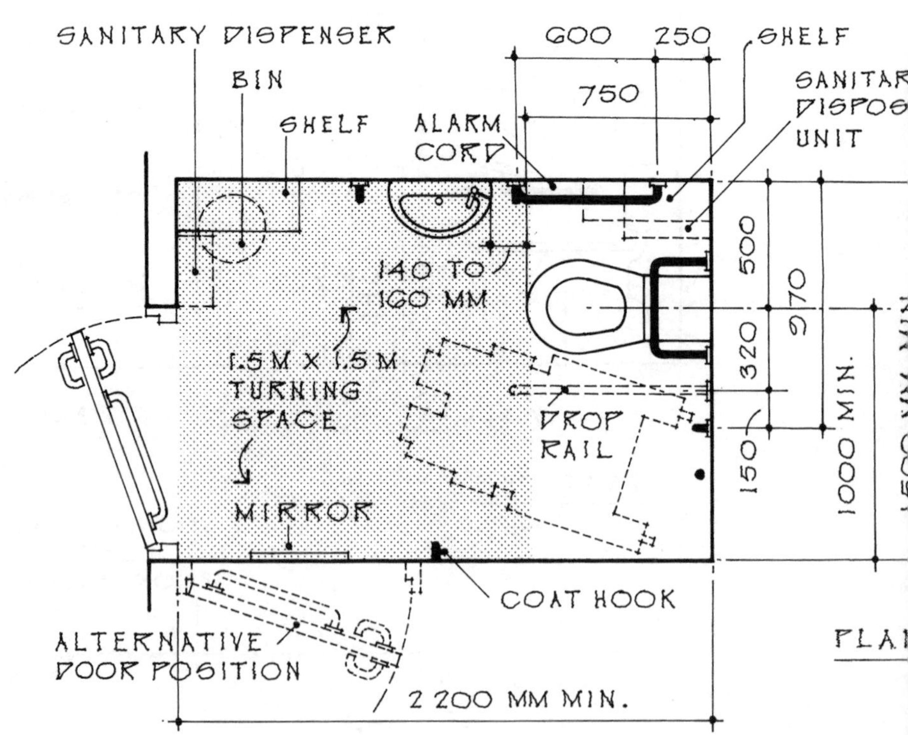

5 As of 2007 useful additional references include the following. For publicat details see the bibliography.

• CHOOSING TOILET EQUIPMENT AND ACCESSORIES, an 18-page factsheet published by the Disabled Living Foundation (qv).

atories: public

1 The admirable Victorians created a generous network of clean, high-quality public lavatories that were the envy of the world. Today Britain's public lavatories are a disgrace. In 10 years they have fallen in number from 15 000 to 5 000, and in July 2006 Mr Phil Woolas, the then Minister for Local Government, told a conference titled 'Public Toilet Provision – The Way Forward' that most were between 50 and 100 years old. He said that the decline had been hastened by rising maintenance costs, poor cleaning, vandalism, and the use of public toilets by drug abusers and by homosexual men for having sex. Mr Richard Chisnell, director of the British Toilet Association (qv), told the same conference that 14 million tourists visit London's South Bank each year but there are no public toilets between Lambeth Bridge and the Tate Modern gallery. Many people make use of the lavatories at the National Theatre, Tate Modern, or McDonald's, and the pressure on these facilities were now so great that McDonald's had told the authorities they could no longer cope with the burden near the London Eye.

2 If the consequences of the above are bad for the public at large, they are much worse for disabled (qv) and elderly people, many of whom cannot leave their homes for fear of a disastrous experience. While Britain is in general *over-regulated* (qv), there is an unanswerable case, advanced for instance by the admirable British Toilet Association, for a new law obliging *local authorities* (qv) to provide and maintain a generous network of free, clean public lavatories with opening hours that suit people's needs.

3 In March 2008 The Department for Communities and Local Government (qv) launched an 'All cisterns go' scheme to foster 'pioneering new approaches' to the problem, but critics strongly attacked the scheme's provision for allowing local authorities to impose a charge for use.

4 As of 2008, useful publications include IMPROVING PUBLIC ACCESS TO BETTER QUALITY TOILETS: A STRATEGIC GUIDE – see the bibliography.

5 Useful websites include:

- British Toilet Association www.britloos.co.uk/
- Changing Places www.changing-places.org/
- Incontact www.incontact.org/
- RADAR www.radar.org.uk/radarwebsite/

See also: *National Key Scheme.*

ws

See: *Statutes.*

yers' charter

See: *Consultants' charter.*

F

See: *London Disability Arts Forum.*

an-on seats

See: *Chairs: sit-stand.*

arning aids

1 These are one category of *aids and equipment* (qv), and are meant to help children or adults with *impaired understanding* (qv).

2 As of 2008, useful websites include the following. They are given in alphabetical order.

- Don Johnston www.donjohnston.co.uk/
- Possum www.possum.co.uk/
- Rompa www.rompa.com/
- Winslow www.winslow-cat.com/

See also: *Reading aids;* and *Writing aids.*

Learning and Skills Council (LSC)

1 As of July 2007, this was the *quango* (qv) that is responsible for adult traini in England. But its future is unclear. Soon after Mr Gordon Brown became Prime Minister in May 2007 the Department for Education and Skills (qv) w closed down and replaced by two new bodies, the Department for Children, Schools and Families (qv) and the Department for Innovation, Universities a Skills (qv), and as part of the same reorganisation heavy hints were dropped that much of the LSC's work might be transferred to *local authorities* (qv). N doubt matters will have clarified by the time this is published.
2 Meanwhile, the LSC is one of a proliferating tangle of organisations fallin over each other's feet in trying to coordinate and oversee the Government's programme of vocational courses. Others include the Qualifications and Curriculum Authority (qv); the Quality Improvement Agency (qv); the Traini and Development Agency for Schools (qv); 25 Sector Skills Councils (qv); a several more. If England's youth unemployment problems could be solved by creating ever more bureaucracies, outcomes in the UK would be exemplary. Alas, they are not – see for instance *New deal: employment.*
3 As of 2008, useful references include the following. For publication details see the bibliography.
• THE COMPETITIVE ECONOMIC PERFORMANCE OF ENGLISH CITIES, a 263-page report by Professor James Simmie of Oxford Brookes University and others, criticises bodies such as the LSC for being expensive and ineffective.

Learning difficulty
Learning disability

1 Both terms are used to describe a difficulty in concentrating, learning, and understanding that is severe enough for the person involved to need addition help and support with his or her everyday life. More specifically:
• The term *learning difficulty* is appropriate for moderate difficulties that can be largely overcome with good care and good schooling. Given such help, people in this category (of whom there may be roughly 1 million in the UK) can often blend into society very successfully.
• The term *learning disability* is appropriate for a lifelong condition, acquire before, during or soon after birth, the most common example being Down's syndrome. These are people (of whom there may be about 200 000 in the UK with severe and complex difficulties, who are likely to be in need of social services provision. Some need support with everyday skills like getting dress or cooking, or social skills like holding a conversation, while those at the mo severe end of the spectrum may require 24-hour care.
2 It is misleading to apply the single term *learning difficulties* to both of thes very different groups, as some *pressure groups* (qv) are trying to make us do Such pressures should be resisted.
3 As of 2008, useful websites include:
• Mencap www.mencap.org.uk/

See also: *Impaired understanding; Mental Capacity Act*; and *Mental impairment.*

Le Corbusier

1 A dedicated, clever and talented pioneer of modern architecture whose influence on the built environment has been for the most part catastrophic. A examples:
• His vision of urban areas consisting of tall buildings spaced widely apart in parklike settings, seductively presented in a variety of unbuilt projects during the 1920s and 1930s, led directly to countless charmless and dysfunctional N Towns and urban renewal schemes the world over. Traditional housing on the ground, with lively streets and private gardens, were replaced by graceless sl and tower-blocks in lifeless landscapes; unloved by their inhabitants and

looking tattier with every passing year. Communities are still paying the price for this 30-year-long experiment.

• His influential 1952 Unité d'Habitation at Marseilles is a stunningly powerful work of sculpture, but for most people this 140 m (450 ft) long concrete megastructure, which contains around 340 apartments for 1 600 people plus associated shopping and leisure facilities on 15 floors, cannot offer an acceptable setting for family life. The apartments lack privacy; the children's bedrooms are effectively sliding-door closets about 2 m (6 ft) deep; there is nowhere practical for children to play, and the only way parents and offspring can get temporary relief from maddening togetherness is to flee to the outside. As a prototype for future habitats this impressive edifice would undoubtedly suit a minority of enlightened and probably childless people, but for most – and especially for families – it is utterly impractical.

• The deliberately crude concrete forms and surfaces he pioneered in the Unité and at Chandigarh in India, which have been emulated the world over by starry-eyed disciples, are ferociously detested by most of the unfortunate citizens whose townscapes have been blighted by these grim, rain-stained monsters. The style appears to be passing, but huge damage has been done to town and cityscapes.

• The concrete sun-break grilles on his Chandigarh buildings, a typically clever idea applied with no attempt at testing in real-life situations, achieved (according to his colleague the late Jane Drew) the very opposite effect to the one that was intended. Instead of protecting the interior from solar heat gain these thermal reservoirs absorb immense amounts of heat from the blazing Punjab sun and then radiate the heat into the already hot interiors. The present author, who grew up in sub-tropical Pretoria, can personally testify to the hellish effects of massive concrete sun-break grilles upon building interiors in hot climates.

2 A clue to le Corbusier's attitude to the human users of his visionary cities and buildings may be found in the Modulor, a proposed proportioning system for building design which he claimed to be 'based on the human body' and on 'real, bodily facts'. His sketch of the human body on which the system was based is shown below in black. Unfortunately, as the accompanying figures show, most building users are smaller, frailer, and altogether less heroic than the 1 829 mm tall figure of Corb's imagination.

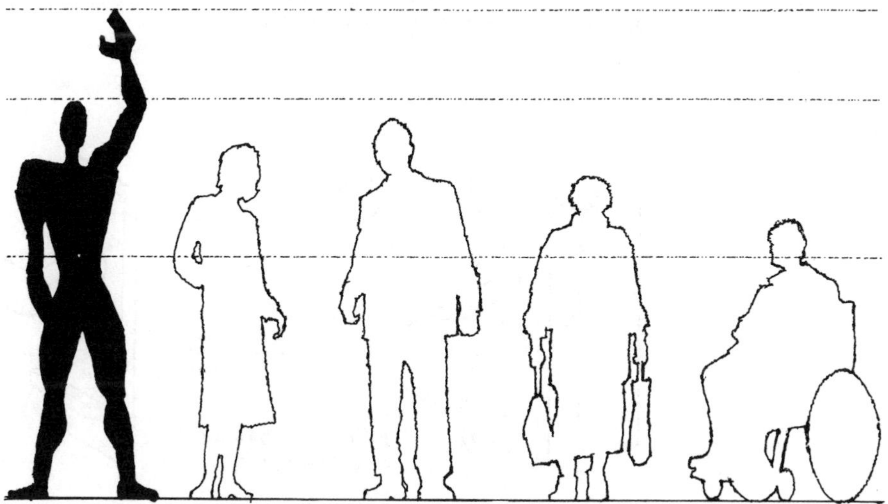

In addition to wanting a new kind of city and a new kind of architecture, le Corbusier clearly wished for a new kind of human being that would be worthy of the environment designed for their use by the master and his disciples. For an example of the latter see Robin Hood Gardens, designed in 1968 by the late

Alison and Peter Smithson, on the corner of Cotton Street and Robin Hood Lane in Poplar, London E14 (while they preferred to live in an elegant period house near The Boltons in Kensington).

3 In sum: le Corbusier was a clever but arrogant designer with an unfortunate addiction to *ideology* (qv) and untested grand theory. The lesson to be drawn from his damaging mistakes is not that we should refrain from trying new idea but that it is wise to move forward in small steps, with plenty of testing along the way; and not to despise the opinions of ordinary people.

4 Useful references include the following. For publication details see the bibliography.

- FORM FOLLOWS FIASCO: WHY MODERN ARCHITECTURE HASN'T WORKED by Peter Blake. See in particular p 33.
- THE IDEAS OF LE CORBUSIER ON ARCHITECTURE AND URBAN PLANNING by Jacques Guiton. See in particular pp 66–69.

See also: *Architecture; Déformation professionelle*; and *New Urbanism.*

Lecture theatres See: *Auditorium design.*

Legal proceedings See: *Civil proceedings.*

Legislation See: *Over-legislation; Statute*; and *Statutory instrument.*

Legislation: access See: *Authoritative practical guides.*

Legislative and Regulatory Reform Act See: *Civil liberties.*

Leisure facilities See: *Sports facilities.*

Level thresholds See: *Thresholds: level.*

Lever handles

1 Para 6.5 of BS 8300 (qv) recommends that doors should be fitted with lever handles, rather than spherical or near-spherical designs. The latter can be difficult to grip, especially for people with weak or *arthritic* (qv) hands.

2 The geometry and dimensions shown below will satisfy BS 8300, and thus make it easier to demonstrate that the requirements of the Disability Discrimination Act (qv) have been met.

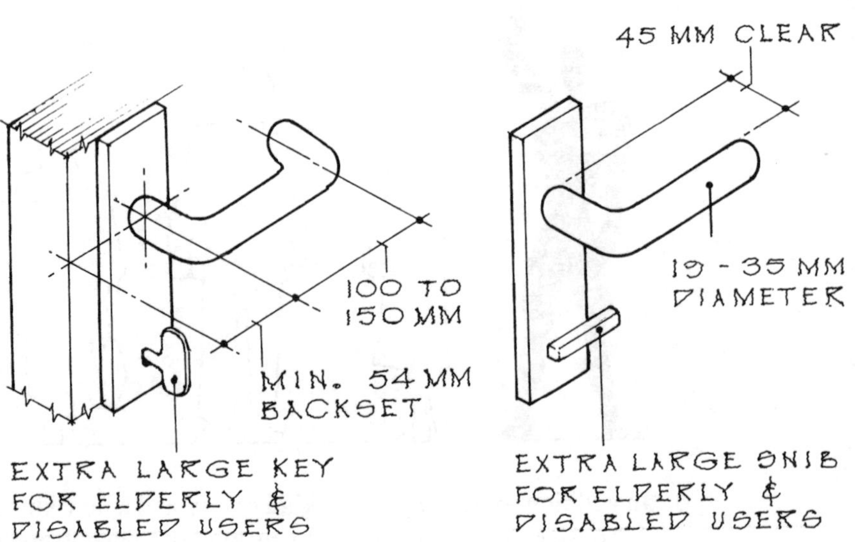

See also: *Authoritative practical guides*; *Doors*; and *Impaired dexterity.*

er taps

1 Para 12.1.6 of BS 8300 (qv) recommends that kitchen sinks be fitted with lever taps, which can be pushed by palm or wrist with no need for gripping. This type of operation is easier for all users, especially those with weak or arthritic hands, and the installation of such fittings will make it easier to demonstrate that the requirements of the Disability Discrimination Act (qv) have been met.

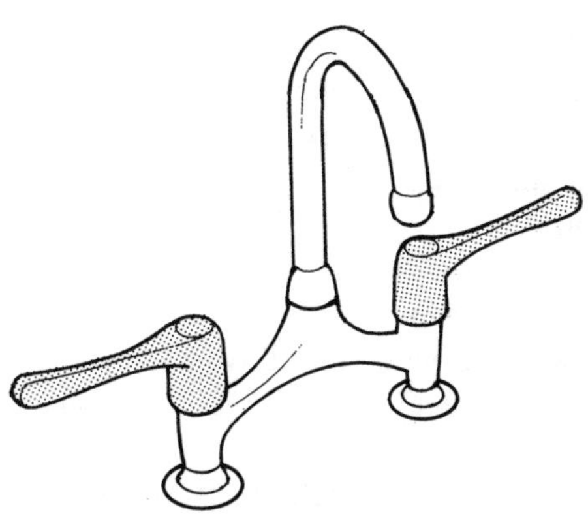

2 Lever taps are also recommended for bathrooms (para 12.2.7) and lavatories (para 12.4.5), though in these cases they will probably be mixer taps with an up-and-down action and not as illustrated above.

See also: *Authoritative practical guides; Impaired dexterity*; and *Taps*.

erty

See: *Civil liberties; Human rights;* and *Speech: freedom of.*

raries: basic nning data

1 Distilled planning data on all the major building types is given in the METRIC HANDBOOK: PLANNING AND DESIGN DATA. For publication details see the bibliography.
2 For additional references, case studies, and commentary on design quality visitwww.cabe.org.uk/default.aspx?contentitemid=40.

raries: inclusive

1 As of 2008, useful references to *inclusive* (qv) library design and management include the following. They are given in alphabetical order. For publication details see the bibliography.

- 21ST CENTURY LIBRARIES: CHANGING FORMS, CHANGING FUTURES shows how public libraries have reinvented themselves in the last decade, reviving their role as beacons for civic pride and social and economic regeneration. It outlines a number of future visions for Britain's public libraries.
- THE DISABILITY PORTFOLIO is a collection of 12 guides on how best to meet the needs of *disabled people* (qv) as users and staff in museums, archives, and libraries. It gives authoritative advice, information and guidance to help overcome barriers and follow good practice.
- LIBRARY SERVICES FOR VISUALLY IMPAIRED PEOPLE: A MANUAL OF BEST PRACTICE is a short website-based guide that gives brief advice plus extensive references to other useful publications and websites.

2 Useful websites include the following. They are given in alphabetical order.

- Chartered Institute of Library and Information Professionals www.cilip.org.uk/
- Department for Culture, Media and Sport www.culture.gov.uk/
- Directgov www.direct.gov.uk/
- Museums, Libraries & Archives Council www.mla.gov.uk/
- People's Network www.peoplesnetwork.gov.uk/

Libraries: selected case examples

1 The Adapt Trust (qv) makes regular awards to arts venues that are considered to be especially meritorious in terms of overall *accessibility*. Award-winning libraries in the four or five years up to 2003 included the following:

- Brixworth Library, Northamptonshire.
- Gateshead Central Library, Prince Consort Road, Gateshead.
- Kidderminster Library.
- Papworth Everard Library, Cambridge.

Highly commended examples include:

- Clocktower Library in Katharine Street, central Croydon, London.
- Rushden Library, Northants.
- Taunton New Library.

Examples of merit include:

- Hastings library, East Sussex.
- Broad Green Library in London Road, West Croydon, London.
- Forest Gate library, Newham, London E7. The library is situated in a converted telephone exchange.

2 In 2006 the Jubilee Library in Brighton, by Bennett's Associates, won that year's Prime Minister's award (qv).

3 For other examples visit www.mla.gov.uk/

See also: *Access awards*.

Library design: complying with the Disability Discrimination Act

See: *Educational, cultural and scientific buildings*.

Licensing Act 2003

1 According to the Department for Culture, Media and Sport (qv) this Act provides 'a balance by promoting freedom of choice while ensuring the necessary safeguards are in place to tackle problem premises' – visit www.culture.gov.uk/

2 Numerous voluntary and community sector organisations (qv) think otherwise. As instances:

- The Act, which came fully into force in 2005, has obliged volunteers running local community halls where alcohol is sold to go on a training cour to obtain an 'accredited personal licence qualification', and to fill in a 20-pag application form which requires 60 pages of explanatory notes. One voluntee told the press that he had been obliged to use a week of his unpaid time, including attending two seminars, to comply with procedures aimed at the prevention of crime and disorder, the prevention of public nuisance, and the protection of children from harm. None of these things had ever been a problem in his small village.
- The Act obliges 'two in a bar' entertainers to obtain a licence before they ca strike up, which was unnecessary before. Warned about the effects this woul have on informal entertainment the licensing Minister, Lord McIntosh denie any such risk and said: 'My view is that there will be an explosion of live music'. Questioned about the expected 'explosion' in 2007, the Government would only say that the impact had been 'broadly neutral'. But an article in t *Daily Telegraph* on 18 June 2007 by Philip Johnston, titled '*Our masters are deaf to our wishes*' (see the bibliography), quotes research from the Departm for Culture, Media and Sport (qv) that shows a big impact on the smallest venues – the very ones that used to operate under the 'two in a bar' exemptio These venues, where a pianist might once have played every night without

the owner having to fill in a form, must now obtain Temporary Event Notices (TENs) which cost £21 each, and in many cases time-consuming *risk assessments* (qv).
3 In the House of Commons Mr Peter Luff MP, who had long warned Ministers of the likely consequences of the Act, to no effect, said on 8 June 2005 (visit www.peterluff.co.uk/hansard.asp): 'The Licensing Act means that villages in every corner of the country will say goodbye to the traditional touring circus, see more village shops go to the wall, watch local sports clubs forgoing much-needed income and lose their village halls, despite no accidents, no anti-social behaviour, and nothing worrying having occurred in, outside, or even remotely close to them'. Malcolm Moss (MP) and Mr Mark Field (MP), who had had helped to steer the Bill into law, told the House of their regret.
4 The workings of the Act will undermine and in many cases kill local community activities, which the Government assures us are the lifeblood of a healthy society. Mr Gordon Brown is committed to cutting the burden of red tape on *charities* (qv) by a quarter by the year 2010. A radical redrafting of the Licensing Act 2003 would be a good start.
5 As of 2008, useful websites include the following. They are given in alphabetical order.

• Department for Culture, Media & Sport	www.culture.gov.uk/
• Directgov	www.direct.gov.uk/
• Office of Public Sector Information	www.opsi.gov.uk/
• Peter Luff MP	www.peterluff.co.uk/hansard.asp

See also: *Better regulation bodies; Over-legislation;* and *Voluntary and community sector*.

---

long learning

Visit www.lifelonglearning.co.uk/

---

time Homes

1 Lifetime Homes are *general needs* (qv) dwellings, but designed so that the dwelling can be easily adapted to *special needs* (qv) as the original occupants become older, increasingly frail, and possibly *disabled* (qv). The aim is to enable people to 'stay put' in their original home for as long as possible before having to 'move on' to special accommodation or an institution.
2 Building all new homes to LH standard is highly recommended now, and will be mandatory after 2013. Useful references include LIFETIME HOMES, LIFETIME NEIGHBOURHOODS: A NATIONAL STRATEGY FOR HOUSING IN AN AGEING SOCIETY, which sets out government policy; and THE HABINTEG HOUSING ASSOCIATION DESIGN GUIDE (qv), which gives design guidance – see the bibliography.

See also: *Housing: accessibility*.

---

ing Burdens k Force (LBTF)

1 The LBTF was launched in autumn 2006 to help cut bureaucracy in the Department for Communities and Local Government (qv) – visit www.communities.gov.uk/index.asp?id=1508116
2 See also: *Better regulation bodies*; and *Over-regulation*.

---

s

1 Lifts are the safest and most comfortable means of vertical travel for able-bodied and *disabled (qv)* people alike, and are therefore very desirable in multi-storey buildings. Note however that (a) a lift of any kind can only augment a stair, never replace it; and (b) while lifts are ideal as a means of *access* (qv) to the upper levels of a building, they are not permissible as a *means of escape* (qv) in the event of fire except if they are *evacuation lifts* (qv).
2 The diagram on the next page suggests a basic taxonomy of lift types.

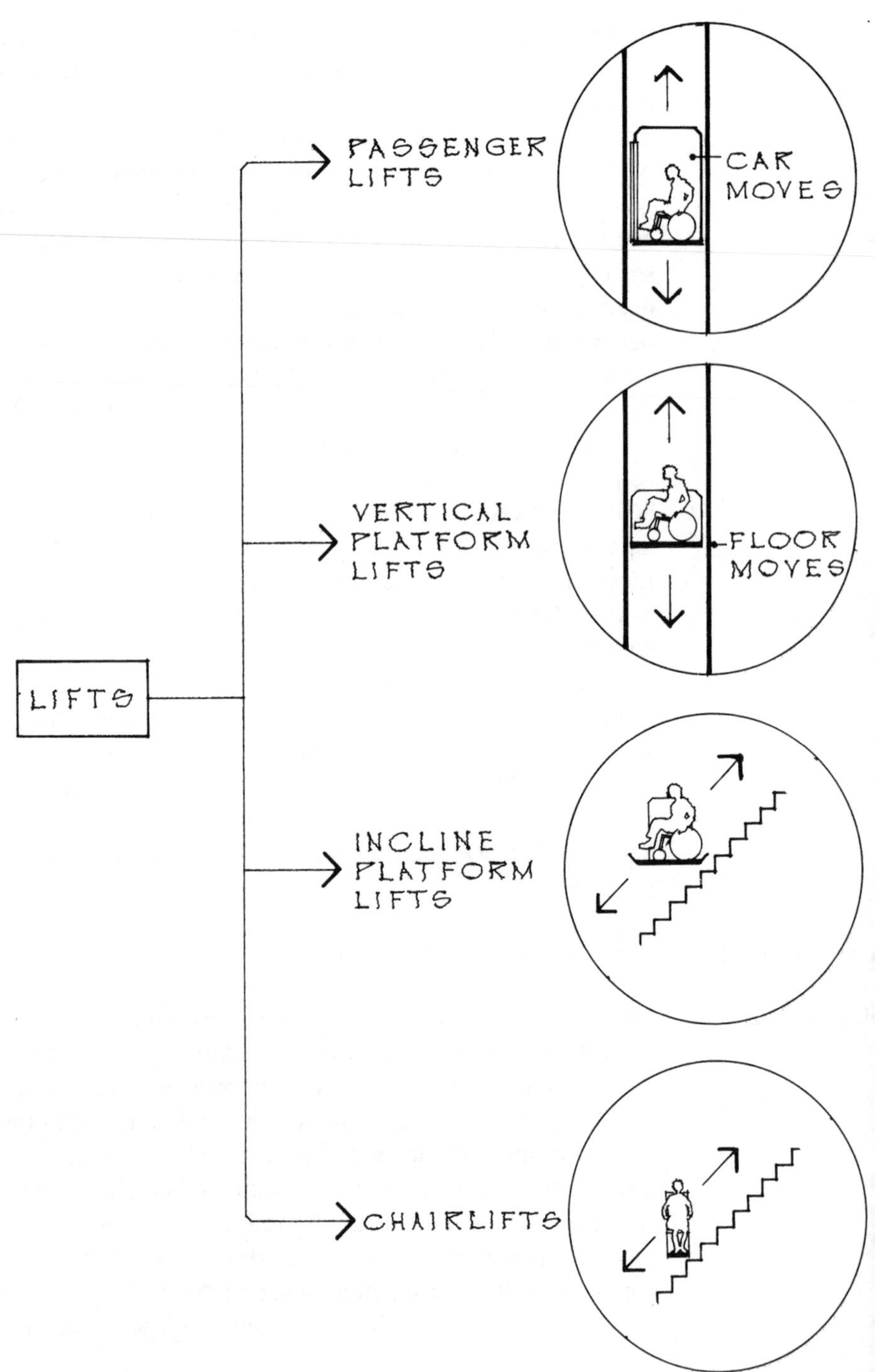

Lifts in public buildings: complying with the Disability Discrimination Act

1 Public buildings must in general comply with the Disability Discrimination Act (qv). In the case of lifts, *authoritative practical guides* (qv) include DESIGNING FOR ACCESSIBILITY (qv) and in addition the following documents. For publication details see the bibliography.

• APPROVED DOCUMENT B (qv), which deals with *fire safety* (qv), and the provisions of which must be heeded in order to comply with the Building Regulations (qv) for England and Wales. As fire safety is a highly specialised subject, it is not dealt with in the present work.

• APPROVED DOCUMENT M (qv), which deals with the use and accessibility of buildings and the provisions of which must be must be heeded in order to comply with the Building Regulations (qv) for England and Wales.

• BS 8300 (qv), which applies to the design of all buildings throughout the U

• INCLUSIVE MOBILITY (qv), which is the most authoritative reference for the *inclusive* (qv) design of *pedestrian infrastructure* (qv) and of *transport-related buildings* (qv) such as bus and coach stations, railway stations, air terminals, and transport interchanges.
• SCHEME DEVELOPMENT STANDARDS, a Housing Corporation (qv) guide that covers all housing in receipt of *Social Housing Grant* (qv).
Except for AD B and AD M the above references do not have the force of law, but conformity with their recommendations will make it easier to demonstrate that the requirements of the Disability Discrimination Act have been met.
2 For places other than England and Wales the following regulatory documents should be consulted:
• In Northern Ireland: TECHNICAL BOOKLET R of the Building Regulations (Northern Ireland) (qv).
• In Scotland: the TECHNICAL HANDBOOKS of the Building (Scotland) Regulations (qv).
• In the USA, where the Americans with Disabilities Act applies: the ADA AND ABA ACCESSIBILITY GUIDELINES (qv).
3 The relevant clauses in AD M, BS 8300, and INCLUSIVE MOBILITY are set out below.

DESIGN ASPECT	DOCUMENT: RELEVANT PARAGRAPHS		
	APPROVED DOCUMENT M	BS 8300	INCLUSIVE MOBILITY
GENERAL NOTES ON LIFTS	3.17 – 3.20	8.4	8.4.5
PROVISION OF LIFTS IN GENERAL	3.21 – 3.24 9.1 – 9.7	8.4.1	8.4.5
DESIGN OF LIFTS IN GENERAL	3.25 – 3.28	8.4.2	8.4.5
PASSENGER LIFTS	3.29 – 3.34	8.4.3	8.4.5
PLATFORM LIFTS (VERTICAL)	3.35 – 3.43	8.4.4	–
PLATFORM LIFTS (INCLINE)	3.44 – 3.49	8.4.5	–
CHAIRLIFTS	–	8.4.5	–

4 Though checks should be carried out before any decisions are finalised, the following notes summarise the order of preference stated in both APPROVED DOCUMENT M and BS 8300:
• The preferred option in all public buildings is a *passenger lift* (qv) that will accommodate a *wheelchair user* (qv). Section 8.4 of BS 8300 gives detailed recommendations on such lifts.
• If a passenger lift is not feasible, and travel is only between two levels, a *vertical platform lift* (qv).
• If that too is not feasible, and travel is only between two levels, an *incline platform lift* (qv) – provided its installation does not conflict with *means of escape* (qv) on staircases.
• If even that is not feasible, a *chairlift* (qv) is the last and least desirable option.
5 For notes on particular lift types see: *Chairlifts; Incline platform lifts*; *Passenger lifts*; and *Vertical platform lifts*.

Lifts: evacuation

1 An evacuation lift is one that may be used for the evacuation of *disabled people* (qv) in case of *fire* (qv). Para 6.39 of APPROVED DOCUMENT B (qv) advises against occupants making use of lifts when there is a fire in a building, but permits provision of an evacuation lift as part of a management plan for the evacuation of disabled persons, provided the lift is appropriately sited, protected, and fitted with safety features.
2 On all these topics, and on the closely associated matter of safe *refuges* (qv) adjacent to lifts, where disabled persons can await assistance, AD B refers designers to BS 5588 part 8 (qv).

See also: *Means of escape*.

Light switches

See: *Controls and switches*; and *Plate and touch switches*.

Lighting design

1 The following definitions and explanations may help non-specialists to understand the guidance on *lighting levels* given a little further down.

LUMINOUS FLUX
(TOTAL OUTPUT)
LUMENS

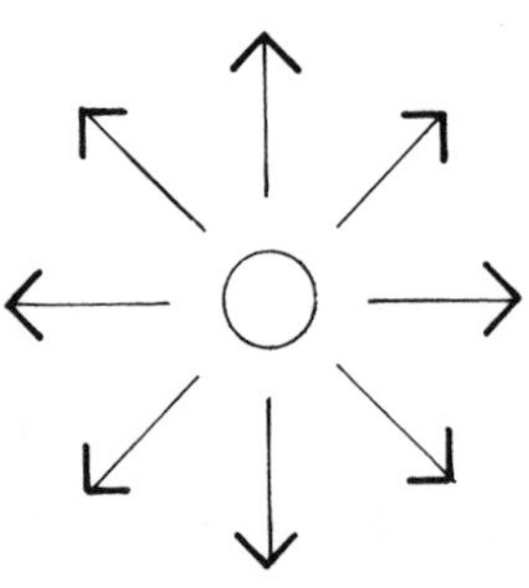

ILLUMINANCE
(INTENSITY OF LIGHT FALLING ON A SURFACE)
LUX

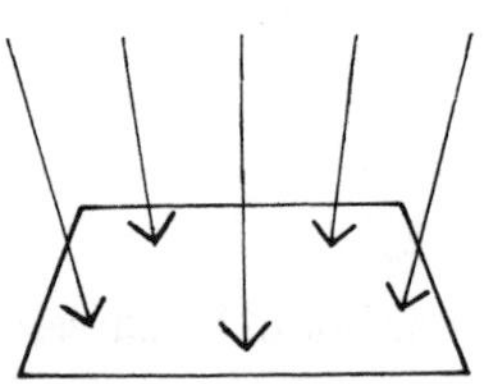

LUMINANCE
(BRIGHTNESS OF A SURFACE)
CANDELAS PER M²

2 The total light output of a source such as a light bulb, in all directions, is called the 'luminous flux' and is measured in *lumens* – see A on the previous page. As approximate examples of the quantities involved:

- Bicycle lamp 10 lumens.
- Domestic spotlight 2 000 lumens.
- Street lamp 13 000 lumens.

3 The intensity of light falling on a horizontal surface such as a desktop or floor is called the 'illuminance' and is measured in *lux* (one lux being one lumen per square metre) – see B on the previous page. As approximate examples:

- Surface illuminated by a candle 1 m away 1 lux.
- Desktop in an average office 500 lux.
- Outdoor surface under an overcast sky 10 000 lux.
- Outdoor surface under a bright sun 100 000 lux.

Recommended lighting levels on floors, stair treads, and worktops, etc., are always stated in terms of illuminance.

4 The apparent brightness of a surface is called 'luminance' and is measured in *candelas* per square metre – see C on the previous page. The luminance of a surface depends on (a) the intensity of light falling upon it, and (b) the light-reflectivity of that surface. Thus a sheet of white paper will appear brighter (i.e. have a higher luminance) than a sheet of black paper receiving the same intensity of light, because it reflects back a higher proportion of the received light. As very approximate examples:

- White paper on an office desk 130 candelas/m^2.
- White paper in strong sunlight 15 000 candelas/m^2.

5 Distilled data on light and lighting design are given in the METRIC HANDBOOK: PLANNING AND DESIGN DATA – see the bibliography.

6 The Society of Light and Lighting sells a complete range of highly authoritative references, both its own and those originating from other sources. Visit www.cibse.org/index.cfm?go=publications.other&L1=164&L2=0

---

ghting design in vellings

1 The following references contain authoritative guidance:

- THE HABINTEG HOUSING ASSOCIATION DESIGN GUIDE (qv) applies most specifically to *housing association* (qv) schemes, but ought to be by the designer's side in the design of all housing developments and individual dwellings.
- THE WHEELCHAIR HOUSING DESIGN GUIDE (qv), also from the Habinteg Housing Association, is another housing design reference whose usefulness extends well beyond the special remit implied by the title.

2 For publication details of the above references see the bibliography.

---

ghting design in blic buildings: mplying with the sability scrimination Act

1 Buildings used by the public must in general comply with the Disability Discrimination Act (qv). In the case of lighting design, *authoritative practical guides* (qv) include DESIGNING FOR ACCESSIBILITY (qv) and in addition the following documents. For publication details see the bibliography.

- APPROVED DOCUMENT M (qv), the provisions of which must be heeded in order to comply with the Building Regulations (qv) for England and Wales.
- BS 8300 (qv), which applies to the design of all buildings throughout the UK.
- INCLUSIVE MOBILITY (qv), which is the most authoritative reference for the *inclusive* (qv) design of *pedestrian infrastructure* (qv) and of *transport-related buildings* (qv) such as bus and coach stations, railway stations, air terminals, and transport interchanges.

Except for AD M the above documents do not have the force of law, but conformity with their recommendations will make it easier to demonstrate that the requirements of the Disability Discrimination Act have been met.

For places other than England and Wales the following regulatory documents should be consulted:

- In Northern Ireland: TECHNICAL BOOKLET R of the Building Regulations (Northern Ireland) (qv).
- In Scotland: the TECHNICAL HANDBOOKS of the Building (Scotland) Regulations (qv).

2 The relevant clauses in AD M, BS 8300, and INCLUSIVE MOBILITY are set out below.

LOCATION	DOCUMENT: RELEVANT PARAGRAPHS		
	APPROVED DOCUMENT M	BS 8300	INCLUSIVE MOBILITY
AT RAMPS AND STEPS	1.25, 1.32,	5.8.8, 5.9.8 8.1.7, 8.2.8	11
IN CORRIDORS AND ROOMS	3.12, 4.9, 4.34, 4.36	9.4, 12.1.13 12.2.15, 12.3.13, 12.4.9, 12.5.11	11

Lighting levels as recommended in BS 8300

1 The following recommendations are stated in terms of minimum *illuminanc*[e] (qv) on the horizontal plane – i.e. floor, stair tread, ramp, or kitchen worktop a[s] the case may be:

LOCATION	SEE PARAGRAPHS	MINIMUM ILLUMINANCE
AT THE TOP AND BOTTOM OF EACH FLIGHT OF A RAMP	5.8.8, 8.2.8	100 LUX
ON THE TREADS OF STEPS AND STAIRS	5.9.8, 8.1.7	
IN LAVATORIES	12.4.9	
IN ACCESSIBLE BEDROOMS	12.5.11	
IN BATHROOMS	12.2.15	100–300 LUX
IN CHANGING AND SHOWER AREAS	12.3.13	
IN KITCHENS	12.1.13	150–300 LUX

2 The foregoing figures should be checked against the relevant references before decisions are finalised.

Lighting levels as recommended in Inclusive Mobility

1 The following recommendations are stated in terms of minimum *illuminanc*[e] of the horizontal plane – e.g. platform, floor, or counter top as the case may b[e]

LOCATION	MIMIMUM ILLUMINANCE
STATION PLATFORMS AND FORECOURTS; UNDERPASSES	50 LUX
LAVATORIES; LIFT INTERIORS	100 LUX
ENTRANCES TO BUILDINGS; PASSAGES AND WALKWAYS	150 LUX
STEPS AND STAIRS AT TREAD LEVEL; RAMPS AT TOP AND BOTTOM; LANDING AREAS OUTSIDE LIFTS; SIGNS, MAPS AND DISPLAYS; TELEPHONES	200 LUX
COUNTER TOPS	250 LUX

2 These figures should be checked against Section 11 of INCLUSIVE MOBILITY (qv) before decisions are finalised.

3 In the USA, where the Americans with Disabilities Act applies, refer to the ADA AND ABA ACCESSIBILITY GUIDELINES (qv).

[Li]ghting: transitional

1 'Ocular adaptation' is the process by which people's vision adapts to changing lighting conditions, eg bright to dim. Persons with *impaired vision* (qv), who include many older people, are likely to have longer adaptation periods than the average, and to find themselves temporarily blinded by sudden and substantial changes in brightness.
2 Designers should therefore strive to avoid:
• Dimly lit entrance lobbies entered from sunlit exteriors. Some intermediate space where peoples' eyes have time to adapt is recommended – see: *Lighting: transitional.*
• Bright lights in otherwise gloomy spaces. Unless there is some valid reason against (for instance, the creation of 'atmosphere' in a nightclub) rooms and spaces in public places should always be evenly lit.
• Incorrectly positioned spotlights shining directly into peoples' eyes – a common nuisance in museums, art galleries, and shops.
• Greatly contrasting lighting levels in adjacent spaces in public buildings such as museums, shops, railway stations, theatres and concert halls. Ideally, brightness levels ought to change gradually as people move from one space to another, not abruptly.
3 These recommendations will sometimes be difficult to achieve. Because designers must try to please many different users, there will often be no single 'correct' answer. In such instances designers can only be advised to think carefully about the kinds of people likely to visit these places; about the level of *risk* (qv) that great lighting contrasts may cause; and then to come to a balanced judgement. By the same token *access officers* (qv), *local access groups* (qv), and other advocates for *ageing* and *disabled people* (qv) must face the fact that designers cannot always please everyone, and that not all spaces can be designed to suit all people. *Inclusive design* (qv) is a matter of give and take, and of reasonable compromise.

[Li]ghtwriters

See: *Speech aids.*

[Li]mb deficiency

See: *Prosthesis.*

[Li]p-readable [te]lephones

See: *Telephones: lip-readable.*

[Li]p-reading

1 Everyone *lip-reads* to some extent, especially in noisy situations; but as people develop a hearing loss they may rely increasingly on this skill. To be able to lip-read effectively they need:
• Reasonably good eyesight – see: *Impaired vision* (qv).
• A good view of the speaker's face. Para 4.34 of APPROVED DOCUMENT M (qv) of the Building Regulations states that where one-to-one communication is important, for instance between a public official and a member of the public, 'lighting should illuminate the face of the person speaking'. In audience and spectator facilities para 4.6 of AD M states that lip-readers should be provided with easily-reachable seats near the front of the seating area.
2 Lip-reading classes are usually run by adult education centres and colleges of further education. The local library, Citizen's Advice Bureau, or adult education centre should be able to advise.

3 As of 2008, useful websites include the following. They are given in alphabetical order.

- Association of Lipspeakers www.lipspeaking.co.uk/
- Directgov www.direct.gov.uk/
- Directgov: disabled people www.disability.gov.uk/
- Royal Association for Deaf People www.royaldeaf.org.uk/
- RNID www.rnid.org.uk/

See also: *British Sign Language; Impaired hearing;* and *Telephones: lip-readable.*

Listeners See: *Hearing aids and listeners.*

Literacy

1 The ability to read and write is essential to a successful life – or indeed any kind of life at all: the present author recalls the heart-wrenching case of two young brothers who had never been out of Brixton because they could not read the signs on buses and trains. To be illiterate is to face a life of struggle, probable poverty, and possibly – owing to the difficulty in getting or keeping a decent wage-earning job – a drift into crime and jail. According to the Home Office four fifths of people in prison lack the literacy skills needed to fill in a job application form.

2 In this respect Britain's state education system started failing its pupils very badly in about the 1980s, and has been doing so ever since (see: *Qualification and Curriculum Authority*). In 2005 only 71% of girls and 56% of boys aged 11 reached level four, which is the standard of writing expected at that age. The respected Royal Literary Fund, which in 1999 started placing professional writers in universities to work one-to-one with students, was staggered by what it found. Students arriving with multiple A grades at A-level could not write anything coherent longer than a paragraph or two, and many thought it unreasonable that they should be asked. For details refer to the RLF's report WRITING MATTERS – for publication details see the bibliography.

3 Recognising the problem, the government launched in 2003 a £2 billion programme titled 'Skills for Life' with the aim of raising the literacy and numeracy skills of 750 000 adults by 2004, and those of 1.5 million adults by 2007 – for details visitwww.lsc.gov.uk/

4 But in its annual report of December 2005 the Adult Learning Inspectorate (ALI) branded the programme 'a depressing failure' despite 'the extraordinary amount of money' invested in it. The main reason, according to ALI, is the poor quality of education provided by state schools, the deficiencies of which are extremely difficult to remedy later in life. For details visitwww.ali.gov.uk/, and for the effect that under-performing state schools are having on British society see: *Social mobility*.

5 The result of these failures is that current levels of illiteracy and innumeracy in the UK are a disgrace. Recent official reports on this situation include the following, listed here in date order:

- SKILLS IN THE UK: THE LONG TERM CHALLENGE, a report written for the Treasury by Lord Leitch, chairman of the National Employment Panel. It was published in 2005.
- SKILLS FOR LIFE: IMPROVING ADULT LITERACY & NUMERACY' SKILLS FOR LIFE, a report produced by the House of Commons Public Accounts Committee in January 2006, stated that up to 16 million adults – nearly half of England's workforce of 30 million full-time and part-time workers – lacked the reading and writing skills that are expected of children leaving primary school. It looked in depth at the effectiveness of the 'Skills for Life' programme

mentioned above, and concluded that the programme had been largely unsuccessful.

• NATIONAL CURRICULUM ASSESSMENTS AT KEY STAGE 2 IN ENGLAND, 2007 (PROVISIONAL), published in August 2007 by the Department for Children, Schools and Families (qv) showed that 40% of children in England were leaving primary school without the expected levels of attainment in all three of the Rs, and unready for secondary school.

6 There are many reasons for this record of failure. Some of them are beyond government control – for instance, family break-down and the generally distracting environment in which youngsters now grow up. But at least two of the prime causes were squarely within government control, and were in fact inflicted on children by theory-obsessed academics and professionals. They are:

• Abandonment of the long-established 'synthetic phonics' method of teaching children to read.

• Abolition of 'streaming' and 'group-setting' – i.e. teaching pupils in groups that are able to progress at roughly similar speed. Such grouping is particularly important for the least able children in a particular subject, who can thereby be given much more time without holding up their more able fellows.

This history of avoidable disaster is recounted in some detail in the next entry.

7 As of 2008, useful websites include:

• CEM Centre	www.cemcentre.org/
• National Literacy Trust	www.literacytrust.org.uk/
• Read Write Inc	www.ruthmiskinliteracy.com/

See also: *Education; Office for Standards in Education*; and *Qualifications and Curriculum Authority*.

---

teracy: how lucational theorists iled four million ildren

1 Until about 40 years ago, British children were very successfully taught to read by the traditional 'synthetic phonics' method. Under this system pupils learn to read by building up words from basic sounds – for instance the individual units c-a-t spell 'cat'. In this way the young reader learns up to 44 'phonemes' (the smallest units of sound) and their related 'graphemes' (the written symbols for the phoneme), and has an excellent chance of becoming proficient at reading. If taught by this method for 20 minutes a day from the first day of primary school, most children can learn to read within 18 months, and virtually all children (including those with *dyslexia*) within a reasonable period thereafter. The present author is indebted to the commentator John Clare for the following account of what happened next.

2 In the 1960s and 1970s this well-tried teaching method was swept away by academics led in the UK by the late Professor Ted Wragg, and replaced by methods such as 'real books' and 'look and say', which expect children to work out the meaning of words largely by gazing at them in the context of other words and pictures, and to learn to read books more or less by osmosis. According to Frank Green, the Canadian theorist who orginated the movement, the learning of reading is spontaneous and effortless and 'requires no particular attention, conscious motivation, or specific reinforcement'. Such ideas delighted the British educational and teacher-training establishments, who despise 'learning by rote' and are bowled over by teaching methods that are 'creative' and 'child-centred'–particularly if they promise learning by near-effortless absorption.

3 While children who are fortunate enough to be clever, and to grow up in bookish middle-class families, may well appear to learn to read almost effortlessly, youngsters lacking these advantages most emphatically (as subsequent events would demonstrate) do not. Nevertheless the new approach

was enthusiastically adopted and enforced by educationists, and especially by those who trained teachers and monitored what they were doing in the classroom.

4 By the middle 1980s, as the now-retired commentator John Clare noted at the time, many parents and classroom teachers were becoming convinced that unprecedented numbers of children were failing to learn to read. These unfashionable views were resolutely ignored by the educational establishment at all levels.

5 In 1990 the educational psychologist Martin Turner came upon evidence tha the educational establishment had been trying to suppress: tests taken by around 4 000 children in nine local education authority (LEA) areas showed that reading standards had fallen between 1985 and 1990, and that the decline was the sharpest for 40 years.

6 In 1996 the then head of Ofsted (qv), Mr Chris Woodhead, ordered an inspection of the teaching of reading in 45 inner London primary schools, and the results so shook the Education Secretary, Ms Gillian Shephard, that she attempted an overhaul of teaching methods in primary schools, including the reintroduction of phonics. The education establishment continued to resist unti publication in February 2005 of a report titled THE EFFECTS OF SYNTHETIC PHONICS TEACHING ON READING AND SPELLING ATTAINMENT: A SEVEN YEAR LONGITUDINAL STUDY, from the Scottish Executive Education Department. The study had involved over 300 children, was carried out by Professor Rhona Johnston from the Department of Psychology at the University of Hull, and D Joyce Watson from the Department of Psychology at the University of St Andrews, and showed conclusively that children taught by the phonic method were three and a half years ahead of their peers in reading at the end of their time at primary school. The advantages were particularly marked among children from illiterate or inarticulate homes – the very youngsters whom Prof Wragg had presumably been trying to help.

7 In December 2005 came the final act in this extended and catastrophic educational experiment – publication of the interim version of Jim Rose's INDEPENDENT REVIEW OF THE TEACHING OF EARLY READING (for the full text visit www.standards.dfes.gov.uk/rosereview/interimreport.doc). The report unequivocally backed a return to the use of synthetic phonics and was accepte in its entirety in March 2006 by the then Education Secretary, Ms Ruth Kelly. In September of that year her successor, Mr Alan Johnson, announced a full-scale return to the traditional phonic method of teaching children to read, to the memorising of multiplication tables at an early age, and to the long-abandoned 'standard written' method of arithmetic calculation – i.e. all the wel established methods that had been tossed overboard at the behest of fashionab *theorists* (qv) and *ideologues* (qv) in the 1970s and 80s. The DfES published a phonics instruction manual in June 2007, amid speculation that many teachers would simply ignore it and continue to teach literacy by now-discredited methods. Time will tell.

8 It was a complete U-turn, but unfortunately too late to help the estimated 4 million children who could over the past 20 years have been taught to read and multiply at an early age, and were not. Their life chances were seriously diminished, and in many cases wrecked, by the intellectual and professional arrogance of educational theorists, teacher training professionals, Department Education officials, local authority advisors, and the teaching unions; and by th negligence of the two watchdogs that did not bark in the night – Ofsted (qv), an the Qualifications and Curriculum Authority (qv). For the probable ultimate outcome for many of these unfortunate youngsters, see: *Crime; Social mobilit* and *Social exclusion.*

9 As of 2008 useful references, in addition to those already mentioned above, include the following. They are given in alphabetical order. For publication details of all references see the bibliography.

- I'M A TEACHER, GET ME OUT OF HERE by Francis Gilbert. A candid and illuminating account, both amusing and frightening, of one young teacher's roller-coaster journey through the English education system.
- THE NEW SCHOOL RULES: THE PARENT'S GUIDE TO GETTING THE BEST EDUCATION FOR YOUR CHILD by Francis Gilbert. An encyclopedic guide, by the same author, to help parents choose the best school for their child, and help the child get the best results.
- READY TO READ? by Anastasia de Waal and Nicholas Cowen. A free, highly informative article on the phonics system from the *think tank* (qv) Civitas.
- SOCIAL TRENDS (qv), published annually by the Office for National Statistics (qv), gives latest statistics plus a well-informed discussion on educational matters. Refer to Chapter 3.
- TEN YEARS OF BOLD EDUCATION BOASTS NOW LOOK SADLY HOLLOW. IT WILL BE HARD POLTICALLY BUT LABOUR MUST ACCEPT ITS VAUNTED POLICIES ON SCHOOLS HAVEN'T WORKED by Jenni Russell.

See also: *Déformation professionelle; Ideology; Office for Standards in Education; Qualifications and Curriculum Authority; Social exclusion*; and *Social mobility*.

---

ve access audits See: *Access audits: live.*

---

bbies: entrance omplying with the isability iscrimination Act

1 A lobby consisting of inner and outer doors, with a space between, may be provided to reduce air loss at a building entrance; to prevent drafts; to increase security; or to provide a *transitional lighting* (qv) zone between outside and inside. The schematic diagram (and it is no more than that) indicates the principle.

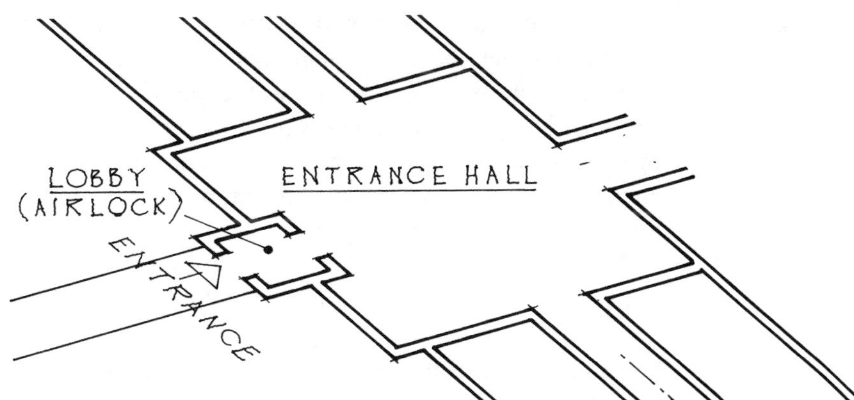

2 Buildings used by the public must in general comply with the Disability Discrimination Act (qv). In the case of entrance lobbies in public buildings, *authoritative practical guides* (qv) include DESIGNING FOR ACCESSIBILITY (qv) and in addition the following documents, publication details of which are given in the bibliography.

- APPROVED DOCUMENT M (qv), the provisions of which must be heeded in order to comply with the Building Regulations (qv) for England and Wales. See paras 3.15–3.16.

For places other than England and Wales the following regulatory documents should be consulted:

- In Northern Ireland: TECHNICAL BOOKLET R of the Building Regulations (Northern Ireland) (qv).

• In Scotland: the TECHNICAL HANDBOOKS of the Building (Scotland) Regulations (qv).
3 In the USA, where the Americans with Disabilities Act applies, refer to the ADA AND ABA ACCESSIBILITY GUIDELINES (qv).

See also: *Entrance halls.*

Lobbies: internal

1 A lobby of the above kind may be provided inside a building for one or both of the following reasons:
• For fire safety: to prevent the spread of fire and smoke, and as part of the provision of *means of escape* (qv) and assistance to fire-fighters as laid down in APPROVED DOCUMENT B (qv) of the Building Regulations.
• For privacy: to shield the view into *lavatories* (qv).
2 In this case relevant clauses in *authoritative practical guides* (qv) include the following:

FUNCTION	DOCUMENT: RELEVANT PARAGRAPHS		
	APPROVED DOCUMENT B	APPROVED DOCUMENT M	BS 8300
FIRE SAFETY DESIGN	3.18, 3.21, 3.44, 4.11 5.12, 5.20, 5.24, 5.25 6.42, 6.43, 6.51, 6.52	–	–
PRIVACY	–	3.15–3.16	6.3.6.2

Local access groups

1 Also referred to as *access groups*, local access groups are groups of *disabled people* (qv) formed to represent the needs of disabled people generally, when local building developments or designs are being proposed or discussed.
2 Such groups must be fully representative of their local communities if they ar to have legitimacy. This may be difficult to achieve. According to a 1994 survey most members of such groups were male *wheelchair-users* (qv) aged 35 to 60, who probably represented well under one in 400 of all building users – see: *Disabled persons: numbers* (qv). Every effort should therefore be made to bring in a wide range of the population, and to ensure that the issues selected fo attention are widely representative and not just the hobby-horses of overly self-assertive individual campaigners.
3 The views of properly representative local access groups are essential in scheme development, and must be sought. But they should be used in a reasonable manner. Many bad schemes have been improved by such dialogue, but many premises that gave much pleasure to the public at large have been closed because they could not offer full equality to a fraction (probably under 1%) of building users. That is not what is meant by *inclusiveness*, which is about breadth and generosity of spirit, not the dog-in-the-manger attitude 'if everyone cannot enjoy it, then no-one shall'.
4 The above problem tends to arise particularly with *historic buildings and site* (qv), which are often difficult to adapt to full accessibility. If unreasonable demands are made by local access groups, then developers and their architects should insist on the following official guidelines being honoured:
• Para 20 (2) of the Disability Discrimination Act (qv) only requires *service providers* (qv) to take such steps as are 'reasonable, in all the circumstances of the case', to ensure that access is not 'impossible' or 'unreasonably difficult' fo disabled persons. The standards of accessibility that are demanded of new

buildings cannot necessarily be applied to existing premises, and the term 'reasonable' must be given full weight.
• Para 0.18 of APPROVED DOCUMENT M (qv) of the Building Regulations states that 'the aim should be to improve accessibility where and to the extent that it is practically possible, always provided that the work does not prejudice the character of the historic building, or increase the risk of long-term deterioration to the building fabric or fittings'.
• Para 13.10 of BS 8300 (qv) states that 'historic buildings should be made accessible to disabled people, wherever possible, without compromising conservation and heritage issues'.
None of these authoritative references speak of complete accessibility to everyone when that proves difficult to achieve.
5 As of 2008, useful references include the following. They are listed in alphabetical order. For publication details see the bibliography.
• PLANNING AND ACCESS FOR DISABLED PEOPLE (qv). See in particular para 8.7.
• PLANNING, BUILDINGS, STREETS AND DISABILITY EQUALITY. A guide to the Disability Equality Duty and Disability Discrimination Act 2005 for local authority departments responsible for planning, design and management of the built environment and streets.
6 Useful websites include:
• Access Group Resources www.accessgroupresources.co.uk/

See also: *Accessible*; and *Inclusion*.

Local authorities

1 As of 2008 there are just under 500 local authorities in England and Wales, representing over 50 million people and spending in 2003 around £78 billion per annum. There are of five types of local authority:

I	County Councils	Typically provide education and social services, leaving housing, environmental health, etc., to District Councils. This system of splitting up the delivery of local government services between County Councils (which cover a wider geographic area) and District councils (which are small and local, and tackle issues such as housing and tourism) is called 'two-tier local government'. For a map of England's counties see: *Regions*.
II	District Councils	See above. Mostly rural.
III	Metropolitan District Councils	Unlike (I) and (II) they do not split up the various services between them but provide the full range of local authority services. They are thus called 'unitary', 'all-purpose' or 'single-tier' councils.
IV	Shire Unitary Councils	As for (III).
V	London Borough Councils	As for (III).

2 It is difficult for a lay person to deduce the category into which any particular local authority falls simply by inspecting its name. For authoritative guidance visit the Local Government Association at www.lga.gov.uk/
3 For the three main levels of local government in England into which the above bodies fit, see: *Government: central, regional, and local.*

See also: *Local government.*

Local authorities: funding

1 According to research by the Halifax, the average council tax per dwelling across the UK in 1997–98 was £564, and in 2007–08 it is £1 078. These figures were published in June 2000. The highest and lowest average bills in 2007 are in adjacent London boroughs – Richmond upon Thames at £1 665 and Wandsworth at £641.
2 On average, council tax pays for about 25% of the costs of local government services, but the figure varies. In a poor borough such as Newham in London it may be only 10%; the London average is around 50%; and in the wealthy borough of Richmond upon Thames it is 80%. The balance is in all cases made up by central government via an annual grant that comes with a huge number of strings attached.
3 There is widespread agreement that local government would be more responsive to local citizens if it received more of its income from them. In Europe, as shown in Figure. 10 of DIRECT DEMOCRACY: AN AGENDA FOR A NEW MODEL PARTY (see the bibliography) the average share of money raised locally is closer to 45%, which probably helps to explain the large degree of autonomy enjoyed by local authorities there. But there is, as yet, no British consensus on how far that principle can be taken – particularly in view of the great disparities of local wealth noted above – or how it is to work. One proposal, in DIRECT DEMOCRACY, is that VAT should be replaced by a variable local sales tax that is levied once at the point of retail. This would make local councils responsible for almost all their local spending, and (judging by overseas examples) become at once more prudent and more imaginative in meeting voters' needs.
4 These matters are too complex to be pursued here. As of 2008, useful references include the following. For publication details see the bibliography:
- DIRECT DEMOCRACY: AN AGENDA FOR A NEW MODEL PARTY, published by the *think tank* Direct Democracy (qv).
- THE LOCALIST PAPERS, published by the same organisation in conjunction with the *Daily Telegraph*.

5 Useful websites include the following. They are given in alphabetical order.
- Direct Democracy www.direct-democracy.co.uk/
- Institute for Fiscal Studies www.ifs.org.uk/
- Local Government Association www.lga.gov.uk/
- Localis www.policyexchange.org.uk/
- New Local Government Network www.nlgn.org.uk/
- New Policy Institute www.npi.org.uk/projects/council%20tax.htm

See also: *Post-democracy.*

Local authority websites

See: *Government websites.*

Local Better Regulation Office (LBRO)

1 One of an array of bodies being created by the Government to help it cut red tape – visit www.dti.gov.uk/consumers/enforcement/lbro/index.html
2 For a comment on other bodies of this kind, and their usefulness, see: *Better regulation bodies.*

cal Education thorities (LEAs) 1 LEAs are *local government* (qv) departments responsible for delivering primary and secondary *education* (qv). The approximately 150 LEAs in England have a duty to improve school performance and tackle failure, pass on schools funding, ensure that *excluded pupils* (qv) are educated, and provide enough school places for children.
2 In December 2006 the Department for Education and Skills' Implementation Review Unit (qv) criticised the role of LEAs, stating that the chain of command between the DfES (qv) and schools had become over-elaborate and was exacerbating the bureaucratic workload on teachers, making it impossible for them to give enough attention to pupils – especially those with *special educational needs* (qv).
3 Quite apart from the above report, the depths to which LEAs have allowed the performance of England's state schools to sink (see for instance: *Literacy*) leads irresistibly to the conclusion that their time is up. A better way forward is suggested under: *Social mobility*.
4 As of 2008, useful websites include:
• Department for Children, Schools and Families
www.dfes.gov.uk/localauthorities/index.cfm

See also: *Better regulation bodies; Education; Education: voucher systems*; and *Social mobility*.

cal government See: *Government; Local authorities*; and *Localism*.

cal Government Act '99 1 The LGA 1999 introduced the *Best Value* (qv) service improvement and inspection regime, replacing the previous Conservative government's Compulsory Competitive Tendering (CCT) regime.
2 As of 2008, useful websites include the following. They are given in alphabetical order.
• IDeA Knowledge www.idea.gov.uk/
• Local Government Association www.lga.gov.uk/
• Office of Public Sector Information www.opsi.gov.uk/

cal Government Act 00 1 The LGA 2000 introduced directly elected mayors and cabinet-style local government, largely scrapping the century-old committee system.
2 As of 2008, useful websites include the following. They are given in alphabetical order.
• IDeA Knowledge www.idea.gov.uk/
• Local Government Association www.lga.gov.uk/
• Office of Public Sector Information www.opsi.gov.uk/

cal Government sociation (LGA) Visit www.lga.gov.uk/

cal government ague table 1 Following Comprehensive Performance Assessment (qv), *local authorities* (qv) are ranked as excellent, good, fair, weak, or poor. Excellent and good councils are allowed more autonomy from central government; weak and poor performing councils are monitored more closely and may, in the worst cases, enjoy ministerial intervention.
2 As of 2008, useful websites include the following. They are given in alphabetical order.
• IDeA Knowledge www.idea.gov.uk/
• Local Government Association www.lga.gov.uk/

Local Government Ombudsman (LGO)	Visit www.lgo.org.uk/
Local Government Performance	1 This is a Department for Communities and Local Government (qv) website that summarises at a glance the performance of all England's *local authorities* (qv), and enables viewers to undertake their own in-depth analysis of any particular authority. 2 For more information visit www.bvpi.gov.uk/pages/Index.asp
Local Involvement Networks (LINks)	See: *NHS structure.*
Local Public Service Agreements (LPSAs)	1 These are agreements between *local authorities* (qv) and central *government* (qv), under which councils are rewarded with extra funding in return for achieving objectives such as – for example – reducing welfare dependency and getting more people into work. Targets are based on a mix of national and local priorities, and in return for achieving them councils get greater freedoms and flexibility, plus financial rewards worth 2.5% of council budgets. 2 For more information visit www.communities.gov.uk/index.asp?id=1134088
Local Strategic Partnership (LSP)	1 An initiative to ensure co-operation between public agencies, *voluntary groups* (qv), and businesses to help achieve the regeneration of deprived neighbourhoods. 2 For more information visitwww.neighbourhood.gov.uk/page.asp?id=531 See also: *Neighbourhood renewal.*
Local tax	See: *Local authorities: funding*; and *Tax: council.*
Localism	1 This is the democratic and *inclusive* (qv) precept that: • As many decisions as possible should be taken by people themselves, not regulated by officials on their behalf; and • The management of those matters that must be regulated should be devolved down to a level as close as possible to the people affected. 2 Britain is at present a highly centralised state. Comparisons are risky, because different countries have different governmental structures; but according to Simon Jenkins, author of BIG BANG LOCALISM, the lowest tier of government in France covers on average a population of around 1 600 people; the lowest tier in Germany and Scandinavia about 5 000; and the lowest tier in Britain 120 000. He estimates that France has an elected representative for every 116 electors; Germany one for every 250; and Britain one for every 2 600. 3 There is a growing feeling in Britain that our centralised state is (a) inefficient, and (b) a generator of cynicism and political apathy because many people have come to feel that – despite freedom of speech; regular elections, and other democratic institutions – government is remote and quite beyond their influence. To quote the *think tank* (qv) Direct Democracy: in Europe and North America matters such as schools admissions policies, health priorities, smoking bans, and housing targets are decided by locally elected officials; in Britain they are settled by a single Minister and then applied uniformly to 60 million people. 4 Even worse, many of the decisions that most tangibly affect our lives are now being taken not by elected representatives of any kind but by unelected *quangos* (qv) such as the Child Support Agency (qv), which largely wasted £50 billion of taxpayers' money before being scrapped; the Learning and Skills Council (qv), the Highways Agency, the National Institute for Health and Clinical Excellence

and several hundred other quangos stretching from Whitehall to the European Union.

5 There are two possible responses to the above.

• The popular call (to which all parties now pay lip-service) for a massive decentralisation of government. This approach has an able advocate in Simon Jenkins, who wants a re-empowerment of the counties and cities to which people feel loyalty, with many services delegated down to municipalities and parishes. Giving people control over services would drive up standards, he argues, and help restore people's faith in British democracy.

• The dissenting view of commentators such as Leo McKinstry, who holds that the British public has come to feel very strongly that all citizens are entitled to receive equal treatment from the state no matter where they live. True localism would inevitably bring variations – perhaps very large ones – in standards of education, healthcare, social protection, and other services across the country. Mr McKinstry argues that most people will find such inequalities quite unacceptable.

Direct Democracy responds to Mr McKinstrys' argument by pointing out that Britain's centralised system has signally failed to deliver nationally consistent outcomes. It states that 'the quality of local schools, hospitals, councils and police forces varies wildly from one area to the next. In the ten best LEAs in the country 60% of pupils obtain five good GCSE passes, whereas in the worst ten half as many do. Hospital mortality figures show patients are twice as likely to die in the worst-performing hospital in England as they are in the best'. And it argues that the major reason why outcomes differ so much – and are at the same time are so often poor – is the Government's insistence on imposing the same state centralist approach everywhere and in all circumstances, regardless of local needs and local conditions.

6 Readers must decide for themselves. As of 2007, useful references include the following. They are given in alphabetical order. For publication details see the bibliography.

• BIG BANG LOCALISM: A RESCUE PLAN FOR BRITISH DEMOCRACY by Simon Jenkins.

• CITIES RENAISSANCE: CREATING LOCAL LEADERSHIP by Michael Heseltine. This is a submission made to the shadow cabinet in June 2008, arguing that the Conservative Party should in the next election be championing a massive transfer of power from central to local government.

• POWER TO THE PEOPLE. THE REPORT OF POWER: AN INDEPENDENT INQUIRY INTO BRITAIN'S DEMOCRACY. This important survey was published by the Joseph Rowntree Foundation (qv) in 2006. See in particular p 46 and p 153 et seq.

• PROSPEROUS COMMUNITIES II: VIVE LA DEVOLUTION. This paper, published in February 2007 by the Local Government Association (qv), proposes a network of about 50 'sub-regions' built around existing structures such as cities, boroughs, counties, and districts. Under this form of devolution, the LGA argues, there would be no need for costly and disruptive reorganisation, merely for collaboration. Different authorities, for instance Hampshire County Council together with Portsmouth and Southampton, both of which are *unitary authorities* (qv), and several neighbouring districts, could form a board made up of members from each to whom the transport and planning powers now wielded by the region would be devolved.

7 Useful websites include the following. They are given in alphabetical order.

• Direct Democracy	www.direct-democracy.co.uk/
• Local Government Association	www.lga.gov.uk/
• Localis	www.policyexchange.org.uk/
• New Local Government Network	www.nlgn.org.uk/

Local transport plans	1 Transport plays a key role in supporting regional and local prosperity, economic growth, and enhancing quality of life. To that end, the Department for Transport (qv) works in partnership with communities at both regional and local level to improve access to jobs and services, particularly for those most in need. 2 For more information on local transport plans visit www.dft.gov.uk/ See also: *Integrated transport*; and *Transport: public*.
London Accessible Housing Register	Visit www.london.gov.uk/mayor/housing/docs/accessible_part_2.pdf
London Assembly	See: *Government: London*.
London borough councils	See: *Local authorities*.
London Development Agency (LDA)	See: *Government: London*.
London Disability Arts Forum (LDAF)	Visit www.ldaf.org/
London government	See: *Government: London*.
London: housing	See: *Housing: London*.
London Index	See: *Areas for regeneration*.
London: mayoral strategies	1 There are eight statutory Mayoral Strategies which together form a blueprint for the future of London. They are: • Air Quality. • Ambient Noise. • Biodiversity. • Cultural and Economic Development. • Spatial Development. • Transport. • Waste Management. 2 See also: *London Plan*.
London Plan	1 The London Plan, which took effect in February 2004, is an overall planning strategy for the 32 London boroughs and the Corporation of London. 2 For more information visit www.london.gov.uk
London Plan: accessibility and inclusiveness	1 The London Plan as published contained the following key provisions for the promotion of greater *accessibility* (qv) in the capital: 2 All London boroughs are required to integrate and adopt certain principles of *inclusive design* (qv) into their Unitary Development Plans (the latter documents have since been superseded by Local Development Frameworks – for an explanation of the latter visit www.planningportal.gov.uk/). This requirement means that buildings must: • Be able to be used safely and easily by as many people as possible without undue effort.

• Offer users the freedom to choose and the ability to participate equally in all the development's mainstream activities, without separation or special treatment.
• Value *diversity* (qv) and difference.
3 All planning applications must include an *Access Statement* (qv) showing how the principles of inclusive design, including the specific needs of *disabled persons* (qv), have been integrated into the proposed development and how inclusion will be maintained and managed.
4 As of 2007:
• All new housing built in London must be built to *Lifetime Home* (qv) standards. This applies to blocks of flats as well as dwellinghouses, both private sector and public sector.
• 10% of new housing should be designed to be *wheelchair accessible* (qv), or to be easily adaptable for residents who are *wheelchair-users* (qv).
5 ACCESSIBLE LONDON: ACHIEVING AN INCLUSIVE ENVIRONMENT is the title of the first Supplementary Planning Guidance to the London Plan. It was issued in April 2004 and gives advice on how to promote and achieve an *inclusive* environment in London. For publication details see the bibliography.
6 For latest information on the above matters visit www.london.gov.uk/

---

ndon Transport See: *Transport for London*.

---

ndon Travelwatch 1 This is the statutory consumer body representing the interest of all travellers, by all means of public transport, in London.
2 For more information visitwww.londontravelwatch.org.uk/

---

ne parent families See: *Familes: single parent.*

---

oked after children See: *Children: looked after.*

---

rd of the Flies See: *Crime: youth.*

---

ttery See: *National Lottery*.

---

uder than words 1 LTW is a best practice charter, developed by the RNID (qv), that will help businesses, *voluntary organisations* (qv) and other bodies to demonstrate that they are in compliance with the Disability Discrimination Act (qv) with respect to customers and employees with *impaired hearing* (qv).
2 For more information visitwww.rnid.org.uk/

See also: *Disability Discrimination Act*; and *Impaired hearing*.

---

w cost home wnership See: *Housing: affordable.*

---

w-rise platform lifts See: *Vertical platform lifts.*

---

w-tech aids and uipment See: *Aids and equipment*; and *Assistive technology*.

T platforms See: *Corduroy hazard warning surfaces*.

---

men x See: *Lighting design*.

M

agic See: *Audioguides*; and *Touch tours*.

ainstream schools See: *Schools: mainstream*.

ainstreaming

1 There are two ways of trying to ensure *equality* (qv) of opportunity for minority groups and disadvantaged people:

- Taking their needs and problems into account from the moment that general policies are framed, and building them into the general framework.

or

- Bolting *special needs* (qv) policies or practices onto the general policies that are already in place.

2 The former is called 'mainstreaming', and is the recommended approach – see: *Inclusion*.

See also: *Bending main programmes; Equality; and Special needs*.

akaton Vocabulary

1 Makaton is a *symbol system* (see: *Speech aids*) comprising a specially selected vocabulary from *British Sign Language* (qv). It is meant for the use of people – especially children – who are *deaf* (qv) or have severe *learning difficulties* (qv), and cannot cope with conventional alphabetical writing.

2 Like Widgit Rebus (qv), the Makaton system consists predominantly of symbols which directly resemble the objects they represent ('pictographs'), rather than more abstract diagrams. Some users therefore find Makaton easier to learn than more sophisticated systems such as Blissymbolics, while on the negative side this simplicity means that Makaton may not be able to provide the linguistic structure and richness which many children and adults need. For some people Makaton therefore provides a useful stepping stone to more formal systems, rather than a complete solution in itself.

3 The figure below shows a selection of simple Makaton symbols just to give the flavour.

PLEASE/ THANK YOU

GOOD MORNING/ HELLO

GOODBYE

4 As of 2008, useful websites include the following. They are given in alphabetical order.

- Directgov www.direct.gov.uk/
- Inclusive technology www.inclusive.co.uk/
- Makaton www.makaton.org/

See also: *Alternative and augmentative communication*; and *Speech aids*.

arket housing See: *Housing: market*.

arket renewal thfinders See: *Pathfinders*.

arriage See: *Families*.

Mayer-Johnson communication

1 This is a *symbol system* (see: *Speech aids*) rather like the one described abo[ve]. But unlike the Makaton (qv) and Widgit Rebus (qv) systems, which are only partly pictographic, the Mayer-Johnson system is almost wholly pictorial – i.e. most of its symbols directly resemble the objects they represent. At the most basic level they represent concepts such as 'I want', 'book', 'hamburger', 'ea[t]' or 'let's talk', which can be assembled to convey coherent messages. A simpl[e] example is shown below:

At later stages of the learning process more complex and subtle concepts are introduced.

2 As of 2008, useful websites include the following. They are given in alphabetical order.

- Directgov www.direct.gov.uk/
- Inclusive technology www.inclusive.co.uk/
- Mayer-Johnson www.mayer-johnson.com/

See also: *Alternative and augmentative communication*; and *Speech aids*.

Mean

See: *Income: median*.

Means of access

See: *Accessible*.

Means of escape

1 There are two complementary aspects to the subject of safe escape from fir[e] in buildings:

- Design precautions, relating to the building layout, construction, and service[s] (lighting, alarms, and sprinkler systems) etc. These are known as *means of escape*.
- Management precautions, known as *evacuation* (qv).

2 These are specialised subjects and it is essential to get expert advice, startin[g] in the first instance with the local *Building Control Body* (qv) and *Fire Authority* (qv).

3 As of 2008, useful initial references include the following. They are given i[n] alphabetical order. For publication details see the bibliography.

- APPROVED DOCUMENT B (qv) of the Building Regulations. The provisions of AD B must be heeded in order to comply with the Building Regulations for England and Wales. In Northern Ireland refer to TECHNICAL BOOKLET E of the Building Regulations (Northern Ireland); and in Scotland to the TECHNICAL HANDBOOKS of the Building (Scotland) Regulations (qv).
- BS 5588–8: FIRE PRECAUTIONS IN THE DESIGN, CONSTRUCTION AND USE OF BUILDINGS – CODE OF PRACTICE FOR MEANS OF ESCAPE FOR DISABLED PEOPLE (qv) deals particularly with means of escape for *disabled people*.
- FIRE SAFETY RISK ASSESSMENT GUIDES. This is a series of individual guides from the Department for Communities and Local Government (qv) on specifi[c] building types including healthcare premises, residential care premises,

educational premises, theatres and cinemas, factories and warehouses, offices and shops, sleeping accommodation, places of assembly, outdoor spaces, and others. They are reasonably priced and aimed at lay people.
4 Useful websites include the following.
• Fire gateway www.fire.gov.uk/

See also: *BS 5588: Part 8; Evacuation; Hazard*; and *Risk*.

Means of escape: basic principles

1 Three matters are crucial to the design of safe escape routes. They are, in brief:
2 The layout of routes. The simplified sketch below, and the notes that follow, summarise some basic principles:

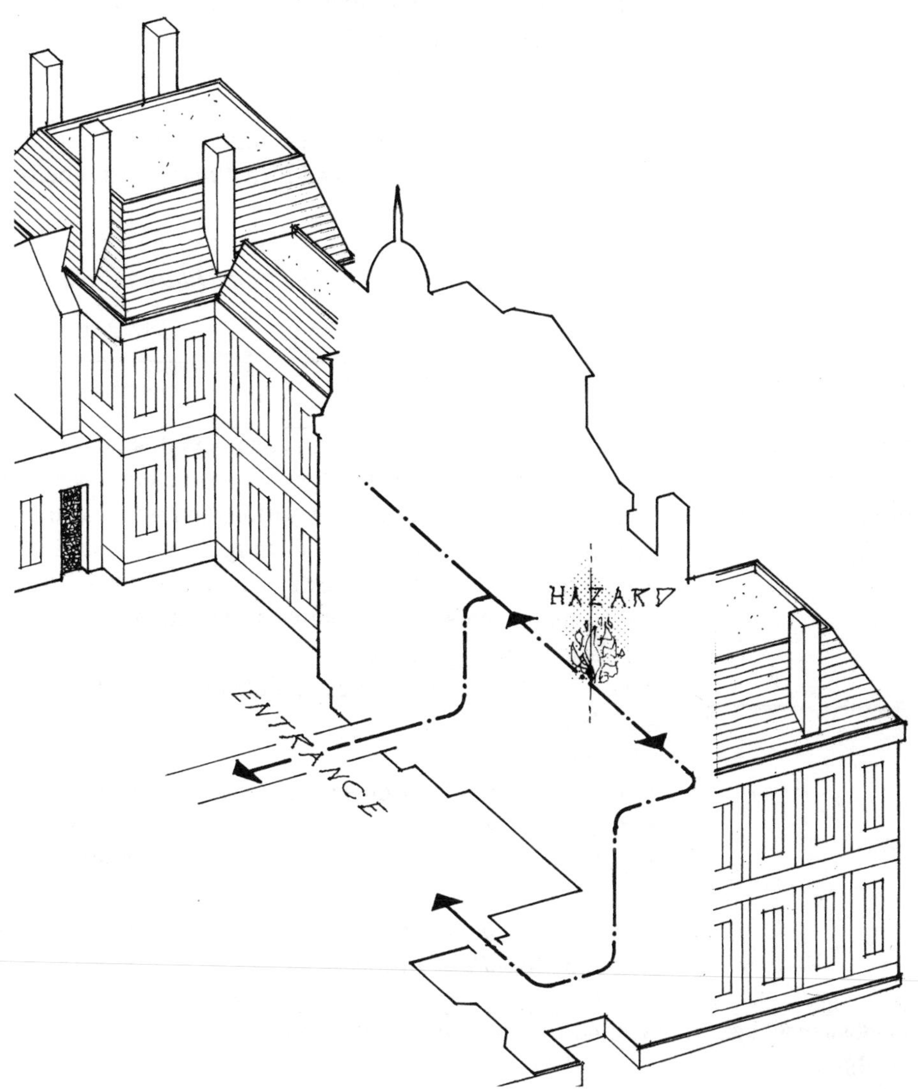

• Because people instinctively try to escape from buildings by the same way as they arrived, it is safest if the same routes are used as far as possible for both access and emergency escape.
• There should in general be two alternative routes leading away from any particular point, to prevent people being trapped if one of the escape routes from the *hazard* (qv) is blocked.
• Escape routes should as far as possible be straight and simple, with daylight visible at each end.
See also *BS 5588: Part 8* (qv), which deals particularly with some of the above arrangements in the case of *disabled* (qv) building occupants.

3 The <u>construction</u> of escape routes. Such routes should, as specified in APPROVED DOCUMENT B (qv) of the Building Regulations and other key references:
• Be 'protected routes', i.e. *corridors* (qv) etc. that are enclosed with fire-resisting floors, ceilings, and walls (the latter including self-closing fire doors) to ensure that the route remains safe from the effects of any adjacent fire for at least 30 minutes.
• Have no changes of level, raised thresholds, or anything else that might impede the swift and safe movement of panicky and confused people fleeing from a fire.
• Have doors that are unobstructed; that open in the direction of escape; and which are easily openable without use of a key.
4 In addition to the above matters, escape routes should also be:
• Clearly signposted. See: *Signage* (qv).
• Well lit. See: *Lighting* (qv).
• Properly ventilated.
5 On all the above matters it is essential to seek the guidance of the local *Building Control Body* (qv) and *Fire Authority* (qv) before any major decisions are taken.

See also: *BS 5588: Part 8; Evacuation; Place of relative safety*; and *Place of safety*.

Median income

See: *Income: median.*

Mediation and the Disability Discrimination Act (DDA)

1 Any *disabled* (qv) person who considers that he or she has been *discriminated* (qv) against without *justification* (qv) by an employer, *service provider* (qv), or landlord may sue for compensation, and/or the issuing of an injunction to prevent the discrimination reoccurring. But many cases can be resolved by negotiation, and this should always be tried first. The Commission for Equality and Human Rights (qv) encourages such attempts, using organisations such as Mediation UK (see below).
2 As of 2008, useful websites include the following. They are given in alphabetical order.

• Directgov	www.direct.gov.uk/
• Equality & Human Rights Commission	www.equalityhumanrights.com/
• Mediation UK	www.mediationuk.org.uk/
• Mediation Works	www.mediation-works.co.uk/

See also: *Community Legal Service Partnerships; Disability Discrimination Act*; and *Disability Law Service*.

Medical model of disability

See: *Disability: other definitions*.

Medical Training and Application Service (MTAS)

1 The MTAS is a computerised system for handling the applications of junior doctors for specialist registrar jobs. It was introduced by the Department of Health (qv) in early 2007 to help get its Modernising Medical Careers (MMC) off to a fast start, but proved to be such a fiasco that it had to be abandoned within two months. First the Government, with its limitless faith in abstract systems in preference to mere fallible human judgement, tried to banish the latter from the process whereby junior doctors are appointed to posts: instead of applicants providing references from their seniors and being interviewed personally by fellow professionals, they found themselves talking to computers

which gave more weight to a crude personality test than to evidence of candidates' medical ability. Then followed the now familiar *IT* (qv) crash: the computer could not cope with the volume of applications; many outstanding candidates failed to get any interviews in the first round; others were shortlisted despite not being qualified for the positions being applied for. Next, in April 2007, came two serious security failures: the supposedly secure website, which contained information on the sexual orientation, previous convictions, addresses, and home phone number of thousands of young doctors, was twice found to be open to anyone with a computer. At that point the system was suspended.
2 For other failures of government computer systems, see: *IT systems*.

---

edicalisation

1 There are genuine mental illnesses just as there are physical illnesses, and the lives of people with severe depression, manic depression (bipolar disorder), or chronic acute anxiety are a constant struggle. In many cases such conditions are covered by the Disability Discrimination Act (qv).
2 There is unfortunately the phenomenon of medicalisation, which promotes ordinary human troubles to the status of impressively named medical disorders that must be written up in textbooks (thus helping publishers sell books), diagnosed by consultants (thus earning them fees at so much per hour), and treated with drugs (thus opening up new markets for pharmaceutical companies). Hot on their trail may come fee-earning psychiatrists and lawyers arguing that the afflicted child or adult is entitled to the full panoply of legal protections that are offered by the Special Educational Needs (qv) legal framework, the Disability Discrimination Act (qv), and so on.
3 In this way shyness may become 'Social Anxiety Disorder'; a tendency to constant worry 'Generalised Anxiety Disorder'; and premenstrual tension 'Premenstrual Dysphoric Disorder'. Some of these newly named conditions may well be genuine mental illnesses, requiring expensive medical treatment, but a healthy degree of scepticism is in order. A useful case example of what goes on behind the scenes is given in the book SELLING SICKNESS: HOW THE WORLD'S BIGGEST PHARMACEUTICAL COMPANIES ARE TURNING US ALL INTO PATIENTS (for publication details see the bibliography). The author Ray Moynihan, an internationally respected health journalist and sometime guest editor at 'The British Medical Journal', describes how a major pharmaceutical company needed to find a new application for its anti-depressant drug, which is named in the book. The company found a brief mention in a psychiatric journal of a little-known nervous condition called Social Anxiety Disorder or SAD, which was characterised by such trivial symptoms as feeling nervous, sweaty, and shy at parties. It hired a PR firm to turn this everyday condition into a disease. The PR firm rounded up patients, experts, and a celebrity sufferer; it presented the SAD story to the press; and in due course the *New York Times* and *American Vogue* ran lengthy features on the ailment. Thousands of people soon decided that they suffered from SAD, doctors prescribed the drug; the pharmaceutical company prospered; and the PR firm won an award for 'Best PR Programme of the year'.
4 Similar stories are told by the more recent works SHYNESS: HOW NORMAL BEHAVIOUR BECAME A SICKNESS by Christopher Lane, and THE LOSS OF SADNESS: HOW PSYCHIATRY TRANSFORMED NORMAL SORROW INTO DEPRESSIVE DISORDER by Allan D Horwitz and Jerome C Wakefield. For publication details see the bibliography.

See also: *Advocacy research; Attention Deficit Hyperactivity Disorder; Dyslexia*; and *Health scares*.

---

encap

Visit www.mencap.org.uk/

Mental Capacity Act 2005	1 This Act aims to protect people with *learning disabilities* (qv) and mental health conditions such as Alzheimer's disease. 2 As of 2008, useful references include the following. For publication details see the bibliography. • MENTAL CAPACITY ACT 2005: CODE OF PRACTICE. 3 Useful websites include: • Directgov www.direct.gov.uk/ See also: *Learning difficulty*; and *Learning disability*.
Mental disability	See: *Mental impairment*.
Mental health care	1 For information on health care policy visit www.dh.gov.uk/en/index.htm 2 As of 2008, useful references include the following. For publication details see the bibliography. • MENTAL HEALTH TEN YEARS ON: PROGRESS ON MENTAL HEALTH CARE REFORM by Louis Appleby. 3 Useful websites include: • Directgov www.direct.gov.uk/
Mental Health Trusts	See: *NHS structure*.
Mental impairment	1 A mental impairment is a reduced functioning of some aspect of the mind. 2 Under the Disability Discrimination Act (qv), which outlaws discrimination against *disabled people* (qv), a mental impairment which has a substantial and long-term adverse effect on a person's ability to carry out normal day-to-day activities is recognised as a *disability* (qv), and a person thus disabled comes under the protection of the Act. As of 2008 the Act does not provide a list of impairments that are covered, but considers the effects of an impairment on a person. Thus someone with a mild form of depression with only minor effect may not be covered, while people suffering from severe depression with substantial effects on their daily lives are likely to be considered as *disabled* under the Act. The *charity* (qv) Mind publishes a list of mental impairments - visit www.mind.org.uk/Information/Booklets/Understanding/index.htm 3 For further details see: *Impaired understanding*. 4 As of 2008, useful websites include: • Directgov: disabled people www.disability.gov.uk/ • Equality & Human Rights Commission www.equalityhumanrights.com/ • National Institute for Mental Health www.nimhe.csip.org.uk/ • Mind www.mind.org.uk/ See also: *Disability: definition in the Disability Discrimination Act; Impaired understanding; Learning disability*; and *Mental Capacity Act*.
Minister for Disabled People	See: *Office for Disability Issues*.
Minister for Women	See: *Government Equalities Office*; and *Women and equality Unit*.
Ministry of Justice	1 This government department (qv) was created in May 2007 to replace the now-defunct Department for Constitutional Affairs. The latter was, during its lifetime, responsible for: • Justice. Responsibilities in this area included running the courts and improving the justice system.

• Rights. Responsibilities in this area included *human rights* (qv) and *freedom of information* (qv) and information rights law.
• Democracy. Responsibilities included law and policy on running elections and modernising the constitution.
The remit of the new Ministry of Justice includes all of the above, plus several areas taken over from the slimmed-down Home Office (qv). The latter include all responsibility for criminal law and sentencing; reducing re-offending; and *prisons* (qv) and *probation* (qv).
2 JUSTICE – A NEW APPROACH by Lord Falconer, Lord Chancellor and Secretary of State for Justice (see the bibliography), sets out in some detail the objectives of this Ministry, which are summarised as being 'to improve the justice system for the public'. This is good news – but only if the judge and jury of what is wanted by 'the public' will be the public itself, and not the usual closed circle of politicians, civil servants, *experts* (qv), *pressure groups* (qv) and *think tanks* (qv). For more on this see: *Post-democracy*.
3 For more information visit www.justice.gov.uk/

See also: *Civil liberties; Crime; Human rights*; and *Privacy*.

---

bile hoists See: *Hoists*.

---

bility aids and uipment 1 Mobility aids and equipment can help two groups of people:
• First, those with *impaired mobility* (qv). Such individuals can be helped to move about more easily by means of a range of aids as shown in the suggested taxonomy below.

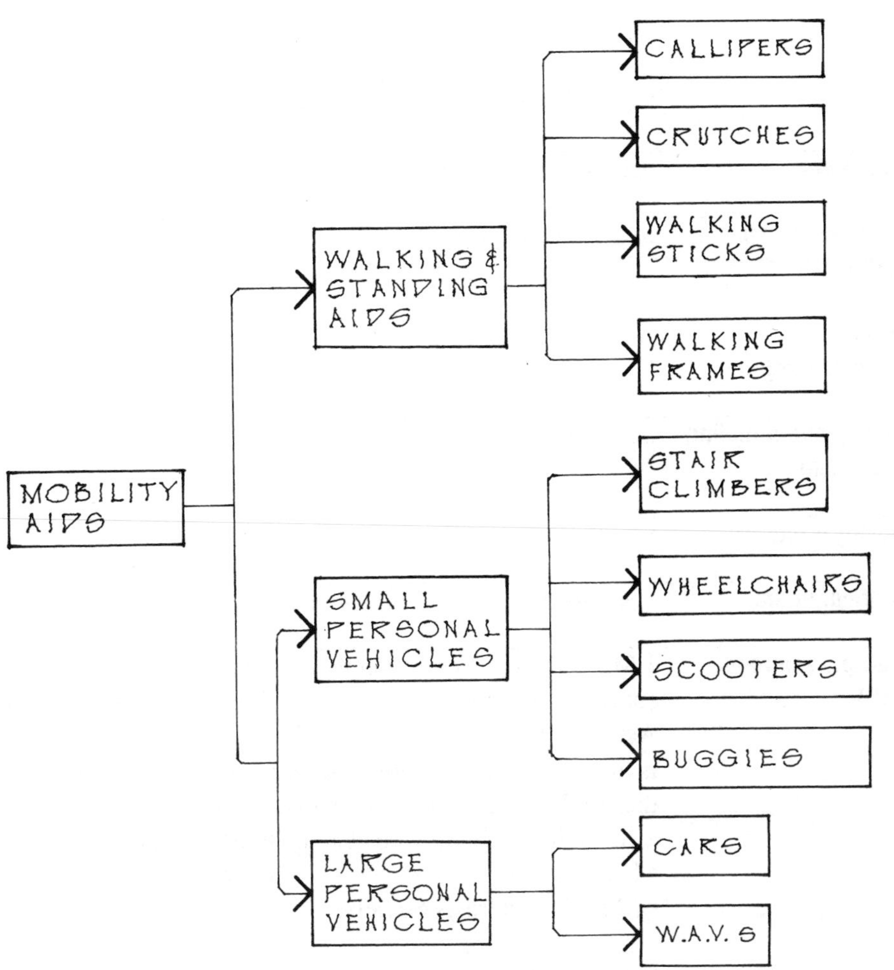

• Second, those with *impaired vision* (qv). Here the function is not to overcome difficulty in movement as above, but to overcome difficulties in *wayfinding* (qv). See therefore entries such as: *Canes for blind people (qv); Guide dogs* (qv); and *Talking signs (qv).*
2 As of 2008, useful websites include the following. They are given in alphabetical order.

• Disabled Living Foundation	www.dlf.org.uk/
• OTDirect	www.otdirect.co.uk/link-equ.html#Platform
• Ricability	www.ricability.org.uk/

See also: *Aids and equipment; Assistive technology*; and *Impaired mobility.*

Mobility and inclusion unit (MIU) See: *Department for Transport.*

Modernising Medical Careers See: *IT systems.*

Moon 1 Like *Braille* (qv), Moon is a system of *tactile* (qv) text that enables *blind* (q people to read by touch. But whereas Braille is based on six raised dots whicl can be used in various combinations to form 63 different symbols, Moon is made up of 14 characters consisting of lines and curves, some of them simplified versions of the alphabet.
2 Moon is generally regarded as useful to those people, usually older persons who lost their sight in later life, who find the more complex dot-based Braille too difficult to master.
3 For more information visit www.rnib.org.uk/

See also: *Blindness*; and *Braille.*

Motorcars See: *Accessible cars.*

Motorway service stations See: *Service stations.*

Move-on accommodation Visit http://england.shelter.org.uk/advice/advice-182.cfm

Moving and handling equipment See: *Hoists.*

Moving walkways See: *Travelators.*

Multiculturism See: *Diversity: alternative models.*

Multiple sclerosis (MS) 1 MS is the most common neurological disorder among young adults, and is specifically recognised as a *disability* (qv) under the Disability Discriminatior Act (qv). This recognition is effective from the time of diagnosis, and not – as is the case with most other conditions – from the point at which the effects of the condition have become 'substantial'.
2 As of 2008, useful websites include the following.

• Directgov	www.direct.gov.uk/
• Multiple Sclerosis Society	www.mssociety.org.uk/

See also: *Disability Discrimination Act.*

lulti-sensory nvironments

1 There are rooms fitted out with a rich variety of soft play equipment, tactile surfaces, visual displays, soft music, and imaginative sound and light effects to provide a safe but stimulating environment for people with profound and multiple *disabilities* (qv). Many of the enjoyable objects and special effects may also have educational and developmental functions. There are special products for particular groups, eg children or older people.
2 As of 2008, useful websites include the following. They are given in alphabetical order.

• Rompa	www.rompa.com/
• SpaceKraft	www.spacekraft.co.uk/

See also: *Children: equipment*.

lulti-sensory npairment

See: *Deafblindness*.

useum design: omplying with the isability iscrimination Act

See: *Exhibition design*; and *Exhibition spaces: accessible*.

useum management: omplying with the isability iscrimination Act

See: *Exhibition management*; and *Exhibition spaces: accessible*.

useums: selected ase examples

1 The Adapt Trust (qv) makes regular awards to arts venues that are considered to be especially meritorious in terms of overall *accessibility* (qv). Award winners in the years up to 2003 included:
• Grosvenor Museum, Chester.
• National Museum of Photography, Film and Cinemas, Bradford.
• Vintage Railway Carriage Museum, Keighly.
• National Railway Museum, York.
• Potteries Museum and Art Gallery, Stoke on Trent.
Other examples of merit include:
• Portland Basin Museum, Ashton under Lyme.
• Slough Museum, High Street, Slough.
• Royal Naval Museum, Portsmouth.
2 The Disability Rights Commission (qv) has recommended the virtual tour at the Tower of London as a good example of how modern technology can make medieval buildings accessible to disabled people.
3 The following museums were shortlisted for the 2004 RIBA Inclusive Design Award (see: *Access awards*):
• Manchester Museum, Manchester.
• Horniman Museum, London SE 23.
For other examples visit www.mla.gov.uk/
4 As of 2008, useful websites include:

• CABE	www.cabe.org.uk/default.aspx?contentitemid=42

See also: *Access awards*.

N

tional Affordable mes Agency	See: *Housing Corporation.*
tional Audit Office AO)	1 The remit of the NAO is to audit the financial statements of all *government departments* (qv) and *quangos* (qv). Visit www.nao.org.uk/ 2 See also: *Audit Commission.*
tional Care andards mmission (NCSC)	1 This is the independent regulatory body responsible for inspecting and regulating almost all forms of residential care and other voluntary and private care services in England. 2 For more information visit www.nice.org.uk/page.aspx?o=32522 See also: *Care Standards Act; Commission for Social Care Inspection; General Social Care Council*; and *Social care.*
tional Centre for lependent Living CIL)	1 An organisation which supports a system of direct payments to disabled (qv) people, thus putting them in charge of their own care and giving them a greater degree of *independence* (qv). Visit www.ncil.org.uk/ 2 For more information visit www.ncil.org.uk/ See also: *Independent Living.*
tional Children's tabase	See: *ContactPoint.*
tional Council for luntary ganisations (NCVO)	Visit www.ncvo-vol.org.uk/ and see: *Voluntary and Community organisations.*
tional Disabled lice Association DPA)	Visit www.ndpa.info/
tional Disability ts Forum (NDAF)	Visit www.ndaf.org/
tional DNA tabase	1 This database, which is a tool in the fight against *crime* (qv), forms part of the Government's dysfunctional *Transformational Government strategy* (qv). As of February 2008 it holds, without their consent and typically for the rest of their lives without remission, the DNA samples of over 4 million people – ie more than 1 in 20 of the British population compared with 1 in 100 for the EU and 1 in 200 for the USA. Around 500 000 of these people have never been convicted of any crime, and 100 000 of the innocents are children. 2 This Orwellian situation has been reached without public consent, and with almost no Parliamentary (qv) debate. Milestones from where we started to where we are now include the following: • Originally police could take DNA samples only from people charged with a crime, and had to destroy the sample if the charge was dropped or the person cleared. • In 2000 the Home Office (qv) scrapped the requirement to erase data. • In April 2004 the Government changed the law to allow the police to forcibly take DNA samples from anyone being arrested for any recordable offence, without requiring permission. The profiles can then be kept permanently, even if the person is acquitted of all charges. This provision includes children above the age of 10 who have committed minor crimes. According to an analysis published in March 2008 by Action on Rights for Children (qv) and

Genewatch, the profiles of more than 1.1 million 10 to 18 year-olds were add to the database between 1995 and 2007, and the figure could rise to 1.5 millio by March 2009. As Ms Terri Dowty, director of ARC, said in 2007: 'We are turning thousands of innocent children into lifelong suspects'.

3 In January 2008 Professor Sir Alec Jeffreys, an inventor of DNA profiling, sharply attacked this constant widening of the net. He protested that a databas that was originally intended to catch convicted criminals if they re-offended now contained the data of 820 000 innocent people.

4 Many people would accept the above measures in the cause of fighting crim if they could be certain (a) that their data was held with complete security, an (b) that conviction on the basis of DNA evidence is completely reliable. On th first point, see: *Privacy*. On the second, DNA profiling is generally accurate, but not completely so. In 1999 Mr Raymond Easton of Swindon, whose DNA had previously been taken after a family dispute and then held on the police database, was arrested for a burglary in Bolton. The case against him was that a DNA sample found at the scene of the crime matched his own on 6 points – what the police called 'a 37 million to one chance'. After the accused proteste that he had never been to Bolton, and demonstrated that he suffered so badly from Parkinson's disease that he could barely dress himself, still less commit a burglary, the samples were compared more rigorously at a further four points of identification. This time they failed to match and the case was thrown out of court. So much for a '37 million to one chance'. Other accused people, who do not have a medical condition that calls a DNA sample into question, may not be lucky enough to have further tests carried out: the words 'DNA', 'latest scientific techniques', and 'expert witness' are so persuasive that DNA evidence tends to be seen by juries (and even judges) as absolutely definitive.

5 As of 2008, useful references include the following. They are given in date order. For publication details see the bibliography:

- DNA FILE ON 100,000 INNOCENT CHILDREN by Philip Johnston, published in th Daily Telegraph on 23 May 2007.
- INNOCENT – BUT ON A CRIMINAL DATABASE by Philip Johnston, published in the Daily Telegraph on 28 May 2007.
- THE FORENSIC USE OF BIOINFORMATION: ETHICAL ISSUES, published in September 2007 by the Nuffield Council on Bioethics. This authoritative paper recommend that the Government must (a) prohibit the police from storing the profiles of innocent people on the national DNA database; and (b) drop plans to extend poli powers to the taking of DNA samples from people suspected of minor offences such as littering or speeding. It argues that there is no evidence that keeping the profiles of people other than active proven criminals 'had any discernable effect' on crime, and that the money would be better spent on recovering DNA traces from crime scenes, which now happens in only 20% of cases.
- MIXED RESULTS – A NATIONAL DNA DATABASE COULD WELL INCREASE, NOT REDUCE, THE NUMBER OF WRONGFUL CONVICTIONS by Professor Allan Jamieson, director of the Forensic Institute in Glasgow.
- The National DNA Database publishes annual reports that may be found at www.homeoffice.gov.uk/science-research/using-science/dna-database/

6 Useful websites include the following:

• GeneWatch UK	www.bbc.co.uk/dna/actionnetwork/A241004
• Liberty	www.liberty-human-rights.org.uk/
• No2ID	www.no2id.net/
• Police National Computer	www.pito.org.uk/products/pnc.php

See also: *Children: fingerprinting by schools; IT systems; Police National Computer; Privacy*; and *Transformational government*.

ational Health Service See: *NHS.*

ational Housing and anning Advice Unit 1 A *quango* (qv) created in June 2007 with the remit of increasing housing supply and affordability – visit www.workfornhpau.co.uk/
2 For more on this subject see: *Housing: affordability.*

ational Housing ederation (NHF) Visit www.housing.org.uk/ and see: *Housing: affordability.*

ational Identity egister 1 The NIR is the database for the National Identity Scheme (qv). As of 2007 it is intended that the NIR will ultimately hold certain personal and biometric details of every adult in the UK. The data will be the stored on three separate, already-existing databases – the Department of Work and Pensions' Customer Information Service, which holds national insurance records; computers in use by the Home Office (qv); and computers in use by the Identity and Passport Service (qv). The records will contain personal information such as name, address, gender, date and place of birth, immigration status, fingerprints, and facial image. It is stated that they will not include information on ethnic origin, medical records, tax records, or religious beliefs.
2 Public and private sector organisations will be able to check the information held on the National Identity Register to help them establish the identity of their customers and staff. For example, people may be asked to prove their identity when opening a bank account or registering with a doctor.
3 For latest information visit www.identitycards.gov.uk/scheme-what-run.asp#nir

See also: *ID cards; Identity Cards Act 2006; Civil rights; IT systems; Privacy*; and *Transformational government.*

ational Identity cheme 1 This highly controversial scheme, which forms part of the Government's ill-starred *Transformational Government strategy* (qv), is intended to introduce ID cards (qv) for cards (qv) for – eventually – all UK residents over the age of 16. Each card will contain the owner's date and place of birth, gender, immigration status, and a digital photograph. It will also have an encrypted microchip containing two fingerprints and a digital facial scan. At checkpoints the chip will be compared against the National Identity Register (qv) which will contain 49 separate items of personal information including all 10 fingerprints. Holders will be expected continually update these 49 items, and (read this sentence slowly) card-holders who fail to notify the authorities each time they change their address will be subject to fines of up to £1 000 per offence.
2 The system is claimed to be 'completely secure' and 'an essential tool in the fight against terrorism'. Author's comments:
• In the three months after October 2007 all claims of government databases being secure were torn to shreds by news that (inter alia) HM Revenue and Customs had 'lost in the post' a CD containing the unencrypted personal details of every child in the country, together with the bank account and NI numbers of their parents or guardians (that's a total of 25 million individuals); that the Dept for Transport had 'lost' a hard disc containing the records of 3 million drivers; and that a Ministry of Defence laptop computer containing the sensitive details of 600 000 military recruits had been stolen from a car – for more on this see: *Privacy*. The chances of the largest database of all being 'completely secure' are nil. For an expert comment refer to ID CARDS ARE THE ULTIMATE IDENTITY THEFT. COMPUTER SYSTEMS ALWAYS FAIL – AND THE NATIONAL DATABASE WILL DO SO BIG TIME by Dr Ian Angell (see the bibliography).

• Instead, therefore, of being an effective tool in the fight against terrorism, a central database could be the very opposite. A store containing all the persona information on every UK citizen in 3 databases, as is proposed, would be a goldmine of information for criminals and terrorists once they had managed t hack into it or once a disc had fallen in their hands.
3 Regular, well-informed commentaries on these matters will be found in Hen Porter's column in the *Observer* on Sundays; and Philip Johnston's column in the *Daily Telegraph* on Mondays.
4 As of 2007, useful websites include the following. They are given in alphabetical order.
• National Identity Scheme www.identitycards.gov.uk/scheme-what.as
• No2ID www.no2id.net/

See also: *ID cards; Identity Cards Act 2006; Civil rights; IT systems; Privacy* and *Transformational government.*

National Information Forum (NIF)

1 An organisation that promotes access to information as a basic right, and as means to empowerment for people who are *disabled* or in any way disadvantaged.
2 For more information visit http://nif.org.uk/

National Institute for Mental Health in England (NIMHE)

1 A body set up to provide research and expertise to help the NHS implement the national service framework on mental health.
2 For more information visit www.nimhe.csip.org.uk/

National Key Scheme (NKS)

See: *RADAR key.*

National Literacy Strategy

See: *Literacy.*

National Lottery

1 The National Lottery was created by Prime Minister John Major in 1994 to generate funds for five 'good causes' – the arts; sport; heritage; the communit (i.e. charities), and – until August 2001 – a variety of millennium projects. Th aim was to foster a golden age of culture and sport for Britain.
2 Many excellent things have been achieved, one of the most important being huge improvement in the *accessibility* (qv) of arts and sports venues, because this has from the start been a precondition for receiving funding. But some commentators do wonder whether these benefits outweigh the fact that many players are poor, unemployed, and/or disabled people desperately hoping to change their luck despite the fact that the odds are stacked heavily against the (the £20 billion that has reportedly been raised for good causes since 1994 sel evidently represents money lost by unsuccessful ticket buyers). That apart, the following criticisms have been made:
• Mr Major assured Parliament in 1994 that all Lottery-funded investments would be additional to normal tax-funded expenditure, a principle that was confirmed by his successor Tony Blair in 1997 when he said: 'We don't believe would be right to use Lottery money to pay for things which are the Governmen responsibilities'. Yet a year later this pledge was broken by Secretary of State Chris Smith with the creation of a 6th category, the New Opportunities Fund (visit www.nof.org.uk/), to which one eighth of Lottery funds were diverted. In 2001 that rose to a third; in 2005 Mr Blair's government passed a Lottery Act giving Ministers direct control over half of Lottery funding; and by 2006 over £3 billion had been diverted from the original good causes. Since then Secretary of State Tessa Jowell has announced that £2.2 billion will be taken from Lottery

funds to support the 2012 Olympic Games, including £112.5 million from Arts Council England, £161.3 million from the Heritage Lottery Fund, and £99.9 from Sport England. The losers will be village halls and small-scale projects to help ordinary people.

• Second, funds have been misspent. £500 million was poured into one of the greatest financial fiascos of our time, the Millennium Dome, and further millions into the Earth Centre and the National Faith Centre, both of which closed soon after opening. £420 000 was given to Peruvian farmers to help them breed fatter guinea pigs for human consumption, £39 200 to a clown to 'tell us what clowning means in the modern world', and £74 000 to a teacher who said he 'wanted to be a sorcerer', while the widely respected Samaritans, who needed an office for 120 volunteers to take telephone calls from deeply distressed and sometimes suicidal people, were told they wanted too much money and did not meet the needs of ethnic minority target groups. Several lifeboat associations have been turned down for the same reason, and a charity hoping to build a house in the Falklands for British veterans visiting the battlefields has had its application rejected.

• Finally, instead of expeditiously investing the funds passed to them, the distributing bodies are sitting on colossal amounts of money (£1.9 billion in 2007, according to Mr Denis Vaughan, the man who helped launch the idea of a national lottery) which they seem unable to spend owing to administrative inefficiency and government-imposed *red tape* (qv).

3 As of 2008, useful references include the following. For publication details see the bibliography.

• THE LARCENY OF THE LOTTERY by Ruth Lea, published in January 2006, looks into some of the above matters more closely.

4 Useful websites include:

• Lottery Funding www.lotteryfunding.org.uk/

See also: *Voluntary and Community Sector.*

---

ational minimum age

See: *Incomes: minimum wage.*

---

ational Offender lanagement Service

See: *IT systems.*

---

ational Probation ervice

1 This service has its origins in the 1907 Probation of Offenders Act, which put the appointment and pay of probation officers onto a statutory footing for the first time. The then Home Office minister Herbert Samuel told MPs that the measure was needed so that offenders 'whom the court did not think fit to imprison on account of their age, character, or antecedents might be placed on probation under the supervision of these officers whose duty it would be to guide, admonish and befriend them'.

2 Those are noble words, and many of the personnel in the present National Probation Service strive to do noble work, but successive governments have so abused the functions of this overburdened service as to make their task impossible. In a book titled THEY CALL IT JUSTICE David Fraser, a former trainer of magistrates, writes: 'When I joined the probation service in the 1960s, the policy was to target for probation people at the beginning of their criminal careers. The purpose was to divert them from crime, and it made sense. In the 1970s all that changed. The new policy was to divert offenders from prison – to save money.' In an article published in the *Guardian* on 5 July 2007 under the title 100 YEARS OLD AND LOOKING FRAIL Mr Eric Goldby, a Methodist minister who worked as a probation officer from 1970 to 2001, gives his own hard-won

views on the way in which successive Governments have distorted the role an ethos of the probation service.
3 One reason for this misuse of the probation service, which has continued unabated for 30 years, is the failure by all recent governments to build the number of *prison* (qv) places that are needed. This policy has had shocking results. On 15 May 2006 the *Daily Telegraph* published a report by its home affairs correspondent Ben Leapman under the title OFFENDERS ON PROBATION CARRY OUT 10,000 CRIMES A MONTH, which quoted Home Office figures showing that over 200 murders were committed by criminals under probation supervision between 1998 and 2004, and well over 100 000 crimes overall eac year. One hundred thousand crimes a year means a hundred thousand victims year – and these are only the crimes for which the police managed to catch a suspect. As most crimes do not result in an arrest, the true figures are almost certainly very much higher.
4 The Government has in recent years increased funding in an effort to impro matters, but most of the money has been spent on hiring more managers and more consultants, and on an overspending, underperforming IT programme called NPSISS (see: *IT systems*). Little of the money has trickled down to overstretched front-line officers, and according a report in February 2007 ther were a thousand vacancies and an exodus of experienced staff. In the same month Mr Andrew Bridges, the head of HM Inspectorate of Probation (qv), tc CRIMINAL JUSTICE MONTHLY that around 6 173 probation officers were trying tc supervise 210 000 criminals, which was an impossible ratio. In the previous two years almost 100 people had been murdered by criminals supposedly und supervision.
5 As of 2008 useful references, in addition to those already mentioned above, include the following. They are given in alphabetical order. For publication details see the bibliography.
• A LAND FIT FOR CRIMINALS by David Fraser. The author, who spent 26 years working in the probation and prison services, demonstrates that successive governments have misled the public on the true nature of their policies on cri and punishment, with traumatic consequences for millions of victims.
• THEY CALL IT JUSTICE by Brian Lawrence.
• CRIMINAL JUSTICE MONTHLY issue 66.
6 Useful websites include the following. They are given in alphabetical order.
• Civitas www.civitas.org.uk/crime/index.php
• Crime & Society Research Association www.csra.org.uk/victims_of_crime.htm
• National Probation Service www.probation.homeoffice.gov.uk/

See also: *Crime; HM Inspectorate of Probation;* and *Home Office*.

National Register of Access Consultants (NRAC) 1 The UK's only officially recognised register of consultants for carrying out access appraisals (qv), access audits (qv), and access statements (qv).
2 For more information visit www.nrac.org.uk/

National Spastics Society Visit www.scope.org.uk/

National statistics See: *Office for National Statistics*.

National Strategy for Neighbourhood Renewal See: *Neighbourhood renewal*.

tional telephone ay service See: *Typetalk.*

vigation See: *Way-finding.*

ets

1 'Neets' is an acronym for young people who are 'Not in Employment, Education, or Training'. The number of Neets aged from 16 to 18 increased from 153 000 in 1997 to 206 000 in 2006. There are also statistics for other age groups (eg 16 to 25), and it is important to establish exactly which age group is being referred to when comparing one set of figures with another, because they tend to differ substantially.

2 Neets tend to be troubled and often unhappy young people. Males typically spend much of their time in *prison* (qv); drink excessively and use drugs; and die prematurely. Females tend to be single mothers with several children and multiple partners (a background that has devastating consequences for the children – see: *Families*), living on benefits, drinking and smoking to excess, and again dying prematurely. THE COST OF EXCLUSION (see below), a study carried out by the London School of Economics for the Prince's Trust, reported in 2007 that around 8% of Britain's 15–19 year-olds are Neets, compared with 3.4% in France and 4.6% in Germany, and predicted that these unfortunate youngsters are likely to spend their entire future lives on *benefit*s (qv) and in *poverty* (qv).

3 Current attempts to deal with the problem are not working. The Government pledged to cut the number of Neets by 20% by 2010, but between 1997 and 2006 the number of young male Neets rose by 27% to around 575 000, and the number of young female Neets by 6% to 669 000 – a total of 1.24 million wasted lives. The latest Government plan, as of April 2007, is to reclassify hard-case Neets as criminals: from 2013 those who fail to stay on in education or on-the-job training will be served with Asbo-style attendance orders or fined £50. Perhaps this forms part of the answer, but it cannot be all of it. The real challenges include:

• Finding ways of improving the general quality of our state school system, which completely fails to prepare many youngsters for work and employment – see: *Education; Literacy;* and *Social mobility.*

• Offering young people who don't like, or aren't good at, academic subjects a better alternative way forward. In July 2007 the House of Lords economic affairs select committee, a body boasting two former chancellors, some leading industrialists, and several world-class economists, produced a report proposing that – provided they applied themselves and reached the required standard – all 14-year-olds be offered a guaranteed apprenticeship when they leave school. As the *Daily Telegraph*'s economics editor, Liam Halligan, commented: for more academic youngsters such a promise would be irrelevant, but for countless others it could be a lifeline, inspiring them to achieve just at the time when they might otherwise begin to drift. But such a system will only work if the available places are (unlike too many past schemes) relevant, of high quality, and provide genuine training. The Lords' report therefore proposes that within five years all government money for apprenticeships should go directly to employers. If apprenticeship funding – around £3 250 per person per year – were channelled into work-based training, rather than into ineffective state-run schemes, 'employers would have an incentive to provide more places, and also to become more actively involved in devising innovative and effective apprenticeships'.

4 As of 2008, useful publications include the following. They are listed in alphabetical order. For publication details see the bibliography.

• THE COST OF EXCLUSION: COUNTING THE COST OF YOUTH DISADVANTAGE IN THE UK by Sandra McNally and Shqiponja Telhaj of the Centre for Economic Performance at the London School of Economics.
• TOWARDS A GOLD STANDARD FOR CRAFT: GUARANTEEING PROFESSIONAL APPRENTICESHIPS by John Hayes and Dr Scott Kelly. This study from the *think tank* (qv) the Centre for Policy Studies, published in June 2007, claims that more than a million young people have been condemned to joblessness by Britain's devalued apprenticeship system. Twice as many youngsters in Germany and Australia enrol on apprenticeships as in the UK, because they ar considered sure-fire routes to jobs.
• WASTED EDUCATION. This report from the *think tank* (qv) the Bow Group (qv) forms part of the Group's 'Invisible Nation' series, and comments on the tens of thousands of children who disappear from the school registers each yea and effectively vanish from public view. They include persistent truants, expellees, victims of bullying, children on witness protection programmes, an some who are ill or caring for sick parents. Most of these 'invisible children', of whom there are (according to Professor Stephen Heppell, chairman of the Inclusion Trust, around 100 000) have a grim future.
• WELFARE ISN'T WORKING: THE NEW DEAL FOR YOUNG PEOPLE by Frank Field and Patrick White. This report from the *think tank* (qv) Reform shows that the number of unemployed young people was higher in 2007 than in 1998, and analyses why.
5 Useful websites include:

• Every Child Matters	www.everychildmatters.gov.uk/ete/neet/
• Frank Field	www.frankfield.co.uk/type3.asp?id=20&type=3
• SmartJustice	www.smartjustice.org/

See also: *Education; Employment; New deal; Social exclusion; Social mobilit Underclass*; and *Workfare*.

Neighbourhood Nursery Initiative (NNI)

1 The Neighbourhood Nursery Initiative was launched in 2000 to expand *childcare* (qv) provision in the 20% most disadvantaged areas in England. The original target was for 45 000 new childcare places for children aged 0 to 5 to be opened by March 2004.
2 Visit www.dfes.gov.uk/rsgateway/DB/RRP/u014486/index.shtml

See also: *Childcare; and Education: pre-school.*

Neighbourhood renewal

1 The Government's National Strategy Plan for neighbourhood renewal is contained in A NEW COMMITMENT TO NEIGHBOURHOOD RENEWAL: A NATIONAL ACTION STRATEGY PLAN, a 127 page report prepared by the Social Exclusion Unit (qv) – see the bibliography. Implementation of the Strategy in England is being overseen by the Neighbourhood Renewal Unit (qv).
2 As of 2008, the numerous administrative units, initiatives, and programmes that may be involved include:
• The 'Active Community Unit' (ACU) is a unit in the Home Office which aims to promote the development of the *voluntary and community sector* (qv) and encourage people to become actively involved in their communities, particularly in deprived areas. Visit www.cabinetoffice.gov.uk/third_sector/
• 'Anti Poverty Strategies' (APS) promote a co-ordinated approach to tackling poverty, including programmes to help people claim *benefits* (qv), manage del etc. Visit www.neighbourhood.gov.uk/glossary.asp
• 'Area Based Regeneration' programmes encourage partners to co-operate, targeting their resources, in areas where economic, social, and environmental

problems combine to lock local communities into a vicious cycle of *exclusion* (qv). Visit www.communities.gov.uk/index.asp?id=1128622
• 'Area Investment Frameworks' (AIF) set out the regeneration priorities for an area with the aim of targeting funding from Regional Development Agencies (qv). Visit www.neighbourhood.gov.uk/glossary.asp
• 'Neighbourhood management programmes' encourage *stakeholders* (qv) to work with service providers to help improve the quality of services in deprived neighbourhoods. Visit www.neighbourhood.gov.uk/page.asp?id=577
• The 'Neighbourhood Renewal Fund' used to provide public services and communities in the 88 poorest *local Authority* (qv) districts with extra funds to tackle deprivation. It was replaced in November 2007 by the Working Neighbourhoods Fund. Visit www.communities.gov.uk/publications/communities/workingneighbourhoods
• 'Neighbourhood Renewal Strategies' embody the strategic aims and targets for a Local Strategic Partnership (qv) that is receiving Neighbourhood Renewal Funding (qv). Visit www.neighbourhood.gov.uk/page.asp?id=535
• 'Neighbourhood Support Funds' are government grants of £10 000 or more to community groups, to help them re-engage disaffected young people. Visit www.dfes.gov.uk/nsf
• 'Neighbourhood wardens' provide a uniformed, semi-official presence in residential areas. They can assist with community safety; *housing* (qv) management; and community development. They may patrol, provide concierge duties, or act as super-caretakers and help to support vulnerable residents. Visit www.neighbourhood.gov.uk/page.asp?id=567
• 'New Deal for Communities' is a programme to regenerate 39 very deprived areas across England over a 10-year period. It is implemented by New Deal Partnership Organisations made up of local communities, voluntary organisations, businesses, and public services. For more information visit www.neighbourhood.gov.uk/page.asp?id=617
3 A NEW COMMITMENT TO NEIGHBOURHOOD RENEWAL: A NATIONAL ACTION STRATEGY PLAN (see the bibliography) refers to many other matters and is essential reading for anyone trying to get an overview of the bewildering array of policies that affect local communities in England.
4 As of 2008, other useful references include the following. For publication details see the bibliography:
• THE ENGLISH INDICES OF DEPRIVATION 2004. This is an official tool used by the Government for identifying areas of poverty in England, which require remedial action.
5 Additional useful websites include:
• CABE www.cabe.org.uk/default.aspx?contentitemid=44

See also: *Poverty in the UK; Safer and Stronger Communities Fund*; and *Social exclusion*.

Neighbourhood Renewal Unit (NRU)

1 The NRU is responsible for overseeing the Government's comprehensive neighbourhood renewal strategy – visit www.neighbourhood.gov.uk/page.asp?id=3
2 Ten studies published by the Joseph Rowntree Foundation (qv) in September 2005 suggested that the government was still failing to increase the prospects of poor areas, despite the work of the NRU – see: *Social inequality*.
3 As of 2008, useful websites include:
• NRU www.neighbourhood.gov.uk/page.asp?id=3
• Neighbourhood renewal www.neighbourhood.gov.uk/

See also: *Regeneration: area-based*; and *Social exclusion*.

Neighbourhood statistics

1 This Office of National Statistics (qv) website gives detailed information on every local neighbourhood in the UK. It is extremely useful and deserves t be better known and more frequently visited.
2 For more information visit www.neighbourhood.statistics.gov.uk/

Neighbourhood wardens

1 Neighbourhood wardens provide a uniformed, semi-official presence in residential areas with the aim of improving the quality of life. They can promo community safety, assist with environmental improvements and housing management, and also contribute to community development. They may patro the area, provide concierge duties, or act as super-caretakers and support vulnerable residents.
2 For more information visit www.neighbourhood.gov.uk/page.asp?id=567

See also: *Street wardens.*

Neighbourhood Watch

1 Six million households in the UK are members of 170 000 Neighbourhood Watch anti-crime schemes. The bulk of the costs are met from volunteers' pockets.
2 For more information visit www.neighbourhoodwatch.net/

See also: *Crime.*

Neighbourhoods

1 Official regeneration initiatives try to target deprived areas as defined by local people, rather than by administrative boundaries. A *neighbourhood* of th kind might for instance include a council estate, or an inner city area centred around a communal facility such as a shopping centre.
2 Useful websites include:

- CABE www.cabe.org.uk/default.aspx?contentitemid=44

New Commitment to Neighbourhood Renewal

See: *Neighbourhood renewal.*

New deal for Employers

1 This is a key part of the Government's Welfare to Work strategy for bringing down unemployment. It pays employers to take on unemployed people and ha several strands – 'New deal for young people'; 'New deal for lone parents'; 'New deal for disabled people', 'New deal 50 plus', etc.
2 Results overall have been disappointing. In August 2007 official data showe that the New Deal had become an expensive revolving door instead of the intended path to steady employment. Around 543 000 people had been throug the scheme twice; 171 000 three times; and 11 900 five times.
3 For young people the scheme has been a complete failure, with more young people out of work in 2007 than in 1997, when chancellor Gordon Brown announced that his ambition was 'nothing less than the abolition of youth unemployment'. According to WELFARE ISN'T WORKING: THE NEW DEAL FOR YOUNG PEOPLE by Frank Field (qv) and Patrick White, published in June 2007 (see the bibliography) 14.4% of under-25s were unemployed in 1997, and 14.5% in 2007.
4 As of 2008, useful websites include:

- Directgov www.direct.gov.uk/
- Frank Field www.frankfield.co.uk/type3.asp?id=20&type=3
- Jobcentre Plus www.jobcentreplus.gov.uk/

See also: *Neets;* and *Workfare.*

New Deal for Communities See: *Neighbourhood renewal.*

New Opportunities Fund (NOP) See: *National Lottery.*

New Urbanism See: *Urban design.*

NGOs See: *Quangos.*

NHS 1 The NHS is a tax-funded national health service that is free at the point of use to all citizens. It is not of course free in reality – its cost in 2008 will be over £100 billion, i.e. more than £1 700 per man, woman and child, which for a two-child family equals a median monthly family healthcare expenditure of over £500. But knowing that anyone falling ill or meeting with an accident is eligible for treatment with no questions asked, for as long as it takes, is one of the most highly prized aspects of British life – and rightly so. Private insurance-funded systems have a role in healthcare, but these schemes have little interest in people with long-term health problems and will go to great lengths to avoid taking them on, and to shed responsibility for members who are found to have a terminal or a long-extended illness.

2 Against that, most people also believe that the NHS has in recent years wasted very large sums of money – and again they are right. Key reports include the following. They are given in date order.

- A BETTER WAY published in 2003 by the *think tank* (qv) Reform, contains proposals from Reform's Commission on the Reform of Public Services for improvement in education, healthcare, and crime.
- PLUNDERING THE PUBLIC SECTOR by David Craig, a consultant turned whistleblower, was published in 2006 and estimated that £70 billion had been wasted.
- In January 2007 John Appleby, chief economist of the *think tank* (qv) the King's Fund, told the BBC that while NHS funding had grown from £34 billion to £91 billion since 1997, only 30% of this had been spent directly on patient care. Consultants' pay rose by 70% while their productivity fell by 20%. Visit www.bbc.co.uk/radio4/today/reports/politics/nhs_cake_20070118.shtml

3 Precise figures may be disputed, but not the general message. While there have undoubtedly been major improvements in the NHS over the past decade and more, they have depended on the proportion of the UK's GDP that is spent on health rising from 6.5% in 1997 to an expected 9% or more in 2008 – a huge increase in real terms. But while the money was poured in, the inherited structure was not adequately reformed, and improvements will be very difficult to maintain when money becomes tighter. In January 2007 Sir Michael Rawlins, head of the National Institute for Health and Clinical Excellence (NICE), stated that healthcare spending would have to rise above 9.3% of GDP in the future to deal with medical inflation and the ageing population; while at the same time economists were predicting that the annual growth in the NHS budget after 2008 must inevitably sink back to 3% or less in real terms after six years at 7% or more.

4 The British Medical Association stated in April 2007 that politicians must stop pretending that the NHS can provide the entire population with the entire range of healthcare services free of charge. It suggested that while the NHS must remain free and available to everyone, the particular services offered on this basis would need in future to be prioritised. It called for an open debate on these matters.

5 Meanwhile, while long-term strategies are debated, there should be no delay in tackling the immediate problems of extreme over-regulation, over-management, and over-complexity. As instances:

- In March 2007 a report titled THE BUREAUCRATIC BURDEN IN THE NHS, the NHS Confederation, a body that represents NHS managers in 90% of Trusts, charged that the NHS is 'having the life choked out of it' by a burden of inspection and regulation that is not shrinking, as claimed by the Government, but rapidly growing. See: *Reducing burdens.*
- In April 2007 an investigation by the Picker Institute, a charity which is regularly commissioned by the NHS to survey its services, found that the NHS had become so bafflingly complex that the Institute's team of 'mystery shoppers' – who, unlike most patients, were experienced informatio searchers who knew about the services they were trying to access – found it virtually impossible to discover where to get these services.

6 As of 2008, useful references on NHS reform include the following. They a given in date order. For publication details see the bibliography:

- A BETTER WAY, published in 2002. See item (2) above.
- PLUNDERING THE PUBLIC SECTOR, published in 2006. See item (2) above.
- NHS REFORM: THE EMPIRE STRIKES BACK by Professor Nick Bosanquet, Henry de Zoete, and Andrew Haldenby, published in January 2007 by the *think tank* (qv) Reform.
- THE BUREAUCRATIC BURDEN IN THE NHS, published in March 2007. See item (5) above.
- AN INDEPENDENT NHS: A REVIEW OF THE OPTIONS, by Professor Brian Edwards head of healthcare development at Sheffield University, published in May 200 by the Nuffield Trust (qv)
- A RATIONAL WAY FORWARD FOR THE NHS IN ENGLAND, published in May 2007 b the British Medical Association, suggests that the Government should set the amount of money and direction of the NHS and leave the running of the NHS to a board of governors appointed by and accountable to Parliament. These are sound ideas, but the report has been criticised for aiming to shift power to medical professionals (for whom the BMA acts as trade union) instead of patients.
- REALISING THE BENEFITS? ASSESSING THE IMPLEMENTATION OF 'AGENDA FOR CHANGE' by James Buchan and David Evans published in July 2007, by the *think tank* (qv) the King's Fund, charges that the introduction of the 'Agenda for Change' pay reform programme, which increased the NHS wage bill to £3 billion a year, was rushed through; that it exceeded all cost estimates; and that 'patients will never feel the intended benefits from the biggest NHS pay refor in UK history unless health service leaders secure changes in the working patterns and productivity of more than a million nurses and other staff'.
- OUR FUTURE HEALTH SECURED? A REVIEW OF NHS FUNDING AND PERFORMANCE by Sir Derek Wanless and others, published in September 2007 by the *think tank* (qv) the King's Fund.
- EURO HEALTH CONSUMER INDEX 2007 published in October 2007 by the Health Consumer Powerhouse in Brussels finds that despite recent huge investment in the UK's health service, we have since 2006 slipped from 15th t 17th of 29 European countries in quality of healthcare. We are well behind all of Europe's rich countries, and are also behind many of the poorer ones such as Estonia, the Czech Republic, and Cyprus. The report identifies two types o healthcare system in Europe – 'Bismarckian' (based on social insurance, with a multitude of insurance organisations which are independent of healthcare providers) and 'Beveridgian' (where financing and provision are handled with one organisation). The top 5 countries, grouped quite closely together near the top of the scale, all have Bismarckian systems. They are, in descending order, Austria, Netherlands, France, Switzerland, and Germany. There is then a gap,

followed by the best-performing Beveridgian country, Sweden, in 6th place; and below that Norway, Finland, Denmark, etc. The UK comes 17th, ahead of Italy, Portugal, Slovenia, and Greece.

• WHY THE NHS IS THE SICK MAN OF EUROPE by James Gubb, published in March 2008 by the *think tank* (qv) Civitas, asserts that NHS is overwhelmed by Government targets, and is steadily falling behind other European countries when we should be learning from them. It finds that their social insurance schemes achieve better results than the NHS, with those who can afford it paying for their health care, and those who cannot afford it being subsidised by taxpayers. Two years ago the Dutch were facing similar problems to ours; they responded by, inter alia, cutting state control of health service purse strings. Today they have what some analysts regard as the most streamlined, equitable, and competitive healthcare system in the world.

7 As of 2008, useful websites include the following. They are given in alphabetical order.

• Civitas	www.civitas.org.uk/nhs/index.php
• Institute for Fiscal Studies	www.ifs.org.uk/
• Institute for Public Policy Research	www.ippr.org.uk/
• King's Fund	www.kingsfund.org.uk/
• NHS Confederation	www.nhsconfed.org/
• Reform	www.reform.co.uk/website/health.aspx

See also: *Better regulation bodies; Dental care; Doctors for Reform;* and *Public services: quality.*

[NH]S Care Records [Se]rvice (NHS CRS)

1 The CRS forms a major part of the Government's severely malfunctioning *Transformational Government strategy* (qv). It is a central computerised database that will ultimately – unless it has meanwhile been abandoned as unsalvageable – hold the medical records of every NHS patient in England. These files will contain the most intimate, confidential, and in many cases embarrassing information about individuals and families, and will be accessible to around a quarter of a million officials.

2 This Orwellian project, on which the public was never consulted and for which its consent has not been obtained, has given rise to much disquiet, and indeed vociferous opposition. The problems are threefold:

• First, everyone accepts that our most personal details must be held by our GP, but few people would willingly consent to having those details on a database to which several hundred thousand people may have access. This applies even when every detail is accurate, but the details on huge databases very often are not accurate, particularly if entered by remote inputters working from a variety of information sources of uneven quality. Identities get confused (how many John Smiths are there in England?); notes are misread; etc. It is absolutely certain (a) that a database containing the records of 50 million people will contain millions of errors; (b) that many of these will be damaging to the people concerned; (c) that the patients concerned will often be unaware of the faulty data, and suffer the results without knowing why; and (d) that if patients do discover mistakes, they are likely to get entangled in a Kafkaesque nightmare in trying to get the errors completely erased from all files.

• Second, taxpayers were assured, when work began, that the total cost would be £12 billion. By late 2006 the system was behind schedule, seriously malfunctioning, greatly overspent, and expert commentators were talking of an ultimate cost of £15 or even £20 billion.

• Third, in October of that year 23 leading British computer scientists publicly questioned whether the programme could ever work, and called for an urgent independent inquiry before proceeding any further. In February 2007

Mr Andrew Rollerson, a top executive with one of the system's main suppliers, Fujitsu, told a conference of IT executives that the programme had not been properly thought out; that it 'isn't working, and it isn't going to work'. In April 2007 the House of Commons public accounts committee issued a damning 175 page report. Since then two opinion polls have found that the overwhelming majority of GPs have no confidence that either their own or their patients' records would be safe on the NHS CRS database. The first was carried out by Medix in November 2007; the second (in which 93% of GPs expressed no confidence) by the BMA in January 2008.

3 In short, we have here a multi-billion pound disaster caused by ministerial arrogance, incompetence and impatience. For similar problems with other large government-commissioned computer systems see: *IT systems*.

4 For a gleam of wisdom we must turn to Iceland, where in 2006 the population forced their government to scrap a law empowering officialdom to upload the entire population's medical records onto a database without their consent. The head of Iceland's cancer registry, Laufey Tryggvidottir, stated: 'When data are entered, there are always many errors. Codes are wrongly entered or even misunderstood.' And uttering words which should be memorised by every UK cabinet minister and civil servant he advised that the best way of keeping mistakes few and small, readily detectable, and readily correctable is for medical databases to be kept smallish and as close as possible to the patients whose data is being stored.

5 Writing in the *Guardian* on 14 June 2007 under the title 'The foundations of an NHS IT system are in place: now start building', the IT commentator Michael Cross suggested three steps that might still save this project:

- Replace the present thoroughly undemocratic policy with one that requires patients to 'opt in' rather than 'opt out'.
- Abandon the thoroughly undemocratic policy of assuming patient consent. The British Medical Association wants the system changed so that patients have to opt in rather than opt out.
- Abandon the attempt to replace individual hospital administrative systems with standard packages procured nationally via the disastrous billion-pound 'local service provider' contracts.
- Instead of trying to make the whole convoy move at the same pace, find centres of excellence and help them spread.

6 As of 2008, useful websites include the following. They are given in alphabetical order.

- Foundation for Information Policy Research www.fipr.org/
- Information Commissioner's Office www.ico.gov.uk/
- NHS Confidentiality Campaign www.nhsconfidentiality.org/
- NHS Connecting for Health www.connectingforhealth.nhs.uk/
- NHS Connecting for Health A-Z www.connectingforhealth.nhs.uk/atoz/li
- NHS CRS www.nhscarerecords.nhs.uk/
- No2ID www.no2id.net/

See also: *Connecting for Health; IT systems; Privacy;* and *Transformational government.*

NHS Connecting for Health

1 NHS Connecting for Health is a Department of Health (qv) *quango* (qv) that is responsible for delivering the National Programme for Information Technology (NPIT) in the NHS. The NPIT comprises eight programmes, including the following:

- NHS Care Records Service (qv).
- Choose and Book, which was meant to be rolled out in 2005 but is behind schedule and very variable in performance – quite satisfactory in some areas, dreadful in others.
- Electronic Prescription Service.

• Quality Management and Analysis System.
In addition, NHS Connecting for Health provides services that were delivered by the NHS Information Authority until its closure in April 2005.
2 As of 2008, useful websites include:
• NHS Connecting for Health www.connectingforhealth.nhs.uk/
• NHS Connecting for Health A-Z www.connectingforhealth.nhs.uk/atoz/list

HS Estates

See: *Department of Health Estates and Facilities Management.*

HS structure

1 NHS England is divided into 10 Strategic Health Authorities (SHAs), each of which is responsible for providing the primary and secondary healthcare services in its particular area.
2 Each of the 10 SHAs is in turn split into five types of Trust, all of which are *quangos* (qv). The interrelationship between these bodies is shown graphically on the following diagram.

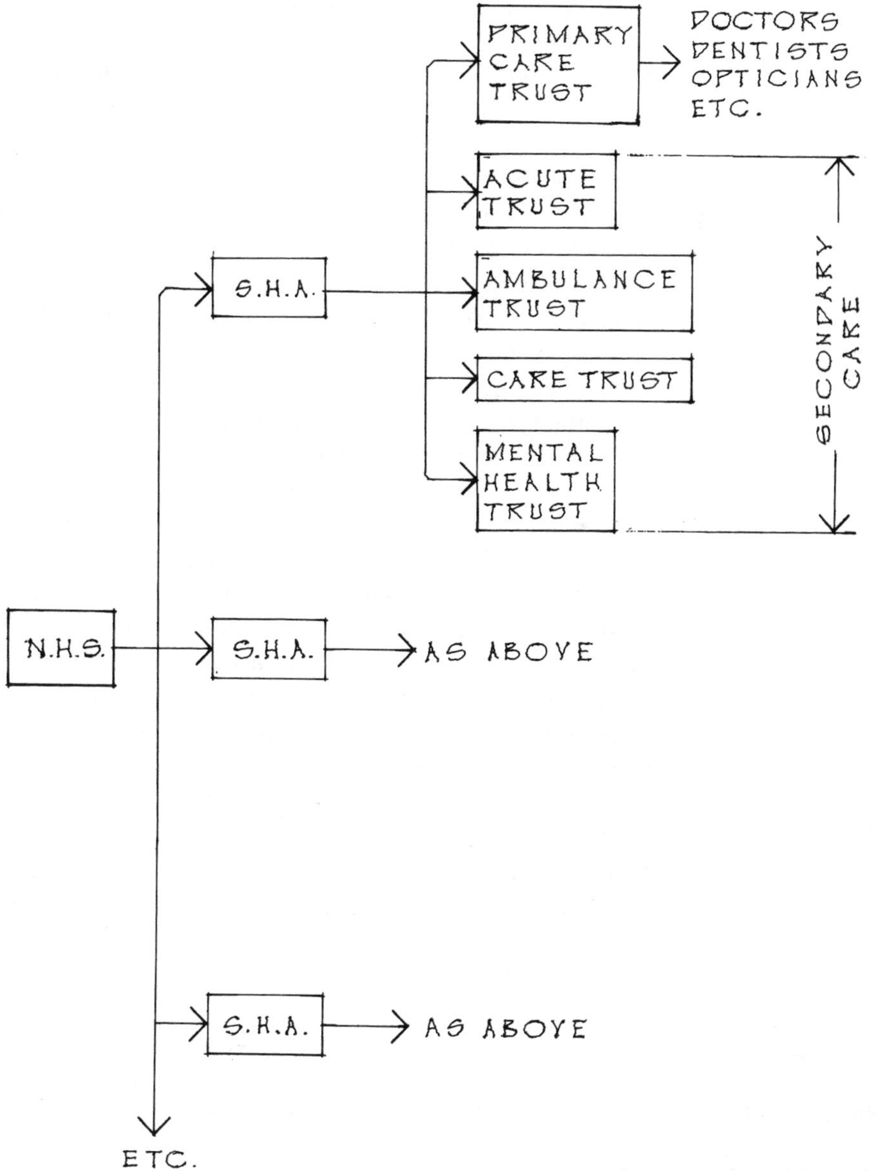

• Primary Care Trusts (PCTs). 'Primary care' is the official term for the care that is provided when people first notice a health problem and visit their local

GP, dentist, optician, or whatever the case may be, and PCTs try to deal with health problems before they become more serious and require 'secondary car for instance in hospitals.

- Acute Trusts, which manage the hospitals in their area.
- Ambulance Trusts, which run the ambulance services in their area.
- Care Trusts, which were introduced in 2002 with the aim of providing bette integrated health and social care. By combining both responsibilities under a single management, it was hoped that they could increase continuity of care ar simplify administration.
- Mental Health Trusts, which provide health and social care services for people with mental health problems.

For more information on the above visit www.nhs.uk/aboutnhs/howtheNHSworks/Pages/HowtheNHSworks.aspx

3 These notes describe the NHS structure as of January 2007, but there can b no assurance that these details will remain valid for long. Since 1997 the NH has undergone at least four (by some counts seven) structural reorganisations. As examples:

- In 2002 the 96 District Health Authorities in England were replaced by 28 Strategic Health Authorities. In 2006 the latter were reduced to 10. These wer huge and extremely costly restructurings.
- In December 2003 the Community Health Councils (which had for 30 years successfully represented patients) were abolished, and thousands of skilled lo volunteers were dismissed. In their place the Government launched with grea fanfare a Commission for Patient and Public Involvement in Health (CPIHH) and a network of 572 Patient and Public Involvement Forums ('Patient Forums') at a cost of millions of pounds.
- Less than three years later, in July 2006, the same Government announced that Patient Forums would be abolished in mid-2007 and replaced by local involvement networks (LINks).

This is an extremely expensive way of running a business. Addressing the *Guardian* newspaper's Public Services Summit in January 2007, Mr Nick Manning, head of public sector management and performance at the Organisation for Economic Cooperation and Development (OECD), accused British governments of a constant and 'mindless promotion of change' despit little proof that the proposed reforms would actually work. He characterised t attitude as 'We know that plan A isn't working, so plan B must be better'. Th speaker was commenting on UK public services in general, but his strictures apply with particular force to the NHS.

4 UK governments should take to heart a quote from Lord Layard, director of the Wellbeing Programme at the LSE Centre for Economic Performance: 'Which country has not restructured its healthcare system since the 1950s? Which country has the highest life expectancy? The answer to both is Japan'.

See also: *Dentistry*.

Noise See: *Acoustics*.

Nolan principles See: *Commission for Public Appointments*.

Non-departmental public bodies
Non governmental bodies See: *Quangos*.

rmal distribution

1 If any human characteristic such as height, weight, or IQ is plotted on a graph, with the characteristic under consideration plotted on the horizontal axis from low (left) to high (right), and the frequency with which that characteristic occurs plotted on the vertical axis, then the result very commonly turns out to be a smooth, symmetrical bell-shaped curve as shown below.

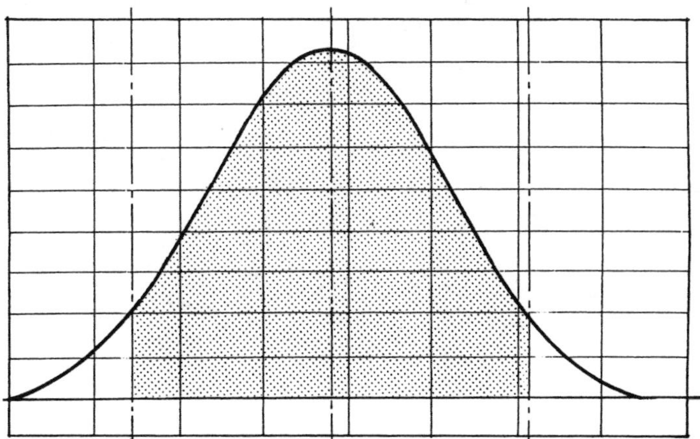

2 This particular curve shows the frequency distribution for the 'stature' (or, in everyday language, height) of adult British men – for more detail see: *Percentiles.* Because this symmetrical bell-shaped curve is so widespread in biology, and particularly in *anthropometry* (qv), it is commonly called the 'normal distribution'. Unfortunately this purely statistical term may be assumed by some to carry hurtful connotations of 'normal' and 'abnormal' people. Some statisticians therefore prefer to use the term 'Gaussian distribution', after the German mathematician Johann Gauss (1777–1855) who first described this kind of distribution.
3 Other curves may be found for other classes of objects or phenomena – curves skewed to the left or right (for an example see: *Income: median*), flat-topped curves, sharply peaked curves, even two-humped curves. They need to be interpreted with care – for more on this see p 19 of bodyspace below.
4 As of 2008, useful references include the following. For publication details see the bibliography.
• BODYSPACE: ANTHROPOMETRY, ERGONOMICS AND THE DESIGN OF WORK by Stephen Pheasant. Refer in particular to Chapter 2.

See also: *Anthropometrics;* and *Percentiles.*

rmal provision

See: *Inclusive design.*

sings

See: *Stairs.*

tification under the ta Protection Act

1 Under the Act every organisation that processes personal information must notify the Information Commissioner's Office (qv), unless they are exempt. Failure to notify is a criminal offence
2 For details of notification requirements and exemptions visit www.ico.gov.uk/what_we_cover/data_protection/notification.aspx

See also: *Data Protection Act;* and *Privacy.*

SISS

See: *IT systems*

AC

See: *National Register of Access Consultants.*

Nuisance tenants	See: *Anti-social behaviour.*
Numeracy	See: *Literacy.*
Nursery care	See: *Childcare;* and *Neighbourhood Nursery Initiative.*
Nursing homes	See: *Care homes for older people.*

esity

1 Obesity is defined in relation to people's Body Mass Index (BMI), and the latter is found by dividing a person's weight (in kilos) by their height squared (in metres). Thus a weight of 69 kilos and a height of 1.75 m gives a Body Mass Index of $69 \div 1.75^2 = 22.54$.

2 Health authorities define a BMI of 18.5 to 24.9 (such as that above) as 'healthy'; 25 to 29.5 as 'overweight'; and over 30 as 'obese'.

3 The graph below, from Figure 7.8 in SOCIAL TRENDS 35 and Figure 7.7 in SOCIAL TRENDS 36 (see the bibliography) shows a steadily growing incidence of obesity among both adults and children in the UK.

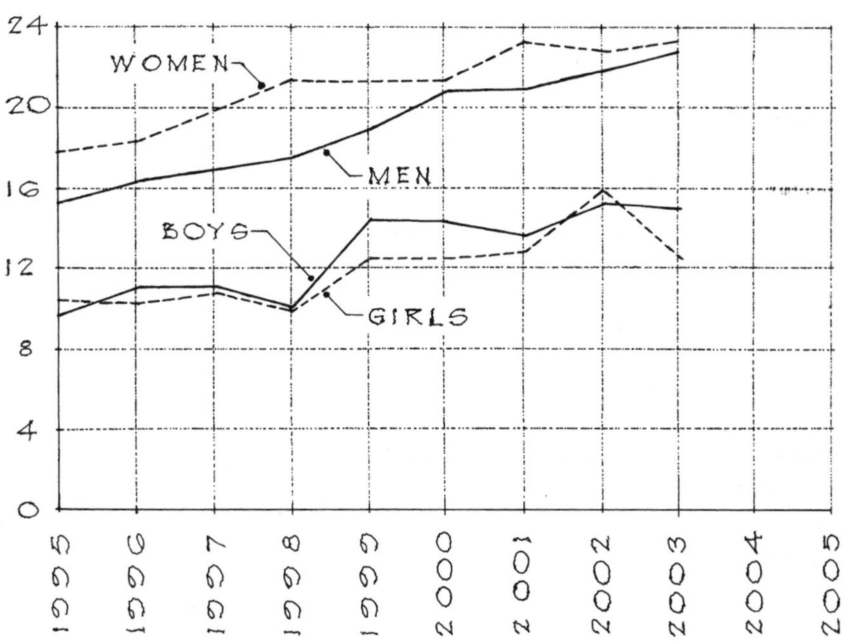

This trend will have two consequences:

- The obese people themselves are likely to suffer a greater incidence of heart disease, diabetes (which is a major cause of blindness), and premature death than would have otherwise have been the case.
- The standard *anthropometric* (qv) tables of body dimensions used by the designers of clothes, chairs, car interiors, and bus, train, and aircraft seats, etc., will need revision as people get larger.

4 As of 2008, useful references include the following. They are given in alphabetical order. For publication details see the bibliography.

- ANALYSIS OF THE NATIONAL CHILDHOOD OBESITY DATABASE 2005–06 by Crowther, Dinsdale, Ruttin and Kyffin.
- CHOOSING EQUIPMENT FOR THE HEAVIER PERSON.

5 Useful websites include the following. They are given in alphabetical order:

- Association for the Study of Obesity www.aso.org.uk/
- Dept of Health statistics www.dh.gov.uk/PublicationsAndStatistics/fs/en
- Dept of Health policy and guidance www.dh.gov.uk/PolicyAndGuidance/HealthAndSocialCareTopics/Obesity/fs/en
- Size UK www.fashion.arts.ac.uk/5913.htm
- Social Issues Research Centre www.sirc.org/
- Social Trends www.statistics.gov.uk/socialtrends

See also: *Anthropometric data*; and *Seating*.

Occupational therapists	1 Occupational therapists or OTs offer practical advice and help to people who are experiencing difficulties with day-to-day tasks such as getting about the house, washing, dressing, cooking, or getting out and about; and more particularly on the most appropriate *aids and equipment* (qv) and *home adaptations* (qv) that may be available. 2 They are distinct from *physiotherapists* (qv) who help people maintain the strength, movement, and function of joints and muscles. Thus an occupational therapist may help an arthritic person get a bathroom *hoist* (qv) and a ch*airlift* (qv), while a physiotherapist may advise that person on helpful exercises and therapy. 3 For more information visit www.cot.co.uk/ See also: *Equipment: entitlement*; and *Workplace assessments*.
Ocular adaptation	See: *Lighting: transitional*.
Office chairs	See: *Chairs: office*.
Office design: basic planning data	1 Distilled planning data on all the major building types is given in the METRIC HANDBOOK: PLANNING AND DESIGN DATA. For publication details see the bibliography. 2 For general design guidance, commentary, and case examples visit: • CABE www.cabe.org.uk/default.aspx?contentitemid=36
Office design: complying with the Disability Discrimination Act	See: *Administrative and commercial buildings*.
Office desks	See: *Worktops*.
Office for Disability Issues (ODI)	1 IMPROVING THE LIFE CHANCES OF DISABLED PEOPLE, a joint report of the Cabinet office (qv), and five other government departments, was published in January 2005 and proposed that the Government should work to improve the lives of *disabled people* (qv) via practical measures in four key areas: • Helping disabled people to achieve *independent living* (qv). • Improving support for families with young disabled children. • Facilitating a smooth transition from childhood into adulthood. • Improving support and incentives for getting and keeping work. 2 The report proposed that this strategy be implemented by a new Office for Disability Issues reporting to the Minister for Disabled People (qv). The ODI was duly created, and in turn set up a new advisory body called 'Equality 2025: the UK Advisory Network on Disability Equality' to help get the *Disability equality duty* (qv) implemented across the UK. 3 For more information, including the ODI's 2007 progress report visit www.officefordisability.gov.uk/ See also: *Independent living*.
Office for Fair Access	1 OFFA is a *quango* (qv) that has responsibility for promoting and safeguarding fair access to higher education for under-represented groups. 2 For more information visit www.offa.org.uk See also: *Aimhigher; Education: tertiary*; and *Social mobility*.

ice for National atistics (ONS)

The ONS has long been the most authoritative single source of information of UK statistics, but the effects of government spin have eroded public confidence in its veracity. An ONS survey conducted in 2005 found that only 17% of people believed government statistics to be free from ministerial interference, and only 14% thought that governments used the data honestly. In September 2007 it was therefore proposed that a new watchdog, the Statistics Board (qv), would oversee the work of the ONS as from April 2008. The new body was duly created, but called instead the UK Statistics Authority (qv).
2 Visit www.statistics.gov.uk//

ice for Standards Education (Ofsted)

1 Here are some of the achievements of Ofsted; the Qualifications and Curriculum Authority (qv); and the Department for Children Schools and Families (qv) and its predecessors, all pulling together. Despite an education budget that is now near the top of the international league a quarter of a million English children will leave primary school in 2008 more or less illiterate; around 25 000 will leave secondary school without a single qualification to show for their 11 years of compulsory education; around 250 schools that are officially judged to be failing will take in a new cohort of pupils to miseducate; and youth unemployment may reach an 11-year record, partly because so many British school-leavers are effectively unemployable.
2 No hint of this disgraceful record appears on Ofsted's self-congratulatory website. How did things get to this point? The commentator John Clare has identified three main factors:

• Ofsted's predecessor, Her Majesty's Inspectorate (HMI), drew its Inspectors from the very top of the teaching profession. They would spend up to a week in the classroom, using their exceptional experience to recognise and encourage good teachers without trying to hold them to any single 'correct' method of teaching. The same spirit infused Ofsted under its first head, Mr Chris Woodhead, whose inspectors spent most of their time in classrooms and reported rigorously and in detail on what they found there. But after 2000, when Chief Inspector Mike Tomlinson took over, these well-tested traditions were abandoned. Now the school being inspected, not the inspector, sets the agenda of the inspection, and the latter has been reduced to a shallow box-ticking survey aimed more at ensuring that the school is precisely implementing centrally-directed government policies than at evaluating the effectiveness of teaching.
• As part of the same process, didactic teaching (ie being taught by a well-informed teacher imparting a lifetime's hard-won knowledge to pupils, and striving to educate, challenge, and stretch them) is now officially frowned upon while 'active learning' (ie children finding things out for themselves) is praised, despite the fact this is a learning style that does not generally suit pupils from homes without books and lacking a lively intellectual ambience. For more on this see: *Literacy*.
• Ofsted's remit has been massively broadened and now includes local education authorities (LEAs); childcare; local authority and private schools; further education; and teacher training. In April 2007 it is expected to be merged with other inspectorates including the Adult Learning Inspectorate and the Children and Family Court Advisory Service.

3 The new type of inspection has done absolutely nothing to create the 'world class education system' promised by Secretary of State for Education David Blunkett in January 1999. One in five 11-year-olds leaves primary school unable to read and write well enough to deal with the secondary school curriculum; and in January 2007 Sir Digby Jones, Chancellor of The Exchequer Gordon Brown's skills troubleshooter, publicly described our secondary school system as having been 'dumbed down to the lowest common denominator' (he should have said 'highest common factor', but probably knew that most of his semi-numerate listeners would not understand that term).

4 In June 2006 Ms Christine Gilbert replaced David Bell as Chief Inspector o Ofsted. We must wait and see whether she proves able to repair the damage that has been done to school inspection standards since 2000, but the events noted under *Jargon* (qv) are not encouraging.
5 In April 2007 Dr Mary Bousted, general secretary of the Association of Teachers and Lecturers, told the union's annual conference in Bournemouth t Ofsted had failed completely in its principal aim of improving school standar and should be stripped of its powers of inspection so that the job of monitorin school performance can be handed to independent advisers.
6 As of 2008, useful references include the following. They are given in date order, with a note on the central argument made by each. For publication details see the bibliography.
• INSPECTION, INSPECTION, INSPECTION ! by Anastasia de Waal was published in July 2006 by the *think tank* (qv) Civitas. Ofsted's inspection methods are actively undermining true education: every box must be ticked, and anything that doesn't have a box is ignored.
• WORKING ON THE THREE RS – EMPLOYERS' PRIORITIES FOR FUNCTIONAL SKILLS MATHS AND ENGLISH was commissioned by the DfES (qv) and published by th Confederation of British Industry in August 2006. Its comments on the state o British education are scathing.
• 2020 VISION: REPORT OF THE TEACHING AND LEARNING IN 2020 REVIEW GROUP was published by the DfES in December 2006. See the comments under *Jargo*
• ARTIFICIAL LEGACY: BLAIR'S SCHOOL ACHIEVEMENT IS A SHAM by Anastasia de Waal and Nicholas Cowen was published in June 2007 by Civitas. The improvement in exam results over the past 10 years have been achieved large by moving goalposts and fiddling results.
• BLAIR'S EDUCATION: AN INTERNATIONAL PERSPECTIVE was written by Professo Alan Smithers and published by the Sutton Trust in June 2007. Mr Blair's educational policies have generally failed, but education statistics are now so corrupted that it is almost impossible to establish the precise truth.
• THE CORRUPTION OF THE CURRICULUM, published in June 2007 by Civitas. Th school curriculum has been degraded by political interference, with tradition subjects being hi-jacked by fashionable causes and teachers being expected to advance social goals instead of imparting a body of academic knowledge tha has been built up over centuries.
• ARE SCHOOLS BEING INSPECTED TO DEATH? by Melanie Read was published in the Times on 31 March 2008.
7 Useful websites include the following. They are given in alphabetical orde

• Campaign for Real Education	www.cre.org.uk/
• Civitas	www.civitas.org.uk/education/index.php
• CEM Centre	www.cemcentre.org/
• Office for Standards in Education	www.ofsted.gov.uk/
• Qualifications and Curriculum Authority	www.qca.org.uk/
• Reform	www.reform.co.uk/website/education.as
• Teachernet	www.teachernet.gov.uk/

See also: *Education; Literacy*; and *Qualifications and Curriculum Authority*.

Office of Government Commerce (OGC)

1 One of the organisations with which the Commission for the Built Environme (CABE) collaborates to help improve design quality in the *built environment* (qv It is an independent Office of the Treasury reporting to the Chief Secretary.

2 For more information visit www.ogc.gov.uk/

See also: *CABE*.

fice of Public Sector ormation
1 The full texts of all recent UK laws can be found on the OPSI website, to which there are frequent references throughout the present work.
2 Visit www.opsi.gov.uk/acts.htm

fice of Rail gulation (ORR)
See: *Railway system*.

fice of the mmissioner for blic Appointments
See: *Commission for Public Appointments*.

fice of the Third ctor (OTS)
1 The OTS gives governmental encouragement and support to *charities* (qv) and *Voluntary and community organisations* (qv).
2 Visit www.cabinetoffice.gov.uk/thirdsector/

See also: *Voluntary and Community Sector*.

fices and mmercial buildings
See: *Administrative and commercial buildings*.

f-street platform ge warning surfaces
1 Off-street platform edge warning surfaces, consisting of rows of flat-topped domes, are one of six types of *tactile paving surface*. Their function is to warn *blind* (qv) people that they are approaching the platform edge at:
• A heavy rail platform (eg National Rail).
• An underground platform (eg London Transport).
• An off-street light rapid transit platform (eg London Docklands Light Railway).
They must not be applied to other situations, as consistency of use is essential if misunderstandings are to be avoided.

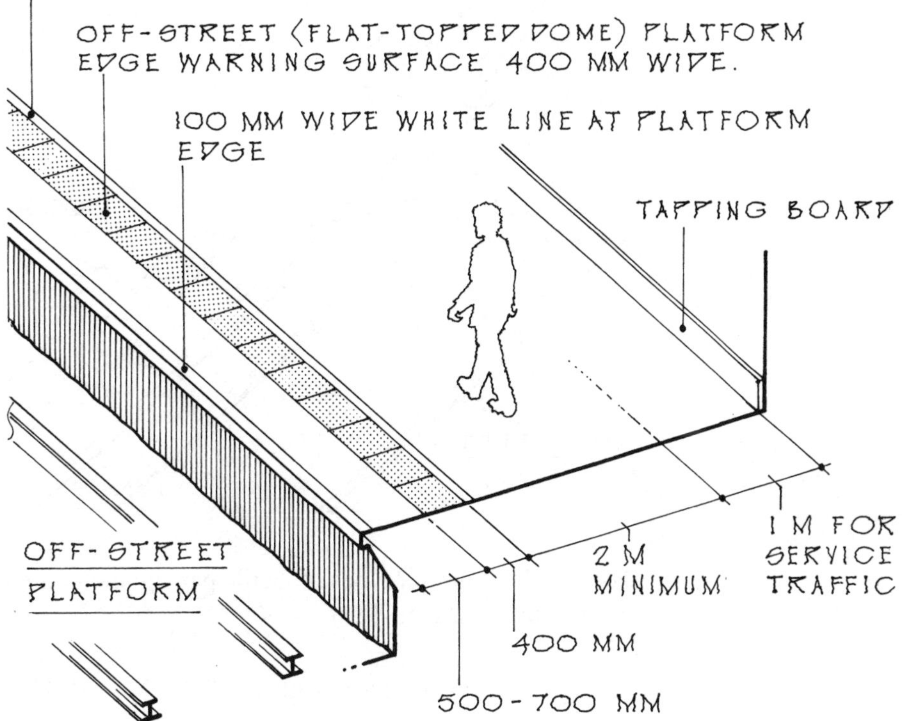

2 The diagram above shows a typical example, but the rules for correct use i particular situations are very precise, and designers must in all cases consult Chapter 3 of GUIDANCE ON THE USE OF TACTILE PAVING SURFACES (qv) and para 4.3 of INCLUSIVE MOBILITY (qv) – see the bibliography.
3 For details of the six types of tactile paving, including corduroy surfaces, s *Tactile pavings*.

See also: *On-street platform edge warning surfaces*.

Ofsted See: *Office for Standards in Education*.

Ombudsman See: *Parliamentary Commissioner for Administration*.

On-street platform edge warning surfaces 1 On-street platform edge warning surfaces, consisting of rows of lozenge shapes, are one of six types of *tactile paving surface*. Their function is to wa *blind* (qv) people that they are approaching the platform edge at an on-street light rapid transit (LRT) platform, eg those for trams. They must not be used other situations, as consistency of use is essential if misunderstandings are tc avoided.
2 The diagram below shows a typical example, but the rules for correct use in particular situations are very precise, and designers must in all cases cons Chapter 4 of GUIDANCE ON THE USE OF TACTILE PAVING SURFACES (qv) and para 4.4 of INCLUSIVE MOBILITY (qv) – see the bibliography.

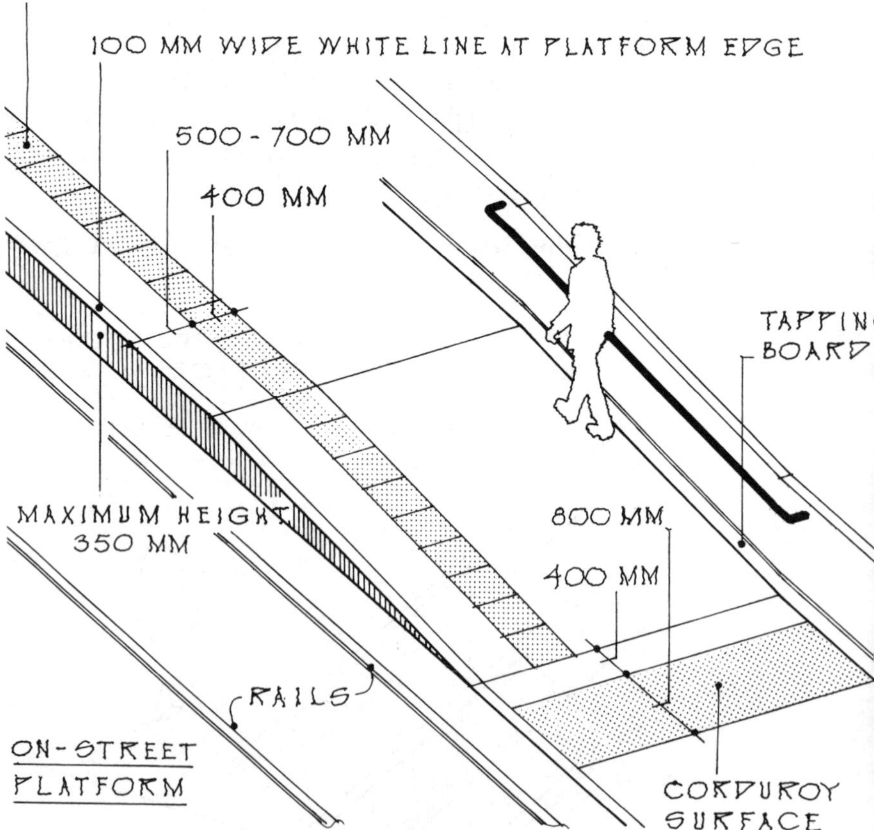

3 For details of the six types of tactile paving, including on-street platform e warning surfaces, see: *Tactile pavings*.

See also: *Off-street platform edge warning surfaces*; and *Tramstops*.

e in, One out rule	See: *Better regulation bodies*.
en for Business	1 A core reference that should be on the shelf of any designer involved with business premises. Read in conjunction with DESIGNING FOR ACCESSIBILITY (qv), the straightforward guidance given in its 80 pages will help building designers get most of their early decisions broadly right. 2 For publication details see the bibliography. See also: *Authoritative practical guides*.
en Market HomeBuy	See: *Housing: affordable*.
en Rights Group RG)	1 The ORG works to preserve and extend traditional civil liberties in the digital world, and to raise awareness of digital rights abuses. Its reports include, for instance, MAY 2000 ELECTION REPORT: FINDINGS OF THE OPEN RIGHTS GROUP ELECTION OBSERVATION MISSION IN SCOTLAND AND ENGLAND – see: *Electoral fraud* and the bibliography. 2 For more information visit www.openrightsgroup.org/ See also: *Civil rights*; and *Privacy*.
era houses	See: *Auditoria*.
portunity Age	1 This is the Government's strategy for an *ageing* (qv) society. First published in March 2005 as OPPORTUNITY AGE: MEETING THE CHALLENGES OF AGEING IN THE 21ST CENTURY (see the bibliography), the strategy focuses on three key areas: • Work and income. • Active ageing. • Services. The strategy is coordinated by the Secretary of State of the Department for Work and Pensions (qv). 2 For more information visit www.dwp.gov.uk/opportunity_age/ See also: *Ageing*.
portunity for All	1 This is the annual Government report about tackling *poverty* (qv) and *social exclusion* (qv). For publication details see the bibliography. 2 For more detail visit www.dwp.gov.uk/ofa/
ticians	See: *NHS structure*.
tdoor surfaces	See: *Surfaces: outdoor*.
erhead hoists	See: *Hoists*.
er-legislation	1 Civilised life depends on the rule of law, but there is a growing feeling that law-making in most modern societies, including the UK, is getting out of hand. There are three problems: • First, laws are being passed on matters that would be better left to people's judgement. Commenting on the American situation, Philip Howard argued in his 1995 book THE DEATH OF COMMON SENSE: HOW LAW IS SUFFOCATING AMERICA (see below) that regulators and lawyers in the USA have become so determined to lay down precise rules for every situation, in order to stamp out

arbitrariness and unfairness, that citizens are in danger of losing the ability to use their common sense and take personal responsibility for their decisions. Paradoxically, the results of excessively rule-governed behaviour are often m arbitrary and unfair than what went before – for UK examples see: *Risk*, and the article MY 7-YEAR-OLD SON, A TINY SPLINTER AND THE FIVE-HOUR FARCE TH SHOWS HOW THE NHS HAS LOST THE PLOT by the BBC's John Humphrys. It was published in the *Daily Mail* on 25 March 2008.

• Second, governments are making far too many laws. According to the Home Affairs commentator Philip Johnston (see below), the Government passed 45 Acts of Parliament between 1997 and 2007, including 6 on immigration, 7 on terrorism, 12 on education, 11 on health and social care, and 25 on criminal justice – the latter creating virtually one new crime per day for ten years running. Has our immigration system been brought under control, are our schools and hospital dramatically better, and our streets safer after this tidal wave of 455 new laws and 32 000 *statutory instruments* (qv)? Were the latter *proportionate* (qv), and thoroughly considered? Or were all too many of them passed mainly for the sake of a good headline?

• Third, laws are over-complex, and very often incomprehensible except to specialists. As one example, the Equality and Diversity Forum (qv) comment in 2004 that anti-*discrimination* (qv) legislation in Britain was embodied in 3 Acts, 52 Statutory instruments, 13 Codes of practice, 3 Codes of Guidance, a 16 EU Directives and recommendations – for the full text visit www.edf.org.uk/news/SEAleafletforwebsite.doc Not only do such constantly accumulating layers of legislation create a nightmare for ordinary citizens an veritable gold-mine for lawyers, but the haphazard process of accretion leads inevitably to bad law. Some people are covered and others are not; of those th are covered the degree of protection is unequal; and key terms have different definitions in different Acts. The Forum called for a Single Equality Act (qv) that (a) brings together in one place all those provisions that we really want a discards all the rest; (b) uses common standards and definitions throughout; a provides simple, effective remedies; and (c) is written in language that everyo can understand. Such an Act is expected to be announced in October 2008, an at time of writing its contents are unclear.

2 Arising from all of the above, the UK badly needs a reforming government that will:

• Stop trying to control every aspect of our lives by law.

• Stop continually adding new layers of legislation to old (see for instance the remarks under item (7) of *Special educational needs*), and concentrate instea on replacing whole swathes of accumulated existing legislation with fewer an simpler new laws, written from scratch.

• Write laws in plain English instead of phrasing Bills (a) in language that is understood only by lawyers, and (b) constantly referring to other statute law without further explanation, thus excluding all readers who lack specialist legal knowledge. In this connection one hopes that publication in June 2006 of the Coroner Reform Draft Bill marked the dawn of a new era. This ground-breaking document has an easy-to-understand interpretation of every clause running alongside the legal text, and spells out all the relevant sections of previous legislation so that lay readers see the full picture. The new drafting style was requested by the Constitutional Affairs Minister Harriet Harman, w said that it was time that the public could read the laws passed in their name. For the full text visit www.dca.gov.uk/legist/coroners_draft.pdf

• Finally, introduce a formal practice of 'post-legislative scrutiny' to verify that laws are actually achieving their intended outcomes. In March 2008, after the above notes were written, the Government yielded to years of pressure anc published proposals for such a scheme.

3 As of 2008, useful references include the following. They are given in alphabetical order. For publication details see the bibliography.
• THE DEATH OF COMMON SENSE: HOW LAW IS SUFFOCATING AMERICA by Philip Howard.
• POST-LEGISLATIVE SCRUTINY: A CONSULTATION PAPER published by the Law Society in 2005.
• WHAT IS THE POINT OF ALL THIS LEGISLATION? by Philip Johnston. *Daily Telegraph*, 29 October 2007.

See also: *Better regulation bodies; and Post-democracy.*

er-regulation

1 Here are three indicators of the degree to which regulation in the UK has grown over the past few years:
• In January 2006 Sir Roger Toulson, the Chairman of the Law Commission, stated that there had been 'a legislative explosion' over the life of the Commission. In 1965, the year in which the Commission was established, there were 7 500 pages of new laws, including both *statutes* (qv) and secondary legislation. In 2003 there were 16 000 pages plus another 11 000 pages of EU (qv) legislation. By 2004 the figure for domestic legislation amounted to almost 18 000 pages. By February 2006 the flood of new laws had become so great that Parliament could no longer debate all of them. The Government therefore decided to abandon the attempt and launched a bill which, if passed, would have enabled it to virtually set aside the need for Parliamentary debate and effectively rule by diktat. (Fortunately a massive rebellion forced it to abandon this idea, and pass instead the more reasonable Legislative and Regulatory Reform Act 2006 – see: *Civil liberties*).
• Later in 2006 the commentator Ross Clark (see below) did a count and found that 'in the 12 months to 31 May 2006 the Government had passed 3 621 separate pieces of legislation … that is more than 10 new sets of rules and regulations for each day of the year'. To get an idea of the overall weight of these regulations he sampled 10% of a year's output, factored it up, and got a total of 72 400 pages of regulations plus 26 200 pages of explanatory notes. That adds up to approximately 100 000 pages of new regulation in one year, and there is no reason to suppose that those particular 12 months were in any way exceptional.
• Between 1997 and 2006, during the tenure of chancellor Gordon Brown, *Tolley's Yellow Tax Handbook*, which is the standard reference work used by accountants, grew from 4 555 to 9 806 pages.
2 Here are four examples of the effect that this mountain of regulation is having on British public services:
• When the £45 billion 'Building schools for the Future' programme was launched in 2004, the Government promised to rebuild or refurbish every secondary school in England over the next 15 years. This was claimed to be the biggest schools investment programme in British history. In January 2007 the Department for Education and Skills (qv) was forced to admit that not a single project had yet been completed; that only 5 building contracts had been signed in 2006 instead of the scheduled 100; that only 14 of the 100 new schools scheduled for opening in 2007 were expected to be completed in that year; and that only 56 of the 200 schools due to open in 2008 were expected to be ready. Schools and construction firms were united in blaming a procurement process that had become 'mired in governmental red tape'.
• In 2006 the Government launched a Pathfinder scheme aiming to place up to 2 000 vulnerable children in some of the country's top private boarding schools, and give them a new start in life. A year later, in September 2007, it was revealed that only 3 children had so far been placed. Those involved complained that the scheme had become 'stifled in red tape'.

• In April 2007 the chief constable of Suffolk, Mr Alastair McWhirter, used his last day in office to complain of the 'phenomenal' amount of paperwork th is now required of the police – for instance, 128 pieces of paper handled by 56 different people before a simple assault case got to court – and that was before the criminal justice unit, the CPS, and the courts started adding to the paper trail. To check his figures the *Daily Mail* staged a simple crime involving two people: a man is punched in a busy street, suffering a broken tooth and bruise nose, and his assailant runs off. With the cooperation of the police, the standar sequence of events was then traced to the point at which an accused person appears in court. It transpired that 131 separate forms had to be written up or filled in by 17 different police officials, culminating in a court appearance involving another 20 people within the court system. All of this followed a simple crime involving two people; one hardly dares ask what would have resulted from an incident involving six or ten people. The investigation was published on 19 April 2007 under the title 'Paper Tigers'.

• In October 2007 the Ministry of Justice (qv) started piloting its Children's Court programme in Preston. It aimed to deal with 300 young offenders in its first 12 months, but 5 months later only 8 hearings had been held. The blame was placed squarely on red tape.

• Precise figures on the amount of red tape under which the Department for Education and Skills (now superseded by the Department for Children, Schoo and Families) buried teachers and other frontline staff are hard to come by, but according to Simon Jenkins, author of THATCHER AND SONS: A REVOLUTION IN THREE ACTS (see the bibliography), the DfES between 1997 and 2006 issued 500 regulations, 350 policy targets, 175 efficiency targets, 700 notes of guidance, 17 plans and 26 separate incentive grant streams. Sir Simon quotes Hansard as recording in 2001 that an average of 3 840 pages of instructions were being sent to schools in England each year, and states that a survey in 2006 showed one in four head teachers considering quitting their jobs because of the volume of red tape.

3 The Government has tried to tackle the problem of over-regulation via a series of measures outlined under *Better regulation bodies* (qv), with negative success – instead of being reduced, as promised, the burden of red tape is very rapidly growing.

4 As of 2008, useful references include the following. They are given in alphabetical order. For publication details see the bibliography:

• AVOIDING REGULATORY CREEP is a report released in October 2004 by the Better Regulation Task Force.

• BUMPER BOOK OF GOVERNMENT WASTE, from the *think tank* (qv) the Taxpayers Alliance gives examples of tax-payers' money poured out on wasteful scheme

• THE BUREAUCRATIC BURDEN IN THE NHS describes the effects of bureaucratic overload in one sorely afflicted organisation.

• THE GREAT DECEPTION by Christopher Booker and Richard North traces the history of the EU and Britain's relationship with it.

• HOW TO LABEL A GOAT: THE SILLY RULES AND REGULATIONS THAT ARE STRANGLING BRITAIN by Ross Clark. In addition to estimating the amount of legislation and regulation now being loaded onto businesses, charities and citizens (see the text above), the author also notes how little of it is ever debat in Parliament. The 29 Acts (qv) that became law in the 12 months to 31 May 2006 are dwarfed by the 3 592 statutory instruments – ie orders and regulations – that came into effect without Parliamentary debate.

See also: *Better regulation bodies; Contract of compliance; Gambling Act; Houses in multiple occupation; and Licensing Act*; and *Post-democracy*

P

ckaging See: *Product packaging.*

ge turners 1 These are *reading aids* (qv) for persons suffering from *paralysis* (qv) or *impaired dexterity* (qv). There are two types:
• Manual page turners are simple and cheap but may not suit severely disabled people. For examples see the catalogues of Sunrise and other suppliers of *Products for easier living* (qv).
• Electric page turners may be the only solution for people who are unable to manipulate reading materials. Examples include the 'Readable' page turner from Quality Enabling Devices, and the 'Gewa' page turner from Possum, both of which enable people to read a book regardless of their position – for instance, lying in bed.
2 As of 2008, useful websites include the following. They are given in alphabetical order.

• Disabled Living Foundation	www.dlf.org.uk/
• Possum	www.possum.co.uk/
• QED	www.qedonline.co.uk/

See also: *Aids and equipment;* and *Reading aids.*

gers See: *Alerting devices.*

get-Gorman signed eech 1 Previously known as Paget-Gorman Systematic Sign Language or PGSS, this is a simultaneous grammatical representation of spoken English that is used to help teach language to children who are *deaf* (qv) or have severe *learning difficulties (qv)*. The signs have been artificially developed within a logical system but do not correspond to *British Sign Language* (qv) signs.
2 For more information visit www.patient.co.uk/showdoc/26739554/

See also: *Alternative and augmentative communication*; and *Speech aids.*

aralympics Visit www.paralympics.org.uk/

aralysis 1 A loss of muscular strength, control, or feeling in any part of the body. Types of paralysis include the following:

• Monoplegia	Paralysis of one limb.
• Hemiplegia	Paralysis of one side of the body.
• Diplegia	Paralysis involving both sides of the body and affecting the legs more severely than the arms.
• Paraplegia	Paralysis of both legs.
• Triplegia	Paralysis of three limbs.
• Quadriplegia	Paralysis affecting all four limbs. Also called *tetraplegia.*

2 For household products that might prove helpful in some of these cases see: *Products for easier living*. For advice on assistive *aids and equipment* (qv) contact the Disabled Living Foundation – see below.
3 As of 2008, useful websites include the following. They are given in alphabetical order.

• Directgov: disabled people	www.disability.gov.uk/
• Disabled Living Foundation	www.dlf.org.uk/
• Disabled Living Online	www.disabledliving.co.uk/
• NHS Direct	www.nhsdirect.nhs.uk/
• OT Direct	www.otdirect.co.uk/

See also: *Aids and equipment; Paralysis;* and *Products for easier living.*

arent power See: *Education: voucher systems.*

Parish See: *Local government.*

Parking See: *Car parking.*

Parks and green spaces

1 As of 2008, useful websites include:

- Cabe Space www.cabe.org.uk/default.aspx?contentitemid=
- Council for National Parks www.cnp.org.uk/
- Countryside Agency www.countryside.gov.uk/
- English Nature www.english-nature.org.uk/
- Forestry Commission www.forestry.gov.uk/
- Royal Parks of London (qv) www.royalparks.gov.uk/

2 See also: *CABE; Countryside facilities;* and *Urban planning.*

Parliament

1 According to official websites such as www.direct.gov.uk/and www.parliament.uk/about/how/role.cfms, Parliament is the 'highest authority the land' with 'supreme authority for government and law-making in the UK a a whole'. Its main roles are described as (a) 'debating and passing all laws'; (b 'examining and challenging the work of the Government'; and (c) 'enabling th Government to raise taxes'.

2 If our elected Parliament truly did enjoy these powers and exercise these responsibilities, Britain would indeed be a politically *inclusive* (qv) society an the people would be sovereign. But except for (c) each of the above statements is misleading, which may help to explain why around a third of the UK electorate no longer bother to cast a vote at general elections.

- First, Parliament is not the 'supreme authority for law-making in the UK as a whole', Brussels is. The European Communities Act 1972 (qv) gives EU decisions automatic primacy over British Acts of Parliament.
- Second, Parliament does not debate and pass 'all laws', it does so for fewer than half. Between 50% and 80% (estimates vary) of our legislation is made in Brussels and nodded through a UK Parliament which, under the European Communities Act 1972, is powerless to block its passage onto the lawbook – see as an example: *Employment Equality (Age) Regulations*. Bodies called 'European scrutiny committees' of the House of Commons can occasionally intervene, but their effect on the incoming legislative flood is marginal.
- Third, the ability of Parliament to 'examine and challenge the work of the Government' has been close to neutered. The guillotine, a Parliamentary device that was invented to curtail obstructive tactics and was rarely applied, is now used as a matter of course to cut off debates in the House of Commons. According to the commentator Mr Henry Porter it was used on 80 occasions in the 94 years between 1881 and 1975, and on 216 occasions in the 6 years between 1997 and 2003.

3 In short, our elected Parliament has lost most of its traditional powers, and th electorate have noticed – see: *Post-democracy*. A reforming government and a new constitutional settlement are urgently needed to restore its role as the voic of the people – see: *Constitution*.

4 Regular commentaries on these matters will be found in Christopher Booker Notebook each week in the *Sunday Telegraph*; Henry Porter's column in the *Observer* on Sundays; and Philip Johnston's column in the *Daily Telegraph* on Mondays.

5 As of 2008, useful references include the following. For publication details see the bibliography.

- ABC: A BALANCED CONSTITUTION FOR THE 21ST CENTURY. This pamphlet by the constitutional lawyer Martin Howe briefly discusses some of the above issues and suggests a way forward.
- THE GOVERNANCE OF BRITAIN. This *green paper* (qv), published soon after Prime Minister Gordon Brown took office in June 2007, states: 'We want to

forge a new relationship between citizen and state and begin the journey towards a new constitutional settlement – a settlement that entrusts parliament and the people with more power'.

6 Useful websites in addition to those mentioned under (1) above include:

• My Society www.mysociety.org

See also: *Constitution; European Communities Act;* and *Post-democracy.*

Parliamentary English See: *West Lothian question.*

Parliamentary Commissioner for Administration 1 Also called the Parliamentary 'Ombudsman' (from the Swedish for 'legal representative'), this is an official with a duty to investigate complaints against *public authorities* (qv), and *government departments*.

2 For more information visit www.ombudsman.org.uk/

Partnerships for Progression See: *Aimhigher.*

Partnerships for Schools See: *Building Schools for the Future.*

Passageways See: *Corridors and passages.*

Passenger conveyors: complying with the Disability Discrimination Act 1 Passenger conveyors are moving surfaces very like escalators (qv) except (a) that the metal plates form a sloping flat surface, not steps; and (b) that they rise at an angle of only 10° or 12°, not 27° or 30° as in the case of escalators as shown on the simplified diagram below.

2 Public buildings and spaces must in general comply with the Disability

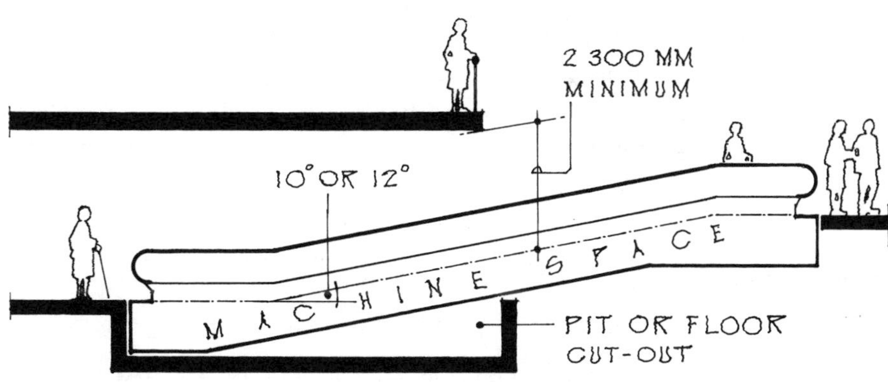

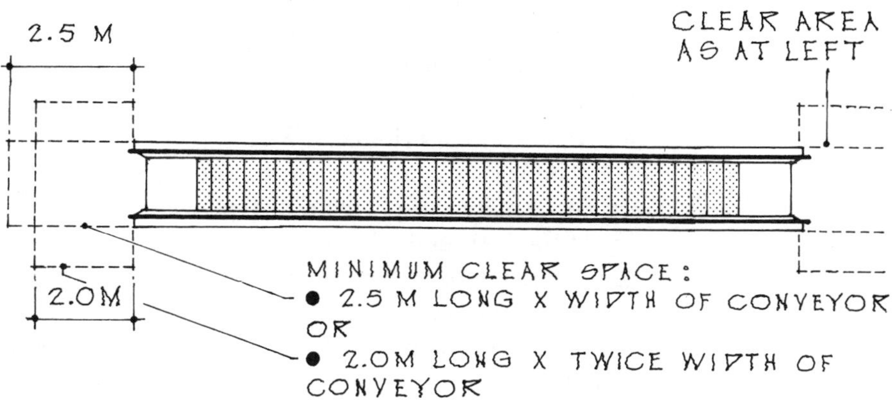

Discrimination Act (qv). In the provision and design of passenger conveyors, relevant *authoritative practical guides* (qv) include the following:
• BS 8300 (qv): the recommendations in para 8.5 apply to all buildings in general.
• INCLUSIVE MOBILITY (qv): the recommendations in para 8.4.4 apply to *transport-related buildings* (qv) such as bus and coach stations, railway stations, air terminals, and transport interchanges. They ought in fact to be considered in all heavily trafficked public buildings and spaces.
The above guides do not have the force of law, but conformity with their recommendations will make it easier to demonstrate that the requirements of the Disability Discrimination Act have been met.
3 In the USA, where the Americans with Disabilities Act applies, refer to the ADA AND ABA ACCESSIBILITY GUIDELINES (qv).
4 As of 2008 useful websites include the following. They are given in alphabetical order.

• Lift and Escalator Industry Association	www.leia.co.uk/
• Kone	www.kone.com/
• Otis	www.otis.com/
• Schindler	www.schindlerlifts.co.uk/

See also: *Escalators;* and *Travelators.*

Passenger lifts

1 *Passenger lifts* are one of four categories of lift, the others being vertical platform lifts, incline platform lifts, and chairlifts – see the taxonomic diagram under: *Lifts.*
2 While passenger lifts are more expensive than the alternatives, they offer the easiest and safest form of vertical travel and should be installed wherever the relatively high cost can be afforded. For more on this see the notes under *Lifts: complying with the Disability Discrimination Act.*
3 As of 2008 useful websites include the following. They are given in alphabetical order.

• Lift and Escalator Industry Association	www.leia.co.uk/
• Kone	www.kone.com/
• Otis	www.otis.com/
• Schindler	www.schindlerlifts.co.uk/

Passenger restraint systems

See: *Wheelchair securing systems.*

Pathfinders

1 Following on from the THE SUSTAINABLE COMMUNITIES PLAN (see the bibliography), which was published on 5 February 2003, the Pathfinder programme is intended to revive run-down areas, particularly in the North of England, by refurbishing large tracts of obsolete housing, or demolishing and replacing them with viable, attractive new housing.
2 In 2005 some Pathfinder programmes began to encounter vehement opposition from local communities and from organisations such as CABE (qv) and SAVE Britain's Heritage. What was at issue was the threat to demolish up to 400 000 houses, many of which were loved by their owners and could, it was argued, be more cheaply renovated (perhaps £20 000 per dwelling) than rebuilt (over £100 000 per dwelling) with less distress to communities. At Nelson in Lancashire it took two public inquiries and the combined opposition of English Heritage, the Prince's Foundation, Save Britain's Heritage, the Victorian Society, and others before the views of local people were respected and the government-imposed plan halted. At Darwen, near Blackburn in Lancashire, the owners of recently refurbished properties were informed that their homes

were unfit for habitation despite the fact that Mr Brian Clancy, a past President of the Institution of Structural Engineers, examined a sample of eight condemned houses and could find nothing wrong with them, stating that one was 'an ideal little first-time buyer house' while others were 'an absolute palace' and 'an absolutely wonderful property'. At Edge Lane in Kensington, Liverpool, where the local council planned to knock down 360 Victorian and Edwardian homes, the authorities were taken aback by the scale of opposition to their demolition plans.
3 In April 2007, after a period of indecision following the departure of Mr John Prescott, his successor, Ms Ruth Kelly, decided to press ahead with the programme.
4 The Pathfinder website www.englishpartnerships.co.uk/pathfinders.htm contains, as of October 2007, no hint of these conflicts or of any public unhappiness whatsoever.
5 As of 2007, useful sources of information including the following. For publication details see the bibliography.
• HOUSING MARKET RENEWAL: REPORT BY THE COMPTROLLER AND AUDITOR GENERAL. This official report, published since the above words were written, finds the effectiveness of Pathfinder schemes to be 'unclear', and while 'intervention has improved housing conditions for some, for others it has led to heightened stress'.

See also: *Housing.*

---

athfinders for chools See: *Building Schools for the Future.*

---

athways to work See: *Incapacity benefit.*

---

atient and Public volvement Forums See: *NHS structure.*

---

atient lifting aids See: *Hoists.*

---

ay gap See: *Incomes: gender pay gap.*

CTs See: *NHS;* and *Public services: quality.*

---

edestrian crossings 1 Pedestrian crossings across vehicular routes fall into two classes, 'controlled' and 'uncontrolled'. Summary note are given below. For latest guidance visit www.dft.gov.uk/pgr/roads/tpm/ltnotes/?version=1
2 Controlled crossings are formal crossing points, controlled by lights. Types include the following:
• Pelican crossings are controlled by pedestrians. To activate the lights, the pedestrian pushes a button on the wait box.
• Puffin crossings are 'intelligent' crossings. They are similar to Pelican crossings, but have built-in electronic sensors which can detect pedestrians on the crossing and delay vehicles until the walkers have safely crossed.
• Toucan (or 'two-can') crossings are for pedestrians and cyclists. They are very like the two above, and are sited where cycle routes cross roads. Cyclists do not have to dismount before crossing.
• Pegasus crossings are for pedestrians, cyclists, and horses, and have an additional push button at a higher level so that a horse rider does not have to dismount.
• Zebra crossings have bold white stripes and a flashing orange beacon at each

end. Pedestrians have automatic priority over vehicles, which must stop if a person is crossing or about to cross. These crossings tend to delay vehicular traffic and are used only where traffic flows are low or moderate – usually in residential areas. They are often used in conjunction with a raised street surface to reduce vehicle speeds at street crossings. They are being gradually phased out (according to an AA report in January 2008, a thousand have been removed in recent year) but are popular with pedestrians, who dislike having to wait for the lights to change even when there is little or no traffic.

3 Uncontrolled crossings are informal facilities where the volumes of pedestrian movement are too low to justify stopping the flow of traffic. They do not have lights and are typically found across side roads, or away from junctions.

4 *Blister surfaces* (qv), consisting of parallel rows of flat-topped blisters, should be used as shown below at all pedestrian crossings where the *kerb* (qv) has been dropped to be flush with the carriageway. The purpose of such surfaces is to warn people who are *blind* or suffer from severely *impaired vision* that they are about to step onto a carriageway.

See also: *Blister surfaces; Kerbs;* and *Traffic advisory leaflets.*

Pedestrian environments: complying with the Disability Discrimination Act

1 The pedestrian environment comprises all walkable areas used by the public parks and precincts; the pavements alongside streets; the approaches to public buildings; the interiors of buildings used by the public, etc. This entire network should be easily negotiable by the whole population, including frail people using *walking frames* (qv); mothers pushing *baby-buggies* (qv); people in *wheelchairs* (qv) or *scooters* (qv); and everyone else.

2 Private dwellings and their sites are not subject to the Disability Discrimination Act (qv), but all pedestrian areas in the public realm must comply with the Act. *Authoritative practical guides* (qv) include DESIGNING FOR ACCESSIBILITY (qv) and in addition the following documents. For publication details see the bibliography.

• APPROVED DOCUMENT M (qv), the provisions of which must be heeded in order to comply with the Building Regulations (qv) for England and Wales. Refer in particular to Section 1.

• BS 8300 (qv), which applies to all buildings in the UK. Refer in particular to Sections 4 and 5.

• INCLUSIVE MOBILITY (qv), which applies to pedestrian precincts, public footways, public footpaths, road crossings, etc. Refer in particular to Sections 1 to 7.

Except for AD M the above references do not have the force of law, but conformity with their recommendations will make it easier to demonstrate that the requirements of the Disability Discrimination Act have been met.

For places other than England and Wales the following regulatory documents should be consulted:

• In Northern Ireland: TECHNICAL BOOKLET R of the Building Regulations (Northern Ireland) (qv).

• In Scotland: the TECHNICAL HANDBOOKS of the Building (Scotland) Regulations (qv).

• In the USA, where the Americans with Disabilities Act applies: the ADA AND ABA ACCESSIBILITY GUIDELINES (qv).

3 The schematic diagram below shows some of the main elements of the pedestrian environment. For the wider context see the diagram under *Built environment.*

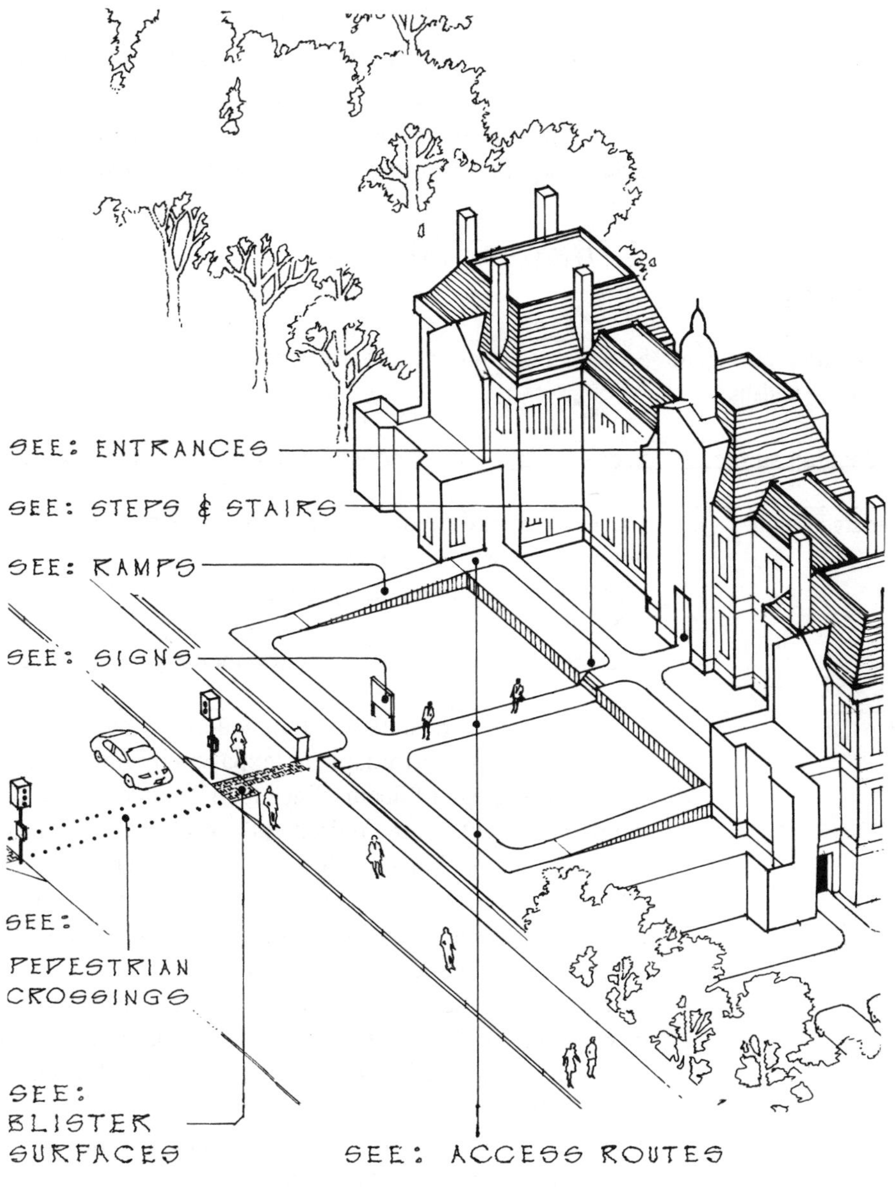

See also: *Access routes;* and *Countryside facilities.*

Pedestrian movement maximum gradients

1 The gradients shown below will in general satisfy the provisions of the documents quoted in the previous entry, though each particular case should be checked in detail before decisions are finalised.

LOCATION	MAXIMUM GRADIENT	SOURCE DOCUMENT
PUBLIC FOOTWAYS[1]	1:100 IDEAL 1:50 GOOD 1:40 FAIR 1:20 ACCEPTABLE 1:12 MAXIMUM (RAMP)[3]	INCLUSIVE MOBILITY
ACCESS ROUTES TO BUILDINGS[2]	1:20 PREFERRED 1:12 MAXIMUM (RAMP)	AD M & BS 8300

[1]These are the pavements alongside streets, and other paths and pedestrian areas in the public realm.
[2]These are the pathways within sites, for instance connecting the site entrance or car park with the building entrance; or connecting one building entrance with another.
[3]A gradient as steep as 1:12 is acceptable only over very short distances, say 1 metre maximum. Like all surfaces that are steeper than 1:20 it must be designed as a *ramp* (qv), with handrails, kerbs, etc

2 See also: *Access routes*; and *Footpaths and footways*.

Pedestrian infrastructure

See: *Pedestrian environment.*

Pedestrian movement: minimum widths

See: *Access routes; Corridors and passages;* and *Countryside facilities.*

Pedestrian surfaces

See: *Access routes; Cobbled surfaces; Gravel* surfaces; and *Ramps.*

Pension credit

1 Everyone aged 60 or over and living in Great Britain is entitled to a certain minimum pension credit every week.
2 The exact sum is regularly adjusted. For latest information visit:
- Directgov www.direct.gov.uk/
- Pension Service www.thepensionservice.gov.uk/pensioncredit/

Pensions

1 When the welfare state was founded in the late 1940s, the state retirement pension was made affordable to taxpayers by setting the retiring age fairly clo to the average age of death. Since then average life expectancy has increased dramatically, but not the retirement age. People born in 1901 could expect to live to 47; people born in 2004 to 79; and it is projected that people born in 2021 will have a life expectancy of 82 – which may turn out an underestimate in view of current advances in medical science. Meanwhile the retirement age h not advanced beyond the upper 60s.
2 The result of the above trends is that while there were more than five people of working age for every pensioner in 1950, there will be just two in 2050, an they will be expected to sustain that pensioner at a standard of living undrean of in 1950.
3 Recently a second problem has emerged: in April 2008 the Office for National Statistics (qv) revealed that more than 60% of adults are not paying into a private pension scheme – that is the highest proportion since records began 12 years ago, and equates to 22 million people of working age. One

expert foresaw 'a generation of pensioners living on baked beans and Rich Tea biscuits at an age when they thought they would be enjoying themselves'.
4 To meet these twin challenges, many commentators argue that the UK Government will have to follow the example of other European countries, where working people and their employers are obliged to save fairly large sums, and either persuade or force the UK population to do the same. As of 2008, despite a dozen government-sponsored pension reports or pieces of legislation since 1988, neither the Government nor the opposition has had the courage to grip this hot potato. If they do not do so fairly soon we are storing up a substantial degree of future pensioner poverty.
5 The latest attempt at achieving a British consensus was the independent Pensions Commission, set up by the Government in December 2002 under Sir Adair Turner. Its report of April 2006, popularly known as the Turner Report, tried to set the agenda for the first half of the 21st century just as Lloyd George's reforms did for the first half of the 20th century, and Lord Beveridge's for the second half. The debate is ongoing: for the latest situation see the references below.
6 As of 2007 useful references include the following. For publication details see the bibliography.
• REFORMING PENSIONS, by Nick Bosanquet, Derek Scott, Andrew Haldenby, and Colin Taylor.
7 Useful websites include the following. They are given in alphabetical order.

• Civitas	www.civitas.org.uk/welfare/index.php?
• Frank Field	www.frankfield.co.uk/type3.asp?id=20&type=3
• National Pensions Debate	www.dwp.gov.uk/pensionsreform/debate/
• New Policy Institute	www.npi.org.uk/projects/pensions.htm
• Pensions Policy Institute	www.pensionspolicyinstitute.org.uk/
• Pensions reform	www.dwp.gov.uk/pensionsreform/
• Pensions Service	www.thepensionservice.gov.uk/
• Reform	www.reform.co.uk/filestore/pdf/Pensions%20reform.pdf
• Turner report	www.pensionscommission.org.uk/

nsions combined recasting service

1 The Pensions Service's *combined forecasting service* allows pension scheme members to see forecasts of both their state pension and private pension together.
2 Visit www.thepensionservice.gov.uk/pensionforecast/home.asp

nsions: Financial sistance Scheme AS)

1 A scheme announced in 2004 to help 125 000 people who had been stripped of their company pensions since 1997, and therefore faced hardship in old age even though they had diligently paid into a pension scheme.
2 By May 2007 the FAS had paid out £5 million to fewer than 1 500 people while spending £10 million on administration; had failed to reach 10 000 people who were already at retirement age and in dire need of immediate action; and seemed ready to abandon them to their fate. As of November 2007 the Government is coming under extreme pressure to announce a fair compensation scheme, and the signs are that it will give way.
3 For more information visit www.dwp.gov.uk/lifeevent/penret/penreform/fas/

nsions: public and ivate

1 SQUANDERED: HOW GORDON BROWN IS WASTING OVER ONE TRILLION POUNDS OF OUR MONEY by David Craig, a government consultant turned whistleblower, is a devastating account of government profligacy with our money. The author alleges, inter alia, that private sector workers are being forced to pay more in taxes to fund public sector pensions than they manage to save for their own retirement (see: *Pensions* above).
2 For publication details see the bibliography.

Percentiles

1 In a collection of statistical data, a *percentile* is the value below which falls a specified percentage of that collection – for instance 5% of the data fall below the 5th percentile, 50% of the data fall below the 50th percentile, and so on. Or, to put it another way: the 5th percentile is the value that has 5% of the collection the 50th percentile is the value that has 50% of the collection; and so forth.
2 Taking as an example the stature (standing height) of adult British men, as shown on the graph on the next page, it is seen that:
- A height of 1 635 mm includes 5% of British adult men, the remaining 95% being taller.
- A height of 1 745 mm includes 50% of British adult men, the remaining 50% being taller.
- A height of 1 860 mm includes 95% of British adult men, the remaining 5% being taller.

3 Some of the matters above are explained in more detail under *Normal distribution* (qv).

Percentiles use in design

1 The considered use of percentiles helps designers to deal with the important problem of how 'broad' an average they should design for. Thus:
- If some product for men were tailored exactly to a stature of 1 745 mm (the standing height of the average British adult male) it would suit the few who ar clustered closely around that dimension, but inconvenience the many who are either significantly shorter or taller than the average.
- If, going to the other extreme, the product were designed in such a way that i suited the entire user population from the shortest man in Britain (say 1 000 mm) to the tallest (say 2 200 mm), the result might well be impractical or impossibly expensive.
- A fairly arbitrary solution that has been found to work well enough for many purposes is to design for the 5th to the 95th percentiles – i.e. for the middle 90% of the user population. The dimensional range is usually not so great as t be unfeasible, yet only 1 in 20 people is excluded at each end of the scale. For more on this refer to p 21 of BODYSPACE below.

2 Before applying this rule one must consider in each particular case the consequences of the mismatch for the 10% of people who are not catered for. Thu
- If a dining table is too high for one in twenty of the user population, and too low for another one in twenty, the consequence is merely some mild discomfo twice or three times a day for one in ten of the population. Designing for the 5th to the 95th percentile may therefore be judged acceptable.
- If an escape hatch from a submarine is too small for one in twenty of the use population the consequences could well be lethal. The mismatch is therefore unacceptable and it is necessary to design for the 100th percentile of the population in terms of body size. This will allow everyone, from the smallest t the largest, to get through.

3 Having sketched the theoretical background, it must be said that designers should in practice be wary of using data from anthropometric tables even after the most careful thought. Buildings, furniture, objects, and spaces are seldom used exclusively by the neatly specified groups described in such tables (eg 'adult men aged 19–45' or 'elderly women aged 65–80'), and far more likely t be used by unpredictable varieties of people – including of course children wit their very different body dimensions. The safest course (and one that will mak it easier to demonstrate compliance with the Disability Discrimination Act) is to use dimensional recommendations such as those given in BS 8300 (qv) and INCLUSIVE MOBILITY (qv), which have been carefully developed, with due consideration of factors such as the above, by expert bodies such as the British Standards Institution (qv).
4 As of 2008, useful references include the following. They are given in alphabetical order. For publication details see the bibliography.

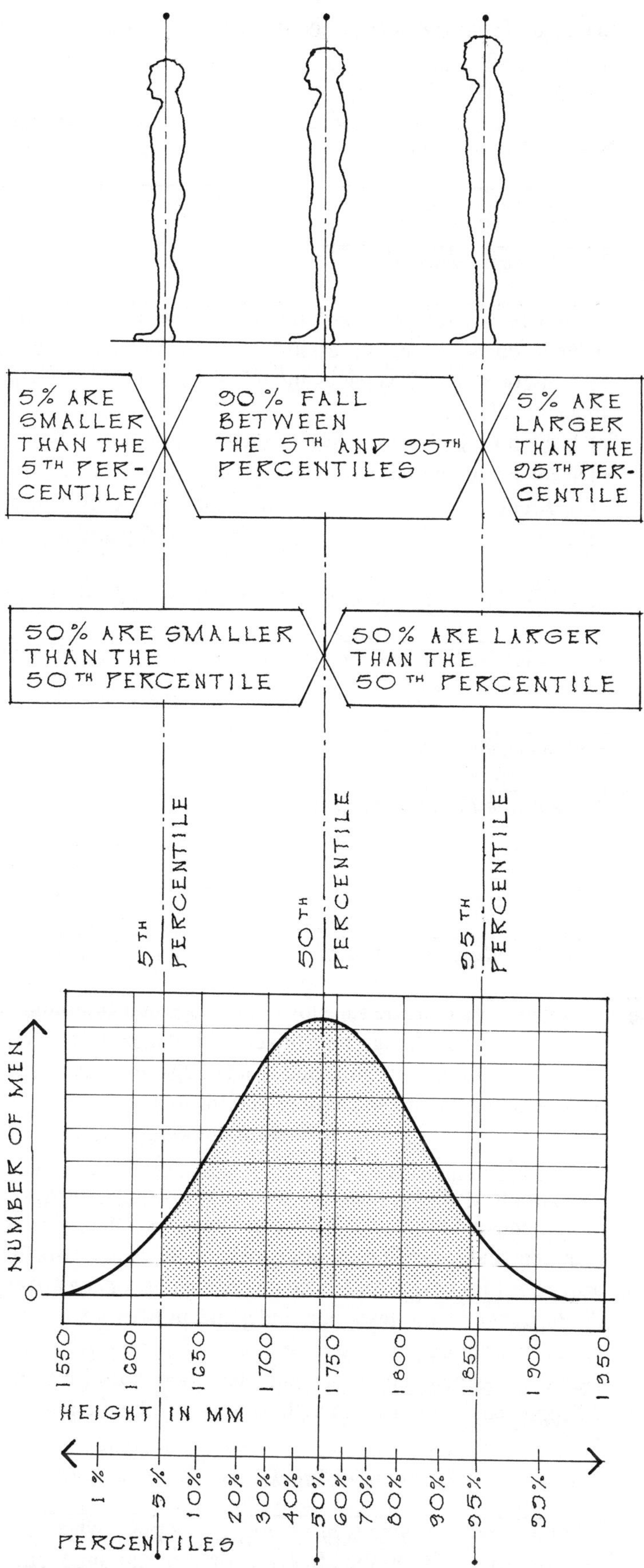

• BODYSPACE deals simply and clearly with all of the above matters – refer especially to pp 15–44.

• THE METRIC HANDBOOK gives distilled dimensional data for building planning and design.

See also: *Anthropometrics;* and *Normal distribution.*

Perception See: *Sensation.*

Perch-type seats See: *Seats: sit-stand.*

Persistent poverty See: *Poverty: persistent.*

Personal alarms See: *Alarms: personal.*

Personal care

1 Personal care refers to activities such as washing and bathing, using the lavatory, dressing and undressing, using medications or cosmetics; etc.
2 As of 2008, useful websites include the following. They are given in alphabetical order.

- Disabled Living Foundation www.dlf.org.uk/
- Disabled Living Online www.disabledliving.co.uk/
- OTDirect www.otdirect.co.uk/link-equ.html#Platform
- Ricability www.ricability.org.uk/

See also: *Aids and equipment; Bathing aids;* and *Products for easier living.*

Personal mobility vehicles See: *Buggies; Powered wheelchairs; Scooters;* and *Stair climbers.*

Petrol stations See: *Service stations.*

PFI See: *Private Finance Initiative.*

Phonemes
Phonics See: *Literacy.*

Physical impairment See: *Disability;* and *Impairment.*

Physical regeneration

1 Policies which tackle run-down buildings and communal areas. The process is different from, though interlinked with, 'economic regeneration' (which ain at creating jobs and wealth) and 'social regeneration' (which tackles the social problems that lead to social deprivation, such as crime and drugs).
2 See also: *Neighbourhood renewal*; and *Pathfinders.*

Physiotherapists

1 Physiotherapists (or physios) specialise in helping people maintain the strengt movement, and function of joints and muscles. Their skills lie in the physical treatment and rehabilitation of people and they are distinct from *occupational therapists* (qv) who offer advice on *aids and equipment* (qv) to help frail or *disabled people* (qv) lead more independent lives. Thus a physiotherapist may advise an arthritic person on exercises and therapy, while an occupational therapist may help that person get a bathroom *hoist* (qv) and a *chairlift* (qv).
2 People can be referred to a physiotherapist by their GP.

See also: *Occupational therapists.*

Place of relative safety

1 In the context of *means of escape* (qv): a place where people can be temporarily safe from the effects of fire, thus giving them time to move on to *place of safety* (probably the great outdoors) in a more leisurely manner shoul this prove to be necessary. Generally it is a *protected lobby* (qv) or *protected stairway* (qv) – see the diagram on the next page.

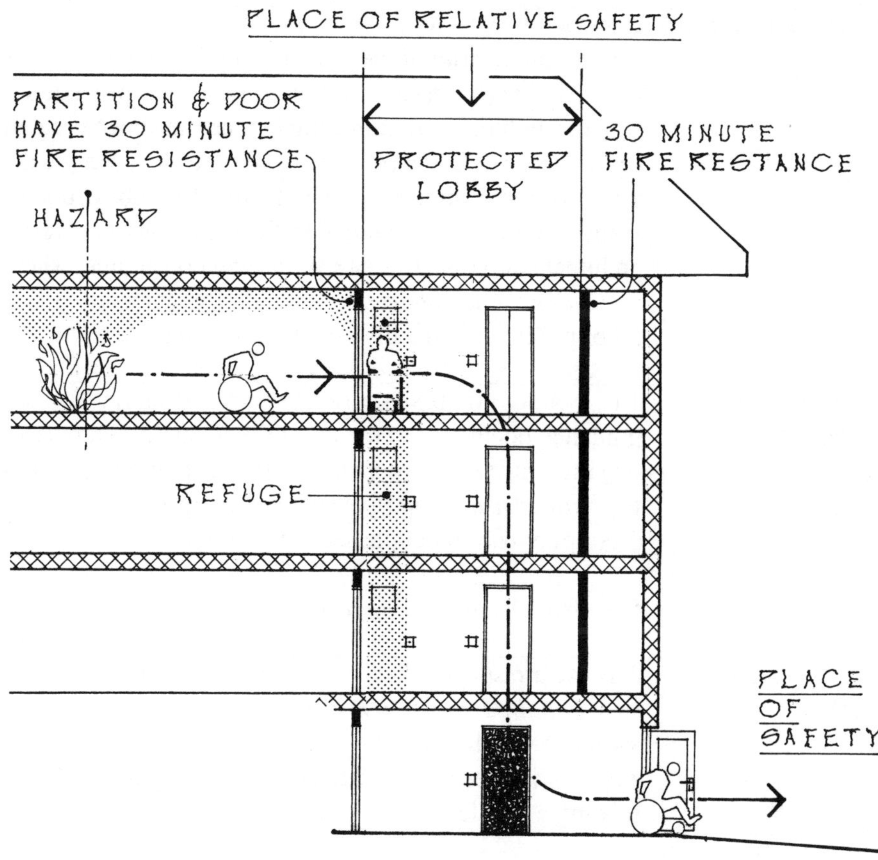

2 See also: *Means of escape.*

---

ace of safety See above.

---

aces of work See: *Workplaces.*

---

aces of worship See: *Religious buildings.*

---

anning 1 The purpose of planning is to protect the character and amenity of localities, for instance by preventing discordant development (eg a noisy factory in a residential area), and ensuring that necessary developments are located where they will do most good for most people and least harm to the environment.
2 Visit www.planningportal.gov.uk/

---

anning and access disabled people 1 PLANNING AND ACCESS FOR DISABLED PEOPLE: A GOOD PRACTICE GUIDE, is an essential reference for planners and architects. Conformity with its contents will make it easier to demonstrate that the requirements of the Disability Discrimination Act (qv) have been met. The document:

- Describes how planners and everyone else involved in the development process can play their part in delivering *inclusive* (qv) physical environments – i.e. built environments that can be readily used by everyone.
- Stresses the importance of early consultation with *disabled people* (qv) when formulating development plans and preparing planning applications.
- Explains all relevant legislation and policies and suggests ways in which these can be implemented and enforced effectively.

2 For publication details see the bibliography.

See also: *Access legislation; Inclusive design;* and *Planning.*

Planning and diversity

1 DIVERSITY AND EQUALITY IN PLANNING: A GOOD PRACTICE GUIDE is an official publication to help local planning authorities deal with *diversity* (qv) and *equality* (qv) issues in their policies and procedures. In order to thread its way through the regulatory thicket that has been created by *over-legislation* (qv), *over-regulation* (qv) this is not the clear practical guide that practical planners and designers need, but a mind-numbing document that will – as in so many other cases mentioned in the present work – force many planners and architects into the arms of fee-earning consultants whether they like it or not – see: *Consultants' charter*.
2 For publication details see the bibliography.

Planning and the Disability Equality Duty

1 PLANNING, BUILDINGS, STREETS AND DISABILITY EQUALITY is an official guidance document to the Disability Equality Duty (qv) for *local authority* (qv departments with responsibility for the planning, design and management of the *built environment* (qv) and *streets* (qv).
2 For publication details see the bibliography.

See also: *Disability Equality Duty.*

Plate and touch switches

1 These are special electrical switches that are more user-friendly for people with *impaired vision* (qv) or *arthritic* (qv) hands than conventional toggle switches which must be gripped.
• Plate switches are operated by depressing or touching a plate instead of moving a toggle.

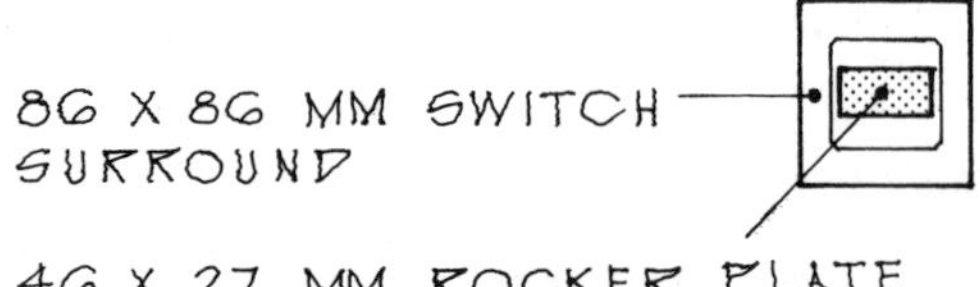

• Touch switches differ from the above in that the entire panel is pressure-sensitive and need only be touched with the fingers, elbow, or any other body part depending on the nature of the impairment. It therefore suits people with extremely poor coordination or hand function.

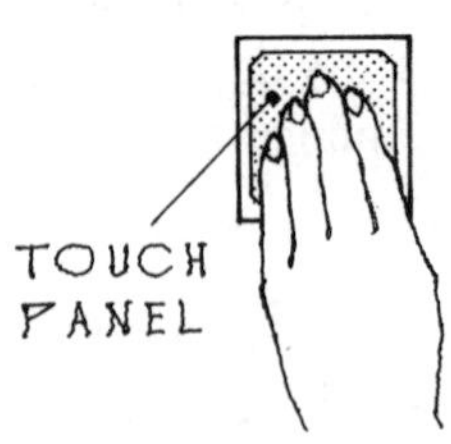

2 For recommended mounting heights see: *Controls and switches*.
3 As of 2008, useful websites include the following. They are given in alphabetical order.

• Disabled Living Foundation	www.dlf.org.uk/
• Ricability	www.ricability.org.uk/

See also: *Products for easier living;* and *Touch and proximity switches*.

Platform lifts

1 As shown on the next page there are two types of platform lift – vertical and incline – which together constitute two of the four basic categories of lift types – see: *Lifts*.

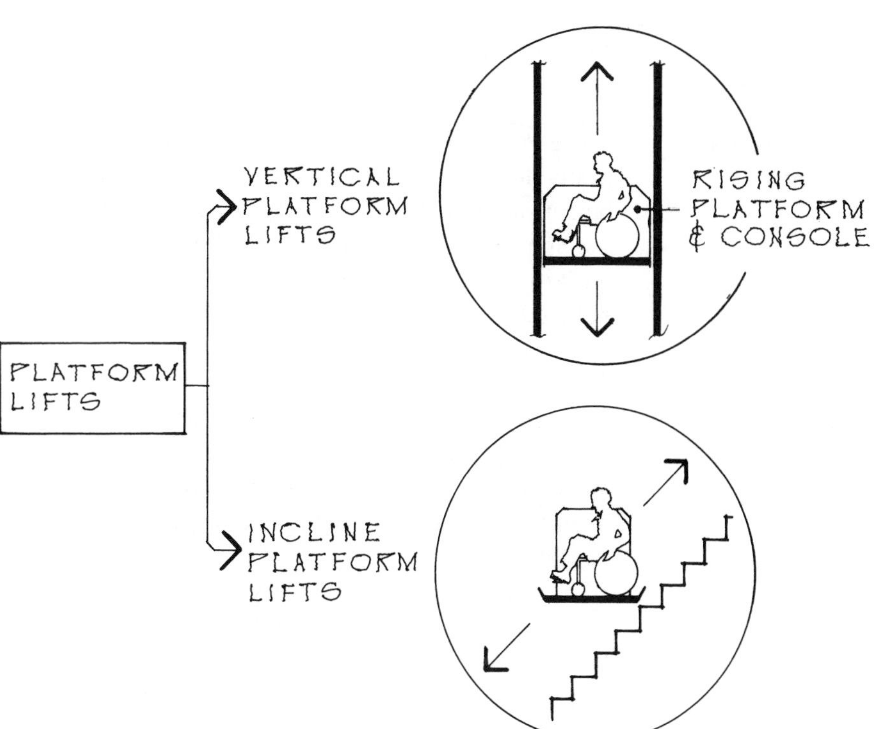

2 For more detail on each see: *Incline platforms lifts*; and *Vertical platform lifts*.

luralism See: *Diversity*.

olice and red tape 1 In April 2007 the chief constable of Suffolk, Mr Alastair McWhirter, used his last day in office to complain of the 'phenomenal' amount of paperwork that is now required of the police – for instance, 128 pieces of paper handled by 56 different people before a simple assault case got to court – and that was before the criminal justice unit, the CPS, and the courts got involved. To check his figures the Daily Mail staged a simple crime involving two people: a man is attacked in a busy street and his assailant runs off. With the cooperation of the police, the standard sequence of events was then traced to the point at which an accused person appears in court, and published on 19 April 2007 under the title 'Paper Tigers'. It found that 131 forms had to be filled in by 17 different police officials, leading to a court appearance involving another 20 people within the court system. This followed a simple crime involving two people. What, it must be asked, would have resulted had the incident involved 6 or 10 people?

2 Subsequent reports in 2007 include the following:

- In October Home Office (qv) figures showed that all promises to cut red tape and move officers from their desks to front-line duties had come to nothing
- visit http://police.homeoffice.gov.uk/performance-and-measurement/performance-assessment/assessments-2006-2007/
- In December it emerged that police officers in England and Wales spent 13.6% of their time on patrol in 2006–07, compared with 15.3% in 2004–05.

3 In April 2008 the Home Office finally admitted that too many officers were concentrating on low-level crimes in order to meet central 'detection' targets, confirming a belief held by 70% of local police commanders.

4 As of 2008, useful references include the following. For publication details see the bibliography.

- THE REVIEW OF POLICING: INTERIM REPORT by Sir Ronnie Flanagan was commissioned by the Home Office and published in September 2007.
- THE PUBLIC AND THE POLICE by Harriet Sergeant

See also: *Over-regulation*.

Police and risk Assessments

See: *Risk assessment.*

Police National Computer

1 The Police National Computer is a national information system that is available to the police and a variety of other non-policing organisations. It can be consulted 24 hours a day by police officers, the Criminal Records Bureau, the Special Branch, and MI5 to vet people who have been stopped for questioning, or have applied for sensitive posts in schools, dealing with the vulnerable, or working in government departments. For more information visit www.pito.org.uk/products/pnc.php

2 The purpose is admirable, but as with *ContactPoint* (qv), the *National DNA Database* (qv), the *NHS Care Records Service* (qv), and a number of other Government databases the PNC poses a serious threat to our *civil liberties* (qv) and *privacy* (qv), and is penetrating our lives ever more deeply despite never having been discussed with, let along agreed to by the electorate. The threat is twofold:

• From childhood onward every trivial misdemeanor we commit is now being recorded and stored in databases to which hundreds of thousands of officials have access, to haunt us for the rest of our lives. To take one recent example, reported in the *Daily Telegraph* on 19 December 2006 under the title 'Model pupil ends up with criminal record for a push in the playground', this database now holds details of the 14-year-old boy, described by his school as 'a model pupil', who pushed over (reportedly without injuring) another pupil whom he suspected of bullying his brother. In May 2006 young master Persaud was issued with a reprimand which resulted in his name being placed on the PNC. I will remain on Criminal Records Bureau (qv) checks for the rest of his life – a fact that will probably prevent him ever getting a job as a teacher, which according to his father is his most fervent ambition.

• This would be bad enough even if all the data held by the PNC were 100% accurate. But they are not. In January 2007 the Government confessed to a failure to log on 27 000 crimes committed by Britons abroad, some of them of these very serious. If errors of this kind are possible, then so are others. Some expert commentators believe that the PNC is not fit for purpose now, and may never be.

3 For notes on the performance of several other large Government databases see: *IT systems.*

Political correctness (PC)

1 Political correctness started with the commendable aim of excising racist an other nastiness from everyday speech, and its achievements in that regard are wholly to be welcomed. No-one can seriously regret the expulsion of words like nigger, faggot, or jewboy from our vocabulary.

2 Unfortunately that virtuous impulse has soured into bossy and intolerant forms of behaviour that are quite inconsistent with a free society. They take various forms, some harmless, some not so harmless, and others seriously malignant.

• Among the moderately harmful varieties of PC is the assiduous attempt by sel appointed representatives of minority groups to stop other people using certain words or expressions which they have decreed to be offensive – see the next ent

• A more malignant variety is the organised attempt to silence those who hold 'mistaken' or 'offensive' views not by rational argument, but by blocking publication of their writings, obstructing their appointment to positions of influence in colleges and universities or in publishing, and trying to prevent their views being heard. Numerous examples could be adduced, but let the hounding of the moderate and courteous philosopher Roger Scruton, recounte in his book GENTLE REGRETS (see below), stand for all. Repeated attempts by academic philosophers to prevent Professor Scruton's views being heard include a letter from an Oxford philosopher to his publisher, Longman, urging the latter to stop publishing his books (see 55 of the above book); and his arrival at the University of Glasgow to deliver an invited lecture, only to find that the Philosophy Department was organising a boycott of his lecture while (by a neat irony) elsewhere on the campus an honorary degree was being conferred on Mr Robert Mugabe – see p 53. These episodes neatly encapsulate the PC attitude to freedom of thought and debate.

3 It is a sign of the times that a group of 60 academics felt it necessary in December 2006 to found an organisation called Academics for Academic Freedom (qv), and call for a change in the law to ensure that scholars are guaranteed the freedom to question and test received wisdom and put forward unpopular views.

4 All strands of PC should be resisted. Truth and the well-being of society are better served by a rich variety of views, freely tested in open debate, than by the prohibitions imposed by thought and language police. As the commentator Henry Porter has put it in the *Observer* newspaper: 'Abuse is the corollary of free speech. I would prefer everyone to be well-mannered and respectful, yet I believe gays have the right to be rude about the church and the church to be rude about gays, without either running to the law'.

5 As of 2008, useful references include the following. They are given in alphabetical order. For publication details see the bibliography.

• CHILDHOOD WELLBEING, a research study published by the Department for Children, Schools and Families, reports the fears of many parents that their children's lives are being 'systematically undermined' by political correctness and the health and safety culture.

• GENTLE REGRETS by Roger Scruton. This is a good-humoured book of recollections, not a political tract, but the author's ruminations on life, the arts, architecture, and other matters provide a humane counterview to the rigidities of political correctness.

• THE RETREAT OF REASON: POLITICAL CORRECTNESS AND THE CORRUPTION OF PUBLIC DEBATE IN MODERN BRITAIN by Anthony Browne. A polemic published by the *think tank* (qv) Civitas.

6 Useful websites include:

• Academics for Academic Freedom www.afaf.org.uk/

See also: *Speech: freedom of.*

---

olitical correctness peech codes

1 The conflict between PC speech codes and clear communication is seen in the following examples from the EQUAL TREATMENT BENCH BOOK, an official guide for courtroom personnel (see the bibliography). Section 5.1.3 of this otherwise admirable publication lays down the following rules on courtroom terminology:

• Avoid the term 'wheelchair bound' and say instead 'wheelchair-user'.

• Avoid using 'suffers from' and say instead 'has', or use other more neutral terminology when speaking of peoples' disabilities.

• Avoid using 'mental illness' and say instead 'mental health problems'

• Avoid using 'mental handicap' and say instead 'learning disabilities or difficulties'.

Taking these injunctions in turn:

• A majority of *wheelchair-users* (qv) are part-time users who can also move about without their wheelchairs. A few are people such as *paraplegics* (qv) who are helpless without their wheelchairs and quite literally 'wheelchair bound'. Wheelchair-users in the second group are very different indeed from the first, and their interests are not served by misusing the English language to pretend otherwise.

• Many people with acute *arthritis* (qv), severe cerebral palsy, or paraplegia do indeed 'suffer' from these distressing conditions; and journalists and others should not be prohibited from speaking accurately. Interviewed by the journal DISABILITY NOW (qv) in August 2005, the BBC journalist Frank Gardner, who became wheelchair bound after being shot while working in Saudi Arabia, was scathing about what he termed such 'censorship'. Defending his right to call himself a 'sufferer' he asked why he should conceal the fact that was in pain. 'I'm first and foremost a journalist', he told his rather shocked interviewer, an 'I like to call a spade a spade.'

• There are mental illnesses just as there are physical illnesses. To pretend otherwise is wrong and irresponsible.

• While 'learning disabilities' or 'learning difficulties' are kindly and acceptable terms in many general contexts, there are occasions when the fact that someone is 'mentally handicapped' or 'mentally retarded' is true, relevant and needs to be clearly stated.

3 The linguistic diktats quoted above are becoming so all-pervasive (even the publishers of the *Oxford Dictionaries of English* have lately succumbed), and are purveyed with such an air of authority, that ordinary people need to be reminded that they are the true owners of the English language and are entitle to think their own thoughts and choose their own words. Here is an admirable extract from an article by Rosa Monckton, President of the charity Kids, published in the *Sunday Telegraph* on 16 May 2004 under the title 'I will talk of my daughter in English, not Newspeak'. Writing about her daughter, she observes: 'Domenica is mentally handicapped ... I do not begin to understand what a 'learning disability' could possibly mean, and a 'learning difficulty' implies that there is some possible cure. This is manifestly not the case … The awful euphemism of 'special needs' means nothing and everything. Everybod on the planet has special needs of one form or another, but we don't all requir the help that is necessary for those who actually are mentally handicapped. A I defend my right, as a mother, to use those words'.

4 As of 2008, useful references include the following. For publication details see the bibliography.

• EQUAL TREATMENT BENCH BOOK: AN OFFICIAL GUIDE FOR COURTROOM PERSONNEL

See also: *Speech: freedom of.*

Polling stations

1 A survey of 1 965 polling stations published in 2001 by the *charity* (qv) Sco found that only a third of polling stations had been fully *accessible* (qv) during the previous general election, and that hundreds of *disabled* (qv) voters had faced major problems. The latter ranged from slippery floors and dangerous ramps to total inaccessibility. Some were forced to complete their ballot papers on the street and others did not vote at all.

2 A similar survey involving more than 1 300 disabled people was carried out i May 2005. Sensory and physical accessibility were tested, and two thirds of polling stations failed the test – visit www.pollsapart.org.uk/

3 An experiment in all-postal voting in June 2004 was widely judged not to offer a solution to the problem. While the option of postal voting is greatly to be welcomed, the parliamentary affairs manager of the Disability Rights Commission (qv) strongly criticised the Government for pressing ahead with all-postal voting despite manifest and widespread opposition.

See also: *Electoral fraud;* and *Voting systems.*

r: core The *core poor* are the poorest 10% of the population. See: *Poverty*.

ulation: ethnic nposition 1 For the latest statistics, plus several pages of authoritative commentary on this topic, refer to Chapter 1 of the annual survey SOCIAL TRENDS (qv). The 2006 edition of this publication also contains an informative special introductory chapter titled 'The different experiences of the United Kingdom's ethnic and religious populations'.

2 There are two levels of classification:

- Level 1 has five main ethnic groups: White, Mixed, Asian or Asian British, Black or Black British, Chinese or other ethnic group.
- Level 2, which is officially preferred, includes White British, White Irish, Other White, Mixed, Indian, Pakistani, Bangladeshi, Other Asian, Black Caribbean, Black African, Other Black, Chinese, and finally Other ethnic groups.

3 Using the more detailed classification, the 2002 census showed that almost 70% of people in Great Britain described their ethnicity as White British and their religion as Christian. Other large faith groups included Pakistani Muslims, Indian Hindus, Black Caribbean Christians, Black African Christians, and Indian Sikhs.

4 BEYOND BLACK AND WHITE: MAPPING NEW IMMIGRANT COMMUNITIES, a report published in 2005 by the think tank (qv) the Institute of Public Policy Research gave what was claimed to be the most detailed and up-to-date analysis of how the balance between Britain's indigenous and foreign-born populations was changing. As instances:

- The proportion of the UK population that was born abroad grew from 5.7% of the total in 1991 to 7.5% in 2001. In absolute numbers this was an increase of 1.1 million, thus accounting for half the 2.2 million overall growth in the UK population over the decade.
- The overwhelming majority of the above people live in London (41% of the total) and in the south-east of England. In London a quarter of the population of the population was born abroad, compared with fewer than 3% in Wales or the northeast of England. In Wembley foreign-born citizens are a majority, and several other parts of London are heading in the same direction, as are some parts of the east and west Midlands.
- The average immigrant earns more, and pays more tax, than the average native-born citizen. In particular, Australians, New Zealanders, and Filipinos do better than the UK average economically, while Iranians, Iraqis, Bangladeshis, and people from the former Yugoslavia are currently doing worse. Angolans top the unemployment list.

5 As of 2008, useful references include the following. They are given in alphabetical order. For publication details see the bibliography.

- BEYOND BLACK AND WHITE: MAPPING NEW IMMIGRANT COMMUNITIES by Sarah Kyambi (editor) was published in 2005 by the think tank (qv) the Institute for Public Policy Research.
- SOCIAL TRENDS (qv) gives latest statistics plus commentary on the above and related matters.

See also: *Diversity*.

ulation: growth 1 For the latest statistics, plus authoritative commentary on this topic, refer to Chapter 1 of SOCIAL TRENDS (qv), with the proviso that immigration figures are now so inaccurate (as conceded by the Government in 2007) that readers would be well-advised to check them against those from the think tank (qv) Migrationwatch at www.migrationwatchuk.org/

2 The UK population grew from 55.9 million people in 1971 to 59.8 million in 2004. In 2006 it was forecast to reach 67 million in 2031 (with most of the increase in England, especially in the South East, and very little in Scotland) thus placing very great pressure on *housing* (qv), the NHS (qv), other public services and *pensions* (qv). In November 2007 the ONS estimated that the population could conceivably reach 108 million (sic) in 2081.
3 In the past, population growth could be explained almost entirely by the balance between birth rates and death rates, but the main cause as of 2007 is immigration. Of the anticipated overall population rise of around 7 million between 2005 and 2030, more than 50% is expected to be the direct result of immigration, and another 30% will be formed by the children of recent immigrants.
4 THE ECONOMIC IMPACT OF IMMIGRATION. 1ST REPORT OF SESSION 2007–08 OF THE HOUSE OF LORDS SELECT COMMITTEE ON ECONOMIC AFFAIRS, published in April 2008 (see the bibliography), gives an authoritative, trenchant and somewhat troubling analysis of the unprecedented wave of immigration now flowing into the UK.
5 As of 2008, useful websites include:

- Civitas www.civitas.org.uk/pubs/immigrationMain.php
- Migrationwatch www.migrationwatch.co.uk/
- Social Trends www.statistics.gov.uk/socialtrends

Population: racial integration

1 MEASURING RESIDENTIAL SEGREGATION, a report written by Dr Ludi Simpson and published in November 2005 by the Cathie Marsh Centre for Census and Survey Research at the University of Manchester, argues that Britain is becoming less racially segregated, contrary to popular belief – the latter expressed for instance in 2005 by the then Director of the Commission for Racial Equality (qv), Mr Trevor Phillips.
2 The researchers compared data from the 1991 and 2001 censuses in 8 850 electoral *wards* (qv) in England and Wales and found that the number of mixed neighbourhoods or wards, where at least 10% are from an ethnic minority, rose from 964 to 1 070 in the decade, and predict that by 2010 the number will rise to 1 300. Dr Simpson said: 'Our research suggests there is a lot of good news. There is more mixing. On balance there is neither retreat nor 'white flight'. The larger populations of black and Asian communities that have been highlighted are there simply due to natural growth, i.e. people having children. There are no black or Asian ghettos anywhere in the UK where families of one colour are trapped. In all parts of Britain, the children of immigrants are moving away from so-called ghettos. After a couple of generations the mixing will be far more noticeable and the population growth of these groups will have slowed and probably stopped'.
3 For publication details of the above reference see the bibliography.

See also: *Diversity;* and *Social exclusion.*

Positive discrimination See: *Discrimination: positive.*

Possum

1 Possum ('Patient Operated Selector Mechanism') is a government-approved supplier of high-tech *aids and equipment* (qv). Visit www.possum.co.uk/
2 See also: Environmental control systems; and Speech aids.

Post-democracy

1 Britain is a democracy, and in most ways a very good one. We enjoy freedom of speech; every citizen has the vote; ordinary people can (and regularly do) eject governments from power after free and open elections; and successive UK governments have created a welfare state which, despite much public

grumbling, cares for British citizens in a way that most of the world's population can only envy.

2 And yet very many British citizens have started to feel that within the outward shell of democratic institutions they are progressively losing political power and influence. Their increasing loss of faith in the effectiveness of representative democracy is seen in the fact that between a third and two thirds of the UK electorate (depending on the kind of election) no longer feel it worth casting their vote. This cynicism is confirmed by several studies, some of which are listed at the end of this entry.

3 Going back to first principles, it is probably safe to say that what most people mean by democracy is:

• First, being free to live as one likes, provided that no harm or serious inconvenience is caused to others.

• Second, having a real say in the formation and application of all laws and regulations that may limit such freedom.

Today both of the above principles are being eroded in Britain (and to some degree in most democracies) not by tyrants but by generally well-meaning governments in pursuit of managerial convenience.

4 On the first point, our freedom to live and speak as we choose is being invaded by misused laws (see: *Civil liberties*); too many laws (see: *Over-legislation*); busy-body regulations (see: *Over-regulation*); and prying databases (see: *Privacy*).

5 On the second point, examples of our growing inability to influence the way we are governed include the following:

• Unlike their counterparts in Europe or the USA, British *local authorities* (qv) now have so little autonomy that voters increasingly cannot be bothered to turn out for local elections – see: *Local authorities: funding;* and *Localism.*

• The few responsibilities that do still rest with local authorities are being progressively passed upward to new regional bodies that are further away from the people they serve, and – even worse – unelected – see: *Local authorities;* and *Regions.*

• At national level, our elected Parliament now makes less than half of the legislation going through Westminster. Between 50% and 80% (estimates vary) of our legislation comes from Brussels and cannot be blocked by MPs – see: *Parliament.*

• Unelected and virtually unaccountable *quangos* (qv) have come to run huge swathes of what we like to think of as the 'public' sector. The quango empire now includes, for instance, much of the NHS, including England's 28 Strategic Health Authorities and 303 Primary Care Trusts. A full list is given on pp 134–178 of PUBLIC BODIES 2006 (see below).

• Finally, an increasing number of decisions that ought to rest with Britain's democratically elected government are being taken by judges who are beyond the reach of the electorate. We refer here not to strictly legal decisions, which must clearly rest with judges acting independently, but to those cases where two legal principles clash and the conflict is resolved not by Parliament, which ought to make law, but by judges, who ought only to interpret law. A topical example is the conflict between (a) the right of the UK Government to protect the safety of its citizens and (b) the right of an illegal immigrant to be protected from inhuman or degrading treatment if he or she is returned to their country of origin. Both are legal rights. Sometimes a choice must be made between the two, and a decision taken on where the balance lies. Existing laws do not give the answer, and a more fundamental judgement must be made – one that broadly reflects the wishes of the people, as all laws must ultimately do. Such fundamental decisions properly belong with the UK's elected Parliament, but are regularly being made by judges who possess neither the training nor the political legitimacy for such a task. For more on this see: *Constitution;* and *Human Rights Act.*

6 In a word, political power is in many (though certainly not all) ways drainin away from the electorate. Citizens who ought to be participants are becoming onlookers. All the forms of democracy remain in place, but the substance is lessening. Except for the power to decide once every 4 or 5 years whether to keep or eject the current government, British citizens (unlike for instance the Swiss) have almost no part to play in decisions which deeply affect their lives The result can fairly be termed 'post-democracy'.

7 As of 2008, useful references include the following. They are given in alphabetical order. For publication details see the bibliography.

- COPING WITH POST DEMOCRACY by Colin Crouch. Professor Crouch argues that the democratic moment has passed and that we are moving into a post-democratic age in which power becomes increasingly concentrated in the han of a professionalised elite and the ordinary people are less and less able to participate in public life.
- DIRECT DEMOCRACY: AN AGENDA FOR A NEW MODEL PARTY is a proposal by a group of MPs, MEPs, and others for a new, truly democratic, bottom-up rathe than top-down political structure for the UK.
- THE GOVERNANCE OF BRITAIN. This *green paper* (qv) was published within days after Prime Minister Gordon Brown took office in June 2007 and calls f 'a journey towards a new constitutional settlement – a settlement that entrusts parliament and the people with more power'.
- PUBLIC BODIES 2006 lists all of the approximately 900 unelected and virtuall unaccountable *quangos* (qv) which now run many (perhaps a majority) of the UK's public services.
- POWER TO THE PEOPLE. THE REPORT OF POWER: AN INDEPENDENT INQUIRY INTO BRITAIN'S DEMOCRACY, published by the Joseph Rowntree Foundation (qv). The Power inquiry into Britain's democracy, which spent 18 months gathering evidence from thousands of people across the country and publishe the results in 2006, found no evidence of public apathy towards community a public affairs, but a steep decline of belief in formal democracy.
- SUPPLY SIDE POLITICS: HOW CITIZENS' INITIATIVES COULD REVITALISE BRITISH POLITICS by Matt Qvortrup. Professor Qvortrup, acknowledged as one of the world's leading authorities on referenda, suggests that direct democrac does not result in populist legislation and ill-considered laws, but the opposit and that it could help to re-engage the British electorate.
- A SURVEY OF PUBLIC ATTITUDES TOWARDS STANDARDS OF CONDUCT IN PUBLIC LIFE REPORT. The 2006 edition of this ongoing, long-term official investigation revealed findings very similar to those above. The Committee's Chairman, Sir Alistair Graham, expresses alarm at the corrosive effects of thi low and deteriorating degree of trust upon democracy itself, and calls for a 'sea change' in governmental behaviour.
- WHAT DEMOCRACY IS FOR: ON FREEDOM AND MORAL GOVERNMENT by Stein Ringen. The Norwegian Dr Ringen is Professor of Sociology and Social Polic at Oxford University, and one of Europe's leading social scientists. His book criticises the state of British democracy and offers useful proposals.

8 Useful websites include the following. They are given in alphabetical order.

- Committee on Standards in public Life www.public-standards.gov.uk/
- Direct Democracy www.direct-democracy.co.uk/
- openDemocracy www.opendemocracy.net/
- My Society www.mysociety.org
- Power Inquiry www.powerinquiry.org/report/index.php

See also: *Parliament*.

st-legislative rutiny See: *Over-legislation.*

verty 1 There are two measures of poverty – absolute and relative.

• Absolute poverty is defined in terms of people's access to a number of basic needs. The latter are usually taken to be (in alphabetical order) clean water, education, food, health, information, sanitation, and shelter. People lacking access to any one of these goods may be defined as 'deprived'; and if they are deprived of two or more then they may be defined as being in 'absolute poverty'. These are terms and yardsticks that have for been used for instance by UNICEF – for more information on this visit www.unicef.org/

• Relative poverty is a term applied to people whose income falls below a certain percentage of the national *median income* (qv). In the UK this threshold is officially set at 60% of the median – a figure that some commentators find unduly high, but one that is widely accepted and seems reasonable to the present author. The graph below, from Figure 5.12 in SOCIAL TRENDS 36 (see the bibliography), illustrates the concept. It shows weekly household disposable incomes in Great Britain in 2003–04.

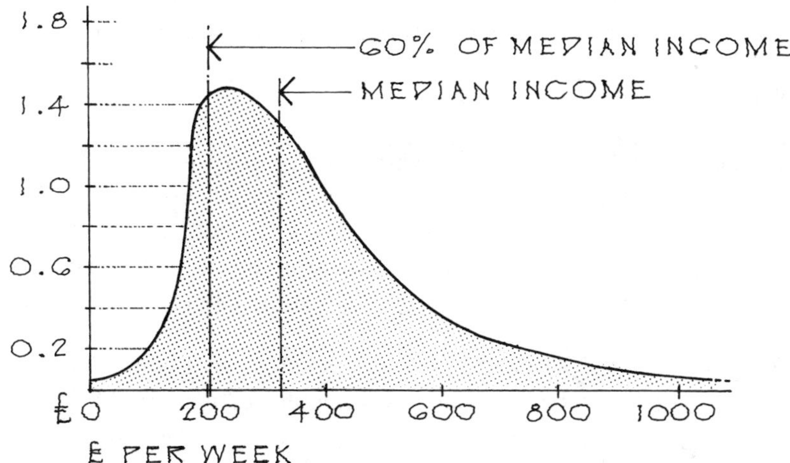

2 The next graph, from Figure 5.15 of SOCIAL TRENDS 37, shows how the percentage of people in Great Britain whose incomes are below 60% and 50% of median household disposable income changed from 1961 to 2004–05.

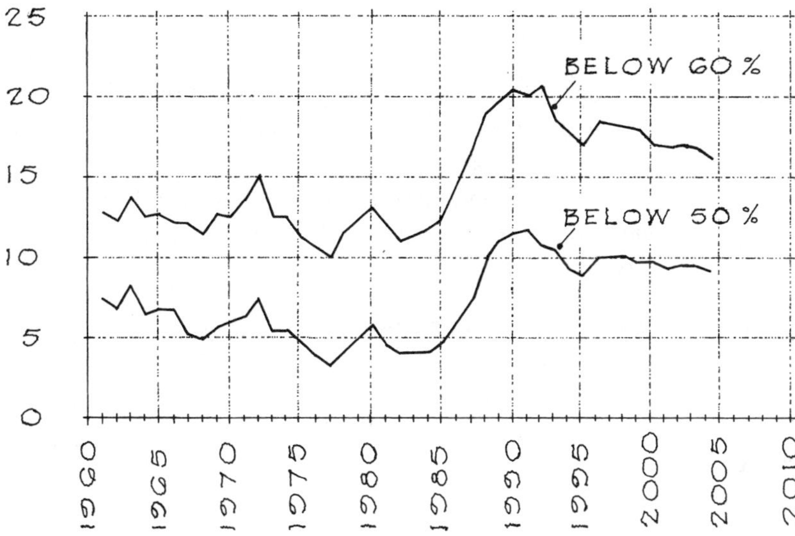

In April 2007, after the above graph was published, the Department for Work and Pensions (qv) announced that between 2004–05 and 2005–06 the number of people in poverty rose from 12.1 to 12.8 million, and the number of children in low-income households from 3.6 to 3.8 million – the first such increases in almost 10 years – see: *Child poverty*.

3 As of 2007, useful references include the following. They are given in alphabetical order. For publication details see the bibliography.

• THE ENGLISH INDICES OF DEPRIVATION 2004. This is an official tool used by the Government for identifying areas of poverty in England, which require remedial action.

• OPPORTUNITY FOR ALL: EIGHTH ANNUAL REPORT 2006. This is the annual Government report about tackling poverty and social exclusion.

4 Useful websites include the following. They are given in alphabetical order.

• Centre for Social Justice www.centreforsocialjustice.org.uk/
• Dept for Communities & Local Government www.communities.gov.uk/
• Joseph Rowntree Foundation www.jrf.org.uk/
• New Policy Institute www.npi.org.uk/projects/rfa.htm
• Office for National Statistics www.statistics.gov.uk/
• Opportunity for all www.dwp.gov.uk/ofa/

See also: *Income: median; and Incomes: inequality.*

Poverty: child See: *Child poverty.*

Poverty: disabled people 1 As of 2008 the poverty rate for *disabled* (qv) adults is twice as high as that for others, and higher than it was10 years ago.

2 MONITORING POVERTY AND SOCIAL EXCLUSION IN THE UK 2006 by G Palmer, T MacInnes and P Kenway, from the Joseph Rowntree Foundation (qv) gives more information. For publication details see the bibliography.

Poverty: ethnic minorities 1 POVERTY AND ETHNICITY IN THE UK by Lucinda Platt, a study from the Joseph Rowntree Foundation (qv) published in April 2007, states that the poverty rate among *ethnic minorities* (qv) in the UK is around 40%, which is twice as high as for white people.

2 For other data refer to Chapter 5 of the official annual publication SOCIAL TRENDS (qv).

3 For publication details of the above references see the bibliography.

See also: *Incomes: inequality.*

Poverty: pensioner See: *Pensions.*

Poverty: persistent 1 People in persistent poverty are officially defined as those who have lived below the 'poverty threshold' (see*: Poverty*) for at least three out of four consecutive years.

2 This rate is holding fairly steady around 11% to 12% of the population. For more detail refer to Chapter 5 of the latest edition of SOCIAL TRENDS (qv) – see the bibliography.

Poverty: rural See: Commission for Rural Communities.

Poverty trap See: *Tax Credits.*

Power to the people See: *Post-democracy.*

[Po]wered doors

1 The installation of powered entrance doors can do more to make a building *accessible* (qv) than almost any single other provision. While self-opening doors are particularly associated with entrances, they can also be useful inside the building. They are of course particularly helpful and appropriate when the door is heavy, as is the case with many historic buildings – see: *Historic buildings and sites.*

2 As of 2007, useful websites include the following,

• Automatic Door Suppliers' Association www.adsa.org.uk/

See also: *Doors;* and *Entrance: accessible.*

[Po]wered doors in public [bu]ildings: complying [w]ith the Disability [Di]scrimination Act

1 Buildings used by the public must in general comply with the Disability Discrimination Act (qv). In the case of powered doors in public buildings, *authoritative practical guides* (qv) include DESIGNING FOR ACCESSIBILITY (qv) and in addition the following documents. For publication details see the bibliography.

• APPROVED DOCUMENT M (qv), the provisions of which must be heeded in order to comply with the Building Regulations (qv) for England and Wales.

• BS 8300 (qv), which applies to all buildings in the UK.

• INCLUSIVE MOBILITY (qv), which applies to *pedestrian infrastructure* and to transport-related facilities such as bus and coach stations, railway stations, air terminals, and transport interchanges; and is an advisory document.

• AUTOMATIC DOOR SYSTEMS, published by the Centre for Accessible Environments (qv), gives impartial and well-informed advice on all aspects of selection and installation.

Except for AD M the above references do not have the force of law, but conformity with their recommendations will make it easier to demonstrate that the requirements of the Disability Discrimination Act have been met.

For places other than England and Wales the following regulatory documents should be consulted:

• In Northern Ireland: TECHNICAL BOOKLET R of the Building Regulations (Northern Ireland) (qv).

• In Scotland: the TECHNICAL HANDBOOKS of the Building (Scotland) Regulations (qv).

2 The relevant clauses in AD M, BS 8300, and INCLUSIVE MOBILITY are set out below.

DESIGN ASPECT	DOCUMENT: RELEVANT PARAGRAPHS		
	APPROVED DOCUMENT M	BS 8300	INCLUSIVE MOBILITY
GENERAL NOTES	2.8 – 2.13 2.18 – 2.24 3.7 – 3.10; 4.18	6.3.3	8.2
REVOLVING DOORS	2.20	6.3.5	8.2
LOW ENERGY DOORS	3.7 – 3.10	6.3.4	8.2
CONTROLS	2.18 – 2.21 4.23	6.3.3	8.2

Powered curtain drawers	See: *Environmental controls.*
Powered lifters	1 *Worktops* (qv) such as desks, tables, sinks, and hobs can be mechanically raised or lowered to suit people with special needs. The action can be controlled by switch or by remote control via a handset or some other device. 2 As of 2008, useful websites include: • Disabled Living Foundation www.dlf.org.uk/ • Ricability www.ricability.org.uk/ See also: *Adjustable furniture; Assistive technology; Products for easier living;* and *Worktops.*
Powered wheelchairs	See: *Wheelchairs: powered.*
Powered window openers/closers	See: *Windows: controls.*
Precautionary principle	1 The principle that where there are threats of serious or irreversible damage, lack of full scientific certainty should not be used as a reason for postponing cost-effective measures to deal with the situation. 2 It is relevant for instance to *risk assessments* (qv). For more on this visit www.hse.gov.uk/
Premature babies	See: *Babies: premature.*
Premises: letting of	See: *Landlords: complying with the Disability Discrimination Act.*
Prenatal diagnosis	See: *Abortion.*
Preschool learning	See: *Education: pre-school.*
Pressure groups	1 These are organisations that try to influence public policy in the interest of a particular cause by gathering and disseminating information; mounting campaigns to draw attention to their views and gain the support of the public; and lobbying governments to change the law. 2 They fall broadly into three groups: • First, 'bottom-up' voluntary organisations that were created by concerned citizens. Examples include Age Concern; Arthritis Care, Barnardo's, RADAR (the Royal Association for Disability and Rehabilitation); the RNIB (Royal National Institution of the Blind); the RNID (Royal National Institution for Deaf People); Scope, which concerns itself with cerebral palsy; and Sense, which works for *deafblind* (qv) people. • Second, charitable trusts that were created by single large benefactors. The Peabody Trust and the Joseph Rowntree Foundation (qv) are two well-known examples. • Third, 'top-down' *quangos* (qv) that were created and are fully funded by the Government. They include for instance the Commission for Equality and Human Rights (qv). The existence of bodies of the first two types is vital to a healthy society, but some of them have started to change in disquieting ways. 3 First, many of these organisations are no longer run by *volunteers* (qv) selflessly trying to make the world a better place, but by highly-paid executiv who measure their success in terms of media headlines. Campaigns of the

traditional type used to win the respect and support of people across the political spectrum; those of the new kind are often strident and politicised, and may alienate almost as many people as they win over.

4 Second, the new style of campaigning is often associated with a dogmatism and intolerance that leads to constant attempts to shut down debate instead of opening it up. ASH and other anti-smoking groups may be right about the dangers of environmental tobacco smoke (ETS), but their near-libellous assaults on the integrity of any researcher who dares to come to other conclusions, and their assiduous efforts to deny opponents the right to be heard, belong in a totalitarian society. See: *Political correctness*; and *Speech: freedom of.*

5 Third, some pressure groups are developing an unhealthily close relationship with government. This relationship operates both upward and downward.

• In the upward direction, the more pushy pressure groups exert a powerful influence upon government policy, and are now playing a greater role in the shaping of public policy than the theoretically sovereign electorate, who often hold a very different view. In matters to do with *children* (qv), for instance, the opinions of parents now count for almost nothing compared with those of unrepresentative *experts* (qv) and *theorists* (qv) acting through single-issue pressure groups (the Government's decision in October 2007 not to criminalise parents who smack the bottom of a naughty child, after polling found that 70% of parents opposed such a law, is an exceptional but welcome step in the right direction). In some cases these spokespersons do not represent the views even of the groups they claim to speak for. When, for instance, Lord Joffe's carefully drafted Assisted Dying for the Terminally Ill bill came before Parliament in late 2005 the pressure group Scope (qv) lobbied vigorously against it. When Scope polled the views of the disabled people it claimed to represent, 93% favoured the bill and 7% were against (see DISABILITY NOW, December 2005, p 17), but the pressure group nevertheless continued its vehement assaults upon the bill. See also: *Self-deliverance and disabled people.*

• In the downward direction, *charities* (qv) of all kinds, including pressure groups, are beginning to receive a lot of government funding, and in consequence run a distinct risk of becoming obedient government tools instead of the awkward, dissenting bodies they ought to be. For more on this see: *Charities and the state.*

See also: *Advocacy research; Charities and the state; Think tanks;* and *Voluntary and community sector.*

[Pri]me Minister's [Aw]ard

1 The Prime Minister's Award for Better Public Building was launched in 2000 as part of the Better Public Buildings Initiative, to mark – it was stated – the government's commitment to a well-designed, high-quality built environment in the 21st century. It is jointly sponsored by CABE (qv) and the Office of Government Commerce (OGC). Past winners include:

2001 The Tate Modern, London.
2002 The City Learning Centre at Brislington School, Bristol.
2003 Bournemouth Library.
2004 The A650 Bingley Relief Road, Yorkshire.
2005 The Jubilee Library, Brighton.
2006 City of London Academy, Bermondsey, London.

2 As of 2008, useful websites include:

• Better Public Buildings	www.betterpublicbuildings.gov.uk/
• CABE	www.cabe.org.uk/
• Office of Government Commerce	www.ogc.gov.uk/

See also: *Access awards.*

Prime Minister's Strategy Unit

1 The Strategy Unit was set up in 2002, bringing together the then Performan and Innovation Unit, the Prime Minister's Forward Strategy Unit, and parts o the Centre for Management and Policy Studies. It reports to the Prime Minis through the Cabinet Secretary, and is stated to have three key roles:

- To support the development of strategies and policies in key areas of government in line with the Prime Minister's priorities.
- To carry out occasional strategic audits to identify opportunities and challenges facing the UK and the UK Government.
- To develop as a centre of excellence in order to enhance strategy across government.

2 One of the Unit's key areas of activity has been the fight (so far wholly unsuccessful) to reduce the burden of *red tape* (qv) that is smothering businesses, public sector bodies, *charities* (qv), and *voluntary societies* (qv), and the creation of a series of *better regulation bodies* (qv) to further that purpose. It was also responsible for the creation of the Office for Disability Issues (qv).

3 For more detail visit www.strategy.gov.uk/

See also: *Better Regulation Bodies; Cabinet Office;* and *Government.*

The Prince's Charities

1 This is a group of 18 charities of which the Prince of Wales is President. Sixteen of them were founded by the Prince personally, and the group adds u to the largest charitable enterprise in the UK, raising and spending over £100 million annually. Those who have worked with these charities and/or benefite from them, are almost unanimous in their praise for the combination of high ideals and good sense that characterise these enterprises.

2 For one example of the group's work see: *Neets.*

3 Another telling example is the work being done by the Prince's Foundation for the built environment, which is closer to what millions of ordinary people feel about buildings and cities than the professional institutions, schools of architecture, government departments, and other bodies who claim to be interpreting and responding to the needs of the public in these matters – see: *Architecture*.

4 As of 2008, useful websites include the following. They are given in alphabetical order.

• Prince's Charities	www.princeofwales.gov.uk/
• Prince's Foundation of the Built Environment	www.princes-foundation.or

Principal entrance

1 With respect to dwellings:

- Para 0.30 of APPROVED DOCUMENT M (qv) of the Building Regulations defines the principal entrance as 'the entrance which a visitor not familiar wi the building would normally expect to approach, or the common entrance to a block of flats'.
- Paras 6.2 and 6.5 of AD M state that on plots which are reasonably level, *wheelchair-users* (qv) should be able to approach the principal entrance or, if that is impossible, to approach a suitable alternative entrance.

2 With respect to non-domestic buildings:

- Para 0.29 of APPROVED DOCUMENT M (qv) of the Building Regulations defines the principal entrance as 'the entrance which a visitor not familiar wi the building would normally expect to approach'.
- Paras 1.17 and 1.18 of AD M state that people with *impaired mobility* (qv) who arrive as passengers should be able to alight close to the principal entranc Refer to the original document for details.

3 With respect to buildings in general: paras 6.1 to 6.3 of BS 8300 (qv) give additional recommendations to those above. While BS 8300 is an advisory document and does not have the force of law, conformity with its recommendations will make it easier to demonstrate that the requirements of the Disability Discrimination Act (qv) have been met.
4 The schematic diagram below notes some of the key terms.

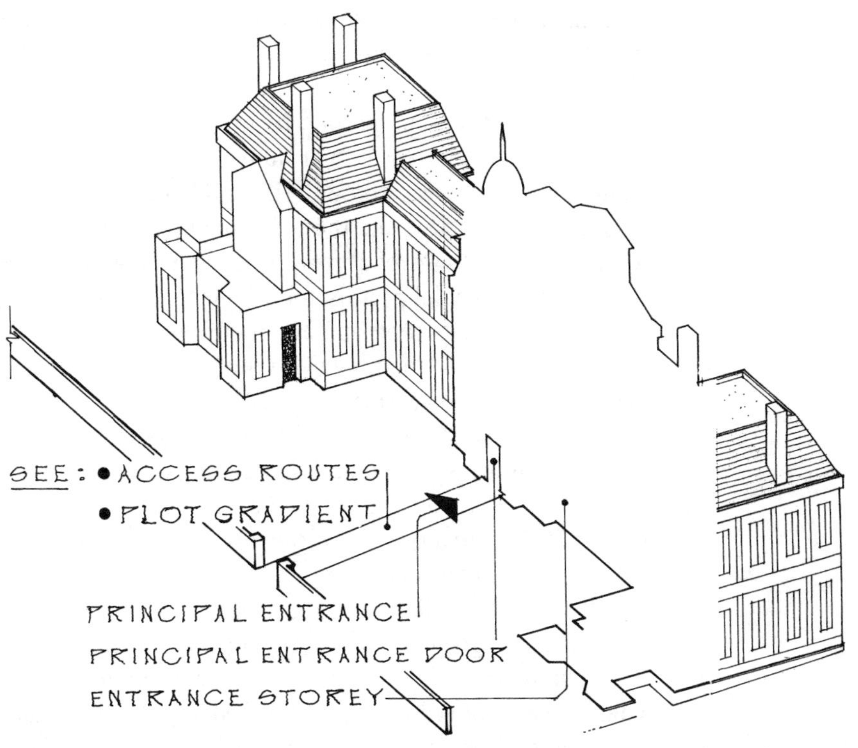

See also: *Access routes;* and *Entrance halls and reception areas.*

sons

1 Prisons are secure institutions where the relatively few people who commit the majority of serious crimes may be kept so that everyone else can be safe. According to p 133 of A BETTER WAY (see the bibliography) the Government believes that half of all crime is committed by approximately 100 000 persistent offenders, of whom about 20 000 are in prison at any one time. The remaining 80 000 are out on the streets, making the lives of law-abiding citizens a misery. If these 80 000 were behind bars the crime rate might be around half of what it is today (see the graph under: *Crime*), and we would be closer to a society in which women could walk the streets at night without fear; and parents could let their children freely roam about without worrying about drug dealers or juvenile gangs. For a demonstration of the crime-reducing effects of incarcerating more criminals, see the graphs on p 129 of A BETTER WAY.
2 We need to build very many new prisons, not only to lower the crime rate but also to gradually transform British prisons from slums to civilised places that offer some possibility of rehabilitation. Of the 141 prisons in England and Wales more than 30 are over 100 years old, and 1 in 6 are more than 150 years old. Cells are overcrowded; there are few workshops; and most inmates have no purposeful activity. Prisoner numbers cannot be much reduced by greater use of community sentences, as most prisoners now are serious, violent and persistent offenders. The time has come to start closing our sordid existing prisons and to build as many new, civilised institutions of various kinds as may be required. A total of 125 000 places would bring our prison places/crime ratio up to the EU average – visit www.civitas.org.uk/data/prisonPopEU2003.htm
3 For more on the above matters see: *Crime*.

Privacy

1 The right to privacy is an essential human liberty. As of 2007 it is threatened by the rapid development of information technologies that make it much easie than in the past to surreptitiously gather and store vast amounts of personal information, combined with a ruthless governmental determination to apply those technologies. In some cases the Government justifies these violations of our privacy by arguing that they are necessary in order to fight terrorism. But the then leader of the Liberal Democratic party, Sir Ming Campbell, correctly put it in July 2007: 'Of course the public has a right to security, but that includes security from the power of the state'.

2 The scale of change that has come over government/citizen relations in Britai in recent decades is staggering. In a famous passage, the great 20th-century historian AJP Taylor wrote that 'in August 1914, a sensible, law-abiding Englishman could pass through life and hardly notice the existence of the Stat beyond the post office and the policeman. He could live where he liked and as he liked. He could travel abroad or leave his country for ever without a passpo or any sort of official permission'. Ninety years later the state has pushed its tentacles into every nook and cranny of our lives. Over the past 10 years it has become determined to know everything about us, and is storing what it discovers in vast (and worryingly error-prone) databases of which we know little and over which we have virtually no control. Shamefully, on the Privacy International website (see below) Britain is the only western country to be coloured black, alongside China and Russia.

3 As of 2007 here are some of the most important of these invasions, given in alphabetical sequence:

• CCTV (qv) surveillance now permeates every aspect of our lives. Unlike the other instances given below, most people are well aware of this particular form of intrusion upon their privacy and – according to testimony given to the Hom Affairs select committee on 1 May 2007 by the Information Commissioner (q Mr Richard Thomas – not only accept the presence of 4.5 million CCTV cameras, but feel protected rather than threatened by them, believing that they help fight crime without unduly affecting the lives of law-abiding citizens. Reading A REPORT ON THE SURVEILLANCE SOCIETY (see below) may change their minds.

• ContactPoint (qv) is intended to hold the details of all 11 million children in England and Wales, whether their parents agree or not. The envisaged degree of state intrusion into the private lives of families is so great that Terri Dowty the children's rights group Action Rights for Children (ARCH) asked: 'Who i bringing children up? Are parents effectively nannies for the state's children, c are children born to families and the state just helps families when they ask fc it?'. A report written by a panel of experts on child protection, law, and computers, and published in December 2006 by Mr Richard Thomas, Parliament's Information Commissioner, charged that the database would shatter family privacy; undermine parental authority; violate the law; would n be secure; would inevitably contain inaccurate information (with potentially traumatic consequences for falsely accused parents); would waste millions of pounds; and would actually put children in greater danger because the huge amount of information would make it much harder to spot those in genuine danger.

• The Criminal Records Bureau (qv), which is used for checking the backgrounds of all people whose work might bring them into contact with children, carried out almost 9 million such checks in the three years prior to 2006, and in future up to 9.5 million adults in England and Wales – more thar third of the adult working population – could be subject to ongoing criminal checks to establish their suitability to work with children. The people obliged to undergo such checks would include lollipop ladies, youth club workers, 16-year-old boys teaching sport to younger children at weekends, and a father who offers to coach his son's team on a Sunday morning. In all of these cases failure to have undergone a formal criminal records check would result in a criminal record and possibly a £5 000 fine.

• Fingerprinting of children by schools. In July 2007 we learnt that schools had been given ministerial permission to take fingerprints from children as young as five without any requirement for parental permission. According to a survey by Mr Greg Mulholland MP, this was already being done and parental permission had been sought in fewer than 20% of cases.

• Identity cards. Under the Identity cards Act 2006 (qv) every citizen will in due course be obliged to inform the authorities each time he or she changes their address, or be fined up to £1 000 per offence.

• The National DNA Database (qv) is the largest in the world, holding the profiles of 5% of the UK population, compared with 1% in the EU and 0.5% in the USA. The UK figure includes several hundred thousand children, an estimated 100 000 of whom have never been charged with, let alone convicted of, any crime. Current law allows the police to take samples (eg a swab from the mouth) without consent, and by force if necessary, on arrest for most offences. This includes children above the age of 10 who have committed some minor indiscretion.

• The NHS Care Records Service (qv) is a central database that is intended in due course to hold the medical records of every NHS patient in England. It will contain the most intimate and in many cases embarrassing information about individuals and families; and will be accessible to around 200 000 officials. It is absolutely certain (a) that a database containing the complex records of around 50 million people will contain millions of errors; (b) that many of these will be damaging to the people concerned; (c) that patients may often be unaware of the faulty data, and suffer the results without knowing why; and (d) that if patients do happen to find mistakes, they are very likely to find themselves getting entangled in a Kafkaesque nightmare if they try ensure that the errors are completely erased from all files.

4 The above developments are disturbing in four respects:

• First, the extent of the net that is being thrown over us. State databases now under development are intended to contain personal information on (a) all children; (b) all NHS patients; (c) one in three of the working population; (d) everyone who has had any kind of brush with the police; and (e) some people who are merely the victims of crime.

• Second, the fact that most of these databases were created without any public consultation, often without Parliamentary scrutiny, and in some cases (eg ContactPoint) in conditions of virtual secrecy.

• Third, the factual unreliability of such databases. In May 2006 the Criminal Records Bureau (qv) found that it had wrongly labelled 2 700 innocent people as paedophiles, pornographers and violent criminals. In January 2007 the Home Secretary conceded that over half of 540 serious criminals (including rapists and murderers) had not been put on the Police National Computer (qv).

• And fourth, their lack of security. In November 2007 HM Revenue and Customs admitted that it had 'lost in the post' a CD containing the unencrypted personal details of every child in the country, together with the bank account and NI numbers of their parents or guardians – that's 25 million ultra-sensitive personal records. In December 2007 the Department for Transport admitted that a hard disc containing the records of 3 million drivers had 'gone missing'. In January 2008 the Department of Work and Pensions admitted that hundreds of benefit files, bank statements, and passport forms had been found scattered on a roundabout near Exeter. In the same month the Ministry of Defence admitted that a laptop computer containing the sensitive details of 600 000 military recruits had been stolen from a car (the 3rd such incident since 2005), and in March 2008 it admitted that 11 000 military ID cards been lost or stolen in the past 2 years. In December 2007 the Ministry of Justice admitted that 4 compute discs containing confidential details of court cases had been 'lost in the post'.

5 As of 2008, useful references include the following. They are given in alphabetical order. For publication details see the bibliography.

• CROSSING THE THRESHOLD: 266 WAYS IN WHICH THE STATE CAN ENTER YOUR HOME by Harry Snook. In the 1950s there were 10 legal powers of entry; by the 1990s there were more than 60, and in most cases force can be used. Any questioning of an official could be deemed to be an obstruction, resulting in a fine of up to several thousand pounds or in some cases imprisonment – fc instance, a £1 000 fine for obstructing an inspector of hedge heights.

• DILEMMAS OF PRIVACY AND SURVEILLANCE – CHALLENGES OF TECHNOLOGICAL CHANGE by Professor Nigel Gilbert and others, and published by the Royal Academy of Engineering in March 2007.

• PROTECTION OF PRIVATE DATA, the first report of the Justice Committee of the House of Commons 2007–08, is scathing about the way companies and government departments handle people's private data, and calls for the reckles use of such data to be made a criminal offence.

• A REPORT ON THE SURVEILLANCE SOCIETY. This study, produced by the Surveillance Studies Network for the Information Commissioner (qv), suggests that the UK is more intensely surveilled than any other advanced Western society and makes brief recommendations.

• THE ROAD TO SOUTHEND PIER: ONE MAN'S STRUGGLE AGAINST THE SURVEILLANCE SOCIETY by Ross Clark.

• WHAT PRICE PRIVACY? a report produced for Parliament by the Information Commissioner's Office (qv) in May 2006 exposes the degree to which our mos private details can now be readily bought from private agencies of various kinds, and calls for prison sentences for offenders.

• WHAT PRICE PRIVACY NOW? a follow-up report to the above, reviewing progress up to December 2006.

6 Useful websites include:

• British Institute of Human Rights www.bihr.org.uk/
• Foundation for Information Policy Research www.fipr.org/
• Freedominfo.org www.freedominfo.org/countries/united_kingdom.htm
• Information Commissioner's Office www.ico.gov.uk/
• Liberty www.liberty-human-rights.org.uk/
• Manifesto Club www.manifestoclub.com/
• openDemocracy www.opendemocracy.net/
• Open Rights Group www.openrightsgroup.org/
• Privacy International www.privacyinternational.org/
• Your Rights www.yourrights.org.uk/

See also: *Civil liberties; Data Protection Act; Freedom of Information Act; Open Rights Group; Post-democracy;* and *Transformational government.*

Probation

See: *National Probation Service.*

Product design

1 Difficult-to-use products inconvenience most people to some degree, but fo those with physical or mental *impairments* (qv) the difficulties can be capplin The latter include not only those people normally thought of as 'disabled' but also most older people, and many normally able-bodied adults during periods of illness or while recovering from an injury. These individuals add up to a substantial proportion of the population.

2 While there are special products suitable for older, disabled, or temporarily unfit people (see: *Products for easier living*), such non-standard products are expensive, and the people who need them tend to be on lower than average incomes. The real need is for all products to be designed for easy use by all or nearly all people – see: *Inclusive design*. In view of the number of users who are non-standard in some way – either physically or mentally – such a policy would make business as well as ethical sense.
3 As of 2008, useful references include the following. For publication details see the bibliography.
• INCLUSIVE DESIGN: PRODUCTS FOR ALL CONSUMERS by Lindsey Etchell and David Yelding.
4 Useful websites include:
• Ricability www.ricability.org.uk

See also: *Inclusive design;* and *Products for easier living.*

---

roduct packaging

1 Research by the Occupational Ergonomics Department of Nottingham University, commissioned by the Department of Trade and Industry (qv) and published in February 2003, suggested that over 60 000 people require hospital treatment each year after injuring themselves while trying to open difficult packaging. Such mishaps occur when people give up trying to open a container in the manner intended and resort to using a knife, scissors, or teeth. Injuries requiring hospital treatment included:
• Cuts caused by trying to opening cans of corned beef requiring a special key (9 000 cases).
• Cuts caused by trying to cut open other types of can with a knife (18 000 cases, including amputated fingers).
• Cuts to the hands, or broken teeth, caused by trying to twist or prise the lids off glass jars.
• Eye injuries caused by people accidentally spraying the contents of aerosol bottle into their eyes while trying to open them.
2 Among the most serious hazards are glass bottles, which caused a large number of cuts to the hands when the bottles broke while being opened, injuries to feet when bottles were dropped, and damage to teeth as people tried to grip bottles while trying to open them.
3 According to another survey a third of people over fifty do not buy certain items because they have difficulty getting them open. Such products include bleach bottles; jars of jam, marmalade, pickles or beetroot; tins of sardines; tins of corned beef; sauce bottles; plastic milk bottles; and ring-pull cans.
4 By all historic standards life today is extremely safe, and 60 000 recorded injuries a year in a population of around 50 million adults do not add up to a major problem in the overall scheme of things. But they do represent much unnecessary suffering that could be readily prevented by more thoughtful design.

---

roducts for easier ving (PELs)

1 PELs are aids such as tap turners, easy-grip cutlery, powered can openers, etc., which make *independent living* (qv) very much easier for older and less able people.
2 As of 2008, useful general websites include the following. They are given in alphabetical order.
• Disabled Living Foundation www.dlf.org.uk/
• Disabled Living Online www.disabledliving.co.uk/
• OTDirect www.otdirect.co.uk/link-equ.html#Platform
• Ricability www.ricability.org.uk/

3 Websites of product suppliers include:

• Ability	www.redcross.org.uk/
• AKW Medicare	www.akw-medicare.co.uk/
• Chestercare	www.benefitsnowshop.co.uk/
• Coopers Healthcare Services	www.sunrisemedical.co.uk/
• DCS Joncare (children)	www.dcsjoncare.com/
• General Medical Supplies	www.genmedical.com/
• Keep Able	www.keepable.co.uk/
• Nottingham Rehab	www.nrs-uk.co.uk/
• Ways and Means	www.patient.co.uk/

4 Specific websites for people with *impaired vision* (qv) and *impaired hearing* (qv) include:

• RNIB	www.rnib.org.uk/
• RNID	www.rnidshop.com/

Proportionateness

1 It is a basic tenet of good law that the measures taken to deal with a problem should be proportionate to the magnitude or complexity of that problem – or, putting it in everyday language, don't use a sledgehammer to crack a nut. The principle is confirmed in the five 'Principles of good regulation' laid down in 2004 by Prime Minister Tony Blair's Better Regulation Task Force (see: *Better regulation bodies*), and still officially accepted. The five principles are:

- Proportionality.
- Accountability.
- Consistency.
- Targeting.
- Transparency.

In the words of the Task Force: 'Regulators should only intervene when necessary. Remedies should be appropriate to the risk posed and costs identified and minimised;' and 'Policy solutions must be proportionate to the perceived problem or risk and justify the compliance costs imposed'.

2 Despite having the Prime Minister's backing these sound precepts have been ignored ever since – see: *Over-legislation* and *Over-regulation*.

Prosthesis

1 Prostheses are artificial devices used to substitute for body parts which are defective or missing. They include artificial limbs, hearing aids, false teeth and eyes, plastic heart valves, and blood vessels.

2 The term refers most commonly to limb replacement. Artificial limbs have a long history, but they have lately become more natural-looking and more comfortable to wear. The comparatively new field of *bionics* has developed myoelectric, or bionic, arms, which are electronically operated and worked by minute electrical impulses from body muscles.

3 As of 2008, useful websites include the following. They are given in alphabetical order.

• Blatchford	www.blatchford.co.uk/
• Dorset	www.dorset-ortho.co.uk/
• Limbless Association	www.limbless-association.org/
• Ossur	www.ossur.com/
• Ortho Europe	www.ortholite.co.uk/
• Otto Bock	www.ottobock.co.uk/
• RSL Steeper	www.rslsteeper.com/

Protected corridors, lobbies, and stairways

See: *BS 5588: part 8; Means of escape;* and *Place of relative safety.*

otection from om Harassment Act 977 See: *Civil liberties.*

ovider of services See: *Service providers.*

SD See: *Public Sector Duty.*

SV See: *Public service vehicles.*

ublic address stems: complying ith the Disability iscrimination Act

1 Buildings used by the public must in general comply with the Disability Discrimination Act (qv), which means that they should be readily usable by people with *impaired hearing* (qv).
2 Para 9.3.1 of BS 8300 (qv) states that 'public address systems should be clearly audible and, wherever practicable, supplemental information'.
Guidance is given in *authoritative practical guides* (qv), which include DESIGNING FOR ACCESSIBILITY (qv) plus the following:
• APPROVED DOCUMENT M (qv): the provisions in paras 0.14, 3.6, and 4.31–4.36 must be heeded in order to comply with the Building Regulations for England and Wales.
• BS 8300 (qv): the recommendations in section 9.3 apply to all buildings in the UK.
• INCLUSIVE MOBILITY (qv): the recommendations in sections 9.1–9.4 apply to *transport-related buildings* such as bus and coach stations, railway stations, air terminals, and transport interchanges.
Except for AD M, the above guides do not have the force of law, but conformity with their recommendations will make it easier to demonstrate that the requirements of the Disability Discrimination Act have been met.
3 In places other than England and Wales the following regulatory documents should be consulted:
• In Northern Ireland: TECHNICAL BOOKLET R of the Building Regulations (Northern Ireland) (qv).
• In Scotland: the TECHNICAL HANDBOOKS of the Building (Scotland) Regulations (qv).
• In the USA, where the Americans with Disabilities Act applies: the ADA AND ABA ACCESSIBILITY GUIDELINES (qv).

See also: *Hearing enhancement systems.*

ublic authorities

1 As of 2007 there is no precise definition of the term 'public authorities', and no definitive list of such bodies. But they are broadly defined as organisations whose functions (or certain of whose functions) are functions of a public nature, and they include *government* (qv) departments, *local authorities* (qv), colleges and schools, *NHS* (qv) trusts, and all *quangos* (qv). It has been suggested that there may be around 45 000 public authorities, but the figure is difficult to verify.
2 All such organisations have a '*duty to promote equality*' (qv).
3 For more information see Chapter 9 of DISABILITY DISCRIMINATION: LAW AND PRACTICE By Brian Doyle – see the bibliography.

See also: *Commission for Equality and Human Rights; Duty to promote;* and *Equality.*

ublic bodies See: *Quangos.*

Public interest companies (PIC)	1 These are organisations usually set up to deliver a public service with public money, but run along the lines of a business with operational independence from government. They are also known as 'public benefit organisations' and 'mutuals'. They are often accountable instead to local service users, staff, or commissioners. 2 Foundation hospitals are cited by official websites as one example.
Public Interest Research Group (PIRU)	1 PIRU is a group of researchers in universities and elsewhere who do empiric research to provide a sound basis for public policy, and to uncover the spin and evasion that clog up much public debate. 2 For more information visit www.publicinterest.ac.uk/
Public Order Act 1986	See: *Civil liberties.*
Public realm	1 The spaces between and within buildings that are publicly accessible. They include streets, squares, forecourts, parks, and open spaces. The Disability Discrimination Act (qv) applies to all such spaces. 2 See also: *Built environment; Pedestrian environments;* and *Spaces: public.*
Public Service Duty	See: *Duty to promote.*
Public Service Vehicles (PSVs)	See: *Buses and coaches.*
Public Service Vehicle Accessibility Regulations 2000	1 The Disability Discrimination Act (qv) effectively outlaws discrimination against *disabled people* (qv) in public life, and to this end Part 5 of the Act makes provision for anti-discrimination regulations to be passed for three sets of transport vehicles: • Taxis. • Public service vehicles (i.e. buses and coaches). • Rail vehicles. 2 In the case of public service vehicles (which are defined as 'vehicles in public service which can carry more than eight passengers') the Act allows the government to make 'PSV regulations' on both the design and operation of such vehicles. The aim is that *disabled people* should be able to get in and out safely, and be carried safely and in reasonable comfort. For *wheelchair-users* (qv) the aim is that these things should be possible while they are sitting in the *wheelchairs* (qv). 3 In England and Wales the Public Service Vehicles Accessibility Regulations 2000 are the only regulations that have so far been enacted. They apply to the design of buses and coaches carrying over 22 passengers – see: *Bus and coach design: complying with the Disability Discrimination Act.* 4 Requirements to do with wheelchair accessibility include: • *Wheelchair* (qv) spaces and provisions. • Boarding lifts and ramps. • Entrances and exits. • Gangways. • *Signs* (qv) and markings. • Communication devices. • Lighting. 5 Requirements of a more general nature include: • Floors and gangways. • *Seats* (qv) and priority seats. • Steps.

• Handrails, *grab rails* (qv) and handholds.
• Communication devices.
• Kneeling systems.
• Route and destination displays.
6 The regulations are being phased in gradually between 2000 and 2017.
7 As of 2008, useful websites include:

• Department for Transport	www.dft.gov.uk/
• DPTAC	www.dptac.gov.uk/pubs.htm#06
• Office of Public Sector Information	www.opsi.gov.uk/
• Vehicle and Operator Services Agency	www.vosa.gov.uk/

See also: *Buses and coaches;* and *Transport: public.*

---

ublic services: uality

1 Between 1997 and 2007 the UK Government has in effect been conducting a nationwide experiment to test whether public services can be improved by a huge infusion of taxpayers' money without accompanying fundamental reform. The following figures, among others, demonstrate beyond reasonable doubt that the experiment has failed.
• In *education* (qv), expenditure in England grew in real terms from £28 billion in 2001–2 to £41 billion in 2005–6. Yet in December 2007 the OECD's Programme for International Student Assessment (Pisa) reported that, between 2000 and 2006, UK 15-year-olds had fallen from 7th to 17th place internationally in reading skills, and from 8th to 24th place in mathematics. A few weeks earlier the Progress in International Reading Literacy (Pirls) survey reported that between 2001 and 2006 English pupils had fallen from 3rd to 19th place internationally in reading, writing, and comprehension skills. Faced with these figures, Secretary of State Ed Balls admitted in December 2007 that so far from improving Britain was – in his words – 'going backwards'.
• In *social mobility* (qv), which is heavily influenced by the quality of education that is received by the population as a whole, Britain is near the bottom of the league for the developed world, and according to most published studies getting comparatively worse.
• In healthcare, massive increases in state expenditure have resulted in some improvement, but according to state of healthcare 2007 (see the bibliography), which is an annual report from the Healthcare Commission (qv), the performance of Primary Care Trusts or PCTs (qv) – which act as gatekeepers to all other NHS services – has worsened. As in education, Britain is actually going backwards.
• In *dental care* (qv), a survey conducted on behalf of Citizens Advice revealed in March 2007 that one year after the Government's reforms of NHS dentistry were introduced, two million people could not get treatment. Two thirds said they were going without dental treatment owing to 'huge problems accessing NHS dentistry and not being able to pay for private treatments'. Some were making round trips of up to 120 miles to reach a NHS dentist.
2 It is fair to conclude that the time has come to call off this extremely expensive failed experiment, and to decisively move away from centralised public services run by government diktat, to decentralised ones in which state funding follows patient or parent choice.

See also: *Education: voucher systems; Localism;* and *Post-democracy.*

---

ublic spaces

See: *Public realm;* and *Spaces: public.*

---

ublic telephones

See: *Telephones: public.*

Public toilets	See: *Lavatories: public.*
Public transport	See: *Transport: public.*
Pubs and restaurants: complying with the Disability Discrimination Act	See: *Refreshment facilities.*
Pull cords	See: *Controls and switches.*
Pupils: exclusion	See: *Schools: exclusion.*
Push-and-go doors	See: *Powered doors.*
Push buttons	1 For the heights of push buttons in general see: *Controls.* 2 For the heights of lift call buttons and lift control buttons see: *Passenger lifts*
Pushchairs	1 *Disabled* (qv) parents and carers may require special *baby buggies* (qv) and other types of pushchair. 2 As of 2008, useful websites include: • Ricability www.ricability.org.uk/

uadriplegia See: *Paralysis*.

ualifications and urriculum Authority 1 The QCA is a *quango* (qv) that until June 2007 was sponsored by the now-abolished Department for Education and Skills (qv), and since that date by the newly created Department for Children, Schools and Families (qv). In September 2007 the Minister for Schools, Mr Ed Balls, announced that the QCA would soon be replaced by a new and more independent regulatory body reporting to Parliament (qv) and not, as before, to a Minister. The details have not yet been clarified as this is written, and readers should visit www.dcsf.gov.uk/ for latest information.

2 The shocking failures of the existing QCA (of which its glowingly self-congratulatory website gives no hint) are however very likely to be perpetuated by its supposedly 'independent' successor. While reporting to Parliament rather than a Minister is a welcome improvement, it seems, in November 2007, that the new organisation will be funded by the Government; that its remit will be set by Ministers; and that its senior members will be appointed by the very Ministers whose departments are being scrutinised – in short, a triple lock to ensure its ultimate obedience to what the Government wants.

3 This is the very reverse of what is required. The following paragraphs summarise – as a cautionary tale – how we got to where we are, and conclude with a proposal for a better future.

4 The main tasks of the Qualifications and Curriculum Authority during its lifetime were (a) to 'maintain and develop the national curriculum', and (b) to 'regulate awarding bodies and maintain standards in tests and examinations'. Its record on these matters may be judged by the following facts:

• As of 2007, one in five adults in Britain is functionally illiterate – see: *Literacy*.

• In August 2006 a report titled WORKING ON THE THREE R'S, commissioned by the DfES and published by the CBI, complained that a third of businesses were having to send employees for remedial lessons in basic literacy and numeracy that should have been learned at school. The CBI publicly described the performance of the QCA as a 'disgrace'. Its report was based on a survey of companies employing in total almost a million full-time staff.

• In the same month Dr Tony Gardiner, who speaks for the UK Mathematics Foundation, accused curriculum and exam authorities of having 'torn the guts out of A-level maths' and making the exam easier than ever.

• In the same month Dr Frank Close, professor of physics and fellow of Exeter College, Oxford, said that 'every physics department has been aware that A-level students do not have the same knowledge base they did even ten years ago.'

• In the same month it was reported that a third of nurses with GCSE English and maths qualifications failed a basic English and maths test including questions at the level of: 'How many minutes are there in half an hour?' and 'Is 8pm the same as 18.00, 19.00, 20.00 or 21.00 hours'? As patient safety depends upon nurses being able to make calculations and keep accurate notes, this failure rate greatly alarmed the hospital and the Nursing and Midwifery Council. But when similar concerns were expressed two months earlier, in mid-June, a QCA spokesman was quoted by the BBC as saying that GCSEs were 'not seen as a preparation for work'. For some of the consequences of an education that does not prepare young people for work see: *Literacy; Social exclusion; Social mobility*; and *Underclass*.

• Since then it has emerged that candidates sitting English BTEC, equivalent to a GCSE, could pass this exam without ever reading a novel, poem or play (news reports, December 2007), and that as from 2009 pupils would be able to pass

French, German, and Spanish GCSE exams without being required either to read or write these languages (news reports, December 2007).

5 Quite clearly the Qualifications and Curriculum Authority has failed to 'maintain standards in tests and examinations'. Research carried out by Robert Coe at the University of Durham (see below) has shown that an A-level candidate who gained an A grade in 2006 would have got a B in 1996 and a C in 1988.

6 The above decline in exam standards has been accompanied by a deterioration in the quality of exam marking. The *Guardian* has published several accounts by newly-appointed markers who were disturbed, and in some cases shocked, by what they found. These include:

- 'In capable hands?' by Felicity Carus. 21 August 2007.
- 'Exam board under fire over marking claims' by Polly Curtis. 26 August 2005.
- 'It really is that bad' by Tom Smith. 25 August 2005.

7 The above is a deplorable record, yet in August 2007 it was reported that the head of the QCA in 2006, Mr Ken Boston, received £273 000 in annual pay and benefits compared with £43 563 for his predecessor in 1998.

8 When considering how the above deeply damaging trends can be reversed, a good starting point is the figure below. It is from p 9 of STANDARDS OF PUBLIC EXAMINATIONS IN ENGLAND AND WALES (see below) and shows the percentage of 15-year-olds achieving five or more GCSEs at grades A*–C or equivalent. The graph reveals that grade inflation took off in 1988 the year in which Secretary of State Kenneth Baker transferred responsibility for regulating the then Department for Education and Science from an independent watchdog to the Department itself, the latter acting via a succession of *quangos* (qv) or quango like bodies of which the QCA is the latest.

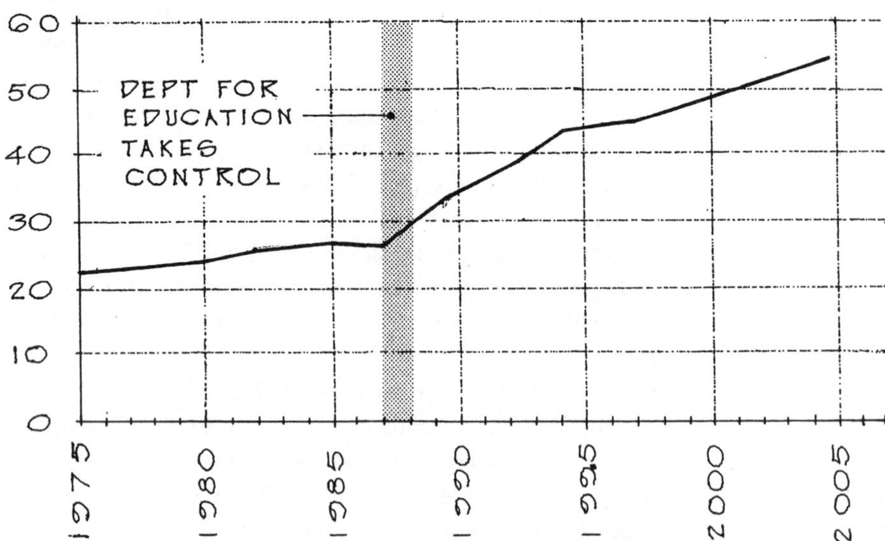

9 The way forward is clear. Self-policing has been a disaster. Responsibility for maintaining curriculum and exam standards should now be vested in a new body that is (a) not funded by the Government; (b) has no connections with the Government, and (c) is dominated by the universities and employers. A good example is set by the Swiss non-profit International Baccalaureate Organisation which administers the International Baccalaureate qualification. The IBO is completely independent from government and maintains robust standards. Between 1993 and 2004 the A-level pass rate rocketed from 81% to 96%, while the IB Diploma pass rate rose slightly from 81% to 83%. It is hardly surprising

that a growing number of British schools are switching from A-levels to the IB qualification.

10 As of 2008, useful references include the following. They are given in date order. For publication details see the bibliography.

• STANDARDS OF PUBLIC EXAMINATIONS IN ENGLAND AND WALES, published in 2005, is the source of the graph and some of the statistics quoted above. It draws heavily on the work of the Curriculum, Evaluation and Management Centre at Durham University.

• WORKING ON THE THREE RS – EMPLOYERS' PRIORITIES FOR FUNCTIONAL SKILLS IN MATHS AND ENGLISH was commissioned by the Department for Education and Skills (qv) and published by the Confederation of British Industry in 2006.

• ARTIFICIAL LEGACY: BLAIR'S SCHOOL ACHIEVEMENT IS A SHAM by Anastasia de Waal and Nicholas Cowen, was published in June 2007 by the *think tank* (qv) Civitas.

• BLAIR'S EDUCATION: AN INTERNATIONAL PERSPECTIVE by Alan Smithers, was published in June 2007. It argues that Mr Blair's policies have not delivered results commensurate with a near doubling in education spending, but that British education statistics have been interfered with so much that it is difficult to establish the truth.

• THE CORRUPTION OF THE CURRICULUM was published in 2007 by the *think tank* (qv) Civitas.

• NATIONAL CURRICULUM ASSESSMENTS AT KEY STAGE 2 IN ENGLAND, 2007 (PROVISIONAL), published in August 2007 by the Department for Children, Schools and Families (qv), shows that 40% of children in England were leaving primary school without the expected levels of attainment in all three of the Rs, and are unready for secondary school.

• SYSTEM REDESIGN – 2: ASSESSMENT REDESIGN by David Hargreaves, Chris Gerry and Tim Oates, published in November 2007 by the body that runs the Government's flagship scheme for specialist schools, accuses Sats (including GCSE tests) of having been so dumbed down that they should now be scrapped. Professor Hargreaves is a former head of the QCA.

• MATHS FEARS OVER NEW SCHOOLS DIPLOMAS by Anthea Lipsett was published in February 2008.

11 Useful references include the following. They are given in alphabetical order.

• CEM Centre	www.cemcentre.org/
• Civitas	www.civitas.org.uk/education/index.php
• Dept for Children, Schools and Families	www.dfes.gov.uk/
• Office for Standards in Education	www.ofsted.gov.uk/
• Reform	www.reform.co.uk/website/education.aspx
• Teachernet	www.teachernet.gov.uk/

See also: *Aimhigher; Accessible; Department for Children, Schools and Families; Literacy; National curriculum; Office for Standards in Education*; and *Social exclusion*.

---

uality and Outcomes amework See: *General Practitioners*.

---

uality Assurance gency for Higher lucation (QAA)

1 The QAA is reponsible for safeguarding and improving the academic standards and quality of higher education in the UK – visit www.qaa.ac.uk

2 THE ENGLISH QUESTION by Professor Thomas Docherty (see the bibliography) contains a devastating attack on the QAA for being a useless – indeed, positively damaging – generator of *red tape* (qv).

Quality Improvement Agency (QIA)

1 The QIA is one of a proliferating tangle of government departments and *quangos* (qv) coordinating and overseeing the Government's programme of vocational courses for young people. Others include the Department for Children, Schools and Families (qv); the Qualifications and Curriculum Authority (qv); the Training and Development Agency for Schools (qv); the Learning and Skills Council (qv); sector skills councils; and more. For a comment see: *Learning and Skills Council*.
2 For more information visit www.qia.org.uk/

Quality Standards Task Group (QSTG)

1 Established by the National Council for Voluntary Organisations (NCVO), but overseen by an independent board, the QSTG works with *charities* (qv) an *voluntary organisations* (qv) to improve quality standards in all aspects of the work.
2 For more information visit www.ncvo-vol.org.uk/index.asp?id=2055

Quangos

1 Quangos – from 'quasi-autonomous non-governmental organisations' – may be loosely defined as bodies which (a) have a role in the processes of national government, and are appointed and funded by central government; but (b) are not government departments or parts of such, and operate to a greater or lesse extent at arm's length from Ministers. They are officially known as 'Non-departmental Public Bodies' (NDPBs), or in some cases as 'Public Bodies' (PBs), depending upon type.
2 Quangos are distinct from, but sometimes confused with, two other types of organisation:
• Executive Agencies, which are part of their parent departments and make no claim to being at arm's length. Examples include the Child Support Agency (qv), the Criminal Records Bureau (qv), the Identity and Passport Service (qv Jobcentre Plus (qv), and the Office for National Statistics (qv). For a full list visit www.cabinetoffice.gov.uk/ministerial_responsibilities/executive_agencie
• Non-ministerial Departments, which are departments in their own right. Examples include the Charity Commission (qv) and the Office for Standards i Education (qv). There is talk of the Office for National Statistics (qv) being converted into such a department, but we await a decision. For a full list visit www.cabinetoffice.gov.uk/ministerial_responsibilities/departments/
3 Quangos themselves may be broadly divided into the following four categories:
• 'Executive bodies', which carry out operational and regulatory functions. There are 199 of these including Arts Council England (qv), the British Museum, English Heritage, the Housing Corporation (qv), and the National Lottery Commission (qv).
• 'Advisory bodies', which usually advise the Government on specific issues. There are 447 of these including the Better Regulation Commission (qv) and t Building Regulations Advisory Committee.
• 'Tribunal bodies'. There are 41, including Asylum Support Adjudicators and the Office of Surveillance Commissioners.
• 'Public Broadcasting Authorities' and 'Corporations'. There are 23, including the BBC and Channel Four Television Corporation.
4 Much disquiet has been expressed about the legitimacy of these unelected organisations. There are certainly far too many of them (see below), but that does not necessarily mean that there should be none. It seems better, for instance, for the Royal Opera House, the National Theatre, and the Royal Ball etc., to be funded via an arm's length quango such as the Arts Council, than requiring them to submit their artistic plans directly to a Minister. (Though

against this it must be said that the latter arrangement was adopted in Wales in 2004. The outcome may hold useful lessons.)

5 Those comments apart, the role of quangos in our public services has become so large as to demand a major public debate. Causes for concern include the following:

• These bodies are unelected.

• There is a huge number of them – according to Table 17.1 of PUBLIC BODIES 2006 (see below), just short of 900, but possibly many more.

• They are proliferating alarmingly and apparently uncontrollably. At the Labour Party conference in 1996 Mr Tony Blair pledged to 'put the unaccountable quango state and its patronage … in the dustbin of history where it belongs'. But according to PUBLIC BODIES 2006 around 300 of the 882 quangos then in existence were created by Mr Blair's government; and we know from other sources that state spending on quangos has risen from £24 billion when Mr Blair took office in 1997, to £123 billion in 2007.

• While quangos go through money like monkeys through bananas (£123 billion represents over £2 000 for every man, woman, and child in Britain) their achievements are often near-invisible. The failures of the Qualifications and Curriculum Authority (qv) are reported above; the Learning and Skills Council (qv) spends over £8 billion a year, but British employers complain vociferously that workplace skills are low and failing to improve.

• Senior appointments to these bodies are drawn disproportionately from a pool of well-connected people. In November 2006 the shadow Culture Secretary Mr Hugo Swire MP published a list of 20 individuals with close government connections who have between them been appointed to 83 influential public posts at salaries of up to £120 000 for a two- or three-day working week. In October 2007 the Commissioner for Public Appointments (qv) disclosed that 394 Labour Activists had been appointed to quangos over the previous year, compared with 96 Conservatives and 78 Liberal Democrats.

• It is usual for Board members to be appointed by the Minister for the parent department or on his recommendation; for funding to be controlled by him; and for their chief executive to report to him. These powers enable Ministers to ensure that desired policies are implemented while the term 'arms-length' enables them to avoid accountability. In short: a sector that spends well over £100 billion a year, employs 1.4 million people, and runs most of what we like to think of as the 'public' sector (for instance much of the NHS, including England's 28 Strategic Health Authorities and 303 Primary Care Trusts) is beyond reach of the electorate even though it has major impacts upon their everyday lives. For a comment see: *Post-democracy*.

6 In 2007 Lord Heseltine, who has twice been Secretary of State for the Environment, suggested a wholesale transfer of power from unelected quangos such as Learning and Skills Councils (qv), the Housing Corporation (qv), and Regional development agencies (qv) to elected *local authorities* (qv).

7 As of 2008, useful references include the following. They are given in alphabetical order. For publication details see the bibliography.

• THE ESSENTIAL GUIDE TO BRITISH QUANGOS 2005 by Dan Lewis, published by the Centre for Policy Studies.

• PUBLIC BODIES 2006, published by the Cabinet Office.

8 Useful websites include the following. They are given in alphabetical order.

• Directgov www.direct.gov.uk/

• Economic Research Council http://quangos.ercouncil.org/home/

• Public Bodies Directory
www.civilservice.gov.uk/other/agencies/public_bodies/index.asp

See also: *Parliament*; and *Post-democracy*.

Quotas

1 A quota system is one whereby the Government or some other agency specifies, or imposes by law, the percentages of people from particular population groups who must be taken on as students, employees, etc.
2 Most people are strongly antipathetic to quota systems (see: *Equality*), and UK governments have hesitated to openly introduce them in employment and other areas. They have, instead, encouraged bodies such as *Arts Council England* (qv) and the distributors of *National Lottery* (qv) funds to achieve a broadly similar effect by imposing upon theatres, museums, and other applicants for funding 'targets' that are quotas in all but name – for an exampl see: *Race* Relations *Act 1976*. This camouflaged approach may soon change. When launching FAIRNESS AND FREEDOM: THE FINAL REPORT OF THE EQUALITIES REVIEW (qv) in February 2007, Mr Trevor Phillips, chairman of the newly-established Commission for Equality and Human Rights (qv), strongly suggested that the time had come for the introduction of numerical quotas in order to end 'immovable, persistent disadvantage'. The Government would be wise to think twice and to consult the public as a whole – not just a closed circ of *experts* (qv), *pressure groups* (qv) and *think tanks* (qv) – before initiating a moves in that direction.

See also: *Discrimination; Diversity*; and *Equality*.

R

ace discrimination 1 *Discrimination* (qv) on the basis of race is outlawed in the UK by anti-discrimination legislation – for details see: *Discrimination.*
2 See also: *Equality*; and *Equality and Human Rights Commission.*

ace Equality Duty 1 For details of this Duty (qv) visit www.cre.gov.uk/duty/index.html
2 See also: *Discrimination; Duty to promote; Equality*; and *Single Equality Act.*

ace Relations Act 76 1 Under this Act, taken together with the Race Relations (Amendment) Act 2000, it is unlawful for anyone to *discriminate* (qv) against another on grounds of race, a term that is taken to include race, colour, nationality, and/or ethnic or national origins.
2 The law specifies four main types of discrimination:
• Direct discrimination occurs if anyone can show (a) that they have been treated less favourably than others in similar circumstances, and (b) that the reason for such treatment was their race. Racist abuse and harassment are forms of direct discrimination.
• Indirect discrimination may occur if an apparently non-discriminatory requirement or condition which applies equally to everyone, has a disproportionately large negative effect on people belonging to a particular racial, ethnic, or national group. An instance might be a rule that pupils must not wear headgear, if this has the effect of forbidding the wearing of a turban by Sikhs or a yarmulka by Jews.
• Victimisation occurs if anyone can show (a) that they have been treated less favourably than others in similar circumstances, and (b) that the reason for this treatment was that they had complained about racial discrimination, or supported someone else who had.
• Harassment, broadly speaking, means violating the dignity of someone from the groups mentioned above, or creating an intimidating, hostile, degrading, humiliating, or offensive environment for them.
3 Given the shameful discriminatory treatment that racial and ethnic minorities have suffered in the past, and to some degree still suffer, some form of legal protection against hurtful discrimination is to be welcomed, and the research report TALKING EQUALITY (see below) indicates that most people in Britain agree with that principle.
4 But that same survey, reinforced by informal evidence, suggests that current legislation and practices do not always get the balance right. The following instances were in the news as this was written:
• In 2005 the Arts Council England (qv) wrote to over 1 000 grant recipients warning that funds would depend on their ability to serve the needs of ethnic minority audiences, adding that: 'We will closely monitor the development of your action plan and your progress in meeting your race equality objectives, and future funding may include considerations on your ability to meet race equality targets'. The letter was accompanied by a guidance pack setting out targets – the latter bearing a marked resemblance to *quotas* (qv) – that might be achievable by many metropolitan organisations, but only by them. It is a fact that over 90% of the UK's population is 'white British' (to use the official term), and in much of the country close to 100%. In these areas such targets are not only near-impossible to meet, but would, if achieved by various coercive devices, deny the will of local people. Commenting on the Arts Council circular, Jude Orange, a spokesperson for the Norwich Puppet Theatre, said that the theatre tried hard to serve minority groups, but 97% of the local population was white and 'I think a show which caters for black and Asian people would be unlikely to find an audience in Norwich. If every year we had to put on plays for them we would be happy to do that – but we would be playing to

empty houses'. There were similar comments from several other perfectly fai minded arts organisations in non-metropolitan and rural areas.
5 The report TALKING EQUALITY: WHAT MEN AND WOMEN THINK ABOUT EQUALITY IN BRITAIN TODAY indicates that target-setting of the crude and coercive kind noted above, which has now become very widespread, does not have the support of mainstream Britain. See: *Equality*.
6 As of 2007, useful references include the following. They are given in alphabetical order. For publication details see the bibliography.

- FAIRNESS AND FREEDOM: THE FINAL REPORT OF THE EQUALITIES REVIEW published by the Cabinet Office.
- TALKING EQUALITY: WHAT MEN AND WOMEN THINK ABOUT EQUALITY IN BRITAIN TODAY by Melanie Howard and Sue Tiballs and published by the Equal Opportunities Commission.

7 Readers are reminded that the present work only gives simplified summarie of original documents and does not purport to provide full, authoritative, and up-to-date statements of the law. As of 2007, useful websites for checking the latest situation include the following. They are given in alphabetical order.

- Directgov www.direct.gov.uk/
- Equalities Review www.theequalitiesreview.org.uk
- Equality & Human Rights Commission www.equalityhumanrights.com
- Office of Public Sector Information www.opsi.gov.uk/acts.htm

See also: *Commission for Equality and Human Rights; Diversity; Equality*; a *Race Equality Duty*.

Racial and Religious Hatred Act 2006

1 This Act makes it an offence for anyone to incite (or 'stir up') hatred agains others on the grounds of their religion.
2 If passed as first proposed, the Act would have made it illegal to express provocative views on anyone's religion, thus making deep inroads into our *civ liberties* (qv). The original Bill was defeated largely by a campaign mounted outside Parliament by the writers' organisation Pen, and especially by the acto Rowan Atkinson.
3 As of 2007, useful websites include the following. They are given in alphabetical order.

- Directgov www.direct.gov.uk/
- Equalities Review www.theequalitiesreview.org.uk
- Equality & Human Rights Commission www.equalityhumanrights.com
- Office of Public Sector Information www.opsi.gov.uk/acts.htm

See also: *Civil Liberties*.

Racial integration

See: *Population: racial integration*.

RADAR

1 The Royal Association for Disability and Rehabilitation (RADAR) works fo the *inclusion* (qv) of *disabled people* (qv) in mainstream society.
2 The RADAR key is not an electronic door-opening device, as might be supposed, but a large metal key issued by RADAR to disabled people to give them access to over 4 000 *public lavatories* (qv) which are normally kept locked as a precaution against misuse or vandalism.

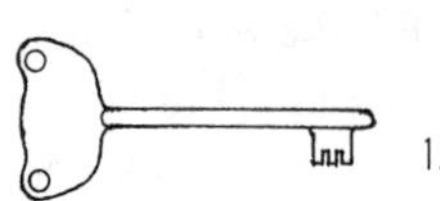

RADAR KEY

3 For more information visit www.radar.org.uk/

adiators See: *Heat emitters*.

ail Group See: *Railway system*.

ail platforms 1 For design guidance refer to paras 8.4.7–8.4.9 of INCLUSIVE MOBILITY (qv) – see the bibliography. This is an *authoritative practical guide* (qv), and conformity with its recommendations will make it easier to demonstrate that the requirements of the Disability Discrimination Act (qv) have been met.
2 See also: *Off-street platforms; On-street platforms*; and *Tramstops*.

ail safety See: *Office of Rail Regulation*; and *Transport safety*.

ail travel 1 The figure below shows that the number of passenger journeys made by train in Great Britain is somewhat higher than in 1970, and increasing. At date 'A' council-run bus services were privatised by the 1985 Transport Act, and at date 'B' the railways were split up and privatised by the 1993 Railways Act.

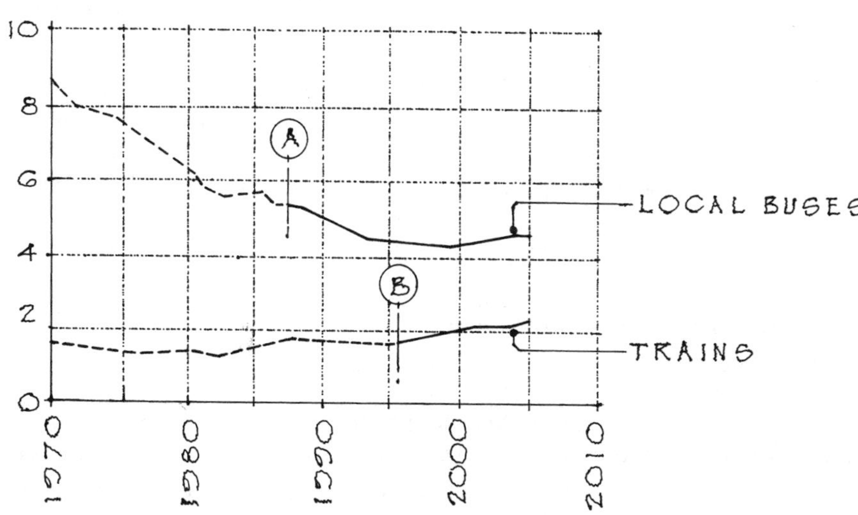

2 The graph is from www.statistics.gov.uk/cci/nugget_print.asp?ID=1094
For more detailed data refer to Table 12.16 in SOCIAL TRENDS 37 (qv), which was published after this was written.

ail travel: cessibility 1 Under the twin influences of the Disability Discrimination Act (qv) and growing consumer power, the *accessibility* (qv) of rail travel is steadily improving. As examples:
• Stations are becoming better equipped, and the UK's train operating companies are increasingly willing to make special arrangements for *disabled* (qv) passengers. Especially if given advance notice they claim to be able to arrange for staff to meet disabled passengers at their departure station, accompany them to the train, and see them safely on board. Similar arrangements can be made at the destination station and at transfer stations along the way.
• Information at stations is becoming more *accessible*. The Rail Passengers' Council has produced *tactile maps* (qv) and a web-based guide (visit www.describe-online.com) to London's Euston and Paddington stations for users with *impaired vision*, and more will follow.
2 Against that, *footbridges* (qv) across railway tracks present a serious problem. Many *disabled* (qv) and older people cannot climb these stairs, and are therefore partly excluded from traveling by rail. Equipping such bridges with reliable,

well-maintained lifts is very expensive (perhaps in the region of £5 million, compared with £1 million for a conventional bridge) and it was disclosed in 2006 that Network Rail was replacing its old footbridges with new ones on a 'like for like' basis, meaning that the new bridges are just as unusable by disabled people as the old.
3 As of 2007, the 'Access for All' scheme is a £370 million Department for Transport (qv) fund that is ring-fenced for improving the accessibility of rail stations up until 2015. This is the 3rd round of such funding.
4 NATIONAL RAIL MAP FOR PEOPLE WITH MOBILITY DIFFICULTIES shows all the railway stations in Britain, with information on matters such as which stations have platforms than can be accessed without having to use steps. Visit www.nationalrail.co.uk/
5 As of 2007 useful references include the following. For publication details se the bibliography.
• RAILWAYS FOR ALL: THE ACCESSIBILITY STRATEGY FOR GREAT BRITAIN'S RAILWAYS. Among other things this document describes the 'Access for All' funding scheme.
6 As of 2007, useful websites include the following. They are given in alphabetical order.
• Access for All stations www.dft.gov.uk/transportforyou/access/rail/railstations/access
• Department for Transport www.dft.gov.uk/transportforyou/access/rail/
• Describe Online www.describe-online.com/

See also: *Door to Door*.

Rail travel: guidance for disabled people

See: *Door to Door*.

Rail vehicles: complying with the Disability Discrimination Act

1 The Disability Discrimination Act (qv) effectively outlaws *discrimination* (qv) against *disabled people* (qv) public life, and to this end Part 5 of the Act makes provision for anti-discrimination regulations to be passed for three sets of public transport vehicles:
• Taxis (qv).
• Public service vehicles, i.e. buses and coaches (qv).
• Rail vehicles (see below).
2 As of 2007, all rail vehicles are subject to the Rail Vehicle Accessibility Regulations (RVAR) 1998 and the Rail Vehicle Accessibility (Amendment) Regulations (RVAAR) 2000, which lay down detailed design requirements for doors, steps, floors; *seats* (qv), controls, interior transparent surfaces, handrai (qv), door handles, passenger information, and toilets. For publication details these documents see the bibliography.
3 Readers are reminded that the present work only gives simplified summarie of original documents and does not purport to provide full, authoritative and u to-date statements of the law. The situation will be constantly developing in coming years, and designers and managers must keep up with current information.
4 As of 2007 useful websites include the following. They are listed in alphabetical order.
• Department for Transport www.dft.gov.uk/transportforyou/access/rai
• DPTAC Access Directory www.dptac.gov.uk/adnotes.htm
• DPTAC Publications www.dptac.gov.uk/pubs.htm
• DPTAC Trains www.dptac.gov.uk/trains.htm
5 Useful references include the following. For publication details see the bibliography.

• CODE OF PRACTICE. PROVISION AND USE OF TRANSPORT VEHICLES: STATUTORY CODE OF PRACTICE: SUPPLEMENT TO PART 3 CODE OF PRACTICE.
This is an essential reference in the application of the Disability Discrimination Act.

See also: *Transport: complying with the Disability Discrimination Act.*

---

ailway safety See: *Office for Rail Regulation.*

---

ailway stations See: *Transport-related buildings.*

---

ailway system 1 As of 2008 the UK's railway system is run by the following organisations:
• Network Rail owns, operates, and maintains the railway infrastructure – what the Railway Act calls 'the permanent way'. The latter comprises the tracks, the signalling, and most of the stations. Network Rail is a not-for-dividend private company that is wholly owned by the state.
• The Office of Rail Regulation (ORR) is the independent regulator of the railways. It has two areas of responsibility – (a) economic regulation; and (b) health and safety regulation.
• The Rail Group in the *Department for Transport* (qv) looks after the strategic direction of the railways, and is in many ways the successor to the Strategic Rail Authority (qv), which existed from 2000 to December 2006. The Rail Group has an overview of the system and tries to suggest to Network Rail and other parties what they ought to be doing.
• The trains are owned by private equity companies known as RoSCos (Rolling Stock Companies).
2 For more information visit the following websites:

• Network Rail	www.networkrail.co.uk/
• Office of Rail Regulation	www.rail-reg.gov.uk/
• Rail Group	www.dft.gov.uk/pgr/rail/

See also: *Rail travel*; and *Transport: safety*.

---

aised kerbs See: *Kerbs.*

---

amps 1 Any pedestrian surface such as an *access route* (qv) or public *footpath or footway* (qv) that is steeper than 1:20, but shallower than 1:12, must be designed as a ramp, with a *slip-resistant surface* (qv), *kerbs* (qv); *guardings* (qv) *handrails* (qv), and regular level landings as resting places. If site gradients are steeper than 1:12 then s*tairs* (qv) must be used.
2 Ramps present a difficult design problem, and most are frankly hideous – particularly if they consist of more than one flight, which is inevitable if the rise is greater than 500 mm. The few reasonably pleasing entrance ramps known to the present author include the following, most of which are illustrated in EASY ACCESS TO HISTORIC BUILDINGS (see the bibliography):
• The entrance to the RIBA building at 66 Portland Place, London W1. The rise is only a few inches, therefore the design challenge was modest, but the solution is a delight.
• The Cromwell Road entrance to the Victoria and Albert Museum, London SW7. An existing flight of steps has been replaced by a new flight of much shallower steps, with a pair of symmetrical ramps behind. The new composition is entirely modern and functional, yet pleases the eye and harmonises well with Sir Aston Webb's late 19th century building façade behind.

• The same generic idea of a central flight of shallow steps with a pair of symmetrical ramps behind has been applied at the western entrance, overlooking St James Park, of the Government Offices in Great George Street, London SW1. The detailing is however less elegant, and the handrail – always a problem in these situations – is disappointing.
• At the Royal Academy in Piccadilly, London SW1, a pair of symmetrical ramps have been formed within the entrance portico where they are not conspicuous – a generic solution that can be applied to many classical buildings.
3 The prefabricated ramps now on the market tend to be particularly hideous, and one lives in hope that an enlightened manufacturer and talented designer may soon get together and develop a range of products that would enable building owners to meet the provisions of APPROVED DOCUMENT M (qv) of the Building Regulations and BS 8300 (qv) without wrecking the appearance of ou buildings and public spaces.

Ramps: dwellings

1 The following references contain authoritative recommendations:
• Section 6 of APPROVED DOCUMENT M (qv). The provisions in AD M must be heeded in order to comply with the Building Regulations (qv) in England and Wales.
• THE HABINTEG HOUSING ASSOCIATION DESIGN GUIDE (qv) applies most specifically to *housing association* (qv) schemes, but ought to be by the designer's side in the design of all housing developments and individual dwellings.
• THE WHEELCHAIR HOUSING DESIGN GUIDE (qv), also from the Habinteg Housing Association, is another illustrated manual whose usefulness extends fa beyond the special remit implied by the title.
• THE HOUSING DESIGN GUIDE applies specifically to housing in London.
2 For publication details of the above references see the bibliography.

Ramps: public buildings: complying with the Disability Discrimination Act

1 Buildings used by the public must in general comply with the Disability Discrimination Act (qv). In the case of ramps in public buildings, *authoritative practical guides* (qv) include DESIGNING FOR ACCESSIBILITY (qv) and in addition the following documents. For publication details see the bibliography.
• APPROVED DOCUMENT B (qv), which deals with *fire safety*. Its provisions must be heeded in order to comply with the Building Regulations (qv) in England and Wales.
• APPROVED DOCUMENT K (qv) which deals, inter alia, with stairs and ramps and protection against falling. Its provisions must be heeded in order to compl with the Building Regulations in England and Wales.
• APPROVED DOCUMENT M (qv), which deals with access to and use of buildings. Its provisions must be heeded in order to comply with the Building Regulations in England and Wales.
• BS 8300 (qv), which applies to all buildings in the UK.
• INCLUSIVE MOBILITY (qv), which is the most authoritative reference for the *inclusive* (qv) design of *pedestrian infrastructure* (qv) and of *transport-related buildings* (qv).
• GUIDE TO SAFETY AT SPORTS GROUNDS (qv), which is the most authoritative reference for the *inclusive design* of sports facilities.
Only AD B, AD K, and AD M above have the force of law, but conformity with the rest will help to demonstrate that the requirements of the Disability Discrimination Act have been met. The relevant clauses in each are set out below. Note that in the matter of *guarding* (qv) and *handrail* (qv) heights, AD M takes precedence over AD K, which as of 2007 is due for revision.

EXTERNAL RAMPS:

DOCUMENT	RELEVANT PARAGRAPHS
APPROVED DOCUMENT K	2.1–2.7 and 3.1–3.3
APPROVED DOCUMENT M	1.19–1.26 and 1.34–1.37
BS 8300	5.8
INCLUSIVE MOBILITY	3.6 and 8.4
GUIDE TO SAFETY AT SPORTS GROUNDS	7.1–7.12 and 13.8–13.18

INTERNAL RAMPS:

DOCUMENT	RELEVANT PARAGRAPHS
APPROVED DOCUMENT B	6.27–6.30
APPROVED DOCUMENT K	2.1–2.7 and 3.1–3.3
APPROVED DOCUMENT M	3.14, 3.20; 3.52–3.53 and 4.12
BS 8300	8.2
INCLUSIVE MOBILITY	3.6 and 8.4
GUIDE TO SAFETY AT SPORTS GROUNDS	7.1–7.12 and 13.8–13.18

For external ramps to public buildings, the design features and dimensions shown below will in general satisfy the provisions in the above references, though each particular case should be checked in detail before decisions are finalised.

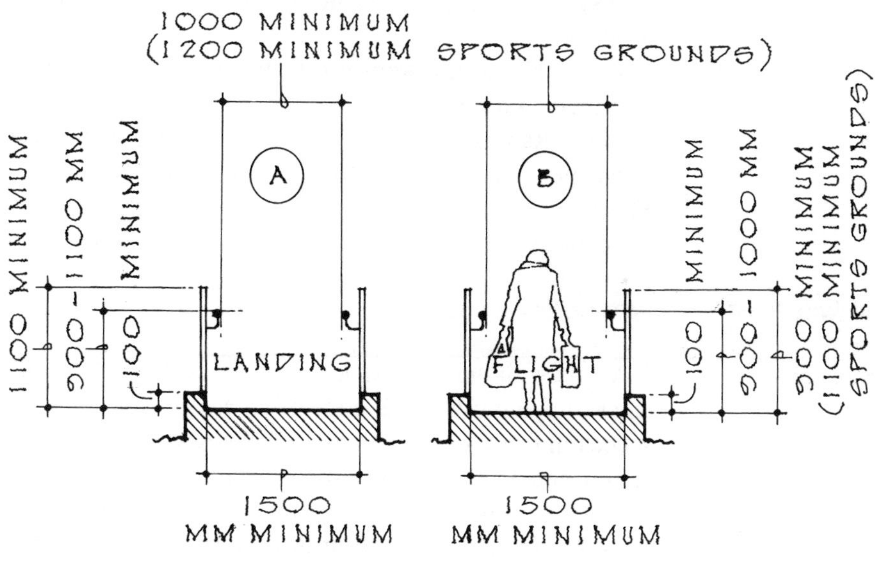

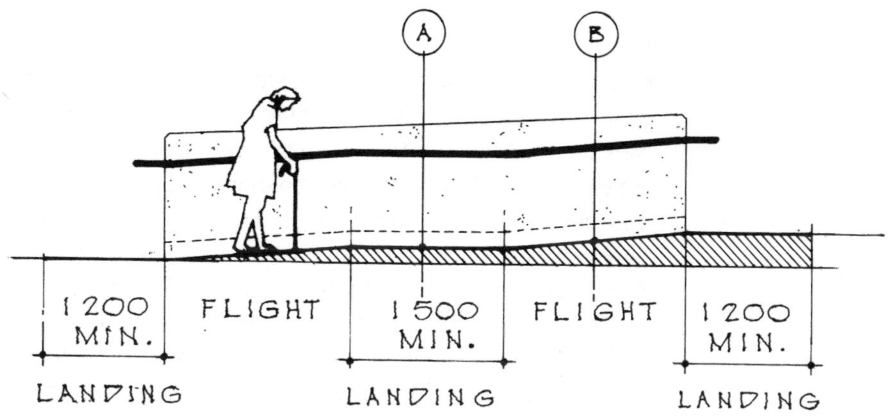

For places other than England and Wales the following regulatory documents should be consulted:
2 In Northern Ireland: TECHNICAL BOOKLETS E, H, and R of the Building Regulations (Northern Ireland) (qv).
3 In Scotland: the BUILDING (SCOTLAND) REGULATIONS and the associated TECHNICAL HANDBOOKS (qv).
4 In the USA, where the Americans with Disabilities Act applies, refer to the ADA AND ABA ACCESSIBILITY GUIDELINES (qv).

See also: *Guardings*; and *Handrails*.

Ramps: portable

1 Lightweight *ramps* that can be transported in a car (or even a suitcase) and used as the need arises can be of great help to *wheelchair* (qv) and *scooter* (qv) users. There are models for use at stepped building entrances, for getting onto platforms in auditoria, and for entering wheelchair accessible vehicles (WAVs) taxis, and caravans etc. For ease of storage and transportation some can be folded, some are telescopic, while others can be rolled up.
2 Para 5.8.9 of BS 8300 (qv) states that portable or temporary ramps:
• May be acceptable in existing situations where limited space precludes a more satisfactory solution, but should not be used as design solutions in new buildings.
• Should have a surface width of at least 800 mm; a drainable, *slip-resistant* (qv) surface; and upstands along both sides to prevent wheelchair tyres veering off the edge.
3 As of 2008, useful websites include the following. They are given in alphabetical order.

• Access Ramps	www.accessramps.co.uk/
• Disabled Living Foundation	www.dlf.org.uk/
• Portaramp	www.portaramp.uk.com/
• Ricability	www.ricability.org.uk/

Read Write Plus

See: *Skills for Life*.

Reading aids

1 People with *impaired vision* (qv) can be helped to lead more *independent* (qv) and satisfying lives by a range of *aids and equipment* (qv) covering the full spectrum from *low-tech* (simple devices such as large-print books) to *high-tech* (electronic devices such as speech synthesisers). Currently available aids include the following, listed in low-tech to high-tech sequence:
• Books printed in large print, *Braille* (qv), or *Moon* (qv).
• Magnifiers and other optical aids. Professional advice should be sought from opticians before purchase, and it is best to buy from organisations such as the RNIB.
• Talking books, magazines, and newspapers. These are printed publications which have been recorded for playback. As one example, the Talking Newspaper Association of the UK provides over 200 national and approximately 1 100 local newspapers and magazines on audio tape, computer disk, email, internet download, and CD-ROM – visit www.tnauk.org.uk/ Similarly the RNIB works in partnership with the text-to-speech company Rhetorical to produce talking versions of newspapers and magazines. See: *Right to Read Alliance*; and visit www.rnib.org.uk/
• Closed-circuit television (CCTV). The printed material is placed on a reading table under the CCTV camera and the latter produces a magnified image on a monitor or TV screen.
• Computers. Standard computers enable users to enlarge text on the screen to a size that suits their eyesight. When this facility is combined with the

enormous amount of information that is now available on the internet and on CD roms, people with low vision have at their disposal a very powerful tool for a richer life – indeed, the internet probably represents the largest advance for visually impaired people since the invention of Braille. If a standard computer set-up is not able to meet a particular user's needs then there are various specialised magnification programmes or 'speech-back' options available. Organisations such as AbilityNet (qv) offer assistance.

• Electronic reading machines, also called optical character readers (OCR). These devices scan and translate printed text into a text file which can then be (a) read by a speech synthethiser and presented as synthesised speech; or (b) read by a computer and presented as enlarged text on the screen.

2 Turning from people with *impaired vision* to ones with *impaired dexterity* (qv) or *paralysis* (qv), the latter can be helped by *page turners* (qv).

3 As of 2008, useful websites include the following. They are given in alphabetical order.

• AbilityNet	www.abilitynet.org.uk/
• Disabled Living Foundation	www.dlf.org.uk/
• RNIB	www.rnib.org.uk/
• Talking Newspaper Association UK	www.tnauk.org.uk/

See also: *Aids and equipment; Blindness; Impaired vision*; and *Right to Read Alliance*.

ebus symbols See: *Widgit Rebus*.

eception areas See: *Entrance halls and reception areas*.

eception desks See: *Counters and reception desks*.

eclining chairs See: *Chairs: special*.

ecreation and sports ctivities for disabled eople

1 The principal organisation in this field is Disability Sport Events (qv), which works to ensure that *disabled people* (qv) can participate in sport at all levels from grass roots to international standard. It does so through the network of clubs, schools, and individuals who constitute its membership. Visit www.disabilitysport.org.uk/

Other organisations include the following:

2 The British Wheelchair Sports Foundation (BWSF) develops sports opportunities for disabled children and adults at both the recreational and competitive levels. It organises events at the Stoke Mandeville Stadium and elsewhere, and works closely with Britain's fourteen Wheelchair Sports Associations. Visit www.britishwheelchairsports.org/

3 British Blind Sport (BBS) is the co-ordinating body of sport for *blind* (qv) and *partially sighted* people in the UK. It has committees for archery, athletics, bowls, cricket, football, goalball (a sport devised for the blind and for which BBS is the Governing Body), judo, martial arts, swimming, and ten-pin bowling. There is also considerable interest in chess, golf, riding, sailing and skiing. Visit www.britishblindsport.org.uk/

4 The British Deaf Sports Council (BDSC) organises inter-club leagues, inter-regional tournaments, and national championships for people with *impaired hearing* (qv). Sports include athletics, badminton, basketball, bowls, cricket, football, golf, indoor games, netball, swimming, table tennis, tennis, ten pin bowling, and adventure sports. The BDSC also has links with squash and rugby. Visit www.britishdeafsportscouncil.org.uk/

5 The Dwarf Athletic Association UK (DAAUK), which promotes sport for very short people, to help them compete on an equal footing all the way up to the World Dwarf Games, the Paralympic, and the Olympic games. Visit www.daauk.org/
6 The English Federation for Disability Sport (EFDS) is the national body for developing sport for disabled people in England. Visit www.efds.net/
7 Scottish Disability Sport is the governing and co-ordinating body for sport fo disabled people in Scotland. Visit www.sportscotland.org.uk/
8 The UK Sports Association for People with Learning Disability (UKSA) develops sport for people with *learning disabilities* (qv) at UK and internationa levels
9 Level Playing Field is a national campaign that celebrates the opportunities for disabled people to be actively involved in football at all levels. Visit www.nads.org.uk/
10 The London Sports Forum for Disabled People promotes the sporting interests of the one million disabled young people and adults in the Greater London area. It tries to ensure that all disabled people in London have greater opportunities to get involved in a sport at a level of their choice. Visit www.londonsportsforum.org.uk/
11 As of 2008, other useful websites include the following. They are given in alphabetical order:

- Directgov www.direct.gov.uk/
- Directgov: disabled people www.disability.gov.uk/
- IDeA Knowledge www.idea.gov.uk/
- OTDirect www.otdirect.co.uk/link-equ.html#Platform

Recreation and sports aids

1 Special equipment is available to help *disabled people* (qv) participate in a wide variety of sports, including the following:

- Angling. Aids include buggies (qv) and *powered wheelchairs* (qv) to take disabled people over rough terrain; *wheelyboat* (qv) wheelchair-accessible boats; one-handed fishing rods and reels; grasping cuffs; and hook-holders which securely hold the hook while it is being baited.
- Athletics. Aids includes *wheelchairs* (qv) that are specially designed for racing or for playing court games such as badminton, basketball, lawn tennis, racquetball, wheelchair rugby (quad rugby) or wheelchair tennis.
- Biking and triking. For an indication of the range of adapted bikes and trikes etc. that are available visit the National Association of Bikers with a Disability (NABD) at www.nabd.org.uk
- Bird-watching. Aids include buggies and powered wheelchairs to take disabled people over rough terrain to viewing sites.
- Bowls. Aids include buggies with wide wheels, which enable disabled bowls enthusiasts to play from a sitting position and move about without damaging the bowling green.
- Exercise. Aids include *exercise and rehabilitation* aids (qv) for use in gymnasia, etc. Some of these can be used from a wheelchair.
- Go-karting. Karts are available with all the controls mounted on the steering column, or otherwise adapted to suit particular forms of impairment.
- Golf. Aids include buggies which allow disabled people to play Handigolf from a sitting position.
- Horse riding. Aids include special saddles and saddle support cushions for disabled horse riders; and hoists which enable disabled people to get onto a hors
- Rowing. Aids include boats with modifications such as special seats to suit people with particular *impairments* (qv), or increased width for greater stabilit
- Yachting: Aids include boats with weighted keels, or with floats fitted on bot sides to give greater stability.

2 As of 2008, useful websites include the following. They are given in alphabetical order.
• Disabled Living Foundation (qv) www.dlf.org.uk/
• OTDirect www.otdirect.co.uk/link-equ.html#Platform

See also: *Aids and equipment*.

---

ecreation and sports uildings See: *Sports-related buildings*.

---

ed tape See: *Better regulation bodies*; *Over-regulation*; and *Police and red tape*.

---

educing burdens 1 A programme within the Department of Health (qv) with the stated aims of achieving 'less red tape, better healthcare, and less NHS bureaucracy'.
2 In March 2007 its website claimed that 'we have already made a substantial amount of progress in reducing burdens on frontline staff', but in the same month the NHS Confederation, a body that represents NHS managers in 90% of Trusts, alleged that despite governmental promises to ease the burden of regulation, and a promise by the Healthcare Commission that it would adopt a 'light touch' in inspection, things were getting worse, not better. Its 20-page report, titled THE BUREAUCRATIC BURDEN IN THE NHS (see below) complained that the NHS was having the life choked out of it by the ever-growing burden of inspection and regulation. At least 56 (and possibly more) bodies were entitled to visit and inspect NHS hospitals and Trusts, and the annual health check by the Healthcare Commission (qv), which is the principal inspection body, required 500 separate information topics to be addressed, a task the NHS Confederation considered well-nigh impossible. The report described the position as intolerable.
3 As of 2008, useful references include the following. For publication details see the bibliography.
• THE BUREAUCRATIC BURDEN IN THE NHS, published by the NHS Confederation.
4 Useful websites include the following. They are given in alphabetical order.
• NHS Confederation www.nhsconfed.org/
• Reducing Burdens www.dh.gov.uk/

See also: *Better regulation bodies*; and *Over-regulation*.

---

eferenda See: *Treaty Establishing a Constitution for Europe*.

---

eform Treaty See: *Treaty Establishing a Constitution for Europe*.

---

efreshment facilities: asic planning data 1 Distilled planning data for most of the major building types is given in the METRIC HANDBOOK: PLANNING AND DESIGN DATA.
2 For publication details see the bibliography.

---

efreshment facilities: esign: complying with ne Disability iscrimination Act 1 Buildings used by the public must in general comply with the Disability Discrimination Act (qv). In the case of refreshment facilities, *authoritative practical guides* (qv) include DESIGNING FOR ACCESSIBILITY (qv) plus the following:
2 APPROVED DOCUMENT M (qv), the provisions of which must be heeded in order to comply with the Building Regulations (qv) for England and Wales. Paras 4.13–4.16 deal specifically with refreshment facilities. They include the following provisions:

• In shared kitchens used by *wheelchair-users* (qv) and others for tea-making, etc., para 4.16 recommends a worktop surface height of 850 mm, with a clear height of at least 700 mm underneath, as shown below.

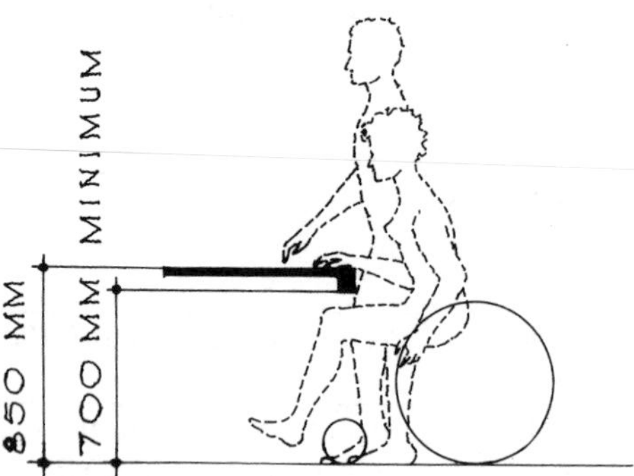

• In pubs, restaurants, cafes, and other refreshment facilities, para 4.16 recommends that one part of every serving counter should have a surface heig of not more than 850 mm, to suit people in wheelchairs. The remainder should be higher, to suit people standing.
For table heights see: *Tables*.
3 BS 8300, the provisions of which apply to all buildings in the UK. Para 12.1 gives extensive advice on worktop heights and layouts in *kitchens* (qv); and para 13.5 gives brief design recommendations for all refreshment facilities, including public houses, restaurants, and cafes. Para 13.5.3 recommends the following table heights:
• In general, clear height from floor surface to underside of table should be 700 mm.
• For some tables in each refreshment space, clear height from floor surface to underside of table should be 750 mm to accommodate *wheelchairs* with armrests.
BS 8300 is an advisory document and does not have the force of law, but conformity with its recommendations will make it easier to demonstrate that the requirements of the Disability Discrimination Act have been met.
4 For places other than England and Wales the following regulatory documents should be consulted:
• In Northern Ireland: TECHNICAL BOOKLET R of the Building Regulations (Northern Ireland) (qv).
• In Scotland: the TECHNICAL HANDBOOKS of the Building (Scotland) Regulations (qv).
• In the USA, where the Americans with Disabilities Act applies, refer to the ADA AND ABA ACCESSIBILITY GUIDELINES (qv).

See also: *Tables*; and *Worktops in kitchens: public buildings*.

Refreshment facilities: management: complying with the Disability Discrimination Act See: *Service providers*.

Refuge See: *BS 5588 part 8; Means of escape*; and *Place of relative safety*.

Regeneration 1 Government regeneration policies may be of three types:
• Social regeneration policies, which tackle the problems that lead to social deprivation, such as crime and drugs.

• Physical regeneration policies, which tackle run-down buildings and communal areas.
• Economic regeneration policies, which have the aim of creating jobs and wealth.
The three policy strands should always be seen as interlinked.
2 See also: *Neighbourhood renewal.*

egeneration: area-sed See: *Neighbourhood renewal.*

egeneration Areas See: *Areas for Regeneration.*

egional Assemblies See: next entry.
egional Development encies

egions 1 In accordance with the EU vision of a Europe of regions rather than of nation states, the Government put six Acts through Parliament in 1998:
• Three of these gave devolved governments to Scotland, Wales, and Northern Ireland.
• A fourth gave the same to London.
• The fifth set up eight regional development agencies and assemblies in the rest of England, the latter intended to be the embryos of future regional governments.
• The last Act, explaining all the rest, divided the UK into 12 vast regional constituencies used for elections to the European Parliament.
Governmental maps thereafter ceased to show a country called England, only an entity called the UK divided into twelve areas of roughly comparable populations – see the map on the next page.
2 The list below gives, on the left, the recently-created English *regions*, and on the right the historic *counties* of which the regions are made up. The websites given on the left are for the Government Office (see below) that represents each individual region.

North East www.go-ne.gov.uk	01 Northumberland 02 Tyne and Wear 03 Durham 04 Teesside
Yorkshire and the Humber www.goyh.gov.uk	05 North Yorkshire 06 West Yorkshire 07 South Yorkshire 08 The Humber
East Midlands www.go-em.gov.uk	09 Derbyshire 10 Nottinghamshire 11 Lancashire 12 Leicestershire 13 Northamptonshire
East of England www.go-east.gov.uk	14 Norfolk 15 Suffolk 16 Cambridgeshire 17 Hertfordshire 18 Bedfordshire 19 Essex

SOUTH EAST www.go-se.gov.uk	20 Oxfordshire 21 Buckinghamshire 22 Berkshire 23 Surrey 24 Hampshire 25 Kent 26 East Sussex 27 West Sussex 28 Isle of Wight
LONDON www.go-london.gov.uk	29 London

South West www.gosw.gov.uk	30 Gloucestershire 31 Avon 32 Wiltshire 33 Somerset 34 Dorset 35 Devon 36 Cornwall (including Isles of Scilly)
West Midlands www.go-wm.gov.uk	37 Staffordshire 38 Shropshire 39 West Midlands 40 Warwickshire 41 Hereford and Worcester
North West www.go-nw.gov.uk	42 Cumbria 43 Lancashire 44 Greater Manchester 45 Cheshire 46 Merseyside

3 Each of the English regions shown on the map (though London is in some respects different – see: *Government: London*) has been given the following main institutions:

• A Regional Development Agency (RDA). These unelected bodies were set up in the 8 English regions in April 1999. The 9th, in London, was created in July 2000 after establishment of the Greater London Authority. They are supposed to stimulate economic development and regeneration, but according to investigations by the commentator Leo McKinstry 'their vast annual budget of £2.3 billion is largely wasted on marketing, seminars, internal job creation, monitoring, action plans, and all the other exercises so cherished by [modern] officialdom'. Visit www.englandsrdas.com/home.aspx

• A Regional Assembly. These bodies, also unelected, were set up with the intention that they would gradually evolve into elected regional parliaments. This idea met with so much popular resistance that the newly installed Prime Minister, Mr Gordon Brown, announced in July 2007 that they would be scrapped and their powers split between elected local councils (qv) and the unelected Regional Development Agencies – but mainly the latter.

• A Government Office for each region. Their roles are to co-ordinate the delivery of central government services in their regions. They are playing an increasingly pivotal role in the government of England, and their individual websites are given in the listing of regions above. For an overview visit www.communities.gov.uk/index.asp?id=1139549

• The Regional Co-ordination Unit (RCU) co-ordinates the work of the nine Government Offices of the Regions. For more information visit the website above.

4 The wholesale reshaping of English government outlined above was put in place without reference to the electorate. Only when the process had been underway for five years did the Government try a first experiment in retrospective consultation. Selecting the region that was considered most likely to favour the new structures, it launched a referendum in the North East of England in November 2004, asking the electorate to approve the creation of an elected Regional Assembly. The proposal was rejected by 78% to 22%. England's eight unelected regional assemblies outside London nevertheless remained in place until July 2007, when (as noted above) Prime Minister Gordon Brown announced their abolition. But he did not abolish the masterplan of an England broken up into regions run by unelected and unaccountable bodies. Each region still exists, has its own Minister, and has its own unelected

Regional Development Agency and extremely expensive Regional Government Office (see above).
5 As of 2007, useful websites include the following. They are given in alphabetical order.
• Department for Communities and Local Government www.communities.gov.uk/index.asp?id=1139476
• Direct Democracy www.direct-democracy.co.uk/
• Local Government Association www.lga.gov.uk/
• Localis www.policyexchange.org.uk/

See: *Government; Localism*; and *Post-democracy*.

Registered Social Landlords (RSLs)

1 RSLs are the new landlords of *social housing* (qv), increasingly replacing *local authorities* as managers of the existing social housing stock and the providers of new dwellings. Social landlords need to register with the Housing Corporation (qv) to qualify for grants to build new homes. Most are *Housing Associations* (qv) but they also include trusts, co-operatives, and companies. They are run as not-for-profit businesses, using financial surpluses to maintain existing homes and to help finance new ones.
2 Space and design standards for new RSL dwellings are controlled by the Housing Corporation through a set of Scheme Development Standards (qv) which recommend minimum standards of layout and design.
3 Visit www.housingcorp.gov.uk/server.php?show=nav.489

See also: *Housing Associations; Housing Corporation*; and *Scheme Development Standards*.

Regulations

See: *Statutes*.

Regulatory Impact Assessment (RIA)

1 In August 1998 Prime Minister Tony Blair announced a new anti-red tape measure: no regulation that would have an impact on businesses or *voluntary organisations* (qv) should be considered by Ministers without a Regulatory Impact Assessment (RIA) being carried out. New regulations should be introduced 'with a light touch' and only after a full RIA has balanced the benefits of the proposal against the wider economic, social and environmental impacts; other alternatives have been considered and rejected; and the benefits have been shown to justify the costs.
2 RIAs proved to be ineffective, and in April 2007 the Cabinet Office (qv) announced that as from the following month they would be replaced by *Impact Assessments* (qv) which, it was claimed, would be simpler and more transparent.
3 We must await developments, but there is no sign of IAs being any more effective than RIAs. The hard fact is that without the 'One in, One out' rule (qv) that was recommended to the Government in March 2005 by its own Better Regulation Task Force, and then quietly buried by Tony Blair, essentially bureaucratic initiatives such as these will not succeed.
4 For more information visit:
• www.cabinetoffice.gov.uk/regulation/news/2007/070402_ia.asp
• www.cabinetoffice.gov.uk/regulation/ria/index.asp
• www.cabinetoffice.gov.uk/regulation/

See also: *Better regulation bodies*; and *Over-regulation*.

Relative poverty

See: *Poverty*.

ligious buildings: sic planning data	1 Distilled planning data for most of the major building types is given in the METRIC HANDBOOK: PLANNING AND DESIGN DATA. 2 For publication details see the bibliography.
ligious buildings: sign: complying with Disability scrimination Act	1 Buildings used by the public must in general comply with the Disability Discrimination Act (qv). In the case of places of worship and crematoria *authoritative practical guides* (qv) include DESIGNING FOR ACCESSIBILITY (qv) plus the following: • APPROVED DOCUMENT M (qv), the provisions of which must be heeded in order to comply with the Building Regulations for England and Wales. • BS 8300, the provisions of which apply to all buildings in the UK. Para 13.8 gives brief design recommendations for the above building types. BS 8300 is an advisory document and does not have the force of law, but conformity with its recommendations will make it easier to demonstrate that the requirements of the Disability Discrimination Act have been met. 2 For places other than England and Wales the following regulatory documents should be consulted: • In Northern Ireland: TECHNICAL BOOKLET R of the Building Regulations (Northern Ireland) (qv). • In Scotland: the TECHNICAL HANDBOOKS of the Building (Scotland) Regulations (qv). • In the USA, where the Americans with Disabilities Act applies, refer to the ADA AND ABA ACCESSIBILITY GUIDELINES (qv). 3 As of 2008, additional references include the following. For publication details see the bibliography: • WIDENING THE EYE OF THE NEEDLE: ACCESS TO CHURCH BUILDINGS FOR PEOPLE WITH DISABILITIES. Includes a Code of Practice (qv). 4 Useful websites include: • Through the Roof www.throughtheroof.org/?p=29
eligious scrimination	1 *Discrimination* (qv) on the basis of religion is outlawed in the UK by anti-discrimination legislation – for details see: *Discrimination*. 2 See also: *Equality*.
elocation grant	See: *Home improvement grants*.
emote control	See: *Environmental control systems*.
epair grants for aces of worship	See: *Heritage Lottery Fund*.
esearch: advocacy	See: *Advocacy research*.
esidential homes	See: *Care homes for older people*.
esidential family entres	1 Family centres are therapeutic settings where one or both parents attend with their children to help a family under stress stay together. Residential family centres are ones in which a family lives for a set period, with children remaining under their parents' care while living in the centre. 2 As of 2008, useful websites include: • Directgov www.direct.gov.uk/ • NSPCC www.nspcc.org.uk/

Residential segregation	See: *Population: racial integration.*
Restraint systems in vehicles	See: *Wheelchair securing systems.*
Resin-bound gravel	See: *Gravel: bonded.*
Restaurants	See: *Refreshment facilities.*

Resuscitation

1 'Do not resuscitate' policies in hospitals, especially in the case of *disabled* (qv) and very old people, have become a matter of intense debate. Such cases fall into two quite distinct classes:

• Those instances in which the person involved has asked for such action.

• Those instances in which the patient is not deemed mentally capable of making any decisions, and others must do so on his or her behalf.

2 In the first case the matter is simple: the wishes of the person in question should be obeyed. The law in Britain (unlike those of the Netherlands, Belgium, and the state of Oregon in the USA) does not currently allow this. That is a clear breach of the individual's democratic right of self-determination, and British law is in urgent need of change.

3 In the second case the matter is less clear-cut, and difficult decisions arise. For a well-informed and thoughtful discussion refer to pp 118–121 of DISABILITY RIGHTS AND WRONGS by the *disabled* writer Tom Shakespeare – see the bibliography.

4 In taking these decisions it is important to understand that resuscitation is not necessarily, or even probably, the kindly act that some people assume. There is a risk of internal fractures and ruptures and long-term brain damage; it may be a hellish experience for the patient; it will probably prolong life for only a limited period; and it may well have to be repeated, possibly to the suffering patient's great distress. These facts should be truthfully discussed with people for whom resuscitation may be a prospect, and their views elicited and respected. In two studies where this was done the proportion who opted for resuscitation dropped from 31% before being acquainted with the facts to 22% after. In the case of those whose life expectancy was a year or less, assuming successful resuscitation, it dropped from 11% to 5%.

5 In short, medical technology would be better concentrated on making patients near the end of their natural lives as comfortable and happy as possible for their remaining days, than on prolonging their distress for what are becoming increasingly unreasonable periods as our technological capabilities advance.

See also: *Independent living*; and *Self-deliverance.*

Reversibility

1 Where new ramps, handrails, support rails, signs, or other physical features must be installed in *historic properties* (qv) it is very desirable to do this in such a manner that the additions can later be removed, leaving the original building fabric exactly as it was before. The *adjustment* (qv) is therefore 'reversible'. Examples might include:

• Making a period balustrade that has an ungrippable handrail at the wrong height more *accessible* (qv) by attaching a removable smooth, grippable handrail of the correct diameter at a convenient height – see: *Stairs*; and *Handrails.*

• Making an entrance to a period building accessible by fitting a removable ramp – see: *Ramps.*

- Mounting signs in historic buildings on well-designed and attractive free-standing frames, rather than fixing them to walls – see: *Signs*.

2 Such temporary features should be designed to be as elegant as though intended to be permanent – which generally is their destiny. Many museum and art gallery entrances are disfigured by hideous 'temporary' ramps which received scant design attention in the first place, and then became even shabbier because 'temporary' materials and finishes were thought to be good enough.
3 EASY ACCESS TO HISTORIC BUILDINGS is a free, well-illustrated publication that gives excellent guidance on these matters – see the bibliography.

See also: *Historic buildings and sites*.

BA

1 The Royal Institute of British Architects promotes architects and architecture in Britain. It has around 30 000 members.
2 For more information visit www.riba.org/

BA inclusive design ward

See: *Access awards*.

---

icability

1 The Research Institute for Consumer Affairs publishes consumer reports on products for *disabled* and older people. All Ricability reports are based on rigorous research and are key references in their field. They cover subjects such as domestic appliances, *products for easier living* (qv), telecoms and alarms, etc. Most are free.
2 For more information and a list of publications visit www.ricability.org.uk/

See also: *Disabled Living Foundation*, and *Disability Information Trust*.

---

ight to Know

See: *Your Right to Know*.

---

ight to read alliance

1 The RNIB estimates that over 95% of books are never made available in *Braille* (qv), audio, or large print. It therefore collaborates with 14 other organisations in the Right to Read alliance, which works for more books to be published in the above formats.
2 For more information visit www.rnib.org.uk/

See also: *Reading aids; Talking books*; and *Talking newspapers*.

---

ights

See: *Human rights*.

---

isk

1 A *hazard* (qv) is something with the potential to cause harm – for instance a *fire* (qv). A *risk* is:

- The likelihood of that harm occurring.
- The potential severity of that harm.
- The number of people who might be exposed to that harm.

Virtually everything in the environment (for instance: sunshine, a dog, or a doorstep) is a hazard, and creates some degree of risk. Because most such examples have a low score against the three factors above, or because we understand them well enough to deal with them by the exercise of common sense, we are not unduly troubled by them. At the other extreme, some risks score very highly on one or more of the three factors (for instance: high-voltage electrical installations) and highly regulated formal precautions may then be required.
2 Where to draw the line between those risks which people (and especially children) can be trusted to manage themselves, and those risks from which they should be shielded by law and regulation, is becoming a very difficult and contentious matter. In past times the law was often too lax; today we are going

to the opposite extreme as governments succumb increasingly to unreasonable (and largely unrealistic) demands from various sources for ensuring complete safety everywhere.

3 CABE (qv) has published the following reports :

• In a press release titled 'Compensation culture condemns us to live with dreary public spaces', issued on 24 March 2004, CABE stated that bouncy castles, ancient trees, boating lakes, adventure playgrounds, and firework displays on windy days were disappearing from British parks, because they were considered by the authorities to pose unacceptable risks to park users. It cited as specific examples the decision by Suffolk county council to ban hanging baskets because they might fall on somebody's head (even though this had never happened); the felling of beautiful horse chestnut trees because falling conkers might in theory hurt children; the regular removal of three-in-a-row swings in parks because the outer swings could hit the one in the middle and the removal of a swing from a playing field because it faced the sun and could harm childrens' eyes. Visit www.cabe.org.uk/news/press/

• WHAT ARE WE SCARED OF? THE VALUE OF RISK IN PUBLIC SPACE is the title of a subsequent CABE report, published in February 2005, which argued that the current climate of risk aversion and threat of compensation claims, is leaving architects, designers and their clients little opportunity to create the kind of vibrant public space which help a neighbourhood to thrive.

4 Comparable examples are reported in the media virtually every week. Taking risks of any kind is coming to be seen as positively anti-social and – given Britain's burgeoning *compensation culture* (qv) – potentially very expensive. As of 2008, thought-provoking publications include the following. They are given in alphabetical order. For details of these and other references quoted here, see the bibliography:

• CHILDHOOD WELLBEING, a research study published by the Department for Children, Schools and Families, reports the fears of many parents that their children's lives are being 'systematically undermined' by political correctness and the health and safety culture.

• HOW TO LABEL A GOAT: THE SILLY RULES AND REGULATIONS THAT ARE STRANGLING BRITAIN by Ross Clark gives examples of the mountain of excessiv regulation now descending upon us, and estimates the overall quantity.

• THE IMPROVING STATE OF THE WORLD: WHY WE'RE LIVING LONGER, HEALTHIER, MORE COMFORTABLE LIVES ON A CLEANER PLANET by Indur Goklany serves both as an attestation to the value of good regulation, and the irrationality of the fears that lead to unnecessary regulation.

• LITTLEJOHN'S BRITAIN by Richard Littlejohn usefully brings together a rich cro of the 'barmy diktats' (his words) that are coming to characterise British life.

• PLAYING IT SAFE: THE CRAZY WORLD OF BRITAIN'S HEALTH AND SAFETY REGULATIONS by Alan Pearce brings together an astounding anthology of rules, regulations, and official policies.

• SOUR GRIPES by Simon Carr comments particularly upon the compensation culture.

5 Positive references for planners and designers include:

• LIVING WITH RISK: PROMOTING BETTER PUBLIC SPACE DESIGN examines ten public spaces and streets across England with the aim of helping designers challenge excessively risk-averse decisions.

6 Useful websites include:

• Health and Safety Executive	www.hse.gov.uk/
• RoSPA	www.rospa.com/
• The great debate	www.thegreatdebate.org.uk/frankfuredi.html

See also: *Asbestos; Compensation culture; Health scares*; *and Over-regulation*.

sk and Regulatory Ivisory Council

1 The RRAC is a *quango* (qv) that was created in January 2008 to protect society from the 'cottonwool approach' to risk.
2 As with *Better regulation bodies* (qv), the Government is attempting to fight existing bureaucracy with new bureaucracy, instead of attacking the problem at source. There is no reason to believe that this latest attempt will be any more successful than its predecessors. Visit www.nationalcareforum.org.uk/

sk assessments

1 A risk assessment is the process of (a) judging a *risk*, and (b) deciding how best to deal with it. Such assessments range from the simple and informal to the complex and formal. They may be usefully divided into three commonsense categories:

• Risks are so simple that we assess them almost without conscious thought. For instance, the quick mental check we do before crossing a street or stepping off a moving bus.

• Risks that require a degree of conscious thought. For instance, the cautious review that a sensible policeman would make before tackling a dangerous situation, or the more lengthy kind of assessment that teachers should carry out before taking children on an adventure trip.

• Risks that that are so complex, and/or so remote from our everyday experience, that they can only be assessed in a highly formal manner, taking a lot of time and involving a lot of paperwork. For instance, ensuring *fire safety* (qv) in a train, an underground station, or a tall building.

2 Leaving aside the third category, risk assessment by professionals can be helped by a some simple form-filling, provided this is light enough not to interfere with the serious business of managing and taking responsibility, but most paperwork beyond that level effectively turns managers and professionals into box-tickers who spend their time on bureaucratic procedures as a substitute for thinking and taking decisions. Their backs have been covered, but are they achieving sensible results?

3 For an answer to that last question, consider Britain's police forces. Until quite recently policemen called to a scene of violence would rush to the site and, upon arrival, strive to save lives using their professional judgement and commonsense (not to mention courage and the basic human instinct of helpfulness) in deciding how best to act. But no longer. The new and rapidly growing culture of formal risk-assessment debars them from such action, as illustrated by the following examples:

• In 1995, Lorraine Whiting was shot by her husband before killing himself. The wounded woman spent almost an hour on the phone pleading to be taken to hospital while the police stood outside her house carrying out risk assessments. Mrs Whiting told the police 25 times that the gunman was dead, but they thought that he might be pressuring her to say that and refused to come to her aid. She bled to death.

• In November 2003, in the village of Hermitage in Berkshire, a terrified woman, Julia Pemberton, hiding in a cupboard from her husband, called the emergency services saying she was about to be murdered. The operator heard her desperate cries suddenly cut off by her husband shouting 'You f***ing whore'. Thames Valley Police Officers arrived at her home quickly but then waited outside for nearly six hours, carrying out risk assessments before deciding it was safe to enter. By then the woman and her son were dead. The heart-wrenching transcript of Mrs Pemberton's futile emergency call was published in the *Guardian* on 29 September 2004 under the title 'Woman's final minutes captured in 999 call' (visit www.guardian.co.uk/Archive/).

• In June 2004, in the village of Highmoor Cross in Oxfordshire, Stuart Horgan shot his ex-wife, her sister and her mother with a shotgun, then fled. Neighbours telephoned the emergency services approximately 50 times pleading for help and assuring them that the gunman had gone. But for well over an hour (while a 70-year-old neighbour stood guard outside the victims' house with a lump of wood and his wife cradled the dying woman in her arms) armed officers from Thames Valley Police, wearing body-armour and surveille[d] by a helicopter hovering overhead, sat in a parked car four miles away carrying out risk assessments, meanwhile refusing even to enter the village, let alone the house. They also refused to let ambulancemen (who arrived within three minutes of being called) go to the aid of the victims. By the time a detective sergeant finally decided to defy the rules, behave in the old-fashioned way, and approach the house, the wife and her sister had died.

• The Sussex Police similarly refused to attend a woman, Linda Watson, who had dialled 999 to say that her husband had been shot outside their home. Even after a family friend had checked the body and confirmed Mr Watson's death to the police, they carried out risk assessments for another hour before deciding it was safe to approach.

4 The officers above were obediently following official guidelines, and these are – if anything – tightening. In October 2007 two community support officer[s] refused to follow anglers into a pond to try and rescue a 10-year-old boy, and stood on the bank watching the child drown. The Greater Manchester Police subsequently publicly thanked the officers for 'acting correctly' and not putting their own lives at risk.

5 For the views of a policeman on this situation refer to NO MORE BRAVE POLICEMEN – THE POLICE CAN'T DO THEIR JOBS IF THEY ARE BOUND BY HEALTH AND SAFETY LAWS by Chief Constable Bernard Hogan-Howe. This article was published in *The Times* – see the bibliography.

Risk assessments: basic procedure

1 Risk assessments usually involve five fundamental steps:

• Identifying the *hazard* (qv).

• Deciding who might be in danger.

• Evaluating the *risks* (qv) arising from that hazard; deciding whether existing precautions are adequate; if not, formulating additional measures.

• Recording the above findings and the action taken.

• Reviewing and if necessary repeating the assessment at suitable intervals thereafter.

2 For sources of information on some specific types of risk assessment see the next three entries.

Risk assessments: play areas

1 The over-protective 'nanny state' has taken hold with particularly baleful effect in the area of children's play. In depressing contrast with freedom enjoyed by all previous generations of children, including the freedom to take risks (which is an essential part of growing up), many youngsters today are hardly let out of sight of parents and teachers, and hardly allowed to climb a tree. They are in effect being robbed of their childhood.

2 As of 2008, useful references include the following. For publication details see the bibliography:

• NO FEAR: GROWING UP IN A RISK AVERSE SOCIETY by Tim Gill argues that wrapping youngsters in cotton wool does them more harm than good.

• RISK AND CHILDHOOD, by Nichola Madge and John Barker of Brunel University, supports the above thesis.

• SEEN AND HEARD: RECLAIMING THE PUBLIC REALM WITH CHILDREN AND YOUNG PEOPLE discusses the design and management of the public realm.

3 Useful downloads include:

• HSE www.hse.gov.uk/foi/internalops/sectors/cactus/5_04_11.pdf

• RoSPA www.rospa.com/playsafety/publications/orderform.pdf

sk assessments: :hool trips

1 As of 2007, officially recommended sources of guidance for teachers who take pupils outside the school premises include the following. For publication details see the bibliography:

2 From the Royal Society for the Prevention of Accidents:

- HEALTH & SAFETY AT SCHOOL: SCHOOL TRIPS: PART 1
- HEALTH & SAFETY AT SCHOOL: SCHOOL TRIPS: PART 2

3 From the Department for Education and Skills:

- HEALTH AND SAFETY OF PUPILS ON EDUCATIONAL VISITS – a good practice guide.
- STANDARDS FOR LEAS IN OVERSEEING EDUCATIONAL VISITS – Part 1 of a 3-part supplement to health and safety of pupils on educational visits.
- STANDARDS FOR ADVENTURE – Part 2 of a 3-part supplement to health and safety of pupils on educational visits.
- A HANDBOOK FOR GROUP LEADERS – Part 3 of a 3-part supplement to health and safety of pupils on educational visits.

4 The two RosPa guides above add up to 14 pages of instructions, with 6 additional references listed at the end. The four DfES guides add up to 118 pages of instructions, with more than 50 additional references listed at the end. Because indemnity against damages may depend on following such instructions quite literally to the letter, teachers dare not substitute their own hard-won professional judgement or common sense for these mind-deadening checklists, but must dutifully absorb and apply the contents of the above library of references.

5 The amount of paperwork that is needed has led one major teaching union, the NASUWT, to formally advise its members against taking children off school premises, even to a museum. Therefore museum visits are widely being curtailed and children are the losers. A RESEARCH REVIEW OF OUTDOOR LEARNING, a study commissioned by the DfES and published in November 2006 by the National Foundation for Educational Research and the Field Studies Council, reported that school trips are becoming steadily tamer and more useless. The survey of 3 500 teachers found that instead of going to outdoor activities or the countryside, pupils were more likely to be taken on 'outings' to museums or in the school playground. For publication details see the bibliography.

6 As of 2007, useful websites include:

- Teachernet www.teachernet.gov.uk/wholeschool/healthandsafety/visits/

sk assessments: •rkplaces

1 As of 2007, useful websites include:

- Business Link www.businesslink.gov.uk/
- Health and Safety Executive www.hse.gov.uk/msd/risk.htm

2 For risk assessments in the particular context of fire safety visit www.fire.gov.uk/workplace+safety/saferpremises/riskassessment

sk assessments and sabled people

1 WHOSE RISK IS IT ANYWAY? is the title of a discussion document released by the Disability Rights Commission (qv) in September 2005, which makes a pertinent contribution to the matters touched upon above. For details see the bibliography.

2 The document argues that health and safety risk assessments are 'leading to diminished opportunities across life' for *disabled people* (qv), and calls for a 'grown-up debate' to stop organisations and authorities taking 'wholly disproportionate steps' against disabled people and 'limiting their choices to participate fully in society'. Mr Bert Massie, the chairman of the DRC, stated that 'fear of litigation is replacing sensible action and the end result is disabled

people being denied the chance to take decisions and weigh up risks for themselves' in their everyday lives.
3 An example mentioned in the report is that of Lynne Coupe, from Chorley in Lancashire, who went without a shower for three months because her *stairlift* (qv) broke down and the social services department refused to install a new lift arguing that she needed help getting on and off and that 'using the lift in these circumstances would be against health and safety rules' – even though she had been helped on and off her stairlift for eight years. They offered to take her to a hospital once a week to be showered but her husband refused and installed a shower in the sun lounge despite being refused planning permission by the *local authority* (qv). Mr Coupe offered the sensible comment that the application of rules such as this to people at large would mean that 'no-one would ever get out of bed in the morning'.
4 In November 2005 the journal DISABILITY NOW (qv) reported the case of a disabled woman who was told by Kirklees council that they would have to do a full risk assessment of the Lake District each time she proposed to go kayaking with Council support, and that payment could be withheld from her *carers* (qv) if she 'disobeyed orders' by proceeding without such an assessment. She commented: 'I am 34 years old and can calculate risk. I feel it is my right to be supported in activities I choose rather than those chosen for me'. A spokesman for the Disability Rights Commission (qv) supported her, saying that the risk assessment situation was 'getting preposterous. As long as she's aware of the risks, there shouldn't be anything stopping her. She should have the same rights as anyone else'.
5 Until there is a basic change in law that removes the fear of litigation from people who would like to be sensible, but can't afford the risk, cases such as these will unfortunately multiply.

See also: *Compensation culture; Independence*; and *Post-democracy*.

RNIB

1 The Royal National Institute of the Blind is the UK's most authoritative source of information on *blind* (qv) or near-blind people, their needs, and suitable products and services.
2 For more information visit www.rnib.org.uk

See also: *Impaired vision*.

RNID

1 The Royal National Institute of the Deaf is the UK's most authoritative source of information on *deaf* (qv) or near-deaf people, their needs, and suitable products and services.
2 For more information visit www.rnid.org.uk

See also: *Impaired hearing*.

Road crossings See: *Pedestrian crossings*.

Road safety See: *Transport safety*.

Road travel: accessibility guides See: *Travel: accessibility guides*.

Rollators See: *Walking frames*.

Roller-tip cane See: *Canes for blind people*.

ugh sleepers unit	This unit no longer exists. See: *Homelessness.*
wntree Trust	See: *Joseph Rowntree Trust.*
yal Association for sability and habilitation	See: *RADAR.*
yal Institute of itish Architects	See: *RIBA.*
yal National stitute of the Blind	See: *RNIB.*
yal National stitute of the Deaf	See: *RNID.*
yal Society for entally Handicapped ildren and Adults	See: *Mencap.*
yal Society for the evention of cidents (RoSPA)	1 RoSPA is a registered charity which provides information, advice, and training to improve safety in all areas of life – at work, in the home, in schools, at leisure and play, and on the roads. It publishes excellent factsheets on safety and accident prevention and is, together with the Health and Safety Executive (qv) a leading source of statistics on *accidents* (qv). It tends to take a more commonsensical view of risk and safety than the hyperactive HSE. 2 For more information visit: www.rospa.org/ See also: *Accidents; Health and Safety Executive*; and *Risk assessments.*
le of law	See: *Civil liberties*; and *Human rights.*
ral housing	See: *Affordable Rural Housing Commission.*
ral poverty	See: *Commission for Rural Communities.*

S

feguarding lnerable Groups t 2006

1 On Monday 23 October 2006 this Act was nodded through *Parliament* (qv) without opposition. According to the commentator Philip Johnston, writing in the *Daily Telegraph* (13 November 2006), the original bill was so badly drafted that it needed 250 amendments, which almost certainly means that the law is still riddled with imperfections, yet to be discovered. Meanwhile the implications of the Act as passed were so clearly stated by Brendan O'Neill, writing in the *Guardian* on 25 October 2006 under the title 'Heavy Vetting', that the following notes are based upon that report.

2 The Act obliges every adult who works with children to submit him or herself to a criminal records check. Anyone who doesn't comply will be branded a criminal and may be fined £5 000. This means that one third of the adult working population including teachers, lollipop ladies, and even 16-year-olds who teach sport to younger kids at the weekend, will come under the Act. It will be a criminal offence (with all that may follow from that – see: *National DNA database*; and *Police National Computer*) for a father to coach his son's football team on a Sunday morning without first being vetted. In order to give up two hours of his time to help his son and his son's mates improve their skills dad will have to present three forms of identification, pay £36, and wait several weeks for official clearance. His details will then be stored on a database and he will be subject to 'ongoing monitoring'. Who could blame him if he stays at home instead? What will be the long-term effect on society if fathers are effectively encouraged to stay at home rather than go out to play football with their sons and their friends?

3 Already, according to Mr O'Neill, there is evidence that the constant expansion of vetting is threatening the jumble sales, local football teams, school trips, and sports days that are central to children's lives. Increasingly, schools are having to abandon or scale down events such as fetes and sports competitions because they can't get the adult volunteers vetted in time. He quotes the tennis coach Dan Travis as confirming from his own experience that vetting is helping to destroy the *voluntary sector* (qv) in sport. Perhaps even worse is the more general effect of obligatory and repeated vetting upon child/adult relations in general. As Jim Campbell, the mayor of Oxford, put it in THE CASE AGAINST VETTING: HOW THE CHILD PROTECTION INDUSTRY IS POISONING ADULT-CHILD RELATIONS (see the bibliography): 'The important informal ways in which people relate are going to disappear. Everything will be done under contract. We are in danger of creating a generation of children who are encouraged to look at people who want to help them with suspicion'.
Ms Judith Gillespie of the Scottish Parent Teacher Council commented that the growth of vetting would prevent children from making their own judgments, stop them learning about life, and possibly 'leave them unable to look out for their own safety and incapable of judging between risky and safe adults'.

4 Mr O'Neill concludes: 'Communities thrive on informality, on working things out for themselves, and children thrive on free and easy relationships of trust with their parents, teachers, and other adults. This Act will stifle all of this and give rise to a climate of fear, uncertainty and loathing. They used to say that it took a village to raise a child. Today only state-sanctioned individuals will be allowed to raise a child'.

5 Even before this Act organisations such as the Girl Guides and Scouts are chronically short of volunteers (the Guides have a waiting list of 50 000, the Scouts 30 000, and some parents have resorted to signing their children up at birth). The new law will greatly magnify this problem but would have done nothing to prevent the incident that inspired it – the Soham murders, in which Ian Huntley, a school caretaker, killed two young girls. Huntley did not work at their school; it was his partner who did, and it is she who would have undergone the screening process.

6 As of 2008, useful references include the following. For publication details see the bibliography:
• DON'T TOUCH! THE EDUCATIONAL STORY OF A PANIC by Heather Piper and Ian Stronach.
• NO FEAR: GROWING UP IN A RISK AVERSE SOCIETY by Tim Gill. A leading child expert argues that wrapping youngsters in cotton wool does them more harm than good.
• RISK AND CHILDHOOD by Nichola Madge and John Barker. Two researchers at Brunel University support the above thesis.
7 Useful websites include the following:
• Liberty www.liberty-human-rights.org.uk/
• Manifesto Club www.manifestoclub.com/
• Office of Public Sector Information www.opsi.gov.uk/acts.htm

See also: *Civil Liberties*; and *Privacy*.

Safer and Stronger Communities Fund

Visit: www.neighbourhood.gov.uk/page.asp?id = 1304

Safety

1 Safety has been defined by the International Standards Organisation (ISO) a 'a state of freedom from unacceptable risks of personal harm' – a definition tha recognises the fundamental point that no activity can be absolutely safe or free from risk.
2 This sensible principle is not adequately recognised in current law, leading t perverse decisions such as noted under *Compensation culture* and *Risk assessments*. An Act of Parliament that formally recognises and defines the concept of an accident, for which no-one can be charged or sued for damages, appears to the present author to be overdue.

See also: *Accidents; Compensation culture*; and *Risk assessments*.

Safety markings

See: *Glass: safety markings; Hazards: protection*; and *Street works*.

Safety: products

See: *Product design*; and *Product packaging*.

Safety: transport

See: *Transport safety*.

Scheme development standards

See: *Housing accessibility*.

School buildings

1 As of 2008, useful websites include:
• www.cabe.org.uk/default.aspx?contentitemid=37
• www.teachernet.gov.uk/management/resourcesfinanceandbuilding/
• www.teachernet.gov.uk/management/resourcesfinanceandbuilding/schoolbuildings/
2 See also: *Building Schools for the Future*.

School trips: risk

See: *Risk assessment*.

Schools access initiative (SAI)

1 The SAI provides funding to make *mainstream schools* (qv) more *accessibl* (qv) to children with *disabilities* (qv) or *special educational needs* (qv).
2 Visit www.teachernet.gov.uk/wholeschool/sen/schools/accessibility/sai/

hools: design See: *School buildings*.

hools: exclusion of pils 1 Extremely ill-behaved pupils not only make their teachers' lives a misery and wreck classroom teaching, but their presence in many comprehensive schools is one of the main reasons why many striving parents (for instance the pro-comprehensive school MP Diane Abbott) cannot bring themselves to place their children in these schools. An article in the 'Perpetuity' newsletter (see below) from the University of Leicester, published in June 2007, reported that 99% of a survey group of 300 teachers had been verbally abused by pupils in the past year, and 17% had been threatened with weapons.

2 For the good of just about everyone else such pupils must be taken out of the classroom and educated elsewhere. This has long been far too difficult, with suspensions or exclusions by heads being regularly set aside by external authorities. New rules introduced in 2007 have made it even harder for head teachers to suspend troublemakers and left them with only two options – permanent expulsion, or keeping the disruptive pupils in the classroom to the detriment of everyone else. In March 2008 the President of the Association of School and College Leaders, Mr Brian Lightman, told the annual ASCL conference that this policy had backfired and that the problem of classroom discipline was getting worse rather than better.

3 A rational policy should start from the fact that 87% of exclusions in primary schools and 60% of exclusions in secondary schools in 2006 involved children with *Special Educational Needs* (qv). It is clear that the official policy of placing these pupils in *mainstream schools* (qv) which cannot cope with them, instead of giving them the attention they need in dedicated *special schools* (qv), is damaging the pupils concerned and also the rest. This approach, which owes more to *ideology* (qv) than to common sense, should be reversed. If comprehensive schools are ever to become acceptable to conscientious parents such as Ms Abbott, heads must be given complete authority to exclude misbehaving pupils. That necessarily means that very many *special schools* (qv) will have to be created to educate the latter. These will have to be very much more effective than current Pupil Referral Units (PRUs) which reportedly cost £17 000 per place per annum, and are not working. According to the *charity* (qv) Barnado's, which runs three such schools, the prerequisite for success is to have class sizes of 6 or 7, plus constant and close supervision. In the case of severely disturbed youngsters the cost of such provision may be equal to the cost of keeping an offender in prison (at a guess, somewhere between £25 000 and £40 000 a year) but the cost would be justified. Even if some pupils proved impossible to educate, they would at least be unable to wreck the education of 20 or 30 classmates in a mainstream school.

4 Returning to the present situation: the number of pupils who are permanently excluded from primary, secondary and special schools in Great Britain every year is around 9 000, and slowly falling. The most common reason in England is persistent disruptive behaviour (31% of all permanent exclusions and 26% of all fixed period suspensions), followed by physical assault against a pupil (17%), and verbal abuse or physical assault against an adult (12%). The number of pupils in secondary schools who are suspended for a fixed period, usually 2–3 days, is around 340 000 and slowly rising.

5 As of 2008, useful references include the following. For publication details see the bibliography:

• SOCIAL TRENDS (qv). See in particular Chapter 3.

See also: *Education; Social exclusion; Social mobility*; and *Special Educational Needs*.

Schools: mainstream and special

1 A 'mainstream school' is defined by section 316 of the Education Act 1996 a school that is not a special school. A 'special school' is defined by section 337 of the Act as one that is 'specially organised to make special educational provision for pupils with *special educational needs* (qv) and is for the time being approved by the Secretary of State under section 342' – visit www.hmso.gov.uk/acts/acts1996/96056-bd.htm#337

2 As of 2008, other useful websites include:

- Gloucester Special Schools Protection League www.gsspl.org.uk/
- Office for Standards in Education www.ofsted.gov.uk/
- Specialist Schools & Academies Trust www.specialistschools.org.uk/
- DfES Standards www.standards.dfes.gov.uk/

See also: *Special educational needs.*

Schools: reform

See: *Social mobility.*

Scooters

1 Scooters are a *mobility aid* (qv), and comprise one of five categories of *small personal vehicles* (qv) for frail and *disabled* (qv) people:

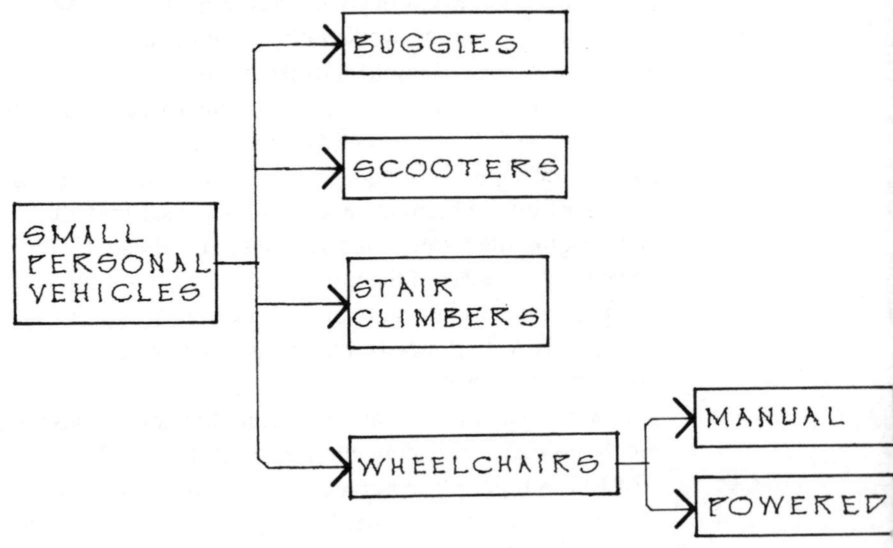

2 They enable people with *impaired mobility* (qv) to get out and about, unhindered by uneven surfaces, kerbs, and ramps, and are run from batteries which are recharged by overnight plugging into a mains socket.

3 The sketch below shows a fairly typical example, but there are many types and specific data should be obtained before any decisions are taken.

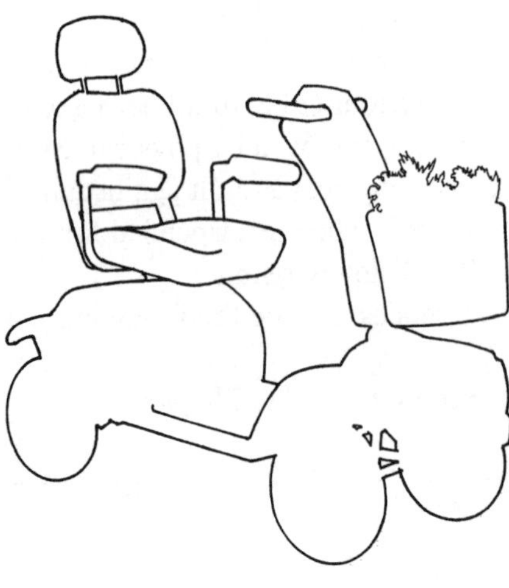

4 As regards scooter storage, many people with *impaired mobility* (particularly older people) live in small dwellings with no indoor space for parking a scooter. Suitable products to meet this need include the Scooter Store, a neat-looking painted steel outdoor mini-garage with automatic door, internal light, and a battery charger. No planning permission is required. For more information visit www.site-safe.co.uk/

5 As of 2008, useful references include the following. They are given in alphabetical order. For publication details see the bibliography:

- CHOOSING A SCOOTER OR BUGGY.
- GET MOBILE – YOUR GUIDE TO BUYING A SCOOTER OR POWERED WHEELCHAIR

6 Useful websites include the following. They are given in alphabetical order.

- Dept for Transport www.dft.gov.uk/transportforyou/access/tipws/
- Disabled Living Foundation www.dlf.org.uk/
- Disabled Living Online www.disabledliving.co.uk/
- KeepAble www.keepable.co.uk/
- OTDirect www.otdirect.co.uk/link-equ.html#Platform
- Patient UK www.patient.co.uk/showdoc/12/
- Ricability www.ricability.org.uk/

7 The annual Mobility Roadshow exhibits scooters and other personal vehicles wheelchairs, etc., and allows test drives. For more information visit www.mobilityroadshow.co.uk/

See also: *Buggies; Mobility aids; Powered wheelchairs*; and *Wheelchairs: space requirements*.

:otland: building gulations

See: *Building (Scotland) Act.*

:otland Office

1 The Scotland Office, headed by the Secretary of State for Scotland, represents the interests of Scotland at Westminster – see: *Government.*

2 Visit www.scotlandoffice.gov.uk/

:ottish Executive

1 The Scottish Executive is the devolved government for Scotland – see: *Government*. It was established in 1999 and is led by a First Minister who is nominated by the Scottish Parliament (qv) and in turn appoints the other Scottish Ministers who make up the Cabinet. It is responsible for most of the issues of day-to-day concern to the people of Scotland, including health, education, justice, rural affairs, and transport.

2 For more information visit www.scotland.gov.uk/

See also: *Constitution; and Government.*

:ottish Parliament

1 The Scottish Parliament exercises several powers that were devolved to it from the UK Parliament at Westminster in 1999. These 'devolved powers' include for education, health, and prisons. Policy areas which remain with the UK Parliament at Westminster are termed 'reserved powers'.

2 The relationship between the Scottish and UK parliaments was ill thought-out and has created the paradoxical situation that English-constituency MPs at Westminster cannot vote on matters devolved to Scotland (eg education), but Scottish-constituency MPs at Westminster are able to vote on these matters for England – a constitutional asymmetry usually referred to as the 'West Lothian question'. This relationship is inequitable and unstable, and cannot survive indefinitely – see *Constitution*.

Scruton, Roger See: *Architecture*; and *Speech, freedom of.*

Seats

1 Owing to the great variety of human shapes, no single seat design can ever satisfy everyone. The problem of matching standardised seats to diverse human need is best met by a combination of three measures:

- Designing all seats to well-established *ergonomic* (qv) standards.
- Where possible, designing seats and chairs with some built-in adjustability – see: *Adjustable furniture*.
- Providing a variety of seat types and sizes in waiting areas etc. so that people can choose one that suits them. Para 11.2.4 of BS 8300 (qv) recommends a mixture of fixed and movable chairs with a variety of seat heights and widths, some with arms and some without.

2 For a discussion of chair design refer to Chapter 4 of BODYSPACE by Stephen Pheasant – see below. Figure 4.6 of that work shows two seat profiles which were found by the eminent French ergonomist E Grandjean to best satisfy a very wide range of users, including (unlike many other studies) people with back trouble as well as ones without. The entire discussion should be studied, but for quick reference the profiles have been redrawn below – one for a multipurpose chair (left) and one for an easy chair (right). Note in particular the bulge to support the *lumbar back* – a notoriously trouble-prone part of the body.

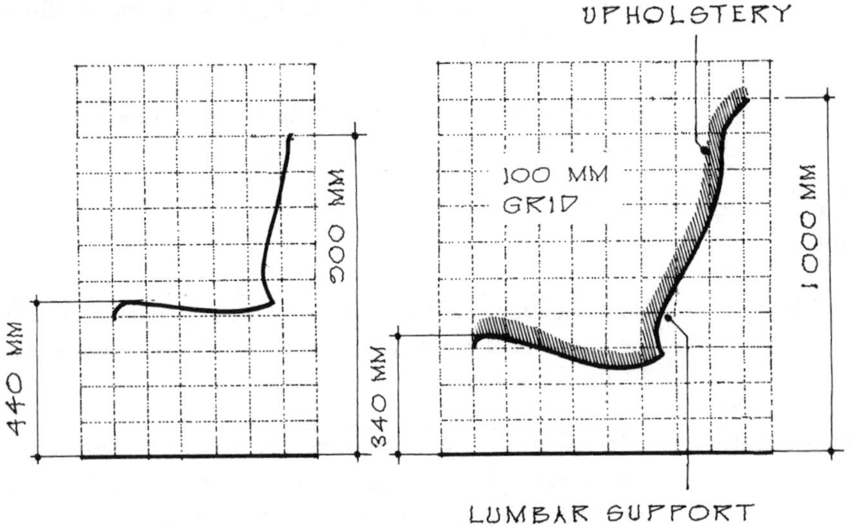

3 The figure below, also from Pheasant, suggests some key seat dimensions that will in many cases offer a reasonable compromise for diverse users. The dimensions were taken in 1996 from recognised UK sources such as BS 5940 and publications from the Health and Safety Executive (qv).

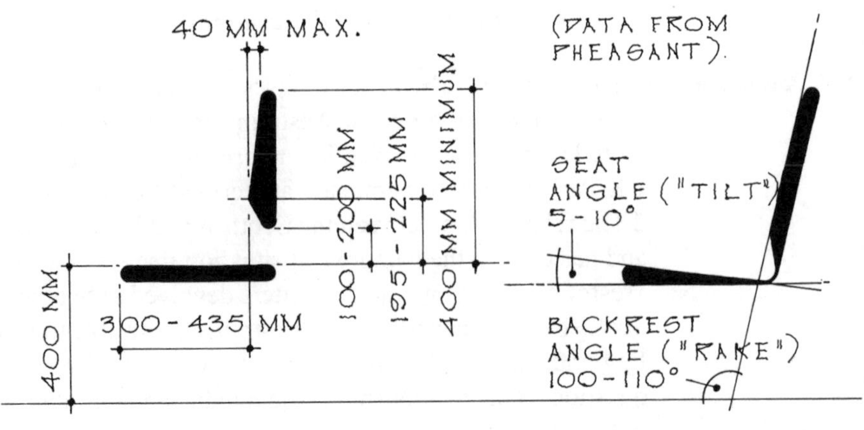

ats in public ldings: complying h the Disability crimination Act

1 Public buildings and spaces must in general comply with the Disability Discrimination Act (qv). In the case of seats in hotels, shops, banks, restaurants, public houses, GP's and dentists' surgeries, airports, railway stations, and theatres, etc., relevant *authoritative practical guides* (qv) include include DESIGNING FOR ACCESSIBILITY (qv) and in addition the following documents. For publication details see the bibliography.

- APPROVED DOCUMENT M (qv), the provisions of which must be heeded in order to comply with the Building Regulations in England and Wales.
- BS 8300 (qv), which applies to all buildings in the UK. It is an advisory document.
- INCLUSIVE MOBILITY (qv), which is the most authoritative reference for the *inclusive design* (qv) of *pedestrian infrastructure* (qv) and of *transport-related* buildings (qv) such as bus and coach stations, railway stations, air terminals, and transport interchanges. It is an advisory document.

While the advisory documents above do not have the force of law, conformity with their recommendations will help to demonstrate that the requirements of the Disability Discrimination Act have been met.

2 The relevant clauses in the above documents are set out below.

SEAT LOCATION	DOCUMENT: RELEVANT PARAGRAPHS		
	AD M	BS 8300	INCLUSIVE MOBILITY
WAITING ROOMS	–	11.2	9.3
LECTURE ROOMS AND AUDITORIA	–	11.3, 11.4, 13.6	–
REFRESHMENT AREAS	4.16	13.5.3	–
TRANSPORT-RELATED BUILDINGS	–	13.1.6	9.3
SPORTS-RELATED BUILDINGS	–	13.7.2	–
LIBRARIES	–	13.9.5	–
PEDESTRIAN ROUTES	–	–	2.4, 3.4
BUS STOPS	–	–	6.1.4

3 Recommendations in the above documents include the following:

- Para 11.2 of BS 8300 suggests a seat height, or a compressed cushion height, of 450 mm to 475 mm for fixed seating. A height slightly greater than the normal 450 mm is likely to be more convenient for people with restricted mobility, who cannot easily raise themselves from a low seat. BS 8300 also recommends a mixture of seating options for customers or visitors to a building – eg fixed or movable, with or without arms, and so on.
- Para 6.1.4 of INCLUSIVE MOBILITY recommends a sit-stand bench at a height of about 580 mm for bus stops (see next entry). Para 9.3 mentions seat heights of 420 mm to 580 mm for waiting areas, with a *median* (qv) height of 470 mm to 480 mm. Armrests should be about 200 mm above seat level, and minimum seat width is 500 mm.

4 The figure on the next page summarises the seat heights recommended in the above documents, and shows Pheasant's recommendation of 400 mm for comparison. Because elderly and DISABLED people often find it difficult to raise themselves from a low seat, BS 8300 and INCLUSIVE MOBILITY recommend greater seat heights than Pheasant. The dimensions for seat depth (front to back) and for backrest height are taken from BODYSPACE, as BS 8300 and INCLUSIVE MOBILITY make no recommendations on these matters.

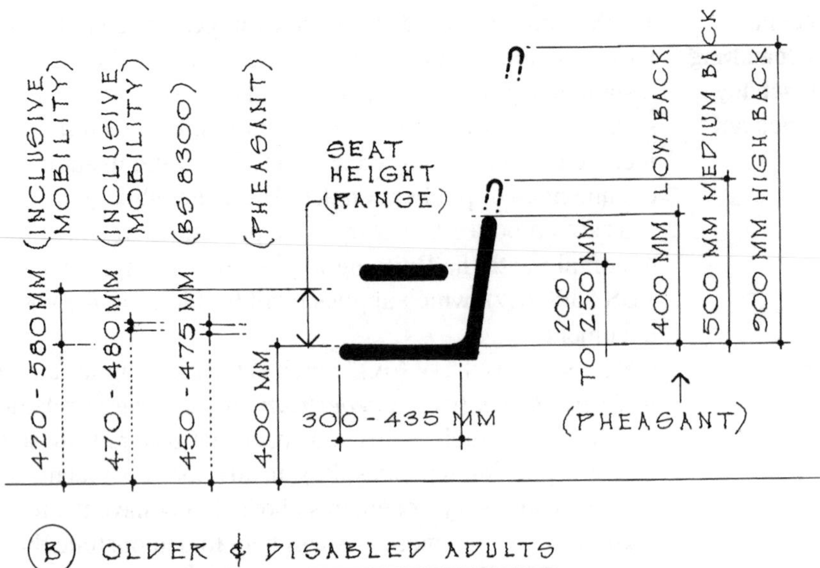

5 BS 8300 makes no recommendations on seat width. The recommendations below are from INCLUSIVE MOBILITY and Pheasant.

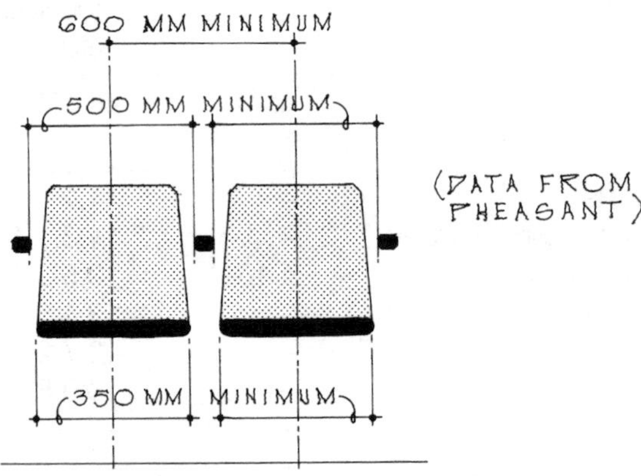

6 The figure below shows a seat design used at several outer London Transport stations, eg on the Wimbledon line. This design is comfortable, and satisfies the authoritative recommendations tabulated at the start of this entry. Note the contoured profile of the backrest.

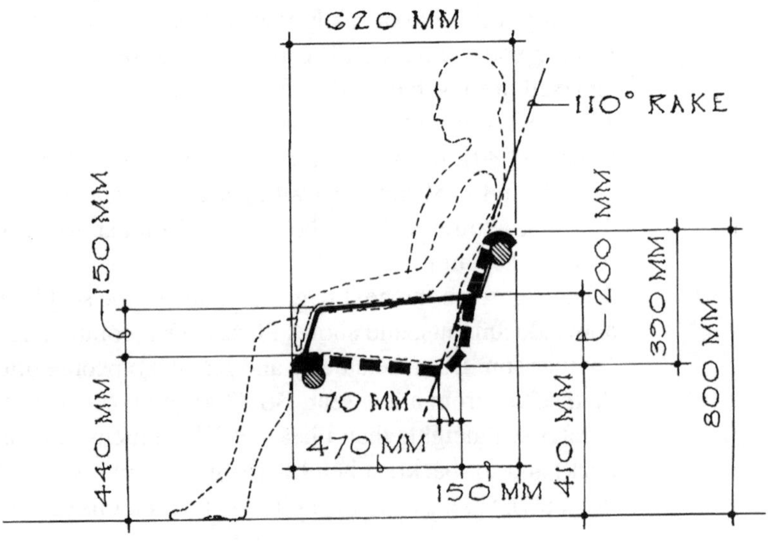

Seats in public vehicles 1 All public transport vehicles (qv) must comply with the Disability Discrimination Act (qv). *Authoritative practical guides* (qv) include the following:

• THE RAIL VEHICLE ACCESSIBILITY REGULATIONS (RVAR) 1998. Para 8 and diagrams B1 to B8 give design data for seats in *rail vehicles* used on railways, tramways, and monorail systems. The figure below summarises some of the key criteria, but the original document must consulted.

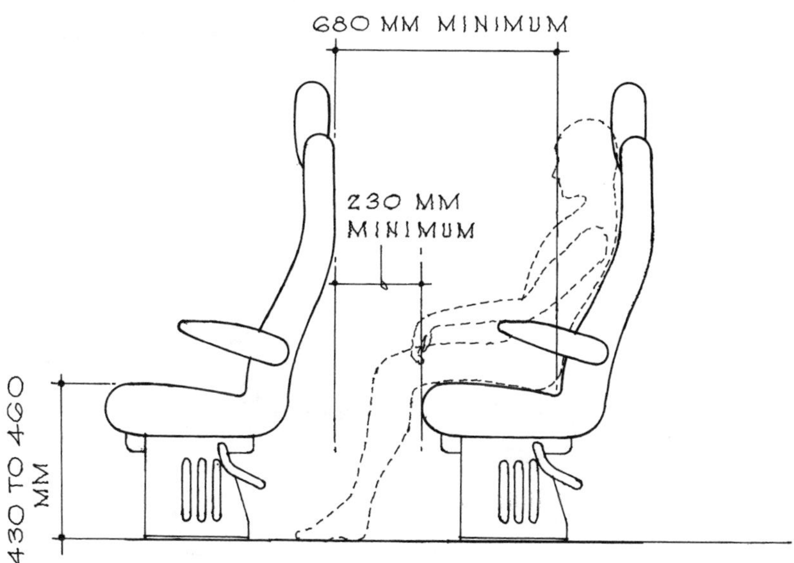

• ACCESSIBILITY SPECIFICATION FOR SMALL BUSES DESIGNED TO CARRY 9 TO 22 PASSENGERS. Paras A5 and B5 give design data for seats in small *buses*. The figure below summarises some of the key criteria, but the original document must consulted.

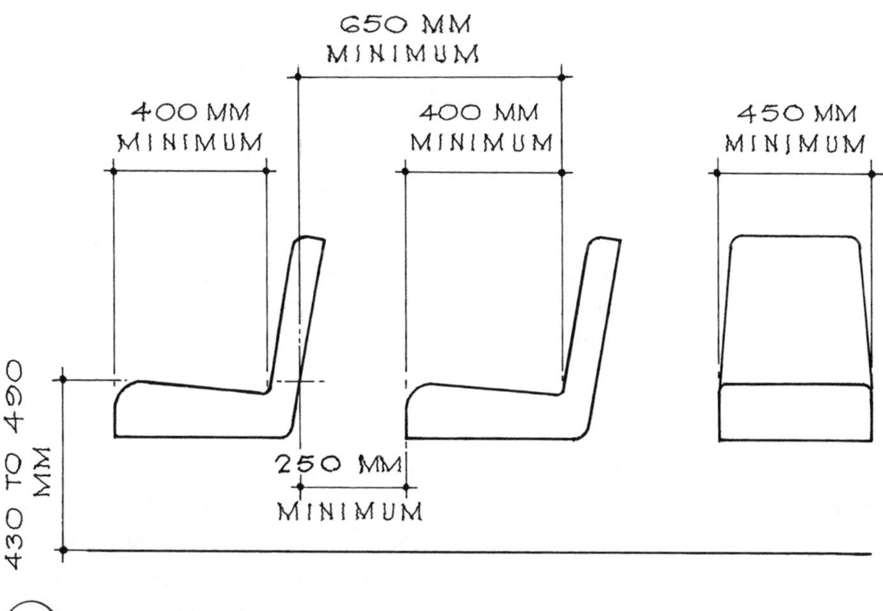

(A) SEATS IN SMALL BUSES

2 For publication details of the above documents see the bibliography.

See also: *Transport: complying with the Disability Discrimination Act.*

Seats: sit-stand

1 Seats which are leant against, not sat upon (also called *lean-on* or *perch-type* seats), may be appropriate in the following situations:

• For people who find sitting down painful owing to sciatic pain, hip replacement problems; stiff knees etc.; or for people working at a high bench or assembly line.

• At bus stops or underground railway stations, where waiting times tend to be short, and where sit-stand seats take up less platform space than conventional ones.

2 Figure A below shows a bench for bus shelters, as recommended in para 6.1 of INCLUSIVE MOBILITY (qv).

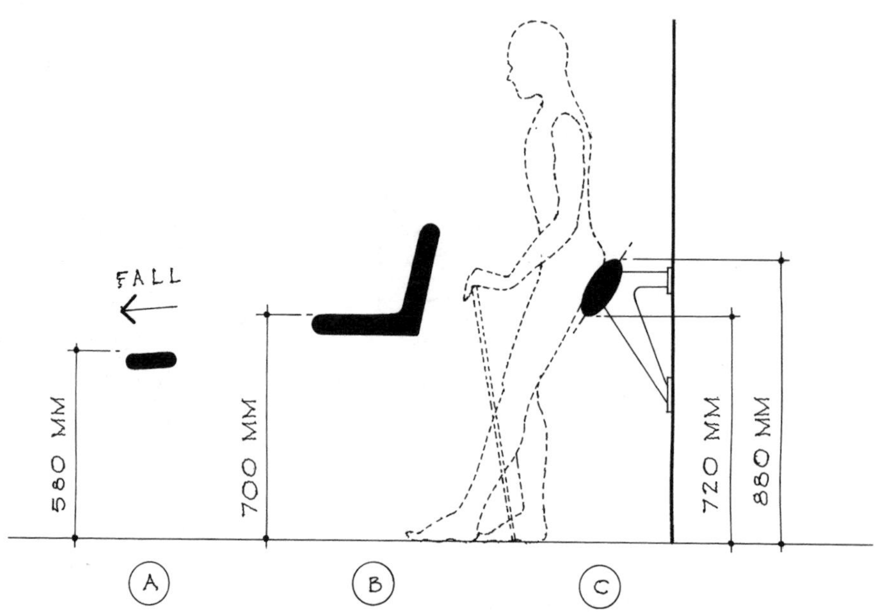

• Figure B above shows a perch-type seat for waiting areas and platforms, as recommended in para 9.3 of INCLUSIVE MOBILITY.

• Figure C shows a case example of a sit-stand seat design used at several inner London Transport stations, for instance the Jubilee line platform at Westminster underground station.

3 As of 2008, useful websites include the following. They are given in alphabetical order.

• HAG Capisco www.backinaction.co.uk/
• RH Support www.posturite.co.uk/
• Stokke Move www.backworks.co.uk/
• Variant XL www.back2.co.uk/

See also: *Bus and coach stops*.

Sector Skills Councils (SSCs)

1 The UK has 25 Sector Skills Councils, each of which is an employer-led, independent organisation that covers a specific business sector. Their goals are to reduce skills gaps and shortages and to improve productivity and performance for that sector. They cover roughly 85% of the UK workforce and are funded and monitored by the Sector Skills Development Agency. For a comment see: *Learning and Skills Council*.

2 For more information visit www.ssda.org.uk/

Secured by design

See: *Crime reduction by design*.

e Security Service t 1989 and 1996 See: *Civil liberties*.

eing aids See: *Reading aids*.

gregated shared cle/footway rface

1 This convoluted phrase is the official term for the *tactile paving surface* (qv) that should be used on all contiguous cyclist/pedestrian lanes, if the two lanes are not physically separated by railings, a difference in level, or some other prominent form of differentiation.

2 The arrangement is shown below:

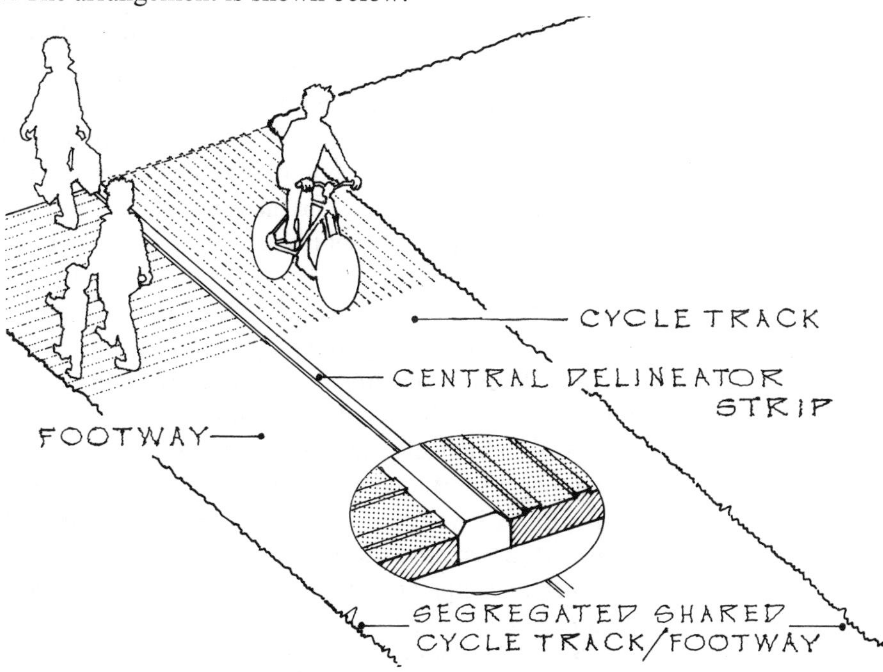

It will be seen that the surface is formed of two types of tactile pavings:

• Parallel rows of flat-topped bars, with the bars laid along the direction of travel on the cycle track, and across the direction of travel on the footway. The purpose is to help people with *impaired vision* distinguish between the footway and the cycle track, and not accidentally use the wrong lane.

• A 150 mm wide raised strip between the two tracks, forming a 'kerb' that stops people from accidentally crossing the divide.

3 The textured surfaces shown on the diagram should be laid at the start and finish of segregated shared cycle track/footways; at junctions with other routes, and at intervals along the route. The central delineator strip should however run along the entire length of the route except at junctions and crossings with other routes.

4 There are no specific recommendations on colour, except that that the central delineator strip should be white.

5 The arrangement shown on the diagram is schematic. The rules for correct use in particular situations are very precise and designers must in all cases consult Chapter 5 of GUIDANCE ON THE USE OF TACTILE PAVING SURFACES (qv) and para 4.5 of INCLUSIVE MOBILITY (qv) – see the bibliography.

6 For details of the six types of tactile paving, including the above, see: *Tactile pavings*.

gregation: racial See: *Population: racial integration*.

Self-deliverance

1 People can have sound reasons for wishing to end their lives – for instance being in pain, suffering from a disability that makes life an intolerable burden (eg *tetraplegia* or advanced *motor neurone disease*), or being old and wanting to pass away with dignity rather than in misery. Clearly this decision rests with the person involved: the fundamental principle of *independent living* (qv), whereby people control their own lives instead of having others decide things for them, must apply in all areas of life, throughout life, and finally in the manner in which we depart from this life. A SOCIAL HISTORY OF DYING (see below), a depressing study of the manner in which increasing numbers of old people actually die, as opposed to the way they fondly imagined they would, adds weight to their right to control these ultimate decisions.

2 BRITISH SOCIAL ATTITUDES SURVEY: THE 23RD REPORT (see below) found, in the words of the National Centre for Social Research, that 'the current law tha prohibits assisted dying is at odds with public opinion – four out of five peopl in Britain say that the law should allow a doctor to end someone's life at the person's request if they have an incurable and painful illness from which they will die'. This figure has held steady since the 1980s. Four fifths of practising Catholics and Protestants support it, as do three quarters of *Church Times* readers. Most *disabled people* (qv) also appear to agree, despite an avalanche of counter-propaganda by various *pressure groups* (qv) which claim to speak for them. When in late 2005 the Assisted Dying for the Terminally Ill bill was brought before Parliament by the humane Lord Joffe (whose record includes being Nelson Mandela's lawyer) the pressure group Scope (qv) asked readers its newspaper DISABILITY NOW (qv), which is written and read by disabled people, for their views. It transpired that 93% favoured the bill and 7% were against – see DISABILITY NOW, December 2005, p 17. Scope nevertheless continued to campaign vociferously against the Bill. For a comment see: *Ideology*; and *Pressure groups*.

3 A 'Death with Dignity Act' has existed in the state of Oregon in the USA since 1997 and allows terminally ill patients to request a lethal injection from their doctors. Compliance with such a request could put a British doctor in prison for 14 years, but in Oregon it has worked well with none of the dreadfu consequences ('legalised murder' etc.) feared by some opponents of self-deliverance. The people of Oregon have repeatedly endorsed the Act in referenda. For more information visit http://egov.oregon.gov/DHS/ph/pas/

4 In Britain the charity Dignity in Dying works to achieve patient choice at th end of life.

5 As of 2008, useful references include the following. They are given in alphabetical order. For publication details see the bibliography.

- BRITISH SOCIAL ATTITUDES SURVEY: THE 23RD REPORT – PERSPECTIVES ON A CHANGING SOCIETY. See the notes above.
- DISABILITY RIGHTS AND WRONGS by Tom Shakespeare. Mr Bert Massie, chairman of the Disability Rights Commission (qv) said of this magisterial work: 'If you read only one book on disability rights this year, make this the book'. Refer in particular to pp 118–119 and 122–32.
- FINAL EXIT by Derek Humphry, also available as a DVD, covers available methods of self-deliverance. After publication it remained in the *New York Times* best seller list for 18 weeks, and is now in its 3rd edition and still sellin briskly. There are evidently many people in the world who wish to avoid a bac death. The DVD, and pp 132–140 of the book, describe a simple method that gentle, quick, certain, and involves no unpleasantnesses of any kind.
- A SOCIAL HISTORY OF DYING by Professor Allan Kellehear, a sociologist at Bath University. See the notes above.

6 Useful websites include the following. They are given in alphabetical order.
- Brunel University www.brunel.ac.uk/news/pressoffice/cdata/euthanasia
- Dignity in Dying www.dignityindying.org.uk/
- Oregon Death with Dignity Act http://egov.oregon.gov/DHS/ph/pas/
- Self-deliverance www.finalexit.org/

See also: *Pressure groups*; and *Resuscitation*.

EN

See: *Special Educational Needs*.

---

ensation

1 The term *sensation* refers to the raw data detected by the *senses* (for instance, red is a colour sensation).
2 The five *senses* possessed by able-bodied people are sight, hearing, taste, smell, and touch, and they use all of these to a greater or lesser degree to apprehend the world and to navigate their way through it. The *built environment* (qv) should be designed so as to make such apprehension and navigation easy, even for people some of whose senses are *impaired* (qv).
3 The term *perception* refers to the interpretation of *sensory* information using the raw data detected by the senses plus previous experience.

---

ensory impairment

1 This term is generally used to refer to people with *impaired hearing* (qv) or *impaired vision* (qv), or a combination of both. In the latter case *deafblind* (qv) is the most readily understood term, and therefore the one to use.
2 For more information see the italicised entries.

---

ensory room

1 A room, especially in schools, designed to facilitate multi-sensory experiences through the use of light, touching/feeling and sound – or, in the *jargon*, 'through the use of light, tactile modifications, and auditory access'.
2 Useful websites include the following. They are given in alphabetical order.
- Disabled Living Online www.disabledliving.co.uk/
- Rompa www.rompa.com/
- Spacekraft www.spacekraft.co.uk/

---

erious Organised rime and Police Act 005

See: *Civil liberties*.

---

ervice providers: omplying with the isability iscrimination Act

1 The Disability Discrimination Act (qv) creates a *civil duty* (qv) for service providers not to *discriminate* (qv) against *disabled* (qv) customers or would-be customers. The term includes shops, pubs, banks, estate agents, and virtually every organisation that provides any kind of service to members of the public, regardless of size, regardless whether it is in the private or public sector, and regardless of whether that service is offered free or for payment.
2 The Act does not cover private clubs with fewer than 25 members. In this context the term 'private club' means clubs such as a golf, cricket or film clubs, or private clubs for particular groups of people such as military or political clubs, which provide their services only to members admitted after a genuine process of selection under club rules. Clubs which virtually anyone can join after payment of a membership fee (as in a health club or video rental shop) are not private and do come within the Act. And if a private club hosts an event that is open to the public, eg a wedding reception, it will on such occasions be subject to the Act. It is clear that very few organisations are truly exempt from the provisions of the Act.
3 Two groups of service providers are treated somewhat differently from the mainstream:

• Providers of educational services. Partly because education is by its very nature a service with special characteristics, the provision of education has developed a complex set of regulations of its own. For more information visit www.teachernet.gov.uk/wholeschool/sen/disabilityandthedda/
• Providers of transport services. While transport buildings (bus, coach and railway stations, ferry and air termini, etc.) are treated like all other buildings, vehicles (taxis, buses, trains, ships, and aircraft, etc.) present a special probler which must be dealt with in its own way. For more on these matters see: *Transport: complying with the Disability Discrimination Act*, and visit www.dft.gov.uk/transportforyou/access/dda2005/
4 As explained under Disability Discrimination Act (qv) service providers, unlike *employers* (qv), do not owe the duty of non-discrimination to particula identified disabled persons, but to all disabled people in society at large – a miscellany of unknown people adding up to perhaps 20% or even on latest definitions 25% of the total population. Service providers can avoid or remov discrimination against disabled people by making reasonable adjustments to their way of doing things, or to their buildings – for a more detailed explanati see: *Adjustments under the Disability Discrimination Act.*
5 The present work only gives simplified summaries of original documents and does not purport to provide full, authoritative and up-to-date statements of the law. For latest information refer to DISABILITY DISCRIMINATION: LAW AND PRACTICE (see the bibliography) and the following websites:
• Directgov: disabled people www.disability.gov.uk/
• Equality & Human Rights Commission www.equalityhumanrights.co
6 Other useful references include the following:
• CODE OF PRACTICE – RIGHTS OF ACCESS: SERVICES TO THE PUBLIC, PUBLIC AUTHORITY FUNCTIONS, PRIVATE CLUBS AND PREMISES. This *code of practice* (q is the most authoritative single reference on the above matters, and must be consulted.

See also: *Adjustments under the Disability Discrimination Act*; and *Disability Discrimination Act.*

Service providers: transport

See: *Transport providers.*

Service stations

1 *Disabled* (qv) drivers may have problems even with manned facilities becau there is often only one person on duty, who is generally too busy to spare them the time. Supermarket filling stations, which usually have two staff on duty, provide a welcome exception to the rule. The Happy Help scheme, run by the organisation Garage Watch, offers helpful advice – visit www.garage-watch.co.uk/
2 GOWRINGS MOBILITY UK ROAD ATLAS (see the bibliography) gives guidance to accessible service stations throughout the UK.

Service stations: complying with the Disability Discrimination Act

1 Buildings used by the public must in general comply with the Disability Discrimination Act (qv). In the case of service stations, *authoritative practical guides* (qv) include DESIGNING FOR ACCESSIBILITY (qv) plus the following:
• BS 8300, the provisions of which apply to all buildings in the UK. Para 13 gives brief design recommendations for *transport-related buildings* (qv) including motorway services.
• CODE OF PRACTICE ON FACILITIES FOR DISABLED MOTORISTS AT FILLING STATIONS is a simple leaflet for operators of filling stations.

• MAKING GOODS AND SERVICES ACCESSIBLE TO DISABLED CUSTOMERS: A GUIDE FOR SERVICE STATIONS was published in 2004 by UKPIA, the trade association representing the main oil refining and marketing companies in the UK, and the Centre for Accessible Environments (qv). The Guide outlines the practical measures for improving access lavatories. In many cases fairly minor changes to layout can make life much easier for disabled customers.
The above documents are advisory and do not have the force of law, but conformity with their recommendations will make it easier to demonstrate that the requirements of the Disability Discrimination Act have been met.
2 For publication details of the above references see the bibliography.
3 As of 2007, useful websites include the following. They are given in alphabetical order.

• DPTAC Access Directory	www.dptac.gov.uk/adnotes.htm
• DPTAC Publications	www.dptac.gov.uk/pubs.htm
• UKPIAm	www.ukpia.com/

See also: *Service providers*;

---

rvices: public

See: *Public services*.

---

tting down points

1 Setting down points are the places where passengers alight from vehicles before entering buildings or other premises.
2 For design recommendations see the next two entries.

---

tting down points: ellings

1 The following references contain authoritative recommendations:
• Section 6 of APPROVED DOCUMENT M (qv). The provisions in AD M must be heeded in order to comply with the Building Regulations (qv) in England and Wales.
• THE HABINTEG HOUSING ASSOCIATION DESIGN GUIDE (qv) applies most specifically to *housing association* (qv) schemes, but ought to be by the designer's side in the design of all housing developments and individual dwellings.
• THE WHEELCHAIR HOUSING DESIGN GUIDE (qv), also from the Habinteg Housing Association, is another illustrated manual whose usefulness extends far beyond the special remit implied by the title.
2 For publication details of the above references see the bibliography.

---

tting down points: blic buildings: mplying with the sability scrimination Act

1 Buildings used by the public must in general comply with the Disability Discrimination Act (qv). In the case of setting down points for public buildings, *authoritative practical guides* (qv) include DESIGNING FOR ACCESSIBILITY (qv) and in addition the following documents. For publication details see the bibliography.
• APPROVED DOCUMENT M (qv), which deals with access to and use of buildings. Its provisions must be heeded in order to comply with the Building Regulations in England and Wales.
• BS 8300 (qv), which applies to all buildings in the UK. It is an advisory document.
• INCLUSIVE MOBILITY (qv), which is the most authoritative reference for the *inclusive* (qv) design of *pedestrian infrastructure* (qv) and of *transport-related buildings* (qv) such as bus and coach stations, railway stations, air terminals, and transport interchanges. It is an advisory document.
While the advisory documents above do not have the force of law, conformity with their recommendations will make it easier to demonstrate that the requirements of the Disability Discrimination Act have been met.

2 The particular clauses in these documents which give guidance on setting down points are tabulated below.

CONTEXT	DOCUMENT: RELEVANT PARAGRAPHS		
	AD M	BS 8300	INCLUSIVE MOBILITY
BUILDINGS IN GENERAL	–	4.2	–
NON-DOMESTIC BUILDINGS	1.18	–	–
TRANSPORT BUILDINGS	–	–	5.1–5.51

See also: *Car parking*; and *Principal entrance*.

Sex discrimination

1 Discrimination on the basis of sex is outlawed in the UK by anti-discrimination legislation – see: *Discrimination*. More specifically:
2 The Equal Pay Act 1970 makes it unlawful for employers to discriminate between men and women in terms of their pay and conditions if the latter are doing the same or similar work; work rated as equivalent in a job evaluation study by the employer; or work of equal value.
3 The subsequent Sex Discrimination Act 1975 prohibits sex discrimination against individuals in the areas of (a) employment; (b) education; (c) the provision of goods, facilities and services; and (d) in the sale or management o premises. The Act covers women and men of any age, including children, and applies in England, Wales, and Scotland. There are however circumstances in which sex discrimination is not unlawful. They include:
• When a charity is providing a benefit to one sex only, in accordance with its charitable instrument.
• When people are competing in a sport in which the average woman is at a disadvantage to the average man because of physical strength, stamina, or physique.
• In insurance, where the discriminatory treatment reasonably relates to actuarial or other relevant data.
• Where the person's sex is a 'genuine occupational qualification' – for instanc some jobs in single-sex schools, or acting jobs that need a particular role to be played by a man or a woman.
4 Readers are reminded that the present work only gives simplified summaries of original documents and does not purport to provide full, authoritative, and up-to-date statements of the law. As of 2007, useful references for checking th latest situation include the following. For publication details see the bibliography.
• CODE OF PRACTICE – SEX DISCRIMINATION gives guidance to employers, trade unions, and employment agencies.
5 Useful websites include the following. They are given in alphabetical order.
• Directgov www.direct.gov.uk/
• Equality & Human Rights Commission www.equalityhumanrights.com/
• Equality Minister www.womenandequalityunit.gov.u
• Office of Public Sector Information www.opsi.gov.uk/acts.htm

See also: *Discrimination; Diversity; Equality; Gender equality duty; Incomes. gender pay gap*; and *Victimhood*.

Sex Equality Duty

See: *Gender Equality Duty*.

Sheltered housing See: *Housing: sheltered.*

Ships and ferries: complying with the Disability Discrimination Act

1 The on-shore terminals which serve ships and ferries are covered by the Disability Discrimination Act (qv) – for details see: *Transport related buildings*.
2 As of 2007, the vessels themselves are not covered by the Act. However, THE DESIGN OF LARGE PASSENGER SHIPS AND PASSENGER INFRASTRUCTURE: GUIDANCE ON MEETING THE NEEDS OF DISABLED PEOPLE, a document prepared and published by DPTAC (qv), which is the Government's advisory body on *accessible* (qv) transport, gives authoritative guidance on:

- Access to and within terminals.
- Shore to vessel transition.
- On-board accommodation.
- Lighting, steps, stairs and ramps on vessels.
- Information and announcements.
- Management and training.

For publication details of this reference see the bibliography.
3 As of 2007, useful websites include the following. They are given in alphabetical order

- Dept for Transport www.dft.gov.uk/transportforyou/access/dda2005/
- DPTAC Access Directory www.dptac.gov.uk/adnotes.htm
- DPTAC maritime www.dptac.gov.uk/ferries.htm
- DPTAC publications www.dptac.gov.uk/pubs.htm

See also: *Transport: complying with the Disability Discrimination Act.*

Shire unitary councils See: *Local authorities.*

Shops: complying with the Disability Discrimination Act See: *Administrative and commercial buildings.*

Shopmobility

1 As of 2007 there are around 250 Shopmobility schemes across the UK. They are highly valued by *disabled* (qv) and elderly frail people, for whom a shopping expedition can be transformed from a trial to a pleasure, and they can play a helpful role in town centre regeneration. Typically, users are met at their car or bus by a *powered wheelchair* (qv) or *scooter* (qv); they transfer onto the latter and can then drive around the shopping precinct freely and independently, loading their purchases into the basket mounted on the vehicle. They are untroubled by *kerbs* (qv) and inclines, and may in town centres (such as Camden Town in London or Kingston upon Thames) have the use of their vehicle for the whole day.
2 In December 2006 DPTAC (qv), the Government's advisory body on *accessible* (qv) transport, commented that current provision was very uneven across the country, and called on *local authorities* (qv), town centre leaders, retail operators, and developers to support such schemes.
3 For more information visit www.justmobility.co.uk/shop/

Showers See: *Bathrooms*; and *Changing rooms and showers.*

Sign and symbol systems

1 These are *communication* systems (qv) for those *deaf* (qv) and hard-of-hearing people who use signs for communication, and *symbols* (qv) to represent words or concepts. See the following entries:

- Blissymbolics.

- British Sign Language.
- Makaton.
- Mayer-Johnson.
- Signalong.
- Signed English.
- Widgit Rebus.

See also: *Alternative and augmentative communication; Communication aids and systems*; and *Speech aids*.

Sign design guide: a guide to inclusive signage

1 This is a core reference for architects, building managers, and others who get involved with the design or maintenance of *signs* (qv) in public buildings or spaces. It was written with the support of the RNIB (qv) and the Sign Design Society, and while it does not have the force of law, conformity with its recommendations will help to demonstrate that the requirements of the Disability Discrimination Act (qv) have been met.
2 For publication details see the bibliography.

See also: *Authoritative practical guides*; and *Signs*.

Sign languages

See: *Sign and symbol systems*.

Signalong

1 Signalong is a *symbol system* (qv) that is designed to help children and adults with *impaired communication* (qv) or *learning disabilities* (qv). It is based on British Sign Language (qv).
2 Visit www.signalong.org.uk/wa/

See also: *Sign and symbol systems*.

Signed English (SE)

Visit www.nas.org.uk/

Signs

1 Signs have functions, and the first step in planning a sign system in a building, a street, a heritage site, or a countryside walk is to thoroughly think through the job that needs to be done. The diagram below suggests a taxonomy of sign types and functions.

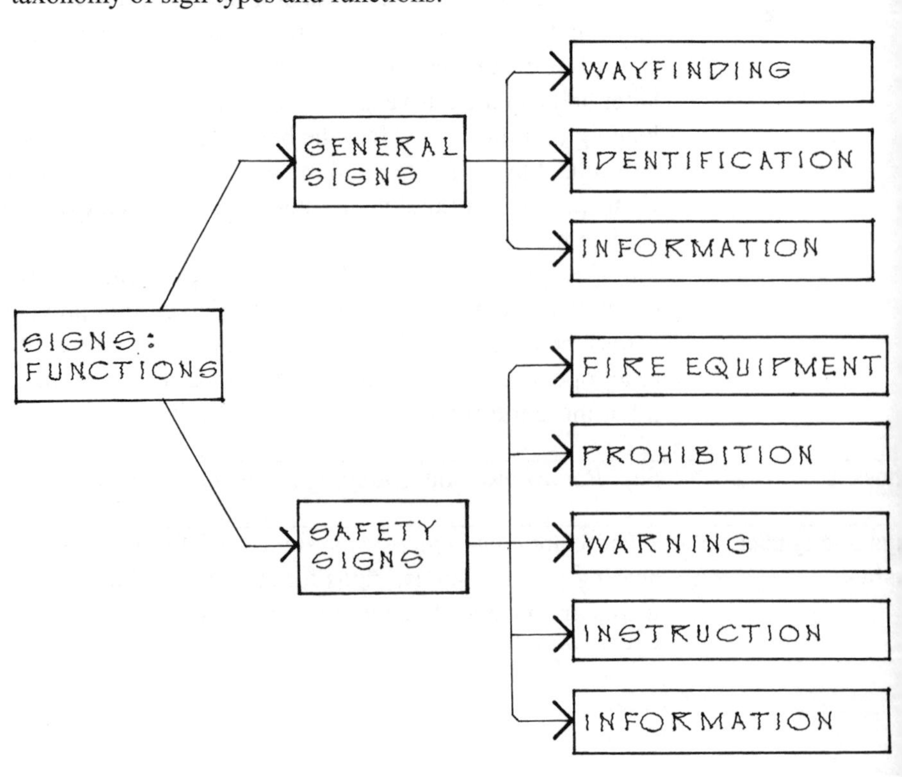

2 'General' signs help people to find their way about in daily life. There are four types:

• Wayfinding signs. These include 'orientation' signs which tell people where they are (i.e. a map with a spot 'you are here'), and 'direction' signs, which show them which way to go to get to their destinations. Wayfinding signs may contain many texts such as 'Turn left for casualties' or 'This way for lifts', plus numerous arrows, and the layout needs careful thought to achieve the greatest possible clarity.

• Identification (also called location) signs. These identify places – eg 'Inquiries', 'Eye department', or 'Dr John Smith'. They tell people when they have arrived at their destination.

• Information signs. These give information – eg 'Open 9 am to 6 pm', or the explanations alongside museum exhibits. If detailed reading is required (eg the texts at museum exhibits) it may be necessary to have two sets of signs – one at the standard height, and another mounted between 1.0 m and 1.1 m above floor level for children and wheelchair-users.

For more detailed notes on general signs see the next entry.

3 'Safety' signs help people to avoid danger and to escape safely in an emergency. There are five types as shown on the previous diagram. They are required under the Health and Safety (Safety Signs & Signals) Regulations 1996 – visit www.legislation.hmso.gov.uk/si/si1996/Uksi_19960341_en_2.htm Symbols and pictograms are standardised, and must use the colours shown in the table below.

TYPE OF MESSAGE	EXAMPLES	COLOUR
FIRE EQUIPMENT	1 FIRE ALARMS 2 FIRE EXTINGUISHERS 3 FIREMAN'S SWITCHES	WHITE TEXT ON RED PANEL
PROHIBITION	1 NO SMOKING 2 NO ENTRY	WHITE TEXT ON RED PANEL
WARNING	1 DANGER: HIGH VOLTAGE 2 DANGER: CONFINED SPACE	BLACK TEXT ON YELLOW PANEL
INSTRUCTION	1 KEEP CLEAR 2 ASSEMBLY POINT 3 LEAVE BUILDING BY NEAREST EXIT	WHITE TEXT ON BLUE PANEL
INFORMATION	1 FIRE EXIT 2 REFUGE POINT 3 EMERGENCY EVACUATION LIFT	WHITE TEXT ON GREEN PANEL

4 As of 2007, useful websites include the following. They are given in alphabetical order.

• British Sign and Graphics Association www.bsga.co.uk/
• Health & Safety Executive www.hse.gov.uk/pubns/indg184.htm
• Safetyshop www.safetyshop.com/
• Stocksigns www.stocksigns.co.uk/

gns: general: omplying with the isability iscrimination Act

1 Buildings and spaces used by the public must in general comply with the Disability Discrimination Act (qv). In the case of general signs, *authoritative practical guides* (qv) include DESIGNING FOR ACCESSIBILITY (qv) and in addition the following documents. For publication details see the bibliography.

• APPROVED DOCUMENT M (qv), the provisions of which must be heeded in order to comply with the Building Regulations in England and Wales.
• BS 8300 (qv), which applies to all buildings in the UK. It is an advisory document.
• INCLUSIVE MOBILITY (qv), which is the most authoritative reference for the *inclusive* (qv) design of *pedestrian infrastructure* (qv) and of *transport-related buildings* (qv) such as bus and coach stations, railway stations, air terminals, and transport interchanges. It is an advisory document.
• SIGN DESIGN GUIDE: A GUIDE TO INCLUSIVE SIGNAGE (qv). This book is published by the RNIB. It is an advisory document.
While the advisory documents above do not have the force of law, conformity with their recommendations will make it easier to demonstrate that the requirements of the Disability Discrimination Act have been met.
2 The relevant clauses in the above documents above are set out below. All sections of the SIGN DESIGN GUIDE are relevant.

DESIGN ASPECT	DOCUMENT: RELEVANT PARAGRAPHS		
	AD M	BS 8300	INCLUSIVE MOBILITY
GENERAL PRINCIPLES	2.5,2.7,3.5,3.18	9.2 TO 9.2.1	10
LOCATION	–	9.2.2	–
MOUNTING HEIGHT	–	9.2.2	10.1.5
WORDING	–	9.2.3.1	–
LAYOUT	–	9.2.3.1	–
TYPEFACES	–	9.2.3.1	10.1.3
TYPE SIZES	–	9.2.3.1	10.1
COLOUR	–	9.2.3.2	10.1.4

3 The mounting heights shown on the figure below will in general satisfy the provisions in the above references, though each case should be checked in detai before decisions are finalised.

Note the possible duplication of *information signs* at two heights if detailed reading is required (for instance the texts at museum exhibits) – one at the standard height for standing adults; the other at a lower level for children and *wheelchair-users* (qv).

See also: *Entrance halls and reception areas*.

Signs: standard symbols

1 For 'general' signs BS 5810 recommends the standard symbols shown on the following pages.

1 Wheelchair. This symbol can be combined with other symbols for wc, lift, parking space, etc.

2 Designated parking space for disabled driver, or for car with disabled passenger.

3 Level access route, avoiding ramps or steps.

4 Wheelchair access via ramp.

5 Number of steps. Shows the location of stairs, and may give the number of steps in numerals.

6 Lift available, of a suitable size for wheelchair-users.

7 Help available to visitors/users on the premises.

8 Unisex toilet (qv) complying with section 5 of APPROVED DOCUMENT M of the Building regulations.

9 Male toilet for use by wheelchair-users, complying with section 5 of APPROVED DOCUMENT M of the Building Regulations.

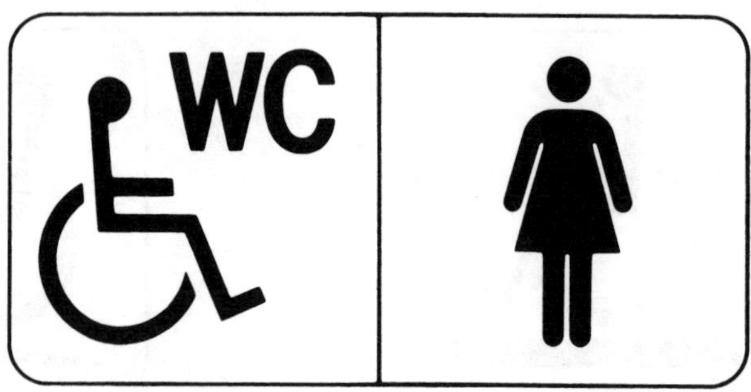

10 Female toilet for use by wheelchair-users, complying with section 5 of APPROVED DOCUMENT M of the Building Regulations.

11 Eating/drinking facility with access for wheelchair-users.

12 Facilities for blind and partially sighted people.

13 Equipment is available to help deaf and hard of hearing people.

14 Facilities are available to help deaf and hard of hearing people.

15 Public telephone usable by people sitting in wheelchairs.

16 Guide dogs are admitted.

Signs: supplemental types

1 The previous pages have dealt with the painted signboards that make up over 90% of most signage systems. There are also three supplemental types:

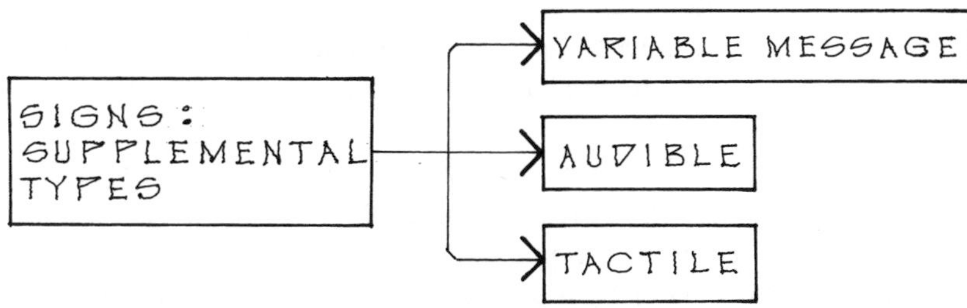

2 Variable message signs, composed for instance of electronic characters formed by luminous dots, are common in railway stations, in underground stations, and at bus stops. They should never have to be viewed against a bright sky behind, and must never be positioned in such a way that direct sunlight can fall upon the characters. In both cases the electronic characters become effectively invisible.

3 Audible information should augment important visual signs in public places, to help people with *impaired vision* (qv). Options include:

• Conventional *public address* (qv) systems. These can be used to convey important information to visually impaired people along with the rest of the public.

• *Induction loop* (qv) systems. These can be used for making announcements in shopping centres, stations, and airports. People wearing suitable reception devices – which could be a *hearing aid* (qv) set to the 'T' position – can then hear the announcement clearly, without background noise.

• *Talking signs* (qv).

4 Tactile signs (qv) should augment important visual signs in public places, where this is sensible and feasible, to help people who have *impaired vision*. Options include raised conventional letters and symbols, or Braille (qv).

5 The relevant clauses in APPROVED DOCUMENT M of the Building Regulations, BS 8300, and INCLUSIVE MOBILITY are tabulated below.

SIGN TYPE	DOCUMENT: RELEVANT PARAGRAPHS		
	AD M	BS 8300	INCLUSIVE MOBILITY
VARIABLE MESSAGE SIGNS	–	–	10.1.6
AUDIBLE INFORMATION	4.35	9.2.5	10.1.8
TACTILE INFORMATION	4.35	9.2.4	10.1.7

Simplification plans See: *Better regulation bodies.*

Single Community Programme Visit www.neighbourhood.gov.uk/page.asp?id=517

Single Equality Act 1 The Government announced in 2007 that it will bring before Parliament, in November 2008, a Single Equality Act to replace the existing patchwork of 116 separate pieces of anti-discrimination legislation covering race, nationality, sex, disability, faith, sexual orientation and age. In addition to bringing together in one place all UK law on *discrimination* (qv) and *equality* (qv), rewritten and hopefully simplified, the Act is expected to move from a negative, anti-discrimination focus saying 'You must not do this, that and the other', to a more positive pro-fairness law saying : 'You can do this'; 'You should do that'; and sometimes 'You must do such and such'. The Equalities and Human Rights Commission (qv) will be in charge of applying the new law.

2 For background understanding, here are some of the key decisions and events that led up this supremely important law:

• In the Queen's speech of November 2004 the Government announced its intention of legislating to establish a new Commission for Equality and Human Rights (qv).

• On 25 February 2005 the Equality Minister Jacqui Smith, together with the Cabinet Office (qv) Minister David Miliband, announced a root and branch review to investigate the causes of persistent discrimination and inequality in British society. It was called the 'Equalities Review' and was led by Mr Trevor Phillips, now chairman of the Commission for Equality and Human Rights – visit www.theequalitiesreview.org.uk/background.aspx

• Also in February 2005, the Government announced a 'Discrimination Law Review' which would work alongside the 'Equalities Review'. Its task would be to review discrimination and legislation in Great Britain, consider the recommendations of the Equalities Review (with which it would maintain close contact) and then make proposals for a clear and streamlined Single Equality Act. The Discrimination Law Review was led by the Women and Equality Unit (qv) – visit www.womenandequalityunit.gov.uk/dlr/index.htm

• In February 2007 the 'Equalities Review' produced its report, titled FAIRNESS AND FREEDOM: THE FINAL REPORT OF THE EQUALITIES REVIEW, which went out for public consultation from 12 June 2007 to 4 September 2007 – visit www.theequalitiesreview.org.uk/publications.aspx. Chapters 1 to 4 review the meaning, incidence, and persistence of inequalities in British society, and Chapter 5, read in conjunction with 'Annex A: Defining and Measuring Equality – a New Framework' proposes a methodology for tackling these problems. The latter is stupefyingly bureaucratic. It includes as part of the proposed toolkit a 'measurement framework' comprising 73 'central and valuable capabilities for adults', with a separate framework for children yet to be published. One must ask whether the authors were aware of the Government's pledge to cut the burden of red tape on government departments, businesses, and charities by one quarter in real terms by 2010 (see: *Better regulation bodies*); whether the Better Regulation Commission (qv) has reviewed these proposals for *proportionality* (qv) and transparency; and what were the BRC's conclusions.

• In June 2007 the 'Discrimination Law Review' produced its final report, titled A FRAMEWORK FOR FAIRNESS: PROPOSALS FOR A SINGLE EQUALITIES BILL FOR GREAT BRITAIN – A CONSULTATION PAPER, which was put out for a consultation period of 12 June 2007 to 4 September 2007 – visit www.communities.gov.uk/index.asp?id=1511211

• In July 2007 the newly-installed Prime Minister Gordon Brown announced the creation of a new Government Equalities Office (qv) which would have responsibility for reviewing the whole of discrimination law, and taking forward work on this Act.

3 It is impossible to know in 2007 how all of the above will work out, but the prospects for a truly simple law, easy to understand and apply (see: *Better regulation bodies*) are unpromising. There is a very real risk that what is being created, despite all governmental promises, is a new *consultant's charter* (qv) that will enrich lawyers and specialist consultants, create work for an army of administrators, and leave the rest of us baffled.

4 As of September 2007, useful websites include the following. They are given in alphabetical order.

• Directgov	www.direct.gov.uk/
• Equality & Human Rights Commission	www.equalityhumanrights.com/
• Equality Minister	www.womenandequalityunit.gov.uk/
• Office of Public Sector Information	www.opsi.gov.uk/acts.htm

See also: *Better regulation bodies; Diversity; Duty to promote; Equality; Government Equalities Office; Over-regulation*; and *Single Equality Act.*

ingle households See: *Housing: single household.*

ingle issue pressure roups See: *Pressure groups.*

ingle parent families See: *Children: single parent.*

it-stand seats See: *Seats: sit-stand.*

itting chairlifts See: *Chairlifts.*

ingle-tier local overnment See: *Local authorities.*

kills councils See: *Sector Skills Councils.*

kills for Life

1 Skills for Life is the national strategy for improving adult *literacy* (qv) and numeracy in England. It was founded in 2003 with the target of helping 1.5 million adults improve their skills by 2007.

2 SKILLS FOR LIFE: IMPROVING ADULT LITERACY & NUMERACY SKILLS FOR LIFE, a report produced by the House of Commons Public Accounts Committee in January 2006, concluded that the programme had been largely unsuccessful – visit www.parliament.uk/ What this tells us is probably not so much that the programme is at fault, but rather that the consequences of poor schooling are extremely difficult to remedy in later life. See in this connection *Education, Literacy*, and *Social mobility.*

3 As of 2007, useful references include the following. For publication details see the bibliography.

• SKILLS FOR LIFE: THE NATIONAL STRATEGY FOR IMPROVING ADULT LITERACY AND NUMERACY SKILLS. This is the launch document of the programme, published in 2001 with a foreword by the then Secretary of State for Education Mr David Blunkett.

• SKILLS FOR LIFE: IMPROVING ADULT LITERACY & NUMERACY SKILLS FOR LIFE. This report by the House of Commons Public Accounts Committee assesses the effectiveness of the programme up to January 2006.

4 As of 2007, useful websites include the following. They are given in alphabetical order.

- Dept for Children, Schools and Families www.dfes.gov.uk/
- Get On campaign www.dfes.gov.uk/readwriteplus/
- National Literacy Trust www.literacytrust.org.uk/

See also: *Education, Learning and Skills Council; Literacy*, and *Office for Standards in Education*.

Sleeping policemen

See: *Traffic calming*.

Slings

1 A sling is the fabric support in which the user of an overhead *hoist* (qv) sits o reclines.

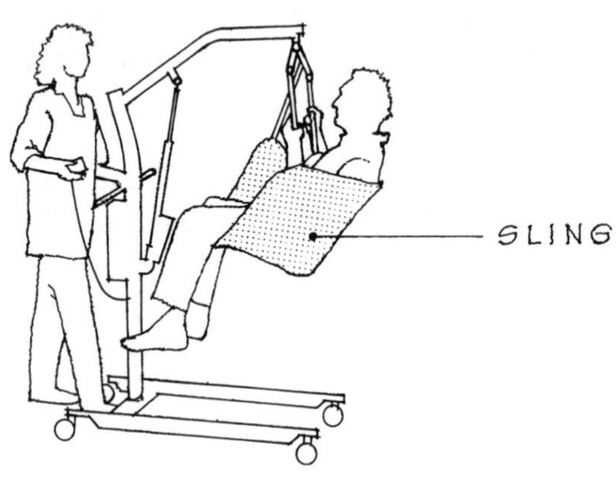

2 Slings must be closely tailored to the needs and posture of the individual use An *occupational therapist* (qv) will be able to advise on the right type for each particular patient.
3 See also: *Hoists*.

Slip-resistance

1 Many accidents are caused by people slipping on wet floors or pavings. It is therefore important for walking surfaces to be *slip-resistant* and give a firm foothold and good wheelgrip, not only when they are dry but also – and especially – when wet.
2 *Authoritative practical guides* (qv) which should be consulted by designers include:

- APPROVED DOCUMENT M (qv), the provisions of which must be heeded in orde to comply with the Building Regulations in England and Wales.
- BS 8300 (qv), which applies to all buildings in the UK. It is an advisory document.
- INCLUSIVE MOBILITY (qv), which is the most authoritative reference for the *inclusive* (qv) design of *pedestrian infrastructure* (qv) and of *transport-related buildings* (qv) such as bus and coach stations, railway stations, air terminals, and transport interchanges. It is an advisory document.

The relevant clauses in the above documents are:

DOCUMENT	RELEVANT PARAGRAPHS
APPROVED DOCUMENT M	1.26, 3.6, 3.14, 5.18, 5.21.
BS 8300	5.8.6, 9.1.3, ANNEX C
INCLUSIVE MOBILITY	3.10.

For places other than England and Wales the following regulatory documents should be consulted:

• In Northern Ireland: TECHNICAL BOOKLETS E, H, and R of the Building Regulations (Northern Ireland) (qv).
• In Scotland: the BUILDING (SCOTLAND) REGULATIONS and the associated TECHNICAL HANDBOOKS (qv).
3 Annex C of BS 8300 lists sixteen commonly used flooring and paving materials and rates them in terms of slip potential. The original document should be consulted, but here is a simple summary. All the evaluations refer to the performance of the surfaces when wet.
• Clay tiles with a carborundum surface finish have the best performance rating of all.
• Next best is a group of materials including carpets, cork tiles, chequer plate GRP profiles, and mastic asphalt.
• Concrete, cast iron, clay pavers, clay tiles, granolithic, and linoleum perform only moderately well.
• Ceramic tiles and float glass have the worst performance rating.
4 In the USA, where the Americans with Disabilities Act applies, refer to the ADA AND ABA ACCESSIBILITY GUIDELINES (qv).

---

nall buses See: *Bus and coach design.*

---

nall personal hicles See: *Mobility aids.*

---

nart Homes 1 Already computers turn our heating systems on and off; control our washing machines and microwave ovens; and provide new ways of monitoring the safety and security of our homes. The smart home of the future is one in which the various *communication systems* (eg telephone, radio, TV, and fire and burglar alarms) and *environmental control* systems (eg central heating, lights, and door and window closers), which normally operate separately, work together as a single integrated system to give the occupants hitherto unknown levels of comfort and security. At present such systems are very expensive, but costs will fall over time.
2 As of 2008, useful references include the following. For publication details see the bibliography.
• SAFETY, SECURITY AND ENVIRONMENTAL CONTROLS: A SPECIFIER'S GUIDE TO ELECTRONIC ASSISTIVE TECHNOLOGY IN THE HOME from the Centre for Accessible Environments (qv) gives a simple introduction to these matters.
3 Useful websites include the following. They are given in alphabetical order.
• AutomatedHome www.automatedhome.co.uk/
• BRE www.bre.co.uk/pdf/events/Smart_controls_EIB_Sep_2004.pdf
• JRF www.jrf.org.uk/housingandcare/smarthomes/default.asp
• Scott Sutherland www.rgu.ac.uk/sss/research/page.cfm?pge=2547
• Siemens www.automation.siemens.co.uk/

See also: *Environmental controls.*

---

nart Justice See: *Crime: youth.*

---

noke alarms
noke detectors See: *Alarms.*

---

ocial attitudes survey See: *British Social Attitudes Survey.*

---

ocial care See: *Social protection; and Social services.*

Social care institute for excellence (Scie)

1 Scie – pronounced 'sky' – is charged with reviewing research and practice in social care. Its task is to produce best practice guidelines for staff and services, setting out which methods do not work as well as effective ones.
2 For more information visit www.scie.org.uk/

Social enterprises

1 These are businesses that trade primarily to achieve social aims, though they may in some cases make a profit. Social aims might include job creation, vocational training, and the provision of local services.
2 The websites below give details of many possible types of social enterprises. One inspiring example, just to show what can be done if a group of energetic citizens band together, is the residents' association called Coin Street Community Builders, which was formed to fight plans to develop a wall of office buildings along their stretch of the Thames riverside. In 1984 the group set up a not-for-profit business with the aim of creating, instead, a mixture of flats for poor people, shops, restaurants, and parks on the south bank between Waterloo and Blackfriars. After approaching 30 banks in vain for a loan to refurbish the disused Oxo Tower, they finally found one that was willing to accept the land as collateral. Today we are able, as a result of their efforts, to wander along the riverside between the London Eye and the Tate Modern, sit in the gardens, have a meal, and visit the shops and cafés at the Oxo Tower and Gabriel's Wharf. And the crucial point is that poor people can live there. This ground-breaking initiative was led by Mr Iain Tuckett.
3 As of 2008, useful websites include the following. They are given in alphabetical order.

- Business Link www.businesslink.gov.uk/
- Cabinet Office www.cabinetoffice.gov.uk/third_sector/social_enterprise
- Coin Street Community Builders www.coinstreet.org/
- Social Enterprise Coalition www.socialenterprise.org.uk/
- Social Enterprise London www.sel.org.uk/

See also: *Neighbourhood renewal*; *Regeneration* and *Voluntary and Community Sector*.

Social entrepreneurs

1 These are the equivalent of commercial entrepreneurs, but operating in the social, not-for-profit sector. They aim to seek new and inventive solutions to social problems.
2 For more information visit www.sse.org.uk/

Social exclusion

1 The Social Exclusion Unit (qv) defines social exclusion as 'a combination of linked problems such as unemployment, poor skills, low incomes, poor housing high crime environments, bad health, poverty and family breakdown'. The term can be applied to people or to areas.
2 There are many reasons why people drift into social exclusion. The major ones include:

- Family problems during childhood. See: *Families*.
- Having been in care as a child. Despite the tiny proportion of the total population that has been in care, a third of Britain's *prison population* (qv) have come through the care system; so have a third of Britain's *homeless* (qv); and so have around half of the women who have the misfortune to be prostitutes. See: *Children in care*.
- Having difficulty with reading and writing. See: *Literacy*.
- Drug addiction.

3 For government programmes aimed at improving matters see: *Social Exclusion Task Force*.

4 As of 2007, useful publications include the following. They are given in alphabetical order. For publication details see the bibliography.

• THE COST OF EXCLUSION: COUNTING THE COST OF YOUTH DISADVANTAGE IN THE UK by Sandra McNally and Shqiponja Telhaj. This study was carried out by the London School of Economics for the Prince's Trust – see: *Prince's Charities*.

• OPPORTUNITY FOR ALL: EIGHTH ANNUAL REPORT 2006 – STRATEGY DOCUMENT 2006. This is the annual Government report about tackling poverty and social exclusion.

5 Useful websites include the following. They are given in alphabetical order.

• Directgov www.direct.gov.uk/
• Frank Field www.frankfield.co.uk/type3.asp?id=20&type=3
• Focus on Social Inequalities www.statistics.gov.uk/focuson/socialinequalities/
• *Guardian* newspaper http://society.guardian.co.uk/socialexclusion/0,11499,630068,00.html
• Opportunity for all www.dwp.gov.uk/ofa/
• Reform www.reform.co.uk/

See also: *Children in care; Literacy; Neets*; and *Social inclusion*.

---

ocial Exclusion Task orce

1 The Social Exclusion Task Force was set up in June 2006 to replace the now-defunct Social Exclusion Unit (see below). It is located in the Cabinet Office (qv) in order to ensure a cross-departmental approach.

2 Visit www.cabinetoffice.gov.uk/social_exclusion_task_force/index.asp

See also: *Child poverty; Children in care; Homelessness; Literacy; Neets; Neighbourhood renewal; Schools: exclusion of pupils*; and *Sure Start*.

---

ocial Exclusion Unit SEU)

1 In the mid-1990s Britain had more children growing up in unemployed households than anywhere else in Europe, and the highest teenage pregnancy rate. Child poverty had trebled between 1979 and 1995; notified drug addicts quadrupled in the decade to 1996, and in the early 1990s there were said to be about 2 000 people sleeping rough in London every night.

2 Shortly after coming to power in 1997 the new Labour Government launched the Social Exclusion Unit to analyse the reasons behind these trends and 'to find joined-up solutions to joined-up problems'. The Unit published reports on five key issues:

• Neighbourhood renewal.
• Rough sleepers.
• Teenage pregnancy.
• Young people not in education, training or employment.
• Truancy and school exclusion.

As a result, the Government set up a dedicated Whitehall unit for each of these problem areas, to improve joint working between departments.

3 In June 2006 the Social Exclusion Unit was replaced by the *Social Exclusion Task Force* (qv).

---

ocial housing

See: *Housing: affordable*; and *Housing: social*.

---

ocial housing grant

1 This is a capital grant provided by the Housing Corporation (qv) to fund *Registered Social Landlords* (qv) to develop *social housing* (qv).

2 For more information visit www.housingcorp.gov.uk/

See also: *Housing: affordable*; and *Housing: social*.

Social inequality

1 As of 2008, useful references include the following. For publication details see the bibliography.

• FOCUS ON SOCIAL INEQUALITIES 2005, published by the Office for National Statistics (qv), sets out the statistical data on six categories of inequality in the UK – education, training and skills; work; income; living standards; health; and participation.

• A MORE EQUAL SOCIETY? NEW LABOUR, INEQUALITY AND EXCLUSION, a study carried out by the Centre for Analysis of Social Exclusion at the London School of Economics, finds that while the new Labour Government of 1997 took *poverty* (qv) and *social exclusion* (qv) seriously, and managed to turn the tide in key areas, Britain remains a very unequal society.

2 As of 2008, useful websites include the following. They are given in alphabetical order.

• Centre for Analysis of Social Exclusion www.sticerd.lse.ac.uk/case/
• Focus on Social Inequalities www.statistics.gov.uk/focuson/socialinequalities/
• Frank Field www.frankfield.co.uk/type3.asp?id=20&type=3
• Joseph Rowntree Foundation www.jrf.org.uk/
• New Policy Institute www.npi.org.uk/
• Office for National Statistics www.statistics.gov.uk/
• Social Trends www.statistics.gov.uk/StatBase/Product.asp?vlnk=13675

See also: *Gini coefficient; Health inequality; Incomes: inequality*; and *Social exclusion*.

Social integration

See: *Population: racial integration*.

Social landlords

See: *Registered Social Landlords*.

Social model of disability

See: *Disability: other definitions*.

Social mobility

1 Social mobility occurs when people move across social class boundaries, or from one occupational level to another. More precisely:

• The term 'intergenerational' mobility refers to changes that occur between generations, i.e. parents to children.

• The term 'intragenerational' mobility refers to career shifts within a particular person's lifetime.

In both cases the mobility can be upward or downward.

2 Recent studies show (a) that the UK and the USA have the lowest rates of social mobility in the advanced world; and (b) that Britain is not only failing to improve, but that we have actually gone backwards in the past 50 years. The references at the end of this entry give more detailed information, but here are the key points.

3 INTERGENERATIONAL MOBILITY IN EUROPE AND NORTH AMERICA by Jo Blanden, Paul Gregg and Stephen Machin, published in 2005 by the Centre for Economic Performance, compared the life chances of children born in the 1950s, 1970s, and 1980s in eight European and North American countries. It found that:

• The UK and USA are substantially less mobile than Canada and the Nordic countries.

• In the USA the degree of social mobility has not changed much; but in the UK it was markedly worse for people born in 1970 than for those born in 1958, and has failed to improve for those born in the 1970s and 1980s.

4 There are two main reasons for Britain's low social mobility:
• The UK's growing *income inequality* (qv). Countries with only a moderate gap between top and bottom incomes, such as Canada and Sweden, have high mobility; and countries with a large gap between top and bottom incomes, such as the USA and the UK, have low mobility. The conventional measure of income inequality is the *Gini coefficient* (qv).
• The abysmal quality of education that is currently available to the children of lower-income families in the UK, thus wrecking their chances of bettering themselves.
5 The danger of the latter development was anticipated a decade ago by Mr Andrew (now Lord) Adonis, who eight years later would become Prime Minister Tony Blair's schools minister. In a prescient book titled A CLASS ACT: THE MYTH OF BRITAIN'S CLASSLESS SOCIETY, he and his co-author Stephen Pollard argued that:
• In the first two thirds of the 20th century, Britain saw a steady improvement in social mobility, largely through the agency of grammar schools which gave bright children from poorer backgrounds an upward ladder.
• The final third of the 20th century saw the gradual disappearance of grammar schools, coupled with a steady decline in the quality of education offered by comprehensive schools – for more on this see: *Literacy*; and *Qualifications and Curriculum Authority*.
The disappearance of grammar schools, and the dumbing down of many or most comprehensive schools, have combined to cut off the upward routes for children whose parents cannot afford private school fees or a house within the catchment area of a state school of excellence. In 1997 Adonis and Pollard were hopeful that the new Government with its emphasis on 'education, education, education' was going to reverse this trend, but they have been disappointed. Figures vary from year to year, but of the 100 top-performing schools in the UK around five are now comprehensives (which represent perhaps 34% of all schools), and 65 are independents (which represent perhaps 7% of all schools). The rest are grammars.
6 What can be done? As there are around 170 grammar schools and 24 000 others, expanding the tiny grammar school sector would benefit a few children and do nothing for the rest – especially the underachieving bottom 25%, who by definition would never get into a grammar school and who in later life, as near-unemployable adults, are at the root of some of Britain's most intractable social problems. The way forward must be a radical improvement of the state school sector as a whole. The available evidence, adduced in relevant entries throughout this book, suggests the following essential measures:
• Removing the Qualifications and Curriculum Authority (qv) from the Department for Children, Schools and Families (qv) and making it a completely independent enforcer of standards. Stabilising curriculum and exam standards would help to put an end to the current thoroughly scandalous situation that permits bad schools to conceal their failings behind ever-improving results from ever-easier exams. For evidence of the latter refer to STANDARDS OF PUBLIC EXAMINATIONS IN ENGLAND AND WALES.
• Radically shrinking Local Education Authorities (qv), and channelling most or all educational funding direct to schools for the latter to spend on teachers and pupils without first losing a large slice to administrators.
• Fostering the growth of the greatest possible variety of independent, self-governing, non-selective schools funded by the state but run by their head teachers. Such schools could be set up by educational groups, charities, businesses, and other organisations with a minimum of red tape, provided they meet certain minimum scholastic standards.

• Having good teachers, and enabling them to teach. That means giving heads in all mainstream schools complete authority (a) to fire bad teachers, and (b) to exclude disruptive pupils – see: *Schools: exclusion*.
• Above all, making the above reforms effective by enabling parents to place their children in a school of their choice, regardless of family income. This almost certainly means some form of voucher scheme – see: *Education: voucher systems*. Genuine parent power is not only a democratic entitlement, but has proven in other countries to be a powerful weapon for fostering good schools and eliminating bad ones.
7 As of 2007, useful references include the following. They are given in date order. For publication details see the bibliography.
• A CLASS ACT: THE MYTH OF BRITAIN'S CLASSLESS SOCIETY by Andrew Adonis and Stephen Pollard was published in 1998. See the text above.
• A BETTER WAY, a 140-page report published by the *think tank* (qv) Reform in 2003, proposes radical reforms in the key areas of education, healthcare, and policing.
• GENERATIONAL INCOME MOBILITY IN NORTH AMERICA AND EUROPE, edited by Miles Corak, was published in 2004.
• INTERGENERATIONAL MOBILITY IN EUROPE AND NORTH AMERICA by Jo Blanden, Paul Gregg, and Stephen Machin is a study that was supported by the Sutton Trust and published by the Centre for Economic Performance at the London School of Economics and Political Science, and published in April 2005. The UK comes bottom of the class.
• STANDARDS OF PUBLIC EXAMINATIONS IN ENGLAND AND WALES, published in 2005 by the *think tank* (qv) Reform, draws on the work of the Curriculum, Evaluation and Management Centre at Durham University to document the sustained fall in examination standards that is causing deep damage to Britain's education system, and the life chances of the pupils passing through that system.
• BLAIR'S EDUCATION: AN INTERNATIONAL PERSPECTIVE written by Professor Alan Smithers and published by the Sutton Trust in June 2007, argues that Mr Blair's educational policies have failed. English independent schools top the international league table; comprehensive schools are far below; and the gap between the two is now greater than in any other advanced country.
• BREAKTHROUGH BRITAIN: ENDING THE COSTS OF SOCIAL BREAKDOWN VOLUME 3: EDUCATIONAL FAILURE, edited by Ryan Robson, was published in 2007. It forms part of the well-researched programme of policy proposals made to the Conservative Party in 2007 by the Rt Hon Iain Duncan Smith's Social Justice Policy Group.
• LIFE CHANCES: ACCOUNTING FOR FALLING INTERGENERATIONAL MOBILITY by Jo Blanden, Paul Gregg, and Lindsey Macmillan was published in the March 2007 issue of the 'Economic Journal'.
• THE MILLENNIUM COHORT STUDY: SECOND SURVEY – A USER'S GUIDE TO INITIAL FINDINGS by Kirstine Hansen and Heather Joshi. This major study, published in 2007 by the Centre for Longitudinal Studies at the Institute of Education, University of London, found that children from disadvantaged families were a full year behind their middle-class contemporaries at the age of three, with little hope of catching up. It was unable to determine how much effect the *Sure Start* (qv) programme, which was launched in 1998 with the aim of remedying this situation, had had in achieving its purpose.
• NEIGHBOURHOOD EDUCATION. This is one of a set of Localist Papers produced by the *think tank* Direct Democracy (qv) and published by the Centre for Policy Studies (qv) in 2007. See also: *Localism*.
• THE NEW SCHOOL RULES by Francis Gilbert, published in 2007. The author, who has taught in comprehensive schools for the past 15 years, holds that the

dreadful conditions in many state schools should not be blamed on parents, pupils, or teachers, but upon 'shockingly poor, incompetent and meddling governance'.

• REDUCING INEQUALITIES: REALISING THE TALENTS OF ALL by Leon Feinstein and others, a report published in September 2007 by the National Children's Bureau, analysed data from a study that tracked the lives of 17 000 people born in 1970. It found that inequalities in British society are entrenched; social mobility is a myth; and despite billions of pounds of government funding to cut child poverty, the gap between the poorest and richest children is probably wider now than it was 30 years ago.

• TEN YEARS OF BOLD EDUCATION BOASTS NOW LOOK SADLY HOLLOW. IT WILL BE HARD POLITICALLY BUT LABOUR MUST ACCEPT ITS VAUNTED POLICIES ON SCHOOLS HAVEN'T WORKED by Jenni Russell, was published in the *Guardian* on 14 November 2007. The title tells the story.

8 Useful websites include the following. They are given in alphabetical order.

• Reform	www.reform.co.uk/website/education.aspx
• Specialist Schools Trust	www.specialistschoolstrust.org.uk/
• Sutton Trust	www.suttontrust.com/

See also: *Aimhigher; Education; Education: voucher systems; Neets*; and *Underclass*.

Social protection

1 The term 'social protection' refers to the assistance given to people who are in need owing to *age* (qv); *disability* (qv); *family* (qv); *poverty* (qv); illness; or other circumstances. Such help and support may come from *central government* (qv), *local authorities* (qv), and from private bodies in the *voluntary and community sector* (qv). It may be in the form of direct cash payments such as social security *benefits* (qv) or *pensions* (qv); payments in kind such as free bus passes or prescriptions, or through the provision of services, for instance via the NHS (qv).

2 The bar chart, from Table 5.31 in SOCIAL TRENDS 33, shows that UK government expenditure on social security benefits in 2001 exceeded the total amount spent on health, education, and defence added together.

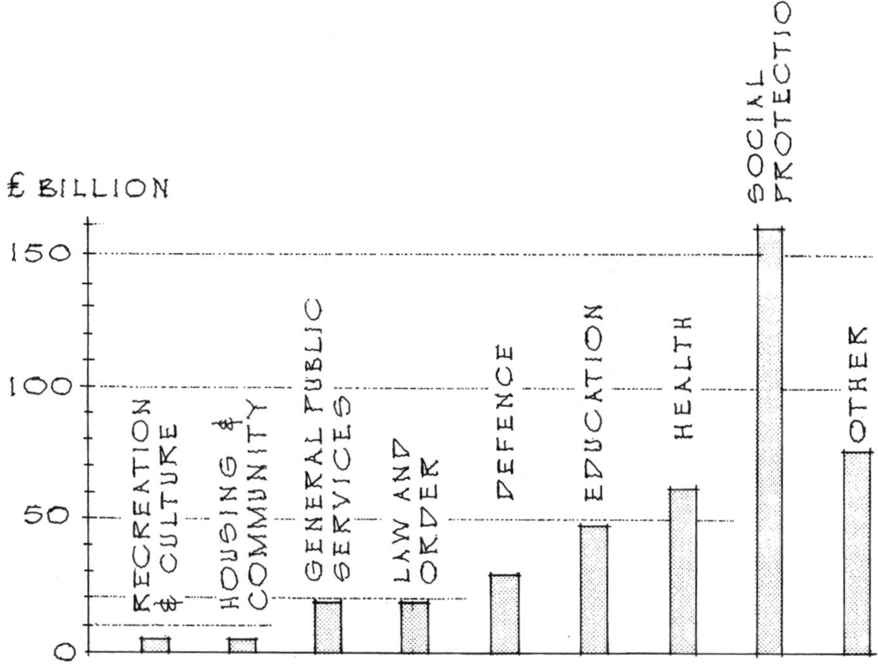

3 The graph below, from Figure 8.3 in SOCIAL TRENDS 37, shows how UK expenditure on social security benefits (including the state retirement pension, disability allowance, income support, and pension credit) climbed in real terms from £57 billion in 1977–8 to £130 billion in 2005–06.

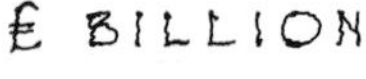

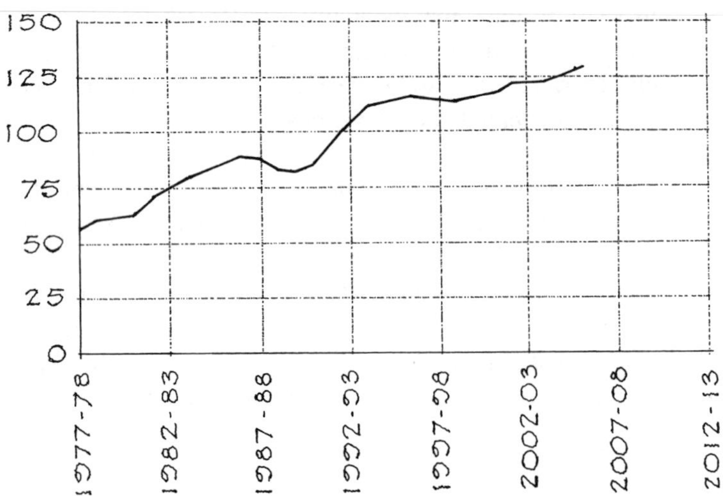

4 Finally, this pie chart from Figure 8.4 in SOCIAL TRENDS 37 shows that in 2003–4 local authority expenditure on social services and social care for older people was almost as much (44%) as the sums spent on all other recipient groups added together. This trend is almost certain to grow.

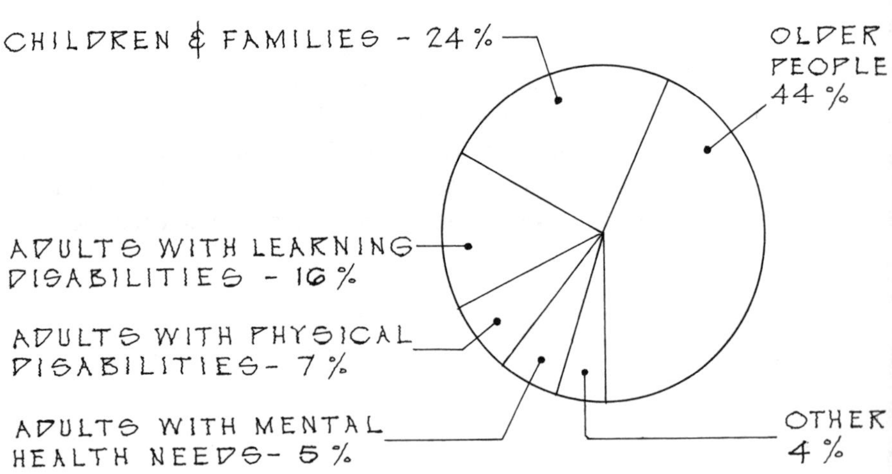

5 As of 2008, useful references include the following. They are given in alphabetical order. For publication details see the bibliography.

- LIVING WELL IN LATER LIFE, a study by the Healthcare Commission (qv), the Audit Commission (qv), and the Commission for Social Care Inspection (qv), argues that old people are being failed by the NHS (qv) and other services.
- SECURING GOOD SOCIAL CARE FOR OLDER PEOPLE: TAKING A LONG TERM VIEW, a study by Sir Derek Wanless for the *think tank* (qv) the King's Fund.
- SOCIAL TRENDS (qv), an annual publication of the Office for National Statistics (qv), gives latest statistical data plus background explanation.

6 Useful websites include the following. They are given in alphabetical order.

- Commission for Social care Inspection www.csci.org.uk/
- Department of Health www.nhs.uk/England/SocialCare/Default.cmsx
- Directgov www.direct.gov.uk/
- Frank Field www.frankfield.co.uk/type3.asp?id=20&type=3

- General Social Care Council www.gscc.org.uk/
- Social Care Association www.socialcaring.co.uk/
- Social Care Institute for Excellence www.scie.org.uk/
- Social Care Online www.scie-socialcareonline.org.uk/about.asp

See also: *Care homes; Care Standards Act; Commission for Social Care Inspection*; and *General Social Care Council.*

Social regeneration

1 The process of tackling the social problems that lead to social deprivation, such as crime and drugs. The process is different from, but interlinked with, *physical regeneration* (which tackles run-down buildings and communal areas) and *economic regeneration* (which is aimed at creating jobs and wealth). The area can range from a housing estate to a whole region.
2 See also: *Neighbourhood renewal.*

Social segregration

See: *Population: racial integration.*

Social Services Departments (SSDs)

1 Social Services Departments are the *local authority* (qv) departments in England and Wales that are responsible for the provision of personal social services. The services provided by them include social work, home care, and community care.
2 For more information contact the local authority concerned. Useful websites include www.adss.org.uk/

See also: *Social protection.*

Social Services Inspectorate (SSI)

1 The SSI's duties were transferred on 1 April 2004 to the Commission for Social Care Inspection (qv), which is now the single inspectorate for social care in England.
2 For historical SSI publications visit www.dh.gov.uk/

Social Trends

1 SOCIAL TRENDS is the source of preference of all social statistics in the present work. It is an annual publication from the Office for National Statistics (qv) and draws together key statistics from all government departments and many other organisations to paint a broad picture of British society today and how it has changed over recent decades. Each edition contains updates of the 13 standard chapters, plus a special extended study of one chosen topic. In the 2006 edition, for instance, this special essay was titled 'The different experiences of the United Kingdom's ethnic and religious populations'.
2 For publication details see the bibliography.

See also: *Office for National Statistics.*

Social work

Visit www.basw.co.uk/ and www.gscc.org.uk/

Space syntax

1 In the 1970s Professor Bill Hillier and colleagues at the Bartlett school of architecture at University College, London, set out to discover why people found some town centres and other urban spaces friendly and easy to navigate, leading to such places being popular and well-used, and why other places were felt to be unfriendly and difficult to use, leading to the dysfunctional and unloved spaces which are the curse of too many modern cities. They subsequently claimed to have developed a way of analysing the layout of spaces, particularly in terms of connectivity, that allows designers to predict whether they are likely to work well or badly in the above terms.

2 As of 2007, useful websites include:
- Space Syntax — www.spacesyntax.com/
- Space Syntax Laboratory — www.spacesyntax.org/
- Space syntax research — www.spacesyntax.net/

See also: *Streets*; and *Urban design*;

Spaces: public

1 As of 2007, useful references include the following. For publication details see the bibliography:
- IT'S OUR SPACE: A GUIDE FOR COMMUNITY GROUPS WORKING TO IMPROVE PUBLIC SPACE. This 114 page booklet from CABE (qv) outlines ways of improving public space, and gives two case examples.

2 Useful websites include the following. They are given in alphabetical order.
- CABE — www.cabe.org.uk/default.aspx?contentitemid=41
  www.cabe.org.uk/default.aspx?contentitemid=42
  www.cabe.org.uk/default.aspx?contentitemid=44
  www.cabe.org.uk/default.aspx?contentitemid=45
- Civic Trust — www.civictrust.org.uk/
- Mayor of London's Architecture & Urbanism Unit
  www.london.gov.uk/mayor/auu/publications.jsp
- New Urbanism — www.newurbanism.org/newurbanism/principles.htm

See also: *Parks*; and *Streets*.

Speaking aids

See: *Speech aids*.

Speaking household appliances

See: *Talking household appliances*.

Speaking lampposts

See: *Talking lampposts*.

Special educational needs (SEN)

1 A minority of children find learning very difficult. The reasons may include emotional difficulties (possibly connected with conditions at home); low intelligence; disorders such as *autism* (qv) or *dyslexia* (qv), or profound brain damage. These children are said to have Special Educational Needs (SEN). Until a few decades ago there was very little provision for such youngsters. They were routinely thought to be subnormal, and mostly dumped in special schools where many were regarded as ineducable and left in conditions of shocking neglect.

2 The 'Committee of Inquiry into the Education of Handicapped Children and Young People', chaired by Mary (now Baroness) Warnock and reporting its conclusions in 1978, addressed this deplorable state of affairs. To see the report visit www.catalogue.nationalarchives.gov.uk/RdLeaflet.asp?sLeafletID=89
Baroness Warnock attacked the dumping of vulnerable children in no-hope segregated institutions and called for more provision to be made for these pupils in *mainstream schools* (qv) where they could be part of normal society. In view of subsequent developments it is worth noting that she stressed two points:
- That mainstream provision was neither a panacea nor a cheap solution, and must be generously funded.
- That the needs of the child must be paramount in whatever decisions were made about mode of education.

3 The 1981 Education Act implemented many of the Warnock recommendations and introduced the Special Education Needs (SEN) framework (qv), including the 'statementing' system that was intended to

ensure that education authorities would accurately identify children with problems and deal effectively with their needs. For latest guidance on SEN visit www.teachernet.gov.uk/wholeschool/sen/

4 Though many good things were achieved by these reforms, three things have gone badly wrong:

• First, in thrall to the *social model* of disability (qv), and probably also hoping that this policy would save money, officials have been placing a growing proportion of SEN pupils in mainstream schools. As shown in the graph, which is from Figure 3.3 in SOCIAL TRENDS 34 (qv), the percentage of SEN pupils in England who are in mainstream schools rose from 56% (110 000 pupils of a total of 195 000) in 1994, to 60% (150 000 pupils of a total of 250 000) in 2003. These policies have brought us to the dire situation that there are around 9 000 fewer places in special schools in 2007 than there were in 1997, but 9 000 more children with special needs in pupil referral units.

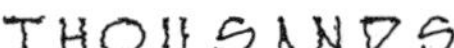

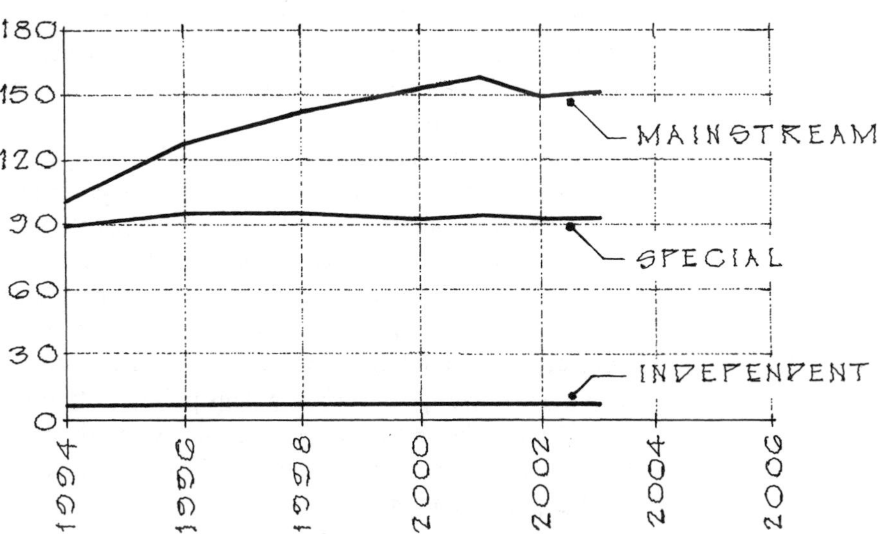

Many parents objected vehemently to the above process, but they were and still are being ignored by Ministers – the latter sometimes acting in a highly hypocritical manner. In 2005, the only full year in which Secretary of State for Education Ruth Kelly ran the DfES, she closed 2 600 places in special schools (more than any other education Secretary since the Labour government came to office in 1997), but two years later, in January 2007, she moved her *dyslexic* (qv) son out of the mainstream school system into a private £17 000 a year special school. Also in 2005 Baroness Warnock published a book titled SPECIAL EDUCATIONAL NEEDS: A NEW LOOK (see below) in which she admitted that things had not worked out as originally intended and called for an end to the closure of special schools, which she considered were better able to provide a good environment for emotionally vulnerable pupils than large mainstream classes.

• Second, the 'statementing' system has never worked well, and by 2006 had come to resemble an impenetrable maze, if not a brick wall, for thousands of desperate parents. Under this system parents can be sure of getting appropriate help for their children only if they can obtain a 'quantified, detailed and specific' statement of (a) the child's condition, and (b) what extra support he or she needs. Until that process is complete the child may be left without suitable teaching, and that is precisely what happens in numerous cases, forcing parents to fight exhausting battles with unresponsive bureaucracies. Thousands of already stressed parents have been spending between £2 000 and £10 000 (and in some cases £18 000 or more) of their own money, and months of precious time, arguing their child's case before appeals tribunals. For more information on these matters visit www.ipsea.org.uk/

• Third, not nearly enough money has been made available for SEN pupils. Putting these youngsters in mainstream schools without spending the very larg sums that are necessary to protect and nurture very difficult pupils in a mainstream setting, has proved disastrous for pupils and teachers alike. The pupils are often mercilessly bullied, and over-stressed teachers cannot cope wi their behavioural problems. It is significant that 87% of exclusions in primary schools and 60% of exclusions in secondary schools in 2006 involved childre with SEN. Instead of getting an *inclusive* (qv) education, as theorists and *ideologues* (qv) like to think, many are getting no education at all, with ruinou implications for their own future.

5 Evidence of teachers' views includes the following:

• A survey of heads and teachers published in the *Times Educational Supplement* on 14 October 2005 under the title 'Thousands better off in specia schools' showed that 80% of teachers wanted an end to further closure of special schools. The survey of 511 classroom teachers and 206 heads in England and Wales was carried out by FDS International on behalf of the TES visit www.tes.co.uk/

• In May 2006 the National Union of Teachers (NUT), which had supported a 'integrated' approach to schooling since 1996, published a report titled THE COSTS OF INCLUSION: A STUDY OF INCLUSION POLICY AND PRACTICE IN ENGLISH PRIMARY, SECONDARY AND SPECIAL SCHOOLS (see below). Quoting this report, the NUT stated that the Government's doctrinaire insistence on inclusion was harming children, which in the case of more severely afflicted youngsters amounted to a form of abuse, and announced that it was joining th campaign to save special schools. This policy shift left the Government's inclusion agenda with no supporters among the six teachers' unions in Englan all of whom had by 2006 condemned the damaging effect it has on the childre on other pupils, and on staff.

• In December 2006 the Department for Education and Skills' Implementation Review Unit (qv) stated that the bureaucratic workload imposed on teachers b DfES policies made it impossible for them to give enough attention to pupils – especially those with SEN.

6 It is clear that the present ideologically driven and undemocratic 'top-down' approach to decision-making must give way to a 'bottom-up' approach in whi the views of teachers take precedence over those of administrators, and parent choice emphatically takes precedence over both of these.

7 As of 2007, useful references include the following. They are given in alphabetical order. For publication details see the bibliography.

• THE COSTS OF INCLUSION: A STUDY OF INCLUSION POLICY AND PRACTICE IN ENGLISH PRIMARY, SECONDARY AND SPECIAL SCHOOLS by Professor John MacBeath (chairman of Educational Leadership at Cambridge University and Director of Leadership for Learning) and Emeritus Professor Maurice Galton (University of Leicester).

• SPECIAL EDUCATIONAL NEEDS AND THE LAW by Simon Oliver. This work brings together the law, principles, and policies of SEN as they stand in February 2007. It is an essential reference, but a brief glance at its 366 pages reveals the extent to which laws which were intended to benefit children have become, instead, a veritable gold-mine for fee-earning specialist lawyers and consultants while children, parents, and teachers look in from the outside. The time is overdue for the monstrous accumulation of SEN laws, regulations, code of practice, and commentary to be swept away and replaced by vastly simpler new legislation written from scratch to achieve the desired result – see: *Over-legislation*. Indeed, voices are beginning to be heard from within the world of SEN to the effect that the statementing system has had its day, and that a new approach is needed.

- DISABILITY DISCRIMINATION: LAW AND PRACTICE by Brian Doyle.
- SPECIAL EDUCATIONAL NEEDS: A NEW LOOK by Mary Warnock. The originator of the present system admits that much of it has gone wrong, and calls for changes.

8 Useful websites include the following. They are given in alphabetical order.

- Audit Commission www.audit-commission.gov.uk/reports/
- Dept for Children, Schools and Families www.dfes.gov.uk/
- Gloucester Special Schools Protection League www.gsspl.org.uk/
- NASUWT www.teachersunion.org.uk/
- National Autistic Society www.nas.org.uk/
- Teachernet www.teachernet.gov.uk/wholeschool/sen/

See also: *Disability Discrimination Act*.

---

ecial educational eeds: co-located hools

1 Co-located (also known as co-sited) schools provide a compromise between *mainstream schools* (qv) and *special schools* (qv) as settings for teaching children with special educational needs. A mainstream school and a special school are located on the same site, thus giving SEN pupils the special facilities and attention they need, while keeping them close to mainstream society.
2 For an instructive example visit www.hadleylearningcommunity.org.uk/

---

ecial needs using

See: *Housing: special needs*.

---

ecial provision

See: *Inclusive design*.

---

ecial schools

See: *Schools: special*.

---

ecialist schools ogramme

See: *Excellence in Cities*.

---

eech aids

1 People with severe speaking difficulties, caused for instance by *stroke*, can be helped by a variety of *augmentative communication aids* (qv) ranging from 'low-tech' to 'high-tech', the latter category also referred to as *Electronic Assistive Technology* or *EAT* (qv). This division suggests the following taxonomy of speech aids, as more fully discussed in the next two entries.

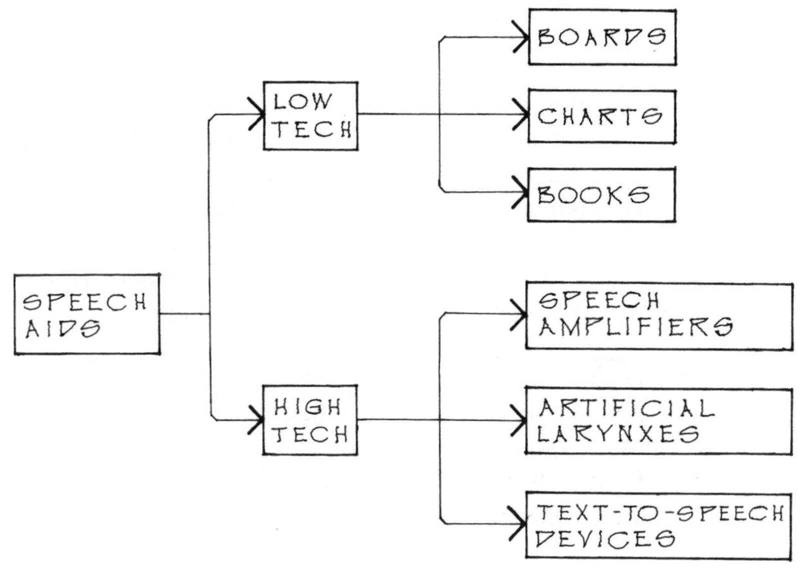

2 Aids of all these types can help people who are non-speaking, or have severely impaired communication, to take part in everyday life and interact wit other people. If professional advice be needed, the Royal College of Speech and Language Therapists is the professional body of and for speech and language therapists (SLTs) in the United Kingdom and Ireland – visit www.rcslt.org/

See also: *Aphasia; Alternative and augmentative communication; Communication aids*; and *Stroke*.

Speech aids: low-tech 1 Low-tech speech aids include alphabet, symbol and picture boards, charts an books. Such aids can usefully be divided into two categories:

• Text-based material, which is appropriate for children or adults who are comfortable with the written word. Typical examples are spelling lists, word lists, and word/phrase lists, which can be presented in the form of display boards, charts, or books.

• Symbol-based material, which is appropriate for people who have difficultie with reading and writing. This can again be presented in the three forms noted above. In this context *symbols* are line drawings that stand for objects, actions, and ideas – for instance, the following simple Makaton symbols:

PLEASE/ THANK YOU

GOOD MORNING/ HELLO

GOODBYE

Such symbols can be built up into large, comprehensive systems; and these ca be used in a *low-tech* way via pen and paper as described here, or in a *high-tec* way via sophisticated electronic devices as noted in the next entry.

2 Three main sources of symbols are currently used in the UK – Makaton (qv) Mayer-Johnson (qv), and Widgit Rebus (qv). Others include Blissymbolics (qv); Signalong (qv), which is based on British Sign Language (qv); and SymbolWorld (qv).

3 As of 2008, useful websites include the following. They are given in alphabetical order.

• Don Johnson www.donjohnston.co.uk/
• Easyinfo www.easyinfo.org.uk/
• Inclusive Technology www.inclusive.co.uk/
• Speechmark www.speechmark.net/

4 For information on specific symbolic systems visit:

• Blissymbolics www.blissymbolics.org/
• Makaton www.makaton.org/
• Mayer-Johnson www.mayer-johnson.com/
• Signalong www.signalong.org.uk/
• SymbolWorld www.symbolworld.org/
• Widgit Rebus www.widgit.com/

Speech aids: high-tech 1 High-tech aids can offer far more sophisticated help than boards and books, but do not replace the latter. Each type of aid has its own essential role and must be used where appropriate.

2 High-tech aids include (a) amplifiers that can improve the audibility of very softly-spoken people simply by making their speech louder; (b) devices that enable people with severely impaired speech to create intelligible sounds by

means of artificial larynxes; and (c) devices that enable wholly speechless people to speak by typing their words, which are then converted by sophisticated computer software into sound.

3 These types, whose place in the overall scheme of things is shown on the earlier taxonomic diagram, are dealt with in turn:

4 Speech amplifiers are portable devices that enable people with weak or whisper voices to speak more loudly. Apart from the beneficial effects of increased volume, these relatively simple devices cannot however improve the clarity of the speech. Examples include the EV3 speech amplifier, comprising a microphone near the mouth, linked to a flat box that can either be hung from the neck or clipped to the belt. Visit www.hear4you.com/

5 Artificial larynxes are hand-held electronic devices which generate a tone that is transmitted into the oral cavity, whereupon the tone is moulded into words by the user. There are two variants:

- Neck type artificial larynxes are pressed against the neck or cheek of the speaker, who presses a button to activate the sound source, and then converts the ensuing tone into speech by mouthing the words.
- Mouth type artificial larynxes are not pressed against the body, but have a thin tube which is placed in the mouth of the user.

Considerable practice is required in the use of both of the above types of speech aid.

6 Text-to-speech communication aids, also known as 'speech generating devices' (SGDs) or 'voice output communication aids' (VOCAs), allow a speechless person to type a word or sentence on a keyboard, whereupon the unit speaks that word or sentence out loud.

7 As shown below, text-to-speech communication aids come in two broad categories – those for *face-to-face communication* (i.e. two people having a conversation in the same room) and those for *distance communication* (i.e. two people having a conversation by telephone).

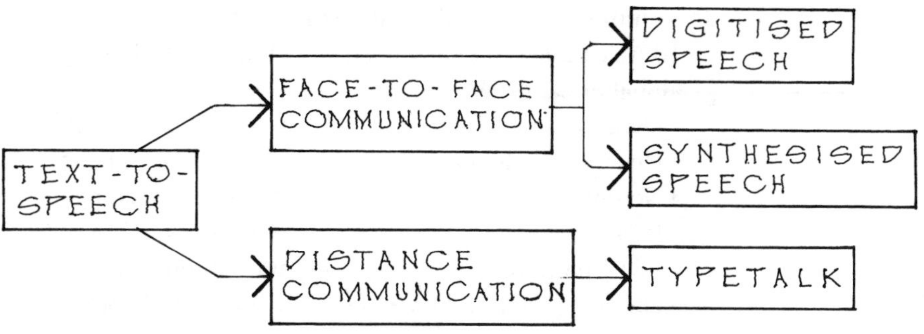

8 Face-to-face communication aids range from desktop models like – as of 2007 – Toby Churchill's 'Lightwriter' (below left) to small hand-held models like Possum's 'Say-it! SAM' (below right).

Models such as those above can – like low-tech aids – be *text-based*, for children or adults who are comfortable with the written word; or *symbol-based* for people who have difficulties with reading and writing.
• Text-based models have conventional computer keyboards. As of 2007, examples include 'Proteor DialO' and 'Cameleon' with EZ Keys – visit www.possum.co.uk/
• Symbol-based models have keyboards using symbols such as PCS, Rebus, or Bliss. Examples include 'MessageMates' and 'Cameleon' with Personal Communicator – visit www.possum.co.uk/
• Some models, for instance Possum's popular 'Say-it! SAM', offer both text and symbol options.
As shown by the figure on the previous page, face-to-face communication aids can produce their sounds by either of the following methods:
• Digitised speech. These are messages that have been spoken into a microphone and recorded, and are then played back when required by, for instance, touching a picture on the keyboard. They tend to sound more natural than the type below.
• Synthesised speech. This is artificial computer-generated speech using software such as DECtalk, InfoVox, or EuroTalk. Such speech is becoming much more natural with advancing technology. For more detail visit www.ace-centre.org.uk/
9 Distance communication aids are in the form of *textphones* (qv). Textphones have a keyboard and display screen, thus enabling speechless people to make telephone calls by typing rather than speaking their words. These devices are very like conventional telephones, but instead of a mouth and earpiece the textphone has a keyboard and display screen so that speechless telephone users can type their words. The typed message can then (a) be read by the person at the other end, or (b) the typed words can be spoken at the other end via the RNID's *typetalk* (qv) service, which uses a human interpreter as a go-between. Most textphones are desktop-sized, but some – eg the 'Nokia Communicator 9210i' – are hand-held.
10 As of 2008, useful websites include the following. They are given in alphabetical order.

• DynaVox	www.dynavox.co.uk/
• Hearing Products	www.hear4you.com/
• Novomed	www.novomed.net/
• Possum	www.possum.co.uk/
• PRI Liberator	www.prentromint.com/
• Quality Enabling Devices	www.qedltd.com/
• Sensory Software	www.sensorysoftware.com/
• Toby Churchill	www.toby-churchill.com/
• Vodafone	www.vodafone.co.uk/textphone

Speech codes

See: *Political correctness*.

Speech-enabled websites

See: *Accessible computers*; and visit www.browsealoud.com/

Speech: freedom of

1 Not only is freedom of speech the essential basis of *democracy* (qv), but a YouGov poll of 3 500 people published in the *Daily Telegraph* on 27 July 200[illegible] under the heading 'What does it mean to be British?' found that the single val[illegible] that the majority of respondents rated most highly was 'the British people's right to say what they think'.

2 UK citizens undoubtedly enjoy a greater degree of freedom of speech than ever before, and more than most of the rest of the world's population, but ominous pressures include the following.

3 First, laws which are passed on governmental assurances that they are aimed only at criminals, and are then used to prosecute ordinary citizens for expressing opinions or even making jokes. For instance:

• The Public Order Act 1986 (qv), which makes it an offence to stir up hatred against groups of persons. In May 2005 Mr Sam Brown, an Oxford student out celebrating the end of his finals, asked a mounted police officer: 'Do you realise your horse is gay?' The policeman radioed for assistance and within minutes two squad cars arrived, disgorging a group of officers who arrested Mr Brown under Section 5 of the Act for making homophobic remarks. A police spokesperson told the student newspaper *Cherwell* that the remarks had been 'not only offensive to the policeman and his horse, but any members of the general public in the area'. The accused spent a night in the cells and was fined £80, which he refused to pay. He was acquitted after a court hearing in January 2006.

• The Terrorism Act 2000 (qv), which according to the Home Office website was passed 'in response to the changing threat from international terrorism'. In September 2005 Mr Walter Wolfgang, an 82-year-old member of the Labour Party, was first frogmarched out of the annual party conference for heckling the Foreign Secretary Jack Straw on the Iraq war, and then prevented from re-entering by police citing the Terrorism Act. After initial denials a Sussex police spokesman admitted that Mr Wolfgang had been issued with a section 44 stop-and-search form under the Terrorism Act.

• The Serious Organised Crime and Police Act 2005 (qv), which was used in October 2005 to arrest and prosecute Miss Maya Anne Evans for standing by London's Cenotaph and reading out the names of 97 British soldiers killed in Iraq, and to prosecute the Mr Neil Goodwin for doing an impersonation of Charlie Chaplin outside Parliament. He was acquitted after a court hearing in January 2006.

For more on the above see: *Civil liberties*.

4 Second, the intolerance of all too many academics and *pressure groups* (qv), who are so convinced of the rightness of their views that they work assiduously to prevent contrary views being aired. A well-known example is the long vendetta carried on by some academic philosophers against the moderate and courteous, but stubbornly free-thinking Roger Scruton. Their attempts to prevent his arguments being heard include a letter from an Oxford philosopher to his publisher, Longman, suggesting that the latter stop publishing his books (see p 55 of his autobiographical GENTLE REGRETS), and Professor Scruton's arrival at the University of Glasgow one day to deliver an invited lecture, only to find that the Philosophy Department had organised a boycott of his talk while elsewhere on the campus – by a particularly neat irony – an honorary degree was being conferred upon Mr Robert Mugabe – see p 53. Other examples include persistent attempts by anti-smoking campaigners such as the *pressure group* (qv) ASH to silence sceptics on the ground that the debate on the dangers of environmental tobacco smoke 'is over', and similar campaigns by supporters of the officially favoured theory on global warming. They are wrong. In a democracy the debate is never over.

5 Third, social pressures (fostered by groups such as the above) that are making people increasingly nervous about expressing honest and reasonable views lest they be branded 'racist', 'sexist', or 'homophobic'. In May 2007 Professor Ted Cantle of the Institute of Community Cohesion (qv), author of the official report on the 2001 race riots in Oldham, told a BBC *Panorama* investigation that *political correctness* (qv) and laws such as the Race Relations Act qv) were

helping to stifle open debate on important matters, and driving people's fears underground.
6 It is both significant and disturbing that a group of 60 British academics felt necessary in December 2006 to found Academics for Academic Freedom (qv) and call for a change in the law to ensure that scholars are given complete freedom to question and test received wisdom and put forward unpopular view Their website argues that one of the most important things students can learn university is how to argue and debate, but laments that 'restrictive legislation, and the bureaucratic rules and regulations of government *quangos* (qv), and o universities themselves, have undermined academic freedom. Many academic are fearful of upsetting managers and politicians by expressing controversial opinions. Afraid to challenge mainstream thought, they pursue self-censorshi AFAF's call is for a change of law that would guarantee all academics the unfettered right to speak out on any issue, both inside and outside the classroom, whether or not it was part of their area of academic expertise, and whether or not these issues are deemed offensive. It is depressing that such a plea had to made more than 200 years after the death of Voltaire.
7 As of 2008, useful websites include the following. They are given in alphabetical order.

- Academics for Academic Freedom www.afaf.org.uk/
- Foundation for Information Policy Research www.fipr.org/
- Freedominfo.org www.freedominfo.org/countries/united_kingdom.ht
- Information Commissioner's Office www.ico.gov.uk/
- Liberty www.liberty-human-rights.org.u
- Office of Public Sector Information www.opsi.gov.uk/acts.htm
- Your right to know www.yrtk.org/

See also: *Human rights; OpenNet Initiative; Political correctness*; and *Pressu groups*.

Speech impairment See: *Impaired speech*.

Speech recognition 1 'Speech recognition' is the technical process that enables a computer to (a) recognise the words that are spoken by a person into a microphone; and (b) convert those spoken words to another form – eg a different language, printed text, British Sign Language (qv), etc.
2 'Voice recognition' is a more advanced variant that can actually recognise a specific voice or speaker.
3 One example of the many potential applications of this technology is 'Tessa' an acronym for Text and Sign Support Assistant. This is a computer programme that uses virtual animation, plus speech-recognition software, to a communication between *deaf* (qv) people and shop assistants. A virtual (i.e. computer-generated) young woman named Tessa appears on a TV screen at th shop counter. If deaf customers can give the shop assistant a rough idea of wh they want to inquire about, the assistant can help them by speaking into a microphone, whereupon Tessa will convert the assistant's spoken words into British Sign Language hand signals. By this means a two-way conversation ca be carried on.
4 Visit www.sys-consulting.co.uk/web/projects/project_view.jsp?code=TESSA

See also: *British Sign Language*; and *Speech aids*.

Speech synthesiser See: *Speech aids*.

Sport Action Zones Visit www.sportengland.org.uk/

port England See: *Sports councils.*

ports councils 1 As of 2008, the official sports councils for the UK are:

- Sport England www.sportengland.org/
- Sport Northern Ireland www.sportni.net/
- Sports Council for Wales www.sports-council-wales.co.uk/
- Sportscotland www.sportscotland.org.uk/
- UK Sport www.uksport.gov.uk/

2 The duties of these organisations include acting as distribution bodies for National Lottery (qv) funds.

See also: *English Federation of Disability Sport.*

ports activities for isabled people See: *Recreation and sports activities for disabled people.*

ports-related uildings: basic lanning data 1 Distilled planning data for most of the major building types is given in the METRIC HANDBOOK: PLANNING AND DESIGN DATA. For publication details see the bibliography.

2 For general design guidance, commentary, and case examples visit:

- CABE www.cabe.org.uk/default.aspx?contentitemid=42

ports-related uildings: design: omplying with the isability iscrimination Act 1 Buildings used by the public must in general comply with the Disability Discrimination Act (qv). In the case of stadia; outdoor and indoor sports centres; club-houses; swimming pools; and fitness suites and exercise rooms *authoritative practical guides* (qv) include DESIGNING FOR ACCESSIBILITY (qv) plus the following. For publication details see the bibliography.

2 APPROVED DOCUMENT M (qv), the provisions of which must be heeded in order to comply with the Building Regulations (qv) for England and Wales. The following paragraphs are directly relevant:

Spectators & seating:	0.14; 4.2; 4.5–4.8; 4.11–4.12; and 4.36.
Sports facilities:	2.13; 4.5; 4.11; and 5.6.
Changing rooms & showers:	5.16–5.18.
Doors:	2.13.
Stadia:	4.11.
Toilets:	5.6.

3 BS 8300, the provisions of which apply to all buildings in the UK. Para 13.7 gives brief design recommendations for all the building types listed above.

4 GUIDE TO SAFETY AT SPORTS GROUNDS (qv). POPULARLY KNOWN AS THE GREEN GUIDE, this work concentrates on designing and managing for safety, particularly (but not only) for disabled viewers.

5 ACCESSIBLE STADIA: A GOOD PRACTICE GUIDE TO THE DESIGN OF FACILITIES TO MEET THE NEEDS OF DISABLED SPECTATORS AND OTHER USERS (qv) is a more recent publication than the Green Guide, and augments the latter with the latest guidance on the requirements of *disabled people*, especially wheelchair-users.

Except for AD M the above documents do not have the force of law, but conformity with their recommendations will make it easier to demonstrate that the requirements of the Disability Discrimination Act have been met.

In places other than England and Wales the following regulatory documents should be consulted:

- In Northern Ireland: TECHNICAL BOOKLET R of the Building Regulations (Northern Ireland) (qv).
- In Scotland: the TECHNICAL HANDBOOKS of the Building (Scotland) Regulations (qv).

6 In the USA, where the Americans with Disabilities Act applies: refer to the ADA AND ABA ACCESSIBILITY GUIDELINES (qv).
7 As of 2008, other useful references include the following. For publication details see the bibliography.
• STADIA: A DESIGN AND DEVELOPMENT GUIDE. This is an excellent general handbook for stadia design.
8 Useful websites include the following. They are given in alphabetical order.
• CABE www.cabe.org.uk/default.aspx?contentitemid=42
• Sport England www.sportengland.org/publications-2003-04.pdf

See also: *Leisure activities for disabled people*.

Sports-related buildings: management: complying with the Disability Discrimination Act

See: *Service providers*.

StageText

1 StageText is a *captioning* (qv) service for *deaf* (qv) and hard-of-hearing visitors to *theatres* (qv) and other arts venues, enabling them to read what is being spoken or sung on the stage, as it happens. Sound effects and off-stage noises are also included.
2 There are two types – open and closed:
• With open captioning the text is displayed on a screen that can be located on above, below, or beside the stage. The caption unit can be placed in position before the performance and removed immediately afterwards. In addition to helping people with *impaired hearing* (qv) the captioning is also found useful by general listeners where the acoustics are poor, and by people in the audience whose first language is not English. In addition the system is relatively inexpensive and fosters a sense of togetherness, with everyone in the audience enjoying the performance together and no-one being labelled as 'disabled' by having to collect and use special equipment, or having to sit in specially-equipped seats.
• With closed captioning the captions are displayed to individual members of audience using one of a number of closed-caption devices. These include hand-held screens which the user holds throughout the performance; small screens installed on the back of the seat in front; and even special spectacles. These systems have the advantage that the captions are invisible to other audience members and do not interfere with the stage aesthetics, but the disadvantages are that deaf audience members have to collect a hand-held device or sit in a designated area of the auditorium, where they may not be able to sit with their hearing friends. In addition, hand-held and seat-back screens require the user to constantly adjust the focus of the eye from viewing screen to stage and back again, which is tiresome.
3 Under the Disability Discrimination Act (qv) StageText and similar services come under the heading of *auxiliary aids and services* (qv). The proper provision of such services will make it easier for theatre owners and managers to demonstrate that the requirements of the Act have been met – see: *Service providers*.
4 A GOOD PRACTICE GUIDE TO OPEN CAPTIONING, a publication funded by Arts Council England (qv), gives authoritative advice on the above matters – see the bibliography.

See also: *Cinemas*.

aircasing 1 A phrase used to describe a process whereby people vary the stake in the ownership of their home according to their circumstances. If a shared owner is struggling to pay his or her mortgage costs, they can 'staircase down' by cutting the share of the home that they own and pay for more of the home in rent. Conversely, if their income rises they could 'staircase up' by buying a bigger share in their home, or buy it outright.

2 See also: *Key worker housing*.

airclimbers 1 Stairclimbers are a *mobility aid* (qv), and comprise one of five categories of *small personal vehicles* (qv) for frail and *disabled people* (qv):

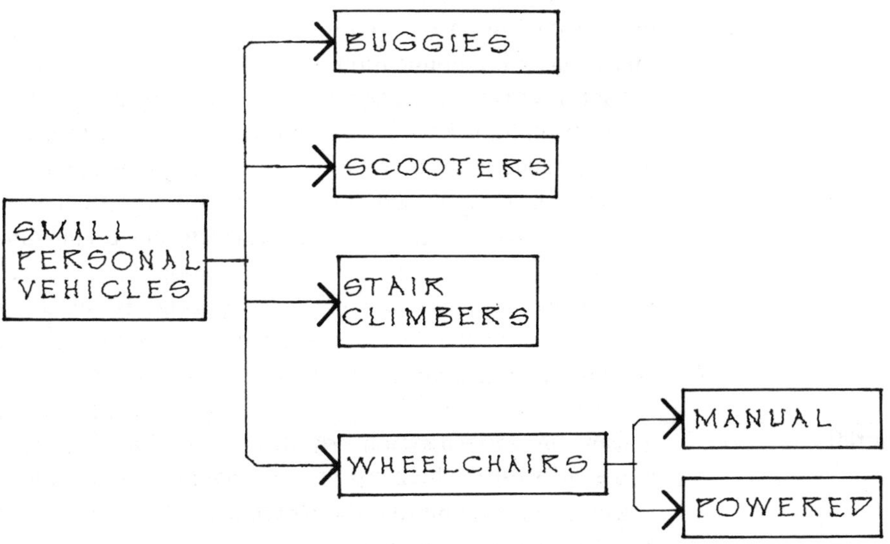

They can carry people with *impaired mobility* (qv) up or down a flight of stairs. Some, like the notional model shown below, move on rubberised caterpillar tracks, others have wheelclusters.

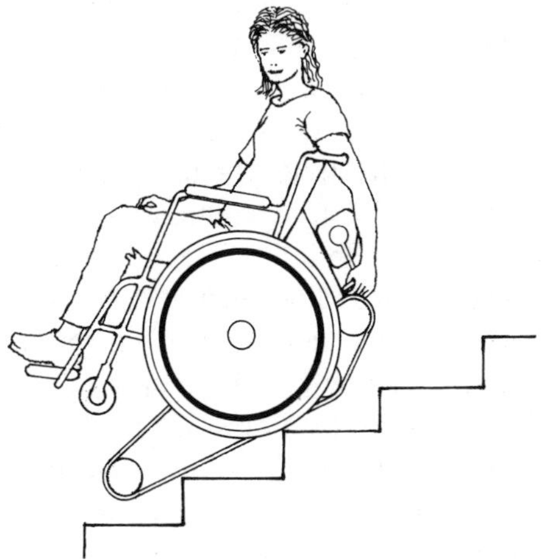

2 Stairclimbers are particularly useful in period buildings that are open to the public, and which cannot be fitted with any kind of permanent installation such as a *platform lift* (qv) either because this is technically unfeasible, or because it would harm the architectural character of the setting – see: *Historic buildings and sites: techniques*. As of 2006 a useful case example may be seen in the National Portrait Gallery at Trafalgar Square, London. As it was impossible to provide lift access to the Royal Landing situated between the ground and first floors, a stairclimber is kept on hand for *wheelchair-users* (qv).

3 Against the many advantages of these ingenious devices must be set the following drawbacks:
• Most stairclimbers require a trained operator, therefore the user does not have full *independence* (qv).
• Many models have the seat raised fairly high above the traction gear, giving a somewhat precarious appearance. Nervous people may be reluctant to use them.
• Some steep or narrow stairs with winders may not be negotiable by stairclimbers.
4 As of 2008, useful references include the following. For publication details see the bibliography.
• CHOOSING EQUIPMENT TO GET UP AND DOWN STAIRS, a factsheet from the Disabled Living Foundation.
5 Useful websites include the following. They are given in alphabetical order.

• Advanced Stairlifts	www.advancedstairlifts.co.uk/
• Baronmead	www.baronmead.com/
• Disabled Living Foundation	www.dlf.org.uk/
• Disabled Living Online	www.disabledliving.co.uk/
• KTP	www.loops.connectfree.co.uk/page16.html
• Mobility	www.mobility-mhs.co.uk/

See also: *Evacuation chairs*.

Stairlifts

1 There are two varieties of stairlift:
• Chairlifts (qv), which require *wheelchair-users* (qv) to transfer from their *wheelchairs*(qv) onto the chairlift, which may be difficult or impossible.
• Incline platform lifts (qv), which carry the passenger while sitting in his or her wheelchair.
Together they constitute two of the four basic categories of lift types – see: *Lifts*.

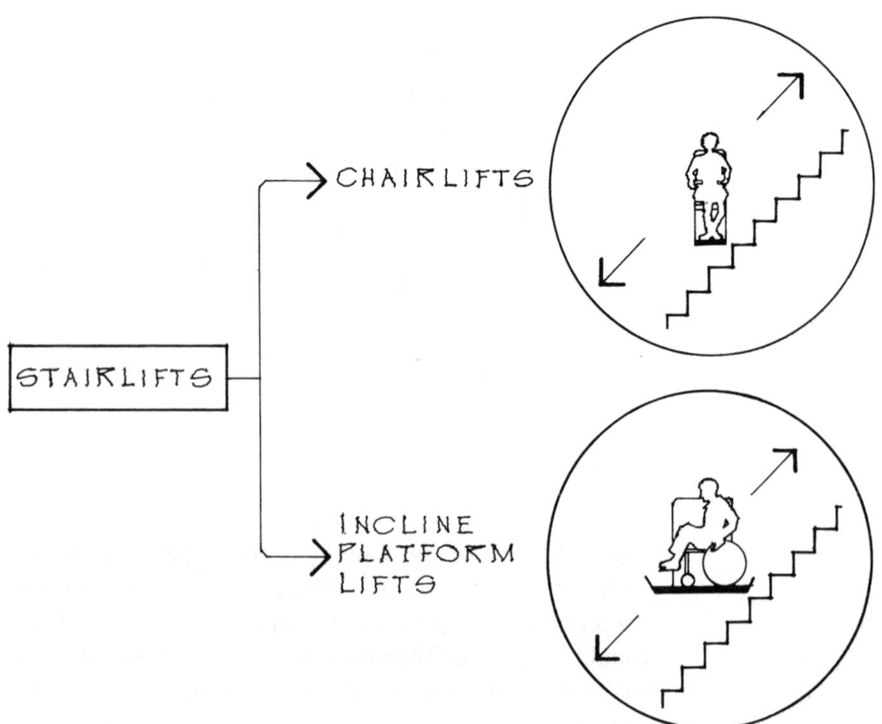

2 For guidance on where and where not to use stairlifts see: *Lifts*.
3 For more detail on the various types of stairlifts see: *Chairlifts*; *Platform lifts* and *Incline platform lifts*.

tairs and steps

1 Any pedestrian surface that is steeper than 1:12 must be designed as one or more flights of steps. For gradients shallower than 1:12 see: *Ramps*.
2 The following entries give guidance.

tairs: dwellings

1 The following references contain authoritative recommendations for the design of both external and internal stairs for dwellings:
• Sections 1–4 of APPROVED DOCUMENT B (qv), which deals with *fire safety*. The provisions in AD B must be heeded in order to comply with the Building Regulations (qv) in England and Wales.
• Sections 6–9 of APPROVED DOCUMENT M (qv), which deals with *access* to and use of buildings. The provisions in AD M must be heeded as above.
• THE HABINTEG HOUSING ASSOCIATION DESIGN GUIDE (qv) applies most specifically to *housing association* (qv) schemes, but ought to be by the designer's side in the design of all housing developments and individual dwellings.
• THE WHEELCHAIR HOUSING DESIGN GUIDE (qv), also from the Habinteg Housing Association, is another illustrated manual whose usefulness extends far beyond the special remit implied by the title.
• THE HOUSING DESIGN GUIDE applies specifically to housing in London.

tairs: public uildings: complying ith the Disability iscrimination Act

1 Buildings used by the public must in general comply with the Disability Discrimination Act (qv). In the case of stairs in public buildings, *authoritative practical guides* (qv) include DESIGNING FOR ACCESSIBILITY (qv) and in addition the following documents. For publication details see the bibliography.
• APPROVED DOCUMENT B (qv), which deals with *fire safety*. Its provisions must be heeded in order to comply with the Building Regulations (qv) in England and Wales.
• APPROVED DOCUMENT K (qv) which deals, inter alia, with stairs and ramps and protection against falling. Its provisions must be heeded in order to comply with the Building Regulations in England and Wales.
• APPROVED DOCUMENT M (qv), which deals with access to and use of buildings. Its provisions must be heeded in order to comply with the Building Regulations in England and Wales.
• BS 8300 (qv), which applies to all buildings in the UK.
• INCLUSIVE MOBILITY (qv), which is the most authoritative reference for the *inclusive design* (qv) of *pedestrian infrastructure* (qv) and *transport-related buildings* (qv).
• GUIDE TO SAFETY AT SPORTS GROUNDS (qv), which is the most authoritative reference for the *inclusive design* of sports facilities.
Only AD B, AD K, and AD M above have the force of law, but conformity with the rest will help to demonstrate that the requirements of the Disability Discrimination Act have been met. The relevant clauses in each are set out below. Note that in the matter of *guarding* (qv) and *handrail* (qv) heights, AD M takes precedence over AD K, which as of 2007 is due for revision.

EXTERNAL STAIRS: DOCUMENT	RELEVANT PARAGRAPHS
APPROVED DOCUMENT B	Consult the index to AD B volume 2
APPROVED DOCUMENT K	1.1–1.32
APPROVED DOCUMENT M	1.27–1.37
BS 8300	5.9–5.10
INCLUSIVE MOBILITY	3.6 and 8.4
GUIDE TO SAFETY AT SPORTS GROUNDS	7.1–7.12 and 13.8–13.18

INTERNAL STAIRS: DOCUMENT	RELEVANT PARAGRAPHS
APPROVED DOCUMENT B	Consult the index to AD B volume 2
APPROVED DOCUMENT K	1.1–1.32
APPROVED DOCUMENT M	3.50–3.51; 3.54–3.55; and 4.12
BS 8300	8.1
INCLUSIVE MOBILITY	3.6 and 8.4
GUIDE TO SAFETY AT SPORTS GROUNDS	7.1–7.12 and 13.8–13.18

The design dimensions shown below will in general satisfy the provisions in the above references, though each particular case should be checked in detail before decisions are finalised.

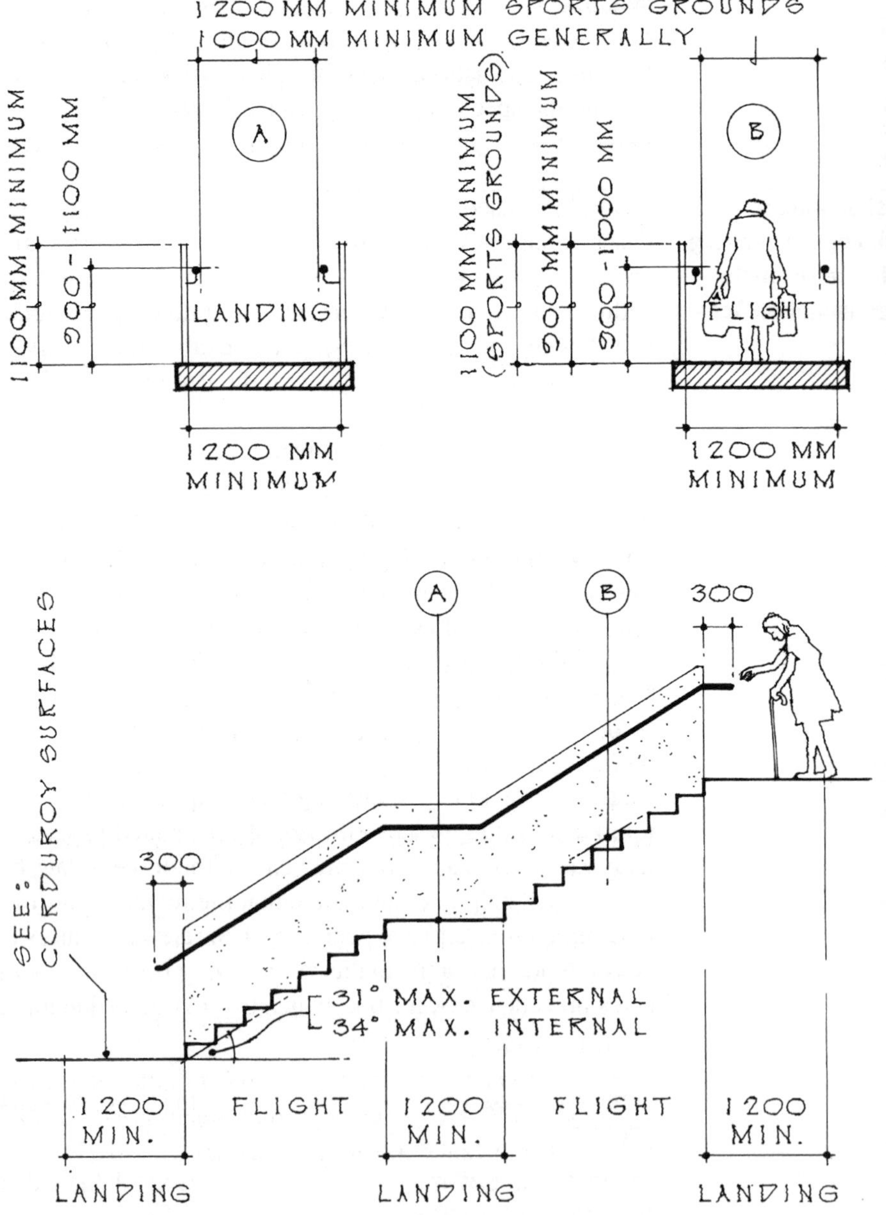

2 For places other than England and Wales the following regulatory documents should be consulted:

- In Northern Ireland: TECHNICAL BOOKLETS E, H, and R of the Building Regulations (Northern Ireland) (qv).

• In Scotland: the BUILDING (SCOTLAND) REGULATIONS and the associated TECHNICAL HANDBOOKS (qv).
3 In the USA, where the Americans with Disabilities Act applies, refer to the ADA AND ABA ACCESSIBILITY GUIDELINES (qv).

See also: *Guardings*; and *Handrails*.

anding and walking ds See: *Walking and standing aids*.

arter home tiative See: *Key workers*.

arter homes See: *Housing: affordable*.

ate retirement nsion See: *Pensions*.

atement of mmunity volvement (SCI) 1 A 'statement of community involvement' sets out the policies of a *Council* for involving its local communities (a) in the preparation, alteration, and review of planning policy documents; and (b) in deciding planning applications.
2 For more information visit www.planningportal.gov.uk/

atementing See: *Special educational needs*.

ations See: *Transport-related buildings*.

atistics See: *Office for National Statistics*.

atistics Board 1 As of September 2007, the Statistics Board is scheduled to replace the Statistics Commission in April 2008. The latter body was a *quango* (qv) with the task of ensuring that official statistics are trustworthy and responsive to public needs, but had lost the confidence of the public. The new Statistics Board is intended to be a more powerful and independent watchdog. Its proposed chairman, Sir Michael Scholar, has told the media that he will report all transgressions to *Parliament* (qv), and that culprits will then have to answer directly to MPs.
2 Unfortunately it would appear that the Board's brief will only cover figures emanating from the Office for National Statistics (qv), which represent about 25% of all official data and cover matters such as employment and economic trends; and that the Board will have no control over the remaining 75% of national statistics which are issued directly by departments such as the Home Office (crime and immigration statistics), the Department for Children, Schools and Families (schools statistics), and the NHS (health statistics). As these tend to be the figures that are least believed by the public, this much-publicised reform must be accounted as more image than substance.
3 In April 2008, as this book was going to press, the Government announced the creation of a UK Statistics Authority (qv), presumably instead of the proposed Statistics Board. For details visit www.statistics.gov.uk/

See also: *Crime: statistics*; and *Office for National Statistics*.

atistics: crime See: *Crime statistics*.

atistics: employment See: *Employment*.

Statistics: health and social care	Visit: www.dh.gov.uk/PublicationsAndStatistics/fs/en
Statistics: misuse	See: *Advocacy research; Health scares; Medicalisation*; and *Tactile pavings*.
Statistics: transport	Visit www.dft.gov.uk/pgr/statistics/
Statistics: workplace accidents and safety	Visit: www.hse.gov.uk/statistics/
Statutes	1 A statute is an Act of Parliament stating the law in broad principles. Matters of detail are customarily covered by *statutory instruments* such as *regulations* - see below. 2 For the full texts of all recent UK statutes and statutory instruments visit www.opsi.gov.uk/acts.htm
Statutory instruments and codes of practice	1 Because legislation is often too complex to be contained in a *statute*, the detail is commonly contained in separate delegated legislation. The latter may come in the form of regulations or orders, which are collectively termed *statutory instruments*. 2 Examples of statutes and the statutory instruments which flow from them include the following: • The Building Act (qv) is a statute, and the Building Regulations (qv) are statutory instruments. • The Disability Discrimination Act 1995 (qv) is a statute, and the Disability Discrimination Act 1995 (Amendment) Regulations 2003 are statutory instruments. 3 Lower down the hierarchy come *codes of practice*, which give guidance on good practice in their particular fields. Unlike statutory instruments these are not the law, or even authoritative explanations of the law; but if a code of practice has been issued by a recognised body such as the British Standards Institution (qv) or the Commission for Equality and Human Rights (qv) then the guidance, recommendations, and 'good practice' examples contained therein must, if relevant, be taken into account by courts of law. Examples of codes of practice that are relevant to the present work include: • BS 8300: 2001 DESIGN OF BUILDINGS AND THEIR APPROACHES TO MEET THE NEEDS OF DISABLED PEOPLE – CODE OF PRACTICE, published by the British Standards Institution (see: *British standards*). • CODE OF PRACTICE – RIGHTS OF ACCESS: SERVICES TO THE PUBLIC, PUBLIC AUTHORITY FUNCTIONS, PRIVATE CLUBS AND PREMISES, published by the Disability Rights Commission (qv). See the bibliography for publication details of the above. 4 As of 2007, useful websites include the following: • For an explanation of Acts of Parliament: www.parliament.uk/about/how/laws/acts.cfm • For the full texts of all recent Acts and Statutory instruments: www.opsi.gov.uk/stat.htm
Staying Put grant	See: *Home improvement grants*.
Step in to Learning programmes	See: *Skills for Life*.
Steplifts	See: *Vertical platform lifts*.

rategic Health thorities See: *NHS structure*.

rategic Rail thority 1 This short-lived *quango* (qv) was created in February 2001 by the then Deputy Prime Minister John Prescott, with a 10-year £60 billion plan to repair Britain's dilapidated network and improve the quality of our rail services. A mandatory requirement of full *accessibility* (qv) of all facilities was an admirable feature of all SRA plans.

2 The SRA did not last much longer than Mr Prescott's £180 billion Transport 2010 (qv) plan and was wound up after passage of the Railways Act 2005 in April 2005.

3 For the current railway setup see: *Railway system*. For latest information on the UK's constantly reorganised railway system visit the Rail Group at www.dft.gov.uk/

rategic road twork 1 These are London's busiest roads. The network includes the 'Transport for London Road Network' (300 miles of red routes and other important streets) and the motorways within Greater London.

2 Visit www.gos.gov.uk/gol/transport/161558/traffic_management_act/

reet calming See: *Traffic calming*.

reet crossings See: *Pedestrian crossings*.

reet furniture: mplying with the isability iscrimination Act 1 Street furniture such as lampposts, signposts, bollards, waste bins, planting boxes, and seats are *hazards* (qv) that create some degree of *risk* (qv) for all pedestrians, and particularly for *wheelchair-users* (qv) and people with *impaired vision* (qv). *Authoritative practical guides* (qv) for dealing with these problems include the following:

- BS 8300 (qv), which applies to all buildings in the UK.
- INCLUSIVE MOBILITY (qv), which applies to pedestrian precincts, public footways, public footpaths, road crossings, etc.

These references do not have the force of law, but conformity with their recommendations will make it easier to demonstrate that the requirements of the Disability Discrimination Act have been met.

2 The relevant clauses in each are set out below.

DESIGN ASPECT	DOCUMENT: RELEVANT PARAGRAPHS	
	BS 8300	INCLUSIVE MOBILITY
BUS STOPS	–	6
PARKING CONTROL EQUIPMENT	4.1.4.2	5.5
PARKING METERS	4.1.5	5.5
COLUMNS, LITTER BINS, SEATS, ETC.	5.7.1	3.7, 3.9
HAZARDS	5.7.2	3.11
FREE-STANDING STAIRS & RAMPS	5.7.3	–
AUTOMATIC TELLER MACHINES	5.7.4	–
OTHER OBSTRUCTIONS	–	3.8, 3.9, 3.11

3 The design features and dimensions shown below will in general satisfy the provisions in the above references, though each particular case should be checked in detail before decisions are finalised.

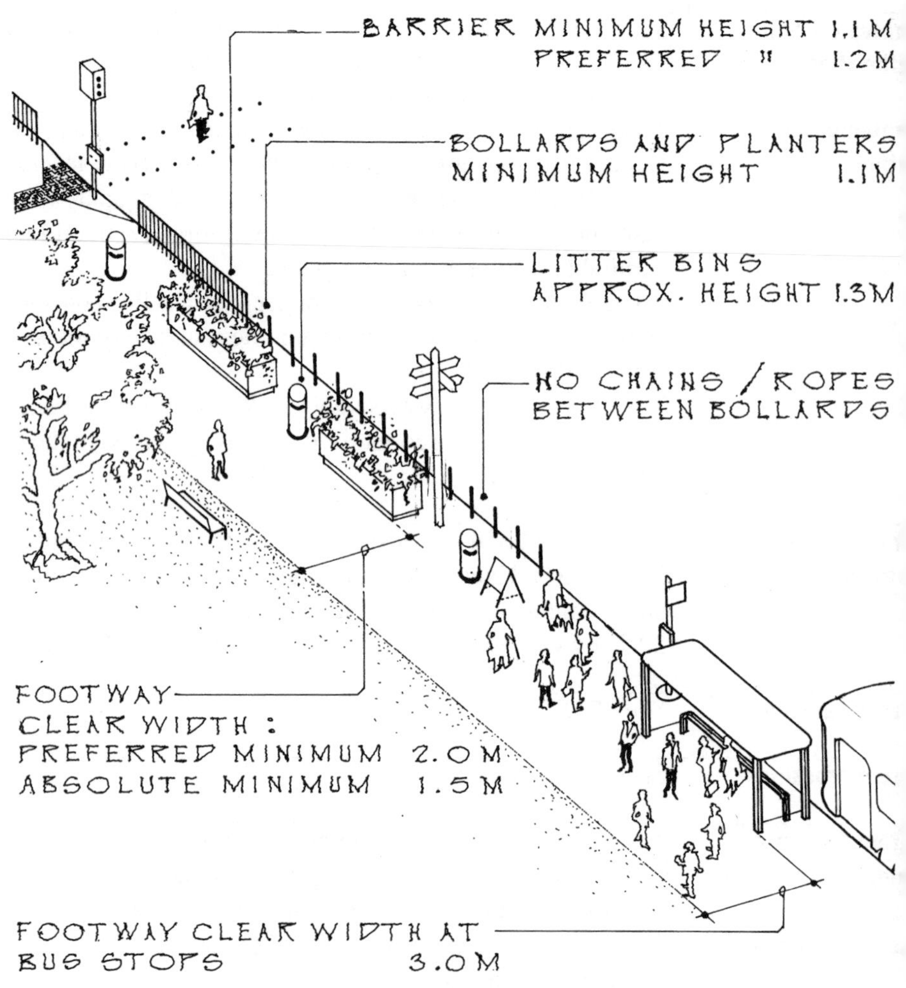

See also: *Hazards: protection*; and *Pedestrian environments*.

Street wardens

1 Street wardens are highly visible uniformed patrols in town and village centres, public areas, and neighbourhoods. They are similar to *neighbourhood wardens* (qv) but their emphasis is on caring for the physical appearance of the area; tackling environmental problems such as litter, graffiti and dog fouling; and helping to deter anti-social behaviour and reduce the fear of crime.
2 Visit www.crimereduction.gov.uk/wardens/wardens23.htm

See also: *Neighbourhood wardens*.

Street works: safety precautions

1 Neither *canes* (qv) nor *guide dogs* (qv) give warning of above-knee obstacles and therefore leave their users vulnerable to injury from high-level obstacles such as overhanging branches; untrimmed hedges (holly can do terrible damage to people's eyes); horizontal or oblique scaffolding poles; open windows; etc. Such hazards should either not exist or, if this cannot be avoided, be surrounded by warning pavings or protective railings.
2 Relevant *authoritative practical guides* (qv) include the following:
• BS 8300 (qv): para 5.7.1.2 refers to free-standing posts and columns; and paras 5.7.2 and 5.7.3 to other hazards.
• INCLUSIVE MOBILITY (qv): para 3.8 refers to scaffolding and street works.
These references do not have the force of law, but conformity with their recommendations will make it easier to demonstrate that the requirements of the Disability Discrimination Act have been met.

3 The markings shown on the diagrams will in general satisfy the provisions in the above references, though each particular case should be checked in detail before decisions are finalised.

The diagram below shows a free-standing post or column, and a pit or some similar hazard.

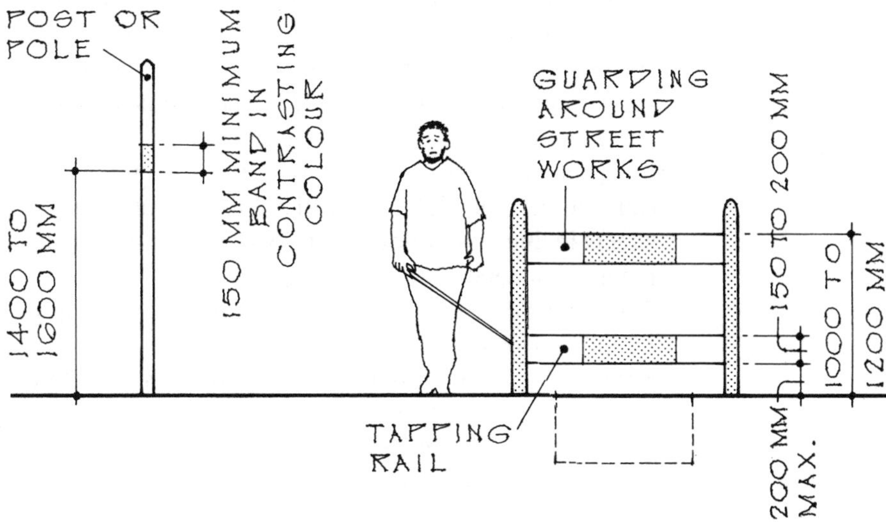

The diagram below shows a door that opens outward into a pedestrian route. For more on this see: *Hazards: protection*.

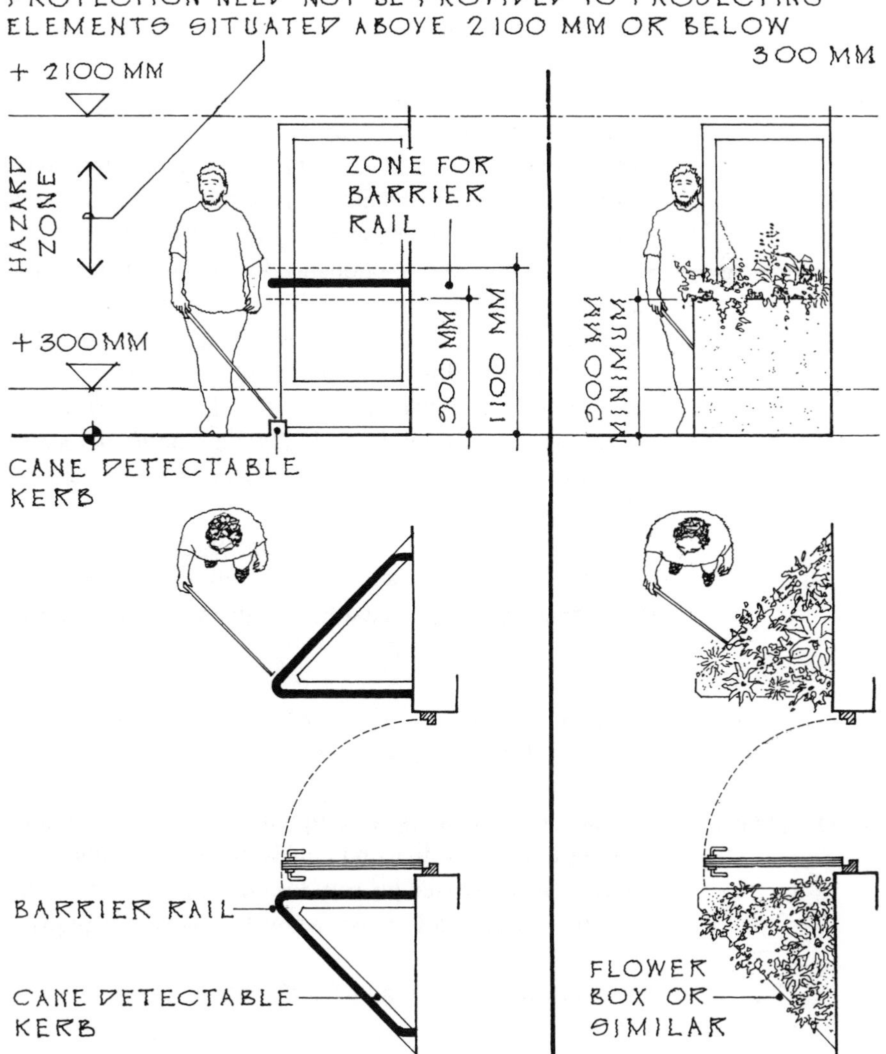

Streets

1 British streets have become less friendly places than they used to be and the effects are damaging, especially for children. In 2007 an opinion poll carried out for Play England (visit www.playengland.org.uk/Page.asp) found that only 21% of children now play regularly in the street or area near their homes, whereas 71% of adults can remember doing so when they were children. They recalled a world where – to quote one respondent – children were 'free, relaxed enjoying themselves'. Writing in the *Guardian* on 1 August 2007 under the title 'Britain has lost the art of socialising the young', Peter Wilby comments on this survey, and on the sad changes that have come about since Richard Hoggart wrote of his childhood in THE USES OF LITERACY.

2 We can't recover the world of the 1950s, but that does not mean that nothing can be done. As of 2008, useful references for improving the quality of our streets include the following. They are given in alphabetical order. For publication details see the bibliography.

3 The following free publications are from CABE (qv):

- CIVILISED STREETS.
- IT'S OUR SPACE is a guide for community groups working to improve public space.
- LIVING WITH RISK: PROMOTING BETTER PUBLIC SPACE DESIGN examines ten public spaces and streets across England with the aim of helping designers challenge excessively risk-averse decisions.
- OUR STREET: LEARNING TO SEE is a teacher's guide.
- TRANSFORMING OUR STREETS reports upon progress since CABE's earlier publication PAVING THE WAY.

4 Other references include:

- BRITAIN HAS LOST THE ART OF SOCIALISING THE YOUNG by Peter Wilby. See the notes above.
- CAR PARKING: WHAT WORKS WHERE, published by English Heritage, gives general advice on car parking provision.
- INCLUSIVE URBAN DESIGN: STREETS FOR LIFE, a book written by Elizabeth Burton and Lynne Mitchell, deals particularly with the design of streets and outdoor spaces to cater for the needs of older people.
- MANUAL FOR STREETS, published by the Department for Transport (qv) with support from CABE (qv) and the Department for Communities and Local Government (qv), aims to improve the quality of life through good street design. It is well illustrated and highly informative.
- SEEN AND HEARD: RECLAIMING THE PUBLIC REALM WITH CHILDREN AND YOUNG PEOPLE was published in 2007 by the *think tank* (qv) Demos.
- THE USES OF LITERACY by Richard Hoggart. See the notes above.

5 Useful websites include:

- CABE www.cabe.org.uk/default.aspx?contentitemid=45
- CABE www.cabe.org.uk/default.aspx?contentitemid=1484
- Play England www.playengland.org.uk/Page.asp

See also: *Architecture; Crime prevention by design; Space syntax; Urban design*; and *Walking distances*.

Streets for people

1 These are areas where a comprehensive package of measures are implemented in order to improve the street environment and enhance a sense of community – typically an increased priority for public transport, walking and cycling.

2 For more information visit www.dft.gov.uk/pgr/roads/tpm/tal

See also: *Home zones*; and *Traffic calming*.

Supplementary provision

See: *Inclusive design.*

Support rails in buildings

1 Support rails (also called *grab rails*) are seen mostly in bathrooms, shower areas, and wcs. But these aids, which come partly within the general category of *walking and standing aids* (qv), can also be useful in other situations. Their functions include:

- Giving frail, elderly or *disabled* (qv) people something to push or pull against when they are trying to get to their feet.
- Offering such people a means of steadying themselves when they are sitting down.
- Offering them a firm grip when they are trying to transfer from one position to another.
- Helping them keep their balance while walking, dressing, or even while standing.

2 Well-positioned support rails at hazardous locations (baths, showers, stairs, and steps) can also save lives, particularly if the users are elderly or frail people. The Royal Society for the Prevention of Accidents (qv) estimates that falls account for over 40% of all injuries in homes.

See also: *Handrails.*

Support rails in public buildings: complying with the Disability Discrimination Act

1 Buildings used by the public must in general comply with the Disability Discrimination Act (qv). In the case of support rails in public buildings, *authoritative practical guides* (qv) include DESIGNING FOR ACCESSIBILITY (qv) and in addition the following documents. For publication details see the bibliography.

- APPROVED DOCUMENT M (qv), the provisions of which must be heeded in order to comply with the Building Regulations for England and Wales.
- BS 8300, the provisions of which apply to all buildings in the UK. It is an advisory document.
- INCLUSIVE MOBILITY (qv), which is the most authoritative reference for the *inclusive design* (qv) of *pedestrian infrastructure* (qv) and *transport-related buildings* (qv) such as bus and coach stations, railway stations, air terminals, and transport interchanges. It is an advisory document.

The advisory documents above do not have the force of law, but conformity with their recommendations will make it easier to demonstrate that the requirements of the Disability Discrimination Act have been met.

2 Relevant clauses in these documents are set out below:

LOCATION	DOCUMENT: RELEVANT PARAGRAPHS		
	AD M	BS 8300	INCLUSIVE MOBILITY
BATHROOMS	5.4,	12.2.10	–
CHANGING & SHOWER AREAS	5.4, 5.18	12.3	–
LAVATORIES	5.4, 5.8, 5.10, 5.11, 5.14	12.4	9.6

3 The profiles shown below will in general satisfy the provisions in the above references, though each particular case should be checked in detail before decisions are finalised. The dimensions for *handrails* (qv) on *stairs* (qv) and *ramps* (qv) are also shown for comparison.

In places other than England and Wales the following regulatory documents should be consulted:
• In Northern Ireland: TECHNICAL BOOKLET R of the Building Regulations (Northern Ireland) (qv).
• In Scotland: the TECHNICAL HANDBOOKS of the Building (Scotland) Regulations (qv).

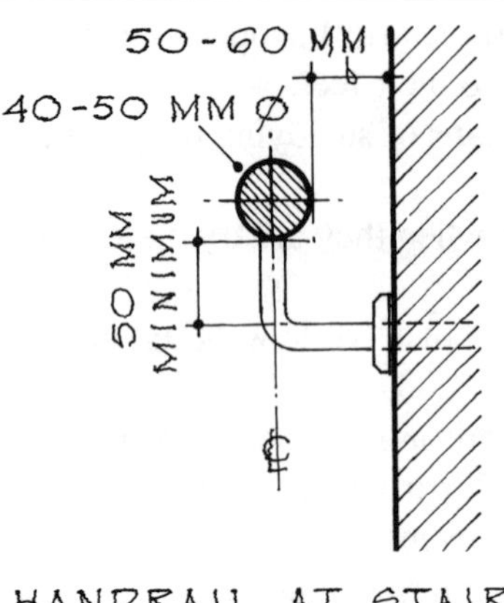

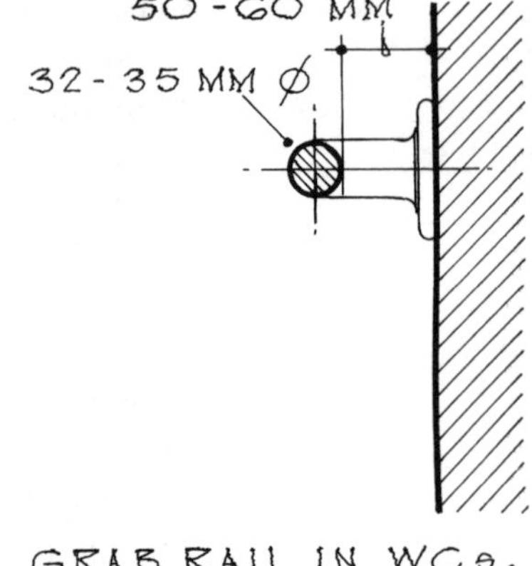

4 As of 2008, other useful references include the following. For publication details see the bibliography.
• CHOOSING AND FITTING GRAB RAILS, a 13-page factsheet from the DLF.
5 Useful websites include the following. They are given in alphabetical order.

• Disabled Living Foundation	www.dlf.org.uk/
• Disabled Living Online	www.disabledliving.co.uk/
• Hewi	www.hewi.com/
• Intrad Wellgrip	www.intrad.com/wellgrip/

See also: *Handrails*.

Support rails in vehicles

1 At time of writing, design recommendations for support rails have been published for the following vehicle types:
• Small buses. ACCESSIBILITY SPECIFICATION FOR SMALL BUSES makes recommendations on the positioning and design of support rails, handrails, and handholds in 'small buses', which are defined as buses and coaches carrying 9 to 22 passengers. These are vehicles thought of by most people as 'minibuses' rather than 'buses' – including for instance factory produced midi-buses and mini-buses; chassis-built minibuses; van conversions, etc. To avoid misunderstanding, an example is shown below.

For publication details see the bibliography.
• Trains. Para 11 of the Rail Vehicle Accessibility Regulations (RVAR) 1998 gives guidance on the positioning and design of support rails, handrails, and handholds in trains and other rail vehicles.
• Ships. THE DESIGN OF LARGE PASSENGER SHIPS makes recommendations on the positioning and design of support rails and handrails in ships.
2 For publication details of the above references see the bibliography. While some of these documents are advisory and do not have the force of law,

conformity with their recommendations will make it easier to demonstrate that the requirements of the Disability Discrimination Act have been met.

Supported housing See: *Special needs housing.*

Supporting People 1 The Supporting People programme by the Department for Communities and Local Government (qv) offers vulnerable people the opportunity to improve their quality of life by providing a stable environment which enables greater *independence* (qv).
2 For more information visit www.spkweb.org.uk/

Sure Start 1 Sure Start is a multi-billion pound government programme to deliver (in the words of its website) 'the best start in life for every child by bringing together early education, childcare, and health and family support' – for more detail visit www.surestart.gov.uk/. It is part of the Government's drive to reduce *child poverty* (qv) and *social exclusion* (qv), and aims to improve the health and well-being of families and children before and from birth so that children are ready to flourish when they go to school. It was launched with the intention of having 500 local Sure Start programmes running by 2004, 2 500 by 2008, and 3 500 by 2010. These figures are for England only; other parts of the UK have separate schemes.
2 The aim is admirable, and the centre nearest to the present author (at 166 Roehampton Lane, London SW15 4HR) is enormously popular, but at this early stage there are mixed views on the success of a scheme that has – together with other early years and childcare services – absorbed £21 billion of taxpayers' money since 1997. As two examples:
• In August 2007 the results of a 6-year study of 35 000 children by Durham University's Curriculum Evaluation and Management Centre showed that after a string of Government policies (including the Sure Start programme) aimed at boosting pre-school children's educational achievement, their development and skills were actually no better than in 2000. While it is not known what proportion, if any, of the 35 000 went through Sure Start, the study's co-author, Christine Merrell, told the BBC that one would have expected such a major government programme to have resulted in some measurable changes in the sample.
• In early 2008 the latest research results on the 2 500 Sure Start Children's Centres then in operation stated that, when compared with equivalent 3 year-olds with no centre in their area, Sure Start children exhibited 'more social behaviour, greater independence and self-regulation', and were less likely to have suffered injuries.
3 The overall picture is unclear. But if future results disappoint, one reason may be that that well-educated parents whose children don't really need Sure Start are using these highly attractive facilities as care centres or handy drop-off places for their kids while they do other things, while the ill-educated and out-of-work parents whose children really do need help are barely using the centres. Perhaps, some have suggested, receipt of benefits should be made conditional on placing children in Sure Start centres.
4 Readers must decide for themselves. As of 2008, useful references include the following. For publication details see the bibliography.
• CHANGES IN CHILDREN'S COGNITIVE DEVELOPMENT AT THE START OF SCHOOL IN ENGLAND 2000–2006 by Christine Merrell, Peter Tymms, and Paul Jones. This study was published in August 2007 by the Curriculum Evaluation and Management (CEM) Centre at Durham University. See the notes above.
5 Useful websites include the following. They are given in alphabetical order.

• CEM Centre	www.cemcentre.org/
• Directgov	www.direct.gov.uk/
• National Evaluation of Sure Start	www.ness.bbk.ac.uk/
• Sure Start	www.surestart.gov.uk/

See also: *Child poverty;* and *Social exclusion*.

Surveillance See: *Privacy*.

Switches See: *Controls and switches*.

Switches: plate and touch See: *Plate and touch switches*.

Switches: touch and proximity See: *Touch and proximity switches*.

Symbols 1 A symbol is a mark that stands for or represents something else – for instance the simple representations of clouds, raindrops, bright sun, hidden sun, etc., that are used on weather maps to convey an idea more quickly and vividly than could be done by words.

2 In the context of an inclusive environment the following types of symbols can have a useful role:

- Symbols used on signage systems in streets, buildings, and parks etc. See: *Signs*.
- Symbols used on maps. See: *Accessible maps*.
- Symbols which help children or adults who have *learning difficulties* (qv) and cannot cope with conventional text. See: *Speech aids*.

See also: *Sign and symbol systems*.

SymbolWorld See: *Widgit Rebus*.

Synface See: *Telephones: lip-readable*.

Syntax: space See: *Space syntax*.

Synthesised speech See: *Speech aids*; and *Talking household appliances*.

Synthetic phonics See: *Literacy*.

T

ables

1 Standard table heights are generally around 700 mm to 750 mm from floor to table top. But people in *wheelchairs* (qv) will need at least 700 mm to the underside of the table, and possibly 750 mm – see the sketch below.
2 Options for providing such non-standard heights include:
• Leg raisers that can be inserted under the feet of a standard table.
• A purpose-made higher table.
• An adjustable-height table – see: *Adjustable furniture*.
3 As of 2008, useful websites include the following. They are given in alphabetical order.

• Coopers Healthcare Services	www.sunrisemedical.co.uk/
• DCS Joncare	www.dcsjoncare.freeserve.co.uk/
• Disabled Living Foundation	www.dlf.org.uk/
• Disabled Living Online	www.disabledliving.co.uk/
• Keep Able	www.keepable.co.uk/
• Nottingham Rehab	www.nrs-uk.co.uk/

DCS Joncare is particularly apposite for products to suit the special needs of children.
4 In pubs, restaurants, cafes, and other refreshment buildings, para 13.5.3 of BS 8300 (qv) recommends the following table heights:
• Some tables in each refreshment space should have a clear height from floor surface to underside of table of 750 mm, to accommodate wheelchairs with armrests.
• For the rest, clear height from floor surface to underside of table should be 700 mm minimum

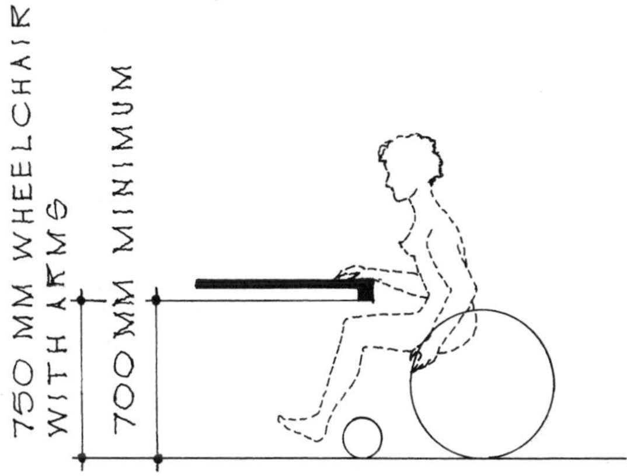

5 The dimensions given above do not apply to serving counters or office desks. For these see: *Counters and reception desks; Refreshment facilities;* and *Worktops*.

actile displays and mages

1 A tactile display is a diagram or some other image that can be 'read' by the fingertips, thereby conveying information to people with *impaired vision* (qv) – see: *Braille; Tactile signs;* and *Tactile text*.
2 As of 2008, useful websites include the following. They are given in alphabetical order.

• Centre for Tactile Images	www.cs.york.ac.uk/tactileimages
• RNIB	www.rnib.org.uk/
• Touchtype	www.touchtypesigns.com/pages/home.htm

See also: *Touch tours*

Tactile pavings

1 A tactile paving is a profiled paving surface that provides warning or guidance to people with *impaired vision* (qv).
2 The most familiar type is the *blister surface* (qv) – also known as 'knobbly bubbles' – that is used at *pedestrian crossings* (qv) to warn *blind* (qv) people that they are about to step onto a carriageway. Britain's town and cityscapes have become blighted by these surfaces and everyone, from the *local authority* (qv) officials who install them to the many pedestrians who are discomfited by stepping on them, assumes that their use is justified by their helpfulness to *disabled people* (qv). But such helpfulness has never been adequately demonstrated. Pages 199–207 and 392–96 of DESIGNING FOR THE DISABLED: THE NEW PARADIGM by Selwyn Goldsmith give a detailed account of how the inability of politicians and government ministers to stand up to *pressure groups* (qv), and their willingness to tolerate flawed *advocacy research* (qv) in the interests of what they believe to be a good cause, triumphed over scientific method. For publication details see the bibliography.
3 Until a future Government takes a new and rational look at the subject (see p. 529), the use of tactile surfaces remains official policy.
As shown below there are six types, each meant for a particular purpose.

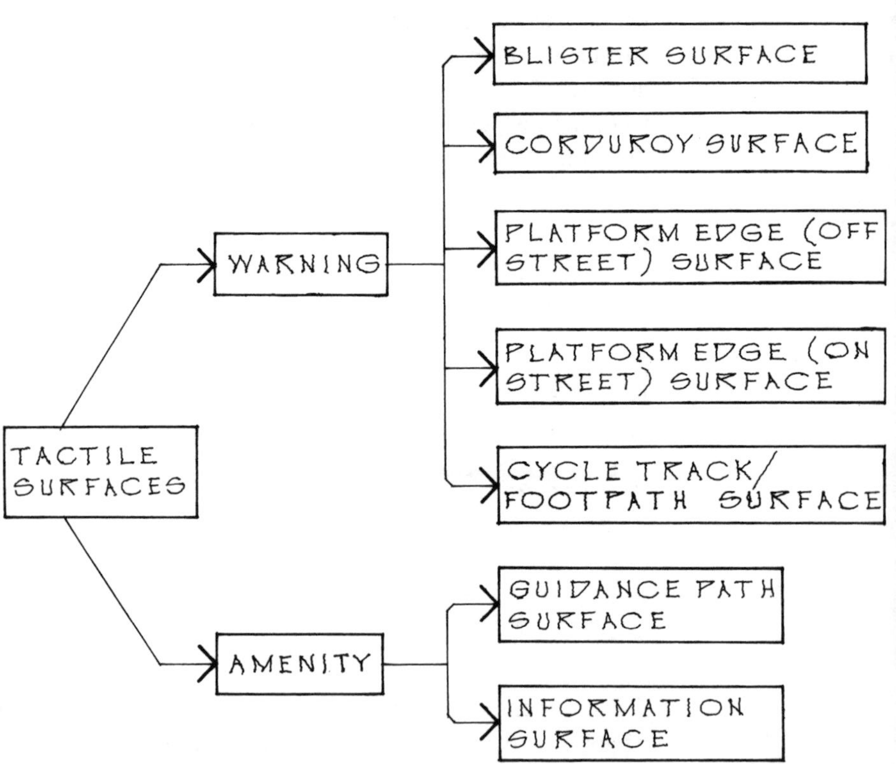

4 Each of the above types must be used in strict accordance with official guidelines in terms of location, layout, orientation, and colour in order to avoid misunderstandings that could have serious consequences. The principal guidance documents on the correct use of tactile pavings include:

- APPROVED DOCUMENT M (qv), the provisions of which must be heeded in order to comply with the Building Regulations for England and Wales.
- BS 8300, the provisions of which apply to all buildings in the UK.
- INCLUSIVE MOBILITY (qv), which is the most authoritative reference for the *inclusive design* (qv) of *pedestrian infrastructure* (qv).
- GUIDANCE ON THE USE OF TACTILE PAVING SURFACES (qv), which sets out the technical specifications and the rules for correct use of tactile pavings, and is an advisory document.

Except for AD M the above documents do not have the force of law, but conformity with their recommendations will make it easier to demonstrate that the requirements of the Disability Discrimination Act (qv) have been met. The most relevant clauses in the three main documents include the following:

DESIGN ASPECT	DOCUMENT & RELEVANT PARAGRAPHS		
	APPROVED DOCUMENT M	BS 8300	INCLUSIVE MOBILITY
PEDESTRIAN CROSSINGS	1.13	–	4.1
FOOTWAYS AND FOOTPATHS	–	5.5.2	4.2, 4.5
CYCLE TRACKS	–	–	4.2, 4.5
EXTERNAL STEPS	1.33	5.9.6	4.2
PEDESTRIAN PRECINCTS	–	–	4.6, 4.7
PLATFORMS	–	–	4.2, 4.3
LEVEL CROSSINGS	–	–	4.2

Tactile pavings: types

1 Blister-surfaced pavings have parallel rows of blisters that are 5 mm high, 25 mm in diameter, and are spaced 65 mm centre to centre. They are used at *pedestrian crossings* (qv) where the *kerb* (qv) has been dropped to be flush with the road surface as shown under *blister surfaces*.

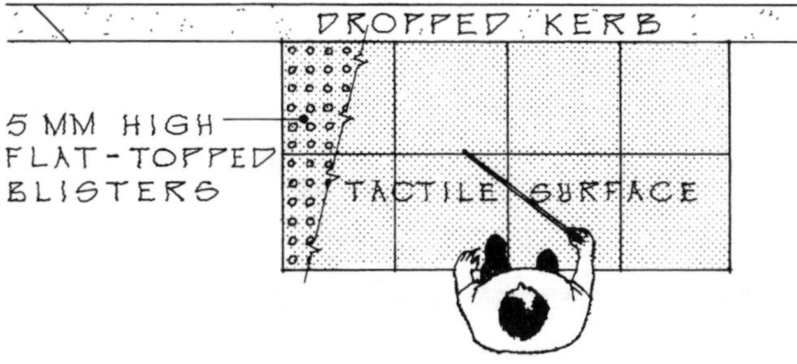

2 Corduroy surfaces consist of rows of rounded bars that are 6 mm high, 20 mm wide, and are spaced 50 mm centre to centre. They are always laid with the bars running at right angles to the direction of travel as shown below.

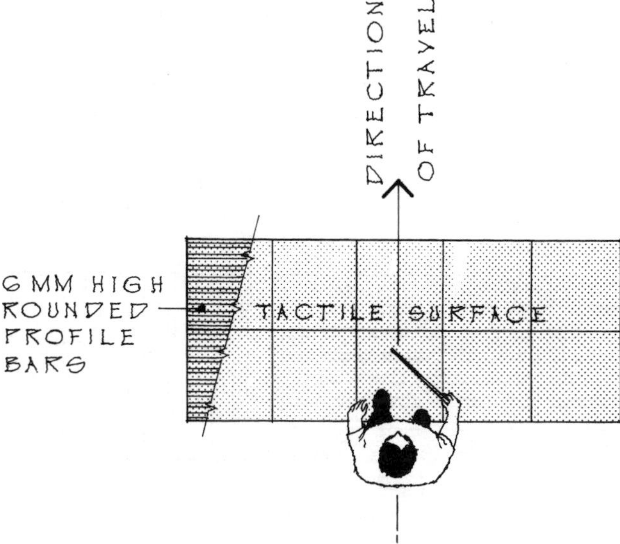

As illustrated under that entry, *corduroy surfaces* are used (a) at the top and bottom of external stairs; and (b) at the foot of a ramp leading to an on-street light rapid transit platform.

3 Off-street platform edge warning surfaces have offset rows of flat-topped domes that are 5 mm high, and spaced 67 mm apart centre to centre as shown below. They are used at London Underground and Network Rail platforms.

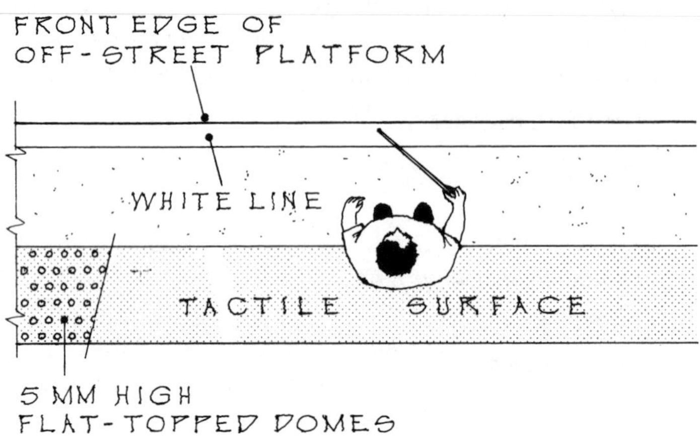

4 On-street platform edge warning surfaces have rows of lozenge shapes that are 6 mm high, with rounded edges, measuring 150 mm x 83 mm in size as shown below. They are used at all on-street light rapid transit platforms such as the tramways in Croydon (London) and Manchester.

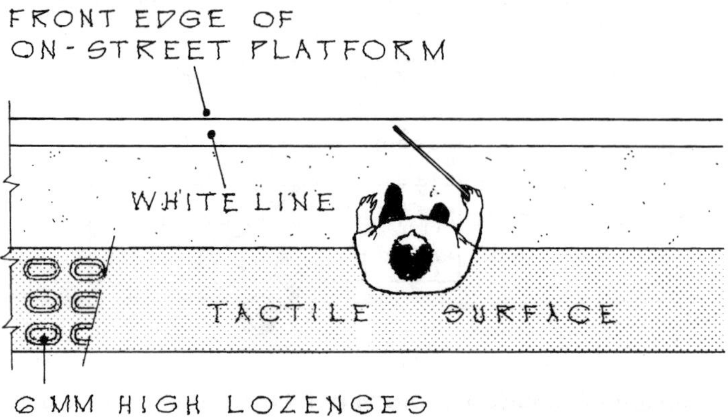

5 Segregated shared cycle track/footway surfaces consist of parallel rows of flat-topped bars, with the bars laid along the direction of travel on the cycle track, and across the direction of travel on the footway.

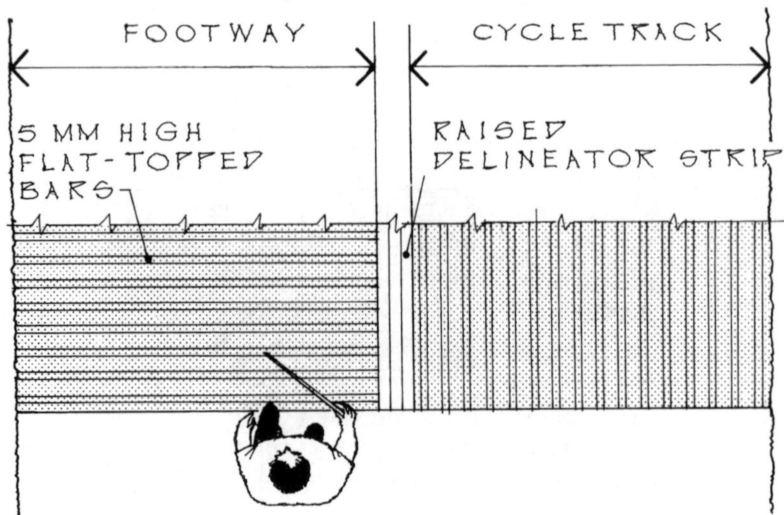

The purpose of segregated shared cycle track/footway surfaces is to help people with *impaired vision* (qv) distinguish between the footway and the cycle track, and not accidentally use the wrong lane.

6 Guidance path surfaces have, unlike all the foregoing ones, an amenity rather than a warning function. They consist of raised flat-topped bars that are 5.5 mm high, 35 mm wide, and are spaced 45 mm apart. Their function is to guide people with *impaired vision* along a route where the traditional clues are not available. For instance:

• In pedestrian precincts, where an easily-recognised footway between the property line and carriageway does not exist.

• In pedestrian precincts, where people need to be guided around obstacles such as cycle stands.

• In pedestrian precincts, where a number of visually-impaired people need to find a specific location.

• In *transport terminals*, to guide people between facilities.

7 Information surfaces again have an amenity rather than a warning function. They do not have tactile features such as blisters or bars, but are detectable because they are made of a yielding material and feel slightly softer underfoot than conventional paving materials. Their function is to help people with *impaired vision* locate amenities such as a bus stop, help point, telephone kiosk, post box, information board, or cash dispenser. Within *transport terminals* (qv) they can be used to indicate ticket offices, help points, waiting rooms, and lavatories.

[T]actile pavings: the [f]uture

1 As mentioned in para (2) of p. 526, the supposed benefits of tactile pavings have never been properly established. Some people have suspected that most of these discomforting surfaces actually cause more nuisance or risk to pedestrians than they are worth.

2 In an article titled THE CHANGING FACE OF TACTILE PAVING (see the bibliography), the *access consultant* (qv) Brian Towers supports the latter view. A study involving a wide range of people and organisations across the UK led him to two conclusions:

• The use of *blister surfaces* (qv) at *pedestrian crossings* (qv) is of considerable value.

• All other uses of tactile pavings, except for platform edge-warning systems (see: *Off-street* and *On-street platforms*) are too confusing to be of any value, even when official guidelines are strictly obeyed. Few of the people for whom these pavings were intended, other that those with *guide dogs* (qv) or trained in the use of long *canes* (qv), seem to know what the various types of pavings are for.

3 The limited usefulness of these pavings is only part of the problem. The study found that 'there are many people for whom tactile paving is more of a curse than a blessing. For example, people who walk with difficulty find all forms of tactile paving not only difficult but sometimes painful to walk over. Blister paving in particular can present a serious tripping hazard'. The irony of the situation is highlighted in the following paragraph from a Health and Safety Laboratories (HSL) tactile paving survey: 'It is … possible that tactile paving may pose a significant trip hazard specifically to those individuals it has been designed to aid'.

4 It seems clear that a rethink of official policy is in order. Mr Towner can be contacted at briantowers@theaccessconsultancy.co.uk

[T]actile plans

See: *Tactile displays and images*.

[T]actile signs

See: *Braille*; and *Signs: supplemental types*.

Tactile text	See: *Braille*; and *Moon*.
Talking books and newspapers	1 These are *reading aids* (qv) for people with severely *impaired vision* (qv). The concept is simple: books or newspapers are read aloud by professional narrators, recorded to a high audio standard, and the readings are then made available to listeners in a variety of formats. 2 As of 2008, useful websites include the following. They are given in alphabetical order. • Calibre www.patient.co.uk/showdoc/26738752/ • RNIB www.rnib.org.uk/ • Talking Book Shop www.talkingbooks.co.uk/ • Talking Newspaper Association www.tnauk.org.uk/ • Travellers Tales www.talkingbookclub.co.uk/ See also: *Impaired vision*; and *Reading aids*.
Talking household appliances	1 Household appliances, especially kitchen equipment, which give spoken warnings or information are becoming readily available. They make life safer and easier for people with severely *impaired vision* (qv). 2 As of 2008, useful websites include: • Cobolt Systems Ltd www.cobolt.co.uk/about.asp • RNIB www.rnib.org.uk/ See also: *Products for easier living*.
Talking signs	1 Talking signs are mounted for instance on lampposts, and give spoken messages to passers-by. The message can be activated manually by push buttons, or automatically by infra-red sensors, or by a combination of both. Their most relevant use is as information points in pedestrian precincts, shops, hospitals, or in other large or complex venues to guide and assist passers-by who have *impaired vision* (qv) or *impaired understanding* (qv), cannot read, or do not have English. 2 As of 2008, useful websites include: • Tiresias www.tiresias.org/index.htm
Tap turners	1 A tap turner is a lever that can be fitted onto a standard tap-handle, enabling people with *arthritis* or weak hands to turn the tap more easily. 2 As of 2008, useful websites include: • Disabled Living Foundation www.dlf.org.uk/ • Disabled Living Online www.disabledliving.co.uk/ • Ricability www.ricability.org.uk/ See also: *Products for easier living*.
Taps	1 Two aspects of tap design and specification are important for ensuring a high degree of convenience and safety for the whole population: • First, taps should be controllable by pushing a lever with wrist or palm, with no need to grip a tap head with the fingers. This helps people with stiff or arthritic hands (see: *Impaired dexterity*); makes it easier for all users to turn a tap with slippery, soapy hands; and avoids users in public lavatories having to grip a handle that is probably heavily contaminated with other people's faecal bacteria. The figure shows a lever-operated mixer tap for kitchen sinks, and there are other types to suit all applications.

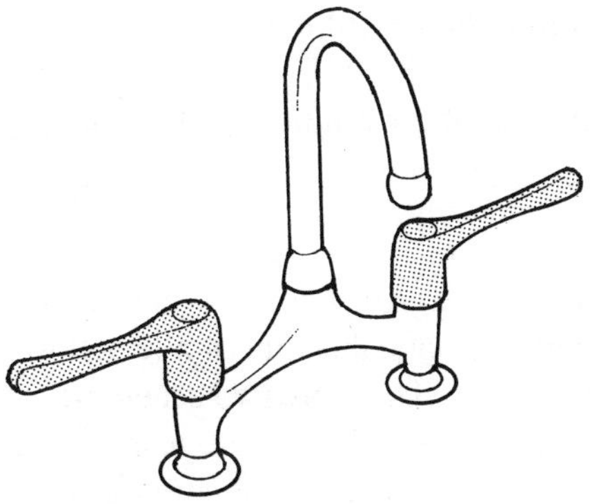

• Second, hot water taps should be temperature-controlled to minimise the risk of accidental scalding. This is particularly important in the case of bath taps, as anyone who has seen a child who has fallen into a bath of near-boiling water can testify.
2 Some of the above matters are already covered by regulations as noted in the table below. Beyond that, proposed amendments to Part G (Hygiene) of the Building Regulations (qv) are likely to require that a thermostatic mixing valve be installed in all new hot taps to prevent scalding.
3 As of 2008, *authoritative practical guides* (qv) for tap specification include the following
• APPROVED DOCUMENT M (qv), the provisions of which must be heeded in order to comply with the Building Regulations for England and Wales.
• BS 8300, the provisions of which apply to all buildings in the UK. BS 8300 does not have the force of law, but conformity with its recommendations will make it easier to demonstrate that the requirements of the Disability Discrimination Act have been met.
Relevant clauses in the above documents include the following:

TAP LOCATION	DOCUMENT: RELEVANT PARAGRAPHS	
	APPROVED DOCUMENT M	BS 8300
TAPS GENERALLY	5.3	–
BATHROOMS	5.3, 5.4(a) & (b)	12.2.7
CHANGING & SHOWER AREAS	5.3, 5.18(p)	12.3.6
KITCHENS	–	12.1.6
LAVATORIES	5.3, 5.4(a) & (b)	12.4.5.1

For places other than England and Wales the following regulatory documents should be consulted:
• In Northern Ireland: TECHNICAL BOOKLETS E, H, and R of the Building Regulations (Northern Ireland) (qv).
• In Scotland: the BUILDING (SCOTLAND) REGULATIONS AND THE ASSOCIATED TECHNICAL HANDBOOKS (qv).
4 In the USA, where the Americans with Disabilities Act applies, refer to the ADA AND ABA ACCESSIBILITY GUIDELINES (qv).

Targets

1 Since 1997 the public sector in the UK has been run on the basis of meeting targets set by central government. Results have been very mixed, and include grotesque practices such as police forces devoting their time to quickly-solved, trivial cases rather than difficult ones; and thousands of seriously ill patients deliberately being kept waiting in ambulances for up to 5 hours because

A&E units will not admit them until it is certain that they can be treated within the target time limit.
2 For other manifestations of government by bureaucracy see: *Overregulation*.

Tax: council

See: *Local authorities: funding*.

Tax credits

1 Tax credits were launched in April 2003 by the Chancellor of the Exchequer, Mr Gordon Brown, with the laudable aim of alleviating hardship among the most vulnerable poor people. There are two schemes:
• Child tax credit, payable to people who are responsible for a child or young person living with them. It is aimed at overcoming *child poverty* (qv) – a term that is defined under that entry.
• Working tax credit, payable to working people on low incomes. It is meant to encourage unemployed people into work by helping them overcome the *poverty trap* – a phrase referring to people on *benefit* (qv) being deterred from taking low-paid jobs because, after paying for the travel and other costs associated with going to work, they could actually be worse off employed than they were unemployed.
2 The problem that has emerged with these well-intentioned subsidies is that instead of encouraging people with low skills or low motivation to move up in life, they tend to trap them in low-paid jobs plus income support – very much as state subsidies tended to preserve unproductive industries in the 1970s. The effect is to disguise poverty instead of curing it. After 5 years of failure there is an emerging consensus that a new approach is needed – one that incentivises and helps low-paid people to improve their skills and progress to more skilled, better-paid work.
3 Meanwhile it is worth noting in some detail the criticisms that have been made of the current tax credit schemes:
• First, like most well-intended government schemes they are grotesquely over-complicated (for another instance see: *Child Support Agency*).
• Second, again like most government schemes, their *IT systems* (qv) were rushed into operation without adequate preparation. A few months after starting they collapsed and the helplines went into meltdown, leaving callers distraught or desperate. Despite massive funding from taxpayers these systems have malfunctioned ever since.
• Third, payments are based on predictions of future earnings, and if these turn out to be wrong (which is in many cases unavoidable) then beneficiaries must repay the excess money. Thus it came about that in June 2005 almost 2 million families, including some of the poorest, least numerate and least literate people in the land, were instructed to repay money they had accepted in good faith, and had long since spent. According to the Parliamentary *Ombudsman* (qv) many had been forced to run up debts and even live off food parcels as a result. Prime Minister Tony Blair apologised to those who had suffered, but in 2006 the fiasco was repeated.
4 Despite assurances that these problems were being resolved, Ms Ann Abraham, the then Parliamentary Ombudsman, delivered a damning report in October 2007 which found that government action to force some of the most most vulnerable people in the land to return moneys paid to them in error, and long since spent, was traumatising many families. According to a Citizen's Advice report released in the same week, some families had been forced to sell or remortgage their homes; others had suffered family break-up. David Harker, chief executive of Citizen's Advice, said that some claimants were so afraid of finding themselves with another large sum to repay that they were opting out of the tax credit system altogether even though it was an important part of their family income.

4 The mountain of wasted money, and the agonies suffered by poor people being chased to repay money they no longer possessed and could not possibly raise, might just about be forgiven if these schemes were fulfilling their stated aims of reducing poverty and encouraging poor people back into work, but here too the news is bad.
5 The following reports are given in date order.
• THE POVERTY TRADE-OFF: WORK INCENTIVES AND INCOME REDISTRIBUTION IN BRITAIN, a report published in October 2006 by the Joseph Rowntree Foundation (qv) and the Institute for Fiscal Studies (qv), stated that owing to the extension of means-testing, 'Labour's reforms to date have acted to weaken both incentives to be in work at all, and incentives to those in work to increase their earnings'. Because half of their extra earnings would be confiscated by the Government, millions of people have elected to remain poor on benefit rather than try to become self-supporting families earning a decent living.
• WELFARE ISN'T WORKING: CHILD POVERTY, a report on the Government's drive to end child poverty, co-authored by the Labour MP Frank Field and published in June 2007, states (a) that 'severe child poverty has been untouched and persistent child poverty remains high'; (b) that hitting the poverty-reduction target for 2010 will require additional government expenditure of £4 billion a year, which is extremely unlikely; and (c) that the tax credit scheme 'brutally discriminates against two-parent families', with damaging consequences for the children concerned and for society – see in this connection: *Child poverty*; and *Children: single-parent.*
6 Clearly this is not the way to do things. It would be far better to start raising the threshold at which people start paying income tax and to keep raising it – as circumstances permit – until everyone under 60% of median earnings (see*: Poverty*) is automatically exempt. For instance, a first step of raising the personal allowance to £7 185 would take 2.5 million people out of income tax. Relieving poor people of the need to pay taxes would be more just, and make more practical sense from almost every point of view, than first taxing their pitiful earnings and then offering to compensate them via hideously complicated subsidies.
7 As of 2008, useful websites include the following. They are given in alphabetical order.

• Direct Gov	www.direct.gov.uk/en/MoneyTaxAndBenefits/index.htm
• Frank Field	www.frankfield.co.uk/type3.asp?id=20&type=3
• HM Revenue and Customs	www.hmrc.gov.uk/leaflets/wtc2.htm
• Institute for Fiscal Studies	www.ifs.org.uk/
• Joseph Rowntree Foundation	www.jrf.org.uk/
• Reform	www.reform.co.uk/

See also: *Child poverty; Incomes: inequality; IT systems; Poverty*; and *Welfare reform*.

axation

1 Taxes are of two kinds:
• Direct taxes, comprising income tax and national insurance. According to Table 5.9 of SOCIAL TRENDS 37 (qv) the poorest tenth of taxpayers lose 11% of their earnings to income tax, the middle tenth 16%, and the richest tenth 22%. NIC contributions average around 7% for all groups.
• Indirect taxes, comprising VAT and the customs and excise duties that are included in the prices of consumer goods. According to Figure 5.10 of SOCIAL TRENDS 37 the poorest fifth of UK households lose around 29% of their disposable household income to indirect taxes, the middle fifth around 21%, and the top fifth around 14%.

2 Looking at the effects of both forms of taxation together, the Liberal Democrat MP Mr Vince Cable elicited from the Treasury in September 2007 that in 2005–06 the poorest 10% of households had average household incom of £8 366 of which 44.2% was taken in direct and indirect taxes; households o the *median* (qv) income of £24 700 paid 35.3.% in taxes; and the top 10% of households had an average income of £88 334 and paid 35% in taxes. The top 1% of households earned £92 300 and paid 31%.

3 These disparities were worsened by tax changes announced by chancellor Gordon Brown in 2007, and introduced in April 2008. These reduced the standard rate of income tax (which applies to incomes up to £36 000) from 22 to 20p, and made up the resultant £7 billion loss in tax revenues by scrapping the special 10p rate that had hitherto applied to the first £2 230 of earnings above the personal allowance of £5 225. The effect is to cut income tax on the well-off and increase it on 5 million of Britain's poorest people, and then to rel on the fiendishly complex *tax credit* (qv) system to compensate the latter. The shift caused bafflement across the political spectrum. As James Buchan notes i ADAM SMITH AND THE PURSUIT OF PERFECT LIBERTY (see the bibliography) even the 18th century patron saint of economic freedom argued that 'the subjects of every state ought to contribute towards the support of the government, as nearly as possible, in proportion to their respective abilities'. Or, in the words o Kingsley Amis's Lucky Jim: 'If one man's got ten buns and another's got two, and a bun has to be given up by one of them, then surely you take it from the man with ten buns'. See also: *Incomes: inequality.*

4 Turning from the impact of taxation upon taxpayers to the overall amounts raised, and how these sums are spent:

• Overall taxation as a proportion of the UK's GDP has for the past few decade fluctuated between a low of 33% or 34% (1979 and again in 1994) and a high of 38% or 39% (now). OECD figures have suggested that government spendin as a percentage of GDP will rise from 37.5% in 2000 to 45.1% in 2008.

• The figure below, from Table 5.31 in SOCIAL TRENDS 33, shows the proportions of its taxation and other income spent by the Government on the provision of goods and services to its citizens. Exact definitions are crucial, an readers should consult the original text in SOCIAL TRENDS.

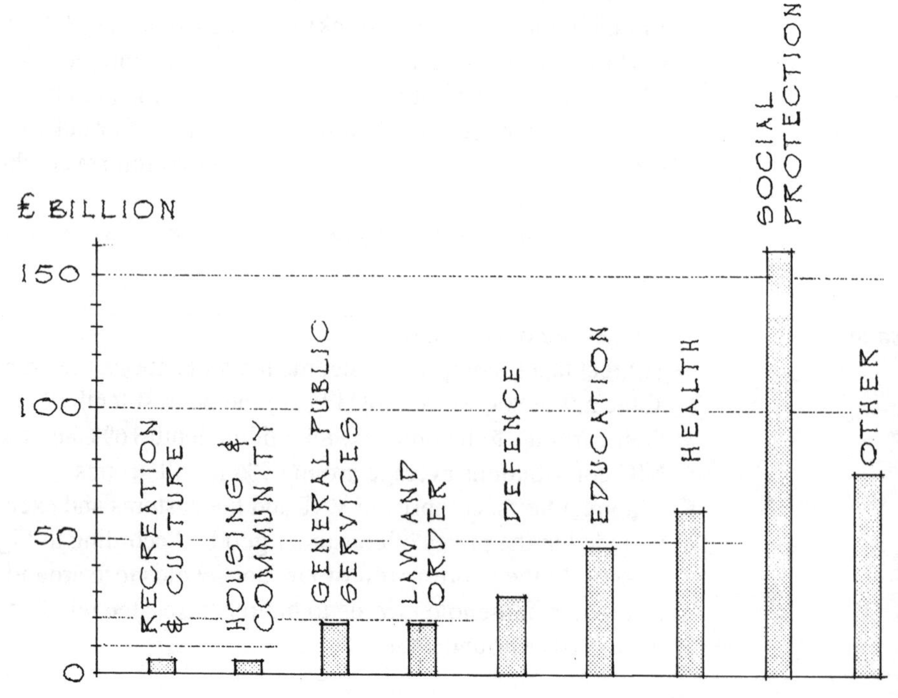

5 As of 2008, useful references include the following. For publication details see the bibliography.
• SOCIAL TRENDS (qv) gives latest statistical data.
• SQUANDERED: HOW GORDON BROWN IS WASTING OVER ONE TRILLION POUNDS OF OUR MONEY comments on whether our taxes are well spent.
6 Useful websites include:
• HM Treasury www.hm-treasury.gov.uk/
• Office for National Statistics www.statistics.gov.uk/
• Reform www.reform.co.uk/website/economy/taxation.aspx

See also: *Incomes: inequality*.

---

xi ranks: complying th the Disability scrimination Act

1 As of 2008, *authoritative practical guides* (qv) for the location and design of taxi ranks include the following:
• Section 7 of INCLUSIVE MOBILITY (qv) – see the bibliography. While this is an advisory document and does not have the force of law, conformity with its recommendations will make it easier to demonstrate that the requirements of the Disability Discrimination Act (qv) have been met.
2 Useful websites include:
• DPTAC Access Directory www.dptac.gov.uk/adnotes.htm
• DPTAC Publications www.dptac.gov.uk/pubs.htm

---

xis: complying with e Disability scrimination Act

1 The Disability Discrimination Act (qv) effectively outlaws discrimination against *disabled people* (qv) in public life, and to this end Part 5 of the Act makes provision for anti-discrimination *regulations* (qv) to be passed for three sets of public transport vehicles:
• Taxis.
• Public service vehicles (i.e. buses and coaches).
• Rail vehicles.
2 In the case of taxis the Act allows the Government to make regulations on both the design (for instance minimum door opening sizes and floor areas, and the provision of ramps) and the operation of taxis. The aim is that disabled people should be able to get in and out of the taxi safely, and be carried safely and in reasonable comfort. For *wheelchair-users* (qv) the aim is that these things should be possible while they are sitting in their wheelchairs.
3 From 2000 all licensed London taxis have had to be wheelchair-accessible; outside London the situation varies. The only national regulation that applies as of 2008 is that taxi drivers cannot refuse to carry *guide dogs* (qv) without good reason (for instance that the driver suffers from asthma and is allergic to dogs), and there seems little prospect of other major measures coming into effect before 2010.
4 Useful references include the following. They are listed in alphabetical order. For publication details see the bibliography.
• CODE OF PRACTICE. PROVISION AND USE OF TRANSPORT VEHICLES: STATUTORY CODE OF PRACTICE: SUPPLEMENT TO PART 3 CODE OF PRACTICE. This is an essential reference in the application of the Disability Discrimination Act.
• WHEELS WITHIN WHEELS: A GUIDE TO USING A WHEELCHAIR ON PUBLIC TRANSPORT published by Ricability (qv). This very informative 29-page guide was published in 2005.
5 The situation will be constantly developing in coming years, and designers and managers must keep up with current information. As of 2007, useful websites for checking the latest situation include:
• Department for Transport
www.dft.gov.uk/transportforyou/access/taxis/usefulcontacts

- DPTAC Access Directory www.dptac.gov.uk/adnotes.htm
- DPTAC Publications www.dptac.gov.uk/pubs.htm
- DPTAC taxis www.dptac.gov.uk/taxis.htm
- Licensed Private Hire Car Association www.lphca.co.uk/
- Licensed Taxi Drivers' Association www.ltda.co.uk/
- Ricability www.ricability.org.uk/

See also: *Transport: complying with the Disability Discrimination Act.*

Telecare

1 A technology that uses sensors to link people at home to *alarms* (qv) at control centres, thus helping older and other vulnerable people to live *independently* (qv) and safely in their own homes rather than go into care. Suc systems can bring great peace of mind not only to the vulnerable people they serve, but also to their *carers* (qv), who usually cannot be with them 24 hours a day.

2 As of 2008, useful references include the following. For publication details see the bibliography:

- CHOOSING EQUIPMENT TO MAINTAIN SAFETY AND INDEPENDENCE AT HOME – INTRODUCING TELECARE. This is a factsheet from the Disabled Living Foundation (qv).

3 Useful websites include the following. They are given in alphabetical order.

- Care Services Improvement Partnership www.changeagentteam.org.uk/
- Department of Health www.dh.gov.uk/
- Disabled Living Foundation www.dlf.org.uk/
- Disabled Living Online www.disabledliving.co.uk/

See also: *Smart homes.*

Telemedicine

1 The use of communications such as video and computers to provide remote diagnosis and healthcare, allowing more care to be provided in the home.

2 As of 2008, useful websites include:

- Telemedicine www.cee.hw.ac.uk/Databases/telemed.html
- Telemedicine and E-Health Information Service www.tis.bl.uk/

See also: *Smart homes.*

Telephone outlets: heights above floor

See: *Controls and switches.*

Telephone services for deaf people

1 BT and other telecommunications service providers offer several special services to help *deaf* (qv) and non-speaking people conduct conversations wit hearing people, and with each other, via the telephone.

2 As of 2008, useful websites include the following. They are given in alphabetical order.

- BT www.btplc.com/age_disability/index.htm
- Ricability www.ricability.org.uk/
- RNID www.rnid-typetalk.org.uk/
- Sign Community www.signcommunity.org.uk/

See also: *Speech aids.*

Telephones: lip-readable

1 Lip-readable telephones have a computer screen linked to the telephone. Incoming speech is converted to realistic lip and facial movements on the face which a deaf person can *lip-read* (qv).

2 For more information visit www.rnid.org.uk/

lephones: private

1 BT and other companies offer a wide range of *accessible* (qv) products and services for people whose needs are not met by the standard range. Of products, the 'big button telephone' shown below is one of the simplest and most popular.

2 As of 2008, useful websites include:

- BT www.btplc.com/age_disability/index.htm
- Ricability www.ricability.org.uk/
- RNID www.rnid.org.uk/shop

lephones: public ildings: complying th the Disability scrimination Act

1 Facilities used by the public must in general comply with the Disability Discrimination Act (qv). In the case of telephones in public buildings, *authoritative practical guides* (qv) include designing for accessibility (qv) and in addition the following documents. For publication details see the bibliography.

- APPROVED DOCUMENT M (qv), the provisions of which must be heeded in order to comply with the Building Regulations for England and Wales.
- BS 8300, the provisions of which apply to all buildings in the UK.
- INCLUSIVE MOBILITY (qv), which is the most authoritative reference for the *inclusive design* (qv) of *pedestrian infrastructure* (qv) and *transport-related buildings* (qv) such as bus and coach stations, railway stations, air terminals, and transport interchanges. It is an advisory document.

Except for AD M the above documents do not have the force of law, but conformity with their recommendations will make it easier to demonstrate that the requirements of the Disability Discrimination Act have been met.

The most relevant clauses in the above documents include those tabulated below:

ASPECT	DOCUMENT: RELEVANT PARAGRAPHS		
	APPROVED DOCUMENT M	BS 8300	INCLUSIVE MOBILITY
GENERAL	0.14(e); 4.14; 4.36(d)(e)	10.4.1	9.2
PROVISION & LOCATION	–	10.4.1	9.2
CONTROLS	–	10.4.2	9.2
BOOTHS	–	10.4.3	9.2

For places other than England and Wales the following regulatory documents should be consulted:

- In Northern Ireland: TECHNICAL BOOKLETS E, H, and R of the Building Regulations (Northern Ireland) (qv).
- In Scotland: the BUILDING (SCOTLAND) REGULATIONS and the associated TECHNICAL HANDBOOKS (qv).

The design features and dimensions shown below will in general satisfy the provisions in the above references, though each particular case should be checked in detail before decisions are finalised.

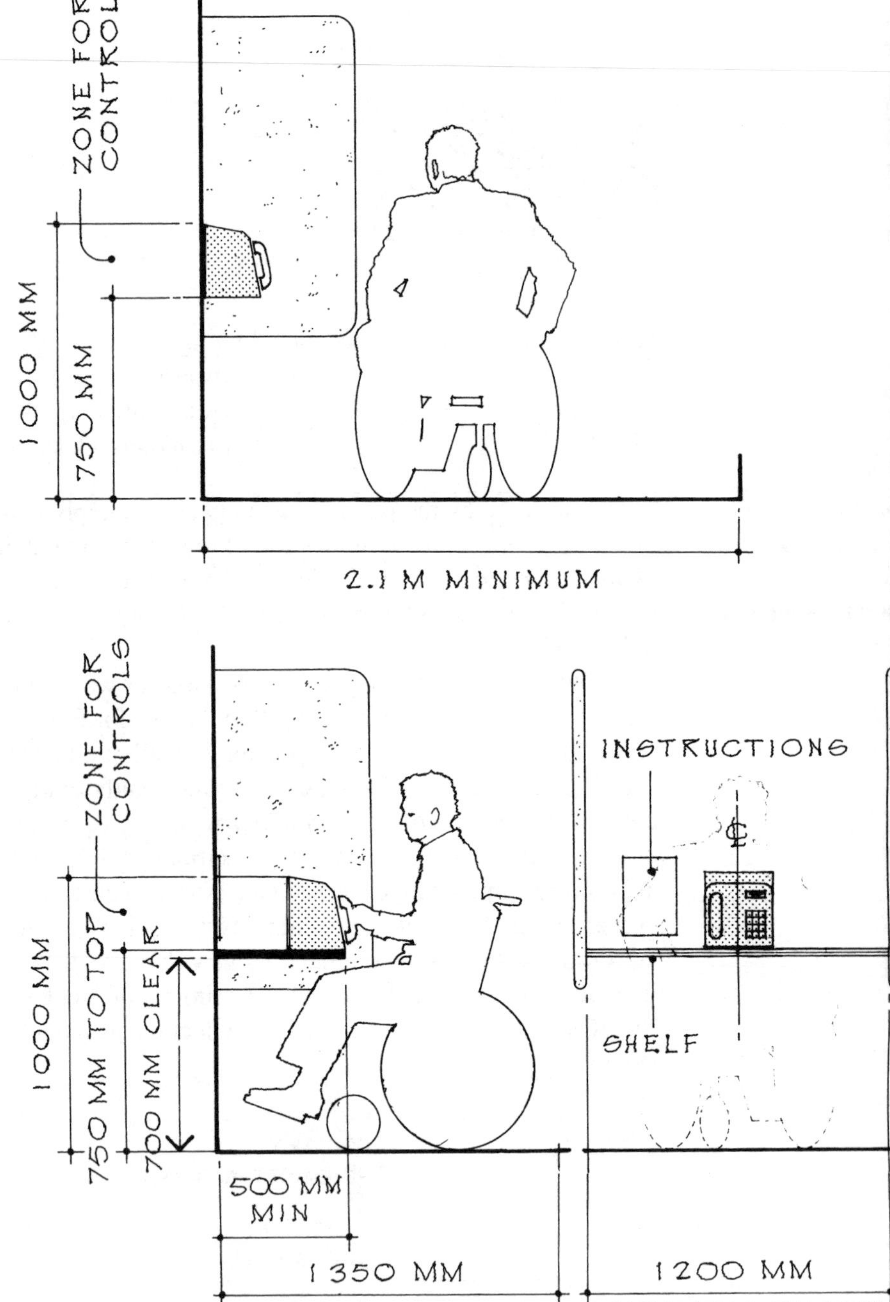

2 In the USA, where the Americans with Disabilities Act applies, refer to the ADA AND ABA ACCESSIBILITY GUIDELINES (qv).

Terminals and transport interchanges See: *Transport-related buildings.*

Terminology See: *Political correctness: speech codes.*

rrorism Act 2000	See: *Civil liberties*.
ssa	See: *Speech recognition*.
traplegia	See: *Paralysis*.
trapods	See: *Standing and walking frames*.
xt-to-speech	See: *Speech aids*.
xtDirect	See: *Telephone services for disabled people*.
xtphones	1 These are special telephones (sometimes called Minicom, though that is the brand name of a particular product) which enable people who are *deaf* (qv) or have *impaired speech* (qv) to communicate by typing and reading, rather than talking and listening. The appliance, which must be used by both participants, consists of a keyboard above which is mounted a small 'scrolling' display screen. Users dial each other just as with a conventional telephone. When a call is answered each participant types in what he/she wishes to say, and sees both outgoing and incoming text displayed on the screen. 2 For more detail visit the websites given under: *Telephone services for disabled people*.
xtured pavings	See: *Tactile pavings*.
eatres: complying ith the Disability iscrimination Act	See: *Auditorium design*; and *Service providers*.
eatres: elected case xamples	1 The Adapt Trust (qv) makes regular awards to arts venues that are considered to be especially meritorious in terms of overall *accessibility* (qv). Examples in the past few years have included the following. 2 Award-winning theatres: • Cambridge Arts Theatre, Cambridge www.cambridgeartstheatre.com/ • Mayflower Theatre, Southampton www.the-mayflower.com/ • Orange Tree Theatre, Richmond www.orangetreetheatre.co.uk/ 3 Highly commended theatres: • Norwich Playhouse, Norwich www.norwichplayhouse.co.uk/ • Plowright Theatre, Scunthorpe www.northlincs.gov.uk/plowright • Royal Lyceum Theatre, Edinburgh www.lyceum.org.uk/ • Theatre Royal, York www.theatre-royal-york.co.uk/ 4 For brief notes on selected cases see the following entries.
heatres: case xample 1	1 When Sadler's Wells, the famous theatre for opera and ballet in Finsbury, London EC1, was redeveloped in 1996–97 the owners grasped the opportunity to aim for full *accessibility* (qv) for all users, including those – such as *wheelchair-users* (qv) and *blind* people (qv) – for whom it is often thought impossible to offer equality in this particular building type. The briefing/design process, and the successes and failures of the final outcome, form an instructive case study for clients and designers contemplating a theatre project. They are usefully summarised on pp 12–15 of issue 74 of ACCESS BY DESIGN, the journal of the Centre for Accessible Environments – see the bibliography. 2 For more information visit www.sadlerswells.com/

Theatres: case example 2	1 While many theatres now try to ensure *accessibility* (qv) for *disabled* (qv) members of the audience, the 230-seat Rhoda McGaw Theatre at the Peacocks Centre in Woking is one of the relatively few to provide also for disabled performers. 2 For more information visit www.theambassadors.com/rhodamcgaw/
Theatres: case example 3	1 Despite replicating a 17th century original, Shakespeare's Globe Theatre on Bankside in Southwark, London SE1, is commendably wheelchair accessible. *Wheelchair-users* (qv) have level access all the way to box office and information desk, and have access to all parts of the auditorium. The standees-only 'yard' at ground level gives wheelchair users a very poor view, but views from the accessible Gentlemen's gallery are excellent. 2 For more information visit www.shakespeares-globe.org/
Theatres: case example 4	1 The Regent's Park Open Air Theatre in London NW1 has existed since 1932 but was completely redesigned to modern standards in the late 1990s and reopened in 2000. It is now a delightful and completely *accessible* (qv) venue, and an excellent case example of what can be achieved by thoughtful clients and designers. The scheme is described and evaluated on pp 12–16 of issue 84 of ACCESS BY DESIGN, the journal of the Centre for Accessible Environments – see the bibliography. 2 For more information visit www.openairtheatre.org/
Theory	See: *Déformation professionelle; Experts*; and *Ideology*
Therapeutic exercisers	See: *Exercise and rehabilitation aids.*
Thermoform displays	See: *Tactile displays.*
Think tanks	1 All political movements, policies, and legislation – including those behind the drive for a more *inclusive* (qv) society – start with ideas, and the past four decades have seen a proliferation of organisations created specifically for the purpose of breeding and publicising ideas. They are popularly known as 'think tanks'. 2 The first think tank – though it did not call itself that, and the term would not be coined until 1959 – was probably the Fabian Society, an intellectual movement founded in Britain in 1884 by Beatrice and Sidney Webb, George Bernard Shaw, and HG Wells for the purpose of researching, discussing, and publishing non-Marxist (i.e. evolutionary, not revolutionary) socialist ideas. Its founders believed in the power of patient argument, hoping to gradually permeate their ideas into the circles of those with power – a project that proved in the long run to be markedly successful. 3 This example was not followed on any substantial scale until the 1960s, whe many well-funded organisations began to be formed in the United States, whe they now play a major role in the development particularly of right-of centre ideas. The trend-setting example in Britain was the Centre for Policy Studies c CPS, which was founded in 1974 to champion economic liberalism, now more popularly known as 'Thatcherism'. 4 More such organisations are springing up every year and it is quite possible that the formative ideas, as yet unknown, which will shape major government policies and legislative programmes in Britain in the mid-21st century are bei hatched at this moment in some think tank. We should take note of what they are doing. The following are some of the better-known UK organisations as of

2008. Some of the descriptions are based on their own websites – caveat emptor!

• 18doughtystreet.com – Politics for Adults is one of the newest, but also (at time of writing) one of the best-resourced examples. It is centre-right, lively, and very ambitious. It is funded by Stefan Shakespeare, founder of the YouGov polling organisation. Visit http://18doughtystreet.com/

• The Adam Smith Institute concentrates mainly on the promotion of free market policies. It pioneered privatisation first in Britain under Mrs Margaret Thatcher, and then elsewhere. Visit www.adamsmith.org/

• The Bow Group exists to publish research and stimulate debate within the Conservative Party, and claims to represent all strands of conservative opinion. Notable recent publications include WASTED EDUCATION – see: *Neets*. Visit www.bowgroup.org/

• Catalyst is a think tank of the left which focuses on developing practical policies for the redistribution of wealth, power, and opportunity. It claims to be the only public policy think tank that is 'unequivocally committed to labour movement values and objectives', and in October 2003 won the 'One To Watch' prize at *Prospect* magazine's Think Tank of the Year Awards. Visit www.catalystforum.org.uk/

• The Centre for Policy Studies (CPS) is a centre-right think tank which – together with the Adam Smith Institute – helped convert the Tory Party to the principles of liberal economics (see above). The stated aim of its founders was to end what they regarded as thirty years of quasi-socialism, good intentions, and disappointment. No British think tank has come close to the influence exerted by the CPS between 1974 and 1979 under Keith Joseph and Alfred Sherman. Recent reports include HANDLE WITH CARE – see: *Children in Care*. Visit www.cps.org.uk/

• Civitas is an offshoot of the Institute for Economic Affairs (qv), and aims to deepen public understanding of the legal, institutional, and moral framework that makes a free and democratic society possible. A particular goal of its studies is the achievement of a better division of responsibilities between government and civil society. In 2006 Civitas was reported to have had the 4th highest number of media mentions of UK think tanks at 226, just behind Reform (qv) at 230. Visit www.civitas.org.uk/

• The Crime & Society Research Association does excellent work in highlighting the injustices suffered by the victims of crime, explaining the reasons behind the staggering rise in crime since the 1950s (see: *Crime*), and proposing sensible countermeasures. Visit www.csra.org.uk/

• Demos was launched in 1993 with the aim of helping to reinvigorate public policy and political thinking, and to develop radical solutions to long term problems. It has promoted a radical agenda of eye-catching, media-friendly policies, and its many reports and studies have looked into a notably wide range of topics. Visit www.demos.co.uk/

• Direct democracy calls for more local decision-making – see: *Localism*. The publication DIRECT DEMOCRACY: AN AGENDA FOR A NEW MODEL PARTY (see the bibliography) sets out its views on these matters. Visit www.direct-democracy.co.uk/

• The Fabian Society (see above) remains one of the UK's important left-of-centre think tanks. It is affiliated to the Labour Party, but is organisationally and editorially independent and welcomes everyone interested in progressive ideas. In 2005 most government ministers are members, and so are over 200 MPs. Visit www.fabian-society.org.uk/

• The Institute for Fiscal Studies (IFS) aims to provide independent economic analysis on matters of public concern (everything from NHS spending to pensioner poverty). Much of its good reputation was built up by its former director Mr Andrew Dilnot, who was willing to spell out truths that were inconvenient to governments. The IFS has been a winner of *Prospect* magazine's 'Think tank of the year' award. Visit www.ifs.org.uk/

• The Institute for Public Policy Research (IPPR) arose from the ashes of the Labour Party's defeat in the 1987 election and aimed to reinvigorate leftwing thinking. The IPPR was so close to Tony Blair that its director, Mr Matthew Taylor, became Prime Minister Blair's head of policy; two employees become cabinet ministers; and it played a key role in setting the Blair Government's direction. The IPPR has been a winner of *Prospect* magazine's 'Think tank of the year' award. Visit www.ippr.org.uk/

• The Institute of Economic Affairs (IEA) was founded in 1955 and is the UK's original free-market think-tank. Its goal is to explain free-market ideas to the public, including politicians, students, journalists, businessmen, academics and anyone interested in public policy. Visit www.iea.org.uk/

• The Institute of Race Relations (IRR) was established in 1958 as an educational charity to carry out research and to collect and publish information on race relations throughout the world. In 1972 it converted itself from being an information-collecting and disseminating organisation to an anti-racist think tank. Visit www.irr.org.uk/

• The Joseph Rowntree Foundation (qv) is the UK's largest independent social policy research and development charity (though some critics allege that its very substantial government funding is beginning to cast doubt on its independence – see: *Charities*). It supports a wide programme of research in housing, area regeneration, social care, disability, families, young people, work, incomes, and poverty. It also carries out practical, innovative projects in housing and care through the Joseph Rowntree Housing Trust. Visit www.jrf.org.uk/

• Localis forms part of Policy Exchange (qv), and is an independent research organisation that was created to develop new ideas for local government. www.policyexchange.org.uk/

• The King's Fund is a respected authority on hospitals and healthcare services, and its numerous reports on these matters are required reading – see for instance: NHS. Its publication SECURING GOOD SOCIAL CARE FOR OLDER PEOPLE (see the bibliography) was named Think Tank Publication of the Year in 2006 by *Prospect* magazine. Visit www.kingsfund.org.uk/

• Migrationwatch is a non-political body that is concerned about the present scale of immigration into the UK. It was noticeably shunned by the media until the Government was forced to admit that its immigration estimates were more accurate than the official figures. Since then it has come in from the cold, and in 2006 was reported to have had 310 media mentions, second only to the Taxpayer's Alliance at 529. Visit www.migrationwatchuk.org/

• The New Local Government Network (NLGN), named 'Think tank of the year 2004' by *Prospect* magazine, seeks to revitalise local political leadership and empower local communities. Visit www.nlgn.org.uk/

• The New Policy Institute (NPI), supported by the Rowntree Foundation (qv), produces authoritative statistics on *poverty* (qv) and *social exclusion* (qv) in the UK. Visit www.npi.org.uk/

• The Nuffield Trust researches policies for improving the quality of health care in the UK, and improving the health of the population. Its study AN INDEPENDENT NHS: A REVIEW OF THE OPTIONS (see: NHS) suggests that control of primary care and hospitals should be passed to an independent corporation similar to the BBC. Visit www.nuffieldtrust.org.uk/

• The Policy Exchange promotes ideas such as small-government and *localism* (qv). In 2005 its study UNAFFORDABLE HOUSING was named 'Pamphlet of the Year' by *Prospect* magazine, and in 2006 the organisation itself won the magazine's 'Think Tank of the Year' award. In 2007 the PE's founding chairman Michael Gove MP became Shadow Secretary of State for Children, Schools and Families, and its deputy director James O'Shaughnessy was hired as Leader of the Opposition David Cameron's policy adviser. It is a think tank to watch. Visit www.policyexchange.org.uk/

- The Policy Studies Institute (PSI) works to promote economic well-being and a better quality of life. Visit www.psi.org.uk/
- Politiea is especially concerned with the place of the state in people's daily lives. It deals with issues in the areas of pension provision and long-term care; strategies for high employment, labour regulation and law; the respective roles of state and parents in education; constitutional change; and the economic and political choices facing the country with regard to globalisation and the EU. Visit www.politeia.co.uk/
- Reform is a liberal think tank that was founded by the Conservative MP Nick Herbert, who as of 2007 is Shadow Secretary of State for Justice, but it is a cross-party organisation, with the Labour MP Frank Field and the Liberal Democrat MP Jeremy Browne serving on its advisory board. It plays a major role in the debate about the modernisation of the criminal justice, education and health services, and produces a daily media summary which the *Guardian*'s Polly Toynbee describes as 'an extremely useful resource for those with an interest in domestic politics'. In 2006 it was reported to have had the 3rd highest number of media mentions of UK think tanks at 230, after the Taxpayer's Alliance (qv) and Migrationwatch (qv). Visit www.reform.co.uk/
- The Smith Institute states that its current work focuses especially on 'the policy implications arising from the interactions of equality, enterprise, and equity'. Recent publications include TOWARDS A NEW CONSTITUTIONAL SETTLEMENT – see: *Constitution*. Visit www.smith-institute.org.uk/
- The Social Issues Research Centre (SIRC) aims to conduct research on positive aspects of social behaviour as well as the more problematic aspects that are the focus of most social science. Visit www.sirc.org/
- The Social Market Foundation (SMF) takes a particular interest in the market-based reform of state healthcare, education, and welfare provision. Visit www.smf.co.uk/
- The Taxpayer's Alliance campaigns for lower taxes (qv). In 2006 it was reported to have had the highest number of media mentions (529) of all UK think tanks. Visit www.taxpayersalliance.com/
- The Young Foundation is named after the late Michael Young, a remarkable thinker and doer who wrote the 1945 Labour manifesto, initiated the Open University, and founded 'Which?' Magazine. Notable recent publications include THE NEW EAST END: KINSHIP, RACE AND CONFLICT – see: *Human rights and social friction*. Visit www.youngfoundation.org.uk/

5 As of 2008, useful references include the following. For publication details see the bibliography.

- THE TOP TWELVE THINK TANKS IN BRITAIN by Toby Helm and Christopher Hope.

6 Useful websites

- The *Guardian* http://politics.guardian.co.uk/thinktanks/
- Sourcewatch www.sourcewatch.org/index.php?title=Think_tanks
- World Think Tank Directory www.policyjobs.net/World_Think_Tanks/

See also: *Ideology*; and *Pressure groups*.

---

hird sector See: *Voluntary and Community sector*.

---

hird sector: Office of See: *Office of the Third Sector*.

Thresholds: level

1 The traditional doorstep forms an awkward barrier for people in *wheelchairs* (qv), and tends to make buildings inaccessible to them. Current building regulations therefore prohibit stepped entrances in new buildings. Para 2.7(e) c APPROVED DOCUMENT M (qv) of the Building Regulations and para 6.2 of BS 8300 (qv) require building entrances to have level thresholds as illustrated below.

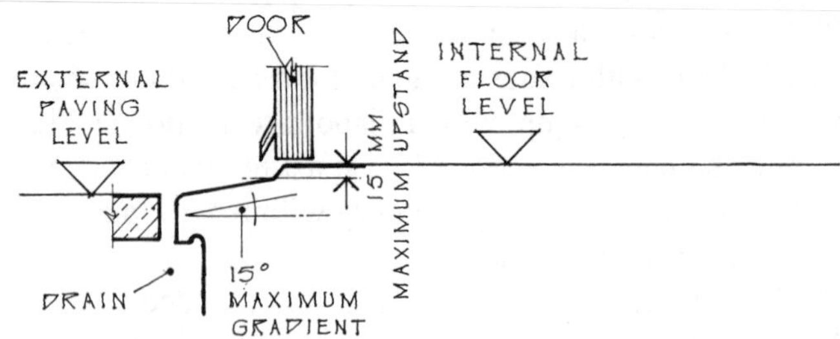

2 While level thresholds ensure greater *accessibility* (qv), they seriously reduce the defences of a building against flood waters. For more on this see the next entry.
3 As of 2008, useful references include the following. They are given in date order. For publication details see the bibliography.
• ACCESSIBLE THRESHOLDS IN NEW HOUSING: GUIDANCE FOR HOUSE BUILDERS AND DESIGNERS. This official guide was published in 1999.
• APPROVED DOCUMENT M (qv) of the Building Regulations was published in 2004.

Thresholds: level, and flood risk

1 A threshold detail such as that shown above almost certainly requires the internal floor level to be about the same as that of the ground outside, or at any rate no more than 40 mm or perhaps 50 mm above it. This departure from the traditional rule of raising the internal floor at least 150 mm or 200 mm above ground level coincides with two other developments:
• The onset in Britain of rising water levels and more frequent floods, both of which are likely to persist and perhaps worsen for as far ahead as we can foresee.
• Official proposals to meet the shortage of housing in southern England (see: *Houses: affordability*) by building large numbers of new houses on floodplains such as the Thames Gateway. In the latter case more than 90% of proposed development land lies in designated flood-plain areas.
2 This combination of factors is fairly alarming. How can the contradictory requirements of accessible entrances and safety from flooding be satisfactorily resolved? Official thinking appears to place great reliance on embankments and drainage systems to keep floodwaters out of entire sites, which is sensible, but such works will have to be proof against terrorist attacks or even bombardment in a war, both of which are possible in the next century, if they are to be relied upon.
3 Useful references include the following. For publication details see the bibliography.
• DEVELOPMENT AND FLOOD RISK: A PRACTICE GUIDE COMPANION TO PPS25 'LIVING DRAFT'. A CONSULTATION PAPER. This consultation paper was put out by the Department for Communities and Local Government (qv) in February 2007 for response by the end of August.
4 Useful websites include:

- Department for Communities www.communities.gov.uk/
- Planning Portal www.planningportal.gov.uk/

rough-floor lifts See: *Vertical platform lifts*.

asters See: *Appliances: domestic*; and *Products for easier living*.

ilets See: *Lavatories*.

ileting and bathing ds See: *Activities of daily living*; and *Products for easier living*.

lerance See: *Diversity; Political correctness*; and *Speech: freedom of*.

nal contrast See: *Colour in buildings*.

ucan crossings See: *Pedestrian crossings*.

uch

1 Touch is one of the five human *senses* (qv).
2 For *aids* (qv) that exploit the sense of touch to help *blind* (qv) or near-blind people, see *Tactile displays and images*; and *Touch tours*.

uch and proximity vitches

1 These are switches which the hand need only touch (or, in the case of proximity switches, approach within an inch or so) to switch on an appliance. They are more convenient for people with *impaired vision* (qv) or *arthritic* (qv) hands than conventional toggle switches which must be gripped.
2 As of 2008, useful websites include the following. They are given in alphabetical order.

- Disabled Living Foundation www.dlf.org.uk/
- Disabled Living Online www.disabledliving.co.uk/
- Let's automate www.letsautomate.com/
- QED www.qedltd.com/wolfson

See also: *Products for easier living*; and *Plate switches*.

uch lamps

1 These are lamps that are switched on and off by touching the lamp itself, not a separate switch, thus helping to solve the problem of people with poor eyesight being unable to turn on the light because they cannot locate the switch. Several brightness settings are possible, a common range being dim, medium, and bright.
2 As of 2008, useful websites include the following. They are given in alphabetical order.

- Asco Lights www.ascolights.co.uk/
- DwealTime www.dealtime.co.uk/
- Lighting4Uk www.lighting4uk.co.uk/
- The Lighting Superstore www.thelightingsuperstore.co.uk/

See also: *Products for easier living*; and *Plate switches*.

uch tours

1 There is a demand at *art galleries* and *museums* (qv) for touch tours and handling sessions, to give people who are *blind* (qv) or have severely *impaired vision* (qv) an opportunity to appreciate the displayed works.
2 In art galleries such as the Tate Modern in London, such a tour will typically include a sculpture that can be explored through direct handling, and a number of other two and three dimensional works that are explored using a combination

of *tactile images* (qv), handling objects, description, and discussion. In museums such as the National Maritime Museum in London, the tours will typically use a variety of real artefacts, authentic replica objects, and tactile displays such as raised (relief) drawings. Some also feature interesting food and smells.
3 Under the Disability Discrimination Act (qv) touch tours and similar services come under the heading of *auxiliary aids and services* (qv). The proper provision of such services will make it easier for gallery owners and managers, in their role as *service providers* (qv), to demonstrate that the requirements of the DDA have been met.
4 For commendable case examples visit the websites given under: *Exhibition spaces: complying with the Disability Discrimination Act.*

See also: *Audioguides; Virtual tours*; and *Vocaleyes.*

Tourism

See: *Travel and tourism-related buildings.*

Trade organisations and the Disability Discrimination Act

See: *Disability Discrimination Act.*

Traffic advisory leaflets

1 Traffic advisory leaflets from the Department for Transport give authoritative guidance on buses, cycling, home zones, intelligent transport services, parking, pedestrian crossings, traffic calming, traffic management, traffic lights, and traffic signs.
2 For more information visit www.dft.gov.uk/pgr/roads/tpm/tal

Traffic calming

1 This term refers to self-enforcing measures to reduce vehicle speeds – for instance the installation of road humps ('sleeping policemen'), chicanes, or rumble surfaces.
2 Proponents claim that such schemes can improve the environment and reduce accidents, but like many good ideas this one has become unhealthily sacrosanct. Filling up roads with humps is not always a good idea, and is sometimes a very bad one – for instance, the effect of humps upon a patient in a speeding ambulance can be excruciating – see THE IMPACT OF ROAD HUMPS ON VEHICLES AND THEIR OCCUPANTS.
3 As of 2008, useful references include the following. For publication details see the bibliography.

- THE IMPACT OF ROAD HUMPS ON VEHICLES AND THEIR OCCUPANTS.
- LIVING WITH RISK: PROMOTING BETTER PUBLIC SPACE DESIGN, PUBLISHED BY CABE (qv), studies ten public spaces and streets across England with the aim of helping designers challenge risk-averse decisions.

4 Useful websites include:

- Department for Transport www.dft.gov.uk/pgr/roads/tpm/tal/trafficmanagement/
- Seconds Count www.secondscount.org/

See also: *Home zones*; and *Streets for people.*

Traffic lights

See: *Pedestrian crossings*; and *Traffic advisory leaflets.*

Traffic management

Visit: www.dft.gov.uk/pgr/roads/tpm/

Traffic signs

1 THE TRAFFIC SIGNS MANUAL is an authoritative guide to the use of traffic signs and road markings prescribed by the Traffic Signs Regulations and covers the

whole of the UK. It is published in several parts – for more detail visit www.tsoshop.co.uk/bookstore.asp?FO=1167696
2 For other publications contact the Traffic Advisory Unit (qv).

See also: *Pedestrian crossings*; and *Traffic advisory leaflets*.

ain travel: guidance or disabled people See: *Door to Door*.

ain travel: statistics 1. The graph below shows that the number of train journeys in Great Britain declined slightly from 1970 to 1983, and has since been modestly growing. At date 'A' council-run bus services were privatised by the 1985 Transport Act, and at date 'B' the railways were split up and privatised by the 1993 Railways Act.

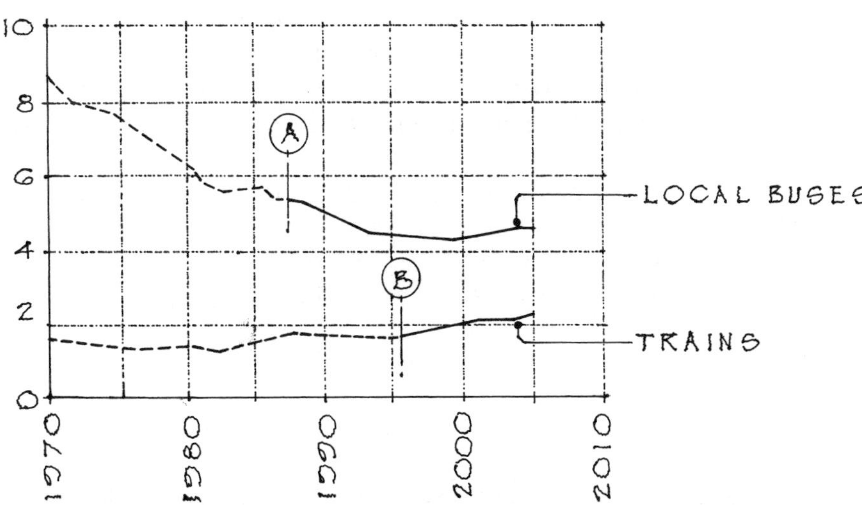

For the source visit http://www.statistics.gov.uk/cci/nugget_print.asp?ID=1094
Figure 12.14 in SOCIAL TRENDS 37 (qv) was published after this was written, and gives more recent data.

See also: *Transport statistics*.

raining and Development Agency for Schools (TDA) 1. The TDA is one of a proliferating tangle of government departments and *quangos* (qv) coordinating and overseeing the Government's programme of vocational courses for young people. Others include the Department for Children, Schools and Families (qv): the Qualifications and Curriculum Authority (qv); the Quality Improvement Agency (qv); the Learning and Skills Council (qv); sector skills councils; and more. For a comment see: *Learning and Skills Council*.
2. For more information visit www.tda.gov.uk/

rains, trams See: *Rail vehicles*.

ramstops 1 For design guidance refer to para 4.4 and section 10 of INCLUSIVE MOBILITY (qv) – see the bibliography.
2 As of 2008, useful websites include:
- DPTAC Access Directory www.dptac.gov.uk/adnotes.htm
- DPTAC Publications www.dptac.gov.uk/pubs.htm

See also: *On-street platforms*; and *Transport: complying with the Disability Discrimination Act*.

Transformational government strategy

1 This ambitious programme, announced in 2005, was intended to make all government services much more personalised via the use of *IT* (qv).
To this end the Government started creating a web of databases such as *ContactPoint* (qv), the *National DNA Database* (qv), the *National Identity Register* (qv), and the *NHS Care Records Service* (qv) to acquire and store every detail of citizens' personal lives. These databases massively invade our right to control information about ourselves. Yet we were never consulted, or even told much, about their creation; are not being asked for permission to have our personal details entered; and have in most cases no right to opt out – affronts made all the worse by the gross insecurity of these information stores (see: *Privacy*). In short, an initiative that promised to expand civil liberties has in some important respects become an Orwellian machine for destroying them.
3 As of 2008, useful references on these matters include the following. For publication details see the bibliography.
• TRANSFORMATIONAL GOVERNMENT: ENABLED BY TECHNOLOGY. This is the original vision as outlined by the Cabinet Office (qv) in 2005.
• DATA PROTECTION WON'T HELP ONCE ALL THE DATA IS GONE by Christina Zaba was published in the *Guardian* on 27 November 2007.
• WHO LOST OUR DATA EXPERTISE? by Michael Cross was published in the *Guardian* on 29 November 2007.
• WHO DO THEY THINK WE ARE? by Jill Kirby.

See also: *IT systems*; *Post-democracy* and *Privacy*.

Transport 2010

A failed and now-abandoned scheme – for details visit www.dft.gov.uk/

Transport: complying with the Disability Discrimination Act

1 The Disability Discrimination Act (qv) effectively outlaws *discrimination* (qv) against *disabled people* (qv) in public life. Private forms of transport such as cars are not subject to the Act. Public transportation is in principle subject to the Act but coverage is partial, and proceeding slowly and incrementally. The following notes briefly summarise a complex situation:
2 The DDA deals with transport-related matters in three broad categories:

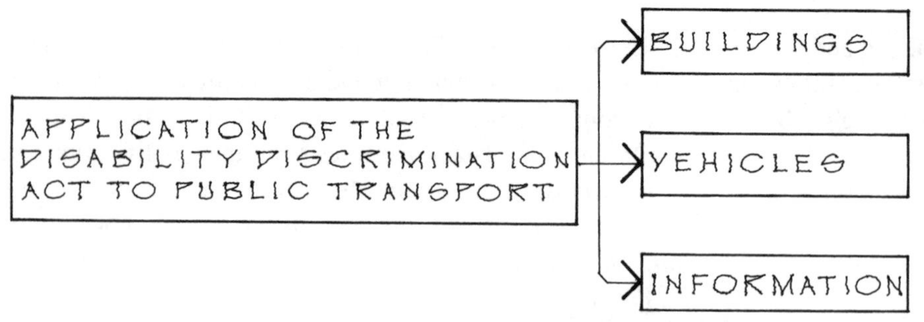

3 Built facilities (bus stops; railway stations; passenger terminals for ships and aircraft; etc.) are fully subject to the provisions of the Disability Discrimination Act. For more detail see: *Transport-related buildings*.
4 Vehicles are, as of 2008, only partly subject to the provisions of Disability Discrimination Act, part 5 of which makes provision for anti-discrimination regulations (qv) to be passed for the following three sets of land-based transport vehicles (but not, at present, aircraft, ships or ferries):
• Taxis (qv).
• Public service vehicles (qv), i.e. buses and coaches.

• Rail vehicles (qv).
The purpose of such regulations is to ensure that disabled people can get on and off vehicles safely and without unreasonable difficulty; and that they are carried safely and in reasonable comfort. Regulations to this effect are coming into effect piecemeal and very slowly, and the end date for compliance by all vehicles may be as late as 2020. For more detail see the following individual entries:

• Taxis.
• Buses and coaches.
• Rail vehicles.
• Public Service Vehicle Regulations 2000.

5 Information provision to travellers is fully subject to the provisions of the Disability Discrimination Act, which requires transport providers to supply disabled persons – for instance people with *impaired hearing* (qv), *impaired vision* (qv), or *impaired understanding* – with information in a format that they can use and understand, at no extra cost and with no delay. Partial guidance will be found in sections 6, 7, 9, and 10 of INCLUSIVE MOBILITY (qv) – see the bibliography.

6 The present work only gives simplified summaries of original documents and does not purport to provide full, authoritative and up-to-date statements of the law. For latest information refer to DISABILITY DISCRIMINATION: LAW AND PRACTICE (see the bibliography) and the following websites:

• Directgov: disabled people www.disability.gov.uk/
• Equality & Human Rights Commission www.equalityhumanrights.com/

7 Other useful references include:

• CODE OF PRACTICE. PROVISION AND USE OF TRANSPORT VEHICLES: STATUTORY CODE OF PRACTICE: SUPPLEMENT TO PART 3 CODE OF PRACTICE. This is an essential reference in the application of the Disability Discrimination Act.

8 Other useful websites include:

• Dept FOR Transport www.dft.gov.uk/transportforyou/access/dda2005/
• Dept for Transport mobility and inclusion unit www.dft.gov.uk/transportforyou/access/miu/
• DPTAC Access Directory www.dptac.gov.uk/adnotes.htm
• DPTAC Publications www.dptac.gov.uk/pubs.htm
• Equality & Human Rights Commission www.equalityhumanrights.com/

See also: *Bus and coach operation; Rail travel; Rail vehicles*; and *Taxis.*

Transport: guidance for disabled people

See: *Door to Door.*

Transport providers

1 'Transport providers' are a special category of *service providers* (qv) under the Disability Discrimination Act. Their duties are set out in CODE OF PRACTICE: PROVISION AND USE OF TRANSPORT VEHICLES: STATUTORY CODE OF PRACTICE: SUPPLEMENT TO PART 3 CODE OF PRACTICE. For publication details see the bibliography.

See also: *Bus and coach operation; Rail travel; Rail vehicles*; and *Taxis*

Transport: public

See: *Transport: complying with the Disability Discrimination Act.*

Transport-related buildings: basic planning data

1 Distilled planning data for most of the major building types is given in the METRIC HANDBOOK: PLANNING AND DESIGN DATA. For publication details see the bibliography.

2 For general design guidance, commentary, and case examples visit:

• CABE www.cabe.org.uk/default.aspx?contentitemid=45

Transport-related buildings: design: complying with the Disability Discrimination Act

1 Buildings used by the public must in general comply with the Disability Discrimination Act (qv). In the case of railway, bus and coach stations; underground railway and rapid transit stations; airports and terminals; sea terminals, and motorway services, *authoritative practical guides* (qv) include DESIGNING FOR ACCESSIBILITY (qv) plus the following:

• BS 8300, the provisions of which apply to all buildings in the UK. Para 13.1 gives brief design recommendations that are relevant to all the building types listed above. It is an advisory document.

• INCLUSIVE MOBILITY (qv), which is the most authoritative reference for the *inclusive design* (qv) of pedestrian infrastructure and transport-related buildings such as bus and coach stations, railway stations, air terminals, and transport interchanges. It is an advisory document.

These references do not have the force of law, but conformity with their recommendations will make it easier to demonstrate that the requirements of the Disability Discrimination Act have been met.

Transport safety

1 The figure below, from Table 12.18 in SOCIAL TRENDS 37 (qv), gives comparative passenger death rates for Great Britain.

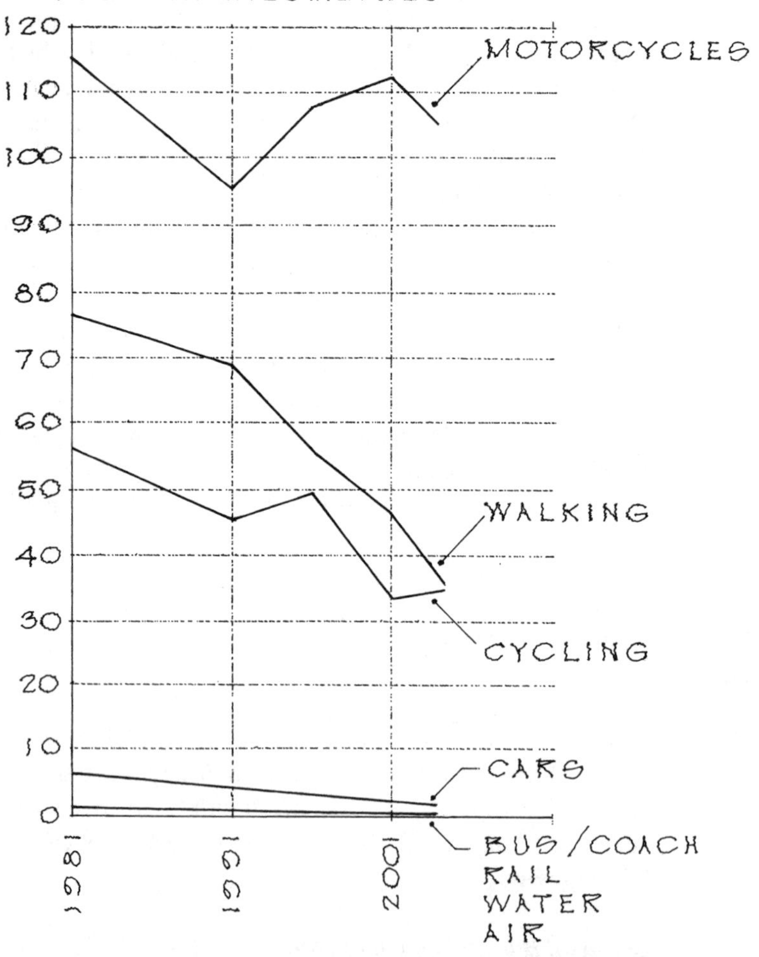

2 When interpreting death rates per passenger kilometre it should be borne in mind that in one hour a walker might travel 5 km, a cyclist 20 km, a motorcycle rider 100 km, and an aircraft passenger 800 km. The death rate per hour of travel, which may be the consideration that is uppermost in travellers' minds, would therefore be very different from the figures in the graph.

3 As of 2007 useful websites include:

• HSE www.hse.gov.uk/workplacetransport/

• Office for National Statistics	www.hse.gov.uk/workplacetransport/
• Pacts	www.pacts.org.uk/
• Rail Safety and Standards Board	www.rssb.co.uk/
• RoSPA	www.rospa.org.uk/roadsafety/index.htm

See also: *Accidents; Office of Rail Regulation*; and *Safety*.

Transport statistics

Visit www.dft.gov.uk/pgr/statistics/datatablespublications/tsgb/

Transport strategy for London

Visit www.london.gov.uk/approot/mayor/strategies/transport/index.jsp

Travel

1 According to Table 12.1 of SOCIAL TRENDS 37 (qv), passenger journeys within Great Britain in 2005 came to the following totals per mode, in billions of passenger kilometres:

• Car and van	678
• Rail	52
• Bus and coach	48
• Air	10
• Motorcycle	6
• Bicycle	4

2 The data in the graph below, from Figure 12.1 of social trends 37, have been converted from passenger kilometres travelled to an index, in order to illustrate the different relative growths of particular modes of transport. Thus, although air travel shows the greatest percentage growth since 1961, the 10 billion passenger kilometres travelled by air in 2004 represent only 1% of all passenger kilometres travelled within Great Britain in that year.

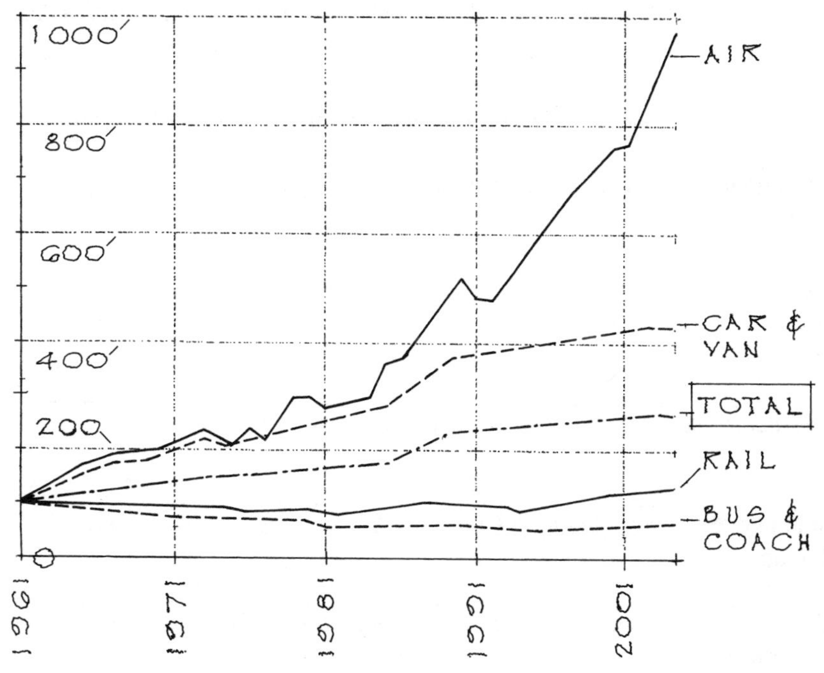

3 For precise definitions on the above matters, and further information, refer to the original documents, publication details of which are given in the bibliography.

Travel: accessibility guides

1 As of 2008, useful references include the following. For publication details see the bibliography.

• THERE AND BACK: A TRAVEL GUIDE FOR DISABLED PEOPLE.

2 For related information see: *Accessible roads*; and *Door to door*.

Travel and tourism-related buildings: basic planning data	1 Distilled planning data for most of the major building types are given in the METRIC HANDBOOK: PLANNING AND DESIGN DATA. For publication details see the bibliography. 2 For general design guidance, commentary, and case examples visit: • CABE www.cabe.org.uk/search.aspx?type=0
Travel and tourism-related buildings: planning: complying with the Disability Discrimination Act	1 Buildings used by the public must in general comply with the Disability Discrimination Act (qv). In the case of hotels, motels, hostels and residential clubs; bed and breakfast guest accommodation; self-catering holiday accommodation; and accommodation providing holiday care, *authoritative practical guides* (qv) include DESIGNING FOR ACCESSIBILITY (qv) plus the following: • APPROVED DOCUMENT M (qv), the provisions of which must be heeded in order to comply with the Building Regulations for England and Wales. Paras 4.4 and 4.17–24 refer specifically to bedrooms in hotels, motels, and student accommodation, etc. • BS 8300, the provisions of which apply to all buildings in the UK. Para 13.11 gives brief design recommendations for all the building types listed above. BS 8300 does not have the force of law, but conformity with its recommendations will make it easier to demonstrate that the requirements of the Disability Discrimination Act have been met. In places other than England and Wales the following regulatory documents should be consulted: • In Northern Ireland: TECHNICAL BOOKLET R of the Building Regulations (Northern Ireland) (qv). • In Scotland: the TECHNICAL HANDBOOKS of the Building (Scotland) Regulations (qv). 2 In the USA, where the Americans with Disabilities Act applies: refer to the ADA AND ABA ACCESSIBILITY GUIDELINES (qv). See also: *Hotels and motels; Service providers; and Tourism.*
Travel and tourism-related buildings: management: complying with the Disability Discrimination Act	See: *Service providers.*
Travel and wheelchair-users	See: *Wheelchair-users and public transport.*
Travel: guidance for disabled people	See*: Door to Door.*
Travel information: complying with the Disability Discrimination Act	See: *Transport: complying with the Disability Discrimination Act.*
Travelators	1 Travelators, also called *moving walkways* or *autowalks*, are moving surfaces which are very like *escalators* (qv) except that they have no steps, and remain flat and level. Their purpose is to reduce walking distances, particularly for

people burdened with luggage, and their most common use is in *airports* (qv) and other transport terminals.
2 INCLUSIVE MOBILITY (qv) comments that travelators can be very helpful where there are long distances to be traversed, with the proviso that a parallel walkway should always be provided adjacent to the moving surface.
3 For some people, particularly older people who are a little unsteady on their feet, stepping onto a moving walkway is an uncomfortable experience. The speed of movement should therefore be kept low – preferably 0.5 m/second, and certainly not more than 0.75 m/second. The surface should be non-slip, and there should be clearly visible emergency stop switches that can be reached and operated by disabled people. An audible warning at the beginning of the travelator, and another prior to the end, are essential for people with *impaired vision* (qv). There should be an unobstructed level 'run-off' at each end of the travelator, extending for at least 6 metres.
4 Travelators' direction of travel should be clearly indicated. The footway at both ends should be marked by *colour contrast* (qv) and by a change in floor finish. The travelator must be well lit, particularly at its entrance and exit. Moving *handrails* (qv) should be rounded in section, and have a colour contrast with the background, and they should extend approximately 700 mm beyond the beginning of the walkway.
5 Recommended travelator width is 1 500 mm, and minimum overhead clearance is 2 300 mm.The side panels of the travelator channel should be finished in a non-reflective surface, and back illuminated side panels can be very disorientating.
6 Useful websites include the following. They are given in alphabetical order.

- Lift and Escalator Industry Association www.leia.co.uk/
- Kone www.kone.com/
- Otis www.otis.com/
- Schindler www.schindlerlifts.co.uk/

See also: *Escalators*; and *Passenger conveyors*.

Treaty Establishing A Constitution for Europe (TECE)

1 This Treaty, which is meant to bring together and replace the many overlapping previous EU treaties and take the EU one further step along the road of ever-closer union, has gone though two stages:
2 The Treaty as first drafted was signed in Rome in October 2004 by representatives of the 27 member states, and had then to be ratified by those states following consultations with their electorates. Fifteen states ratified the TECE, but French voters rejected it in a referendum May 2005, and Dutch voters in June 2005, whereupon the matter was put on hold.
3 In 2007 a Reform Treaty was formulated, the nature of which was described thus by various European leaders:

- Angela Merkel, the German Chancellor: 'The substance of the Constitution is preserved. That is a fact'.
- The Spanish Prime Minister Jose Zapatero: 'We have not let a single substantial point of the Constitutional treaty go'.
- Vaclav Klaus, the Czech President: 'Only cosmetic changes have been made and the basic document remains the same'.
- Margot Wallstrom, EU Commissioner: 'It's essentially the same proposal as the old constitution'.

4 The problem then arose of how to persuade electorates to vote for a virtually unmodified regurgitation of a document they had earlier rejected. Senior EU figures were refreshingly honest about their tactics:

- On 17 July 2007 Mr Valéry Giscard d'Estaing, architect of the original constitution, explained to a London conference how this would be done: 'All

the earlier proposals will be in the new text, but will be hidden and disguised in some way'. Further: 'What was [already] difficult to understand will become utterly incomprehensible, but the substance has been retained'.
• Jean Claude Juncker, Prime Minister of Luxembourg, said: 'Of course there will be transfers of sovereignty. But would I be intellligent to draw the attention of public opinion to this?'
5 As of October 2007, it appears that the Treaty will now be imposed upon us without our consent. This is seriously wrong. In his COMPARATIVE STUDY OF REFERENDUMS: GOVERNMENT BY THE PEOPLE (see the bibliography) Professor Mads Qvortrup quotes the great English constitutional lawyer A V Dicey (1835–1922), whose works are accepted as a cornerstone of the British *constitution* (qv), as arguing that referenda are 'the people's veto' to check Parliament when significant constitutional changes are proposed. This principle was confirmed when the question of devolution was put to the electors in Scotland and Wales before being implemented by Parliament, and again when on 20 April 2004 Prime Minister Tony Blair told the House of Commons that it should debate the European constitutional question and 'then let the people have the final say'. The following year he inserted a promise of a referendum on the European constitutional treaty in the Labour Party's 2005 manifesto – a commitment without qualifications or escape clauses. The constitutional treaty has since been replaced by the reform treaty, which is almost identical. The promised referendum should be held.
6 Useful references include the following:
• Christopher Booker's Notebook, published each week in the *Sunday Telegraph*, provides well-informed regular commentaries.
• The old and the new Treaty texts are displayed side-by-side for easy comparison at www.openeurope.org.uk/research/comparative.pdf
• Open Europe www.openeurope.org.uk/
• Open Europe www.openeurope.org.uk/research/rhetoric&reality.pdf
• Wikipedia http://en.wikipedia.org/wiki/Reform_Treaty

See: also: *Constitution*; and *Post-democracy*.

Tribunals

1 Tribunals are bodies with the function of deciding claims and disputes arising in connection with the administration of *laws* (qv), *regulations* (qv), and the like.
2 Examples of tribunals that may be relevant to users of the present work include:
• Employment tribunals may be called upon to hear disputes between employers and *disabled* (qv) employees, or trade organisations and *disabled* members, under the Disability Discrimination Act (qv). Visit www.employmenttribunals.gov.uk/default.asp
• Special educational needs and disability tribunals (SENDISTs) may be called upon to hear disputes between schools and *disabled* pupils – see: *Special educational needs*. Visit www.sendist.gov.uk/
3 The workings of all tribunals in England, Scotland and Wales are supervised by a *quango* (qv) called the Council on Tribunals – visit www.council-on-tribunals.gov.uk/

See also: *Civil proceedings; Disability Discrimination Act*; and *Special educational needs*.

Triplegia

See: *Paralysis*.

Tunnels and underpasses: complying with the Disability Discrimination Act

1 Facilities used by the public must in general comply with the Disability Discrimination Act (qv). In the case of tunnels and underpasses, *authoritative practical guides* (qv) include the following documents. For publication details see the bibliography.

• INCLUSIVE MOBILITY (qv), which is the most authoritative reference for the *inclusive design* (qv) of *pedestrian infrastructure* (qv) and *transport-related buildings* (qv) such as bus and coach stations, railway stations, air terminals, and transport interchanges. It does not have the force of law, but conformity with the recommendations in section 8.4.6 will make it easier to demonstrate that the requirements of the Act have been met.

2 A wider problem is the inability of many *disabled people* (qv), including all *wheelchair-users* (qv), to negotiate the steps down to an underpass – for instance to cross a road or reach another platform at a rail station. This problem makes it impossible for many of them to undertake journeys by public transport. The ideal solution is to provide a *vertical platform lift* (qv), but these are expensive – see: *Footbridges*.

See also: *Footbridges; Pedestrian environments; Rail travel: accessibility*; and *Transport-related buildings*.

Turning circle

See: *Wheelchair: turning circle*.

TV outlets: heights above floor

See: *Controls and switches*.

Two-tier local government

See: *Local authorities*.

Typetalk

See: *Telephone services for deaf people*.

U

UK Home Care Association (UKHCA)

1 UKHCA is the professional association of home care providers from the independent, voluntary, not-for-profit, and statutory sectors.
2 For more information visit www.ukhca.co.uk/

See also: *Carers*.

UK Statistics Authority

1 This body was created in April 2008 to restore confidence in official statistics by (a) overseeing the Office for National Statistics (qv); and (b) monitoring and assessing all official statistics. Visit www.statistics.gov.uk/
2 See also: *Statistics Board*.

UltraCane

See: *Canes for blind people*.

Underclass

1 This term refers not to people who live in *poverty* (qv) – though many of them undoubtedly do – but to ones who essentially live outside mainstream society, sharing few of those common values such as feeling one ought if possible to behave well, to be educated, to have a job, and to nurture a well-educated and responsible family, that bind most of society together. They form only a small minority of the population overall, but loom large in particular areas, for instance some *housing estates* (qv), where their lack of civility wrecks the peace of mind and quality of life of everyone around. As they tend constantly to be in trouble, their own lives are probably no happier.
2 The existence of an underclass has become a truly major social problem only since about the 1980s, and while there have been many contributing factors, everyone who has looked into the matter agrees that *family* (qv) breakdown is at the root. As this phenomenon, in turn, depends on private decisions taken by millions of individual people, governments cannot directly change the situation. But that is not to say that nothing can be done. Here are three matters that come directly under government control, could be readily implemented, and might help to nudge things in the right direction:
• First, change the bizarre tax rules that actually penalise couples for living together. All published studies conclude that children thrive best, and are most effectively socialised, if brought up in stable two-parent families. The report WELFARE ISN'T WORKING: CHILD POVERTY, co-written by the Labour MP Frank Field, charges that the current *tax credit* (qv) and benefit strategy 'brutally discriminates against two-parent families' and actively discourages single parents from forming stable relationships. He calls for the tax system to be equalised for all parents. For more on this see: *Children: single parent*.
• Second, give school heads absolute authority to exclude intolerably disruptive pupils. Such children, many of them from the current underclass, effectively wreck the education of their class-mates, thus ensuring that many of the latter will form part of the next generation of an uneducated, unemployable, crime-prone layer of society. For more on this see: *Schools: exclusion of pupils*.
• Third, start applying the zero-tolerance policies that are described in THE GREAT AMERICAN CRIME DECLINE: STUDIES IN CRIME AND PUBLIC POLICY by Franklin E Zimring, in order to bring lawlessness under control. Such policies helped to cause a three-quarters drop in youth violence in New York after 1990, and in Britain similar policies cut stranger killings in Manchester from 37 in 1999 to 5 in 2005 For more on this see: *Crime*.
3 As of 2008, useful references on this important and difficult subject include the following. They are given in date order. For publication details of these, and of the references mentioned above, see the bibliography:
• THE DREAM AND THE NIGHTMARE: THE SIXTIES' LEGACY TO THE UNDERCLASS by Myron Magnet. An instructive and fundamentally optimistic book.
• THE EMERGING BRITISH UNDERCLASS by Charles Murray was published in 1990 by the *think tank* (qv) the Institute of Economic Affairs.

• CHARLES MURRAY AND THE UNDERCLASS: THE DEVELOPING DEBATE by Charles Murray with commentaries by Ruth Lister, Frank Field, and others. This report was published in 1996 by the think tank Civitas.
• UNDERCLASS + 10: CHARLES MURRAY AND THE BRITISH UNDERCLASS 1990–2000 by Charles Murray with a comment by Melanie Phillips. This is an update of the work above and was published in 2001.
• SOCIAL TRENDS (qv) is an annual publication from the Office for National Statistics (qv) that draws together key statistics from all government departments and many other organisations to paint a broad picture of British society. It is an essential reference.
4 Useful websites include:
• Civitas www.civitas.org.uk/
• Frank Field www.frankfield.co.uk/type3.asp?id=20&type=3
• Office for National Statistics www.statistics.gov.uk/
• Reform www.reform.co.uk/site/search.aspx?q=underclass

See also: *Children: single parent; Crime; Families; Housing estates; Literacy; Neets*; and *Workfare*.

Underpasses See: *Tunnels and underpasses*.

Underground stations See: *Transport-related buildings*.

Understanding See: *Impaired understanding*.

Unemployment See: *Employment; Incapacity benefit; Jobseeker's allowance; Literacy; Social exclusion*; and *Workfare*.

Unemployment benefit See: *Jobseeker's allowance; New deal*; and *Tax credit*.

Unemployment: youth See: *Neets; New deal*; and *Workfare*.

Unitary councils See: *Local authorities*.

Universal Declaration of Human Rights (UDHR)

1 The UDHR is an advisory declaration adopted on 10 December 1948 at Palaise de Chaillot. The main text starts by proclaiming 'this Universal Declaration of Human Rights is a common standard of achievement for all peoples and all nations, to the end that every individual and every organ of society, keeping this Declaration constantly in mind, shall strive by teaching and education to promote respect for these rights and freedoms and by progressive measures, national and international, to secure their universal and effective recognition and observance, both among the peoples of Member States themselves and among the peoples of territories under their jurisdiction'. This general opening statement is followed by 30 specific articles.
2 For the full text visit www.unhchr.ch/udhr/lang/eng.htm
3 Useful websites include:
• Wikipedia
http://en.wikipedia.org/wiki/Universal_Declaration_of_Human_Rights

See also: *Constitution; European Convention on Human Rights; Human rights*; and *Human Rights Act*.

Universal design See: *Inclusive design*.

Unisex wcs See: *Lavatories*.

Upstand kerbs See: *Kerbs*.

Urban design 1 Urban design is the art of designing buildings, streets, parks, and squares to form a habitat that is safe, convenient, and delightful. At its best, as seen in many of the surviving town and city centres of old Europe and in a few modern places such as New York's Rockefeller Plaza, such design creates a 'theatre of public life', in which existence is raised from everyday practicality to conscious delight. And good urban design aims, of course, at *inclusion* (qv) – ie at creating such environments for the whole of the population, not just a favoured elite.

2 The observation made under *Architecture* (qv), that safe, comfortable buildings without beauty sell the public short, applies with even greater force to the assemblages of buildings that make up our *streets* (qv) and other public spaces. So does the suggestion, in the earlier entry, that the way forward is a revival of the thoroughly proven and much-loved classical vernacular style for most of our buildings, leaving experimental modern styles for those particular clients and users who actually want them. The following notes form a brief supplement to the earlier remarks.

3 The most important part of every building is the façade. What happens behind the façade affects only the occupants and need not please anyone else. Facades are seen by everyone and ought, in a civilised townscape, to give pleasure to everyone. The classical tradition recognises these distinct functions and deals with the façade and the spaces behind the façade as separate entities – for instance, in his Regent Park (1812-28) and Carlton House terraces (1827–32) John Nash designed the street frontages and was content to leave the interiors to others. The modernist dogma that the exterior of the building must 'express' the interior, and on no account be independently composed, can find little justification in history and none in common sense; and its application to building design has done dreadful damage to our townscapes.

4 Turning, then, to façades, the classical vernacular style enables all building designers, from talented architects to ordinary builders, to compose building exteriors that please the eye both individually and as groups. As one example, the series of neo-classical façades on either side of the Royal Institute of Painters in Watercolour (1881) on the south side of Piccadilly, London, demonstrate the application of a visual language that permits buildings of powerful self-expression and almost infinite diversity to stand side-by-side harmoniously, rather as good manners permit people with hugely diverse interests and personalities to co-exist socially.

5 Not only has modernism discarded this visual language, it has also made good street architecture all but impossible by replacing the verticality of classical design with the horizontality that is typical of current buildings. Buildings composed of repetitive, closely-spaced verticals can fluidly follow curved or irregular streetlines, and can be neatly tucked into small or oddly-shaped spaces. Horizontal composition can be highly satisfactory in the case of pavilion-like buildings on open land, but does not suit most urban sites, where such buildings ignore or obliterate street curves instead of respecting them, and either collide awkwardly with their neighbours or stand away from them, thus creating those rubbish-collecting left-over spaces that so disfigure modern cities. As a final point, traditional vertically-composed buildings encourage graceful, expressive skylines as the verticals of the façade are continued upwards into the spires, finials, and other graceful features that create the eye-pleasing skylines of classical architecture. Horizontally-composed buildings most naturally terminate with flat roofs that stand dumbly against the sky, or brutally punch upward like a fist, producing skylines that no-one loves.

6 The New Urbanism, a movement associated particularly with the Swiss architect Leon Kriér, represents one attempt to improve matters. Its practitioners in the USA include Andrés Duany, the architect behind the Seaside settlement in Florida; and in Britain Prince Charles, who has repeatedly shown himself to be a more reliable spokesman for millions of ordinary citizens than the high priests of modern architecture who – with no democratic mandate whatever – advise ministers on government policy in respect of planning and design. Early experimental examples of the New Urbanism in Britain include.

• Poundbury, a development that was commissioned in 1988 by Prince Charles, and is currently taking shape on 400 acres of Duchy of Cornwall land on the western edge of Dorchester. It has mixed-use neighbourhoods, a village square, winding streets, traditional houses built from local stone and timber, and obeys urban design codes to ensure a harmonious style. It is roughly a third complete, with a total of 2 4000 homes planned; and forms an interesting contrast with modernist experiments such as Harlow new town (started 1947), Cumbernauld (started 1955), and Milton Keynes (started 1970).

• Tornagrain, near Inverness in Scotland, where the development of a new town of 5 000 homes was announced in July 2006.

• Coed Darcy Urban Village, a proposed new town that is projected to have 4 000 homes constructed on the site of what was once a BP oil refinery.

7 As of 2008, websites in relation to the New Urbanism include the following. They are given in alphabetical order.

• Coed Darcy www.neath-porttalbot.gov.uk/
• The Congress for the New Urbanism www.cnu.org/
• The New Urbanism www.newurbanism.org/
• Poundbury www.poundbury.info/
• The Prince's Foundation www.princeofwales.gov.uk/
• Tornagrain new town www.tornagrain-newtown.co.uk

8 Websites on urban design in general include:

• Academy of Urbanism www.academyofurbanism.org.uk/
• CABE www.cabe.org.uk/default.aspx?contentitemid=42
www.cabe.org.uk/default.aspx?contentitemid=44
www.cabe.org.uk/default.aspx?contentitemid=45

Department for Communities etc. www.communities.gov.uk/
Guardian http://society.guardian.co.uk/urbandesign/page/0,,602114,00.html
Joint Centre for Urban Design at Oxford Brookes University
www.brookes.ac.uk/schools/be/jcud/
Prince's Foundation for the Built Environment
www.princes-foundation.org/
RIBA Urbanism & Planning Group www.riba.org/
Royal Town Planning Institute www.rtpi.org.uk/
Urban age www.urban-age.net/
Urban Design Alliance www.udal.org.uk/
Urban Design Group www.udg.org.uk/
Urban Forum www.urbanforum.org.uk/
Urban Policy Unit www.neighbourhood.gov.uk/page.asp?id=643
Urban Task Force www.lga.gov.uk/lga/planning/urbantaskforce.htm
Urbed www.urbed.com/

See also: *Architecture; Déformation professionelle; Ideology*; and *Le Corbusier*.

Urban development corporations	See: *English Partnerships*.
Urban green spaces	See: *CABE Space*; and *Greenspace*.
Urban regeneration	See: *Neighbourhood renewal*.
Urban Regeneration Agency	See: *English Partnerships*.
Urban spaces: live auditing	See: *Access audits: live*.
Urban White Paper	Visit: www.communities.gov.uk/index.asp?id=1127168

V

aluing people	Visit: http://valuingpeople.gov.uk/index.jsp
ariable-height urfaces	See: *Tables*; and *Worktops*.
COs	See: *Voluntary and community sector*.
ehicle barriers:	See: *Car parking*.
ehicles: complying ith the Disability iscrimination Act	1 The Disability Discrimination Act (qv) does not apply to privately used vehicles such as private motor cars. For guidance on the latter see: *Accessible cars*. 2 The Act does in principle apply to vehicles used by the public. For details see: *Transport: complying with the Disability Discrimination Act.*
ertical circulation in uildings	See the following entries: • Ramps. • Stairs and steps. • Lifts. • Escalators. • Passenger conveyors. • Travelators.
ertical platform lifts	1 Vertical platform lifts are one of two types of vertical lift, as shown below.

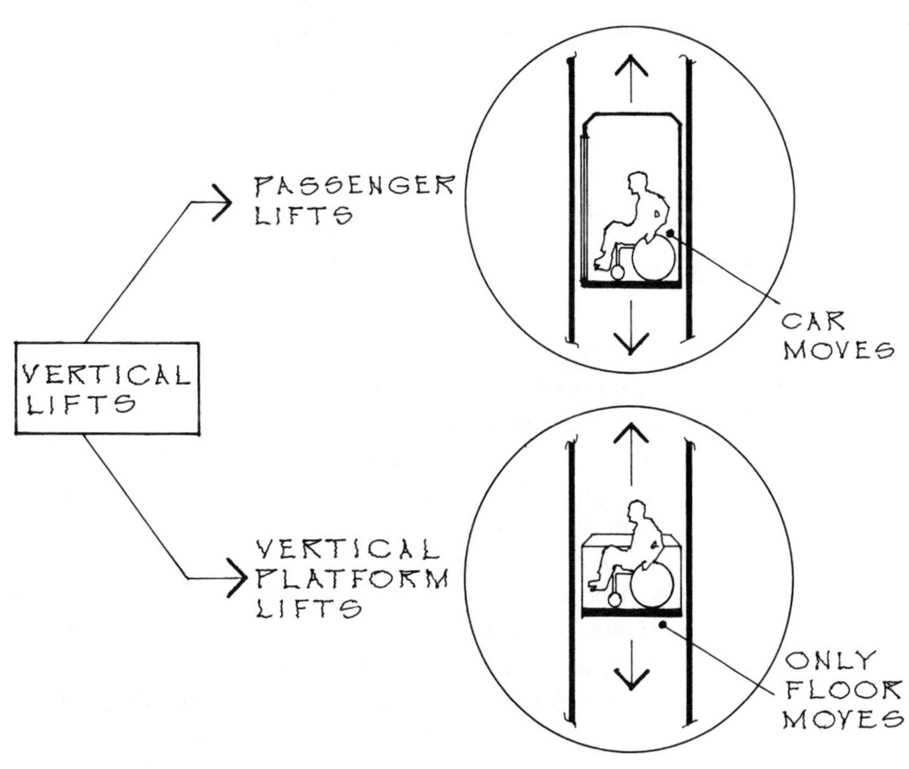

2 The advantages of vertical platform lifts over *passenger lifts* (qv) include the following:
• They take up less floor space.
• They require only a shallow pit.
• They generally do not require a lift motor room.
• They are cheaper.

Their comparative disadvantages include the following:
• They are very much slower.
• They are suitable only for intermittent and relatively light use, not for frequent or continuous use.
• They are suitable only for travel distances of two or a maximum of three storeys.
It should be noted that a *lift* (qv) of any kind can only augment a stair, never replace it; and that lifts are not permissible as a *means of escape* (qv) unless they are *evacuation lifts* (qv).
3 PLATFORM LIFTS, a 32-page publication from the Centre for Accessible Environments (qv), is an *authoritative practical guide* (qv).

Vertical platform lifts types

1 The diagram below suggests a taxonomy of vertical platform lift types, arranged from shortest rise to highest rise:

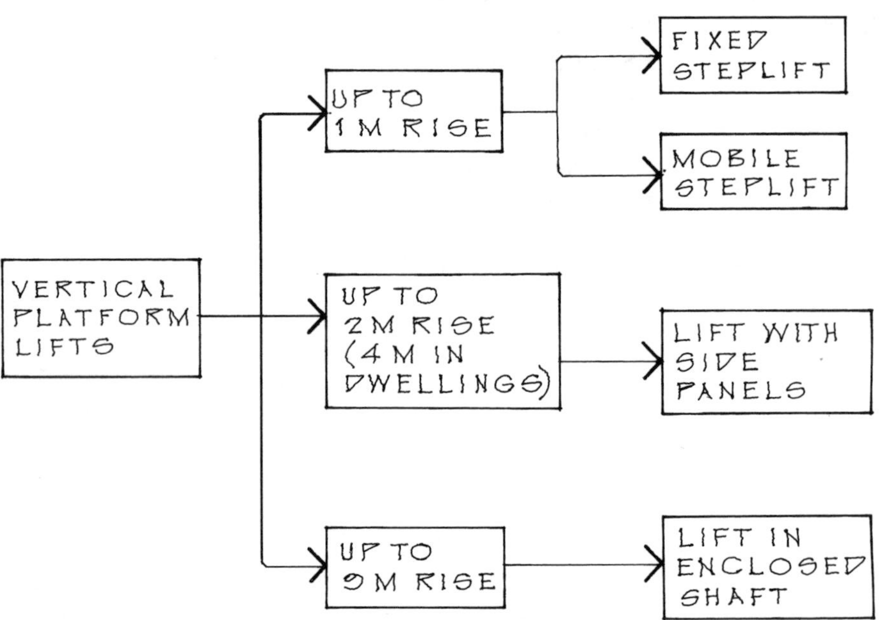

2 Steplifts rise no more than about 1 m, possibly using a scissor-type lifting mechanism. They need only minimal safety precautions owing to the short rise. Controls therefore are simple and the platform does not require any form of enclosure, only a guard rail and a 100 mm high safety flap at floor level to prevent a wheelchair accidentally rolling off the moving platform. The flap folds down flat when the lift has reached its destination.
There are two sub-types:
• Fixed steplifts are commonly used at a doorstep, a split-level floor, an auditorium platform, or a speaker's dais, if a *ramp* (qv) proves to be unfeasible for reasons of space or appearance. Wheelchair users are given complete independence, as they can ride directly onto the platform, and travel up or down, with no need for assistance. For flush entry, which is very desirable, the lift must be sunk into a shallow pit, which is normally between 50 mm and 250 mm deep. A fairly intricate custom-designed installation at the Merchant Taylor's Hall in Threadneedle Street, London EC2, is described under *Vertical platform lifts: case examples* below.
• Mobile steplifts (also called 'portable platform lifts') are broadly similar to the above but can be moved around on castors and parked where required – for instance at a building entrance, a stage, or a vehicle door to assist a wheelchair-user overcome differences of level of up to 1 m. They are best used where

access is needed occasionally or infrequently; in all other cases a permanent solution is preferable.

3 Low-rise platform lifts, with a permitted travel distance of up to 2 m in public buildings or 4 m in private dwellings, must be protected by light metal and glass at the sides.

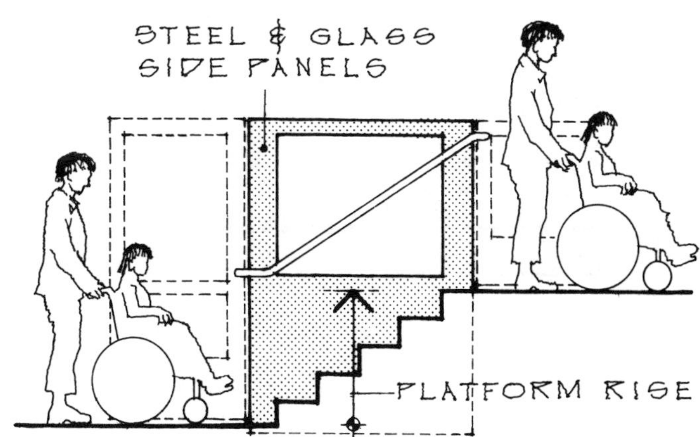

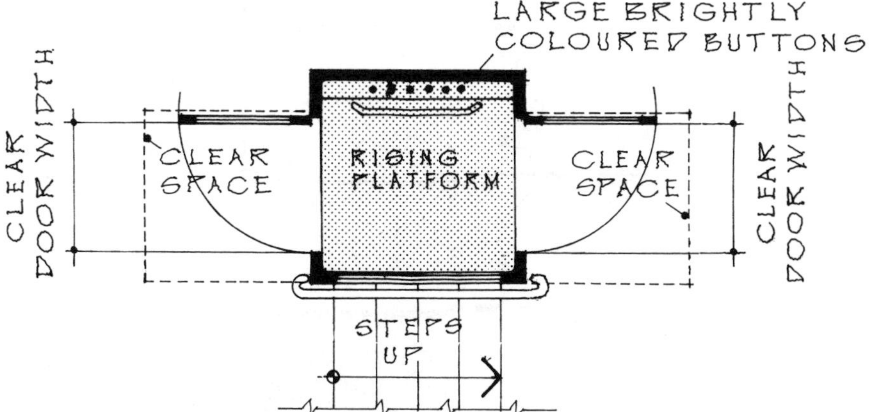

Appropriate uses for low-rise platform lifts include the following:

• In public buildings they are useful for overcoming a difference of level within a storey, eg giving access to a mezzanine floor in a shop, a raised area in a restaurant, etc.

• In private dwellings they can be used for travel between storeys. 'Through-floor lifts', as shown on the next page, are a sub-type specifically intended for such use, allowing *disabled* (qv) residents to ride effortlessly between the ground and first floor of their home (a maximum rise of 4 m) without leaving their wheelchair. They are neat and unobtrusive and can be safely operated by a disabled user without assistance. The system is as follows. Until needed the lift remains parked out of sight upstairs, the only evidence of its existence in the downstairs living area being a pair of vertical rails against the wall and a neat trapdoor panel in the ceiling. When the lift is needed downstairs the user presses the CALL button, and the platform descends through the ceiling trapdoor. The user enters the lift, presses the UP button, and the platform rises smoothly to the upper storey, re-closing the ceiling trapdoor behind it as it passes through. Through-floor lifts can easily be installed in existing dwellings, the main work that is required being the cutting of a floor opening for the lift to travel through, and the fixing of twin wall-mounted rails. In *Lifetime Homes* (qv) a floor area must be designated where a through-floor lift from ground to first floor can be installed in future.

People on low incomes may qualify for financial help towards the cost of installing such a lift in their homes – see: *Home improvement grants*.

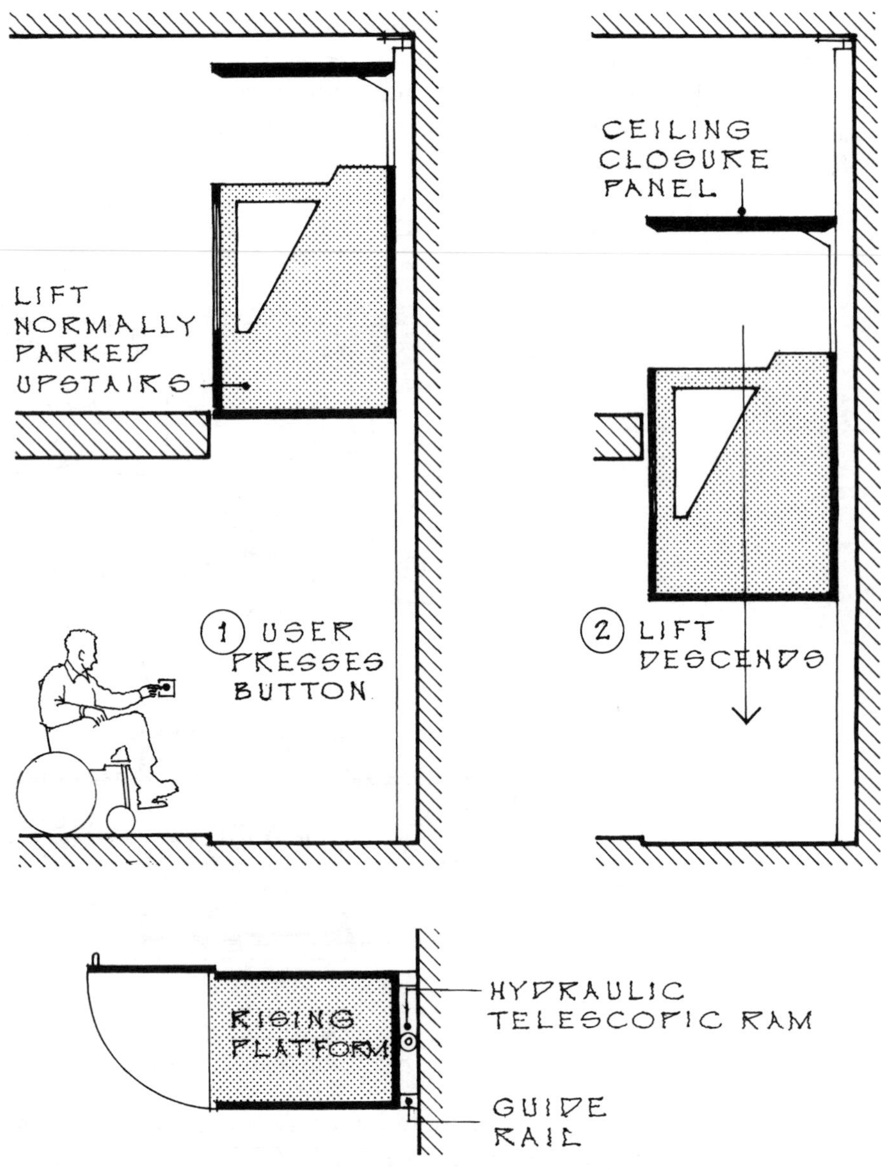

4 High-rise platform lifts, illustrated diagrammatically on the next page, have a permitted travel distance of up to 9 m in public buildings. Unlike short-rise lifts they must be fully enclosed within a shaft, and are permitted to penetrate floors and travel between storeys in public buildings. But they should only be installed if the provision of a *passenger lift* (qv) is quite unfeasible, the latter always being the preferred option (see para 8.4.1 of BS 8300). They are suitable only where traffic is light and travel distances no more than two or three storeys.

As of 2008, useful websites include the following. They are given in alphabetical order.

- Axess4All www.axess4all.com/
- Disabled Living Foundation www.dlf.org.uk/public/factsheets.html
- Patient UK www.patient.co.uk/showdoc/12/
- OTDirect www.otdirect.co.uk/link-equ.html#Platform
- Ricability www.ricability.org.uk/

In the opinion of the present author, Axess4All is a particularly neat and attractive product.

See also: *Lifts*.

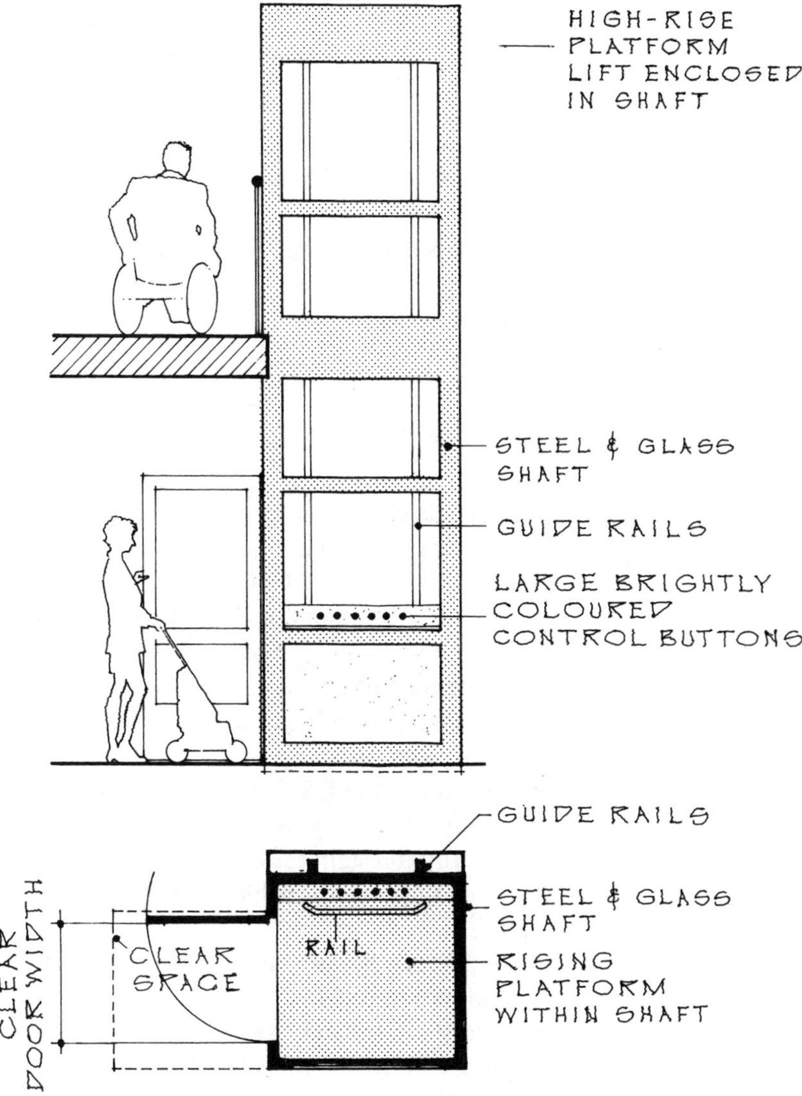

Vertical platform lifts: public buildings: complying with the Disability Discrimination Act

1 Buildings used by the public must in general comply with the Disability Discrimination Act (qv). In the case of vertical platform lifts in public buildings, *authoritative practical guides* (qv) include DESIGNING FOR ACCESSIBILITY (qv) and in addition the following documents. For publication details see the bibliography.

- APPROVED DOCUMENT M (qv), the provisions of which must be heeded in order to comply with the Building Regulations for England and Wales.
- BS 8300, the provisions of which apply to all buildings in the UK. It is an advisory document.
- INCLUSIVE MOBILITY (qv), which is the most authoritative reference for the *inclusive design* (qv) of *pedestrian infrastructure* (qv) and *transport-related buildings* (qv) such as bus and coach stations, railway stations, air terminals, and transport interchanges. It is an advisory document.

While the advisory documents above do not have the force of law, conformity with their recommendations will make it easier to demonstrate that the requirements of the Disability Discrimination Act have been met.

The particular clauses in the above documents which give guidance on the provision of vertical platform lifts are set out below.

BUILDING TYPE	DOCUMENT: RELEVANT PARAGRAPHS		
	AD M	BS 8300	INCLUSIVE MOBILITY
LIFTS IN BLOCKS OF FLATS	–	8.4 to 8.4.2; 8.4.4	–
LIFTS IN NON-DOMESTIC BUILDINGS	3.22; 3.24 to 3.28	8.4 to 8.42; 8.44	
LIFTS IN TRANSPORT-RELATED BUILDINGS	–	–	8.4.5

Though checks should be carried out before any decisions are finalised, the following notes summarise the order of preference stated in both APPROVED DOCUMENT M and BS 8300:

• The preferred option in all public buildings is a *passenger lift* (qv) that will accommodate a *wheelchair-user* (qv). Section 8.4 of BS 8300 gives detailed recommendations on such lifts.

• If a passenger lift is not feasible, and travel is only between two levels, a *vertical platform lift* (qv).

• If that too is not feasible, and travel is only between two levels, an *incline platform lift* (qv) – provided its installation does not conflict with *means of escape* (qv) on staircases.

• If even that is not feasible, a *chairlift* (qv) is the last and least desirable option.

2 For places other than England and Wales the following regulatory documents should be consulted:

• In Northern Ireland: TECHNICAL BOOKLETS E, H, and R of the Building Regulations (Northern Ireland) (qv).

• In Scotland: the BUILDING (SCOTLAND) REGULATIONS and the associated TECHNICAL HANDBOOKS (qv).

3 In the USA, where the Americans with Disabilities Act applies, refer to the ADA AND ABA ACCESSIBILITY GUIDELINES (qv).

Vertical platform lifts: case examples

1 Examples of external platform lifts include:

• The main entrance to the British Museum in London WC1. The museum has a level entrance from Montague Place in the north, but the main entrance from Great Russell Street on the south is situated on a high podium, reached via an imposing flight of steps that used to effectively make the museum inaccessible to *wheelchair-users* (qv) and some other *disabled people* (qv). As part of the ambitious renovation project that created the Great Court at the time of the millennium, an unobtrusive low-rise platform lift was installed to the left of the steps, thus enabling disabled visitors to reach podium level swiftly and easily, with no need to ask anyone for help.

• The glass-enclosed platform lift at London's Millennium Bridge across the Thames. The bridge links the Tate Modern Gallery on the south bank (SE1) with St Pauls Cathedral on the north (EC4), and the lift is situated at the St Paul's end. Being subject to heavy pedestrian traffic, and exposed to a high risk of vandalism, the lift was designed to withstand heavy wear and considerable misuse.

2 Examples of internal platform lifts include:
• The steplift at the entrance to the Merchant Taylor's Hall, Threadneedle Street, London EC2. This grade 1 listed entrance with its short flight of steps was originally inaccessible to wheelchair-users. But a steplift has recently been installed beneath the left half of the flight of steps, where it is completely concealed by the steps. If a disabled person arrives and needs to mount the stairs, he or she presses a button, whereupon the left half of the stair slides back and exposes the steplift, the upper surface of which is flush with the floor. The wheelchair-user rides onto the steplift, is raised to the upper level, and enters the building. The steplift then sinks back into its pit, the left half of the flight of steps slides back into position, and the steplift is again invisible beneath the steps. The action is summarised on the diagram below, and a full description of the lift is given on p 22 of PLATFORM LIFTS – see the bibliography.

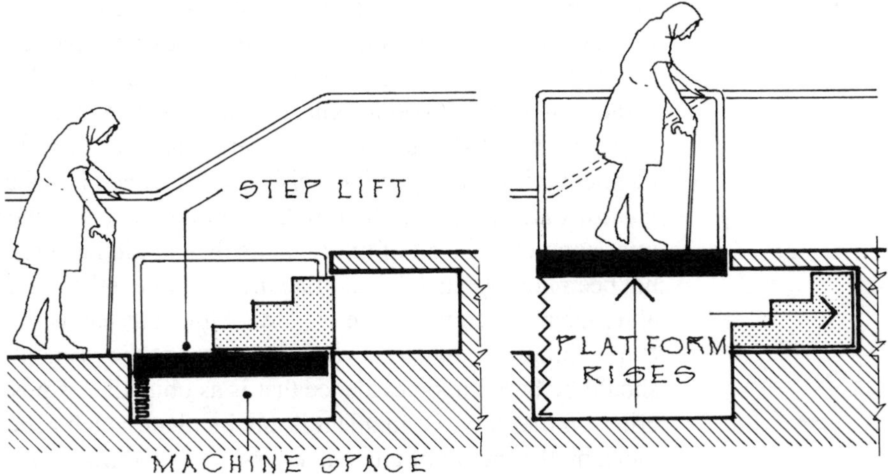

• The very neatly designed low-rise platform lift at the western end of Lancelot's Link, an internal street in Southwark Cathedral in London SE1. The lift was installed as part of a major refurbishment of the cathedral at the time of the millennium. It is briefly illustrated on p 21 of the CAE publication PLATFORM LIFTS (see above), and the project as a whole is illustrated on pp 13–21 of ACCESS BY DESIGN, the journal of the Centre for Accessible Environments – see the bibliography.

Vetting See: *Childcare: vetting*

Vibrating alarms See: *Alerting devices.*

Victimhood 1 Many commentators, including the disabled author Mr Tom Shakespeare, are concerned that, well-intentioned though anti-discrimination laws and regulations are, the labelling of specified groups of people as being in need of legal protection, whereas the rest of the population is not, may subtly lead members of the protected group into the trap of building their identity around an oppressed status. Offered the possibility of interpreting every setback as resulting from oppression, never from any fault of their own, it becomes all too easy to take refuge in victimhood and effectively stop trying. Other commentators have expressed a more fundamental concern: that bestowing a special legal status on some members of society but not on others promotes social division rather than unity – a view that appears to be supported by many or perhaps most of the general public – see: *Equality*.
2 Do these arguments mean that the concept of 'group justice' is flawed, and that society would be better served by a system of 'individual justice' with its

emphasis on justice and redress for individual victims of discrimination? There are two schools of thought on this:

• First, the current official view, which holds that certain groups of people suffer forms of discrimination that are so closely identified with the nature of that group that those people can only be effectively protected by group rights. In line with this approach the Equality and Human Rights Commission (qv), which started work in late 2007, has been given responsibility for protecting six specific groups of people. They are defined in terms of race, religion, disability sexual orientation, gender, and age.

• There are three counter-arguments to the above. First (as Tom Shakespeare has suggested), that the very concept of 'group' membership which underpins the above legislation is unreal. The most important attribute of a person may at one moment be the fact that she is woman, at another her dyslexia, at another the fact that she is a mother, at another the fact that she is Asian, at another the fact that she is a devout Catholic, and so on. We are complex individuals and are best treated as individuals. Second, that grotesque legal consequences have resulted from the attempt to embody group victimhood in law. One of the worst is the principle that people who commit crimes against members of a protected group are now liable to far more severe sentences than if they had inflicted the identical crime upon someone else. Thus in June 2006 the murderers of a *homosexual* (qv) man, Mr Jody Dobrowksi, were sentenced to 28 years imprisonment instead of the 14 years that would have resulted had the victim not been homosexual. What price the fundamental and hard-won principle that all citizens are equal before the law? Almost as bad is the replacement (following the Stephen Lawrence inquiry) of the principle that court findings should be based upon evidence that is as objective as possible, with the new criterion that 'a racist incident is any incident which is perceived (sic) to be racist by the victim or any other person'. The third counter-argument, which arises in part from the previous two, is simply that the concept of group rights encourages the splitting of British society into distinct groups competing with each other for rights and entitlements, and therefore divides society instead of bringing it together.

3 Tuning from principles to numbers, there is the interesting question of how many people enjoy the protection of group rights and how many do not. On pp 5–8 of his book WE'RE (NEARLY) ALL VICTIMS NOW Mr David Green, director of the *think tank* Civitas (qv), has added up the number of people who belong to legally defined 'victimhood groups' and found the answer to be 109% of the population. This is of course impossible, and for a more realistic estimate the calculation must make allowance for overlapping categories – for instance, the fact that the same person may be both black and disabled, or both Jewish and homosexual. An estimate of this kind suggests that around 73% of the total population are members of the six protected groups, while the 27% of the population who are able-bodied, Christian, heterosexual white males of working age are legally out in the cold – see the diagram below.

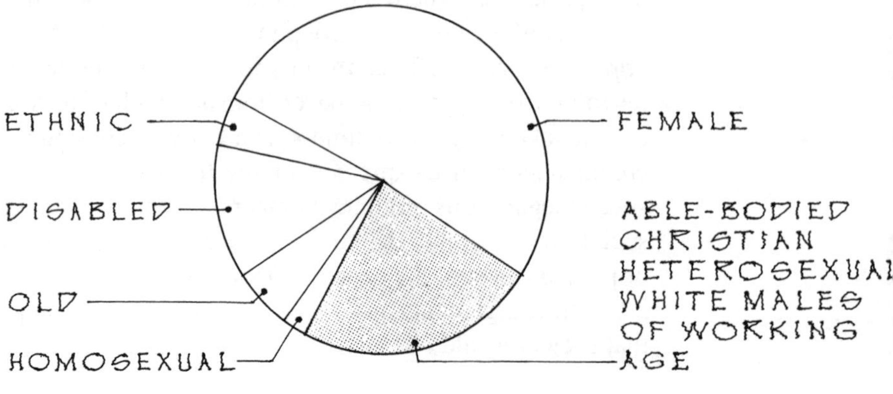

4 As of 2008, useful references include the following. For publication details see the bibliography.
• DISABILITY RIGHTS AND WRONGS by Tom Shakespeare. This is a well-informed and thoughtful book of which Mr Bert Massie, chairman of the Disability Rights Commission said: 'If you read only one book on disability rights this year, make this the book'.
• GAY-BASHING SHOULD NOT BE A HATE CRIME by Johann Hari, a gay author.
• THE STEPHEN LAWRENCE INQUIRY: REPORT OF AN INQUIRY BY SIR WILLIAM MACPHERSON OF CLUNY. For the formulation that 'a racist incident is any incident which is perceived to be racist by the victim or any other person' see pp 313–14, volume 1.
• TALKING EQUALITY is a research report commissioned and published by the Equal Opportunities Commission (qv). It investigated attitudes to equality expressed by the general public, and found that while most people were aware that social inequality and discrimination were widespread, and had in many cases experienced these things themselves, they disliked measures such as positive discrimination, which were discriminatory and to create artificially engineered outcomes, and were sceptical about group rights, except perhaps for *disabled* (qv) and older people. See: *Equality*.
• WE'RE (NEARLY) ALL VICTIMS NOW by David G Green was published by the *think tank* (qv) Civitas. The case of Mr Jody Dobrowksi and others of a similar nature are discussed on pp vii and 48.

See also: *Discrimination; Equality; Equality and Human Rights Commission*; and *Hate crime*.

Victoria Climbié

1 Victoria Climbié was a little girl who was sent from her home in a shanty town in the Ivory Coast to live with her great-aunt in London, in the hope that she would get a good education and enjoy a better life. Instead she was beaten with bicycle chains, kept trussed up in a plastic sack, burnt, and tortured in other ways. After suffering months of horrific abuse she died of her injuries on 15 February 2000, aged just eight.
2 The failings of the great array of social services whose duty it is to prevent such happenings, make an instructive story. In July 1999 the child was admitted to a hospital casualty unit with scalding to her face and head, but released back into the care of the woman who was torturing her. She was on the records of four social services departments, two housing authorities, two police child protection teams, two hospitals, and the NSPCC. According to the later official inquiry, care workers missed twelve chances to save Victoria; and her social worker never once spoke to her without her abusers being present. As the great majority of officials most directly responsible for Victoria were black, there was no question of race prejudice or institutionalised race discrimination. Senior managers were retained and promoted during this period. Indeed, Mr Gurbux Singh, chief executive of Haringey council when the above events were taking place, denied all responsibility and went on to become chairman of the Commission for Racial Equality (qv).
3 A detailed account of the catastrophic incompetence of virtually every official who came into contact with the murdered child was set out under the heading 'Key figures in the Climbié case' by David Batty in the *Guardian* of Monday February 4, 2002. It can be looked up in the *Guardian* archive at http://society.guardian.co.uk/climbie/story/0,,616539,00.html
4 It is sad to report that in February 2007, soon after another dreadful and very similar case had been heard in court (see: *ContactPoint*), Lord Laming, who had chaired the above inquiry, complained in his Victoria Climbié Memorial Lecture that social workers and managers were still not following

straightforward, intellectually undemanding procedures that would help protect children. In an article published in the *Guardian* on 28 February 2007 under the title 'Our children have less protection now than did Victoria Climbié', Ms Liz Davies, a senior lecturer in social work at London Metropolitan University, went further and charged that the system to prevent child abuse had 'for five years … been vanishing before our eyes'.

5 As of 2007, useful references include the following. They are given in date order. For publication details see the bibliography.

- THE VICTORIA CLIMBIÉ INQUIRY.
- THE VICTORIA CLIMBIÉ INQUIRY: REPORT OF AN INQUIRY by Lord Laming.
- KEY FIGURES IN THE CLIMBIÉ CASE by David Batty.
- CLIMBIÉ INQUIRY: THE ISSUE EXPLAINED by David Batty.
- AFTER CLIMBIÉ, CHILDREN ARE AT EVEN MORE RISK. An article by Jill Kirby, published in 2007 after the Kimberly Harte/Samuel Duncan child abuse case – see: *ContactPoint*.
- CLIMBIÉ LESSONS NOT LEARNED, SAYS RESEARCH by John Carvel.

See also: *Children: looked-after*; and *ContactPoint*.

Village halls

1 A popular village hall, that is liked and used by the whole community, can greatly benefit the social well-being of rural areas.

2 Village halls, like many other *social enterprises* (qv) run by the *voluntary and community sector* (qv) are being hard hit by the insupportable weight of pettifogging regulation that is now descending upon even the most modest organisations, making the lives of people who are trying to do good things for the community a misery and resulting in the closure of highly valued facilities – see: *Over-regulation*. Examples of the latter include the following:

- The Licensing Act 2003, which came fully into force in 2005, has obliged volunteers running local community halls where alcohol is sold to go on a training course to obtain an 'accredited personal licence qualification', and to fill in a twenty page application form which requires sixty pages of explanatory notes. One volunteer told the press that he had been obliged to use a week of his unpaid time, including attending two seminars, to comply with procedures aimed at the prevention of crime and disorder, the prevention of public nuisance, and the protection of children from harm, none of which had ever been problems in his small village. In the House of Commons Mr Peter Luff, who had long warned ministers of the likely consequences of the Act, to no effect, said on 8 June 2005 (visit www.peterluff.co.uk/hansard.asp): 'The Licensing Act means that villages in every corner of the country will say goodbye to the traditional touring circus, see more village shops go to the wall, watch local sports clubs forgoing much-needed income and lose their village halls, despite no accidents, no anti-social behaviour, and nothing worrying having occurred in, outside, or even remotely close to them'. Two of the MPs who had helped to steer the Bill into law (Mr Malcolm Moss and Mr Mark Field) were gracious enough to admit to the House that they were not proud of what they had helped to bring about.
- The Gambling Act 2005 (qv) is also closing down modest community events such as bingo evenings for older people. In one example known to the present author the volunteers could not face the onerous record-keeping that is now required, or afford the application and annual licence fees of £1 900 and £1 500 respectively. These things were happening as Secretary of State Tessa Jowell was striving under the same Act to set up around 40 supercasinos around the UK, including several 'resort casinos' which are specifically aimed at luring more people into big-time gambling.

• There have also been press reports of small church halls having to close because they cannot afford to install the wheelchair *ramps* (qv) and/or accessible *lavatories* (qv) which they have been told to provide in order to comply with the Disability Discrimination Act (qv). The closure of facilities that could have given pleasure to 99% of the population because they could not be adapted to cater for the remaining 1% (in many cases fewer) does not strike most people as a sensible use of the law.
3 The organisation ACRE (Action with Communities in Rural England) is an excellent source of good practical guides on the design and management of village halls – see below.
4 As of 2007, useful references include the following. They are given in alphabetical order. For publication details see the bibliography.
• PLAN, DESIGN AND BUILD: PARTS 1 AND 2. Part 1 gives practical guidance and Part 2 shows a number of case examples.
• VILLAGE HALL GOOD MANAGEMENT TOOL KIT. This 32-page A4 booklet plus loose insert and CD-ROM is a complete guide, answering all the important questions or indicating where authoritative information can be found.
5 Useful websites include:
• Action with Communities in Rural England www.acre.org.uk/

See also: *Community Centres*.

Virtual tours

1 Virtual tours of places or buildings are pictorial or multi-media representations of those places which are realistic enough to offer a reasonable substitute for an actual visit. The simplest type is a sequence of colour photographs in a book, or a sequence of colour images on a TV screen. A more realistic experience is offered by a cinematic tour, especially if projected onto a large or wrap-around screen. At the cutting edge of technology are computer-generated simulations of three-dimensional environments with which the viewer can interact in a seemingly real or physical way by using special electronic equipment.
2 Such tours have many uses. Estate agents and architects use walk-through simulations to promote buildings and proposed building designs to clients; the owners of historic properties and cultural venues use them to demonstrate their attractions to tourists other potential visitors; and – the application that is most pertinent to the present work – if an *historic building or site* (qv), or a *museum* (qv) or art gallery, cannot be made physically *accessible* (qv) to *disabled people* (qv), then virtual tours may offer them the next best thing.
3 Under the Disability Discrimination Act (qv) virtual tours and similar services come under the heading of *auxiliary aids and services* (qv). The proper provision of such services will make it easier for gallery and museum owners, in their role as *service providers* (qv), to demonstrate that the requirements of the DDA have been met.
4 For commendable case examples visit the websites given under: *Exhibition spaces: accessible*.

See also: *Audioguides*; and *Touch tours*.

Visibility markings See: *Hazards: protection*; and *Street works*.

Visibility panels See: *Doors*.

Visual contrast See: *Colour in Buildings*.

Visual impairment See: *Impaired vision*.

VocalEyes

1 VocalEyes is an *audio description* (qv) service for *blind* (qv) and partially-sighted visitors to art galleries, museums, heritage sites, and even to theatres and opera houses, enabling them to listen to recorded descriptions of what a painting or sculpture, a heritage site, or a theatre or opera set looks like. In addition to making it possible for them to enjoy visits to such venues on their own, the service also helps them enjoy the social experience of attending such events in the company of sighted friends and relatives.
2 In the case of museums and art galleries the provision of a spoken commentary is fairly straightforward. In the case of theatrical and musical performances it is more problematical. Great care is taken to confine the spoken descriptions to gaps in the dialogue, so that the describer is silent while actors are speaking; and in the case of operatic performances spoken comment is generally given during the overture, or during passages without singing, not over arias or other key passages of music.
3 In addition to providing audioguides VocalEyes sends out pre-show cassette tapes to subscribers in advance of performances, providing them with detailed information on the visual aspects of the production so that they can properly prepare themselves. The tapes additionally contain information on the venue, its location and layout, and its accessibility. This information can also be downloaded as sound files from www.vocaleyes.co.uk/
4 Under the Disability Discrimination Act (qv) virtual tours and similar services come under the heading of *auxiliary aids and services* (qv). The proper provision of such services will make it easier for gallery and museum owners, in their role as *service providers* (qv), to demonstrate that the requirements of the DDA have been met.
5 For commendable case examples visit the websites given under: *Exhibition spaces: accessible.*

See also: *Audioguides*; and *Touch tours.*

Vocational training

See: *Education*; and *Learning and Skills Council.*

Voice-activated controls

1 Voice-activated systems are a form of remote control that enables people who cannot use their hands to operate *environmental control systems* (qv) and other mechanisms by speaking their commands into a microphone. By this method tetraplegics (qv) and other severely *disabled people* (qv) can operate the tv, switch lights on and off, open or close doors and windows, and draw curtains from the comfort of their chair, bed, or wheelchair.
2 As of 2007, useful websites include the following. They are given in alphabetical order.

• AutomatedHome	www.automatedhome.co.uk/
• Possum Controls Limited	www.possum.co.uk/
• Ricability	www.ricability.org.uk/
• RehabTeq Limited	www.rteq.co.uk/
• SmartaSystems	www.smartasystems.co.uk/

See also: *Assistive technology; Environmental control systems*; and *Smart homes.*

Voice recognition

See: *Speech recognition.*

Voluntary and Community Sector

1 BETTER REGULATION FOR CIVIL SOCIETY, an 82-page report published in November 2005 by the Better Regulation Task Force (qv), estimated that there were several million volunteers in the UK, of whom over 400 000 were doing

social care work, and commented that this sector (also known as the 'third sector') was in many cases innovating in areas that the public and private sectors could not reach.

2 Political parties now vie with each other in their expressions of admiration and support for the voluntary sector, and in promises to give it a major role in the provision of future social services. But three problem areas need to be thought through before pinning too much hope on a new golden age of citizen-driven public services.

3 The first is that voluntary organisations are unlikely to be equal to the roles that Governments may wish to transfer to them. The public has come to expect a scale and reliability of public services that can come only from the state. Enthusiastic amateurs can very usefully fill gaps or do topping-up, but they cannot take over. Voluntarism works because it's voluntary, and what it does cannot be multiplied by a hundred.

4 The second is that many voluntary bodies are driven by values and beliefs that do not sit happily with state policies. They may want to cater only for Muslims; only for white boys; only for Anglican women; only for homosexual men; only for elderly tourists; etc. In a diverse democracy these are perfectly legitimate and indeed thoroughly admirable aims, but their exclusivity jars with official policy. Unfortunately the state has at its disposal highly effective methods, both direct and indirect, for imposing *ideological* (qv) uniformity on the natural *diversity* (qv) of views that is characteristic of democracies. As of 2007 it is ruthlessly using both. Here are two examples:

• The first method is control by direct diktat. Britain is fortunate enough to have a dozen Roman Catholic adoption agencies which handle up to a third of all adoptions of older and troubled children, who are generally very difficult to place. These bodies do excellent work, and their disruption rates are among the lowest in the field. In January 2007 the Government decided that as from April of that year all adoption agencies must be willing to place children with *homosexual* (qv) couples. The Roman Catholic church stated that it accepted the legal entitlement of homosexual couples to adopt children, and would refer any such couples who approached it to a more suitable agency, but that its theological principles prevented it from making such placements itself. In this way the church hoped to continue placing hundreds of children in loving homes, while other agencies filled their own particular niches. The Government immediately ordained that no exemptions whatever to the Sexual Orientation Regulations would be permitted, and that Roman Catholic adoption agencies could either conform or cease to be recognised. In 2007 negotiations are underway, but if this decision stands (which seems likely) the losers will be the most hard-to-place children. Thus does the official obsession with uniformity trump *diversity* (qv) and the welfare of vulnerable children.

• A more indirect way of making voluntary organisations implement state policies is via government funding with strings attached. Britain's charities now receive more money from the Government than from voluntary donations. The UK VOLUNTARY SECTOR ALMANAC (see below) reported in June 2007 that 62% of such government money now comes in the form of contracts to carry out specific tasks. The method of execution is prescribed in the finest detail, and charities must produce lengthy monitoring reports to prove that they have not deviated an inch from official procedures. A survey by the Charities Commission (qv) titled STAND AND DELIVER reveals that only 26% of charities providing services now feel free to make decisions without pressure to conform with the wishes of their funders. Mr Stuart Etherington, chief executive of the National Council for Voluntary Organisations (qv) called such pressures on charities to conform 'very dangerous'. Increasingly, he said, voluntary organisations are having to make a choice between being driven by their original mission, or doing what state authorities from whom they receive funds tell them to do – a neat example of the emerging phenomenon of *Post-democracy* (qv). What can be done? In WHO CARES?, a report for the *think tank* (qv) Civitas, Mr Nick Seddon proposes that charities should in future be classed in three groups. Those receiving less than 30% of their income from the state should continue to benefit from charitable status, while those receiving between 30% and 70% should be called state-funded charities and receive more modest benefits. Those receiving over 70% of their income from the state are in his view already de facto state agencies, and should be forced either to reduce their dependency on statutory funding or lose their charitable status.

5 The third problem area is that voluntary organisations are now struggling under a mountain of *over-regulation* (qv) that cripples them even in their current modest activities. A massive cut in red tape will be required just to help them survive, let alone take on major additional duties. The following notes augment those under *Gambling Act* (qv), *Licensing Act* (qv), and *Village Halls* (qv):

• BETTER REGULATION FOR CIVIL SOCIETY, an 82-page report published in November 2005 by the Government's own Better Regulation Task Force (qv), asserted that over-regulation threatened not merely the effectiveness, but in some cases even the survival of valuable charities. Among the examples quoted were recently introduced regulations which required 40 000 existing bell ringers in Church of England churches to undergo criminal record checks every three years, and new bell ringers to go through an eight-point application process that included providing two referees, and being interviewed and vetted by the Criminal Records Bureau (qv). The study found that charity managers spent hours each week on supplying different sets of similar information for central and local government regulators, and that some managers of small charities were spending up to 40% of the working week accounting for funding from central and local government when they should be getting on with their good works.

• Closely related to the above problems is another, not mentioned in BETTER REGULATION FOR CIVIL SOCIETY – volunteers' growing worry about the *risk* (qv) of litigation should something go wrong in the course of their public-spirited activity. In July 2004 the Conservative MP Julian Brazier introduced a private member's bill in the House of Commons, which was aimed at protecting volunteers from unnecessary litigation and red tape. He argued that 'too many adults are deterred from volunteering by the threat of litigation' and was supported by the former Labour health secretary, Mr Frank Dobson, and the former Labour minister for sport, Ms Kate Hoey, who said that the bill had the support of a number of government ministers, particularly in the Department for Education and Skills. The bill was nevertheless talked out.

6 These problems and pressures come at a time when voluntary organisations are about to have their incomes from sources other than government contracts very severely squeezed. In early 2007 Secretary of State Tessa Jowell announced that she proposed to divert £2.2 billion from the 'good causes' budget of the National Lottery (qv) to help meet 2012 Olympic Games costs that jumped from her original estimate of £2.4 billion in 2005 to £10.4 billion (and rising) in 2007. Writing in the *Guardian* on 6 June 2007 under the title 'Under the thumb of the state' Mr Ben Wittenberg, director of policy and research at the Directory of Social Change, warned of potentially devastating consequences for the 156 000 cash-strapped small charities that survive on less than £10 000 a year.

7 The National Coalition for Independent Action (see below) is an alliance of organisations formed in 2007 to mobilise opposition to the above threats to the independence of voluntary organisations.

• BETTER REGULATION FOR CIVIL SOCIETY: MAKING LIFE EASIER FOR THOSE WHO HELP OTHERS. This report was published in 2005 by the Better Regulation Task Force (qv).
• CRUNCH TIME FOR SOCIAL ENTERPRISES: MINISTERS NEED TO DO MORE TO ENCOURAGE SOCIAL ENTERPRISES by Patrick Butler.
• GOVERNMENT'S WAY WITH WORDS IMPACTS THE LIFE-CHANCES OF ALL by Charles Moore. A depressing (though amusingly presented) picture of the growing substitution of jargon and bureaucracy for effective work in voluntary organisations.
• STAND AND DELIVER: THE FUTURE FOR CHARITIES PROVIDING PUBLIC SERVICES presents the results of a survey of registered charities carried by the Charity Commission, concerning their participation in delivering public services.
• UK VOLUNTARY SECTOR ALMANAC 2007. This annual publication contains a wealth of data including the types and sources of the voluntary sector's income; how the money is spent; and the changing nature of the sector.
• WHO CARES? HOW STATE FUNDING AND POLITICAL ACTIVISM CHANGE CHARITY, by Nick Seddon of the *think tank* Civitas (qv), argues that the time has come to make a distinction between charities that are genuinely part of civil society, and those that have effectively become an arm of the state.
8 Useful websites include the following. They are given in alphabetical order.
• Association of Chief Executives of Voluntary Organisations www.acevo.org.uk/
• The Charity Commission www.charity-commission.gov.uk/
• Civitas www.civil-society.org.uk/
• Commission for the Compact www.thecompact.org.uk/
• Futurebuilders www.futurebuilders-england.org.uk/
• *Guardian* http://society.guardian.co.uk/voluntary/0,7886,374546,00.html
• National Coalition for Independent Action www.independentaction.net/
• New Policy Institute www.npi.org.uk/projects/islington.htm

See also: *Charities; Office of the Third Sector; Over-regulation; Risk assessments*; and *Social enterprises*.

Volunteering code

1 A compact between government and voluntary organisations whereby the government agrees to promote and support volunteering, while voluntary organisations agree to maintain high standards of volunteer management.
2 Useful websites include the following. They are given in alphabetical order.
• Compact www.thecompact.org.uk/
• Directgov www.direct.gov.uk/
• Volunteering England www.volunteering.org.uk/

Voting fraud See: *Electoral fraud.*

Voting booths See: *Polling stations.*

Vouchers: education See: *Education: voucher systems.*

Vulnerable children See: *Children: vulnerable*; and *Safeguarding Vulnerable groups Bill.*

W–Z

W3C — See: *Accessible websites.*

Wage: average — See: *Income: median.*

Waiting areas: complying with the Disability Discrimination Act

1 Waiting areas in buildings are most commonly located near the principal entrance, but in large complexes such as *hospitals* (qv), and in *transport-related buildings* (qv) such as airport passenger terminals, bus and coach stations, and railway stations there may be additional waiting areas elsewhere in the building.

2 Buildings used by the public must in general comply with the Disability Discrimination Act (qv). In the case of waiting areas, *authoritative practical guides* (qv) include DESIGNING FOR ACCESSIBILITY (qv) and in addition the following documents. For publication details see the bibliography.

- APPROVED DOCUMENT M (qv), the provisions of which must be heeded in order to comply with the Building Regulations for England and Wales.
- BS 8300, the provisions of which apply to all buildings in the UK. It is an advisory document.
- INCLUSIVE MOBILITY (qv), which is the most authoritative reference for the *inclusive design* (qv) of *pedestrian infrastructure* (qv) and *transport-related buildings* (qv) such as bus and coach stations, railway stations, air terminals, and transport interchanges. It is an advisory document.

While the advisory documents above do not have the force of law, conformity with their recommendations will make it easier to demonstrate that the requirements of the Disability Discrimination Act have been met.

3 The particular clauses in the above documents which give guidance on the provision of vertical platform lifts are set out below.

BUILDING TYPE	DOCUMENT: RELEVANT PARAGRAPHS		
	AD M	BS 8300	INCLUSIVE MOBILITY
BUILDINGS IN GENERAL	3.2–3.6	7.1 and 11–11.2	--
TRANSPORT-RELATED BUILDINGS	--	--	9.3–9.4

For places other than England and Wales the following regulatory documents should be consulted:

- In Northern Ireland: TECHNICAL BOOKLETS E, H, and R of the Building Regulations (Northern Ireland) (qv).
- In Scotland: the BUILDING (SCOTLAND) REGULATIONS and the associated TECHNICAL HANDBOOKS (qv).

4 In the USA, where the Americans with Disabilities Act applies, refer to the ADA AND ABA ACCESSIBILITY GUIDELINES (qv).

See also: *Entrance halls and reception areas; Lobbies: entrance*; and *Seats in public buildings*.

Wales — See: *Government: Wales.*

Walk-in baths — See: *Bathing aids.*

Walking — See: *Pedestrian movement*; and *Traffic management.*

Walking and standing aids

1 Walking and standing aids are one of several classes of *mobility aids* (qv) for frail or *disabled people* (qv). The diagram suggests a taxonomy of types:

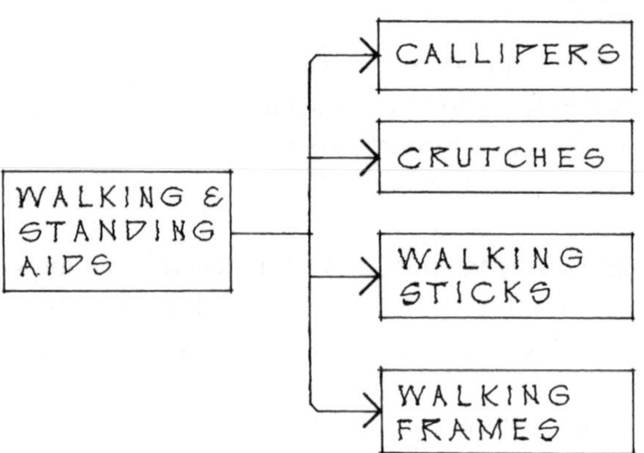

2 Callipers consist of splints fitted to the leg so that the patient can walk without any pressure on the foot, and come in two forms – (a) short, to give support to weak feet; or (long), to give support to weak knees, thighs or calf muscles.
3 Crutches are of three types – (a) Underarm (also called axillary) crutches are generally used when the user's lower limb cannot bear weight; (b) elbow crutches are for people whose lower limbs are able to bear full or partial weight; and (c) forearm trough (also called gutter) crutches are for people who, owing to pain or deformity, cannot take weight through their hands or wrists. They can only be safely used for partial weight-bearing.
4 Walking stick types include (a) conventional walking sticks made of wood or metal; (b) lightweight folding sticks which can be kept in a brief case or handbag; and (c) tripod or tetrapod sticks which can be leant on more heavily than conventional sticks.
5 Walking frames comprise two basic categories – (a) 'non-wheeled' and 'wheeled' – as shown below.

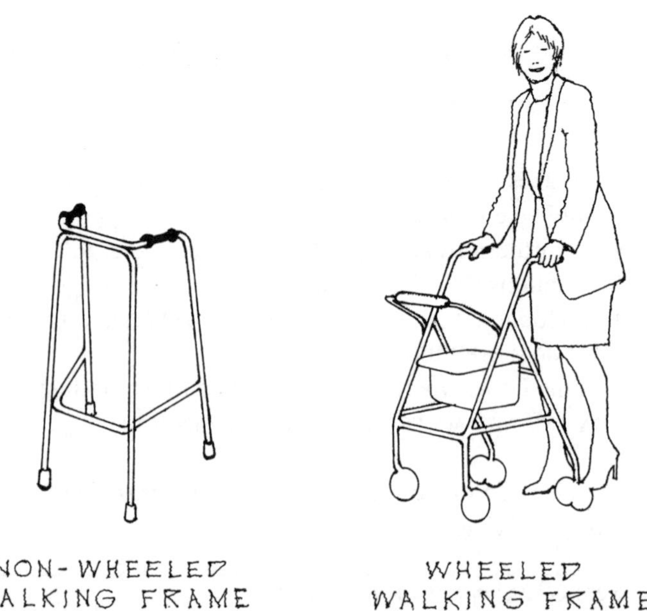

NON-WHEELED WALKING FRAME

WHEELED WALKING FRAME

6 As of 2007, useful references include the following. For publication details see the bibliography.
• CHOOSING WALKING EQUIPMENT, a 25-page factsheet published by the Disabled Living Foundation (qv).

7 Useful websites include the following. They are given in alphabetical order.

• Assist UK	www.assist-uk.org/
• Disabled Living Foundation	www.dlf.org.uk/
• Disabled Living Online	www.disabledliving.co.uk/
• OTDirect	www.otdirect.co.uk/link-equ.html#Platform
• Ricability	www.ricability.org.uk/

See also: *Impaired mobility; Mobility aids*; and *Products for easier living.*

Walking distances

1 In the planning of *pedestrian environments* (qv), para 2.4 of INCLUSIVE MOBILITY (qv) gives the following recommendations on maximum walking distances without a rest. These recommendations affect matters such as the spacing of *bus stops* (qv) sitting areas, etc.

• People with *impaired mobility* (qv) using a *walking stick*	50 m
• People with impaired mobility not needing a walking aid	100 m
• People with normal mobility but *impaired vision* (qv)	150 m

When planning pedestrian environments, conformity with the above recommendations will make it easier to demonstrate that the requirements of the Disability Discrimination Act (qv) have been met.
2 For more detail refer to INCLUSIVE MOBILITY – see the bibliography.

See also: *Pedestrian movement.*

Walking into Health

1 A Government scheme launched in January 2008 with the aim of getting 'a third of England walking at least 1 000 more steps daily by 2012' (sic).
2 For more information visit www.dh.gov.uk/

Walkways

See: *Footpaths and footways.*

Warden alarms

See: *Alarms.*

Wards

1 Every *local authority* (qv) in the country is divided into a number of smaller districts known as 'wards', of which there are 8 414 in England. These form the constituencies for local elections, and return up to three councillors each.
2 The government focuses much of its *regeneration* (qv) work on wards, as they are compact enough to allow money to be targeted at the worst pockets of deprivation.
3 The Office for National Statistics (qv) publishes data on employment, crime, and many other aspects of local life on a ward-by-ward basis, on its website www.neighbourhood.statistics.gov.uk/

See also: *Government: local; Neighbourhoods*; and *Regeneration.*

Warm Front

See: *Home improvement grants*; and visit www.warmfront.co.uk/.

Warning signs

See: *Signage: safety.*

Warning surfaces

See: *Hazards: protection; Street works and scaffolding*; and *Tactile pavings.*

Warnock, Mary

See: *Babies: premature*; and *Special educational needs.*

Wash basins

See: *Lavatories.*

Water sports

See: *Recreational and sports activities for disabled people*; and *Recreation and sports aids.*

WAVs

See: *Wheelchair accessible vehicles.*

Wayfinding aids See: *Signs*; and *Talking signs*.

Wcs See: *Lavatories*.

Websites: accessible See: *Accessible websites*.

Welfare

1 For more detail see the specific entries: *Child benefit; Council tax benefit; Housing benefit; Incapacity benefit;* and *Unemployment benefit*.
2 As of 2008, useful websites include:
• Directgov www.direct.gov.uk/

Welfare dependency

1 In February 2008 the Government admitted that one in five British children are growing up in households with no parent at work. In addition to spending their early years in poverty, these children are unlikely ever to acquire the habit of work and will probably always be poor. This is shocking, but for many of their parents living off *benefits* (qv) is entirely logical, as they are better off doing nothing than working. Should they get a legitimate job, the poorest could suffer a marginal tax rate of 90%, which is more than twice as high as that for city millionaires.
2 What can be done to remedy this huge waste of human potential, which is not good either for the people involved or for the rest of society? Here are two views, the first based on the American experience, and the second from a thoughtful and well-informed British MP.
3 In 1993 President Bill Clinton said: 'This country would be much better off if our children were born into two-parent families. Once a really poor woman has a child out of wedlock, it almost locks her and the child into a cycle of poverty which then spins out of control'. Three years later, in August 1996, he managed after much opposition to get his Personal Responsibility and Work Opportunity Reconciliation Act into law. It established that there would henceforth be no more than five years' worth of state income support for any family in the country, regardless of how many children the mother had; and at the same time a new system to fund childcare for working mothers was inaugurated. The new law caused outrage. Writing in the *Independent on Sunday* on 29 August 2006, under the title 'The way to reduce welfare dependency', the commentator Dominic Lawson quoted the following reactions from eminent Americans:
• Senator Pat Moynihan: 'The people who do this will go to their graves in disgrace'.
• Senator Edward Kennedy: 'This will condemn millions of innocent children to poverty in the name of welfare reform'.
• Marian Edelman, President of the Children Defense Fund: 'This will be a moral blot on our nation that will never be forgotten'.
• Peter Edelman, an assistant secretary at the US Department of Health, who resigned his post in protest: This law will lead to 'more malnutrition, more crime, increased infant mortality, increased drug and alcohol abuse, increased family violence and abuse against women and children'.
• The *New York Times*: 'This is not reform, it is punishment'.
But exactly ten years after the law was passed, in August 2006, President Clinton could write in the *New York Times* that welfare rolls had dropped from 12.2 million to 4.5 million; that the percentage of American children born out of wedlock had shown the first sustained fall since 1946; that the employment rate of single mothers and never-married mothers had doubled; that the poverty rate among single mothers (an index-linked $17 000 for a woman with two children) had fallen from over 50% to under 42%; and that the poverty rate among families with black children had fallen from over 41% to under 33%. There had also been a very dramatic fall in crime, and while it is too early to claim a direct

link between falling crime and President Clinton's welfare reforms, it is certainly true that critics' predictions of more crime following his reforms were proven wrong. A Department of Health official who resigned in 1996 in protest at the new scheme, Mr Wendell Primus, had already reconsidered his original views, telling the *New York Times* in 2001 that 'in many ways, welfare reform is working better than I thought it would. The sky is not falling in. Whatever we have been doing for the past five years we ought to keep going'. While British solutions cannot be identical to ones that worked in the USA, there may nevertheless be lessons to be learnt from the American experience. One is that the 1996 Personal Responsibility and Work Opportunity Reconciliation Act decisively shifted the running of social security from the federal government down to individual states. The latter seized their newfound freedom to experiment with enthusiasm, and best practice soon spread as states copied what was working elsewhere. For more on this refer to LOCAL WELFARE (see below).

4 The thinker behind the Clinton reforms was Willam Galston, a theorist who identified 3 steps to escaping poverty: First, finish school; second, marry before having children; and third, avoid teenage pregnancy. Among those Americans who did all three only 8% were poor; of those who did none, 79% were poor. Visit http://en.wikipedia.org/wiki/William_Galston

5 Be that as it may, here is a British view from the widely respected Labour MP Frank Field, who also serves on the advisory board of the *think tank* (qv) Reform. Writing in the *Daily Telegraph* of 17 June 2007 under the title 'Central control of welfare simply doesn't work' and in other documents, Mr Field has argued that what he describes as 'the failure of the Government's £84 billion welfare to work programme' should be tackled by a 4-part plan.

• First, the Department for Work and Pensions (qv) should put a time-limit on benefits as President Clinton did in the USA.

• Second, it 'should get serious about building partnerships with the private sector', with the latter providing the necessary funding in order to protect the taxpayer. In other countries charitable and private firms have been very successful in helping claimants land decent jobs.

• Third and most importantly, it should pilot independence for some Jobcentreplus offices. While the Department for Work and Pensions should set the budget, these offices should be freed from the wearisome box-ticking and other *red tape* (qv) that currently burdens their operations, and concentrate all their energies on getting claimants into work. Local staff generally know quite a lot about the personal circumstances of claimants. They know who genuinely needs a lot of support and encouragement and who is swinging the lead and they will act accordingly. They are best placed to know which private suppliers can deliver, and which know how to impress bureaucrats and win contracts, but fail to get claimants into work. They should be encouraged to use this knowledge.

• Finally, some of the taxpayers' money that is saved by these efficiencies should be used to reward staff for success.

6 Meanwhile, current government policies to reform the British welfare system are outlined at www.dwp.gov.uk/welfarereform/ This site includes details of the Welfare Reform Act 2007, which aims to take one million people off benefits and into work. To that end it introduced reforms to *Incapacity Benefit* (qv) and *Housing Benefit* (qv), and made a number of other amendments to the benefit system.

7 As of 2007, useful references include the following. For publication details see the bibliography.

• THE DREAM AND THE NIGHTMARE: THE SIXTIES' LEGACY TO THE UNDERCLASS by Myron Magnet. An instructive and fundamentally optimistic book.

• THE WAY TO REDUCE WELFARE DEPENDENCY by Dominic Lawson. The *Independent on Sunday*, 29 August 2006.

• BILL CLINTON SAVED PEOPLE FROM A LIFE OF CRIME. IS GORDON BROWN BRAVE ENOUGH TO FOLLOW HIM? by Dominic Lawson.
The *Independent*, 28 August 2007.
• CENTRAL CONTROL OF WELFARE SIMPLY DOESN'T WORK by Frank Field.
The *Daily Telegraph*, 17 June, 2007.
• LOCAL WELFARE. This is one of a set of Localist Papers produced by the *think tank* (qv) Direct Democracy.
• REFORMING WELFARE by Nicholas Boys Smith. This paper was written before the Welfare Reform Act 2007 came into force but remains well worth reading.
• WELFARE ISN'T WORKING: THE NEW DEAL FOR YOUNG PEOPLE by Frank Field and Patrick White, published in June 2007 by the *think tank* (qv) Reform, reminds us that the aim of the New Deal in 1997 was 'nothing less than the abolition of youth unemployment', instead of which the unemployment rate for under-25s rose from 14.4% in 1997 to 14.5% ten years later.
8 Useful websites include the following. They are given in alphabetical order.
• Civitas www.civitas.org.uk/welfare/index.php
• Directgov www.direct.gov.uk/
• Frank Field www.frankfield.co.uk/type3.asp?id=20&type=:
• Office of Public Sector Information www.opsi.gov.uk/acts.htm
• Reform www.reform.co.uk/website/home.aspx

See also: *Child benefit; Council tax benefit; Housing benefit; Incapacity benefit*; and *Unemployment benefit*.

Welfare policies leading to social friction See: *Housing policy and social friction*.

Welfare reform green paper 1 IN WORK, BETTER OFF: NEXT STEPS TO FULL EMPLOYMENT, published in July 2007, contains proposals to get people into work and off benefits.
2 For publication details see the bibliography.

Welfare to Work See: *New Deal: unemployment*.

West Lothian question 1 This phrase, coined by the late Tom Dalyell MP, refers to the paradox that Westminster MPs with Scottish seats are able to vote on matters affecting English constituencies, but MPs with English constituencies have no corresponding right to vote on key Scottish questions. Thus were tuition fees imposed upon English students with the help of the votes of Scottish MPs at Westminster, whereas English MPs had no say on whether tuition fees ought to be imposed in Scotland. An Ipsos Mori poll published in March 2008 showed that the West Lothian anomaly has become the constitutional issue causing most public dissatisfaction in England.
2 This imbalance cannot prevail indefinitely. A just solution would be the creation of an English Parliament for English laws, to form a counterpart to the present Scottish Parliament (qv) for Scottish laws and Welsh Assembly for Welsh laws. No additional MPs would be required. The arrangement would simply be that on some days Westminster would deal with UK-wide matters such as finance, defence, and foreign affairs, with MPs from all three countries sitting (as at present) and the Union Flag flying from the Victoria tower. On other days it would deal with matters that solely affect England, for instance education, with only the 529 MPs for English constituencies taking part and the Cross of St George flying from the Victoria tower.

2 See also: *Constitution*; and *Government: Scotland*.

Wheelchair-accessible 1 In building design the term 'wheelchair-accessible' means that it is possible for people in *wheelchairs* (qv) to approach, enter, and use a building. For sources of guidance see: *Authoritative practical guides*.
2 In vehicle design the term means that people in wheelchairs should (a) be able to get in and out of such vehicles safely and without unreasonable difficulty; and (b) to travel in them safely and in reasonable comfort. For more detail see: *Transport: complying with the Disability Discrimination Act*.

Wheelchair-accessible housing See: *Special needs housing*.

Wheelchair-accessible vehicles See: *Accessible cars*.

Wheelchair housing design guide See: *Habinteg*.

Wheelchair platform lifts See: *Incline platform lifts*; and *Vertical platform lifts*.

Wheelchair stairlifts See: *Incline platform lifts*.

Wheelchair securing systems 1 When *wheelchair-users* (qv) travel in vehicles, two separate securing systems are required for their safety:
• Wheelchair securing systems. These are metal clamps that secure the wheelchair to the vehicle floor to prevent the chair from moving about during travel.

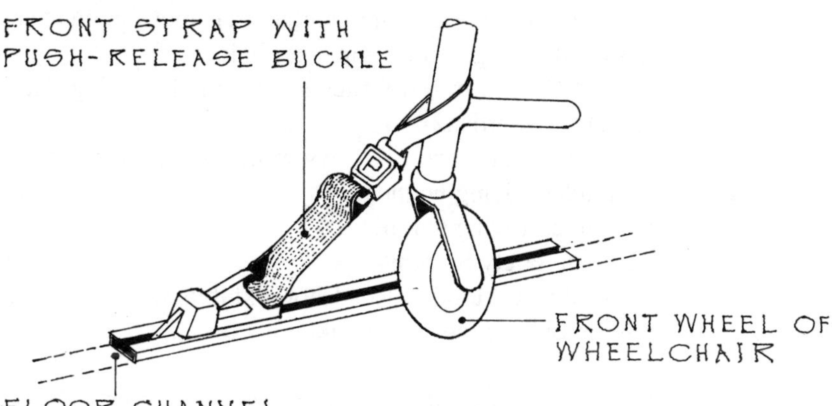

They normally comprise a pair of metal tracks bolted to the vehicle floor, into which two wheelchair clamps are inserted once the wheelchair is in position, and then locked to firmly keep it there. Such securing clamps must in all cases (a) be easy to apply; (b) hold the wheelchair very securely in place during travel; and (c) be easy and quick to release.
• Passenger restraining systems. These are belts, harnesses, or other devices to keep wheelchair-users firmly in their wheelchairs in the event of sudden braking or an accident, thus helping to protect them against injury. Passenger restraining systems must by law be separate from the *wheelchair securing system*. They are quite similar to conventional car seat belts, but of more specialised design, and include the following types:
Safety belts, which consist of a lap belt plus a diagonal belt, are almost identical to conventional car seat belts. The belts must by law be anchored to the vehicle floor.
Harnesses, which have two functions – (a) to provide restraint as above, and (b) to provide postural support for people who need it. Harnesses generally have two vertical shoulder straps joined to a waist belt (not just one diagonal strap as

above), thus keeping the wearer in an upright position. They may also incorporate a crotch strap and other special features. Again all belts and straps must be anchored to the vehicle floor.

Headrests, which help to protect passengers from whiplash injury in the event of an accident. Unlike the restraints described above they are fixed directly to the wheelchair.

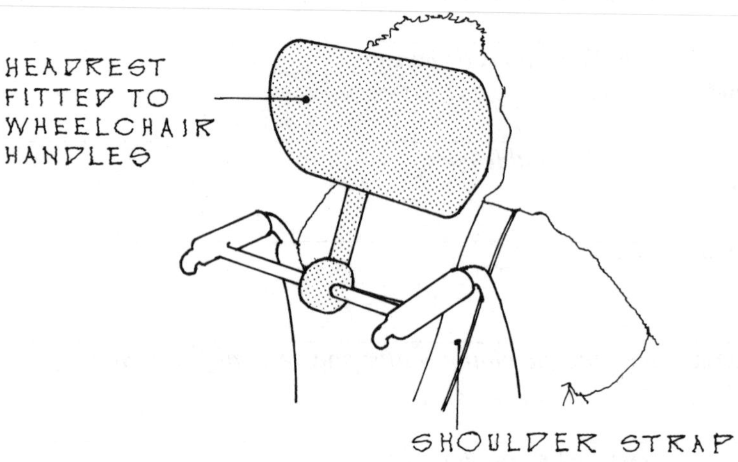

3 The above products are available in many varieties to meet the needs of particular users (including children) and to suit particular types of vehicle including cars, vans, taxis, minibuses, etc. It is essential that reputable products be used; that the matching of product to user be done by experienced experts; and that installation be carried out by specialist firms with a known track record. Ill-informed decisions or faulty installation could have fatal consequences.

4 Comprehensive lists of manufacturers and suppliers are obtainable from the Disabled Living Foundation – see below.

5 As of 2007, useful websites include the following. They are given in alphabetical order.

- Dept for Transport www.dft.gov.uk/transportforyou/access/tipws/
- Disabled Living Foundation www.dlf.org.uk/
- Disabled Living Online www.disabledliving.co.uk/
- Unwin Safety Systems www.unwin-safety.com/

6 The annual Mobility Roadshow exhibits wheelchair securing systems – visit www.mobilityroadshow.co.uk/

See also: *Wheelchair-users and public transport.*

Wheelchair-users

1 Wheelchair-users are people with *impaired mobility* (qv) who, unlike ambulant disabled persons, need a *wheelchair* (qv) in order to get about.

2 This dependence is in most cases partial. Of an estimated three quarters of a million wheelchair-users in the UK, perhaps half a million need a wheelchair for travelling out-of-doors, but have a limited ability to move short distances (for instance around the home) without the wheelchair. A smaller number, including for instance people who are *paraplegic* (qv) are quite literally wheelchair-bound.

3 These realities of life suggest the following terminology when writing or speaking of people who use wheelchairs:

- 'Wheelchair-user' correctly describes (a) the generality of people in wheelchairs; or (b) more specifically, the many people who use wheelchairs but are not completely dependent upon them.
- 'Wheelchair-bound' correctly describes a smaller number of people who are completely dependent on their wheelchairs and helpless without.

Some references, including for instance the EQUAL TREATMENT BENCH BOOK, a guide for courtroom personnel (visit www.jsboard.co.uk/), and HACKED OFF, a guide for journalists (visit www.disabilitynow.org.uk/campaigns/hackedoff/), suggest that the term 'wheelchair-bound' should never be used. This advice is demonstrably unsound, and readers should feel free to decide for themselves the most appropriate language in particular circumstances – see: *Political correctness: speech codes*.

4 In addition to the above distinction between 'wheelchair-users' and people who are 'wheelchair-bound', it may also on occasion be necessary to distinguish between the following groups:

- 'Independent users', who operate their manual or powered wheelchairs themselves and need no assistance.
- 'Assisted users', who need their manual wheelchair to be pushed by a helper.

A helper can generally manoeuvre a wheelchair through slightly narrower doorways than self-operators, but as against that a wheelchair plus helper needs more floorspace in *lifts* (qv) etc.

5 As of 2008, useful websites include:

- Wheelchair Users' Group www.wheelchairusers.org.uk/

See also: *Disability; Impaired mobility; Mobility aids*; and *Wheelchairs*.

Wheelchair-users and housing

See: *Housing: special needs*.

Wheelchair-users and public transport

1 WHEELS WITHIN WHEELS: A GUIDE TO USING A WHEELCHAIR ON PUBLIC TRANSPORT tells wheelchair-users what they can expect from trains, coaches, buses, and taxis. It has information on how they can find out where accessible services are running, gives advice on travelling in a wheelchair, and contains a wealth of other useful information. For publication details see the bibliography.

2 The Department for Transport website 'Transport for You' has a section dedicated to the needs of wheelchair-users, which is kept constantly up to date. Visit www.dft.gov.uk/transportforyou/access/tipws/

3 As of 2008, other useful websites include:

- Directgov: disabled people www.disability.gov.uk/
- Ricability www.ricability.org.uk/

See also: *Air travel*; and *Flight Rights*.

Wheelchairs: manual

1 Manual wheelchairs are a *mobility aid* (qv), and comprise one of five categories of *small personal vehicles* (qv) for frail and *disabled* (qv) people:

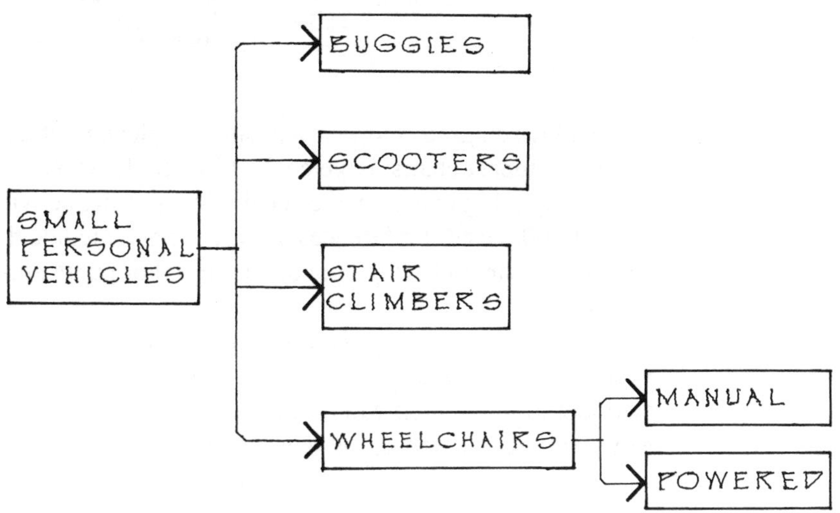

2 Wheelchair dimensions vary considerably. According to para 2.3 of INCLUSIVE MOBILITY (qv) chair lengths range from 707 mm to 1 357 mm, and widths from 511 mm to 741 mm. There is no indisputable average size, but dimensions of 1 100 mm long × 600 mm wide (reducing to 400 mm when the chair is folded), or 1 075 mm long × 630 mm wide, are ones that are commonly quoted.

3 For the purposes of *accessible* design a 'reference wheelchair' size of 1 200 mm × 700 mm has been officially adopted, especially in the design and operation of public transport vehicles – see for instance page 12 of the Rail Vehicle Accessibility Regulations 1998 (qv). The key dimensions are illustrated below, and should be used by designers wherever no better information is available. The dimensions in brackets have been added in case designers find them useful; they do not form part of the official standard.

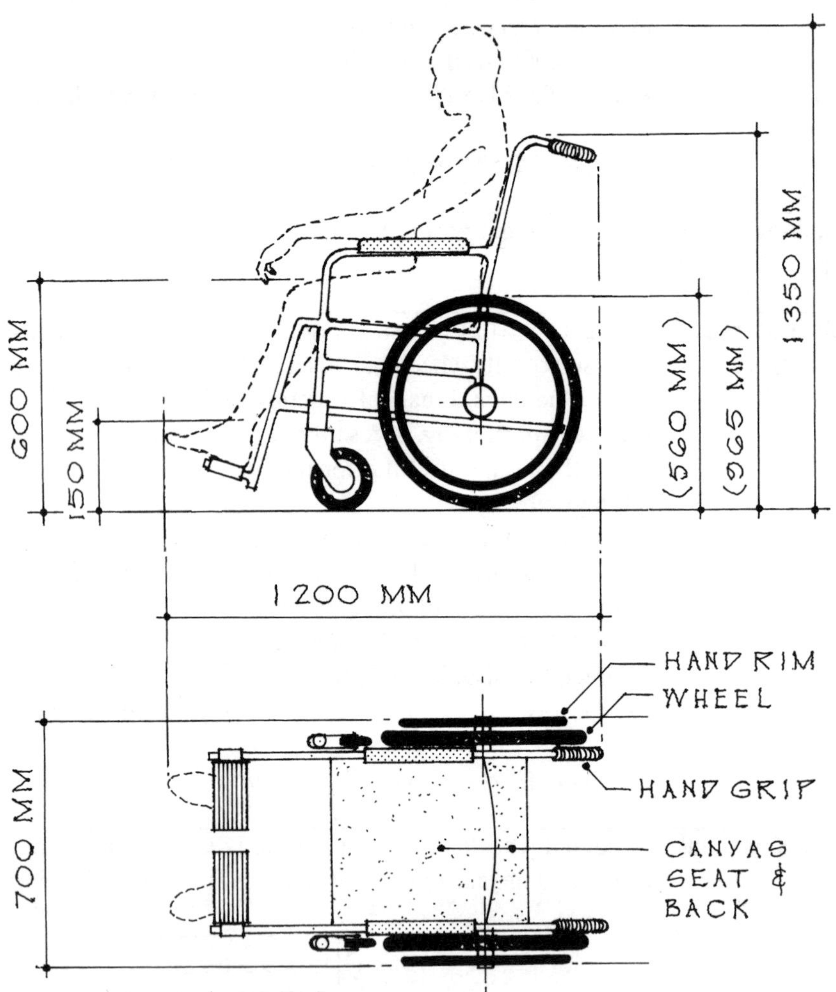

4 The above diagram shows a standard wheelchair. Other wheelchair types include compact chairs, large chairs (for people who are larger or heavier than the average); high performance chairs for sporting activities; etc.

5 As of 2008, useful references on wheelchairs include the following factsheets from the Disabled Living Foundation (qv). For publication details see the bibliography.

- CHOOSING A STANDARD SELF-PROPELLED WHEELCHAIR.
- CHOOSING AN ATTENDANT-PROPELLED WHEELCHAIR.
- CHOOSING A POWERED WHEELCHAIR.
- OUT AND ABOUT WITH YOUR WHEELCHAIR.

6 Useful websites include the following. They are given in alphabetical order.

- Dept for Transport www.dft.gov.uk/transportforyou/access/tipws/
- Disabled Living Foundation www.dlf.org.uk/
- Disabled Living Online www.disabledliving.co.uk/
- KeepAble www.keepable.co.uk/
- OTDirect www.otdirect.co.uk/
- Patient UK www.patient.co.uk/showdoc/12/
- Ricability www.ricability.org.uk/

7 The annual Mobility Roadshow exhibits wheelchairs and other personal vehicles, and allows test drives. For more information visit www.mobilityroadshow.co.uk/

See also: *Impaired mobility; Mobility aids*; and *Products for easier living*.

Wheelchairs: powered 1 Battery-powered wheelchairs, also called powerchairs or electric wheelchairs, are a *mobility aid* (qv), and comprise one of five categories of *small personal vehicles* (qv) for frail and *disabled* (qv) as shown in the previous entry. They are an excellent example of *assistive technology* (qv), helping disabled people to lead more *independent* (qv) lives. The sketch below shows a fairly common example.

2 Simple models such as the one above are basically mobility aids, helping people with *impaired mobility* (qv) to move about. There are also highly sophisticated (and expensive) tilt-in-space models as shown on the next page. These can perform many additional functions:

- They can contribute greatly to the comfort, health and happiness of severely disabled people by enabling them to lie down, sit up, stand erect, recline, or change position whenever they wish and without leaving their chair. These actions may be necessary to maintain comfort or for social, personal hygiene, or medical reasons, and being unable to perform them at will can be very distressing.
- They can help both patients ands *carers* (qv) by reducing the need to move severely disabled people from chair to bed, bed to wc, etc. Such transfers are tiring and awkward for all concerned, and also to some degree risky – see: *Hoists*.

3 The two models illustrated here are fairly arbitrary selections from a very wide range of available models. Shapes and sizes vary greatly, and specific data should be obtained from manufacturers/suppliers before any decisions are taken.

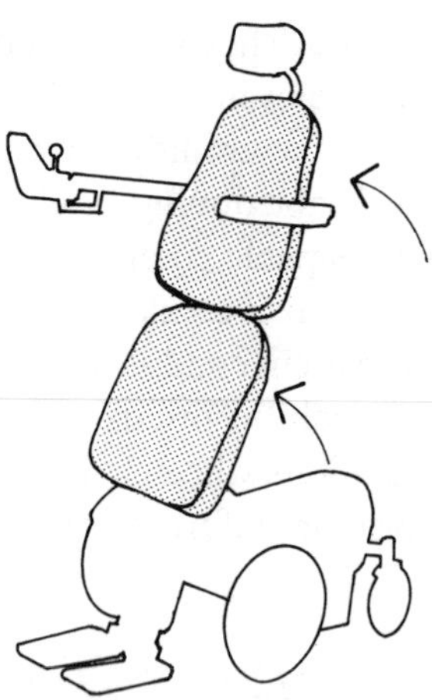

4 As of 2008, useful references include the following. For publication details see the bibliography:

- GET MOBILE – YOUR GUIDE TO BUYING A SCOOTER OR POWERED WHEELCHAIR

5 Useful websites include the following. They are given in alphabetical order.

- Dept for Transport www.dft.gov.uk/transportforyou/access/tipws/
- Disabled Living Foundation www.dlf.org.uk/
- Disabled Living Online www.disabledliving.co.uk/
- KeepAble www.keepable.co.uk/
- OTDirect www.otdirect.co.uk/
- Patient UK www.patient.co.uk/showdoc/12/
- Ricability www.ricability.org.uk/

6 The annual Mobility Roadshow exhibits scooters and other personal vehicles wheelchairs, etc, and allows test drives. For more information visit www.mobilityroadshow.co.uk/

See also: *Buggies; Mobility aids*; and *Scooters*.

Wheelchairs: space requirements

1 Buildings used by the public must in general comply with the Disability Discrimination Act (qv). For guidance on providing enough space for wheelchair use, *authoritative practical guides* (qv) include DESIGNING FOR ACCESSIBILITY (qv) and in addition the following documents. For publication details see the bibliography.

- APPROVED DOCUMENT M (qv), the provisions of which must be heeded in order to comply with the Building Regulations for England and Wales.
- BS 8300, the provisions of which apply to all buildings in the UK. It is an advisory document.
- INCLUSIVE MOBILITY: A GUIDE TO BEST PRACTICE ON ACCESS TO PEDESTRIAN AND TRANSPORT INFRASTRUCTURE is, as of 2008, the most authoritative reference for the *inclusive* design of *pedestrian environments* (qv) and of *transport-related buildings* (qv). It does not have the force of law, but conformity with its recommendations will help to demonstrate that the requirements of the Disability Discrimination Act (qv) have been met. It is a core reference for planners and designers of the built environment.
- GUIDE TO SAFETY AT SPORTS GROUNDS (qv), which is an essential reference for the design of sports facilities, and is an advisory document.

• ACCESSIBLE STADIA: a good practice guide to the design of facilities to meet the needs of disabled spectators and other users (qv), which is a more recent publication than the above guide and gives the latest recommendations on the requirements of wheelchair-users. It is an advisory document.
While the advisory documents above do not have the force of law, conformity with their recommendations will make it easier to demonstrate that the requirements of the Disability Discrimination Act have been met.
• THE WHEELCHAIR HOUSING DESIGN GUIDE (qv) is an invaluable additional reference even though *dwellings* are not, on the whole, subject to the Disability Discrimination Act.
2 The most relevant clauses of the three principal guidance documents are listed below, starting with general *ergonomic* and spatial data, and going on to specific recommendations for particular elements of the built environment.

DESIGN ASPECT	DOCUMENT: RELEVANT PARAGRAPHS		
	APPROVED DOCUMENT M	BS 8300	INCLUSIVE MOBILITY
ERGONOMIC DATA	–	ANNEX D	–
SPACE ALLOWANCES	–	ANNEX E	2.3
EXTERNAL ACCESS ROUTES	1.13; 6.9 to10	5.2 to 5.3	3.1; 12
EXTERNAL RAMPS	1.26; 6.15	5.8.3 to 5.8.4	8.4.2
LOBBIES	2.29; 3.16	6.3.6; ANNEX B	–
CORRIDORS & PASSAGEWAYS	3.14; 7.5	7.2.2	8.3
INTERNAL RAMPS	3.5.3	8.2.3 to 8.2.4	8.4.2
COUNTERS/RECEPTION DESKS	–	11.1	–
WAITING AREAS	–	11.2	9.3 to 9.4
AUDIENCE AREAS	4.12	11.3 to 11.4	–
BEDROOM AREAS	4.24	12.5	–
REFRESHMENT AREAS	4.16	13.5.3	9.4
KITCHEN AREAS	–	12.1	–
BATHROOM AREAS	5.21	12.2	–
CHANGING & SHOWER AREAS	5.18	12.3	–
LAVATORY AREAS	5.1 to 5.14; 10.3	12.4	9.6

3 The following dimensions will in most situations satisfy the above provisions, but no design should be finalised before it has been checked against the relevant paragraphs above.
• Stationary wheelchairs, such as those parked in the seating area of a theatre or sports stadium, require a minimum clear space measuring 1 400 mm long × 900 mm wide – see para 4.12 of APPROVED DOCUMENT M. But with the increasing use of *scooters* a more sensible bay size would be 1 600 mm long × 900 mm wide – see annex E to BS 8300.
• Moving wheelchairs, whether travelling in a straight line or turning 90° or 180°, require the minimum clear widths shown on the next page – for more detail refer to table 2 of APPROVED DOCUMENT M, and to the WHEELCHAIR HOUSING DESIGN GUIDE. These dimensions apply to manual wheelchairs, not to *scooters* or *buggies*.
4 For places other than England and Wales the following regulatory documents should be consulted:
• In Northern Ireland: TECHNICAL BOOKLETS E, H, and R of the Building Regulations (Northern Ireland) (qv).
• In Scotland: the BUILDING (SCOTLAND) REGULATIONS and the associated TECHNICAL HANDBOOKS (qv).

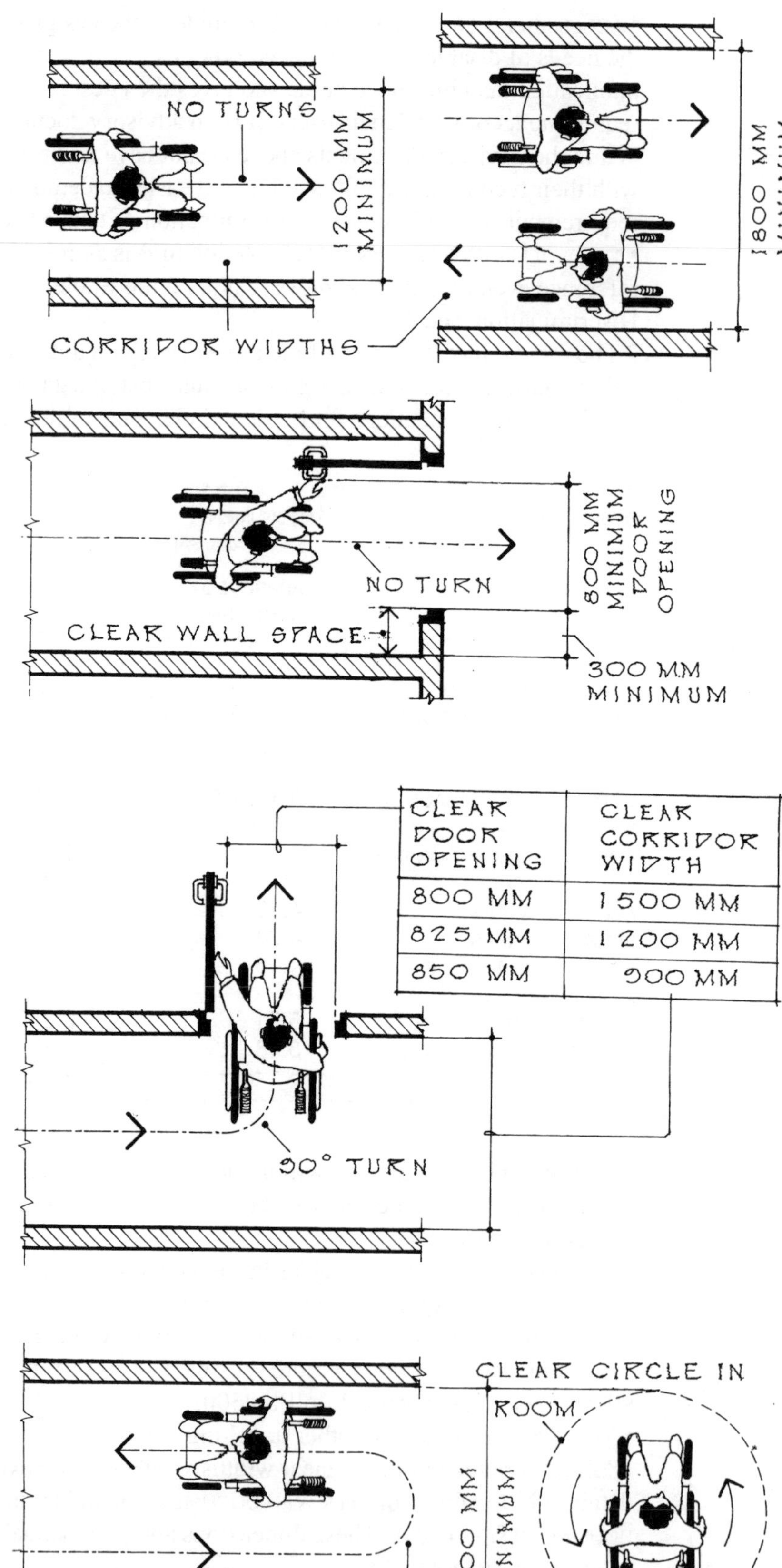

CLEAR DOOR OPENING	CLEAR CORRIDOR WIDTH
800 MM	1500 MM
825 MM	1200 MM
850 MM	900 MM

5 In the USA, where the Americans with Disabilities Act applies, refer to the ADA AND ABA ACCESSIBILITY GUIDELINES (qv).

Wheeled walking frames See: *Walking and standing aids.*

Wheelyboats 1 These are wheelchair-accessible boats that provide *disabled* (qv) anglers with *independent* (qv) access to water. Since their launch in 1985 around one hundred wheelyboats have been supplied. Visit www.wheelyboats.org/
2 For other specialised sports aids see: *Recreation and sports activities for disabled people.*

White asbestos See: *Asbestos.*

White canes See: *Canes for blind people.*

White paper 1 A statement of policy issued by the government, often forming the basis for new *legislation.*
2 A white paper is usually preceded by a consultative *green paper* (qv).

Widgit Rebus 1 Widgit Rebus is a *symbol system* (see: *Speech aids*) designed to help children and adults with *impaired communication* (qv). This includes people – especially children – who are *deaf* or have severe *learning difficulties* (qv), and cannot cope with conventional alphabetical writing.
2 Like Makaton (qv), the Rebus system consists predominantly of symbols which directly resemble the objects they represent ('pictographs'), rather than more abstract diagrams ('ideographs' etc). Some users therefore find Rebus symbols easier to learn than more sophisticated systems such as Blissymbolics. But this simplicity also means that Rebus symbols may not be able to provide the linguistic structure and richness which many children and adults need. For some people the Rebus system therefore provides a useful stepping stone to more formal systems, rather than a satisfactory solution in itself.
3 There are about 7 000 Rebus Symbols. Purely to convey the flavour, a few simple examples are shown below.

SAFETY PIN

SAILOR

SAME

4 SymbolWorld is a website created by Widgit (qv) for adults and older children who have *learning difficulties*, and prefer *symbols* to the written word. It is a screen-based magazine that includes world news, personal contributions, stories, and learning materials. It is designed for easy navigation, with large buttons for links.

5 As of 2007, useful websites include:

- Inclusive Technology www.inclusive.co.uk/
- SymbolWorld www.symbolworld.org/
- Widgit Symbols Development Project www.widgit.com/widgitrebus/

See also: *Alternative and augmentative communication; Makaton vocabulary; Mayer-Johnson communication system*; and *Speech aids*.

Windows: controls

1 Window handles should be easily grippable, and should be operable by a clenched fist with no need for gripping. Para 10.3.2 of BS 8300 (qv) recommends *lever handles* in preference to knobs.
2 Manual controls can be mounted remotely, in the case of high and other awkwardly located windows. A common type uses a turn handle as shown below. The handle can be linked to a maximum of about six individual window operators by means of a cable running within a conduit, up to a maximum conduit length of approximately 18 metres. When the handle is turned the inner cable is driven forward or backwards, thus opening or closing the windows. For simplicity the diagram shows only pivoted windows, but such gear can also operate top-hung, bottom-hung, and side-hung windows.

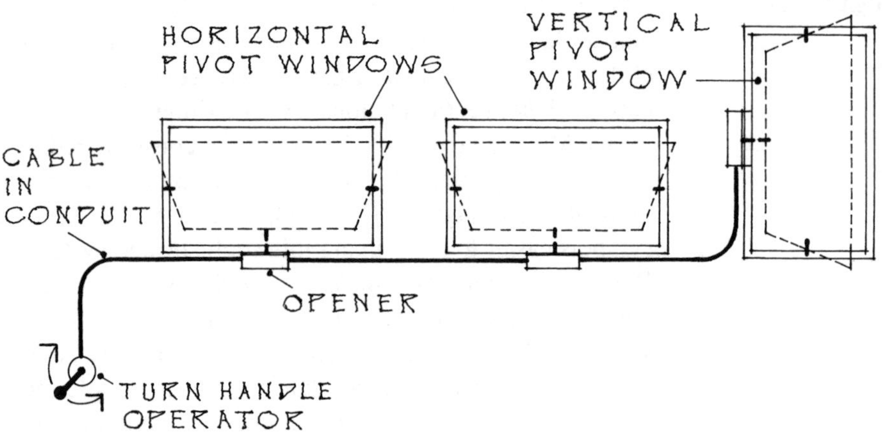

3 Powered versions of the above system are available to assist *disabled people* (qv) and others, requiring only the pressing of a switch rather than having to turn a handle.
4 Electronic controls and operating systems are the most sophisticated of all. There are two categories:

- Remotely-controlled openers/closers are operated by a person using a keypad or some similar device, from a distance – for instance sitting in bed or in a wheelchair. See: *Environmental control systems*.
- Automatic openers/closers activate and control themselves without human intervention. See: *Smart homes*.

5 Powered openers/closers can interface with burglar alarms, smoke alarms, fire alarms, timing devices, thermostats, and rain detectors. The window can also be fitted with remotely-controlled curtain or blind openers/closers. All these options are currently expensive, but costs may be expected to come down over time.
6 Useful websites include:

- Disabled Living Foundation www.dlf.org.uk/
- Disabled Living Online www.disabledliving.co.uk//
- Possum Controls Limited www.possum.co.uk/

See also: *Environmental control systems; Remote control*; and *Smart homes*.

Windows: dwellings

1 The following references contain authoritative recommendations for window design in dwellings:

• APPROVED DOCUMENT K (qv) of the Building Regulations, which deals with safety. Its provisions must be heeded in order to comply with the Building Regulations (qv). Pages 14–15 give guidance on protection against collision with open windows.

• THE HABINTEG HOUSING ASSOCIATION DESIGN GUIDE (qv) applies most specifically to *housing association* (qv) schemes, but ought to be by the designer's side in the design of all housing developments and individual dwellings.

• THE WHEELCHAIR HOUSING DESIGN GUIDE (qv), also from the Habinteg Housing Association, is another illustrated manual whose usefulness extends far beyond the special remit implied by the title.

• THE HOUSING DESIGN GUIDE applies specifically to housing in London.

For publication details of the above references see the bibliography.

Windows: public buildings: complying with the Disability Discrimination Act

1 Buildings used by the public must in general comply with the Disability Discrimination Act (qv). In the case of windows in public buildings, *authoritative practical guides* (qv) include DESIGNING FOR ACCESSIBILITY (qv) and in addition the following documents. For publication details see the bibliography.

• APPROVED DOCUMENT K (qv) of the Building Regulations, which deals with safety. Its provisions must be heeded in order to comply with the Building Regulations (qv).

• BS 8300, the provisions of which apply to all buildings in the UK. BS 8300 does not have the force of law, but conformity with its recommendations will make it easier to demonstrate that the requirements of the Disability Discrimination Act have been met.

• The two documents listed in the previous entry are useful additional references even though they refer specifically to dwellings, which – except for the public spaces in blocks of flats – do not generally come within the ambit of the Disability Discrimination Act.

• The most relevant clauses of these documents are tabulated below.

DESIGN ASPECT	DOCUMENT: RELEVANT PARAGRAPHS			
	AD K	BS 8300	DESIGN GUIDE	WHEELCHAIR HOUSING DESIGN GUIDE
PROTECTION AGAINST COLLISION	K4	–	–	–
GENERAL REQUIREMENTS	–	–	3.11.1	14
HEIGHTS OF SILLS & TRANSOMS	K2	10.3.1	3.11.2	14
CONTROLS	–	10.3.2	3.11.3	14
FINISHES	–	–	3.11.3	–
GLAZING	K2	–	3.11.4	14

The dimensions on the diagram on the next page will in most situations satisfy the above provisions, though no design should be finalised before it has been checked against the relevant paragraphs above.

• Figure A shows precautions against the risk of falling, and applies only to ground floor windows. For more detail refer to APPROVED DOCUMENT K of the Building Regulations.

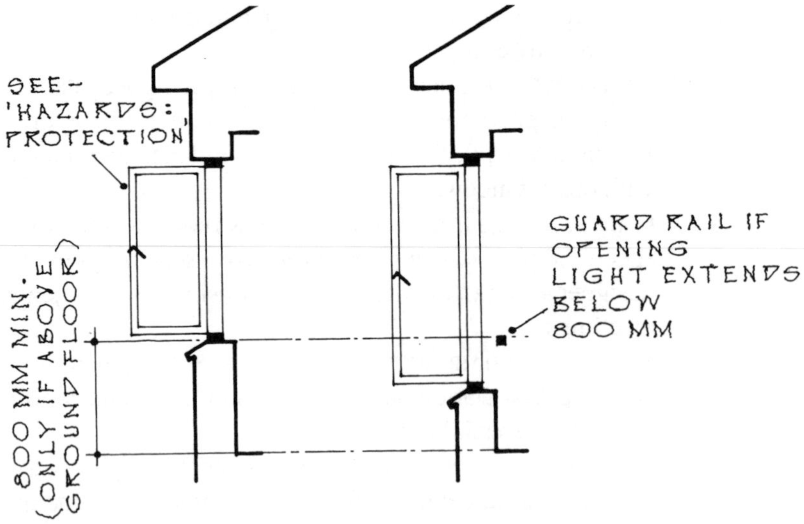

SEE – 'HAZARDS: PROTECTION'
GUARD RAIL IF OPENING LIGHT EXTENDS BELOW 800 MM
800 MM MIN. (ONLY IF ABOVE GROUND FLOOR)

(A) SAFETY

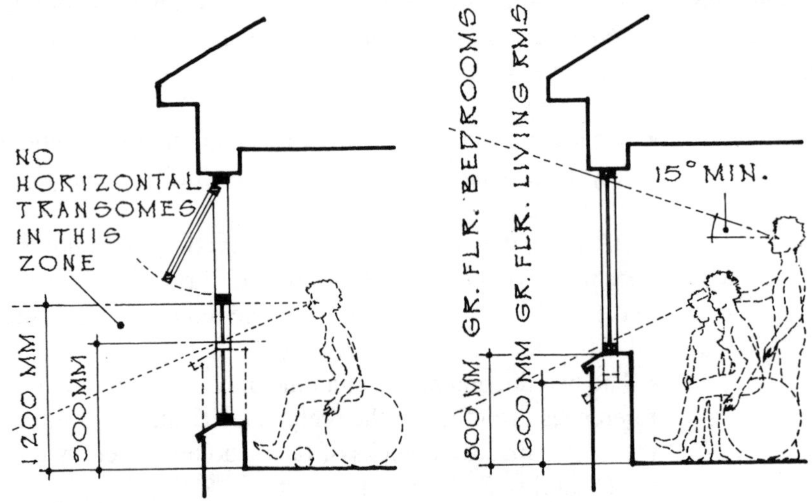

NO HORIZONTAL TRANSOMES IN THIS ZONE
1200 MM
900 MM
800 MM GR. FLR. BEDROOMS
600 MM GR. FLR. LIVING RMS
15° MIN.

(B) VIEW OUT

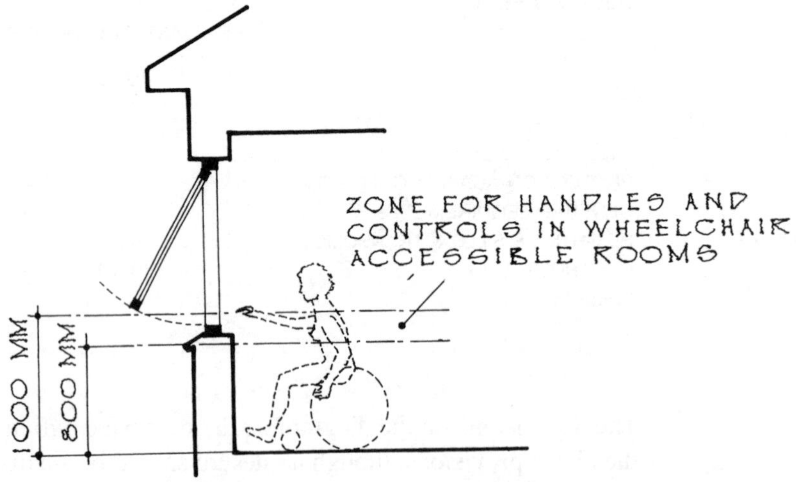

ZONE FOR HANDLES AND CONTROLS IN WHEELCHAIR ACCESSIBLE ROOMS
1000 MM
800 MM

(C) CONTROLS

• Figure B shows suitable dimensions to ensure a good view out. For more detail on the diagram at the left refer to BS 8300. For more detail on the diagram at the right refer to Habinteg's DESIGN GUIDE for housing. Unless their main purpose is ventilation, windows should in general be vertical so that people looking out can see the immediate foreground (for instance the flower beds and gardens right outside the window); the middle distance and horizon; and the sky. Horizontal slot shapes generally do not give a satisfactory view out and should be used with caution.
• Figure C shows suitable heights for all window controls, both manual and powered, and is taken from BS 8300.

Women Unit (WEU) See: *Discrimination; Equality; Incomes: gender pay gap*; and *Sex discrimination.*

Women and Equality Visit www.womenandequalityunit.gov.uk/

Women and Work Commission See: *Incomes: gender pay gap.*

Women: Minister for See: *Minister for Women.*

Women's National Commission. Visit http://www.thewnc.org.uk/ and see: *Government Equalities Office.*

Work Foundation See: *Incomes: inequality: what can be done?*

Workfare 1 Workfare schemes, which operate in one form or another in the USA, Australia, and India, typically require able-bodied young people to work or study for up to 40 hours a week in exchange for *benefit* (qv). If they fail to find work within a certain period (eg one week) they are told to turn up at an allotted time and place to spend eight hours under supervision clearing litter, removing graffiti, or doing some other useful task. In Britain such schemes could in principle be run by *Jobcentres* (qv) or by approved contractors.
2 The rationale behind such schemes is threefold:
• If young people are able to stay at home and do no work for a fair length of time, for whatever good reason, they will probably find it almost impossible thereafter to get back into work (a) because they have lost the habit, and (b) because they can demonstrate no useful experience to prospective employers. A lifetime of unemployment is bad for them, their future families, and the rest of society.
• If taxpayers see out-of-work young people do useful things for the community they are less likely to resent their taxes going into public welfare schemes.
• More fundamentally: it is simply wrong for people who can work to receive money from others without giving something in return.
3 For more information visit http://en.wikipedia.org/wiki/Workfare

See also: *Jobseeker's Allowance; Neets;* and *Welfare dependency.*

Working dogs See: *Assistance dogs.*

Working Neighbourhoods Fund See: *Neighbourhood renewal.*

Workplace assessments	1 Under the Disability Discrimination Act (qv) employers may be required to make *adjustments* (qv) to the workplaces of *disabled* (qv) employees, or applicants for employment. The required measures are best identified by a workplace assessment carried out by an *occupational therapist* (qv) 2 For more information visit www.cot.co.uk/ See also: *Adjustments under the Disability Discrimination Act*; and *Occupational Therapists.*
Workplaces and risk assessments	See: *Risk assessment.*
Workplaces: safety	Visit www.hse.gov.uk/
Worktops: kitchens in dwellings	See: *Kitchens.*

Worktops: kitchens in public buildings

1 For kitchen design in general see: *Kitchens: complying with the Disability Discrimination Act*. For kitchen worktop heights, *authoritative practical guides* (qv) which will make it easier for building designers to demonstrate that the requirements of the Disability Discrimination Act have been met include the following:

2 APPROVED DOCUMENT M (qv), the provisions of which must be heeded in order to comply with the Building Regulations in England and Wales. The following provisions apply:

• In shared kitchens used by *wheelchair-users* (qv) and others for tea-making, etc, para 4.16 recommends a worktop height of 850 mm, with a clear height of at least 700 mm underneath.

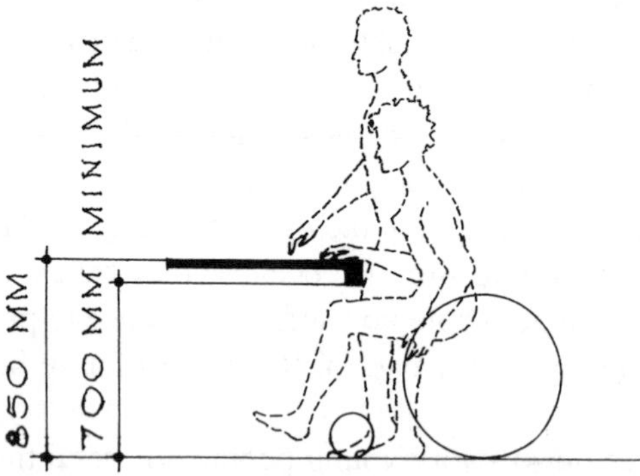

• In pubs, restaurants, cafes, and other refreshment facilities, para 4.16 recommends that one part of every serving counter should be at a height of not more than 850 mm, to suit people in wheelchairs. The remainder should be higher, to suit people standing.

3 BS 8300 (qv), which applies to all buildings in the UK and is an advisory document. The following provisions apply:

• Para 12.1 gives extensive advice on worktop heights and layouts in kitchens. Where kitchens are used by people in wheelchair and others, worktops should be provided at heights of 760 mm for those in wheelchairs, and 900 mm for people standing. Where a single worktop height must suffice for both, 850 mm is recommended. The diagram on the next page summarises these recommendations.

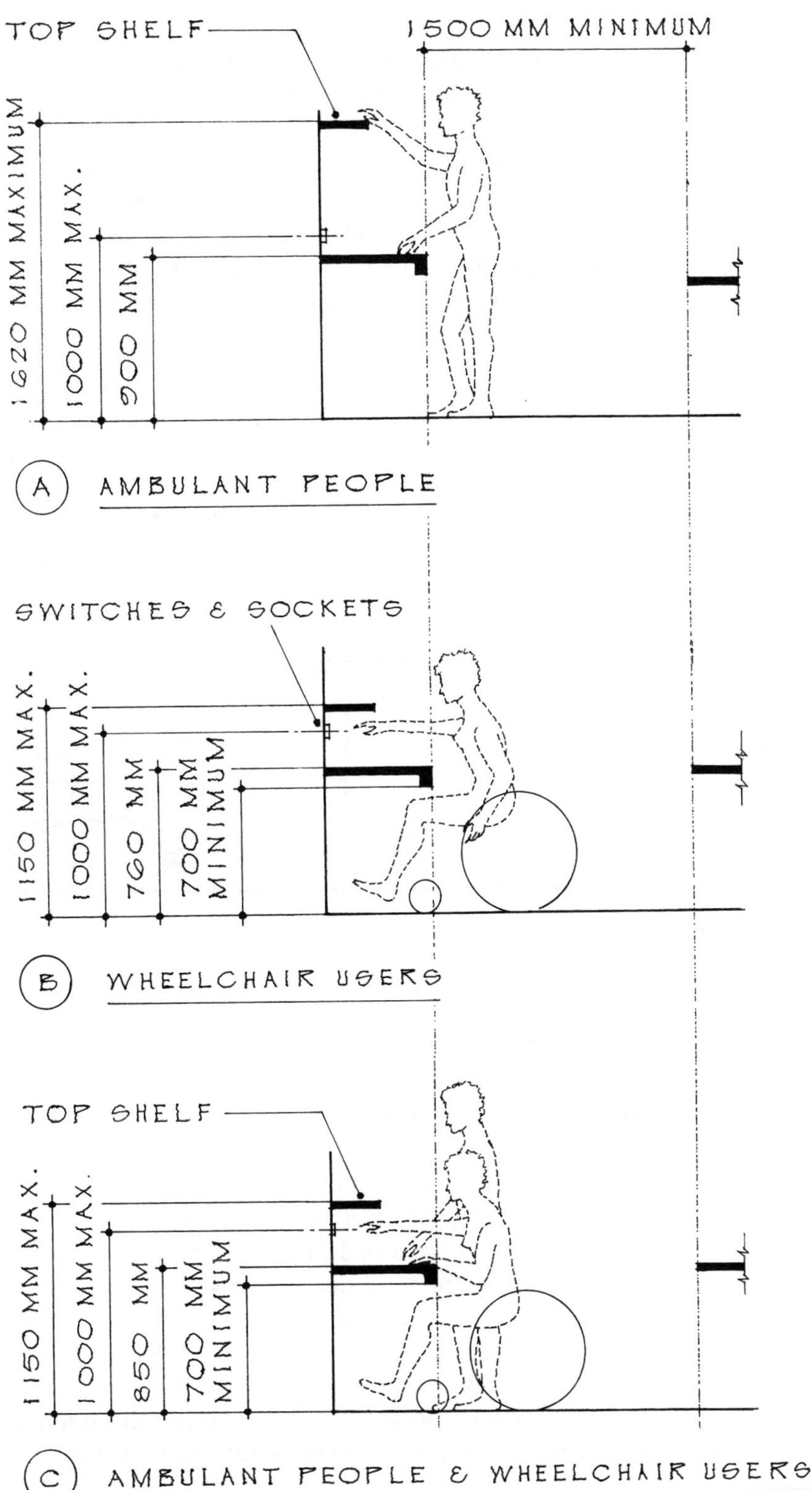

• Para 13.5.3 gives advice on worktop heights in pubs, restaurants, cafes, and other refreshment buildings – for details see: *Refreshment facilities: complying with the Disability Discrimination Act*.

See also: *Counters and reception desks; Kitchens; Powered lifters; Refreshment facilities*; and *Tables*.

Worktops: office

1 BODYSPACE by Stephen Pheasant (see below) reviews the available evidence on suitable desk and worktop heights and makes the following general recommendations:

• Adjustable-height worktops are the ideal, especially if the work is intensively screen-based. All things considered, including the need for adequate kneeroom beneath the worktop, a range of adjustment between 660 mm and 770 mm is probably about right.
• If an adjustable desktop cannot be provided, then recommended fixed worktop heights are 735 mm for men; 705 mm for women; and 720 mm for working populations comprising both.
Chair height must always be adjustable. The diagram summarises some of the key dimensions suggested in the above reference.

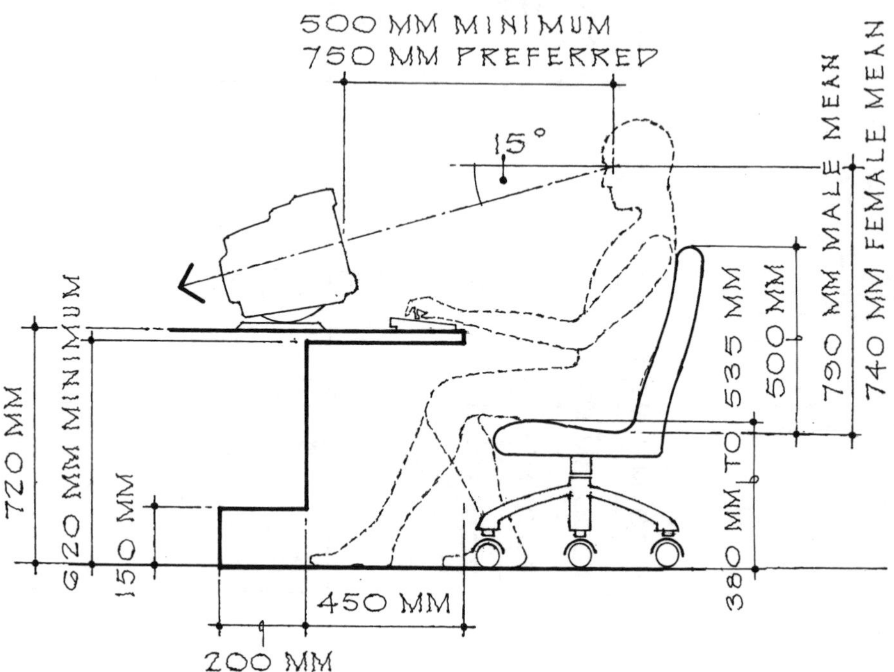

2 As of 2008, useful references include the following. They are given in alphabetical order. For publication details see the bibliography.
• BODYSPACE: ANTHROPOMETRY, ERGONOMICS AND THE DESIGN OF WORK by Stephen Pheasant is a respected and widely used reference, written in a clear and lively style. It gives guidance on long-term basic principles, not on the latest official recommendations, which are subject to change. The main respect in which some of its basic data may need revisiting is that both children and adults have become fatter since it was written – see: *Obesity*.
• BS EN ISO 9241–5: 1999 ERGONOMIC REQUIREMENTS FOR OFFICE WORK WITH VISUAL DISPLAY TERMINALS: WORKSTATION LAYOUT AND POSTURAL REQUIREMENTS is an authoritative official reference.
• THE METRIC HANDBOOK, which contains distilled information on a wide range of design-related matters, gives useful summary data.
3 Useful websites include:
• University of Nottingham www.virart.nott.ac.uk/

See also: *Anthropometrics; Ergonomics*; and *Powered lifters*.

World Health Organisation (WHO) Visit www.who.int/en/

Wrapping rage See: *Product packaging*.

Wrist-operated taps See: *Taps*.

Writing aids

1 People with *impaired dexterity* (qv), caused for instance by arthritis, can be helped to lead more *independent* (qv) and satisfying lives by a range of writing *aids and equipment* (qv) ranging from *low-tech* (simple manual devices) to *high-tech* (sophisticated electronic equipment).

2 Low-tech aids include the following:

• Large diameter grips, in the shape of a cylinder or ball that can be fitted to pens and pencils to help people who find a standard diameter pen too slim to grasp firmly. They are generally made of soft pvc so as to fit onto any common pen thickness.

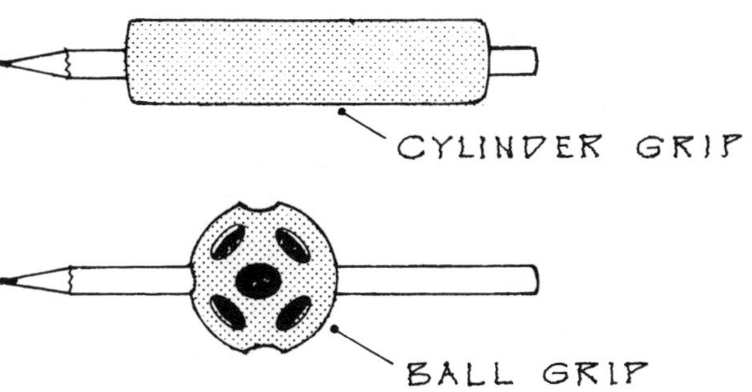

• Small diameter grips, in the shape of textured or knobbly sleeves that can be slid onto pens and pencils to give a more readily grippable surface. Again the usual material is soft pvc.

• Loops which are attached to the pen, and fitted round the hand, to help people with a weak grip maintain the pen in position.

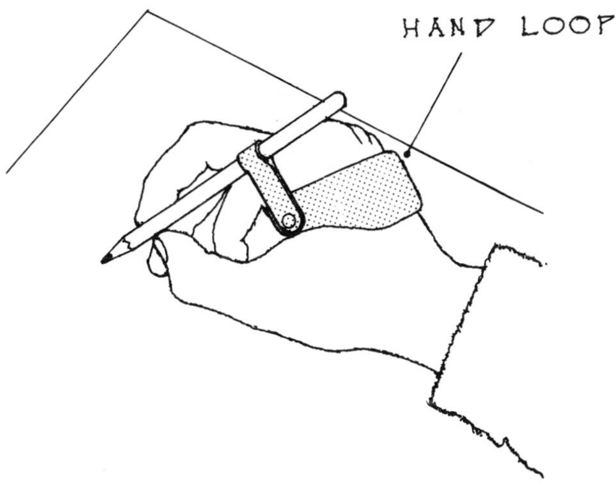

• Weighted pens etc. to reduce tremor, caused for instance by Parkinson's disease.

• Special writing slopes which can be placed on a table or desktop to support the paper at the most convenient angle for writing. They are usually made of clear plastic.

• Wheelchair trays which can be placed across the arms of a *wheelchair*, and comprise a writing surface plus moulded compartments for pens, pencils, and rubbers etc. They enable *wheelchair-users* (qv) to write while sitting in their wheelchairs.

3 High-tech aids include the *computer* (qv). If a standard computer set-up is not able to meet a particular user's needs then organisations such as AbilityNet (qv) can advise on modifications.

4 As of 2008, useful websites include the following. They are given in alphabetical order.

• AbilityNet www.abilitynet.org.uk/

- Disabled Living Foundation www.dlf.org.uk/
- Disabled Living Online www.disabledliving.co.uk/
- Homecraft Rolyan www.homecraftabilityone.com/
- KeepAble www.keepable.co.uk/

See also: *Aids and equipment; Impaired dexterity; Products for easier living*; and *Reading aids*

Yachting See: *Recreation and sports aids.*

Your Right to Know See: *Freedom of Information Act.*

Youth crime See: *Crime: youth.*

Zebra crossings See: *Pedestrian crossings.*

Zimmer frames See: *Walking frames.*

BIBLIOGRAPHY

The following titles are referred to in the main text. Readers are urged to buy books through local bookshops wherever possible: without customer support these low-profit businesses are at risk of disappearing from our high streets and town centres, to the great detriment of civilised life. In cases of difficulty all books, including ones published abroad, can be bought online from www.amazon.co.uk/. Most out-of-print (o/p), secondhand, and rare titles can be ordered online via www.abebooks.co.uk/.

A list of core references for building designers wishing to comply with the all-important Disability Discrimination Act (qv) is suggested in the main text under: *Authoritative practical guides*.

100 YEARS OLD – AND LOOKING FRAIL by Eric Goldby.
Guardian, 5 July 2007. Visit http://society.guardian.co.uk/crimeandpunishment/comment/0,,2116880,00.html

THE 2008 LEXICON: A GUIDE TO CONTEMPORARY NEWSPEAK.
2007, 30pp, £5.00. Published by and obtainable from the Centre for Policy Studies. May be viewed free of charge at www.cps.org.uk/

2020 VISION: REPORT OF THE TEACHING AND LEARNING IN 2020 REVIEW GROUP.
2006, 60pp, free. May be downloaded from www.publications.teachernet.gov.uk/eOrderingDownload/6856-DfES-Teaching%20and%20Learning.pdf

21ST CENTURY LIBRARIES: CHANGING FORMS, CHANGING FUTURES.
2004, 23pp, free. Published by Building Futures and may be downloaded from www.buildingfutures.org.uk/pdfs/pdffile_31.pdf

ABC: A BALANCED CONSTITUTION FOR THE 21ST CENTURY by Martin Howe.
2006, 23pp, £5.00. Published by and obtainable from Politeia. May be downloaded free of charge from www.politeia.co.uk/

ABSENT MINDS: INTELLECTUALS IN BRITAIN by Stefan Collini.
2006, 526pp, £25. ISBN: 0 199 29105 5. Published by Oxford University Press.

THE ACADEMIC EXPERIENCE OF STUDENTS IN ENGLISH UNIVERSITIES by Tom Sastry and Bahram Bekhradnia.
2007, 175pp, free. Published by the Higher Education Policy Institute and may be downloaded from www.hepi.ac.uk/downloads/33TheacademicexperienceofstudentsinEnglishuniversities2007.pdf

ACCESS AND THE DDA: A SURVEYOR'S HANDBOOK by Mark Ratcliff.
2007, £39.95. ISBN-13: 9781842193211. Published by and obtainable from the RICS at www.ricsbooks.com/productInfo.asp?product_id=16546

ACCESS AUDIT HANDBOOK by Alison Grant.
2005, 132pp, £35. Published by the Centre for Accessible Environments (qv) and the RIBA (qv), and obtainable from www.cae.org.uk/ or www.ribabookshops.com/

ACCESS AUDIT PRICE GUIDE.
2005, 300pp, £90. Published by and obtainable from the Building Cost Information Service at 020 7695 1500 or sales@bcis.co.uk

ACCESS AUDITS: A PLANNING TOOL FOR BUSINESS.
2003, DVD/CD-ROM, £12.00. Published by and obtainable from the Centre for Accessible Environments (qv) at www.cae.org.uk/

ACCESS AUDITS OF PRIMARY HEALTHCARE FACILITIES. Health Facilities Note 20.
1997, 66pp, £75.00. ISBN: 0 11 322056 1. Obtainable from the Centre for Accessible Environments (qv) at www.cae.org.uk/

ACCESS BY DESIGN.
Quarterly journal of the Centre for Accessible Environments (qv). For subscription details visit www.cae.org.uk/

ACCESS CONTROL USER GUIDES.
Free. Published by and obtainable from the British Security Industry Association at www.bsia.co.uk/

ACCESS FOR DISABLED PEOPLE – DESIGN GUIDANCE NOTE.
2002, 72pp, £10.50. ISBN: 1 86078 149 7. Published by Sport England and obtainable from the Centre for Accessible Environment (qv) at www.cae.org.uk/

ACCESS FOR DISABLED PEOPLE TO ARTS PREMISES TODAY: THE JOURNEY SEQUENCE by C Wycliffe Noble and Geoffrey Lord.
2003, 192pp, £50.00. ISBN: 0 7506 5779 0. Published by Architectural Press, an imprint of Elsevier at http://books.elsevier.com/

ACCESS FOR DISABLED PEOPLE TO SCHOOL BUILDINGS.
See: BUILDING BULLETIN 91.

ACCESS JOURNAL.
Quarterly journal of the Access Association (qv) and the JMU. For subscription details visit www.jmuaccess.org.uk/28.asp

THE ACCESS MANUAL: AUDITING AND MANAGING INCLUSIVE BUILT ENVIRONMENTS by Ann Sawyer and Keith Bright.
2005, 256pp, £39.99. ISBN: 1 4051 4626 5. Published by Blackwell Publishing at http://bookshop.blackwell.co.uk/

ACCESS STATEMENTS: ACHIEVING AN INCLUSIVE ENVIRONMENT BY ENSURING CONTINUITY THROUGHOUT THE PLANNING, DESIGN AND MANAGEMENT OF BUILDINGS AND SPACES.
2004, 40pp. Published by the Disability Rights Commission and obtainable from the Construction Information Service at http://products.ihs.com/cis/Doc.aspx?AuthCode=&DocNum=267006

ACCESS TO AIR TRAVEL FOR DISABLED PEOPLE: CODE OF PRACTICE.
2006, 73pp, free. May be downloaded from www.dft.gov.uk/transportforyou/access/aviationshipping/

ACCESS TO AIR TRAVEL: GUIDANCE FOR DISABLED AND LESS MOBILE PASSENGERS.
2004, 65pp, free. May be downloaded from www.dft.gov.uk/transportforyou/access/aviationshipping/

ACCESS TO BUILDINGS by Stephen Garvin.
2008, 12pp, £15.00. ISBN-13: 978 1 84806 031 9. Published by and obtainable from IHS BRE Press at www.ihsbrepress.com/

ACCESS TO ATMS: UK DESIGN GUIDELINES by Robert Feeney.
2002, 72pp, £20.00. Published by and obtainable from the Centre for Accessible Environments(qv) at www.cae.org.uk/

ACCESSIBILITY SPECIFICATION FOR SMALL BUSES DESIGNED TO CARRY 9 TO 22 PASSENGERS.
2001, 39pp, free. Published by and obtainable from DPTAC (qv). May be downloaded from www.dptac.gov.uk/buses.htm

ACCESSIBLE HOUSING: QUALITY, DISABILITY AND DESIGN by Rob Imrie.
2005, 264pp, £37.00. ISBN: 0 415 31892 0. Published by Taylor and Francis.

ACCESSIBLE INFORMATION – RNIB CLEAR PRINT GUIDELINES.
2006, 3pp, free. May be downloaded from www.rnib.org.uk/

ACCESSIBLE LONDON: ACHIEVING AN INCLUSIVE ENVIRONMENT.
2004, 82pp, free. May be downloaded from www.london.gov.uk/mayor/strategies/sds/docs/spg_accessible_london.pdf

ACCESSIBLE STADIA.
2003, 106pp, £25.00. ISBN: 0 9546293 0 2. Published by the Football Stadia Improvement Fund and the Football Licensing Authority, and obtainable from www.flaweb.org.uk/docs/specsafe/pubs/acssstad.php

ACCESSIBLE THRESHOLDS IN NEW HOUSING: GUIDANCE FOR HOUSE BUILDERS AND DESIGNERS.
1999, 22pp, £8.00. ISBN: 0 11 702333 7. Obtainable from the Centre for Accessible Environments (qv) at www.cae.org.uk/

ACCOMMODATING PUPILS WITH SPECIAL EDUCATIONAL NEEDS AND DISABILITIES IN MAINSTREAM SCHOOLS.
See: BUILDING BULLETIN 94.

ACOUSTIC GUIDES.
2006, free. Published by and obtainable from Armstrong on telephone 0800 371 849 at www.armstrong-ceilings.co.uk/

ADA AND ABA ACCESSIBILITY GUIDELINES FOR BUILDINGS AND FACILITIES.
visit www.access-board.gov/ada-aba/final.htm

ADAM SMITH AND THE PURSUIT OF PERFECT LIBERTY by James Buchan.
2007, 208pp, £7.99. ISBN: 1 861 97940 1. Published by Profile Books.

AFTER CLIMBIÉ, CHILDREN ARE AT EVEN MORE RISK by Jill Kirby.
2007, 2pp, free. Published by the Centre for Policy Studies and may be downloaded from www.cps.org.uk/search/default.asp

AGAINST THE ODDS: AN INVESTIGATION COMPARING THE LIVES OF CHILDREN ON EITHER SIDE OF BRITAIN'S HOUSING DIVIDE.
2006, 48pp, free. Published by Shelter and may be downloaded from http://england.shelter.org.uk/

AGE DISCRIMINATION: THE NEW LAW by Simon Cheetham with a chapter by Esther White. 2006, £35.00. ISBN-13: 978 1 84661 026 4. Published by Jordans at www.jordanpublishing.co.uk/

AM I MAKING MYSELF CLEAR?
2002, 31pp, free. Published by Mencap and may be downloaded from www.mencap.org.uk/download/making_myself_clear.pdf

AN AGE EQUALITY DUTY: THE TIME HAS COME.
2004, 47pp, £10. Published by and obtainable from Age Concern on telephone 020 8765 7200 or at www.ace.org.uk/

AN ANALYSIS OF THE VALUE ADDED BY SECONDARY SCHOOLS IN ENGLAND: IS THE VALUE ADDED INDICATOR OF ANY VALUE? by Jim Taylor and Anh Ngoc Nguyen.
2008, Oxford Bulletin of Economics and Statistics. For more details contact Professor Jim Taylor at jim.taylor@lancaster.ac.uk

ANALYSIS OF THE NATIONAL CHILDHOOD OBESITY DATABASE 2005–06 by Crowther, Dinsdale, Ruttin and Kyffin.
2007, 23pp, free. May be downloaded from www.dh.gov.uk/assetRoot/04/14/21/56/04142156.pdf

ANTI-SOCIAL BEHAVIOUR LAW by Paul Greatorex and Damian Falkowski.
2006, £49.00. ISBN-13: 978 1 84661 002 8. Published by Jordans at www.jordanpublishing.co.uk/

ANTI-SOCIAL BEHAVIOUR ORDERS: A SUMMARY OF RESEARCH INTO ANTI-SOCIAL BEHAVIOUR ORDERS GIVEN TO YOUNG PEOPLE BETWEEN JANUARY 2004 AND JANUARY 2005.
2006, 20pp, £8.00. Published by and obtainable from the Youth Justice Board at telephone 0870 120 700, or may be downloaded free of charge from www.yjb.gov.uk/

APPROVED DOCUMENTS.
The following Approved Documents (qv) of the Building Regulations for England and Wales are referred to in this work.
APPROVED DOCUMENT B: FIRE SAFETY.
APPROVED DOCUMENT E: RESISTANCE TO THE PASSAGE OF SOUND.
APPROVED DOCUMENT K: PROTECTION FROM FALLING ETC.
APPROVED DOCUMENT M: ACCESS TO AND USE OF BUILDINGS.
APPROVED DOCUMENT N: GLAZING SAFETY.
All Approved Documents can be bought from The Stationery Office at www.tsoshop.co.uk/bookstore.asp?FO=1167765, or downloaded free of charge from www.planningportal.gov.uk/

ARCHITECTURAL IRONMONGERY by Alison Grant.
2005, 40pp, £15.00. ISBN: 1 85946 170 0. Published by the Centre for Accessible Environments (qv) and the RIBA (qv), and obtainable from www.cae.org.uk/ or www.ribabookshops.com/

THE ARCHITECTURE OF DEMOCRACY by Allan Greenberg.
2006, 204pp, $50.00. ISBN: 0 8478 2793 3. Published by Rizzoli International Publications in New York and obtainable from Amazon UK at www.amazon.co.uk/

ARE SCHOOLS BEING INSPECTED TO DEATH? By Melanie Read.
The *Times*, 31 March 2008. Visit www.timesonline.co.uk/tol/comment/columnists/melanie_reid/article3648821.ece

ARTIFICIAL LEGACY: BLAIR'S SCHOOL ACHIEVEMENT IS A SHAM by Anastasia de Waal and Nicholas Cowen.
2007, 9pp, free. Published by Civitas (qv) and may be downloaded from www.civitas.org.uk/pdf/deWaal_artificialachievement_June07.pdf

THE ASBO: WRONG TURNING, DEAD END by Neil Wain and Elizabeth Burney.
2007, 141pp, £12.95. ISBN-13: 978 0 903683 99 9. Published by and obtainable from the Howard League for Penal reform at www.howardleague.org/

ASPIRING TO EXCELLENCE: FINAL REPORT OF THE INDEPENDENT INQUIRY INTO MODERNISING MEDICAL CAREERS led by Professor Sir John Took.
2007, 75pp, free. May be downloaded from www.mmcinquiry.org.uk/MMC_FINAL_REPORT_REVD_4jan.pdf

ASSESSING SECONDARY SCHOOL DESIGN QUALITY: RESEARCH REPORT.
2006, 101pp, free. Published by CABE and may be downloaded from www.cabe.org.uk/AssetLibrary/8736.pdf

AUTOMATIC DOOR SYSTEMS by Alison Grant.
2005, 36pp, £15.00. ISBN: 1 85946 171 9. Published by the Centre for Accessible Environments (qv) and the RIBA (qv), and obtainable from www.cae.org.uk/ or www.ribabookshops.com/

AVOIDING DISABILITY DISCRIMINATION IN TRANSPORT: A PRACTICAL GUIDE FOR BUSES AND SCHEDULED COACHES.
2007, 38pp, free. May be downloaded from www.equalityhumanrights.com/pages/eocdrccre.aspx

AVOIDING DISABILITY DISCRIMINATION IN TRANSPORT: A PRACTICAL GUIDE FOR TOUR COACH OPERATORS.
2007, 40pp, free. May be downloaded from www.equalityhumanrights.com/pages/eocdrccre.aspx

AVOIDING REGULATORY CREEP.
2004, 50pp, free. Published by the Better Regulation Task Force (qv) and may be downloaded from www.brc.gov.uk/upload/assets/www.brc.gov.uk/hiddenmenace.pdf

BABY CARRIERS.
2002, 24pp, 2002, free. Published by Ricability and may be downloaded from www.ricability.org.uk/reports/report-childcare.htm

BARKER REVIEW: DELIVERING STABILITY – SECURING OUR FUTURE HOUSING NEEDS.
2004, 149pp, free. May be downloaded from www.hm-treasury.gov.uk/media/E/3/barker_review_report_494.pdf

BCIS ACCESS AUDIT PRICE GUIDE.
Regularly updated. For latest information visit www.bcis.co.uk/

BENEFITS SIMPLIFICATION: 7TH REPORT OF SESSION 2006–2007, HOUSE OF COMMONS WORK AND PENSIONS COMMITTEE.
2007, 117pp, free. May be downloaded from www.publications.parliament.uk/pa/cm200607/cmselect/cmworpen/463/463i.pdf

BEST PRACTICE GUIDANCE: WHEELCHAIR ACCESSIBLE HOUSING: DESIGNING HOMES THAT CAN BE EASILY ADAPTED FOR RESIDENTS WHO ARE WHEELCHAIR USERS.
see: WHEELCHAIR ACCESSIBLE HOUSING.

BETTER HOMES, GREENER CITIES by Alan W Evans and Oliver Marc Hartwich.
2006, 59pp, free. May be downloaded from www.policyexchange.org.uk/

BETTER REGULATION FOR CIVIL SOCIETY: MAKING LIFE EASIER FOR THOSE WHO HELP OTHERS.
2005, 82pp, free. Published by the Better Regulation Task Force (qv) and may be downloaded from www.brc.gov.uk/upload/assets/www.brc.gov.uk/betregforcivil.pdf.

A BETTER WAY.
2003, 140pp, free. Published by Reform (qv) and may be downloaded from www.reform.co.uk/filestore/pdf/030405%20A%20Better%20Way.pdf

BEVERIDGE REPORT.
See: REPORT ON SOCIAL INSURANCE AND ALLIED SERVICES by W H Beveridge.

BEYOND BLACK AND WHITE: MAPPING NEW IMMIGRANT COMMUNITIES by Sarah Kyambi (editor).
2005, 131pp, £14.95. May be downloaded from www.ippr.org.uk/publicationsandreports/publication.asp?id=308

BIG BANG LOCALISM: A RESCUE PLAN FOR BRITISH DEMOCRACY by Simon Jenkins.
2004, 137pp, free. May be downloaded from www.policyexchange.org.uk/

BILL CLINTON SAVED PEOPLE FROM A LIFE OF CRIME. IS GORDON BROWN BRAVE ENOUGH TO FOLLOW HIM? by Dominic Lawson.
The *Independent*, 28 August 2007. Visit
http://comment.independent.co.uk/columnists_a_l/dominic_lawson/article2900978.ece

BIP 2111:2008 – A COMPREHENSIVE GUIDE TO FIRE SAFETY.
2008, 374pp, £45.00. ISBN-13: 978 0 580 50943 8. Published by BSI and obtainable from www.bsi-global.com/

B IS FOR BULLIED.
2006, 8pp, free. Published by and obtainable from the National Autistic Society on telephone 0845 070 4004 or at www.autism.org.uk/

BLAIR'S EDUCATION: AN INTERNATIONAL PERSPECTIVE by Alan Smithers.
2007, 53pp, free. Published by the Sutton Trust and may be downloaded from www.suttontrust.com/reports/SuttonTrust_BlairsEd19June.pdf

BODYSPACE: ANTHROPOMETRY, ERGONOMICS AND THE DESIGN OF WORK by Stephen Pheasant.
1998, 244pp, £26.50. ISBN: 0 7484 0326 4. Published by Taylor & Francis.

BREAK THROUGH BRITAIN: ENDING THE COSTS OF SOCIAL BREAKDOWN:
• VOLUME 1: FAMILY BREAKDOWN edited by Dr Samantha Callan.
July 2007, 116pp, free. May be downloaded from www.centreforsocialjustice.org.uk/client/downloads/family%20breakdown.pdf
• VOLUME 3: EDUCATIONAL FAILURE EDITED by Ryan Robson.
July 2007, 83pp, free. May be downloaded from www.centreforsocialjustice.org.uk/client/downloads/education.pdf

BRIEFING PAPERS.
Various. Published by the Employers' Forum on Disability and may be downloaded free of charge from www.employers-forum.co.uk/

BRITAIN HAS LOST THE ART OF SOCIALISING THE YOUNG by Peter Wilby.
The *Guardian*, 1 August 2007. Visit http://society.guardian.co.uk/children/comment/0,,2138843,00.html

A BRITISH BILL OF RIGHTS: INFORMING THE DEBATE. THE REPORT OF THE JUSTICE CONSTITUTION COMMITTEE.
2007, 168pp, £9.99. ISBN-13: 978 0 907247 43 2. Published by and obtainable from Justice on telephone 020 7329 5100 or at www.justice.org.uk/

THE BRITISH CONSTITUTION by Anthony King.
2007, 428pp, £25.00. ISBN: 0 199 23232 6. Published by Oxford University Press.

BRITISH JUSTICE: A FAMILY RUINED by Camilla Cavendish.
The *Times*, 21 February 2008. Visit www.timesonline.co.uk/tol/comment/columnists/camilla_cavendish/article3406214.ece

BRITISH SOCIAL ATTITUDES SURVEY: THE 23RD REPORT – PERSPECTIVES ON A CHANGING SOCIETY.
2007, 422pp, £45.00. ISBN: 1 4129 3432 X. May be downloaded free of charge from www.esds.ac.uk/government/bsa/

BRITISH STANDARDS.
All British Standards are available from BSI (The British Standards Institution) at 020 7242 6393. Alternatively visit BSI at www.tso.co.uk/ or Toscadoc at www.toscadoc.co.uk/

BROKEN HOMES AND BATTERED CHILDREN by Robert Whelan.
1994, 96pp, £7.50. Published by and obtainable from Family and Youth Concern at www.famyouth.org.uk/publications.php

BS 5588: PART 8 1999, FIRE PRECAUTIONS IN THE DESIGN, CONSTRUCTION AND USE OF BUILDINGS – CODE OF PRACTICE FOR MEANS OF ESCAPE FOR DISABLED PEOPLE.
1999, 28 pp, £91.00. ISBN: 0 580 28262 7. Obtainable from Toscadoc at www.toscadoc.co.uk/

BS 8300: 2001, DESIGN OF BUILDINGS AND THEIR APPROACHES TO MEET THE NEEDS OF DISABLED PEOPLE, CODE OF PRACTICE.
2001, 173pp, £180.00. ISBN: 0 580 38438 1. Obtainable from Toscadoc at www.toscadoc.co.uk/

BT COUNTRYSIDE FOR ALL STANDARDS AND GUIDELINES – A GOOD PRACTICE GUIDE TO DISABLED PEOPLES' ACCESS TO THE COUNTRYSIDE.
2005, CD-ROM, £50.00. Published by and obtainable from the Fieldfare Trust at www.fieldfare.org.uk/

BUILDING A WORLD-CLASS NHS by Ian Smith.
2006, 39pp, free. May be downloaded from www.reform.co.uk/filestore/pdf/Buildingaworldclassnhs.pdf

BUILDING A WORLD-CLASS NHS by Ian Smith.
2007, 232pp, £20.00. ISBN: 0 23 055380 X. Published by Palgrave Macmillan.

BUILDING BULLETIN 91. ACCESS FOR DISABLED PEOPLE TO SCHOOL BUILDINGS.
1999, 88pp, £14.95. ISBN: 0 11 271062 X. Published by the Department for Education and Employment (qv) and may be obtained from the Centre for Accessible Environments (qv) at www.cae.org.uk/ or from the Stationery Office at www.tsoshop.co.uk/bookstore.asp

BUILDING BULLETIN 93. THE ACOUSTIC DESIGN OF SCHOOLS.
2004, 207pp, £30.00. Published by the Department for Education and Employment (qv) and may be obtained from the RIBA (qv) at www.ribabookshops.com/

BUILDING BULLETIN 94. INCLUSIVE SCHOOL DESIGN – ACCOMMODATING PUPILS WITH SPECIAL EDUCATIONAL NEEDS AND DISABILITIES IN MAINSTREAM SCHOOLS.
2001, 102pp, £19.95. ISBN: 0 11 271109 X. Published by the Department for Education and Employment (qv) and may be obtained from the Centre for Accessible Environments (qv) at www.cae.org.uk/ or from the Stationery Office at www.tsoshop.co.uk/bookstore.asp

BUILDING BULLETIN 100. DESIGN FOR FIRE SAFETY IN SCHOOLS.
2007, 160pp, £19.95. ISBN-13: 978 1 8594 6291 1. Published by the Department for Children, Schools and Families (qv) and may be obtained from the Stationery Office at www.tsoshop.co.uk/bookstore.asp

BUILDING FOR THE FUTURE: 2004 UPDATE by Alan Holmans, Sarah Monk and Christine Whitehead.
2004, £14.50. Published by and obtainable from Shelter at http://england.shelter.org.uk/

BUILDING ON PROGRESS: PUBLIC SERVICES.
2007, 87pp, free. Published by the Cabinet Office and may be downloaded from www.cabinetoffice.gov.uk/upload/assets/www.cabinetoffice.gov.uk/strategy/building.pdf

BUILDING REGULATIONS AND HISTORIC BUILDINGS: BALANCING THE NEEDS FOR ENERGY CONSERVATION WITH THOSE OF BUILDING CONSERVATION – AN INTERIM GUIDANCE NOTE ON THE APPLICATION OF PART L.
2004, 24pp, free. Published by English Heritage and may be downloaded from www.english-heritage.org.uk/upload/pdf/ign_partl_buildingregs.pdf

THE BUILDING REGULATIONS 2000 FOR ENGLAND AND WALES.
Visit www.planningportal.gov.uk/england/professionals/en/4000000000001.html

THE BUILDING REGULATIONS EXPLAINED AND ILLUSTRATED by J R Waters, Keith Bright and M J Billington. 13th edition.
2007, 944pp, £55.00. ISBN: 1 405 15922 7. Published by Blackwell Publishing at http://bookshop.blackwell.co.uk/

BUILDING REGULATIONS EXPLANATORY BOOOKLET.
2003, 69pp, free. Published by the Department for Communities and Local Government (qv) and may be downloaded from www.communities.gov.uk/documents/planningandbuilding/pdf/explanatorybooklet

BUILDING REGULATIONS (NORTHERN IRELAND) 2000.
To read the Regulations visit www.buildingcontrol-ni.com/sections/?secid=5.
To download the Technical Booklets visit www.dfpni.gov.uk/index/law-and-regulation/building-regulations/br-technical-booklets.htm

BUILDING (SCOTLAND) REGULATIONS 2004.
To read the Regulations visit www.scotland.gov.uk/Topics/Business-Industry/support/15242/buildingregs0426
To download the Technical Handbooks visit www.sbsa.gov.uk/tech_handbooks/tbooks2007.htm

BUILDING FOR THE PERFORMING ARTS: A DESIGN AND DEVELOPMENT GUIDE by Ian Appleton.
2008, 283pp, £39.99. ISBN-13: 978 0 7506 6835 4. Published by Architectural Press, an imprint of Elsevier at http://books.elsevier.com/

BUILDING SIGHT: A HANDBOOK OF BUILDING AND INTERIOR DESIGN SOLUTIONS by Peter Barker, Jon Barrick and Rod Wilson.
1995, 180pp, £25.00. ISBN: 1 85878 074 8. Published by HMSO and the RNIB (qv), and obtainable from the Centre for Accessible Environments (qv) at www.cae.org.uk/

BUMPER BOOK OF GOVERNMENT WASTE by Matthew Elliott and Lee Rotherham.
2006, 340pp, £9.99. ISBN: 1 897 59779 7. Published by and obtainable from the Taxpayers' Alliance at www.harriman-house.com/pages/book.htm?BookCode=22852

THE BUREAUCRATIC BURDEN IN THE NHS.
2007, 28pp, free. Published by the NHS Confederation and may be downloaded from www.nhsconfed.org/tools/search.cfm/res/285942/page/8

BY ALL REASONABLE MEANS: INCLUSIVE ACCESS TO THE OUTDOORS FOR DISABLED PEOPLE.
2005, 52pp, free. Published by the Countryside Agency and may be downloaded from http://naturalengland.twoten.com/naturalenglandshop/docs/CA215.pdf

CABE AND THE HISTORIC ENVIRONMENT.
2004, free. May be downloaded from www.cabe.org.uk/AssetLibrary/2365.pdf

CAR PARKING: WHAT WORKS WHERE.
2006, 134pp, free. Published by and obtainable from English Partnerships on telephone 020 7881 1600 or at www.englishpartnerships.co.uk/

CARE MATTERS: TIME FOR CHANGE.
2007, 144pp, free. This *white paper* (qv) is published by the DfES and may be downloaded from www.dfes.gov.uk/publications/timeforchange/docs/timeforchange.pdf

CARE MATTERS: TRANSFORMING THE LIVES OF CHILDREN AND YOUNG PEOPLE IN CARE.
2006, 124pp, free. This *green paper* (qv) is published by the DfES and may be downloaded from www.dfes.gov.uk/consultations/downloadableDocs/6731-DfES-Care%20Matters.pdf

THE CASE AGAINST VETTING: HOW THE CHILD PROTECTION INDUSTRY IS POISONING ADULT-CHILD RELATIONS.
2006, 27pp, free. May be downloaded from www.manifestoclub.com/

THE CASE FOR COUNCIL HOUSING IN 21ST CENTURY BRITAIN.
2007, 96pp, free. Published by Defend Council Housing and may be downloaded from www.defendcouncilhousing.org.uk/dch/

CASE REPORT 34.
See: ENDS AND MEANS: THE FUTURE ROLES OF SOCIAL HOUSING IN ENGLAND.

CENTRAL CONTROL OF WELFARE SIMPLY DOESN'T WORK by Frank Field.
Daily Telegraph, 17 June 2007. Visit www.frankfield.co.uk/type2.asp?id=21&type=2

CHANGES IN CHILDREN'S COGNITIVE DEVELOPMENT AT THE START OF SCHOOL IN ENGLAND 2000-2006 by Christine Merrell, Peter Tymms, and Paul Jones.
2007, 14pp, free. Published by the Curriculum Evaluation and Management (CEM) Centre at Durham University and may be downloaded from www.cemcentre.org/

CHANGES IN COMMUNAL PROVISION FOR ADULT SOCIAL CARE 1991–2001 by Laura Banks, Philip Haynes, Susan Balloch, and Michael Hill.
2006, 108pp, £16.95. ISBN: 1 85935 485 8. Published by the Joseph Rowntree Foundation (qv) and may be downloaded free of charge from www.jrf.org.uk/bookshop/eBooks/9781859354865.pdf

THE CHANGING FACE OF TACTILE PAVING by Brian Towers.
2007, 5pp. ACCESS BY DESIGN, issue 113, pp 27-31. Published by the Centre for Accessible Environments (qv) at www.cae.org.uk/

CHARGE OR RELEASE: TERRORISM PRE-CHARGE DETENTION – COMPARATIVE LAW STUDY.
2007, 65pp, free. Published by the *pressure group* (qv) Liberty, and may be downloaded from www.liberty-human-rights.org.uk/issues/pdfs/pre-charge-detention-comparative-law-study.pdf

CHARLES MURRAY AND THE UNDERCLASS: THE DEVELOPING DEBATE by Charles Murray with commentaries by Ruth Lister, Frank Field, and others.
1996, 190pp, free. Published by Civitas (qv) and may be downloaded from www.civitas.org.uk/pdf/

CHILD POVERTY IN PERSPECTIVE: AN OVERVIEW OF CHILD WELL-BEING IN RICH COUNTRIES.
2007, 52pp, free. Published by Unicef and may be downloaded from www.unicef-icdc.org/presscentre/presskit/reportcard7/rc7_eng.pdf

CHILDHOOD WELLBEING: QUANTITATIVE RESEARCH STUDY.
2008, free. Published by the Department for Children, Schools and Families and may be downloaded from www.dcsf.gov.uk/rsgateway/DB/RRP/u015387/index.shtml

CHILDREN AS VICTIMS: CHILD-SIZED CRIMES IN A CHILD-SIZED WORLD.
2007, 20pp, £5.00. Published by and obtainable from the Howard League for Penal Reform at www.howardleague.org/index.php?id=8

CHILDREN'S DATABASES – SAFETY AND PRIVACY: A REPORT FOR THE INFORMATION COMMISSIONERS.
2006, 194pp, free. May be downloaded from www.fipr.org/childrens_databases.pdf

CHILDREN'S NURSERIES UK MARKET REPORT 2007.
2007, 168pp, £540.00. Published by and available from Laing & Buisson on telephone 020 7833 9123 or at www.laingbuisson.co.uk/

CHOOSING A MOBILE HOIST and similar titles.
See: DISABLED LIVING FOUNDATION FACT SHEETS.

CHOOSING TO BE DIFFERENT – WOMEN, WORK AND THE FAMILY by Jill Kirby.
2006, 56pp, £7.50. Published by and obtainable from the Centre for Policy Studies. May be downloaded free of charge from www.cps.org.uk/historiccatalogue/default.aspCPS

CIRCULAR 01/06 (COMMUNITIES AND LOCAL GOVERNMENT): GUIDANCE ON CHANGES TO THE DEVELOPMENT CONTROL SYSTEM.
2006, 13pp, free. ISBN-13: 978 0 11 753966 2. May be downloaded from www.communities.gov.uk/publications/planningandbuilding/circularcommunities2

CITIES RENAISSANCE: CREATING LOCAL LEADERSHIP by Michael Heseltine.
2007, 13pp, free. May be downloaded from www.conservatives.com/pdf/CityLeadership.pdf

CIVIL AVIATION AUTHORITY GUIDANCE ON SEATING RESTRICTIONS.
2001, free. May be downloaded from www.dptac.gov.uk/pubs/seatrestrict/

CIVILISED STREETS.
2008, 8pp, free. Published by Cabe (qv) and may be downloaded from www.cabe.org.uk/AssetLibrary/11279.pdf

CIVITAS REVIEW.
Published by Civitas (qv). For particulars visit www.civitas.org.uk/pdf/CivitasReviewJanuary07.pdf

CLAIMS ABOUT PHYSICAL BARRIERS TO ACCESS: A GUIDE TO THE DISABILITY DISCRIMINATION ACT.
2007, 140pp, free. May be downloaded from www.equalityhumanrights.com/pages/eocdrccre.aspx

A CLASS ACT: THE MYTH OF BRITAIN'S CLASSLESS SOCIETY by Andrew Adonis and Stephen Pollard.
1998, 336pp, £18.00. ISBN: 0 140 26100 1. Published by Penguin books.

THE CLASSICAL VERNACULAR: ARCHITECTURAL PRINCIPLES IN AN AGE OF NIHILISM by Roger Scruton.
1994, 160pp, 1994. ISBN: 1 85754 054 9. Published by Carcanet Press Limited.

CLEAR PRINT GUIDELINES.
2006, 3pp, free. May be downloaded from www.rnib.org.uk/

CLIMBIÉ INQUIRY: THE ISSUE EXPLAINED by David Batty.
The *Guardian*, 5 August 2005. Visit http://society.guardian.co.uk/climbie/story/0,,557457,00.html

CLIMBIÉ LESSONS NOT LEARNED, SAYS RESEARCH by John Carvel.
The *Guardian*, 2 February 2008. Visit www.guardian.co.uk/society/2008/feb/01/climbie.childprotection

CODE FOR LIGHTING.
2006, 130pp, £158.64. Book + CD-ROM. Published by and obtainable from the Chartered Institution of Building Services Engineers (CIBSE) at www.cibse.org/index.cfm?go=publications.view&item=183

CODE OF PRACTICE: EMPLOYMENT AND OCCUPATION.
2004, 158pp, £13.50. ISBN: 0 11 703419 3. Obtainable from TSO on telephone 0870 242 2345 or at www.tsoshop.co.uk/bookstore.asp. May also be downloaded free of charge from www.equalityhumanrights.com/pages/eocdrccre.aspx

CODE OF PRACTICE FOR SCHOOLS – DDA 1995: PART 4.
2002, 60pp, free. May be downloaded free of charge from www.equalityhumanrights.com/pages/eocdrccre.aspx

CODE OF PRACTICE ON FACILITIES FOR DISABLED MOTORISTS AT FILLING STATIONS.
2000, 3pp, free. Published by DPTAC (qv). May be downloaded from www.dptac.gov.uk/pubs/fill/index.htm

CODE OF PRACTICE (REVISED) FOR PROVIDERS OF POST-16 EDUCATION AND RELATED SERVICES.
2007, 278pp, £15.00. ISBN: 0 11 703730 3. Obtainable from TSO on telephone 0870 242 2345 or at www.tsoshop.co.uk/bookstore.asp. May also be downloaded free of charge from www.equalityhumanrights.com/pages/eocdrccre.aspx

CODE OF PRACTICE – RIGHTS OF ACCESS: SERVICES TO THE PUBLIC, PUBLIC AUTHORITY FUNCTIONS, PRIVATE CLUBS AND PREMISES.
2006, 360pp, £15.00. ISBN: 0 11 703695 1. Obtainable from TSO on telephone 0870 242 2345 or at www.tsoshop.co.uk/bookstore.asp. May also be downloaded free of charge from www.equalityhumanrights.com/pages/eocdrccre.aspx

CODE OF PRACTICE – SEX DISCRIMINATION.
See: GENDER EQUALITY DUTY CODE OF PRACTICE: ENGLAND AND WALES.

CODE OF PRACTICE. THE DUTY TO PROMOTE DISABILITY EQUALITY: STATUTORY CODE OF PRACTICE, ENGLAND AND WALES.
2005, 181pp, £15.00. ISBN: 0 11 703605 6. Obtainable from TSO on telephone 0870 242 2345 or at www.tsoshop.co.uk/bookstore.asp. May also be downloaded free of charge from www.equalityhumanrights.com/pages/eocdrccre.aspx

CODE OF PRACTICE – TRADE ORGANISATIONS AND QUALIFICATIONS BODIES.
2004, 158pp, £15.00. ISBN: 0 11 703418 5. Obtainable from TSO on telephone 0870 242 2345 or at www.tsoshop.co.uk/bookstore.asp. May also be downloaded free of charge from www.equalityhumanrights.com/pages/eocdrccre.aspx

CODES OF PRACTICE: EMPLOYERS AND SOCIAL CARE WORKERS.
2002, 23pp, free. Published by the General Social Care Council and may be downloaded from www.gscc.org.uk/Good+practice+and+conduct/Get+copies+of+our+codes/

COLLAPSE: HOW SOCIETIES CHOOSE TO FAIL OR SURVIVE by Jared Diamond.
2006, 592pp, £9.99. ISBN-13: 9780140279511. Published by Penguin Books.

COLOUR AND CONTRAST – A DESIGN GUIDE FOR THE USE OF COLOUR AND CONTRAST TO MEET THE NEEDS OF DISABLED PEOPLE.
For more information visit www.extra.rdg.ac.uk/ie/

COLOUR CONTRAST AND PERCEPTION.
For more information visit www.extra.rdg.ac.uk/ie/

A COMPARATIVE STUDY OF REFERENDUMS: GOVERNMENT BY THE PEOPLE by Matt Qvortrup.
2005,192pp, £14.99. ISBN-13: 9780719071812. Published by Manchester University Press at www.manchesteruniversitypress.co.uk/

THE COMPETITIVE ECONOMIC PERFORMANCE OF ENGLISH CITIES by Professor James Simmie and others.
2006, 263pp, free. May be downloaded from www.communities.gov.uk/documents/citiesandregions/pdf/153232

COPING WITH POST DEMOCRACY by Colin Crouch.
2000, 76pp, £6.95. Published by and obtainable from the Fabian Society at bookshop@fabian-society.org.uk

THE CORRUPTION OF THE CURRICULUM. Robert Whelan (Editor).
2007, 156pp, £12.00. Published by and obtainable from Civitas (qv) at www.civitas.org.uk/

THE COST OF EXCLUSION: COUNTING THE COST OF YOUTH DISADVANTAGE IN THE UK by Sandra McNally and Shqiponja Telhaj.
2007, 70pp, free. May be downloaded from www.princes-trust.org.uk/

THE COSTS OF INCLUSION: A STUDY OF INCLUSION POLICY AND PRACTICE IN ENGLISH PRIMARY, SECONDARY AND SPECIAL SCHOOLS by John MacBeath, Maurice Galton, Susan Steward, Andrea MacBeath and Charlotte Page.
2006, 78pp, free. ISBN: 0 9543823 3 1. May be downloaded from www.ocdsb.edu.on.ca/general_Info/Spec_Ed/CostsofInclusion.pdf

CRIME AND CIVILIZED SOCIETY by David G Green, Emma Grove and Nadia A Martin.
2005, 366pp, £14.50. Published by and obtainable from Civitas (qv) at www.civitas.org.uk/

CRIME IN ENGLAND AND WALES: MORE VIOLENCE AND MORE CHRONIC VICTIMS by Graham Farrell and Ken Pease.
2007, 8pp, free. Published by Civitas (qv) and may be downloaded from www.civitas.org.uk/pdf/CivitasReviewJun07.pdf

CRIME OPPORTUNITY PROFILING OF STREETS (COPS): A QUICK CRIME ANALYSIS-RAPID IMPLEMENTATION APPROACH.
2005, 48pp, £47.50. Published by and obtainable from BRE Press on telephone 01344 328038 or at www.bretrust.org.uk/page_1col.jsp?sid=592

CRIME PREVENTION EFFECTS OF CLOSED CIRCUIT TELEVISION: A SYSTEMATIC REVIEW by Brandon C Welsh and David P Farrington.
2002, 68pp, free. Published by the Home Office (qv) and may be downloaded from www.homeoffice.gov.uk/rds/pdfs2/hors252.pdf

CRIME PREVENTION THROUGH ENVIRONMENTAL DESIGN: APPLICATIONS OF ARCHITECTURAL DESIGN AND SPACE MANAGEMENT CONCEPTS by Timothy D Crowe.
2000, 360pp, £43.99. ISBN: 0 7506 7198 X. Published by Butterworth-Heinemann, an imprint of Elsevier Press at http://books.elsevier.com/

CRIME STATISTICS: USER PERSPECTIVES. Report No 30.
2006, 194pp, free. May be downloaded from www.statscom.org/

CRIME STATISTICS: AN INDEPENDENT REVIEW.
2006, 38pp, free. May be downloaded from www.homeoffice.gov.uk/rds/pdfs06/crime-statistics-independent-review-06.pdf

CRIMINAL JUSTICE MONTHLY: Issue 66 (Winter 2006–07).
Published by the Centre for Crime and Justice Studies at www.crimeandjustice.org.uk/cjm/issue66.html. For the article 'Probation, the public, and what is possible' visit www.crimeandjustice.org.uk/opus189/cjm66-bridges.pdf

CROSSING THE THRESHOLD: 266 WAYS IN WHICH THE STATE CAN ENTER YOUR HOME by Harry Snook.
2007, 112pp, £10. Published by and obtainable from the Centre for Policy Studies. May be downloaded free of charge from www.cps.org.uk/

CRUNCH TIME FOR SOCIAL ENTERPRISES: MINISTERS NEED TO DO MORE TO ENCOURAGE SOCIAL ENTERPRISES by Patrick Butler.
The *Guardian*, 5 March 2008. Visit http://blogs.guardian.co.uk/joepublic/2008/03/crunch_time_for_social_enterpr.html

THE CULT OF THE AMATEUR by Andrew Keen.
2007, 228pp, £12.99. ISBN-13: 9 781 857883393 0. Published by Nicholas Brealey.

DATA PROTECTION WON'T HELP ONCE ALL THE DATA IS GONE by Chistina Zaba.
Guardian, 27 November 2007. Visit http://politics.guardian.co.uk/comment/story/0,,2217678,00.html

THE DEATH OF COMMON SENSE: HOW LAW IS SUFFOCATING AMERICA by Philip Howard.
1996, 208pp, o/p. ISBN: 0 446 67228 9. Published by Warner Books.

DELIVERING SUCCESSFUL IT-ENABLED BUSINESS CHANGE.
2007, 41pp, free. 27th report of session 2006–07 of the House of Commons Committee of Public Accounts. May be downloaded from www.publications.parliament.uk/pa/cm200607/cmselect/cmpubacc/113/113.pdf

DENTISTRY WATCH: NATIONAL SURVEY OF THE NHS DENTISTRY SYSTEM WITH VIEWS FROM BOTH PATIENTS AND DENTISTS.
2007, 61pp, free. Published by the Commission for Patient and Public Involvement in Health (qv) and may be downloaded from http://147.29.80.160/portal/csc/genericContentGear/download/Dentistry+watch+national+summary+of+results +-+final3-15-10.pdf?document_id=116400639

DESIGN AND ACCESS STATEMENTS: HOW TO WRITE, READ AND USE THEM.
2006, 34pp, free. Published by and obtainable from CABE (qv). May be downloaded from http://cabe.org.uk/AssetLibrary/8073.pdf

DESIGN AND ACCESS STATEMENTS: REPORT FROM A 'LEARNING GROUP' COMPRISING 16 LOCAL PLANNING AUTHORITIES.
2008, 18pp, free. Published by the Planning Advisory Service and may be downloaded from www.pas.gov.uk/pas/core/page.do?pageId=25956%20

DESIGN CODES: THE ENGLISH PARTNERSHIP EXPERIENCE.
2007, 6pp, free. Published by English Partnerships and may be downloaded from www.englishpartnerships.co.uk/

THE DESIGN OF LARGE PASSENGER SHIPS AND PASSENGER INFRASTRUCTURE: GUIDANCE ON MEETING THE NEEDS OF DISABLED PEOPLE.
2000, 53pp, free. Published by and obtainable from DPTAC (qv). May be downloaded from http://www.dptac.gov.uk/pubs/guideship/index.htm

DESIGN OF RESIDENTIAL CARE AND NURSING HOMES FOR OLDER PEOPLE.
Health Facilities Note 19.
1998, 160pp, £15. ISBN: 0 9534158 0 5. PDF only. May be downloaded from www.cae.org.uk/

DESIGNING FOR ACCESSIBILITY by Andrew Lacey.
2004, 72pp, £22.50. ISBN: 1 85946 143 3. Published by the Centre for Accessible Environments (qv) and the RIBA (qv), and obtainable from www.cae.org.uk/ or www.ribabookshops.com/

DESIGNING FOR SPECIAL NEEDS: AN ARCHITECT'S GUIDE TO BRIEFING AND DESIGNING FOR PEOPLE WITH LEARNING DISABILITIES by Maurice Harker and Nigel King.
2002, 192pp, £20.00. Published by and obtainable from the RIBA (qv) at www.ribabookshops.com/

DESIGNING FOR THE DISABLED: THE NEW PARADIGM by Selwyn Goldsmith.
1997, 426pp, £72.00. ISBN: 0 7506 3442 1. Published by Architectural Press, an imprint of Elsevier at http://books.elsevier.com/

DESIGNING GALLERIES: THE COMPLETE GUIDE TO DEVELOPING AND DESIGNING SPACES AND SERVICES FOR TEMPORARY EXHIBITIONS by Mike Sixsmith.
1999, 204pp, £19.99. ISBN: 0 7287 0780 2. Published by and obtainable from Arts Council England (qv) at www.artscouncil.org.uk/

DEVELOPMENT AND FLOOD RISK: A PRACTICE GUIDE COMPANION TO PPS 25 'LIVING DRAFT'. A CONSULTATION PAPER.
2007, 129pp, free. May be downloaded from www.communities.gov.uk/publications/planningandbuilding/developmentflood

DILEMMAS OF PRIVACY AND SURVEILLANCE – CHALLENGES OF TECHNOLOGICAL CHANGE by Professor Nigel Gilbert and others.
2007, 64pp, free. Published by the Royal College of Engineering and may be downloaded from www.raeng.org.uk/policy/reports/pdf/dilemmas_of_privacy_and_surveillance_report.pdf

DIRECT DEMOCRACY: AN AGENDA FOR A NEW MODEL PARTY.
2005, 106pp, free. Published by Direct Democracy and may be downloaded from www.direct-democracy.co.uk/

DISABILITY DISCRIMINATION ACT 1995.
May be downloaded free of charge from www.opsi.gov.uk/acts.htm

DISABILITY DISCRIMINATION ACT 2005.
May be downloaded free of charge from www.opsi.gov.uk/acts.htm

DISABILITY DISCRIMINATION: LAW AND PRACTICE by Brian Doyle.
2008, 592pp, £49.00. ISBN: 978 1 84661 083 7. Published by Jordans at www.jordanpublishing.co.uk/

DISABILITY IN GREAT BRITAIN.
1999, 163pp, £35.00. ISBN: 1 84 123 119 3. Published by and obtainable from the Department for Work and Pensions (qv) at www.dwp.gov.uk/asd/asd5/rrs1999.asp#g

THE DISABILITY PORTFOLIO.
2003 to 2006, 12 volumes. For particulars visit www.mla.gov.uk/

DISABILITY RIGHTS AND WRONGS by Tom Shakespeare.
2006, 232pp, £19.99. ISBN-13: 978 0 415 34719 8. Published by Routledge at www.routledge.com/

DISABILITY RIGHTS HANDBOOK.
2007, 300pp, £20.00. ISBN-13: 978 1 903335 3. Published annually by, and obtainable from, the Disability Rights Alliance at www.disabilityalliance.org/

DISABLED LIVING FOUNDATION FACTSHEETS. The following factsheets, which are mostly around 16pp in extent, may be downloaded free of charge from www.dlf.org.uk/public/factsheets.html
CHOOSING A MOBILE HOIST.
CHOOSING A POWERED WHEELCHAIR.
CHOOSING A SCOOTER OR BUGGY.
CHOOSING A STANDARD SELF-PROPELLED WHEELCHAIR.
CHOOSING AN ACTIVE USER WHEELCHAIR.
CHOOSING AN ATTENDANT-PROPELLED WHEELCHAIR.
CHOOSING AN OVERHEAD HOIST.
CHOOSING AND FITTING GRAB RAILS.
CHOOSING EQUIPMENT FOR THE HEAVIER PERSON.
CHOOSING EQUIPMENT TO GET UP AND DOWN STAIRS.
CHOOSING EQUIPMENT TO MAINTAIN SAFETY AND INDEPENDENCE AT HOME: INTRODUCING TELECARE.
CHOOSING TOILET EQUIPMENT AND ACCESSORIES.
CHOOSING WALKING EQUIPMENT.

DISCOMFORT OF STRANGERS by David Goodhart.
Guardian, 24 February 2004. Visit www.guardian.co.uk/race/story/0,11374,1154684,00.html

DIVERSITY AND EQUALITY IN PLANNING: A GOOD PRACTICE GUIDE.
2005, 186pp, £18.00. ISBN: 1 85 112755 0. Published by and obtainable from Communities and Local Government Publications at communities@twoten.com, or may be downloaded free of charge from www.communities.gov.uk/index.asp?id=1505969

DNA FILE ON 100,000 INNOCENT CHILDREN by Philip Johnston.
Daily Telegraph, 28 May 2007. Visit www.telegraph.co.uk/news/main.jhtml?xml=/news/2007/05/23/ndna23.xml

DOES MARRIAGE MATTER? by Rebecca O'Neill.
2003, 52pp, £2.50. Published by and obtainable from Civitas (qv) on 020 7799 6677 or may be downloaded free of charge from www.civitas.org.uk/

DOMESTIC TECHNICAL HANDBOOK.
See: TECHNICAL HANDBOOKS.

DON'T TOUCH! THE EDUCATIONAL STORY OF A PANIC by Heather Piper and Ian Stronach.
2008, 184pp, $37.95. ISBN: 978 0 415 42008 2. Published by Routledge at www.routledge.co.uk/

THE DUTY TO PROMOTE DISABILITY EQUALITY: STATUTORY CODE OF PRACTICE, ENGLAND AND WALES.
See: CODE OF PRACTICE. THE DUTY TO PROMOTE DISABILITY EQUALITY.

EASIER LIVING: A GUIDE FOR OLDER AND DISABLED PEOPLE LIVING IN LONDON.
2001, 24pp, free. Published by Ricability (qv) and may be downloaded from www.ricability.org.uk/reports/report-childcare.htm

EASY ACCESS TO HISTORIC BUILDINGS.
2004, 55pp, free. Published by English Heritage and may be downloaded from www.english-heritage.org.uk/

EASY ACCESS TO HISTORIC LANDSCAPES.
2005, 72pp, free. Published by English Heritage and may be downloaded from www.english-heritage.org.uk/

THE ECONOMIC IMPACT OF IMMIGRATION. 1ST REPORT OF SESSION 2007-08 OF THE HOUSE OF LORDS SELECT COMMITTEE ON ECONOMIC AFFAIRS.
2008, 84pp, free. May be downloaded from www.publications.parliament.uk/pa/ld200708/ldselect/ldeconaf/82/82.pdf

ECONOMICS AND POLITICS OF CLIMATE CHANGE: AN APPEAL TO REASON by Nigel Lawson.
2006, 18pp, free. May be read or heard at www.cps.org.uk/historiccatalogue/default.asp

EDUCATION AT A GLANCE 2004.
2006, 465pp, £46.00. ISBN-13: 9789264025318. Published by and obtainable from the Organisation for Economic Co-operation and Development (OECD) at www.oecd.org/document/7/0,2340,en_2649_201185_33712135_1_1_1_1,00.html

EDUCATIONAL FAILURE AND WORKING-CLASS WHITE CHILDREN IN BRITAIN by Gillian Evans.
2007, 224pp, £16.99. ISBN: 0 230 55303 6. Published by Palgrave Macmillan.

THE EFFECTS OF SYNTHETIC PHONICS TEACHING ON READING AND SPELLING ATTAINMENT: A SEVEN YEAR LONGITUDINAL STUDY by Professor Rhona Johnston and Dr Joyce Watson.
2005, 70pp, free. May be downloaded from www.scotland.gov.uk/Resource/Doc/36496/0023582.pdf

E PLURIBUS UNUM: DIVERSITY AND COMMUNITY IN THE TWENTY-FIRST CENTURY by Robert D Putnam.
Scandinavian Political Studies. Volume 30, Issue 2, pp 137–174, June 2007. Visit www.blackwellpublishing.com/journal.asp?ref=0080-6757&site=1

THE EMERGING BRITISH UNDERCLASS by Charles Murray.
1990, 96pp, o/p. Published by the Institute of Economic Affairs. The text is included in CHARLES MURRAY AND THE UNDERCLASS: THE DEVELOPING DEBATE (qv).

ENDS AND MEANS: THE FUTURE ROLES OF SOCIAL HOUSING IN ENGLAND by John Hills.
2007, 228pp, free. Case Report 34 from the ESRC Research Centre for Analysis of Social Exclusion may be downloaded from http://sticerd.lse.ac.uk/dps/case/cr/CASEreport34.pdf

THE ENGLISH INDICES OF DEPRIVATION 2004.
2004, 181pp, free. May be downloaded from www.communities.gov.uk/index.asp?id=1128440

ENGLISH PARTNERSHIPS GUIDANCE NOTE: INCLUSIVE DESIGN.
2007, 30pp, free. Published by English Partnerships, and may be downloaded from www.englishpartnerships.co.uk/

THE ENGLISH QUESTION by Thomas Docherty.
2007, 160pp, £15.95. ISBN: 978 1 84519 133 7. Published by Sussex Academic Press at www.sussex-academic.co.uk/

ENHANCING CARE PROVISION FOR BLIND AND PARTIALLY-SIGHTED PEOPLE IN GP SURGERIES: GUIDELINES FOR BEST PRACTICE.
2005, 25pp, free. May be downloaded from www.guidedogs.org.uk/fileadmin/gdba/downloads/social_research/Guidelinesfinal.doc

EQUAL TREATMENT BENCH BOOK: AN OFFICIAL GUIDE FOR COURTROOM PERSONNEL.
2007, 21pp, free. May be downloaded from www.jsboard.co.uk/etac/etbb/index.htm

EQUALITIES REVIEW.
See: FAIRNESS AND FREEDOM: THE FINAL REPORT OF THE EQUALITIES REVIEW.

EQUALITY by R H Tawney.
1931. New edition 1965, 255pp, o/p. ISBN: 0 043 23014 8. Published by HarperCollins Publishers.

THE ESSENTIAL GUIDE TO BRITISH QUANGOS 2005 by Dan Lewis.
2005, 104pp, £15.00. Published by and obtainable from the Centre for Policy Studies at www.cps.org.uk/

ESTATES: AN INTIMATE HISTORY by Lynsey Hanley.
2007, 256pp, £12.00. ISBN: 1 862 07909 9. Published by Granta.

EURO HEALTH CONSUMER INDEX 2007.
2007, 48pp, free. Published by the Health Consumer Powerhouse. May be downloaded from www.healthpowerhouse.com/media/Rapport_EHCI_2007.pdf

EVALUATING HOUSING PRINCIPLES STEP BY STEP.
2008, 84pp, free. Published by Cabe (qv) and may be downloaded from www.cabe.org.uk/AssetLibrary/11349.pdf

EXPERIMENTS IN LIVING – THE FATHERLESS WAY by Rebecca O'Neill.
2002, 20pp, free. Published by Civitas (qv) and may be downloaded from www.civitas.org.uk/pubs/experiments.php

EXPERT POLITICAL JUDGMENT: HOW GOOD IS IT? HOW CAN WE KNOW? by Philip Tetlock.
2006, 352pp, £11.95. ISBN-13: 978 0 691 12871 9. Published by Princeton University Press.

THE FAILED GENERATION: THE REAL COST OF EDUCATION UNDER LABOUR by Chris Skidmore.
2008, 5pp, free. Published by the Bow Group and may be downloaded from www.bowgroup.org/harriercollectionitems/The%20Failed%20Generation.pdf

THE FAILURE OF BRITAIN'S POLICE: LONDON AND NEW YORK COMPARED by Norman Dennis.
2003, 68pp, free. Published by Civitas (qv) and may be downloaded from www.civitas.org.uk/pdf/cs26.pdf

FAIRNESS AND FREEDOM: THE FINAL REPORT OF THE EQUALITIES REVIEW.
2007, 174pp, £15.00. May be ordered by telephone from the Equalities Review on telephone 020 7276 1200, or downloaded free of charge from www.theequalitiesreview.org.uk/

FALLING BEHIND: HOW RISING INEQUALITY HARMS THE MIDDLE CLASS by Robert Frank.
2007, 160pp, $19.95. ISBN: 0 520 25252 7. Published by the University of California Press and obtainable from Amazon UK at www.amazon.co.uk/

FAMILIES, CHILDREN AND CHILDCARE by Penelope Leach, Kathy Sylva, and Alan Stein.
An ongoing study by the Institute for the Study of Children, Families & Social Issues at www.familieschildrenchildcare.org/fccc_frames_home.html

FAMILIES WITHOUT FATHERHOOD by Norman Dennis and George Erdos. Foreword by A H Halsey.
2000, 133pp, £6.00. Published by and obtainable from Civitas (qv) at www.civitas.org.uk/

FAMILY POLICY, FAMILY CHANGES: SWEDEN, ITALY AND BRITAIN COMPARED by Patricia Morgan.
2006, 160pp, £14.00. Published by and obtainable from Civitas (qv) at www.civitas.org.uk/

FAREWELL TO THE FAMILY? by Patricia Morgan.
1999, 175pp, £8.00. Published by and obtainable from Civitas (qv) at www.civitas.org.uk/

FINAL EXIT: THE PRACTICALITIES OF SELF-DELIVERANCE by Derek Humphry.
2002, 220pp, $13.95. ISBN: 0 385 33653 5. Published by Dell Publishing, an imprint of Random House Inc, and obtainable from Amazon Books UK at www.amazon.co.uk/

FINAL EXIT ON DVD: THE ART OF SELF-DELIVERANCE FROM A TERMINAL ILLNESS by Derek Humphry.
2006, DVD, $20.00. ISBN-13: 978 0 9768283 0 3. Published by Ergo and obtainable from Ergo Bookstore at www.finalexit.org/ Total cost including airmail delivery comes to around £16.00.

FIRE SAFETY IN CARE HOMES FOR OLDER PEOPLE AND CHILDREN.
2004, 143pp, £45.00. Published by and obtainable from the Fire Protection Association at www.thefpa.co.uk/

FIRE SAFETY RISK ASSESSMENT GUIDES.
£12.00 per title. Published by the Department for Communities and Local Government (qv), and obtainable from The Stationery Office at www.tsoshop.co.uk/

FITTING THE BILL: LOCAL POLICING FOR THE 21ST CENTURY by Barry Loveday and Jonathan McClory.
2007, 42pp, free. Published by the Policy Exchange and may be downloaded from www.policyexchange.org.uk/images/libimages/263.pdf

FOCUS ON FAMILIES 2007.
2007, 112pp, £40. ISBN-13: 978 1 4039 9323 6. Published by Palgrave Macmillan for the Office for National Statistics. May be downloaded free of charge from www.statistics.gov.uk/downloads/theme_compendia/fof2007/FO_Families_2007.pdf

FOCUS ON SOCIAL INEQUALITIES.
2004, 124pp, free. Published by the Office for National Statistics (qv) and may be downloaded from www.statistics.gov.uk/focuson/socialinequalities/

THE FORENSIC USE OF BIOINFORMATION: ETHICAL ISSUES.
2007,168pp, £10.00. ISBN-13: 978 1 904384 16 8. Published by the Nuffield Council on Bioethics and obtainable on telephone 020 7681 9619.

FORM FOLLOWS FIASCO: WHY MODERN ARCHITECTURE HASN'T WORKED by Peter Blake.
1977, 169pp, $12.95. ISBN: 0 316 09940 6 (cloth); 0 316 09939 2 (paper). Published by Little, Brown and Company and obtainable via Abebooks at www.abebooks.co.uk/

FREEDOM'S ORPHANS: RAISING YOUTH IN A CHANGING WORLD by Julia Margo.
2007, 211pp, £12.95. Published by and obtainable from the Institute for Public Policy Research at www.ippr.org/

FROM PUNISHMENT TO PROBLEM SOLVING – A NEW APPROACH TO CHILDREN IN TROUBLE by Rob Allen.
2006, 36pp, free. Published by the Centre for Crime and Justice Studies and may be downloaded from www.kcl.ac.uk/depsta/rel/ccjs/2006-punishment-to-problem-solving.pdf

THE FUTURE OF BUILDING CONTROL.
2007, 17pp, free. Published by the Department for Communities and Local Government (qv) and may be downloaded from www.communities.gov.uk/pub/104/TheFutureofBuildingControl_id1509104.pdf

GANGS, ALAS, ARE OFFERING WHAT BOYS NEED by Harriet Sergeant.
Sunday Times, 19 August 2007. Visit www.timesonline.co.uk/tol/comment/columnists/guest_contributors/article2284163.ece

GAY-BASHING SHOULD NOT BE A HATE CRIME by Johann Hari.
The *Independent*, 11 October 2007. Visit www.independent.co.uk/opinion/commentators/johann-hari/johann-hari-gaybashing-should-not-be-a-hate-crime-396532.html

GENDER EQUALITY DUTY CODE OF PRACTICE: ENGLAND AND WALES.
2006, 81pp, free. May be downloaded from www.equalityhumanrights.com/Documents/Gender/Public%20sector/EOC%20Code%20of%20practice%20Equality%20Duty.pdf

GENERATIONAL INCOME MOBILITY IN NORTH AMERICA AND EUROPE, edited by Miles Corak.
2004, 320pp, £50.00. ISBN: 0 521 82760 4. Published by Cambridge University Press.

GENTLE REGRETS: THOUGHTS FROM A LIFE by Roger Scruton.
2005, 248pp, £10.99. ISBN: 0 826 48033 0. Published by Continuum.

GET MOBILE – YOUR GUIDE TO BUYING A SCOOTER OR POWERED WHEELCHAIR.
2006, 28pp, free. Published by RADAR (qv) and may be downloaded from www.radar.org.uk/

GLASS IN BUILDINGS by Ann Alderson.
2006, 58pp, £15.00. ISBN: 1 85946 254 5. Published by the Centre for Accessible Environments (qv) and the RIBA (qv), and obtainable from www.cae.org.uk/ or www.ribabookshops.com/

GOOD CHILDHOOD INQUIRY – A NATIONAL INQUIRY.
2007, 8pp, free. Published by and obtainable from the Children's Society at www.childrenssociety.org.uk/

GOOD LOO DESIGN GUIDE by Andrew Lacey.
2004, 52pp, £20.00. Published by the Centre for Accessible Environments (qv) and the RIBA (qv), and obtainable from www.cae.org.uk/ or www.ribabookshops.com/

GOOD PRACTICE GUIDE ON PLANNING FOR TOURISM.
2006, 48pp, free. May be downloaded from www.planningportal.gov.uk/england/professionals/en/1115314090554.html

GOOD PRACTICE GUIDE TO OPEN CAPTIONING.
2003, 50pp, free. May be downloaded from www.stagetext.org/content.asp?content.id=1024

THE GOVERNANCE OF BRITAIN.
2007, 63pp, £13.50. ISBN-13: 97801011717021. Published by and obtainable from the Stationery Office on telephone 0870 242 2345 or at www.tsoshop.co.uk/bookstore.asp. May be downloaded free of charge from www.official-documents.gov.uk/document/cm71/7170/7170.pdf

THE GOVERNMENT'S RESPONSE TO KATE BARKER'S REVIEW OF HOUSING SUPPLY.
2005, 100pp, free. May be downloaded from www.hm-treasury.gov.uk/media/F59/0D/prb05_barker_553.pdf

GOVERNMENT'S WAY WITH WORDS IMPACTS THE LIFE-CHANCES OF ALL by Charles Moore.
Daily Telegraph, 8 March 2008. Visit www.telegraph.co.uk/opinion/main.jhtml?xml=/opinion/2008/03/08/do0802.xml

GOWRINGS MOBILITY UK ROAD ATLAS.
2006, 186pp, £12.99. Published by and obtainable from Gowrings Mobility at www.gowringsmobility.co.uk/roadatlas.html

THE GREAT AMERICAN CRIME DECLINE: STUDIES IN CRIME AND PUBLIC POLICY by Franklin E Zimring.
2006, 272pp, £17.99. ISBN: 0 19 518115 8. Published by Oxford University Press.

GROUP SAFETY AT WATER MARGINS.
Published by the DfES (qv) and may be downloaded free of charge from www.teachernet.gov.uk/wholeschool/healthandsafety/visits/

GUIDANCE ON THE USE OF TACTILE PAVING SURFACES.
1998, 84pp, free. May be downloaded from www.dft.gov.uk/transportforyou/access/tipws/guidanceontheuseoftactilepav6167

GUIDE TO ASSIST IN COMPLIANCE WITH THE DISABILITY DISCRIMINATION ACT.
Free. Several publications obtainable from the British Security Industry Association at www.bsia.co.uk/

A GUIDE TO THE HUMAN RIGHTS ACT 1998.
2006, 47pp, free. May be downloaded from www.dca.gov.uk/peoples-rights human-rights/pdf/act-studyguide.pdf

GUIDE TO SAFETY AT SPORTS GROUNDS.
1997, 248pp, £12.00. ISBN: 0 11 300095 2. Published by and obtainable from the Stationery Office on telephone 0870 242 2345 or at www.tsoshop.co.uk/bookstore.asp

GUNS, GERMS AND STEEL: A SHORT HISTORY OF EVERYBODY FOR THE LAST 13 000 years by Jared Diamond.
1997, 480pp, £8.99. ISBN: 0 09 930278 0. Published by Vintage.

HABINTEG HOUSING ASSOCIATION DESIGN GUIDE.
2004, 80pp, £30. ISBN: 0 95454 620 2. Published by and obtainable from Habinteg (qv) at www.habinteg.org.uk/

HACKED OFF.
2007, 4pp, free. May be downloaded from www.disabilitynow.org.uk/campaigns/hackedoff

HALIFAX HAPPIEST HOMES REPORT.
2005, 20pp, free. May be downloaded from www.sirc.org/publik/happy_homes.pdf

A HANDBOOK FOR GROUP LEADERS. Part 3 of a 3-part supplement to health and safety of pupils on educational visits.
2002, 16pp, free. Published by the DfES (qv) and may be downloaded from www.teachernet.gov.uk/_doc/2578/HANDBOOK%20FOR%20GROUP%20LEADERS%20FINAL%202002.doc

HANDLE WITH CARE: AN INVESTIGATION INTO THE CARE SYSYTEM by Harriet Sergeant.
2006, £12.50, free. Published by the Centre for Policy Studies and may be downloaded from www.cps.org.uk/

HAPPIEST HOMES REPORT.
See: HALIFAX HAPPIEST HOMES REPORT.

HAPPINESS, ECONOMICS AND PUBLIC POLICY by Helen Johns and Paul Ormerod.
2007, 60pp, £10. Published by and obtainable from the Institute of Economic Affairs at www.iea.org.uk/files/upld-book416pdf?pdf

HAPPINESS: LESSONS FROM A NEW SCIENCE by Richard Layard.
2006, 320pp, £8.99. ISBN-13: 9780141016900. Published by Penguin Books.

HEALTH & SAFETY AT SCHOOL: SCHOOL TRIPS: PART 1.
2001, 4pp, free. Published by RoSPA (qv) and may be downloaded from www.rospa.com/safetyeducation/schooltrips/part1.htm

HEALTH & SAFETY AT SCHOOL: SCHOOL TRIPS: PART 2.
2001, 6pp, free. Published by RoSPA (qv) and may be downloaded from www.rospa.com/safetyeducation/schooltrips/part2.htm

HEALTH AND SAFETY OF PUPILS ON EDUCATIONAL VISITS: A GOOD PRACTICE GUIDE.
1998, 72pp, free. Published by the DfEE and may be downloaded from www.nationalmediamuseum.org.uk/pdfs/HASPEV.pdf

HEALTH FACILITIES NOTE 19.
See: DESIGN OF RESIDENTIAL CARE AND NURSING HOMES FOR OLDER PEOPLE.

HEAVY VETTING by Brendan O'Neill.
Guardian, 25 October 2006. Visit http://commentisfree.guardian.co.uk/brendan_oneill/2006/10/a_paranoiacs_charter.html

HOME TRUTHS: THE CASE FOR 70 000 SOCIAL HOMES A YEAR.
2007, 17pp, free. May be downloaded from www.housing.org.uk/

HOUSE ADAPTATIONS: A WORKING FILE FOR OCCUPATIONAL THERAPISTS by Stephen Thorpe.
1999, £25.00. ISBN: 0 903976 26 3. Published by the Centre for Accessible Environments (qv) and may be downloaded from www.cae.org.uk/

HOUSING AND THE DISABILITY EQUALITY DUTY.
2006, 81pp, free. May be downloaded free of charge from www.equalityhumanrights.com/pages/eocdrccre.aspx

HOUSING FOR A COMPACT CITY.
2003, 144pp, free. ISBN: 1 85261 439 0. Published by and obtainable from the Greater London Authority at www.london.gov.uk/

HOUSING FOR PEOPLE WITH SIGHT LOSS: A THOMAS POCKLINGTON TRUST DESIGN GUIDE.
2008, 120pp, £40.00. ISBN: 978 1 84806 029 6. Published by and obtainable from IHS BRE Press at www.ihsbrepress.com/

HOUSING DESIGN GUIDE.
2008. To be published in May 2008 by Design for London. For further particulars visit www.designforlondon.gov.uk/

HOUSING MARKET RENEWAL: REPORT BY THE COMPTROLLER AND AUDITOR GENERAL.
2007, 46pp, free. Published by the National Audit Office (qv) and may be downloaded from http://www.nao.org.uk/publications/nao_reports/07-08/070820.pdf.

HOW SOCIETIES FAIL – AND SOMETIMES SUCCEED.
See: COLLAPSE: HOW SOCIETIES CHOOSE TO FAIL OR SURVIVE.

HOW TO LABEL A GOAT: THE SILLY RULES AND REGULATIONS THAT ARE STRANGLING BRITAIN by Ross Clark.
2006, 272pp, £9.99. ISBN: 1 897 59795 9. Published by Harriman House.

HUBBUB: FILTH, NOISE AND STENCH IN ENGLAND, 1600–1717 by Emily Cockayne.
2007, 335pp, £25. ISBN: 0 300 11214 9. Published by Yale.

ISC CENSUS 2007.
2007, 32pp, free. Published by the Independent School Council, and can be downloaded from www.isc.co.uk/Publications_ISCCensus.htm

THE IDEAS OF LE CORBUSIER ON ARCHITECTURE AND URBAN PLANNING by Jacques Guiton.
1981, 124pp, o/p. ISBN: 0 8076 1005 4. Published by George Braziller Inc. and obtainable via Abebooks at www.abebooks.co.uk/

THE ILLUSION OF DESTINY by Amartya Sen.
2007, 240pp, £8.99. ISBN-13: 9780141027807. Published by Penguin Books.

I'M A TEACHER, GET ME OUT OF HERE by Francis Gilbert.
2004, 224pp, £8.99. ISBN: 1 904 09568 2. Published by Short Books.

THE IMPACT OF INEQUALITY: HOW TO MAKE SICK SOCIETIES HEALTHIER by Richard R Wilkinson.
2005, 368pp, £70.00. ISBN-13: 978 041 537 2688. Published by Routledge at www.routledge.com/

IMPACT OF ROAD HUMPS UPON VEHICLES AND THEIR OCCUPANTS by J Kennedy and others.
2004, 4pp, free. May be downloaded from www.secondscount.org/

THE IMPLEMENTATION OF THE NATIONAL PROBATION SERVICE INFORMATION SYSTEMS STRATEGY.
2001, 46pp, free. May be downloaded from www.nao.org.uk/pn/00-01/0001401.htm

IMPROVING PUBLIC ACCESS TO BETTER QUALITY TOILETS: A STRATEGIC GUIDE.
2008, 56pp, free. Published by the Department for Communities and Local Government (qv) and may be downloaded from www.communities.gov.uk/documents/localgovernment/pdf/713772

THE IMPROVING STATE OF THE WORLD: WHY WE'RE LIVING LONGER, HEALTHIER, MORE COMFORTABLE LIVES ON A CLEANER PLANET by Indur Goklany.
2007, 516pp, $19.95. ISBN-13: 978 1 930865 98 8. Published by the Cato Institute at www.catostore.org/

IMPROVING THE LIFE CHANCES OF DISABLED PEOPLE.
2005, 244pp, free. May be downloaded from www.cabinetoffice.gov.uk/strategy/downloads/work_areas/disability/disability_report/pdf/disability.pdf

IN WORK, BETTER OFF: NEXT STEPS TO FULL EMPLOYMENT.
2007, 91pp, free. Published by the Department for Work and Pensions (qv) and may be downloaded from www.dwp.gov.uk/welfarereform/in-work-better-off/in-work-better-off.pdf

INCLUSIVE ACCESSIBLE DESIGN by Adrian Cave.
2007, 98pp, £17.50. ISBN: 1 85946 250 2. Published by the Centre for Accessible Environments (qv) and the RIBA (qv), and obtainable from www.cae.org.uk/ or www.ribabookshops.com/

INCLUSIVE DESIGN.
See: ENGLISH PARTNERSHIPS GUIDANCE NOTE: INCLUSIVE DESIGN.

INCLUSIVE DESIGN: PRODUCTS FOR ALL CONSUMERS by Lindsey Etchell and David Yelding.
2004, 9pp, free. Published by Ricability (qv) and may be downloaded from www.ricability.org.uk/

INCLUSIVE MOBILITY.
2002, 164pp, free. Published by the Department for Transport (qv) and may be downloaded from www.dft.gov.uk/transportforyou/access/tipws/inclusivemobility

INCLUSIVE SCHOOL DESIGN – ACCOMMODATING PUPILS WITH SPECIAL EDUCATIONAL NEEDS AND DISABILITIES IN MAINSTREAM SCHOOLS.
See: BUILDING BULLETIN 94.

INCLUSIVE URBAN DESIGN: STREETS FOR LIFE by Elizabeth Burton and Lynne Mitchell.
2006, 224pp, £24.99. ISBN: 0 7506 6458 4. Published by Architectural Press, an imprint of Elsevier Press at http://books.elsevier.com/

INDEPENDENT LIVING: THE RIGHT TO BE EQUAL CITIZENS by Sarah Gillinson, Hannah Green and Paul Miller.
2005, 113pp, £10. Published by Demos and obtainable from Scope at www.scope.org.uk/

INDEPENDENT REVIEW OF THE TEACHING OF EARLY READING: FINAL REPORT by Jim Rose.
2006, 105pp, free. May be downloaded from www.standards.dfes.gov.uk/phonics/report.pdf

AN INDEPENDENT NHS: A REVIEW OF THE OPTIONS by Brian Edwards.
2007, 70pp, £10.00. Published by and obtainable from the Nuffield Trust on telephone 020 7631 8450 or at www.nuffieldtrust.org.uk/publications. May also be downloaded free of charge from www.nuffieldtrust.org.uk/ecomm/files/IndependentNHS.pdf

INDEPENDENT SCHOOLS COUNCIL CENSUS 2007.
2007, 32pp, free. May be downloaded from www.isc.co.uk/Publications_ISCCensus.htm

INNOCENT – BUT ON A CRIMINAL DATABASE by Philip Johnston.
Daily Telegraph, 28 May 2007. Visit www.telegraph.co.uk/opinion/main.jhtml?xml=/opinion/2007/05/28/do2802.xml

INSPECTION, INSPECTION, INSPECTION! by Anastasia de Waal.
2006, 160pp, £9.50. Published by and obtainable from Civitas (qv) at www.civitas.org.uk/

INTERGENERATIONAL MOBILITY IN EUROPE AND NORTH AMERICA by Jo Blanden, Paul Gregg and Stephen Machin.
2005, 20pp, free. Published by the Centre for Economic Performance at the London School of Economics and may be downloaded from http://cep.lse.ac.uk/about/news/IntergenerationalMobility.pdf

INTERNAL FLOOR FINISHES by Ann Alderson.
2006, 56pp, £15.00. ISBN: 1 85946 255 5. Published by the Centre for Accessible Environments (qv) and the RIBA (qv), and obtainable from www.cae.org.uk/ or www.ribabookshops.com/

INTERNATIONAL CLASSIFICATION OF IMPAIRMENTS DISABILITIES AND HANDICAPS.
Visit www.empowermentzone.com/icidh.txt

ISC CENSUS 2007.
See: INDEPENDENT SCHOOLS COUNCIL CENSUS 2007.

ID CARDS ARE THE ULTIMATE IDENTITY THEFT. COMPUTER SYSTEMS ALWAYS FAIL – AND THE NATIONAL DATABASE WILL DO SO BIG TIME by Professor Ian Angell.
The *Times*, 7 March 2008. Visit www.timesonline.co.uk/tol/comment/columnists/guest_contributors/article3499317.ece

IS DEMOCRACY POSSIBLE HERE? by Ronald Dworkin.
2007, 192pp, £11.95. ISBN: 0 691 12653 4. Published by Blackwell Publishing at http://bookshop.blackwell.co.uk/

IT'S OUR SPACE: A GUIDE FOR COMMUNITY GROUPS WORKING TO IMPROVE PUBLIC SPACE.
2007, 114pp, free. Published by CABE (qv) and may be downloaded from www.cabe.org.uk/AssetLibrary/9462.pdf

JOURNAL OF HAPPINESS STUDIES.
Visit www1.eur.nl/fsw/happiness/

JUDGE SLATES 'BANANA REPUBLIC' POSTAL VOTING SYSTEM by Sandra Laville.
The *Guardian*, 5 April 2005. Visit http://politics.guardian.co.uk/voting/story/0,,1460951,00.html

JUSTICE – A NEW APPROACH by Lord Falconer.
2007, 29pp, free. Published by the Ministry of Justice (qv) and may be downloaded from www.justice.gov.uk/docs/Justice-a-new-approach.pdf

JUSTICE – FIRST REPORT OF THE JUSTICE COMMITTEE OF THE HOUSE OF COMMONS 2007-08.
See: PROTECTION OF PRIVATE DATA.

KEN'S GAS GUZZLER.
Building Design, 29 July 2005. Visit www.bdonline.co.uk/story.asp?sectioncode=426&storycode=3054384

KEY FIGURES IN THE CLIMBIÉ CASE by David Batty.
The *Guardian*, 4 February 2002. Visit http://society.guardian.co.uk/climbie/story/0,,616539,00.html

A LAND FIT FOR CRIMINALS by David Fraser.
2006, 458pp, £17.99. ISBN: 1 857 76964 3. Published by the Book Guild.

THE LARCENY OF THE LOTTERY by Ruth Lea.
2006, 36pp, £5.00. Published by and obtainable from the Centre for Policy Studies at www.cps.org.uk/. May be viewed at www.cps.org.uk/search/?pageno=12&productsearch=larceny+ of+the+lottery

LEARNING FROM PLACE.
2007, 150pp, £19.95. ISBN-13: 9 781 859 462 829. Produced by the Academy of Urbanism (qv); published by and obtainable from the RIBA (qv) at www.cae.org.uk/ or www.ribabookshops.com/

LEAVING AN ENVIRONMENTALLY SOUND, ATTRACTIVE LEGACY by Quinlan Terry.
See: RICHARD H DRIEHAUS PRIZE BOOK 2005.

LESS IS MORE.
See: REGULATION – LESS IS MORE: REDUCING BURDENS, IMPROVING OUTCOMES.

LETTERS TO LILY ON HOW THE WORLD WORKS by Alan Macfarlane.
2006, 320pp, £14.999. ISBN: 1 861 97780 0. Published by Profile.

LIBRARY SERVICES FOR VISUALLY IMPAIRED PEOPLE: A MANUAL OF BEST PRACTICE.
2005, free. May be downloaded from www.cilip.org.uk/

LIFE CHANCES: ACCOUNTING FOR FALLING INTERGENERATIONAL MOBILITY by Jo Blanden, Paul Gregg and Lindsey Macmillan.
The Economic Journal, March 2007. Visit www.res.org.uk/society/mediabriefings/pdfs/2007/0703/jb-pg-lm.pdf

LIFETIME HOMES, LIFETIME NEIGHBOURHOODS: A NATIONAL STRATEGY FOR HOUSING IN AN AGEING SOCIETY.
2008, 176pp, free. Published by the Department for Communities and Local Government (qv) and may be downloaded from www.communities.gov.uk/documents/housing/pdf/lifetimehomes

LITTLEJOHN'S BRITAIN by Richard Littlejohn.
2007, 384pp, £7.99. ISBN: 0 099 50944 X. Published by Random House.

LIVING WELL IN LATER LIFE: A REVIEW OF PROGRESS AGAINST THE NATIONAL SERVICE FRAMEWORK FOR OLDER PEOPLE.
2006, 108pp, free. ISBN: 1 84562 081 X. Published by the National Audit Office and may be downloaded from www.audit-commission.gov.uk/

LIVING WITH RISK: PROMOTING BETTER PUBLIC SPACE DESIGN.
2007, 88pp, free. Published by Cabe (qv) and may be downloaded from www.cabe.org.uk/default.aspx?contentitemid=1930

LOCAL WELFARE.
2007, 14pp, free. Published by the Centre for Policy Studies and may be downloaded from www.cps.org.uk/latestpublications/?

LONDON BLUE BADGE PARKING GUIDE.
2007, 190pp, £4.50. Published by and obtainable from Gowrings Mobility at www.gowringsmobility.co.uk/roadatlas.html

LONDON HOUSING DESIGN GUIDE.
See: HOUSING DESIGN GUIDE.

LORD OF THE FLIES by William Golding.
1954. New edition 2004, 224pp, £12.99. ISBN-13: 9780571224524. Published by Faber and Faber.

THE LOSS OF SADNESS: HOW PSYCHIATRY TRANSFORMED NORMAL SORROW INTO DEPRESSIVE DISORDER by Allan D Horwitz and Jerome C Wakefield.
2007, 312pp, £17.99. ISBN: 0 19531 304 6. Published by Oxford University Press at www.oup.com/uk/catalogue/?ci=9780195313048

MACPHERSON REPORT.
See: THE STEPHEN LAWRENCE INQUIRY.

MAKE YOUR CONFERENCE ACCESSIBLE.
2003, 5pp, free. Published by the Centre for Accessible Environments (qv) and may be downloaded from www.cae.org.uk/pdf/venues/check.pdf

MAKING CONNECTIONS – A GUIDE TO ACCESSIBLE GREENSPACE.
2004, 115pp, £15.00. Published by and obtainable from the Sensory Trust at www.sensorytrust.org.uk/

MAKING GOODS AND SERVICES ACCESSIBLE TO DISABLED CUSTOMERS: A GUIDE FOR SERVICE STATIONS.
2004, free. Published by UKPIA and may be ordered on telephone 020 7240 0289.

MAKING THE DUTY WORK: A GUIDE TO THE DISABILITY EQUALITY DUTY FOR DISABLED PEOPLE AND THEIR ORGANISATIONS.
2006, 44pp, free. May be downloaded from www.equalityhumanrights.com/pages/eocdrccre.aspx

THE MASTER BUILDERS: LE CORBUSIER, MIES VAN DER ROHE, FRANK LLOYD WRIGHT by Peter Blake.
1996 (reissue), 448pp, o/p. ISBN: 0 393 31504 5. Published by Norton General Books at www2.wwnorton.com/catalog/fall96/master.htm and obtainable via Abebooks at www.abebooks.co.uk/

MATHS FEARS OVER NEW SCHOOLS DIPLOMAS by Anthea Lipsett.
The *Guardian*, 1 February 2008. Visit http://education.guardian.co.uk/1419education/story/0,,2250820,00.html

MANUAL FOR STREETS.
2007, 146pp, £22.50. Published by and obtainable from Thomas Telford at thomastelford.com/

MAY 2000 ELECTION REPORT: FINDINGS OF THE OPEN RIGHTS GROUP ELECTION OBSERVATION MISSION IN SCOTLAND AND ENGLAND.
2007, 68pp, free. Published by the Open Rights Group (qv) and may be downloaded from www.openrightsgroup.org/wp-content/uploads/org_election_report.pdf

MAYOR OF LONDON'S HOUSING DESIGN GUIDE.
See: HOUSING DESIGN GUIDE.

THE MEANING OF PUBLIC AUTHORITY UNDER THE HUMAN RIGHTS ACT: SEVENTH REPORT OF SESSION 2003-04.
2004, 66pp, free. May be downloaded from www.publications.parliament.uk/pa/jt200304/jtselect/jtrights/39/39.pdf

MEASURING RESIDENTIAL SEGREGATION by Ludi Simpson.
2005, free. May be downloaded from www.ccsr.ac.uk/research/migseg.htm

MENTAL CAPACITY ACT 2005: CODE OF PRACTICE.
2007, 302pp, free. May be downloaded from www.dca.gov.uk/legal-policy/mental-capacity/mca-cp.pdf

MENTAL HEALTH TEN YEARS ON: PROGRESS ON MENTAL HEALTH CARE REFFORM by Louis Appleby.
2007, 8pp, free. May be downloaded from www.dh.gov.uk/en/Publicationsandstatistics/Publications/index.htm

THE METRIC HANDBOOK by David Littlefield.
2007, 856pp, £34.99. ISBN: 0 7506 5281 0. Published by Architectural Press, an imprint of Elsevier at http://books.elsevier.com/

MILLENNIUM COHORT STUDY: SECOND SURVEY – A USER'S GUIDE TO INITIAL FINDINGS by Kirstine Hansen and Heather Joshi.
2007, 198pp, free. Published by the Institute of Education, University of London, and may be downloaded from http://image.guardian.co.uk/sys-files/Society/documents/2007/06/11/MCS2.pdf

MINOR ADAPTATIONS WITHOUT DELAY: A PRACTICAL GUIDE AND TECHNICAL SPECIFICATIONS FOR HOUSING ASSOCIATIONS.
2006, 57pp, free. Published by the Housing Association (qv) and may be downloaded from www.housingcorp.gov.uk/server/show/ConWebDoc.7502

MIXED RESULTS – A NATIONAL DNA DATABASE COULD WELL INCREASE, NOT REDUCE, THE NUMBER OF WRONGFUL CONVICTIONS by Allan Jamieson.
The *Guardian*, 28 February 2008. Visit www.guardian.co.uk/commentisfree/2008/feb/28/ukcrime.forensicscience

MONITORING POVERTY AND SOCIAL EXCLUSION IN THE UK 2006 by Guy Palmer, Tom MacInnes and Peter Kenway.
2006, 104pp, £16.95. Published by the Joseph Rowntree Foundation (qv) and may be downloaded rom www.jrf.org.uk/

A MORE EQUAL SOCIETY? NEW LABOUR, INEQUALITY AND EXCLUSION by John Hills and Kitty Stewart (Editors).
2005, 391pp, £19.99. ISBN: 1 861 34577 1. Published by the Policy Press for the Joseph Rowntree Foundation (qv).

MUSEUMS AND ART GALLERIES: MAKING EXISTING BUILDINGS ACCESSIBLE by Adrian Cave.
2007, 136pp, £35.00. ISBN-13: 978 1 85946 175 4. Published by the Centre for Accessible Environments (qv) and the RIBA (qv), and obtainable from www.cae.org.uk/ or www.ribabookshops.com/

MY 7-YEAR-OLD SON, A TINY SPLINTER AND THE FIVE-HOUR FARCE THAT SHOWS HOW THE NHS HAS LOST THE PLOT by John Humphrys.
Daily Mail, 25 March 2008. Visit www.dailymail.co.uk/pages/live/articles/news/news.html?in_article_id=544115&in_page_id=1770

NARROWING THE GAP: FINAL REPORT OF THE FABIAN COMMISSION ON LIFE CHANCES AND CHILD POVERTY.
2006, 224pp, £10.95. Published by and obtainable from the Fabian Society at www.fabians.org.uk/

NATIONAL CHILD DATABASE WILL INCREASE RISK by Eileen Munro.
The *Guardian*, 6 April 2004. Visit http://society.guardian.co.uk/children/comment/0,,1186315,00.html

NATIONAL CURRICULUM ASSESSMENTS AT KEY STAGE 2 IN ENGLAND, 2007 (PROVISIONAL).
2007, 14pp, free. Published by the DfES (qv) and may be downloaded from www.dfes.gov.uk/rsgateway/DB/SFR/s000737/SFR24-2007.pdf

THE NATIONALISATION OF CHILDHOOD by Jill Kirby.
2006, 11pp, £7.50. Published by and obtainable from the Centre for Policy Studies at www.cps.org.uk/ May be viewed free of charge at www.cps.org.uk/historiccatalogue/default.asp

NATIONAL RAIL MAP FOR PEOPLE WITH MOBILITY DIFFICULTIES
For particulars visit www.nationalrail.co.uk/

NEIGHBOURHOOD EDUCATION.
2007, 16pp, free. Published by the Centre for Policy Studies and may be downloaded from www.cps.org.uk/latestpublications/?

A NEW COMMITMENT TO NEIGHBOURHOOD RENEWAL: A NATIONAL ACTION STRATEGY PLAN.
2001, 127pp, free. May be downloaded from www.neighbourhood.gov.uk/publications.asp?did=85

THE NEW EAST END: KINSHIP, RACE AND CONFLICT by Geoff Dench and Kate Gavron.
2006, 288pp, £15.99. ISBN: 1 861 97928 2. Published by Profile Books and available from the Young Foundation at www.youngfoundation.org.uk/node/134

THE NEW SCHOOL RULES: THE PARENT'S GUIDE TO GETTING THE BEST EDUCATION FOR YOUR CHILD by Francis Gilbert.
2007, 288pp, £10.39. ISBN: 0 7499 5127 3. Published by Piatkus.

NEW THINGS HAPPEN: A GUIDE TO THE FUTURE THAMES GATEWAY.
2006, 35pp, free. Published by CABE (qv) and may be downloaded from http://cabe.org.uk/AssetLibrary/9203.pdf

NHS REFORM: THE EMPIRE STRIKES BACK by Nick Bosanquet, Henry de Zoete, and Andrew Haldenby.
2007, 49pp, free. Published by Reform (qv) and may be downloaded from http://www.fundsis.org/docs_act/79_NHS-reform-the-empire-strikes-back.pdf

NO FEAR: GROWING UP IN A RISK AVERSE SOCIETY by Tim Gill.
2007, 96pp, £8.50. ISBN-13: 978 1 903080 08 5. Published by the Calouste Gulbenkian Foundation. May also be downloaded free of charge from www.gulbenkian.org.uk/media/item/1177/89/No-fear.pdf

NO MORE BRAVE POLICEMEN – THE POLICE CAN'T DO THEIR JOBS IF THEY ARE BOUND BY HEALTH AND SAFETY LAWS by Chief Constable Bernard Hogan-Howe.
Times, 5 December 2007. Visit www.timesonline.co.uk/tol/comment/columnists/guest_contributors/article3000981.ece

NO PLACE LIKE UTOPIA: MODERN ARCHITECTURE AND THE COMPANY WE KEPT by Peter Blake.
1993, 348pp, $27.50. ISBN: 0 394 54896 5. Published by Alfred A Knopf and obtainable via Abebooks at www.abebooks.co.uk/

THE NON-COURSES REPORT 2007.
2007, 30pp, free. Published by the TaxPayers' Alliance. May be downloaded from http://tpa.typepad.com/waste/files/noncourses_report_2007.pdf

NON-DOMESTIC TECHNICAL HANDBOOK.
See: TECHNICAL HANDBOOKS.

NOT FIT FOR PURPOSE: £2 BILLION COST OF GOVERNMENT'S IT BLUNDERS by Bobbie Johnson and David Hencke.
Guardian, 5 January 2008. Visit .www.guardian.co.uk/technology/2008/jan/05/computing.egovernment

OFFENDERS ON PROBATION CARRY OUT 10,000 CRIMES A MONTH by Ben Leapman.
Daily Telegraph, 15 May 2006. Visit www.telegraph.co.uk/news/main.jhtml?xml=/news/2006/05/14/ncrim14.xml

ON REPORTING MENTAL HEALTH ISSUES.
2006. May be downloaded free of charge from the Press Complaints Commission at www.pcc.org.uk/news/index.html?article=NDIwMA

OPCS SURVEYS OF DISABILITY IN GREAT BRITAIN.
1984. ISBN 0 116 91266 9. Published by the Department of Health and Social Security and obtainable from www.statistics.gov.uk/CCI/SearchRes.asp?term=OPCS+SURVEYS+OF+DISABILITY+IN+GREAT+BRITAIN.&x=40&y=11

OPEN FOR BUSINESS.
2004, 80pp, £29.00. Published by and obtainable from the Employers' Forum on Disability at www.employers-forum.co.uk/

OPPORTUNITY AGE: MEETING THE CHALLENGES OF AGEING IN THE 21ST CENTURY.
2005, free. May be downloaded from www.dwp.gov.uk/opportunity_age/first_report.asp

OPPORTUNITY FOR ALL: EIGHTH ANNUAL REPORT 2006 – STRATEGY DOCUMENT.
2006, 172pp, free. May be downloaded from www.dwp.gov.uk/ofa/reports/latest.asp

OUR FUTURE HEALTH SECURED? A REVIEW OF NHS FUNDING AND PERFORMANCE by Sir Derek Wanless, John Appleby, Anthony Harrison and Darshan Patel.
2007, 321pp, £50. Published by and obtainable from the Kings Fund (qv). May also be downloaded free of charge from www.kingsfund.org.uk/publications/kings_fund_publications/index.html

OUR MASTERS ARE DEAF TO OUR WISHES by Philip Johnston.
Daily Telegraph, 18 June 2007. Visit www.telegraph.co.uk/opinion/main.jhtml?xml=/opinion/2007/06/18/do1802.xml

OVER-PARENTING IS THE CURSE OF OUR TIME by Johann Hari.
The *Independent*, 4 February 2008. Visit www.independent.co.uk/opinion/commentators/johann-hari/johann-hari-overparenting--is-the-curse-of-our-time-777702.html

OUR STREET: LEARNING TO SEE: A TEACHER'S GUIDE TO USING THE BUILT ENVIRONMENT AT KEY STAGE 2.
2007, 64pp, free. Published by CABE (qv) and may be downloaded from http://cabe.org.uk/AssetLibrary/9789.pdf

PANIC NATION: UNPICKING THE MYTHS WE'RE TOLD ABOUT FOOD AND HEALTH by Stanley Feldman and Vincent Marks.
2005, 272pp, £9.99. ISBN: 1844541223. Published by John Blake Publishing Ltd.

PARANOID PARENTING by Frank Furedi.
2002, 234pp, $14.95. ISBN:1 556 52464 1. Published by the Chicago Review Press and obtainable via Abebooks at www.abebooks.co.uk/

PAS 78 – A GUIDE TO GOOD PRACTICE IN COMMISSIONING ACCESSIBLE WEBSITES.
2006, 66pp, £30.00. Published by and obtainable from BSI at www.bsi-global.com/. May also be downloaded free of charge from www.equalityhumanrights.com/pages/eocdrccre.aspx

PERMANENT DIFFERENCES? INCOME AND EXPENDITURE INEQUALITY IN THE 1990S AND 2000S by Alissa Goodman and Zoë Oldfield.
2004, 39pp, £50. Published by and obtainable from from the Institute for Fiscal Studies on telephone 020 7291 4800. May also be downloaded free of charge from www.ifs.org.uk/publications.php?publication_id=2117

PERSONAL EMERGENCY EGRESS PLANS.
1993, 30pp, £6.00. ISBN: 0 952179 20 2. Published by and obtainable from the Centre for Accessible Environments (qv) at www.cae.org.uk/

PISA – SHOW'S OVER: INTERNATIONAL STUDY EXPOSES GOVERNMENT STANDARDS CHARADE.
Civitas blog, 5 December 2007. Visit www.civitas.org.uk/blog/2007/12/pisa_shows_over_international.html

PLAN, DESIGN AND BUILD:
Part 1: 1997, 120pp, £13.00.
Part 2: 2001, 32pp, £6.00.
Published by and obtainable from ACRE at www.acre.org.uk/

PLANNING AND ACCESS FOR DISABLED PEOPLE: A GOOD PRACTICE GUIDE by Driver Jonas.
2003, 112pp, £15.00. ISBN: 1 851126 04 X. Published by and obtainable from the Department for Communities and Local Government (qv) at www.communities.gov.uk/ May also be downloaded free of charge from www.communities.gov.uk/publications/planningandbuilding/planningaccess

PLANNING, BUILDINGS, STREETS AND DISABILITY EQUALITY: A GUIDE TO THE DISABILITY EQUALITY DUTY AND DISABILITY DISCRIMINATION ACT 2005.
2006, 56pp, free. May be downloaded from www.equalityhumanrights.com/pages/eocdrccre.aspx

PLANNING YOUR HOME FOR SAFETY AND CONVENIENCE – PRACTICAL ADVICE FOR DISABLED AND OLDER PEOPLE by Andrew Lacey.
2002, 40pp, £7.50. ISBN: 0 903976 34 X. Published by the Centre for Accessible Environments (qv) and the RIBA (qv), and obtainable from www.cae.org.uk/ or www.ribabookshops.com/

PLATFORM LIFTS by Alison Grant.
2005, 32pp, £15.00. ISBN: 1 85946 172 7. Published by the Centre for Accessible Environments (qv) and the RIBA (qv), and obtainable from www.cae.org.uk/http://www.cae.org.uk/ or www.ribabookshops.com/

PLAYING IT SAFE: THE CRAZY WORLD OF BRITAIN'S HEALTH AND SAFETY REGULATIONS by Alan Pearce.
2007, 128pp, ISBN: 1905548850. Published by The Friday Project at http://www.thefridayproject.co.uk/

PLAYING IT SAFE: TODAY'S CHILDREN AT PLAY by Diana McNeish and Helen Roberts.
1995, 30pp, £2.50. Published by and obtainable from Barnardos at www.barnardos.org.uk/

PLUNDERING THE PUBLIC SECTOR by David Craig and Richard Brooks.
2006, 240pp, £9.99. ISBN: 1845293746. Published by Constable & Robinson.

POST-LEGISLATIVE SCRUTINY: A CONSULTATION PAPER.
2005, 64pp, free. May be downloaded from www.lawcom.gov.uk/docs/cp178.pdf

POVERTY AND ETHNICITY IN THE UK by Lucinda Platt.
2007, 152pp, £15.95. ISBN-13: 978 1 86134 989 7. Published by the Polity Press. A 4-page summary of the findings may be downloaded free of charge from www.jrf.org.uk/knowledge/findings/socialpolicy/pdf/2059.pdf

THE POVERTY TRADE-OFF: WORK INCENTIVES AND INCOME REDISTRIBUTION IN BRITAIN by Stuart Adam, Mike Brewer and Andrew Shephard.
2006, 64pp, free. Published by the Joseph Rowntree Foundation (qv) and may be downloaded from www.jrf.org.uk/bookshop/eBooks/1590-poverty-benefits-taxation.pdf

POVERTY, WEALTH AND PLACE IN BRITAIN 1968–2005.
2007, 125pp, free. Published by the Joseph Rowntree Foundation (qv) and may be downloaded from www.jrf.org.uk/bookshop/eBooks/2019-poverty-wealth-place.pdf

POWER TO THE PEOPLE. THE REPORT OF POWER: AN INDEPENDENT INQUIRY INTO BRITAIN'S DEMOCRACY.
2006, 312pp, £12.00. Published by the Joseph Rowntree Foundation (qv) and obtainable from York Publishing Distribution on telephone 01904 431213 or orders@yps-publishing.co.uk

THE PRICE OF PARENTHOOD by Jill Kirby.
2005, 12pp, £7.50. Published by and obtainable from the Centre for Policy Studies. May be viewed free of charge at www.cps.org.uk/historiccatalogue/default.aspCPS

THE PRINCIPLES OF INCLUSIVE DESIGN.
2006, 20pp, free. Published by CABE (qv) and may be downloaded from http://cabe.org.uk/AssetLibrary/8853.pdf

PROGRESS IN INTERNATIONAL READING LITERACY by Liz Twist, Ian Schagen and Claire Hodgson.
2007, 172pp, free. Published by the National Foundation for Educational Research and may be downloaded from www.nfer.ac.uk/publications/pdfs/downloadable/PIRLSreport.pdf

THE PROGRESSIVE PATRIOT: A SEARCH FOR BELONGING by Billy Bragg.
2006, 296pp, £17.99. ISBN: 0 593 05343 45. Published by Random House.

PROSPEROUS COMMUNITIES II: VIVE LA DEVOLUTION!
2006, 76pp, free. Published by and obtainable from the Local Government Association on telephone 020 7664 3131. May also be downloaded from www.lga.gov.uk/Documents/Publication/prosperouscommunitiesII.pdf

PROTECTION OF PRIVATE DATA: HC154. FIRST REPORT OF SESSION 2007-08 OF THE JUSTICE COMMITTEE OF THE HOUSE OF COMMONS.
2008, 32pp, £8.50. ISBN: 0215037928. Obtainable from TSO on telephone 0870 242 2345 or at www.tsoshop.co.uk/bookstore.asp May also be also be downloaded free of charge from www.publications.parliament.uk/pa/cm200708/cmselect/cmjust/154/15402.htm

PROVISION AND USE OF TRANSPORT VEHICLES, STATUTORY CODE OF PRACTICE, SUPPLEMENT TO PART 3 CODE OF PRACTICE.
2006, 93pp, £15.00. ISBN: 0 11 703632 3. Obtainable from the Stationery Office on telephone 0870 242 2345 or at www.tsoshop.co.uk/bookstore.asp May also be downloaded free of charge from www.equalityhumanrights.com/pages/eocdrccre.aspx

THE PUBLIC AND THE POLICE by Harriet Sergeant.
To be published in May 2008 by Civitas at www.civitas.org.uk/

PUBLIC BODIES 2006.
2006, 372pp, free. Published by the Cabinet Office (qv) and may be downloaded from www.civilservice.gov.uk/other/agencies/public_bodies/index.asp

PUSHCHAIRS.
2004, 24pp, 2002, free. Published by Ricability (qv) and may be downloaded from www.ricability.org.uk/reports/report-childcare.htm

A QUESTION OF STANDARDS: PRESCOTT'S TOWN HALL MADNESS by Owen Paterson MP and Gerald Howarth MP.
2006, 26pp, free. Published by the Cornerstone Group and may be downloaded from www.cornerstonegroup.org.uk/

RACE EQUALITY DUTY.
See: REGENERATION AND THE RACE EQUALITY DUTY.

RADICAL CLASSICISM: THE ARCHITECTURE OF QUINLAN TERRY by David Watkin.
2006, 256pp, $60.00. ISBN: 0 8478 2806 9. Published by Rizzoli International Publications in New York and obtainable from Amazon UK at www.amazon.co.uk/

RAIL VEHICLE ACCESSIBILITY REGULATIONS (RVAR) 1998.
1998. Obtainable from the Stationery Office on telephone 0870 242 2345 or at www.tsoshop.co.uk/bookstore.asp. May also be downloaded free of charge from www.hmso.gov.uk/si/si1998/98245602.htm#end

RAIL VEHICLE ACCESSIBILITY (AMENDMENT) REGULATIONS (RVAAR) 2000.
2000. Obtainable from the Stationery Office on telephone 0870 242 2345 or at www.tsoshop.co.uk/bookstore.asp. May also be downloaded free of charge from www.hmso.gov.uk/si/si2000/20003215.htm24

RAILWAYS FOR ALL: THE ACCESSIBILITY STRATEGY FOR GREAT BRITAIN'S RAILWAYS.
2006, 45pp, free. May be downloaded from www.dft.gov.uk/transportforyou/access/rail/rfa/railwaysforallstrategy1

A RATIONAL WAY FORWARD FOR THE NHS IN ENGLAND: A DISCUSSION PAPER OUTLINING AN ALTERNATIVE APPROACH TO HEALTH REFORM.
2007, 82pp, free. Published by the British Medical Association and may be downloaded from http://image.guardian.co.uk/sys-files/Society/documents/2007/05/08/rationalwayforward.pdf

READERS AND READING: NATIONAL REPORT FOR ENGLAND 2006.
See: PROGRESS IN INTERNATIONAL READING LITERACY.

READY TO READ? by Anastasia de Waal and Nicholas Cowen.
2007, 14pp, free. Published by Civitas (qv) and may be downloaded from www.civitas.org.uk/pdf/readytoread.pdf

THE REAL LEVEL OF UNEMPLOYMENT 2007 by Professor Steve Fothergill and others.
2007, 61pp, free. Published by Sheffield Hallam University and may be downloaded from www.shu.ac.uk/research/cresr/downloads/The%20Real%20Level%20of%20Unemployment%202007-.pdf

REALISING THE BENEFITS? ASSESSING THE IMPLEMENTATION OF 'AGENDA FOR CHANGE' by James Buchan and David Evans.
2007, 50pp, £10.00. Published by and obtainable from the King's Fund (qv) at http://www.kingsfund.org.uk/ May also be downloaded free of charge from http://www.kingsfund.org.uk/publications/kings_fund_publications/realising_the.html

THE RED TAPE ECONOMY by Graeme Leach.
2005, 38pp, £4.90. Published by and obtainable from Civitas (qv) at www.civitas.org.uk/

REDUCING INEQUALITIES: REALISING THE TALENTS OF ALL by Leon Feinstein and others.
2007, 64pp, £10.00. ISBN-13: 9781905818204. Published by the National Children's Bureau and obtainable from Central Books at www.ncb-books.org.uk/

REFERRALS, ASSESSMENTS, AND CHILDREN AND YOUNG PEOPLE ON CHILD PROTECTION REGISTERS, ENGLAND – YEAR ENDING 31 MARCH 2005.
2006, 132pp, free. May be downloaded from www.dfes.gov.uk/rsgateway/DB/VOL/v000632/VOL01-2006textv1.pdf

REFORMING PENSIONS by Nick Bosanquet, Derek Scott, Andrew Haldenby and Colin Taylor.
2005, 14pp, free. Published by Reform (qv) and may be downloaded from www.reform.co.uk/filestore/pdf/Pensions%20reform.pdf

REFORMING WELFARE by Nicholas Boys Smith.
2006, 143pp, free. Published by Reform (qv) and may be downloaded from www.reform.co.uk/filestore/pdf/Reforming%20welfare.pdf

REGENERATION AND THE RACE EQUALITY DUTY.
2007, 185pp, free. Published by the Equality and Human Rights Commission (qv) and may be downloaded from www.equalityhumanrights.com/Documents/Race/Formal%20investigations/CRE%20Regeneration%20Formal%20investigation.pdf

REGULATION – LESS IS MORE: REDUCING BURDENS, IMPROVING OUTCOMES.
2005, 69pp, free. Published by the Better Regulation Task Force (qv) and may be downloaded from www.brc.gov.uk/downloads/pdf/lessismore.pdf

REPORT CARD 7.
See: CHILD POVERTY IN PERSPECTIVE: AN OVERVIEW OF CHILD WELL-BEING IN RICH COUNTRIES.

REPORT ON SOCIAL INSURANCE AND ALLIED SERVICES by W H Beveridge.
1942, 300pp. For summaries visit www.sochealth.co.uk/history/beveridge.htm and www.fordham.edu/halsall/mod/1942beveridge.html

A REPORT ON THE SURVEILLANCE SOCIETY by Richard Thomas.
2006, 102pp, free. May be downloaded from www.ico.gov.uk/upload/documents/library/data_protection/practical_application/surveillance_society_full_report_2006.pdf

THE RETREAT OF REASON: POLITICAL CORRECTNESS AND THE CORRUPTION OF PUBLIC DEBATE IN MODERN BRITAIN by Anthony Browne.
2006, 121pp, £9.50. Published by and obtainable from Civitas (qv) at www.civitas.org.uk/

THE REVIEW OF POLICING: INTERIM REPORT by Sir Ronnie Flanagan.
2007, 59pp, free. Published by the Home Office. May be downloaded from http://police.homeoffice.gov.uk/news-and-publications/publication/police-reform/The_review_of_policing_inte1.pdf?view=Binary

A REVIEW OF RESEARCH ON OUTDOOR LEARNING by Mark Rickinson, Justin Dillion and others.
2004, 6pp, free. Published by the Field Studies Council and may be downloaded from www.field-studies-council.org/documents/general/NFER/NFER%20Exec%20Summary.pdf
For the full document, costing £8.00, contact the Field Studies Council on telephone 01743 852140, or by email at publications@field-studies-council.org

REVIEW OF THE ELECTORAL COMMISSION.
2007, 115pp, free. Published by the Committee on Standards in Public Life (qv) and may be downloaded from www.public-standards.gov.uk/

RICABILITY CONSUMER REPORTS. The following illustrated reports from the Royal Institute for Consumer Affairs (Ricability) are free, and may be ordered on telephone 020 7427 2460 or downloaded from www.ricability.org.uk/ They provide independent assessments of products and services particularly for older and disabled users.
BABY CARRIERS.
CALLING FOR HELP: A GUIDE TO COMMUNITY ALARMS.
CAR CONTROLS.
GETTING A WHEELCHAIR INTO A CAR.
THE INS AND OUTS OF CHOOSING A CAR.
MOTORING AFTER A STROKE.
MOTORING WITH ARTHRITIS.
MOTORING WITH MULTIPLE SCLEROSIS.
PEOPLE LIFTERS: A GUIDE TO DEVICES WHICH HELP WHEELCHAIR USERS GET INTO A CAR.
PUSHCHAIRS.
STAY IN TOUCH: A GUIDE TO TELEPHONES AND SERVICES FOR OLDER AND DISABLED PEOPLE.
TAKING CONTROL: A GUIDE TO BUYING OR UPGRADING CENTRAL HEATING CONTROLS.
WHAT'S NEW? IN PRODUCTS FOR EASIER AND SAFER LIVING.
WHEELS WITHIN WHEELS: A GUIDE TO USING A WHEELCHAIR ON PUBLIC TRANSPORT.

RICHARD H DRIEHAUS PRIZE BOOK 2005 by Quinlan Terry and others.
2005, 80pp, $20.00. ISBN: 0 9670548 3 4. Published by and obtainable from the University of Notre Dame School of Architecture at www.architecture.nd.edu/

RICHARD NIXON AND THE RISE OF AFFIRMATIVE ACTION: THE PURSUIT OF RACIAL EQUALITY IN AN ERA OF LIMITS by Kevin Yuill.
2006, 265pp, £21.98. ISBN: 0 742 54998 4. Published by Rowman & Littlefield.

RICHISTAN: A JOURNEY THROUGH THE 21ST CENTURY WEALTH BOOM AND THE LIVES OF THE RICH by Robert Frank.
2007, 263pp, £12.99. ISBN: 0 749 92823 9. Published by Piatkus.

RISK AND CHILDHOOD by Nichola Madge and John Barker.
2007, 68pp, free. Published by the Royal Society of Arts and may be downloaded from www.rsa.org.uk/acrobat/risk&childhood_nov07.pdf

THE ROAD TO SOUTHEND PIER: ONE MAN'S STRUGGLE AGAINST THE SURVEILLANCE SOCIETY by Ross Clark.
2007, 176pp, £9.99. ISBN-13: 978 19056441444. Published by Harriman House at www.harriman-house.com/

SAFETY FEARS OVER NEW REGISTER OF ALL CHILDREN by Francis Elliott.
The Times, 27 August 2007. Visit www.timesonline.co.uk/tol/news/politics/article2332307.ece

THE SAFETY OF WHEELCHAIR OCCUPANTS IN ROAD PASSENGER VEHICLES.
2007, 75pp, free. May be downloaded from www.dft.gov.uk/transportforyou/access/tipws/thesafetyofwheelchairoccupan6168?page=4

SAFETY, SECURITY AND ENVIRONMENTAL CONTROLS: A SPECIFIER'S GUIDE TO ELECTRONIC ASSISTIVE TECHNOLOGY IN THE HOME by Andrew Lacey.
2002, 44pp, £10.50. ISBN: 0 903976 35 8. Published by the Centre for Accessible Environments (qv) and the RIBA (qv), and obtainable from www.cae.org.uk/ or www.ribabookshops.com/

SAFETY SIGNS AND SIGNALS: GUIDANCE ON REGULATIONS.
1996, £8.50. ISBN: 0 717 60870 0. Published by and obtainable from the Health and Safety Executive. For further particulars visit www.hse.gov.uk/pubns/indg184.htm

SCARED TO DEATH: FROM BSE TO GLOBAL WARMING – HOW SCARES ARE COSTING US THE EARTH by Christopher Booker and Richard North.
2007, 256pp, £16.99. ISBN: 0 826 48614 2. Published by Continuum Books at www.continuumbooks.com/

SCHEME DEVELOPMENT STANDARDS.
2003, 44pp, free. Published by the Housing Corporation (qv) and may be downloaded from http://www.housingcorp.gov.uk/upload/pdf/SDSv5.pdf

SECURING GOOD SOCIAL CARE FOR OLDER PEOPLE: TAKING A LONG TERM VIEW by Sir Derek Wanless.
2006, 345pp, free. ISBN: 1 85717 544 1. Published by the Kings Fund (qv) and may be downloaded from www.kingsfund.org.uk/resources/publications/securing_good.html

SECURITY WITHOUT THE SPIKES? A PRACTICAL RESOURCE PACK FOR CRIME PREVENTION IN THE PUBLIC REALM.
2004, 21pp, free. May be downloaded from www.gate-it.org.uk/ideas/security-without-the-spikes.doc

SEE IT RIGHT: MAKING INFORMATION ACCESSIBLE FOR PEOPLE WITH SIGHT PROBLEMS.
2007, 140pp, £30.00. Published by and obtainable from the RNIB (qv) at www.rnib.org.uk/

SEEN AND HEARD: RECLAIMING THE PUBLIC REALM WITH CHILDREN AND YOUNG PEOPLE by Joost Beunderman, Peter Bradwell, and Celia Haddon.
2007, 68pp, £10.00. Published by and obtainable from Demos at. www.demos.co.uk/ May also be downloaded free of charge from www.demos.co.uk/files/070928_DEMOS_S&H_Pamphlet.pdf

SELLING SICKNESS: HOW THE WORLD'S BIGGEST PHARMACEUTICAL COMPANIES ARE TURNING US ALL INTO PATIENTS by Ray Moynihan and Alan Cassels.
2005, 254pp, £7.99. ISBN: 1 560 25697 4. Published by Nation Books.

SEND FOR THE SHERIFF.
2007, 20pp, free. Published by the Centre for Policy Studies and may be downloaded from www.cps.org.uk/latestpublications/?

SENSE AND ACCESSIBILITY – HOW TO IMPROVE ACCESS ON COUNTRYSIDE PATHS, ROUTES AND TRAILS FOR PEOPLE WITH MOBILITY IMPAIRMENTS.
2000, 36pp, free. Published by Natural England and may be downloaded from www.countryside.gov.uk/

SEXUAL BEHAVIOUR IN BRITAIN: THE NATIONAL SURVEY OF SEXUAL ATTITUDES AND LIFESTYLES by K Wellings, J Field, A M Johnson, and J Wadsworth.
1994, 480pp, o/p. Published by Penguin Books and obtainable via Abebooks at www.abebooks.co.uk/

SHARED CARE: ESTABLISHING A BALANCE BETWEEN HOME AND CHILD CARE SETTINGS by Lieselotte Ahnert and Michael E Lamb.
'Child Development' vol 74, issue 4, p 1044. July 2003. For particulars visit www.blackwellpublishing.com/journal.asp?ref=0009-3920&site=1

SHYNESS: HOW NORMAL BEHAVIOR BECAME A SICKNESS by Christopher Lane.
2007, 288pp, £18.99. ISBN: 0 30012 446 5. Published by Yale University Press at www.yalebooks.co.uk/

SIGN DESIGN GUIDE: A GUIDE TO INCLUSIVE SIGNAGE by Peter Barker, June Fraser and the Sign Design Society.
2000, 92pp, £20. ISBN: 185878 412 3. Obtainable from the Centre for Accessible Environments (qv) at www.cae.org.uk/

SIMPLE JUSTICE by Charles Murray and others.
2005, 113pp, £9.00. Published by and obtainable from Civitas (qv) at www.civitas.org.uk/

SIMPLIFICATION PLAN 2007: PROMOTING BUSINESS AND ENTERPRISE THROUGH BETTER REGULATION.
2007, 79pp, free. Published by the Department for Business, Enterprise and Regulatory Reform (qv) and may be downloaded from www.berr.gov.uk/files/file42767.pdf

SIMPLIFICATION PLANS: A SUMMARY.
2004, 35pp, free. Published by the Cabinet Office (qv) and may be downloaded from www.cabinetoffice.gov.uk/regulation/documents/simplification/summary.pdf

SKILLS FOR LIFE: IMPROVING ADULT LITERACY & NUMERACY SKILLS FOR LIFE. A REPORT PRODUCED BY THE HOUSE OF COMMONS PUBLIC ACCOUNTS COMMITTEE IN JANUARY 2006.
2006, 37pp, free. May be downloaded from www.publications.parliament.uk/pa/cm200506/cmselect/cmpubacc/792/792.pdf

SKILLS FOR LIFE: THE NATIONAL STRATEGY FOR IMPROVING ADULT LITERACY AND NUMERACY SKILLS.
2001, 67pp, free. May be downloaded from www.dfes.gov.uk/readwriteplus/bank/ABS_Strategy_Doc_Final.pdf

SKILLS IN THE UK: THE LONG TERM CHALLENGE by Lord Leitch.
2005, 18pp, free. Published by HM Treasury and may be downloaded from www.hm-treasury.gov.uk/media/2/3/pbr05_leitchreviewexecsummary_255.pdf

SMART PFI: RIBA POSITION PAPER.
2006, 6pp, free. May be downloaded from www.architecture.com/Files/RIBAHoldings/PolicyandInternationalRelations/Policy/SmartPFI/SmartPFIPositionPaper.pdf

THE SOCIAL ENTREPRENEUR: MAKING COMMUNITIES WORK by Andrew Mawson.
2008, 192pp, £9.99. ISBN 1 843 54661 2. Published by Atlantic Books at www.groveatlantic.co.uk/

A SOCIAL HISTORY OF DYING by Allan Kellehear.
2007, 310pp, £15.99. ISBN: 0 521 69429 2. Published by Cambridge University Press.

SOCIAL TRENDS by the Office for National Statistics (qv).
Annual, 250pp, £49.50. Published by and obtainable from Palgrave Macmillan at www.palgrave.com/products/title.aspx?is=9781403993946. May also be downloaded free of charge from www.statistics.gov.uk/statbase/Product.asp?vlnk=5748&More=N

SOLUTIONS 14.
2007, 79pp, free. Published by and obtainable from the RNID (qv) on telephone 01733 361 199 or at www.rnid.org.uk/bookshop

SOUR GRIPES by Simon Carr.
2006, 216pp, £9.99. ISBN: 0 7499 5119 2. Published by Portrait Press.

SPECIAL EDUCATIONAL NEEDS AND THE LAW by Simon Oliver.
2007, 366pp, £35. ISBN-13: 978 0 85308 575 1. Published by Jordans at www.jordanpublishing.co.uk/

SPECIAL EDUCATIONAL NEEDS: A NEW LOOK by Mary Warnock.
2005, ISBN: 0 902227 12 2. Published by the Philosophy of Education Society of Great Britain and obtainable from www.philosophy-of-education.org/impact/impact_11.asp

SQUANDERED: HOW GORDON BROWN IS WASTING OVER ONE TRILLION POUNDS OF OUR MONEY by David Craig.
2008, 320pp, £8.99. ISBN 1 8452 9832 2. Published by Constable at http://constablerobinson.co.uk/

STAND AND DELIVER: THE FUTURE FOR CHARITIES PROVIDING PUBLIC SERVICES.
2007, 36pp, free. Published by the Charity Commission (qv) and may be downloaded from www.charity-commission.gov.uk/Library/publications/pdfs/RS15text.pdf

STANDARDS FOR ADVENTURE – Part 2 of a 3-part supplement.
2002, 17pp, free. Published by the DfES (qv) and may be downloaded from www.teachernet.gov.uk/wholeschool/healthandsafety/visits/

STANDARDS FOR LEAS IN OVERSEEING EDUCATIONAL VISITS – Part 1 of a 3-part supplement.
2002, 28pp, free. Published by the DfES (qv) and may be downloaded from www.teachernet.gov.uk/wholeschool/healthandsafety/visits/

STANDARDS OF PUBLIC EXAMINATIONS IN ENGLAND AND WALES.
2005, 15pp, free. Published by Reform (qv) and may be downloaded from http://reform.co.uk/website/education.aspx

STATE OF HEALTHCARE 2007.
2007, 139pp, free. Published by the Healthcare Commission (qv) and may be downloaded from www.healthcarecommission.org.uk/_db/_documents/State_of_Healthcare-2007.pdf

THE STATE OF SOCIAL CARE IN ENGLAND 2005–06.
2007, 28pp, free. Published by and obtainable from the Commission for Social Care Inspection at www.csci.org.uk/

THE STEPHEN LAWRENCE INQUIRY: REPORT OF AN INQUIRY BY SIR WILLIAM MACPHERSON OF CLUNY.
1999, £26.00. Published by and obtainable from TSO Parliamentary and Legal Bookshop at www.tsoshop.co.uk/parliament. May also be downloaded free of charge from www.archive.official-documents.co.uk/document/cm42/4262/sli-00.htm/

STOLEN INNOCENCE: A MOTHER'S FIGHT FOR JUSTICE – THE AUTHORISED STORY OF SALLY CLARK by John Batt.
2005, 336pp, £6.99. ISBN-13: 9780091905699. Published by Ebury Press.

STRUGGLING TO PAY COUNCIL TAX: A NEW PERSPECTIVE ON THE DEBATE ABOUT LOCAL TAXATION by Michael Orton.
2006, 60pp, £13.95. Published by the Joseph Rowntree Foundation. May be downloaded free of charge from www.jrf.org.uk/bookshop/eBooks/1577-local-tax-benefits.pdf

A STUDY OF THE DIFFICULTIES DISABLED PEOPLE HAVE WHEN USING EVERYDAY CONSUMER PRODUCTS.
2000, 90pp, free. May be downloaded from www.dti.gov.uk/files/file21460.pdf

THE SUBVERSIVE FAMILY: AN ALTERNATIVE HISTORY OF LOVE AND MARRIAGE by Ferdinand Mount.
1982, 282pp, o/p. ISBN: 0 22 401999 6. Published by Jonathan Cape and obtainable via Abebooks at www.abebooks.co.uk/

SUPPLY SIDE POLITICS: HOW CITIZENS' INITIATIVES COULD REVITALISE BRITISH POLITICS by Matt Qvortrup.
2007, 44pp, £7.50. Published by and obtainable from the Centre for Policy Studies at www.cps.org.uk/ May also be viewed free of charge at www.cps.org.uk/cpsfile.asp?id=677

SURE START AND BLACK AND MINORITY ETHNIC POPULATIONS.
2007, 107pp, free. May be downloaded from www.surestart.gov.uk/_doc/P0002437.pdf

SURVEY OF PUBLIC ATTITUDES TOWARDS STANDARDS OF CONDUCT IN PUBLIC LIFE REPORT.
2006, 104pp, free. May be downloaded from www.public-standards.gov.uk/upload/assets/www.public_standards.gov.uk/public_attitude_survey.pdf

SUSTAINABLE COMMUNITIES: BUILDING FOR THE FUTURE.
For particulars visit www.communities.gov.uk/publications/communities/sustainablecommunitiesbuilding/

SYSTEM REDESIGN – 2: ASSESSMENT REDESIGN by David Hargreaves, Chris Gerry and Tim Oates.
2007, 49pp, £10.00. Published by and obtainable from the Specialist Schools and Academies Trust at http://secure.ssatrust.org.uk/eshop/

TACKLING ANTI-SOCIAL BEHAVIOUR.
2006, 47pp, £13.50. ISBN: 0 102 94291 5. Published by and obtainable from the Stationery Office at www.tsoshop.co.uk/bookstore.asp May also be downloaded free of charge from www.nao.org.uk/publications/nao_reports/06-07/060799.pdf

TACKLING HEALTH INEQUALITIES: 2007 STATUS REPORT ON THE PROGRAMME FOR ACTION.
2008, 111pp, free. Published by the Department of Health and may be downloaded from www.dh.gov.uk/en/Publicationsandstatistics/Publications/DH_083471

TALKING EQUALITY: WHAT MEN AND WOMEN THINK ABOUT EQUALITY IN BRITAIN TODAY by Melanie Howard and Sue Tiballs.
2003, 57pp, free. Published by the Equal Opportunities Commission (qv) and may be downloaded from www.equalityhumanrights.com/pages.eocdrccre.aspx

TAXATION OF MARRIED FAMILIES: HOW THE UK COMPARES INTERNATIONALLY by Don Draper and Leonard Beighton, with a foreword by Professor Robert Rowthorn
2008, 40pp, free. Published by the charity Care, and may be downloaded from www.care.org.uk/ Group/Group.aspx?ID=69135

TEN YEARS OF BOLD EDUCATION BOASTS NOW LOOK SADLY HOLLOW. IT WILL BE HARD POLITICALLY BUT LABOUR MUST ACCEPT ITS VAUNTED POLICIES ON SCHOOLS HAVEN'T WORKED by Jenni Russell.
The *Guardian*, 14 November 2007. Visit www.guardian.co.uk/commentisfree/story/0,,2210600,00.html

THE THAMES GATEWAY: LAYING THE FOUNDATIONS – A REPORT BY THE COMPTROLLER AND AUDITOR GENERAL.
2007, 72pp, free. May be downloaded from www.nao.org.uk/publications/nao_reports/06-07/0607526.pdf

THATCHER AND SONS: A REVOLUTION IN THREE ACTS by Simon Jenkins.
2006, 384pp, £9.99. ISBN-13: 9780141006246. Published by Penguin Books.

THERE AND BACK: A TRAVEL GUIDE FOR DISABLED PEOPLE.
2006, pp, £7.25. Published by RADAR (qv) and may be downloaded from www.radar.org.uk/radarwebsite/tabid/84.default.aspx

THEY CALL IT JUSTICE by Brian Lawrence.
1999, 428pp, £25.00. ISBN-13: 978-1857763720. Published by the Book Guild Ltd and obtainable via Abebooks at www.abebooks.co.uk/

THE TIGER THAT ISN'T: SEEING THROUGH A WORLD OF NUMBERS by Michael Blastland and Andrew Dilnot.
2007, 256pp, £12.99. ISBN: 1 861 97839 1. Published by Profile Books at www.profilebooks.co.uk/

TOWARDS A GOLD STANDARD FOR CRAFT: GUARANTEEING PROFESSIONAL APPRENTICESHIPS by John Hayes and Dr Scott Kelly.
2007, 32pp, £5.00. Published by the Centre for Policy Studies and may be viewed free of charge at www.cps.org.uk/search/default.asp

TOWARDS A NEW CONSTITUTIONAL SETTLEMENT edited by Chris Bryant.
2007, 177pp, £9.95. ISBN: 1 905370 20 2. Published by and obtainable from the Smith Institute. May also be downloaded free of charge from www.smith-institute.org.uk/pdfs/constitution_full.pdf

TRANSFORMATIONAL GOVERNMENT: ENABLED BY TECHNOLOGY.
2005, 25pp, free. Published by the Cabinet Office (qv) and may be downloaded from www.cio.gov.uk/documents/pdf/transgov/transgov-strategy.pdf

TRANSFORMING OUR STREETS.
2006, 8pp, free. Published by CABE (qv) and may be downloaded from http://cabe.org.uk/AssetLibrary/8784.pdf

THREE CHEERS FOR SELECTION: HOW GRAMMAR SCHOOLS HELP THE POOR by Norman Blackwell.
2007, 28pp, £7.00. Published by and obtainable from the Centre for Policy Studies at www.cps.org.uk/

THE TOP TWELVE THINK TANKS IN BRITAIN by Toby Helm and Christopher Hope.
The Daily Telegraph, 24 January 2008. Visit www.telegraph.co.uk/news/main.jhtml;jsessionid=D0PJTGHR3DJ0TQFIQMFCFFOAVCBQYIV0?xml=/news/2008/01/24/nttank224.xml

TOWARDS THE LIGHT: THE STORY OF THE STRUGGLES FOR LIBERTY & RIGHTS THAT MADE THE MODERN WEST by A C Grayling.
2007, 337pp, £20.00. ISBN-13: 9780747583868. Published by Bloomsbury Publishing at www.bloomsbury.com/

TOXIC CHILDHOOD: HOW MODERN LIFE IS DAMAGING OUR CHILDREN AND WHAT WE CAN DO ABOUT IT by Sue Palmer.
2007, 384pp, £7.99. ISBN 13: 9780752880914. Published by Orion Books at www.orionbooks.co.uk/

TRANSFORMING OUR STREETS.
2006, 8pp, free. Published by CABE (qv) and may be downloaded from www.cabe.org.uk/AssetLibrary/8784.pdf

THE TRAJECTORY AND IMPACT OF NATIONAL REFORM: CURRICULUM AND ASSESSMENT IN ENGLISH PRIMARY SCHOOLS (Research Survey 3/2) by Dominic Wyse, Elaine McCreery and Harry Torrance.
2008, 34pp, free. ISBN-13: 978 1 906478 22 3. Published by the University of Cambridge Faculty of Education, and may be downloaded from www.primaryreview.org.uk/Publications/Interimreports.html

THE TWO FACES OF LIBERALISM by John Gray.
2000, 161pp, £13.99. ISBN-13: 9780745622590. Published by and obtainable from Polity at www.polity.co.uk/

UK PASSPORT SERVICE: BIOMETRICS ENROLMENT TRIAL.
2005, 301pp, free. Published by the UK Passport Service and may be downloaded from http://dematerialisedid.com/PDFs/UKPSBiometrics_Enrolment_Trial_Report.pdf

UK VOLUNTARY SECTOR ALMANAC 2007.
2007, 108pp, £25.00. Published by and obtainable from the NCVO (qv) on telephone 0800 2 798 798 or at www.ncvo-vol.org.uk/

UNAFFORDABLE HOUSING: FABLES AND MYTHS by Alan W Evans and Oliver Marc Hartwich.
2005, 41pp, free. Published by the Policy Exchange and may be downloaded from www.policyexchange.org.uk/

UNDERCLASS + 10: CHARLES MURRAY AND THE BRITISH UNDERCLASS 1990–2000 by Charles Murray and Melanie Phillips.
2001, 39pp, free. Published by Civitas (qv) and may be downloaded from www.civitas.org.uk/pdf/cs10.pdf

UNDERSTANDING THE DISABILITY DISCRIMINATION ACT: A GUIDE FOR COLLEGES, UNIVERSITIES, AND ADULT COMMUNITY LEARNING PROVIDERS IN GREAT BRITAIN.
2007, 166pp, free. Published by the Disability Rights Commission (qv) and may be downloaded from www.equalityhumanrights.com/pages/eocdrccre.aspx

UNIVERSAL DESIGN by Selwyn Goldsmith.
1997, 426pp, £38.99. ISBN: 0 7506 4785 X. Published by Architectural Press, an imprint of Elsevier at http://books.elsevier.com/

UNSTOPPABLE GLOBAL WARMING: EVERY 1,500 YEARS by S Fred Singer and Dennis T Avery.
2006, 276pp, £21.72. ISBN: 0742551172. Published by Rowman & Littlefield and obtainable from Blackwells at http://bookshop.blackwell.co.uk/

THE USES OF LITERACY by Richard Hoggart.
1957. New edition 1963, 326pp, o/p. Published by Penguin Books and obtainable via Abebooks at www.abebooks.co.uk/

THE VICTORIA CLIMBIÉ INQUIRY.
2003, free. May be downloaded from www.victoria-climbie-inquiry.org.uk/finreport/reportoverview.htm

THE VICTORIA CLIMBIÉ INQUIRY: REPORT OF AN INQUIRY by Lord Laming.
2003, free. May be downloaded from www.victoria-climbie-inquiry.org.uk/finreport/finreport.htm

VILLAGE HALL GOOD MANAGEMENT TOOL KIT.
2003, 32pp plus loose insert and CD-ROM, £15.00. ISBN: 1871157773. Published by and obtainable from Action with Communities in Rural England (ACRE) on telephone 01285 653 477, or visit www.acre.org.uk/

THE WAY TO REDUCE WELFARE DEPENDENCY by Dominic Lawson.
The *Independent on Sunday*, 29 August 2006. Visit http://comment.independent.co.uk/columnists_a_l/dominic_lawson/article1222369.ece

THE WEALTH AND POVERTY OF NATIONS by David Landes.
1998, 650pp, £20.00. ISBN: 0 316 90867 3. Published by Little, Brown and obtainable via Abebooks at www.abebooks.co.uk/

WELFARE ISN'T WORKING: CHILD POVERTY by Frank Field and Ben Cackett.
2007, 28pp, free. Published by Reform (qv) and may be downloaded from http://reform.moodia.com/filestore/pdf/070611%20Welfare%20isn't%20working%20-%20child%20poverty.pdf?

WELFARE ISN'T WORKING: THE NEW DEAL FOR YOUNG PEOPLE by Frank Field and Patrick White.
2007, 29pp, free. Published by Reform (qv) and may be downloaded from www.reform.co.uk/filestore/pdf/070511%20Welfare%20isn't%20Working%20-%20NDYP.pdf

WE'RE (NEARLY) ALL VICTIMS NOW! HOW POLITICAL CORRECTNESS IS UNDERMINING OUR LIBERAL CULTURE by David G Green.
2006, 86pp, £7.50. ISBN: 1 903386 53 5. Published by and obtainable from Civitas (qv) at www.civitas.org.uk/

WHAT DEMOCRACY IS FOR: ON FREEDOM AND MORAL GOVERNMENT by Stein Ringen.
2007, 334pp, £23.95. ISBN 13: 978 0 691 12984 6. Published by Princeton University Press at http://press.princeton.edu/

WHAT ARE WE SCARED OF? THE VALUE OF RISK IN PUBLIC SPACE.
2005, 48pp, free. Published by Cabe (qv) and may be downloaded from www.cabe.org.uk/default.aspx?contentitemid=477

WHAT HOME BUYERS WANT: ATTITUDES AND DECISION MAKING AMONG CONSUMERS.
2005, 28pp, free. Published by Cabe (qv) and may be downloaded from www.cabe.org.uk/AssetLibrary/2217.pdf

WHAT IS THE POINT OF ALL THIS LEGISLATION? by Philip Johnston.
Daily Telegraph, 29 October 2007. Visit www.telegraph.co.uk/opinion/main.jhtml?xml=/opinion/2007/10/29/do2903.xml

WHAT PRICE PRIVACY?
2006, 45pp, £13.50. Published by and obtainable from the Stationery Office on telephone 0870 242 2345 or at www.tsoshop.co.uk/bookstore.asp. May also be downloaded free of charge from www.informationcommissioner.gov.uk/

WHAT PRICE PRIVACY NOW? THE FIRST SIX MONTHS PROGRESS IN HALTING THE UNLAWFUL TRADE IN PERSONAL INFORMATION.
2006, 32pp, £13.50. Published by and obtainable from the Stationery Office on telephone 0870 242 2345 or at www.tsoshop.co.uk/bookstore.asp May also be downloaded free of charge from www.informationcommissioner.gov.uk/

WE WANT LAWS THAT ACTUALLY WORK by Matthew Parris.
The *Times*, 23 February 2008. Visit www.timesonline.co.uk/tol/comment/columnists/matthew_parris/article3419812.ece

WHEELCHAIR ACCESSIBLE HOUSING: DESIGNING HOMES THAT CAN BE EASILY ADAPTED FOR RESIDENTS WHO ARE WHEELCHAIR USERS. BEST PRACTICE GUIDANCE.
2007, 51pp, free. ISBN-13: 978 1 84781 072 4. Published by the Greater London Authority and may be downloaded from www.london.gov.uk/mayor/strategies/sds/docs/bpg-wheelchair-acc-housing.pdf

WHEELCHAIR HOUSING DESIGN GUIDE.
2006, 124pp, £40. ISBN: 1 86081 897 8. Published by Habinteg (qv) and obtainable from BRE Press at www.brepress.com/ or from the Centre for Accessible Environments (qv) at www.cae.org.uk/

WHEELS WITHIN WHEELS: A GUIDE TO USING A WHEELCHAIR ON PUBLIC TRANSPORT.
See: Ricability.

WHITAKER'S ALMANAC.
Annual, 1408pp, £40.00. ISBN-13: 9780713685541. Published by A&C Black at www.whitakersalmanack.co.uk/

WIDENING THE EYE OF THE NEEDLE: ACCESS TO CHURCH BUILDINGS FOR PEOPLE WITH DISABILITIES by John Penton.
2001, 80pp, £12.95. ISBN-13: 9780715175897. Published by the Council for the Care of Churches at www.cofe.anglican.org/about/cathandchurchbuild/

THE WISDOM OF CROWDS by James Surowiecki.
2005, 320pp£7.99. ISBN-13: 978-0349116051. Published by Abacus Books, and imprint of Little, Brown.

WHO CARES? HOW STATE FUNDING AND POLITICAL ACTIVISM CHANGE CHARITY by Nick Seddon.
2007, 200pp, £9.50. ISBN: 1 903 38656 X. Published by and obtainable from Civitas (qv) at www.civitas.org.uk/

WHO DO THEY THINK WE ARE? By Jill Kirby.
2008, 38pp, £5.00. Published by and obtainable from the Centre for Policy Studies. May be viewed free of charge at www.cps.org.uk/

WHO LOST OUR DATA EXPERTISE? CARELESS USE OF PERSONAL DATA HIGHLIGHTED THE MALAISE AT THE HEART OF THE GOVERNMENT'S IT INFRASTRUCTURE, BUT WILL IT CHANGE THE INTERNAL CULTURE? by Michael Cross.
The *Guardian*, 29 November 2007. Visit www.guardian.co.uk/technology/2007/nov/29/comment.politics

WHOSE RISK IS IT ANYWAY?
2005, free. Published by the Disability Rights Commission (qv) and may be downloaded from www.equalityhumanrights.com/pages/eocdrccre.aspx

WHY THE NHS IS THE SICK MAN OF EUROPE by James Gubb.
2008, 11pp, free. Published by the *think tank* (qv) Civitas and may be downloaded from www.civitas.org.uk/pdf/CivitasReviewMar08.pdf

WORKING ON THE THREE RS – EMPLOYERS' PRIORITIES FOR FUNCTIONAL SKILLS IN MATHS AND ENGLISH.
2006, 116pp, £50.00. ISBN: 0852016387. Published by and obtainable from the Stationery Office on 0870 242 2345 or at www.tsoshop.co.uk/bookstore.asp. May also be free of charge from www.cbi.org.uk/pdf/functionalskills0906.pdf

THE WORLD MAP OF HAPPINESS by Adrian White.
2006, free. May be viewed at www.le.ac.uk/pc/aw57/world/sample.html

WOULDN'T YOU FEEL SAFER WITH A GUN? BRITISH ATTITUDES ARE SUPERCILIOUS AND MISGUIDED by Richard Munday.
The Times, 8 September 2007. Visit www.timesonline.co.uk/tol/comment/columnists/guest_contributors/article2409817.ece

WRITING MATTERS: THE ROYAL LITERARY FUND REPORT ON STUDENT WRITING IN HIGHER EDUCATION FROM 2006.
2006, 108pp, free. May be downloaded from www.rlf.org.uk/fellowshipscheme/research.cfm

YOUNG PEOPLE AND CRIME: FINDINGS FROM THE 2005 OFFENDING, CRIME AND JUSTICE SURVEY by Debbie Wilson, Clare Sharp and Alison Patterson.
2006, 108pp, free. May be downloaded from www.homeoffice.gov.uk/rds/pdfs06/hosb1706.pdf

YOUR RIGHT TO KNOW: A CITIZEN'S GUIDE TO THE FREEDOM OF INFORMATION ACT by Heather Brooke.
2007, 309pp, £13.99. Published by Pluto Press at www.yrtk.org/

YOUTH JUSTICE 2004: A REVIEW OF THE REFORMED YOUTH JUSTICE SYSTEM.
2004, 132pp, free. Published by the Audit Commission and may be downloaded from www.audit-commission.gov.uk/Products/NATIONAL-REPORT/7C75C6C3-DFAE-472d-A820-262DD49580BF/Youth%20Justice_report_web.pdf